ARCHITECT AND ENGINEER

ARCHITECT AND ENGINEER

A Study in Sibling Rivalry

Andrew Saint

YALE UNIVERSITY PRESS
NEW HAVEN AND LONDON

aan IWMJ

Copyright © 2007 Yale University

Designed by Sally Salvesen
Printed in China through WorldPrint

Library of Congress Cataloging-in-Publication Data

Saint, Andrew.
Architect and engineer: a study in sibling rivalry / Andrew Saint.
p.   cm.
Includes bibliographical references and index.
ISBN 978-0-300-12443-9 (alk. paper)
1. Architects and engineers. 1. Title.
NA2543.E54S25 2008
720–dc22

2007016892

FRONTISPIECE: detail of Fig. 262
Coffer-dam in construction for the Pont Royal, Paris,
drawing by Lievin Cruyl, 1687

CONTENTS

Two contrary thoughts on writing about buildings:

The use of history mechanical is of all others the most radical
and fundamental towards natural philosophy; such natural
philosophy as shall not varnish the sume of subtle, sublime
or delectable speculation, but such as shall be *operative* to the
endowment and benefit of man's life.

Francis Bacon, *The Advancement of Learning*

Sind wir vielleicht *hier* um zu sagen: Haus,
Brücke, Brunnen, Tor, Krug, Obstbaum, Fenster,
höchstens: Säule, Turm … aber zu *sagen*
oh zu sagen *so*, wie selber die Dinge niemals
innig meinten zu sein.

Rainer Maria Rilke, *Ninth Duino Elegy*

PREFACE AND ACKNOWLEDGEMENTS

This book is a very long footnote to another I wrote a generation ago, *The Image of the Architect* (1983). That book came out of the first full course I taught at a school of architecture, the Architectural Association. By way of anecdote and example, it sought to offer historical insights about the nature of the profession the students had elected to join. Twelve years later I was appointed to a full-time post in the Department of Architecture at the University of Cambridge. Faced anew, as a non-architect, with the challenge of making history pertinent to would-be professionals, I chose a similar method to explore the tangled thickets of relations between architects and engineers. The course proved not specially popular. No doubt there were deficiencies in its presentation. But the main reason, I now see, was that when you are learning to design, you need to have your confidence built up, not broken down. That is often the effect of studies that treat architecture as part of a larger whole. Like human kind in general, budding architects cannot bear very much reality.

Still, friends and colleagues encouraged me in the belief that the topic was worth pursuing further, not least in view of excitements and controversies about relations between modern-day architects and engineers. Though much has been written about this last issue, it is seldom informed by deeper understanding; indeed there has never been, so far as I am aware, a broad and sustained historical enquiry into the architect-engineer relationship. So I dropped the course but persevered with the book.

The mass of material was overwhelming. I soon realized that I should have to choose, cut and shape severely. Rather than a continuous narrative or theoretical analysis, the notion of six discursive essays or case-studies which could be individually read or dipped into according to taste has been basic to the enterprise. The first is about the impact of military construction and organization, often underestimated. The second and third tackle the so-called 'new materials' of iron and concrete, since I believe (unfashionably) that a materialist slant helps to illuminate professional relations in construction. The fourth takes a building-type, the bridge, in which aesthetics play a part yet the nature of the architect-engineer relationship as commonly conceived tends to be inverted. The fifth looks into the seemingly closer relations between architects and engineers in the making of major projects since about 1930. Finally, the sixth study examines the training of the two professions. That essay is deliberately, some readers may find provocatively, placed last. It is often assumed, particularly in continental Europe, that the outlook of architects and engineers flows from what the French call their *formation*. I have chosen to look through the other end of the telescope, and taken institutions, educational and professional, as the outcome of wider forces.

A reader persistent enough to work right through the chapters may feel in want of a clear thread. So I have appended a conclusion that tries to draw out the pattern I believe I have found. It is not a simple pattern, but one exists. Anyone looking for answers and philosophies is counselled to jump straight from the introduction to the conclusion.

All the essays are about the western tradition in construction. In the main they address patterns of development since about 1660. There are some references back to the Renaissance in Chapter 1 and indeed to the Romans in Chapter 4, but these are brief escapades. In general the book moves gradually forward. The first two chapters run from about 1660 to 1900; the third covers 1750 to 1939; the fifth runs from 1920 to slightly short of the present, while the other two essays range across the full chronology of the book. The other unifying principle is that I have consistently investigated just three countries: Britain, France and the United States. Those three states, I judged, had enough points of similarity and difference to warrant concentrating my efforts on them. Nevertheless episodes in other countries are drawn in as and when they seemed helpful: Italy and Germany feature episodically in this way, as also, less often, do Holland, Russia and Spain; Belgium and Portugal figure only in footnotes.

Set out like that, the omissions seem flagrant. It is the same with individual architects and engineers. Modernists may be disappointed. Why is there so little on Mies, and almost nothing on Nervi or Calatrava? My selections have naturally been personal. This is a long book. It has been said that the trouble with Bruckner's symphonies is not that they are too long but that they are not long enough, because he never has the space to develop his gradually unwinding themes. I have been very aware of that. There could easily have been more chapters. Two themes I specially regret having to omit: the distribution of jobs between architects and engineers in the making, extending and maintaining of cities, on which I have confined my remarks to a few sentences and footnotes in Chapter 1; and their respective roles in the design and manufacture of public structures in series, such as electricity pylons, underground stations, phone-boxes and street furniture.

Though it has been long and hard work writing this book, it has also been fun. Best have been the trips to see things, bridges above all. Unavoidably, I have written often about buildings I have not seen. When I did see things, they invariably appeared in a fresh light and changed my perceptions. If there is one piece of advice I would tender to students of architecture and engineering at every level, it is to limit their study of photographs, drawings and documents, and to go and look.

In gratefully recording debts, I would like to begin by thanking the many scholars and enthusiasts on to whose backs I have crawled. This book is not the outcome of primary research, in the way that is understood today. I have done little work in archives but a lot in libraries, filleting other people's discoveries. In such synthetic work covering a broad field there will be misinterpretations and mistakes, for which I am of course wholly responsible. Of the libraries I have used, I should like to single out the London Library and the Faculty Library for the Departments of Architecture and History of Art at Cambridge under Maddie Brown; also those of the Institution of Civil Engineers under the beneficent and learned Mike Chrimes, and of the RIBA. Little about the history of the built environment cannot be found in one of those four places.

Among friends who have been outstandingly generous and forbearing to me over this project's long life I should first mention Charlotte Ellis, Martin Meade and Anne-Marie Châtelet in Paris, and Mosette Broderick in New York. In London, Robert Thorne has patiently encouraged and gently tested me. So has Jules Lubbock, who read a good deal of the text. Other readers and chasteners of chapters or sections included Tim Benton, Simon Bradley, Steven Brindle, Nick Bullock, Alan Crawford, Gillian Darley, Ed Diestelkamp, Adrian Forty, Elain Harwood, Neil Jackson, Sophie Le Bourva, Robin Middleton, Alan Powers, Sam Price, Dmitry Shvidkovsky, Robin Spence, Gavin Stamp and David Yeomans. I am specially grateful to Hubert Murray for first pointing out to me the importance of the US Corps of Engineers. The upshot was a surreal adventure to their archives outside Washington. Once I had penetrated the bunker, I was welcomed there with warmth and helpfulness. A bracing discussion with Antoine Picon led me to clarify my conclusions. At an early stage Giles Oliver suggested the subtitle, which I have clung to through thick and thin. Anthony Alofsin has been a regular source of friendship and support, notably over issues to do with American architectural education and the role of Frank Lloyd Wright.

I should also like to thank Bill Addis, John Allan, Andrew Barlow, Susie Barson, Peter Blundell Jones, Karen Bowie, Tim Brittain-Catlin, Bob Bruegmann, Neil Burton, Ian Campbell, James Campbell, Peter Carl, Peter Carolin, Amy Chamier, Martin Charles, Mike Chrimes, Myke Clifford, Phil Cooper, Dan Cruickshank, Sophie Descat, James Edgar, Julia Elton, William Fawcett, Ben Foo, Ed Ford, Max Fordham, John Greenacombe, Peter Guillery, Di Haigh, Richard Hill, Deborah Howard, Peter Howell, Nick Jacobs, Helen Jones, Frank Kelsall, Tom Killian, Eda Kranakis, Catherine Leopold, Ellen Leopold, Tony McIntyre, Sebastian Macmillan, John Nicoll, Aart Oxenaar, Nicholas Penny, Michael Port, Bob Proctor, Nick Ray, Liz Robinson, Lily Saint, Frank Salmon, Irénée Scalbert, Katya Shorban, Teresa Sladen, Phil Steadman, James Sutherland, Lynne Walker and Mary Wall, as well as to remember with gratitude and affection three friends who died during the later stages of my work, Tony Baggs and Colin and Rosemary Boyne.

Picture-research has become ever more of a practical and financial burden for the author of any book on architecture and art written without a financial return in view. This one could not have been published without the generous support of the Graham Foundation in Chicago, and the Paul Mellon Centre and the Ove Arup Foundation in London. My thanks are due to all these three institutions for their faith in an unusual project. Even with that help and the resource of the internet, the unwary author looking for pictures across many countries needs all the support he can get. Julia Brown, a queen among picture-researchers, took me under her wing and gave me much support and advice. Claudia Marx did a brilliant job getting German, Swiss and Russian pictures for me. Many others who helped me through the *terra incognita* of high-resolution scans and rights are inadequately acknowledged by the photographic credits.

At Yale University Press, I am grateful to Catherine Bowe and more particularly to my editor Sally Salvesen, a robust and constant support and friend. My deepest thanks are reserved for the dedicatee.

INTRODUCTION

The Department of Architecture at the University of Cambridge, where I went to teach in the autumn of 1995, occupies one end of an early Victorian terrace at the edge of the town centre. There is the usual ramble of extensions behind. Behind again stands an altogether different affair: the Department of Engineering. A blunderbuss of a building housing the university's largest department, it has internal courts and roads of its own, and much-contested parking space along its periphery.

To get to Engineering you must take a slip road beside the flank of the terrace, running the gauntlet, so to speak, of the architects. During term it is especially busy around the hour, as students of structural, mechanical, electrical and other subdivisions of engineering stream in and out of lectures. Their timetable is highly ordered, and it is easy in principle to understand what they learn: maths, science and technology – accuracy, discipline and precision. In the foyer hangs a painting by Terence Cuneo which summarizes the engineering ideal as it appeared when the building opened in 1952 (ill. 1). Prince Philip, then as now Chancellor of the University of Cambridge, backed by a posse of males in gowns, looks down with boyish curiosity from a gallery upon a trench of giant machinery which must be juddering and humming impressively. Here is progress, as for the nation, so for the world.

From the slip road you cannot see much of what goes on as you pass the Department of Architecture, because the once-generous fenestration in its back extension has been largely covered over to get extra pinboard space. A mischievous thought might be that this had been done to exclude the profane. If they glimpsed inside, the main activity that the engineers might observe would be the mysteries of the architectural 'crit', that daylong blend of ritual and endurance-test that is the central act of the studio-teaching calendar. Students pin up their work or set out their fragile models, dally, listen, disappear for a while, look in again, and finally, often falteringly, one by one present their ideas. Teachers and visiting critics fidget or frown according to their lights, before giving vent to shrewd or arbitrary utterances. Between, there are long pauses. Everything runs late and seldom in the right order. Tension is high and exhaustion great, because many students have been up all night finishing their work. The crit can be formative, devastating, illuminating, initiating, alienating or just plain boring. In the terms of the engineers, as a means of conveying skills or facts, it is neither systematic nor rational. Rather, it is an exercise in rhetoric for a calling that must be groomed to persuade.

The architects at Cambridge also have their icon near the entrance: a tiny unlabelled bronze, wall-mounted above the first half-landing of the staircase (ill. 2). Although it is by Henry Moore, the students take no more notice of it than the nonchalant engineers do of their Cuneo. Its abstraction is mysterious; it conveys nothing precise or useful. Only, like most Moores, it cries out to be touched. It seems to say: this is holy, I too matter even if you pass me by, here is art.

One day, many of these student architects and some of these student engineers (the structural engineers at any rate) will work with one another. Their opposed, yet juxtaposed, cultures and professions have to find ways of coming together. Most buildings of

1. The Duke of Edinburgh (Prince Philip) opening the new building for the Department of Engineering, University of Cambridge, 1952. Painting by Terence Cuneo.

complexity need both architects and engineers, as recruits to both disciplines may be told at the start of their studies. Probably the postulant architects will learn more about engineering, however superficially, than engineers will about architecture. For architecture is the more catholic subject, whose aspirants have to know a little about a lot of things. There are more engineers in the world than architects. In many circumstances of construction the former can do without the latter. An architect in practice, on the other hand, who tries to go much beyond interior decoration will not get far without the help of some engineering calculations. Before it is started, a building must be seen to be able to stand up. The engineer is the guarantor that it will.

There is a deeper reason why architects learn something about engineering. Structure is the basic requirement for any design that is to be built. Because of that, most architects acknowledge that it has in some sense to be confronted, incorporated and, quite possibly, expressed. In those terms, an architectural design that does not address structure (and there are many such) is incomplete or illogical. One of the many tasks of an architectural education is to promote an informed attitude towards structure. It may stand at one remove from the act of design, but it always lurks in the background.

The need to make a building firm, sound or whole is as old as architectural theory. Indeed it is older, since the stone blocks, timber posts, mud walls and thatched roofs of the different ancient architectures were always the ultimate determinants of the way in which they were designed. You may theorize about form and ritual to your heart's content. But in the end the properties of available materials, together with man's ingenuity in extending them, have always done the most to shape buildings.

Nevertheless, the historians agree, around the time of the Enlightenment the attitude of writers about architecture towards structure sharpened. The plain idea that the properties of materials had to be understood and accommodated gave way to a forthright theory that set a priority on 'rationalizing' and 'expressing' the raw necessity of structure. An interest and a competence in matters structural became paradigmatic. Laugier, Pugin and Viollet-le-Duc (all French or having links with French culture; theories about struc-

2. Small untitled bronze by Henry Moore at the Department of Architecture, University of Cambridge. Mounting by Bill Howell.

ture seem closely connected with Cartesianism) stand at the head of a host of writers, teachers and practitioners for whom the visible or distinctive expression of structure had become architecture's fundamental tool by the heyday of the Modern Movement.

Let us listen for a moment to a soldier in this army, pronouncing as recently as 1990 about a prominent new London building designed by Terry Farrell. The author is John Winter, an architect of whom it is enough to say that he spent a formative period during his youth in the Chicago office of Skidmore, Owings and Merrill.

Dear Terry

I very much enjoyed our visit round Embankment Place [ill. 3] last week and congratulate you on bringing this project to near-completion without any sign of the compromises which bedevil most big London projects.

After we had completed our visit I felt rather bad that I had left you with a few flippant comments. Your achievement clearly deserved more than that. So here are my reactions, for what they are worth.

1. I think that you are the most skilful urbanist in England. No one else can sort out immensely complicated architecture as well as you can, nor weave a project so well into the existing urban fabric. I really do have immense respect for you at this level.

2. As a lover of clear structure I obviously have problems enjoying your buildings. The top floor at Embankment Place is miraculous, but the hangers do not feel like hangers – with the sole exception of the diagonal ones across the gable wall; and these are clearly not genuine. For me integrity matters.

I realize that I have puritanical tastes and so you will always be more successful with planners and developers than I shall ever be. I delight in your success, but I do get angry that you produce such marvellous strategies and then detail it in a way that I find a real turn-off. Detailing can tell the story of a structure, the arch and the building process.

Yours ever
John Winter.

3. Embankment Place. Air-rights building over Charing Cross Station, London, by Terry Farrell and Company, architects, 1986–91

Such a provocation will seldom go unanswered. Soon, a reply from Farrell comes winging back:

Dear John

Thank you for your letter – the congratulations on Charing Cross and the compliments about my skill as an urbanist. However I must admit that your letter seemed to me to contain so much of the kind of unfair bias about architecture that I have spent the last 10–15 years trying to counter, that I thought it was worth replying to.

I particularly reacted, as you might have expected, to your statement that 'For me integrity matters'. A statement, as I see it, of disconcern and intolerance for the view of others and the satisfaction with your own point of view. For me integrity matters probably every bit as much, and indeed it could be that for me a real and deeper integrity matters! I see those that set themselves a life task of following a narrow doctrine such as functionalism or an exact expression of 'honest' construction or whatever the choice, as like true believers of a political or religious doctrine . . . Architecture is a widely based art and architectural integrity begins with being inspired by, responding to, and indeed serving the community, the place and the context of the time.

. . . For me, architecture begins with urbanism not construction and herein lies, I think, the difference, as I see tolerance as a virtue not a weakness.

Yours ever

Terry Farrell.

Here are encapsulated the history and psychology of the structural approach to architecture: the crusading zeal; the preoccupation with 'detail', interpreted as about structure, not autonomous ornament; and the shift back and forth between ethical and aesthetic languages. Finally, there is the flat rejection of the whole criticism by an architect who sees structure as one among many tasks.

If the intolerance of Winter's letter is obvious, Farrell's response reeks of vague compromise. Winter and Winter's mentors seem to promise clarity, rigour and discipline – the legacy of engineering to architecture. Lectures timed to the hour; definable knowledge; progressive, ascertainable goals; clear technical criteria. It is an approach that can be grasped by insiders and outsiders alike. Thirteenth-century Gothic great church? Good, because the separate elements of vault, wall, buttress and pinnacle do their separate jobs and are visibly distinguished from one another (Viollet-le-Duc). St Paul's Cathedral? Bad, because the upper walls of the aisles hide buttresses and the true construction of the great dome is concealed (Pugin). 1920s villa by Le Corbusier? Brilliant, because the system of supports breaks free from walls and partitions and, while remaining clear and legible, releases the outline and plan-form of the house (Le Corbusier).

Though old now, these clichés of modernist criticism retain their appeal. They assert that the further architects go in expressing – embellishing, drawing attention to, making a song and dance about – structure, the one thing a building cannot do without, the more virtuous their architecture will be. Since tasks vary, this will not be all that architecture needs to do. It may well not be its highest task. But it is raised to universal applicability.

There is a danger, though. If architecture must depend upon structure, doesn't it run the risk of subordinating itself to engineering? The structural engineers are the people who know how to make things stand up, and who can procure you the maximum span or the greatest load-bearing capacity in the most efficient way and at the least expense. If you also believe, as many have done, that beauty lies in grace of construction, and that this in its turn depends upon harnessing and displaying the forces at work in a structure with such regard to economy of means and ends that form and efficiency coincide, then the engineers are poised to take over aesthetic as well as technical leadership in architecture. When Whitehead defined style as 'the achievement of a foreseen end, simply and without waste', he was epitomizing a stance common among modernists. Potentially it delivered architecture over to the engineers.

In the early twentieth century, many people indeed claimed that had come to pass:

> Our engineers are healthy and virile, active and useful, balanced and happy in their work. Our architects are disillusioned and unemployed, boastful or peevish. This is because there will soon be nothing more for them to do . . . Our engineers produce architecture, for they employ a mathematical calculation which derives from natural law, and their works give us a feeling of harmony . . . Today it is the engineer who *knows*, who knows the best way to construct, to heat, to ventilate, to light. Is it not true?

Thus Le Corbusier, articulating in 1920 what theorists had long been hinting. Earlier, Henry van de Velde had written likewise:

> There exists a class of persons whom we can no longer refuse to call artists . . . The artists I am referring to, who have created a new architecture, are the engineers.

Telford, the Rennies, the Brunels and the Stephensons, Paxton (perhaps), the Roeblings, Eiffel; later, Freyssinet, Maillart, Williams, Nervi, Candela, Calatrava: these are the kinds of people whom Van de Velde meant. Ever since the Industrial Revolution, the engineers seemed to have been architecture's true heroes. A sizeable constituency of architects still feels that way.

ESTHÉTIQUE DE L'INGÉNIEUR ARCHITECTURE

4. Le Corbusier-Saugnier, *Vers une architecture*, Paris 1923

Of the spectacle of one professional group abasing itself and attributing all virtue to another, you could almost say it was neurotic or self-destructive. Like many strong emotional reactions, it paid little heed to the facts. Take two obvious cases. Among the most iconic buildings of the nineteenth century are the Crystal Palace and the Eiffel Tower. Both have been endlessly held out as exemplars for a modern architecture, though neither was primarily the product of an architectural office. Yet both were conceived and built as short-life structures, with all the special conditions and environmental limitations that implies.

Let us briefly examine the facts in each case, leaving fuller discussion for later pages. For the Crystal Palace (pp. 131–4), the original competition entries of the architects were found wanting, in part because they could not be built in the time available. They were set aside in favour of an idea not indeed from a professional engineer but from a brilliant amateur – a businessman-inventor-landscapist with experience of glasshouse construction and railway development. The planning, detailing and execution of Paxton's design were assigned to Fox and Henderson, engineer-contractors. Such were the scale and timetable of the project that only organization-men like they, with a formidable track-record of railway-building, could pull it off. Architects were indeed involved, but at the margins, in refining the details and making a colour scheme for the interior. The building was a resounding success; it did its six-month job perfectly and was a stupendous novelty. That did not mean that everyone thought it beautiful. Apart from conservatives and bigots, there were those who found it monotonous. When the Crystal Palace was taken down and re-erected at Sydenham as a permanent structure, Paxton had more control, took care to make many alterations – and employed a good deal more professional help. The result was unquestionably more satisfactory. Nevertheless it was the original raw 'engineering' concept that stuck in the architectural imagination.

The Eiffel Tower (pp. 162–7) has an architect, but his name is even less known than that of Fox and Henderson, the true heroes of the Crystal Palace. He was Stephen Sauvestre. Without Sauvestre it is improbable that the tower would have been built. It was he who refined the profiles and enriched the details of a striking but raw idea originally cooked up by Maurice Koechlin and Emile Nouguier, two engineers within Eiffel's firm. The great arches at the base were Sauvestre's addition. Structurally, they do nothing; they are decorative, or rhetorical if you prefer. They were put there to help sell the tower, to make it acceptable. You can call this a concession to conservativism, realism or taste. The fact remains that it was only after Sauvestre had done his work that Eiffel permitted the project that was to take his name to go forward.

Not that the eponymous attribution is fraudulent or unjust. It was Gustave Eiffel who took the risks, supplied the confidence and the contacts, and bore the bad and good publicity – in short, saw the job through. Honourable though the instinct may be, the task of accurately attributing creativity to individuals in complex, collaborative building projects is doubtful in architecture and even more so in engineering. As in politics or business, so in design, the convention of crediting a named individual, be it Eiffel or Frank Lloyd Wright or Norman Foster, with responsibility for a project is a shorthand. Rather than a guarantee of imaginative artistry, it is chiefly a recognition that without the named person the project would not have happened, or would have taken a quite different form.

That leads us back to the modernist architects' admiration for the legacy of the nineteenth-century engineers. Many reasons were given for why types like Brunel and Eiffel should be emulated: their ability to discard the trappings of history and face modern needs and briefs fair and square; their unselfconsciousness; their courage and ruthlessness; their technical know-how; the scale at which they worked; their capacity to cope with new materials and weave new spaces and enclosures out of them. Since architectural ideology had come to lay stress on structural expression, the man who could put structures together in the most original and striking way was bound to win applause. But in the end it boiled down to a sense of envious wonder: the engineers got things done.

How *did* things get done, especially in complex cases? The question begs to be better addressed. It is the main burden of the essays that follow. Even a cursory glance at our two 'temporary' structures, the Crystal Palace and the Eiffel Tower, shows that more was involved than a raw engineer ploughing on, blind to aesthetics. Architects were there on the sidelines, playing their marginal part. When we look into more complex projects, buildings with strong cultural programmes, buildings where people shelter, work or live, buildings that stand next to others, buildings in sensitive contexts, or buildings with different elements for different uses, it becomes apparent that the relationship between architecture and engineering is seldom simple: that compromise and partnership are the rule, not the exception; and that the balance between the two professions shifts from project to project. Time and again, the disjunction between efficiency and expression conjured up by the difference in educational patterns between engineers and architects is gainsaid by the felicitous impurity of what actually happens when a building is built.

How did those stereotypes arise in the first place? Before the Enlightenment and the Industrial Revolution, there can be difficulty in distinguishing between architect and engineer. They seem to shade into one another, and for that matter into the realms of craft. Were mediaeval castles designed by architects or engineers? The question is addressed in this book's conclusion. But we instinctively feel that it is not quite rightly posed, because our modern professional divisions did not yet exist. Was Brunelleschi an artist, a craftsman, an architect or an engineer? Of course he was all four, and his great dome at Florence reflects the indistinguishable wholeness of his ingenuity. Something similar could be said of Wren. Yet by Wren's time there is a sense that the pattern is changing, and that something has begun to insert itself between conception and execution – one that may be explained in terms of class, of growth in demand, of new complexities in technique, or of divisions of labour in the interests of efficient production.

One temptation has been to look upon this apparent division as some kind of disaster – the dissolution of a former golden unity in the building arts or even a fall from metaphysical grace. On this view, from the apple-eating of material and technical progress and the triumph of 'instrumentality' there sprang the spiritual alienation and exile of architecture – of which the alleged division between architect and engineer is one of many symptoms. Alberto Perez-Gomez has put it this way: 'The arrogance, anguish, and correlative responsibility of architects, engineers and technicians of the nineteenth century contrasted with the general tranquility and self-confidence of their predecessors.'

It is not far from here to those blocked-off windows at Cambridge. From that standpoint, the mission of architectural scholarship and indeed of architectural education becomes redemptive. It seeks a return to paradise, or to a lost harmony of the kind that the poet Milton incomparably described:

> That we on earth, with undiscording voice,
> May rightly answer that melodious noise;
> As once we did, till disproportioned sin
> Jarred against nature's chime, and with harsh din
> Broke the fair music that all creatures made
> To their great Lord, whose love their motion swayed
> In perfect diapason, whilst they stood
> In first obedience, and their state of good.
> O, may we soon again renew that song,
> And keep in tune with Heaven, till God ere long
> To his celestial consort us unite,
> To live with Him, and sing in endless morn of light!

By looking into the detailed, material facts of how architects and engineers have related, this book argues otherwise: that the two professions have always been somewhat different, but that they should be valued as equal. Like any true partners, they must be seen as different equals who must perpetually be learning how to live together.

I

'IMPERIAL WORKS AND WORTHY KINGS'

1. France 1660–1789

VAUBAN'S WORLD

Around the year when Milton wrote his 'At a Solemn Musick' was born the military engineer par excellence of the seventeenth century: Sebastian Le Prestre, the Marquis de Vauban (ill. 6).[1] Long famed as a patriot, soldier and engineer, Vauban enjoys an enlarged reputation today as something too of an economist, statesman and proto-*philosophe*. His robust, enduring fortifications attest France's rise to eminence and splendour. Among those who tend the cult of Vauban, some have seen him as a kind of artist in nationhood. 'His hand worked upon the soil of France as the graver of the aged Rembrandt upon copper,' wrote Daniel Halévy. For Halévy the frontier of France, as Vauban left it, could be compared with Versailles: 'more magnificent still, as well thought out, more powerful, almost as beautiful'.[2]

Nowadays Vauban is cast most often as an engineer, but he would have seen himself first and last as a soldier. His fame is inseparable from Louis XIV's territorial ambitions. Were it not for the legacy of his extant forts and towns, bristling along the perimeter of France, he would be as little sung today as other of the Sun King's military commanders. In every stronghold where he left his mark, streets, promenades, monuments, cafés and, above all, walls perpetuate his name.

Not that Vauban was a very original designer of fortifications. What he had was zeal, opportunity, steadiness of judgement and organization, endurance, and a boundless capacity for taking pains. The outline of his career is soon told. Vauban comes from a middling gentry family with property near Vézelay. He enjoys a modest education, drifts into soldiering during the wars of the Fronde, is captured, changes sides, and becomes an aide-de-camp to Clerville, Mazarin's 'commissaire général' for fortifications from 1659. Then Louis XIV assumes personal power and, through the agency of his great ministers Colbert and Louvois, embarks on a plan of expansion and integration bolstered by warfare. Vauban's career as an expert in siege-warfare – the balancing sciences of attack and defence – now takes off.

In the first instance Vauban is Louvois' man: he supervises siege operations in the borderlands of the north and east within the Minister of War's sphere, where conflict is gruelling and persistent.[3] From Vauban's frank correspondence with Louvois we get to know the man and learn the range of tasks that confront the strategic military engineer of a land-hungry nation-state. By the time Clerville dies in 1677, he has half-displaced him and overcome Colbert's initial hostility. He now takes in marine fortifications, which belong to Colbert's sphere of interests, so gradually drawing the defence of the entire kingdom into his hands.

FACING PAGE
5. Neuf-Brisach, the ramparts.

6. The Marquis de Vauban (1633–1706). Drawing attributed to Charles Le Brun.

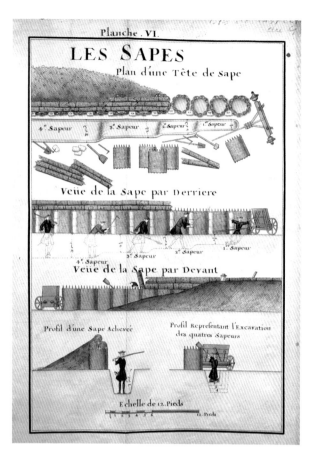

7. Methods of making earthworks and trenches, with exact dispositions of sappers, their tools and materials. From Vauban, *La manière d'attaquer les places*. A reminder that the majority of works built by military engineers were temporary.

Following the deaths of Colbert (1683) and Louvois (1691), Vauban acquires unique prestige, always enjoying the ear of the King. Though France's military fortunes are on the wane and budgets for fortifications are slashed after 1692, he labours on, travels as much as ever and commences his most perfect new 'ville de guerre', Neuf-Brisach on the German border (ills. 5, 8, 9). In the many sieges at which he participates, he never fails to reduce a town or is forced to concede one. As a high-level technician rather than a commander in the field, Vauban is created a marshal of France only late in life, in 1703. By then he is stoutly independent. A prolific writer and correspondent, he branches out from warfare into demography and farming (including a tract on pig-keeping). His antipathy to the revocation in 1685 of the Edict of Nantes, which causes hosts of economically active Protestants to flee France, is blunt and on the record. At the time of his death he has even had printed a subversive pamphlet on taxation, *La Disme Royale*.

THE ROOTS OF SPECIALIZATION

At a Vauban fortification today, one is first impressed by the grandeur of the surviving structures. Maybe too much so: for walls and bastions, gates and barracks of permanent masonry were only a tithe of any castle-builder's tasks. Fortifications are the stage-sets left behind from the drama of siege-warfare, actual or potential. For each specialist in defence, another had to be as adept in attack. Their commanders needed to know about weaponry, artillery trajectories, saps (ill. 7) and scaling methods as well as ravelins, bastions and demi-lunes. Both sides belonged to a science of fortifications that had sprung up and speeded up since artillery and the printed technical treatise put paid to the high crenellated walls of the middle ages. For Vauban, designing was one of many skills whose totality, like those of any army commander, had first and foremost to do with operations: getting things done well, and in due order. He was a critical cog in the machinery of French military expansion. This builder of durable structures was himself part of a structure whose like had not been seen in Europe since the Roman empire. His greatness as an engineer lies in process, not in any single act, art or technique.

If there is a key to the advances made under Vauban, it has to do with continuities: a continuity of time, because the campaigns of Louis XIV's armies were sustained; and of organization, because those armies enjoyed a newly cohesive make-up and political command. These continuities marked a step-change in that recasting of the art of war since the Renaissance known to scholars as the 'military revolution'– one that embraced weapons, manpower and tactics as well as the building-up and breaking-down of new-style strongholds.[4]

The defences raised by Vauban and his subordinates stand midway along a line of development in fortification that spanned some four centuries. The origins of the technique go back to the onset of the Renaissance in Italy. In its primitive manifestations the passionate virtuosity and ideal geometry of those times were to the fore. Exuberance of invention can seldom be disentangled from response to necessity in the military and mechanical drawings of Leonardo and the earliest treatises on fortification. Many such schemes belong, in Janis Langins's phrase, to the realm of 'technical fantasy'.[5] But with the fiercer warfare that convulsed Italy in the early sixteenth century, the new fortification turned into a gravely practical, competitive art. The upshot was the model of multi-angular, bastioned defence known as the 'trace italienne', designed to protect strongholds against flanking fire from ever more potent artillery (ill. 9). A defining moment came in 1534, when Pope Paul III convened experts in Rome to debate its defences, following the sack of the Eternal City seven years before.[6] By then the trace italienne was sweeping Europe. During that decade more than a hundred Italian experts were said to be in France, peddling the new bastions.[7]

Typically, the trace italienne was designed by architect-engineers in partnership with military strategists. 'Architect-engineer' is a handy if clumsy portmanteau-phrase that will often recur in this book.[8] It will generally be used to mean someone who, consciously or otherwise, straddles the two disciplines of construction as they are construed today. The expression cannot be found before 1700. It was unneeded before theoretical and social boundaries became laid upon provinces of construction never perfectly separable in practice. During the Renaissance, there was no border to cross; professional terminology was not yet fixed. Irrespective of status or trade, a man might indifferently be called 'architetto' or 'ingegnere' or indeed something else. The choice of term generally followed not from the skill or numeracy of the individual but from the nature and name of the office he filled at the time.[9]

Not that architect and engineer have ever been wholly indistinguishable. Anyone studying the relationship between the two will soon conclude that though the terms and the jobs are hard to prise apart, they correspond to different facets of the human personality. At the time of the Renaissance, either word retained a nuance of its own. An architect implied someone who imposed ideas upon buildings, someone at least acquainted with the high end of the Vitruvian cluster of values and with art. An engineer meant someone clever with machines, able to harness power, versed above all in the techniques of war. The terminology can be traced back to Roman usage. It is striking that 'architect' has Greek roots and 'engineer' Latin ones. Then as now, the skills they connoted tallied with the strengths of the two great classical cultures, complementary and overlapping in some ways, in others antithetical.[10]

But if the terms 'architect' and 'engineer' had their own resonances, when it came to the arts of practical building it can hardly be denied that they were muddled up together to the point of interchangeability. What began to draw them apart? Scholars of scientific bent like to argue that mathematics is the key. To be educated at the time of the Renaissance meant to be excited by the beauty and potential of mathematics. Naturally, those who proportioned façades applied calculations otherwise than those who surveyed property or worked out ballistic trajectories. But to suppose that this actually divided the 'professions' is to confuse cause and effect. The root of the reason for the incipient parting of ways is surely simpler. It lies in the growth of specialized opportunities among those who designed and managed construction.

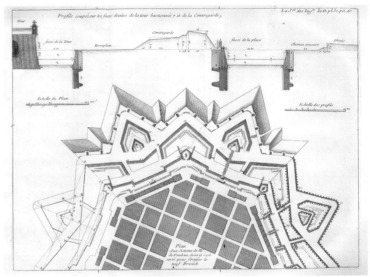

8. Neuf-Brisach, model for the town, c.1705. One of the great collection of models for French fortifications, mostly housed in the Musée des Plans Reliefs at the Invalides. The design is not quite as executed, but shows clearly the triple line of defences, the strict grid with parade square in the centre and the barracks strategically sited behind the main walls.

9. Part-plan of Neuf-Brisach, showing the scale of the 'trace italienne' compared to the town grid, and the difficulty in reconciling their geometries. From Bélidor, *La science des ingénieurs*, 1729, Book VI.

The wealth, scholarship and technology of the Renaissance led not only to more building but to different kinds of building. In the new circumstances an identifiable body of men are able to earn their living from conceiving, designing and running construction projects without dirtying their hands. In a classic division of labour, the next step from this proliferation is the specialization of building experts. But there is a difference from the divisions of labour in the old building guilds or in later, profit-driven industrial processes. Whereas both those are about applying special skills to particular stages during the process of construction or manufacture, the specialized experts in construction who emerge during the later phases of the Renaissance undertake the conception and oversight of the whole of a project – but one of a specific kind. Historically, the demand for the task and its definition come first; the skills and the nomenclature follow.

The specialized military engineer is an instance. The revolutionary scale of sixteenth-century warfare brought to the fore a class of architect-engineers whose talents could confer security, victory and even prosperity. Yet none of the men who developed the trace italienne – great names like Francesco di Giorgio, the Sangallos, Sanmicheli, Michelangelo – was an exclusive specialist in fortification. They and most of their immediate successors among the contending powers of Europe were only episodically employed on matters military; all worked at other things as well, and often for several states. For example, the noble gates and bastions devised by Sanmicheli for Verona and other client cities of the Venetian Republic – not just in the Terrafirma but in Dalmatia, Crete and Cyprus – never monopolized his activity. Even the affluent, far-reaching Serenissima could not keep him permanently at such tasks.[11]

In the later sixteenth century exclusive military specialists start to emerge, like Francesco Paciotto who laid out the citadel of Turin, and Giulio Savorgnan who designed the new town of Palmanova for the Venetians (ill. 10).[12] Such men can well be defined as military engineers, meaning that their pre-eminent focus was the art of war. Not that they did not design buildings or have a spread of skills. Nor were they independent experts working in a strategic or political vacuum. Turin was designed in partnership with the warrior-prince Emanuele Filiberto, Palmanova with the oligarch Marcantonio Barbaro under Venice's formalized committee system, the Provveditori alle Fortezze. Among the reasons why specialists like Paciotto and Savorgnan came into being is that they were soldiers who could be commanded, and owed something like full-time responsibility to the state for which they worked.

Vauban is their successor, in a nation-state with a larger land-mass and a better-defined programme of development. Again, he is classified as an engineer because most things he did were military in nature, not because he could not or did not design decent buildings. In terms of skills he was a generalist. But in terms of focus he was a specialist, sticking largely to one job, which developed and ramified along with the institutional framework that supported him. Unstable though that framework sometimes proved, it offered miracles of continuity and authority compared to the employment any Italian engineer had enjoyed. How had that come about? The answer has to do with advances in the organization of armies.

Often in technical affairs, an enterprise or a state starts by hiring experts as and when they are wanted. If there is much to do, they have plenty of work. In due course management or government decides that by bringing the experts into permanent employment it can do better, learn from experience, get superior results, keep tighter control and perhaps even save money. That happened in France with soldiering in the later seventeenth century. The Spanish in Flanders and Cromwell in England had already deployed what are called standing armies. One definition of a standing army terms it 'the size of a force on which expenditure was accepted as "ordinary"'.[13] The higher you pushed that limit, the more full-time officers and experts you needed to control and direct your force's activities. During the palmiest years of his reign, Louis XIV was spending up to three-quarters of his total revenue on war – a statistic that puts Versailles and other initiatives of French cultural policy into perspective.[14] That entailed a sizeable body of experts at his permanent disposal. France's status as the first full-fledged nation-state depended upon them.

From the military revolution until Napoleon, sieges in Europe normally counted for more than battles. Construction became bound up with the performance of armies. A standing corps of engineers, as opposed to a single salaried officer or experts hired by the campaign or by the task, was the corollary. In France, the Corps du Génie (as it was to be known from Vauban's time onwards) is first dimly recognizable as a small body of native experts around 1600, under Henri IV. The numbers of regularly paid 'ingénieurs du roi' during the first half of the seventeenth century may never have exceeded twenty.[15] When Vauban's predecessor Clerville took command in 1659, they were still of that order. But by 1691 the figure had shot up to 275, divided into three ranks and two distinct departments – Guerre and Marine. At that date we also know that a total of 350 engineers had been recruited over the previous thirty years (in attack, the rate of attrition was grave).[16]

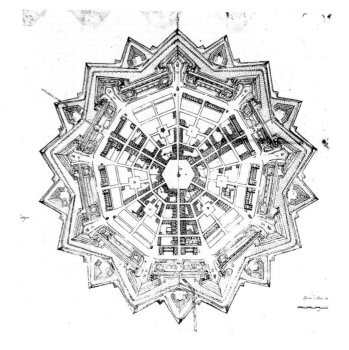

That marks a transformation. Even during the interminable Thirty Years War, campaigns had been too episodic and armies too haphazard for such coherence. The changes brought in by Louis XIV and his administrators were soon copied by other states. When Vauban's lesser-resourced Dutch rival, Menno van Coehoorn, took over his country's defensive strategy in 1697, the first thing he did was to keep on his seventy engineers (many of them French Huguenots) in time of peace.[17]

10. Ideal plan for Palmanova, c.1600, showing attempt to unite the new style of fortification with radial urban planning. Such designs were seldom more than very partly implemented.

THE ART QUESTION

Vauban then was an engineer in the first place by virtue of the job he was hired to perform for a regularly belligerent nation, and the resources that entailed. Naturally he needed technical aptitude along with concentration to get through his mountain of tasks. But his high place in engineering's pedigree is due to breadth of talent and managerial ability, not to an exclusive set of skills. That may help if in contemplating what he left behind we ask, as many have done, is it architecture – even more, is it art?

The constructions for which Vauban and his fellow-engineers were responsible varied along the scale of propriety from hurdles to protect saps and other temporary works connected with attack at one end, through defensive banks of mud and turf, to the design of gates, barracks, churches and complete towns at the other. In between come the battered masonry ramparts, proud and photogenic still today (ill. 11) and affording such a 'grande satisfaction esthétique'[18] – greater satisfaction for current tastes, maybe, than Louis XIV's palaces. Latterly, these structures have given Vauban something of an 'architectural' reputation.

Enthusiasts have struggled to pin down what the expressive power of Briançon or Montdauphin or Neuf-Brisach consists in.[19] Is it the way they fuse with the site, the constraining grip of their periphery, or just some vague quality of space and volume – the *pis aller* of modernist architectural criticism? As with the warehouses and factories of the early Industrial Revolution, applauded now, shunned when they were built, the pleasure we get from castles is retrospective. Like reading about a bygone murder, we relish something that never impinged upon our daily lives, never sheltered, threatened or hindered us, nor cost us money.

There is a larger dimension. Vauban's austere bulwarks inscribe themselves within a tradition to which architects of calibre back to Brunelleschi had contributed. Less

11. At many fortified towns like
Città Castellana, shown here, the
form and texture of the defunct walls
offer an aesthetic pleasure that can
hardly have been available at the time
of their construction.

picturesque than mediaeval castles, they make amends by dint of their order, regularity
and geometry – attributes often easier to read on paper than on site. Contemporaries certainly responded to the appeal of these grand constructions, calling them with whatever
impurity of motive magnificent, fine or beautiful. Louis XIV took as much enjoyment
from the cruel aesthetics of the siege as from anything he ever built, calling the fortifications of Lille 'very beautiful', Racine tells us.[20]

One authority on Renaissance fortifications, John Hale, has devoted a lecture to the
tension between art and engineering they embody. Sensing that the criteria of the times
he is addressing will not help him, Hale's conclusion starts out from a modern
perspective.

We no longer flinch from finding pleasure in works where function predominates. We
can take pleasure in grain stores and cooling towers, admire, indeed, shapes deter-

mined by the mindless flow of air in a wind tunnel. Almost automatically we apply the standards of art to the products of engineering. And we cannot be immune to the human significance of fortifications as an architectural commentary on changing conditions of security and terror. But interest can only be intense if it is accompanied by discrimination. There are fortresses whose lean energy can exhilarate; others are as dull as the ditchwater that surrounds them. Why is this? . . . It is a matter of the relationship of the building to its terrain, of the parts to the mass; the way in which space is sliced into by bastions, cradled in their flanks; gun ports and embrasures can be judged as fenestration can be; the appropriateness of a wall's height and impression of solidity to the shafts, ducts and countermine corridors within it can be sensed and appraised; gates can be valid expressions of the power inherent in the system as a whole or mere façadism. In a word, the question 'art or engineering' can be asked, and can be answered.[21]

Yet it might be asked, can we not have 'both/and' rather than 'either/or'? Why does the disjunction arise at all? That it does so is token of a theory whose origins reach back to the Renaissance and whose stultifying effect has yet to loosen its grip upon western culture.

In brief, it is agreed that before, say, 1400, design and building were looked upon as a continuum. Technical skills, military, civil or ecclesiastical, structural or mechanical, varied according to trade, task and temperament, but just like the titles 'architect' and 'engineer', they overlapped and could not be absolutely divided. After which point, first in Tuscany and Umbria, arose a new idea: that there was a generic set of skills christened 'disegno', scholarly and mathematical by nature, expressible through drawing and separable from execution. These skills entitled the possessor to a gentle status which the plain craftsman or mechanic could not claim. Alberti expounded the theory – largely in his book about painting. In respect of architecture, Serlio and Palladio strengthened it by omission as much as by argument, treating their subject chiefly as a province of design and ducking the developing topic of military engineering. Vasari confirmed things by his selection of material for his artists' lives, so paving the way for centuries of disegno-driven art history. That is why in older accounts of seventeenth-century French architecture, Vauban and his engineers merit scant mention, while the courtly architectes du roi are expatiated upon.

The appropriateness of the disegno theory to construction has been repeatedly challenged. A Platonic intellectualism, it enjoyed no basis in building practice or the building trades. Apart from the overlap between architects and engineers, on a wider scale mathematicians, artists, architects, engineers, technologists, builders and craftsmen were jumbled up throughout the ferment of the Renaissance. Brunelleschi, Francesco di Giorgio, Leonardo, the Sangallos, Sanmicheli, Peruzzi and Michelangelo, to mention only illustrious names, were all involved in hydraulic, military and mechanical projects – and their management (ill. 12). Their prowess as draughtsmen and inventors did not confine them to the court or the bottega. Disegno was an extra weapon in their armoury, not an exclusive one.[22]

Once prosperity and the consequent process of specialization among building experts had gone far enough, it was possible to present one aspect of disegno – the preparation of imaginative designs – as the distinctive skill of architects. Thereafter art, however defined, would come to

12. One of a series of unbuilt designs by Baldassare Peruzzi for a dam and sluice on the River Bruna, c.1530. Renaissance engineering projects presented in this fetching way appeal to modern tastes in a way that elaborate architecture often does not. It is easy to imagine this scheme as having been conceived for reinforced concrete.

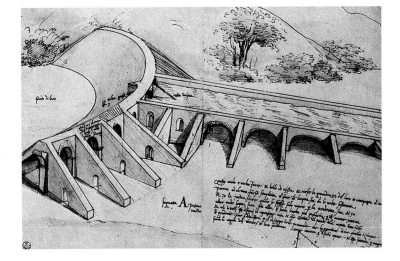

be seen by some who could make a living from it as the chief end of architectural activity. But for other classes of expert and indeed many architects, art has always been one of many aims – sometimes even a by-product, finishing skill or ornamental add-on, as in many crafts and trades.

The technical preoccupations of the early military engineers seldom precluded them from honouring the civic and expressive sides of their craft. There was little need, for instance, to face the whole of the trace italienne in proud and permanent masonry; shock-absorbing earth would often be as effective, if not better. Yet Italian and French engineers (not Dutch ones) saw the architectural expression of the enceinte as an integral part of their job.

In urban form too, art and technique were inextricable. From the Renaissance onwards, new towns often owed their 'trace' to military engineers. They were the experts who tended to be there first. In any exposed situation, defence came first, water supply second – both engineering skills. Within the town, some form of grid was the normal and natural choice of layout, hard though it could be to reconcile with a polygonal enceinte. But the temptations of disegno, with its intimation of a higher order, were never far away. A prime example is the radially planned city, over whose fetching geometries paper artists and armchair strategists enthused alike. In reality it seldom answered to site conditions or efficient internal planning. After a century or so of experiment, radial cities were largely set aside by practical town-planners, civil and military, Vauban included.[23] From such compromises between strategy, technique and the diagrammatics of disegno, the science of town-planning managed its crab-like progress.

By the mid seventeenth century, overall urban form began to be seen as the province of the engineer, not because he was a superior technician or lesser artist but because he was well placed to exercise the broad, strategic side of design. In a famous passage of the Discours de la méthode, Descartes likens the unity of a good book to that of a well-ordered building or well-laid-out town; the design of the imaginary building he ascribes to an architect, but that of the town to an engineer.[24] To this day, that frame of mind survives in the engineering-weighted world of French institutional planning. It implies a hierarchy of aesthetic order, not the presence of art in the planning of an architect and its absence in that of the engineer. On the whole, engineers have dominated modern city-planning. The most famous of all exercises in urban reconstruction, for instance, the so-called Haussmannization of Paris, was controlled by two engineers, Alphand and Belgrand, who have never had their full due.[25] Yet engineers have never quite stamped out the heresy that architects are the broader synthesists, and that the uplifting, the diagrammatic and the purely irrational should control the overall planning or replanning of towns. Hence, for instance, the hiring of art-architects today to regenerate down-at-heel communities.

There is a final reason to be wary of the division between art and technology that has been a legacy of Renaissance art-theory: it was hardly sanctioned by the example of the ancients, and certainly not by that of Rome, the fount of first authority. In Roman culture, where architecture ends and engineering begins is tricky to determine (the question is addressed in respect of bridges on pp. 282–6). The subjects are inextricable in Vitruvius. His muddle of a text was pored over not just to reach consensus about the architectural orders, decode his myths or reconstruct the buildings he describes, but also for his tantalizing hints about process, technique and machinery, many bearing on the arts of war. Julius Caesar's account of bridging the Rhine was studied as intently as the biblical description of Solomon's temple. Then there was Frontinus, neither architect nor engineer but 'curator aquarum' under Trajan – the official in charge of water, that essential medium of technology, welfare and civilization. Frontinus's paean to the aqueducts within his care as preferable to the 'useless' pyramids or the tourist attractions of Greece was to find an echo when Vauban compared the dome of the Invalides to the fifth wheel on a wagon.[26]

13. The Pont du Gard.

Nor were texts all. In the developed world, we enjoy an infrastructure of a scale and quality beyond earlier fantasies. We forget how exemplary and how looming was the physical legacy of Rome to the Renaissance princes and the founders of modern European nationhood. What these rulers sought to be and do, the Roman emperors had once achieved. With its genius for method, its extent of sovereignty, its technical know-how and the military cast of its institutions, Rome was a practical model for those that aimed to advance their city or country through a combination of aggression and construction. As late as the 1720s, some soldiers were still studying ancient military tactics for practical tips;[27] a century later, Rondelet thought the Romans still could not be bettered in building construction. It was not just a matter of prestige, scholarship, style or image, but of sheer technical and administrative catching-up. There were exceptions. Since the Romans lacked gunpowder, their city-defences never had to respond to artillery with the elaboration of the trace italienne. Otherwise, until the Enlightenment and the Industrial Revolution were set fair, Roman remains still reproached the inferiority of modern constructors.

One example. Throughout Louis XIV's reign, the city of Nîmes stagnated for want of water. Nearby lay the long-broken Pont du Gard, costly and vertiginous testimonial to the aqueduct system that had been so persuasive a strand in Roman rule (ill. 13). In emulation of that monument, the King in the 1680s pressed Vauban into building the multi-tiered Aqueduc de Maintenon in order to supplement the famous Machine de Marly on the Seine and bring water all the way from the Eure to Versailles. But the project could not be afforded; many of those employed to build it fell sick and died. It was left unfinished, with a single row of arches. Not until sixty years later did one of Vauban's successors, the Génie engineer Jean-Philippe Mareschal, help revive the textile industry of Nîmes by renewing its Roman water supply with a town-planning scheme that combined hydraulic, archaeological and architectural skills.[28] The Pont du Gard remained

14. Gateway to fortified town designed by Comte Thomassin, 1728, and illustrated in Bélidor, *La science des ingénieurs*, 1729, Book I, Plate 16.

unrepaired. In due course, an insignificant road was built alongside it.

In summary, we should not impose upon the legacy of Vauban or his predecessors categories that belie both our instinctive response and the framework within which the science and art of fortification were pursued between 1500 and 1750. That said, different elements in the Renaissance tradition of construction always enjoyed different weighting, along a continuum from the necessary and the useful to the symbolic.

For a military historian like Anne Blanchard, the doyenne of Vauban scholarship, the 'architecture' in fortifications means just the incidental elements and therefore has minor importance.

Architecture is always an art. Individual elements or practices can be changed or revived as required or according to shifts in fashionable taste. Fortification is different, because it relies upon strict rules aimed at a single and particular end, that of defending a fortified enclosure. If its forms often have aesthetic appeal, that is something extra and not a conscious result of intention, apart from the gates. In the end only the rigorous calculations of the engineer matter.[29]

An older authority, James Fergusson, entertained a more enlarged view of how architecture and fortification related. It might be argued, he quipped,

that there is no cosmoclast like a cannon-ball, and it is absurd to ornament what is sure to be destroyed. This is, however, hardly a fair view of the case: of one hundred bastions that are built, not more than one on an average is ever fired at, and it is a pity that the remaining ninety-nine should disfigure the earth during the whole period of their existence.[30]

In the citadel of Lille or the new town of Neuf-Brisach, the scale and type of construction, from earthworks and walls through arsenals, barracks and hospitals to churches and palaces, switch constantly from the functional to the symbolic, from might to the image of might. They are a totality. We cannot stop at a given point and say: here begins architecture, and there engineering. What we can say is that different notes have been struck in different places, and skills disposed accordingly.

Take the design of gates for fortifications. Here, so long as funds were forthcoming, symbolism and display might make their mark. It depended on the city and gate in question. How much 'architecture' was lavished on a gate depended on the message the authorities wished to convey and therefore on state policy. In the second quarter of the sixteenth century, an expansive Venice favoured magnificent gates for the cities it controlled. In a debate before Charles V in 1535 over the relationship between city-gates and defence, the Ferrarese said that gates ought just to be functional, while the Venetians argued that handsome ones enhanced a city's image and dignity, and attracted visitors. That is why Sanmicheli could fashion gates of such brio, at Verona and elsewhere. Then came retrenchment and a shift in policy; by 1576 Savorgnan was condemning 'pompa et ornamenti impertinenti' and grumbling that a couple of good bastions could have been built with all the money frittered away on the gates of Verona and Zara.[31]

Vauban faced similar conflicts of opinion. A famous case concerned the gates at Strasbourg, which France had memorably conquered in 1681. When Louvois, always on the look-out for economy, ventured that the new gates were extravagant, he earned a tart reply:

They cannot be any narrower or shallower, because they have to be closed securely and people need to be able to pass . . . So the changes for which you ask will have to come down in the end to a mere few triglyphs, metopes and dentils, and also the royal arms and ciphers which are the only ornaments on the gates . . . If you really want to get rid of the ornaments on the gates, just let Tarade [the site engineer] know. It'll save about 400–500 écus, and the result, you can be quite sure, is that they'll be simple enough, and pretty commonplace . . . Everyone in Germany passes through this way and the Germans, who always notice things and are generally good connoisseurs, can very well judge the King's magnificence and the state of the place from the elegance of its gates.[32]

Vauban got his way. As Colbert had said, 'this reign is not about little things, and nothing can be conceived too grandly, even if there must be some proportion.'[33]

Vauban had a system for fortifications, or rather a sequence of systems which his assistants and followers reduced to formulae. That was natural; he had no time to start everything from scratch. So also with architecture: like any designer involved in a sequence or programme of buildings, he repeated and refined what worked well. Even large buildings might be replicated. In 1679 Colbert relayed the King's instructions for Vauban to make a standard barrack design, which was 'engraved and generally distributed'.[34] But though many aspects of the defensive works, for instance the profiles of revetments, were standardized, they were always adapted to sites, materials, budgets and the effect intended. Likewise, gates were racked up and down the scale of grandeur and ornament, as occasion or economy suggested. Most were designed by Vauban's own engineers; some were farmed out to local architects; once in a while the architectes du roi were brought in to buff them up.[35]

That range of choices is echoed in Bélidor's *La science des ingénieurs* (1729), which includes a posthumous record of Vauban's methods for the next generation. Gates used to be more elaborate before the days of detached defence-works, says Bélidor; and he then goes on to illustrate an ideal design for gates (ill. 14),

in order to show that handsome ornament is not incompatible with fortifications; they have been approved by architects of high skill. Perhaps they will be found too rich to be employed in villes de guerre; but let me add that expense has never been something to deter our monarchs, for at Lille, at Maubeuge and in several other places are to be seen gates at least as magnificent as these ones.[36]

There were horses for courses, then, in the architecture of fortification. To change the metaphor, the solos played on various instruments could be embellished to suit the tune in question. Indeed the building of a Vauban town or citadel is like an orchestral composition; interlocking voices are called for at different points in the score. This tallies with what we know of the talents and temperaments within Vauban's band from which the maestro wrung his harmony, and with an era in which sharp boundaries had still to be drawn between the disciplines of construction. 'The Génie is a profession beyond all our capacities,' he remarked; 'it embraces too much for any single man to possess to a sovereign degree of perfection.'[37] It is time to see how it operated.

ARCHIMEDEANS AT WORK

Vauban knew how to lead and to delegate. But he was also collaborative and pedagogic by instinct, and a fierce protector of his 'bande d'Archimède', as he fondly called them. He fostered them, championed them and often badgered the authorities for their salaries to be advanced. In matters of design, though, he was always keen to do the maximum himself. At Dunkirk, an early and favourite fortification of much complexity, the 'lines' were entirely of his devising: 'I traced them all out myself, and there is nothing that I have not measured at least twice. In fact my assistant has not so much as touched the

15. Map of Lille engraved by Isaac Basire for *History of England* by Paul de Rapin-Thoyras, showing the town's fortifications at the time of the War of the Spanish Succession (1700–1713).

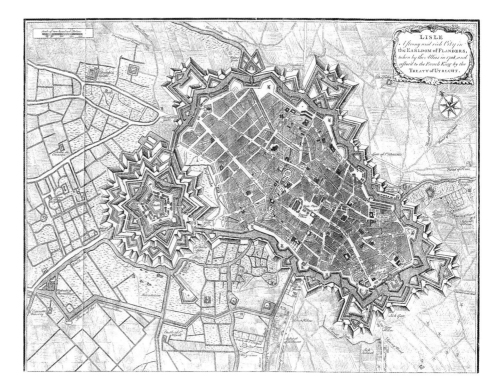

measuring-tape'.[38] At Lille, where there was both collaboration and contention, he asked Louvois to be allowed to get on with his own job 'while the other gentlemen get on with theirs'.[39] Though more was delegated in later years, he went on making detailed plans himself and expected them to be adhered to. His subordinates were to 'lay out the plan settled upon by the Director General', affirms a handbook of 1691.[40]

Procedures and divisions of responsibility were, for their day, lucid. Vauban would arrive at a given site on one of his tours, along with the 'directeur des fortifications de la province', in other words the local head of operations under Louvois, Colbert or their successor, Le Peletier.[41] On site, copious notes would be exchanged with the local ingénieurs du roi. A 'projet des ouvrages' would be concocted, drawn out with different colours for existing and proposed work, and copied: one copy was kept on site, another held by the provincial director and a third sent to headquarters in Paris. Here start the consistency, clarity and clean draughtsmanship of French engineering tradition.

In due course models too began to be made – the origin of the spectacular 'plans-reliefs' to be admired today in the Invalides (ill. 8). If design changes were made locally, normal procedure entailed referring them back to the centre. Contracts were made following tenders based on indicative prices set by the provincial director, engineers having been forbidden from undertaking contracts themselves since 1604. It was all quite different from the sixteenth-century Venetian forts of the Terrafirma, where funds and forced labour had in part to be found locally, compounding the grievances raised by the disruptive art of fortification. Vauban kept his eye on the system, cautioned against accepting over-low tenders, and campaigned for the workforce (which usually included army units) to have proper pay, tools, hours (no Sunday working) and breaks, as well as some provisions for safety. Here were hints of the well-capitalized and managed building procedures of the modern state.

Who were the people that made their mark in Vauban's Génie? Here Anne Blanchard's researches are a gold mine. She has traced the origins of 105 out of the 164 known engineers serving in the Département de la Guerre in 1691.[42] Of these, 64 per cent were of noble origin, reflecting patterns of recruitment for the officer class that persisted throughout the Ancien Régime, and 36 per cent of bourgeois birth. (In the smaller Marine department, the proportions of bourgeois to nobles were inverted, since Colbert took a more meritocratic approach to recruitment than Louvois.) The fathers of thirteen of the war department's engineers are known to have been army officers, while nine are described as architects or entrepreneurs (building craftsmen who undertook contracts) and five as engineers. It is in these last two classes that advancement and leadership are telling.

Jacques Tarade, whom we have met at Strasbourg, is a case in point.[43] The son and grandson of Parisian architect-masons, he had enjoyed civic and royal patronage, having carried out paving for the City of Paris and a carpentry contract for the menagerie at Versailles. From resident engineer at Strasbourg when the clash over the gates took place, he rose to supervising engineer for Alsace and the eastern front in 1696, continuing in that region until 1712. Tarade had an architect brother involved in the fortifications at Dunkirk, and some architectural competence of his own (he published a book about St Peter's, Rome after his retirement). Yet at Neuf-Brisach he was supplemented by J.-H. Mansart, architecte du roi, for the enrichment of the gates – though not for the town's church or barracks. When the architects of the Bâtiments du Roi had to be called in and when they could be circumvented seems to have been fairly arbitrary.

It would be wrong to imagine the Génie during Vauban's lifetime as a self-contained, centralized team of experts managing each detail of every project. Authorship and superintendence of the greater works had to be negotiated between national and local interests, sundry skills and trades, and scrapping, overlapping arms of the state bureaucracy. To get a taste, let us watch his men at work in two cities, Lille and Toulon.[44]

Lille had been captured during the first of Louis' northern encroachments in 1667.[45] The annexing and fortifying of this Flemish border town prompted its growth and modern prosperity. Even before that it was a well-organized commune, with a municipal building service employing a 'maître des oeuvres' to approve or design the façades of all houses.[46] The first step was the citadel, 'une sorte de ville devant la ville'[47] – a common device since Paciotto's innovations at Turin, presenting attackers with two places to reduce instead of one. Lille's (ills. 15, 16) was exceptional because it was so massive and built from scratch, not in expansion of an existing fort. Clerville and Vauban presented rival designs; the latter's, the first major work of his career, were accepted. Construction took place in 1669–74 under the superintendence of the Chevalier de Montgivrault for the army, and Simon Vollant for the locals. The city's engineer, Quillien, also played a role. Michel Nicolas de Lalonde, mathematically minded author of the recent Elémens de la fortification, was Vauban's on-site deputy.[48] The multiplicity of authors should not surprise; the ballistics, hydraulics and logistics of such undertakings needed thinking on many fronts. The Lille citadel was as big a city-planning operation as a new town, and urgent.

In the construction phase Vollant was the key figure. He came from a dynasty of Lille architect-masons, knew how to make best use of the region's men and materials, and had artistic pretensions. Louvois once called Vollant 'un fou et un visionnaire' – an architect at heart, therefore.[49] At any rate, he gave satisfaction enough to be approved as engineer and architect to the Sun King's armies, become treasurer of Lille, and design fortifications also for nearby Menin (where Mansart was brought in to 'correct' the surrounds to the gates).[50] In tandem with the citadel, Lille acquired new walls and a whole new

16. Lille, a corner of the citadel.

17. Double-page spread from the *Atlas de Louis XIV*, showing the town and naval harbour of Toulon with its fortification and protected approaches, according to Vauban's accepted scheme. The new arsenal is above the empty harbour.

18. Cross-section and elevation of the fort at the mouth of the main entrance to Toulon's harbour, 1707, drawn by Claude Masse for his survey of French ports and fortifications. The artist has cautioned: 'The drawings are not to be entirely relied on as they partly follow ideas of M. Verier, ingénieur du Roy.'

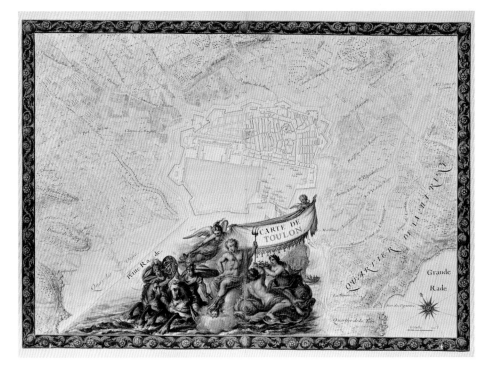

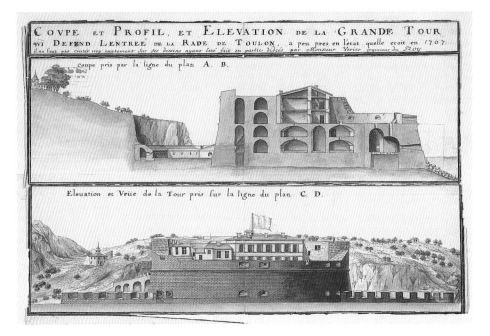

quarter for the town. Civil construction was forbidden until the citadel was advanced. Appointed governor of Lille, Vauban therefore had a house there and often visited, noting in 1674 that construction in the town was in full swing.[51] That year, aesthetic control of street frontages was reaffirmed in a fresh decree. Here Vollant doubtless had free rein; the exuberant Porte de Paris (1685–92), in a loyal, un-Flemish idiom, is known to be his. Who designed the manifold buildings of the citadel, some like the chapel and arsenal of real merit, is less certain.

The endeavours of Vauban, his rivals and his collaborators at Toulon (ills. 17, 18) have been chronicled by Jean Peter.[52] Such vast projects could never be rapid or trouble-free, not least on home territory where vested interests abounded. Proudest of France's naval ports, Toulon reached perhaps 20–25,000 in size during Louis XIV's reign. But its dockyard was outmoded and too small for Colbert's modern Mediterranean fleet, nucleus of the Département de la Marine. Enlarging and enclosing the dockyard lay at the heart of the many schemes put forward from 1665 onwards. Fresh defences were also needed, along with some recasting of the town.

The Marine had its own traditions and technicians, including an engineering service charged with harbour defences and construction at home and abroad. Even after 1678, when Vauban took over coastal fortifications on home territory, the two branches of the Génie were distinct. The bourgeois and technical strand among Colbert's marine engineers, different from the siege-specialists who dominated the inland service, has been mentioned. François Blondel, the leading personality, had been a royal engineer since the 1630s, and then after an interval came back to the Marine side under Colbert.[53] Soldier, diplomat, traveller, mathematician and author as well as architect-engineer, Blondel is best known as the first professor and director of Colbert's Académie d'Architecture. There was friction between him and Louvois, and he is unlikely to have got on well with Vauban. Blondel's built works as an engineer were to be dismissed by the army side as theoretical and dilettantish – 'fortification curieuse'.[54] Under Colbert, whose son (and successor in charge of the Marine) he tutored, Blondel had played an early role in planning the dockyard at Rochefort, precursor to Toulon, and designed the great ropeworks there.[55]

The preparatory phase at Toulon was never confined to the Marine engineers. Some of the early schemes originated with Clerville, others with local officials, high and low, from Nicolas Arnoul, intendant of the galley fleet at Marseilles, to François Gombert, a dockyard clerk and draughtsman who made several plans and rose to become a senior engineer in the reconstruction. Eye-catching submissions also came in from Pierre Puget, master-architect and sculptor in charge of the renowned atelier in the dockyard that made ship's figureheads and other trappings of naval magnificence. Colbert was always susceptible to architecture, as Blondel's designs for Rochefort betrayed. But Puget, he knew, was a 'mégalomane', ignorant of defences and infrastructure. Modelling the new dockyard upon 'an antique palace on the Aegean coast' was a non-starter, so Puget's final utopian project foundered with the rest.[56]

Soon after Vauban inherited the Marine defences in 1678, Colbert turned Toulon over to him. Having not designed a dockyard before, he found the complexity of brief and topography alike heavy going. His plan borrowed the best features of its predecessors. The key was not the enclosing defences, but the drawing-together of multifarious technical facilities for the fleet within a layout which put efficiency before elegance.[57] It took him three exhausting tries, the last in 1682, before everyone was placated. Having spent so much effort, he was loath to see his plan altered. That was inevitable, given the scale of things. Though infrastructural work started in 1680, much remained to be done when Vauban died in 1706.

How and through whom were his designs for Toulon delegated and executed? Vauban's key lieutenant, his ingénieur général and directeur général des travaux, was Antoine Niquet, appointed in 1680. A Parisian who had started life as an instrument-maker and technician at the Académie des Sciences, Niquet owed that first post and his transfer into the Marine engineers to Colbert, who had a penchant for individualists of mathematical or architectural bent. Opinionated and truculent, he hated consulting and could be relied upon to quarrel with colleagues. Colbert knew how to handle him. Early on, Niquet earned a stinging rebuke for 'your ridiculous vanity, which has long rendered you intolerable everywhere you have worked',[58] and had to be threatened with imprisonment. More than once he changed a design by Vauban or substituted a new one of his own without reporting it, to his master's displeasure.

Not that Niquet lacked skills, status or tenacity (he was still employed on the dockyard in 1708). Despite bruising relations, Vauban believed in him. Beyond Toulon, he enjoyed a remit throughout Languedoc and Provence, and also intervened in the Atlantic ports. It was to Niquet that Vauban delegated the improvements (mainly, better aqueducts) made in 1686–93 to the Canal du Midi.[59] That visionary undertaking, linking the Mediterranean and Atlantic, had been financed as to one-third by the state, two-thirds by private enterprise, and completed just five years before. The shortcomings of the original canal were a factor in drawing strategic civil infrastructure into the bosom of the French state. It was the only major civil project that Vauban took under his wing besides the Aqueduc de Maintenon. Niquet waxed rich from his Canal du Midi connection, becoming 'conseiller du roi', tax collector, and mayor of Narbonne.

A team of engineers and inspectors answered to Niquet at Toulon, totalling some sixteen at the start of works and dwindling to about four in the new century. The great site was split between them. Some had been sent to Toulon; others like Gombert were local employees of the Marine. No doubt their skills varied. Longest-serving, and Niquet's eventual deputy, was Jean de Chaumont, an architect-draughtsman linked with the Mansart dynasty. Some fifteen separate building projects (some, like the warehousing sheds, in multiple units) were carried out in the dockyard in 1686–95. Much of the official correspondence over that time concerned alterations to their design, so the strength of the architectural team should not surprise. The contractors were quite separate. Some were local entrepreneurs of standing, not all connected with the building trades. Because of the financial scale of operations, one national figure came in: André Boyer, styled an 'architect' in Paris and involved in royal building works. But though Boyer contracted to build many of the major buildings, including the surviving ropeworks, he sublet most of his 'take' to local masons. In construction, who really builds something can be as elusive as who really designs it.

Three Administrations and Their Purposes

Various attempts were made to set Vauban's 'systems' of fortification down on paper.[60] But the broad constructive achievement of Vauban's Génie was better codified for the next generation in two famous tomes, *La science des ingénieurs* (1729) and *Architecture hydraulique* (1737–53) by Bernard Forêt de Bélidor – among the most thumbed-through and cribbed-from of source-books for architecture and engineering.[61] A mathematician-teacher of pragmatic bent, Bélidor was never himself a member of the Génie, but he had a military background and spent spells on active service. In between, he had worked on the mapping surveys to establish a national meridian line directed from the Paris Observatory. He then taught at the oldest of the technical colleges run by the French military, the school of artillery at La Fère, set up well before the Génie had a school of its own. Bélidor is supposed to have based his books on courses he taught there.

Bélidor's textbooks face two ways. They set out up-to-date mathematical theory in relation to construction, and so are esteemed by historians of scientific engineering. But they also cover practical building of all kinds, running the gamut of infrastructure and architecture over and above the arts of war. In the preface to *La science des ingénieurs*, Bélidor says that Vauban and his successors never had time to lay out what they had done, so he is remedying the omission. As the title *Architecture hydraulique* expresses, he assumes the unity of architecture and engineering. He lays much stress on process, printing whole specifications and estimates for Neuf-Brisach. But he also insists on due use of the orders, picking Vignola as the clearest authority. Competence is the aim, not originality; Bélidor offers techniques and plans for hard-pressed, far-flung engineers who do not aspire to be architects 'du premier ordre'.[62] Barracks and infirmaries from his pages were indeed to turn up little amended as far afield as the Americas, Russia and British India.

As to his scope, Bélidor explains that though the military have made the running, civil works are now coming to the fore. France was waxing richer, less martial, more enlightened. Who now would preside over improving its infrastructure and towns, on which the prosperity of trade and manufactures depended? That was the question for future constructors. Given all that had been put in place in the previous reign, it was bound to be some arm of the State, but which? In a polity with overlapping layers and fractious interests, working that out would be a struggle. The upshot was to be a classic instance of technology transfer – the deflection of military techniques, institutions and management to social and economic ends.

The reforms of state bureaucracy under Henri IV around 1600 had sown the seeds of three administrations for building. Over time these embryos grew into the Génie, which promoted France's military defence and expansion; the Ponts et Chaussées, destined eventually to consolidate and develop the core of national territory; and the Bâtiments du Roi, devoted to the monarch's personal buildings and whatever other jobs he might release them for. For years these organs of state were fluid; their titles, personnel, scope and operations intermingled. By the end of the seventeenth century, under the impress of Colbert and Vauban, they were on course to becoming distinct entities. The Génie had been defined as a corps, with the implications of military discipline attached to that word, though only half its engineers were officers, and uniforms were not yet worn.[63] The Ponts et Chaussées, small as yet, was groping its way towards becoming a corps of engineers on the model of the Génie. Also few in number but prestigious were the architects of the Bâtiments du Roi, quite a different body. It alone was to escape the disciplinary straitjacket of a corps.

19. Jean-Baptiste Colbert (1619–83), portrait by Robert Nanteuil, 1676.

Colbert bought the post of intendant to the Bâtiments du Roi at the start of 1664.[64] Since he was already the finance minister, he could support it generously. As he waxed in power, he forged the service into an instrument aimed at furnishing works of all kinds, enduring or ephemeral, which would house and flatter royal magnificence. 'In lieu of dazzling actions in war, nothing indicates better the greatness and spirit of princes than buildings', Colbert admonished the bellicose young king.[65] As time went on, he wanted more. The feats of French arms, managed from the centre by Louvois and his circle, suggested a model for centralized civil government as well, dedicated to stimulating trade and manufactures until they could dominate the markets of the known world. Such was the dream of this greatest of mercantilists.[66] If Vauban and his engineers could execute military construction to orders, when and where it was needed, might not the whole country benefit from a parallel discipline in civil construction, based on an objective body of architectural knowledge?

Colbert (ill. 19) was devoted to the arts. But his early experiences over the East Wing of the Louvre had taught him about the pride and jealousy of architects. Their output was too discontinuous and too prone to Parisian bickering to be steady. He therefore intended that the Académie d'Architecture, founded in 1671 under the original title of the Académie des Architectes du Roi, should strengthen the Bâtiments by harnessing training and debate with the creation of an exemplary royal architecture, applicable to national tasks. That Colbert dreamt of imposing something like the discipline of a corps on the Académie is suggested by the appointment of François Blondel as its first director and professor. As a veteran marine engineer and an expert on mathematics and artillery, Blondel was expected to establish a fixed body of knowledge for the disputatious topic of architecture, set out royal doctrine and goals for the subject and exercise control over designs. Technique, including 'l'architecture militaire des fortifications', was to the fore in Blondel's lectures.

That dictatorial vision foundered. Colbert and Blondel both died in the 1680s; Louvois, who took over the Bâtiments from Colbert, followed in 1691. With the leading centralizers departed, the economy declined. In hindsight, the impossibility of centralizing architecture is obvious. The diversity of patronage, tasks and tastes, and the stop-go nature of civil construction, always an easy way to regulate an economy, told against it. By 1700 a reaction had set in against focussing architecture upon national endeavours. The last intendant of the Bâtiments du Roi before its organization changed in 1708 was an architect, J.-H. Mansart, eminent in his art, an accomplished technician and diplomat, but no exponent of outward-looking state policy.

Training followed suit. It was the Académie d'Architecture, not the Bâtiments administration itself, which became identified with the small seminary that grew up around the lectures after 1700. Little by little, the school's links with practical tasks and building programmes weakened. In a power-vacuum, its ideals became wedded to the distractions and abstractions of design competitions which, imitated from the fine arts, became a formal part of the educational process in 1720. Thereafter the highest echelon of French architecture eloped in a love-affair with the fine-artists. 'Disegno' in the narrow interpretation had scored its first institutional victory, causing long-term damage to the unity and purposes of the building arts. The educational upshot of that divorce is explored in Chapter 6 (pp. 431–8).

The practical skills of the Bâtiments du Roi were not at first much touched by this dereliction. The service went on drawing in the best architects in France, giving its pupils (often family members – hence dynasties like the Mansarts and the Gabriels) a superb apprenticeship across the range of construction skills. As royal servants in theory, the architectes du roi could be lent out for private or public commissions, subject to favour and politics. One feature of Colbert's vision that survived him was the intendancy system, which he had used to impose his will upon France's provinces and towns.[67] In his commanding role as finance minister, he deployed officials loyal to the centralizing State (often lawyers from Paris) to control the ways in which the Ancien Régime's cumbrous 'generalities' and semi-autonomous 'pays d'état' raised and spent money. When improvement schemes for a town were in the offing, these regional intendants wielded a powerful hand; and where magnificence and prestige were to the point, they were not slow to call in the Bâtiments. The classic example is the campaign to endow each of the major French towns with its 'place royale' – a programme instigated by Louvois in the 1680s, though it took a century to complete.[68] The services of the architectes du roi were intermittently imposed or solicited for the places royales.

Nevertheless the Bâtiments du Roi had no binding or recurrent involvement with national development. Better placed to build up the nation, so it looked still at the time of Vauban's death in 1706, were the military engineers. For a decade thereafter, the Génie remained the one fully constituted corps – the sole building organization with the network and resources to operate throughout France, indeed to grasp it as a whole. Such was the Génie's authority that its construction experts were in demand. Vauban himself, we have seen, had occasionally been siphoned off into civil tasks. The pattern survived him. After a fire ravaged the centre of Rennes in 1720, for instance, the Governor of Brittany's reaction was that he did not trust the local architects: would the government appoint an engineer for the reconstruction? Isaac Robelin, director of fortifications at Brest, was duly dispatched to Rennes, coming up with a master plan for a stolid grid, along with rules for façades and construction standards. He had got some way when he fell out with the townsfolk and resigned in 1724, to be replaced by Jacques Gabriel, architecte du roi and chief engineer of the recently constituted Corps des Ponts et Chaussées.[69]

The Ponts et Chaussées was the construction service of the future. But Génie engineers continued to be called in to embellish French towns, at least in the pays d'état. Jean-Philippe Mareschal, the military engineer who helped extend and revive Nîmes in the

20. The Jardins de la Fontaine at Nîmes owe their design and configuration to the military engineer Jean-Philippe Mareschal, as part of his plans for reviving the city's Roman water system in the 1740s.

1740s with a canal, waterworks and public garden (ill. 20), has already been mentioned. Earlier in his career Mareschal had been posted at Belfort. There he enriched that stronghold's 'équipement urbain' with an arsenal and a new administrator's house and rebuilt one of the main churches. At Nîmes, not a fortified town, he was imposed upon a disgruntled citizenry by the regional intendant, after the ingénieur de la province had designed a modest scheme of improvement, then died. So lavish was Mareschal's revised plan that the town fought tooth and nail against paying for it, in one of the tiffs between municipalities and the national administration that litter the reign of Louis XV. Military urbanism did not always imply solidity without splendour.[70]

It is time to come to the Ponts et Chaussées.[71] The phrase is first found in Colbert's correspondence, and once again it was Colbert who articulated the service, though he did not invent it. In 1662–3 two good architects, François Le Vau and Libéral Bruand, were appointed to posts in this embryonic engineering service, perhaps in hopes of ommissions to follow.[72] Then in 1665 Colbert extended his remit to communications, as part of the national programme of mapping and planning sponsored under his direction. The King, he lectured the regional intendants, wanted his people to work 'unremittingly at re-establishing all the public roads and making all the rivers of his kingdom navigable'. And again: 'commercial advantage and the public weal depend chiefly upon the viability of the roads'.[73] The upshot was less enlightened. Under Louis XIV warmongering came first, buildings in his own honour second, and infrastructure last, despite this advocacy. Those public works that took place – the Canal du Midi is the most famous – were unsystematically procured. A supervisory architect was appointed here, an engineer called in there. 'Their titles mattered little', writes Jean Petot. 'Engineers had to perfect their skills by practical experience, and they were judged by Colbert on their results.'[74] On the rare prestige project, the Bâtiments du Roi intervened.[75]

Identity and coherence had to await the transfer of the brief for roads and bridges to an 'intendant des finances' in the twilight of the old reign, in 1712. Reforms promptly followed in 1713 and 1716, decreed in reaction to the decayed state of internal communications, roads especially. They set up a clearer territorial division of responsibilities, and confirmed the Ponts et Chaussées as a formal corps on the model of the Génie. Even then, the remit of the new corps encompassed neither the regional pays d'état that made up so much of France under the Ancien Régime, nor border areas nor municipalities. Under the 1716 arrangement, the technical staff numbered under thirty; their salaries were low. At first the priority for the small budget was improving roads in the region around Paris. But in time the service would grow, snatching glory from the corps on which it had been modelled.

The infant Ponts et Chaussées differed from the Génie in one telling respect. The top two posts, those of 'inspecteur général' and 'premier ingénieur', were both given the official title 'ingénieur-architecte' – seemingly the first use of such a term.[76] Yet the very need for it is once again a reminder that the roles of architect and engineer continued to be interchangeable, specially so on the civil side of construction. The first premier ingénieur, Jacques Gabriel, was certainly an architect by name and nature: indeed an architecte du roi linked with the Bâtiments administration. Putting an architect at the head of the corps persisted beyond Gabriel. Germain Boffrand, who took his place in 1742, was also an architect, though one with interests in machinery.[77] Even the great Perronet who took over in about 1750 had trained with an architect.

When Gabriel took on the Ponts et Chaussées, he was already awash with private and public commissions. Since national finances still did not permit the corps to take on much new work, his duties were moderate. But his appointment confirms that a cultural dimension was anticipated for its labours. Gabriel did not neglect the road maintenance and improvement that were the nitty-gritty of the Ponts et Chaussées work, and made frequent tours of inspection. But he also took technical and cultural issues arising from the corps' early bridge-building activities to the Académie d'Architecture for its views. Even before Perronet, the Ponts et Chaussées profited from interacting with the Paris-based intelligentsia, as the Génie never had.

The exalted reputation of the Corps des Ponts et Chaussées has its roots in the 1740s, when Daniel Trudaine as its intendant and Jean-Rodolphe Perronet as head of its drawing office, later its first engineer, took the expanding service by the scruff of the neck and converted it into the classic instrument of enlightened construction: a body committed to furthering national progress through building programmes, and to training architect-engineers in breadth and depth as the passport to that end. Behind that leap forward lay Trudaine's ability to deliver sustained funding for public works, and so to set about Colbert's dream of a national infrastructure. With hiccups, the Ponts et Chaussées enjoyed liberal allocations of money under Trudaine's successors up to the Revolution. If the sums never equalled those Louis XIV had lavished on fortifications, the corps' share of public money spent on construction between 1750 and 1789 compared to advantage with what an envious Génie could command. The jibe that the Ponts et Chaussées were the 'ingénieurs de la Finance' gets to the heart of the matter.[78]

Later chapters of this book look at what that development implied for architects and engineers from two angles: one through the bridges the Ponts et Chaussées built (pp. 295–305), the other through the training it offered (pp. 438–40). This chapter concentrates on military constructors and their progeny, so all that must wait. But to put things in perspective, something should be said here about what the Ponts et Chaussées did and did not do, how it related to rival state enterprises, and by what means it managed to present itself by the time of the Revolution as not far short of a national construction service.

Roads and bridges alone were the corps' original remit. But they never made up the whole of communications even in pre-industrial times. Buildings proper were not allotted to the corps, nor could they operate, it has been said, in the pays d'état or within towns. Yet as with the Bâtiments du Roi, if money and technical help were forthcoming from the centre in exchange for an acknowledgement of royal bounty, there could

always be flexibility. Here Gabriel's double status as architecte du roi and premier ingénieur were a boon. In his consuming engagement for the corps, the great Loire bridge outside the town of Blois (1717–26) (ill. 21), his status as a royal architect helped him win a demarcation dispute with an old-established local river service, the Turcies et Levées de la Loire.

In due course that service was taken into the Ponts et Chaussées, as part of a movement from about 1760 to bring rivers, canals and ports within the French generalities under the corps. Beyond its direct control there still remained the pays d'état like Brittany, Burgundy and Languedoc, where many of the major infrastructural projects were planned. Here, before the Revolution swept them away, the states built up regional equivalents to the Ponts et Chaussées.[79] Some, like the Burgundy service, whose biggest task was the Canal de Bourgogne, came to be more or less dependencies, heavily staffed by engineers who had been taught by Perronet in the infant Ecole des Ponts et Chaussées. In the pre-revolutionary years there were enough salaried state engineers to permit some movement between the services. It became not unusual for engineers to transfer from the Génie to the livelier Ponts et Chaussées, and common for them to shift from a junior post in the national service to a higher post in a pays d'état.

The Languedoc directorate of Travaux Publics was the proudest of these regional services.[80] Some of its technicians, like Henri Pitot, had started out in the pre-Perronet Ponts et Chaussées, others like François and Bertrand Garipuy of Toulouse were high-class local architect-engineers. In the Garipuys' bridges of the 1770s at Mirepoix and Gignac or their graceful épanchoirs (feeder locks) with tumbling steps along the Canal de la Robine de Narbonne (ill. 22), the equation between architecture, engineering and hydraulics is as serene and complete as anything coming out of Paris during those halcyon decades for masonry construction.

During the final, tax-conscious years of the Ancien Régime, Languedoc in fact had a reputation for extravagance in public works. When Arthur Young made his famous tours of France in 1787–9, nowhere was he more astounded by the magnificence of main roads and bridges. But he injected a note of English parsimony into his praise.

In this journey through Languedoc, I have passed an incredible number of splendid bridges, and many superb causeways. But this only proves the absurdity and

22. Epanchoir de Gailhousty on the Canal de la Robine de Narbonne, showing bridge with flight of steps, lock, and sluice to right. Designed by the architect-engineers François and Bertrand Garipuy, working for the Travaux Publics of the Languedoc, c.1780.

oppression of the government. Bridges that cost £70,000 or £80,000 and immense causeways to connect towns, that have no better inns than such as I have described, appear to be gross absurdities. They cannot be made for the mere use of the inhabitants, because one-fourth of the expense would answer the purpose of real utility. They are therefore objects of public magnificence, and consequently for the eye of travellers. But what traveller, with his person surrounded by the beggarly filth of an inn, and with all his senses offended, will not condemn such inconsistencies as folly, and will not wish for more comfort and less appearance of splendour.[81]

The main roads which Young admired not just in Languedoc but all over France were the core of what the Ponts et Chaussées and its offshoots constructed.[82] They were not only the corps' initial priority but the practical testing ground for every engineer. Perronet made his mark as a road-builder, first in the outskirts of Paris, then around Alençon. Roads were unglamorous; during his long service in the Limousin, the top technician among Ponts et Chaussées road-making engineers, P.-M.-J. Trésaguet, carried out not a single 'grand ouvrage d'art'.[83] Humdrum though such work might be, roads – like military operations but unlike buildings – fostered an awareness of strategic connection, means to ends, and the values of method and patience. Laying out, making and maintaining a road was both an apprenticeship and a continuum. If you could not cope with the cumbersome forced-labour process applied to French roads, you could hardly aspire to the delicacy of supervising structures.

What about buildings? One should not be seduced by the plethora of imaginary schemes that gushed out of the Ecole des Ponts et Chaussées from the 1760s, once that school's system of educating its pupils through design competitions was up and running. Apart from its prestigious masonry bridges, most of the above-ground structures officially procured through the pre-revolutionary Ponts et Chaussées were small buildings attached to roads, waterways, harbours or bridges. The corps remained primarily an engineering service, not an architectural one, although its entrants were schooled across the disciplines.

The prize schemes mainly stand for the types of exemplary public project that Perronet's expansionist service ached to build. But they do tally with a small number of smart provincial buildings actually constructed to designs by Ponts et Chaussées engineers. These owe less to the ambitions of the centre than to the corps' territorial network. There was usually an expert not far from the spot, Paris-trained after 1760 in technical and architectural competences, and possessed of fair social standing. What more natural than that town-councillors, intendants or governors should prefer such men to the local architect-mason or the chic Parisian architect, whose fees would be high and who would delegate supervision anyway?

One focus of this enlarged activity was the Atlantic seaboard. Much construction took place in France's harbour towns during the pre-revolutionary years, as naval and merchant ports alike were drawn into strategic planning from the centre. Now for the first time open friction developed between the civil and military engineering corps, as trade was pitted against defence in a symbolic duel over social priorities. The Génie was the almost invariable loser. In 1761 responsibility for building in most of France's ports was handed over to the Ponts et Chaussées; others followed, notably Le Havre, whose extension was humiliatingly snatched from the Génie in 1775. Even at Cherbourg, one of the few places where large sums were spent on fortification in the twilight of the Ancien Régime, the actual harbour works fell to the Ponts et Chaussées.[84]

In tandem, a number of handsome buildings rose to the designs of architect-engineers. The ways in which they might be allocated can be followed from Monique Moulin's study of what went on at Rochefort and La Rochelle, neighbouring naval ports both inured to engineers directing their development. [85] At La Rochelle, interventions by the Ponts et Chaussées in the town's architecture went back to 1737, when a local engineer altered the intendant's hôtel, and continued up until the Revolution. In 1760–5 Mathieu Hue, the regional ingénieur en chef, was commissioned by the merchant community to build La Rochelle's bourse (ill. 23), while in 1785–9 his successor Duchesne reconstructed the Palais de Justice – essays both in Parisian neoclassicism. There was an instructive row over Hue's fees. Technically the bourse was a private job. But the syndics expected to get Hue cheap at 5 per cent of construction costs instead of the 10 per cent a Paris architect would have charged, since he was on the spot and receiving a state salary. Hue thought otherwise, in token of the building's ambition. The threat that state engineers could undercut private architects was still worrying Victor Louis in Bordeaux some years later.

23. The Bourse at La Rochelle. A correctly neoclassical design by Mathieu Hue, the regional ingénieur en chef of the Ponts et Chaussées, 1760–5.

Rochefort too acquired some fine buildings. But here the Ponts et Chaussées was excluded in favour of 'ingénieurs de la marine' – a small, resuscitated service built up independently of the Génie after 1750. Its most striking monument was a grandiose pavilion-plan hospital with centralized chapel erected by Pierre Toufaire in the 1780s. Toufaire's career shows the flexibility still feasible then at the margins of the major engineering corps. More architect than engineer, he boasted a house in Paris and a private practice centred upon his native Châteaudun before he was nominated 'ingénieur en chef de la marine' and sent to Rochefort.[86] Besides the hospital, some barracks and a plan for extending the town, Toufaire was also involved, with the controversial help of the English ironmasters John and William Wilkinson, in designing the first foundries at Indret and Le Creusot, promoted to improve the French navy's cast-iron cannon (p.78).

So the Ponts et Chaussées did not have public construction in the ports to itself, nor did things always go its way. Though a decree passed in 1780 stipulating that all public works carried out at the costs of local towns and communities, 'such as presbyteries, prisons, law courts, barracks, dykes, canals and other constructions intended for public utility' should be designed by the corps, it proved a dead letter from the start.[87] There were just too many bodies and interests involved. Nor was the pre-revolutionary Ponts et Chaussées as much engaged with 'social purpose in architecture', to use Helen Rosenau's phrase, as with projects for economic improvement and urban embellishment. Most of the plethoric designs for hospitals, prisons and so forth that poured from Paris in these years came from private architects.[88] Nearly every one of those was an essay in enlightenment aspiration, not a practical project – just like the prize competition designs of the Ponts et Chaussées school.

The decree of 1780 at least revealed how far the wind had blown. Predominant in the ports at the start of the century, Vauban's Génie had been marginalized as constructors to the nation. Not that it was a spent force. Less absorbed now by practical construction, the military corps refocussed upon the arts of war. Fresh disputes about fortification broke out after France's defeats in the Seven Years War. Developments in artillery began to swing the balance back towards open battles, eventually with dramatic results. In the feats of French revolutionary arms, superiority in strategic intelligence played a decisive role: first on the part of Lazare Carnot, the Génie officer whose military planning saved the beleaguered republic, and then of Napoleon, the artillerist whose victories extended France's power across the face of Europe – taking along with it the model of its technical institutions.

Nor did building vanish from the military agenda. There were always forts to update and maintain if seldom to build. Meanwhile, a few individual Génie officers found time in their outposts to think and theorize in ways that had a long-term impact on construction. Such was Amédée-François Frézier, a many-sided Génie engineer who started out as an infantry officer, explored in South America and ended up in charge of fortifications in Brittany. When Frézier died at an advanced age in about 1774, it was said of him that his head was better furnished than his study. From that well-stocked head proceeded a three-volume treatise on stereotomy, and the first contribution from an experienced constructor to the incipient debate about 'rationalist' expression in architecture.[89] Such too was Charles-Augustin de Coulomb, father of modern soil mechanics, who endured nine years (1764–73) entrenching the island of Martinique, supervising 'wretched works under a burning sky' and deducing from them principles for earthworks that were to immortalize him through his *Essai sur une application des règles de maximis et minimis à quelques problèmes de statique, relatifs à l'architecture*.[90]

Coulomb was young enough to have studied at Mézières, the school of military engineering fostered in that town by the Génie from 1749. For those who measure engineering by mathematical and scientific progress, Mézières counts for more than the Ponts et Chaussées school in Paris, then developing in parallel.[91] Its pedagogy was certainly the more fundamental. Gaspard Monge, the intellectual founding-father of the post-revolutionary Ecole Polytechnique, began as a teacher at Mézières. The Polytechnique started from theoretical principle and then applied what had been learnt to a range of engineering disciplines, among which military ends came first. Such had been Monge's approach at Mézières. On the face of things, that left little room for practical architecture. But architectural teaching always survived on the margins of the Mézières curriculum, and was to be retained and reformulated by Durand at the Polytechnique. More of this story and its ramifications in Chapter 6 (pp.440–5) Here it need only be said that the new pedagogy carried French military engineering's traditions over into the nineteenth-century understanding of construction. Architecture, a fellow-traveller clinging to the juggernaut of engineering's onslaught, never quite fell off.

In summary, between 1660 and 1789 the Génie invented and the Ponts et Chaussées developed the model of a military or quasi-military corps of experts, directly employed by government and charged with carrying through a rolling programme of national construction tasks. The scope and the continuities implied by these arrangements were new – unheard of, at any rate, since the Romans. Without the enlightened despotism of French rule and the compacted landmass of the national territory, they would have been impracticable. The tasks involved and the aim of development attached to the programmes made engineering skills dominant in the corps. But architecture, if less securely tied to the furthering of material and technical progress, was always honoured as the expressive partner in the arts of construction and treated as an essential skill. It enjoyed an integral place in both corps.

Codified anew after the Revolution, that model spread by conquest or imitation from France across the western world. In all countries it touched, the élite among the senior building professions, both engineers and architects, came to be cast as experts equipped with a definite institutional 'formation', rank and authority, bent upon disinterested tasks, with the public realm to the fore and private jobs some way behind. Theirs were the lights by which civilized countries, old or young, expected to progress in construction. Progress indeed they did. But it was not the only model. Some nations managed to fumble forward with less authoritative direction.

2.   *Britain 1660–1730*

THE MILITARY SIDE

'They order, said I, this matter better in France.' Sterne's famous dictum applies well to seventeenth-century military engineering as practised in the British Isles. Three nations sharing one monarch, England, Scotland and Ireland had all suffered major disruptions since the 1630s, coupled with fitful land warfare on the soil of each. For a time it looked as though the three kingdoms' instability might be forcibly resolved by Cromwell's standing army, which though small by continental standards forged them for the first time into something like a single political entity. That proved an aberration. When the Stuarts returned in 1660, the English army shrank back to a miniscule 3,000 men.[92] Along with it dwindled all hope of union, not to return until 1707.

National aversion to a standing army is a mantra still inculcated into British schoolchildren. It is one of several principles or myths that go some way to explain the want of concerted support for military construction at the heart of Stuart government. Resistance to taxation without representation is another; the strength of localism another. There is also the old island faith that the navy, or failing that the people, will be defence enough: 'it is not England's profession to trust in lime and stone', bragged army commanders back in 1547.[93]

Yet at that earlier date, the official attitude had been different. Following an invasion scare in 1538, Henry VIII set about building 'divers and many fortifications' in what has been called 'the one scheme of comprehensive coastal defence ever attempted in England before modern times.' The cost of Henry's plethoric, eccentric string of forts, together with his defences of the remaining English enclave in France around Calais, amounted to a figure 'far exceeding the total expenditure on all non-military works in the entire lifetime of this giant among embellishers and palace-builders', records John Hale. The engineering community that devised and built the Henrician forts comprised a handful of the itinerant Italians and other 'strangers' then peddling modern fashions of fortification across Europe, generally on short-term contracts, plus some competent, longer-lasting nationals. Military specialists operating beneath the broad umbrella of the King's Works, the royal building administration, all were under the eye of a watchful tyrant who intervened unremittingly in details of fortification.[94]

Under Elizabethan parsimony there was maintenance but not much new building; the same held true under the early Stuarts. Then from 1636 fortification began to drift away from the King's Works and towards the Ordnance Office.[95] That parting of the ways led to a sharper distinction than obtained in other European countries between official architecture and official engineering, consigning the latter to a cultural limbo. The consequences have been profound. Whereas most students of English architecture know of the hegemonic role played by the King's Works through Inigo Jones, John Webb, Christopher Wren and their successors, the Ordnance Office long languished in obscurity. The King's Works has been commemorated in a six-volume history by eminent authors; the Ordnance Office and the military engineers who laboured for it can show nothing comparable. Likewise, Sir Howard Colvin's dictionary of British architects embraces only a few military engineers, yet includes many lowly civil craftsmen. Even Sir Alec Skempton's recent sequel on British engineers, which has done much to make up the deficiency, confines itself in theory to 'civil engineers'.[96]

The neglect is pardonable. The Ordnance Office was always a pragmatic rather than creative institution. As its name suggests, it was geared in the first place to organizing storage and supply – of weapons, ammunition and sundry military necessities. After 1636 it subcontracted the maintenance of forts to commissions, but did little building itself. The reverses of the Second Dutch War, culminating in de Ruyter's calamitous raid on the Medway and Thames in June 1667, changed all that. Already pledged to overhauling the defences of Portsmouth and Plymouth, Charles II's Government now 'committed the Care of all Fortifications and Repairs to the Ordnance'.[97]

Alas: for 1667 might have been a new start. A few months before, London had succumbed to fire and awaited reconstruction. The Thames, Channel and North Sea ports stood in equal urgency of protection. On both civil and military fronts Charles II was obliged and keen to build. Could he and his advisers have planned better? The array of intellectual talent eager to serve him in the circle of the Royal Society was equal to that available to Louis XIV. It is not far-fetched to imagine an English Bâtiments du Roi or even a Génie operating felicitously under Wren. But might-have-beens must not be long indulged. As the setting aside of radical rebuilding plans after the Great Fire had already shown, Charles II lacked the income and authority to cut through the thickets of opposition. Where consent must be negotiated and budgets eked out, public construction will seldom stir the heart.

Though the Ordnance Office after 1667 was not heroic, it was not without resource.[98] On one side of its technical staff stood the surveyors and accountants, headed from 1669 by the practical mathematician Sir Jonas Moore, who had made his name surveying the Fens. In 1673 Moore published an up-to-date manual on fortification which ran to a second edition. On the other side were the handful of engineers. Even after years of civil war, the best of these were foreign experts. Charles II, who like Henry VIII claimed a personal interest in the topic, told Pepys 'that England has never bred an able engineer of its own, no Englishmen having given their mind to it nor have we had occasion enough to invite any to the study of it'.[99] That position was not to change quickly.

Sir Bernard de Gomme, Charles II's chief engineer, has now emerged from the shadows thanks to a minute study of his career.[100] A Dutch Protestant from Zeeland who had learnt his trade in Low Countries sieges of the 1630s, de Gomme came over to England at the outbreak of the Civil War along with Prince Rupert. He soon became the royalists' most valued engineer and quartermaster. After the interregnum, he applied for arrears of pay and the position of 'surveyor-general' of fortifications. That title he received only in 1682, three years before his death. But from 1661 de Gomme was in charge of updating most of the major fortifications, building a new citadel at Plymouth, a fort at Tilbury, and much at Sheerness and Portsmouth, besides defensive works for England's troubled, ephemeral outposts at Dunkirk and Tangier.

De Gomme can be acquitted of the incompetence that spiteful contemporaries levelled against him. He was a hard worker and all-rounder. Like all first-rank military engineers of his day, his talents ran across surveying, waterworks, architectural design, cartography, tactics and supply. But not even his biographer claims that he stood out in any one branch of his art;

24. Gate to the Royal Citadel, Plymouth, by Bernard de Gomme of the Ordnance Office, 1670.

and unlike his fellow Ordnance officer, Jonas Moore, he enjoyed scant contact with Royal Society members. On the civil side, he possessed the skills of a Dutch polder engineer. As for architecture, the best surviving expression of de Gomme's endeavours is the main gate to the Citadel at Plymouth (1670): a hearty performance with a hint of Flanders about it, but some way off the best French or even English design (ill. 24).

Though the professionals of Charles II's Ordnance Office did as best they could, they betrayed the frustration and quarrelsomeness that go with mismanagement and under-resourcing. De Gomme was notoriously irascible, in part because he enjoyed weak support. No sweeter in temper was his second-in-command and successor Martin Beckman, a Swedish artillery captain by origin. In 1680 the Government awarded a rash monopoly for building all defence works to Thomas and John Fitch of London, the leading engineering contractors of the day. The consequences embroiled Beckman in a running battle with the insouciant Fitches over delays deriving from shortages of labour and materials at the strategic citadel of Hull (1681–4). Maddened to distraction, Beckman liked to resolve disputes by resorting to violent language, as thus: 'the engineer that have [sic] given such direction ought to have his ears cut off, to be nailed to his great toes to cure him of his gout.'[101] Good construction was unlikely in such an atmosphere, let alone convincing architecture.

After the 'Glorious Revolution' of 1688, a Dutch invasion in disguise, things sharpened up. William III, the new king, soldierly to his fingertips and steeped in continental military methodology, bolstered local deficiencies with so much fresh foreign expertise that the House of Lords pleaded that 'for the encouragement of the English, there may not be so

25. Royal Barracks, Dublin, by Thomas Burgh, architect-engineer, 1704–8. This austere building, now the National Museum of Ireland, was the lynchpin of the barrack-building campaign after William III's subjugation of Ireland.

many strangers employed in the Office of Ordnance'.[102] Samples of William's soldier-engineers are Wolfgang Romer, a diplomat's son from the Palatinate, who came over with William III's Dutch–German army, did much work in Ireland and then fortified New York and Boston; and Christian Lilly, also German and also active in Ireland, going on from there to the West Indies.[103] More creative were two engineers from the Huguenot diaspora, rich in technical talent: Nicholas Dubois and Jean de Bott (Johann von Bodt). Both arrived with William and may have been trained under his chief engineer Coehoorn, the so-called 'Dutch Vauban'. Dubois continued in the army till 1713, then branched out into entrepreneurial activities including speculative building and a translation of Palladio; Bott spent a decade in England and Ireland before being taken up by Frederick I of Prussia as one of his architects for Potsdam. Here is the pattern of apprenticeship in military construction leading on to a career in civil architecture. It had been pursued in the previous generation by Captain William Winde, a gentleman-soldier who having failed to advance in Charles II's Ordnance turned to designing country houses.[104]

As in France, it was on the contested fringes of Britain (to be drawn into one nation by the Act of Union in 1707) that the impact of William III's firmer military organization made itself most felt. Since insurgency was perceived as a bigger threat than the set-piece siege, the main form it took in bricks and mortar was not advanced fortification but barrack-building. Most prolific was the Barracks Board set up for Ireland in 1700. Leading from the centre, Dublin rapidly acquired its vast Royal Barracks (1704–8), now the National Museum of Ireland – the largest single building commenced anywhere under William III's aegis. Its architect was Thomas Burgh, who had served as an engineer in Flanders and was once described as a disciple of Coehoorn. Burgh's very plain design has a certain granite nobility (ill. 25). Of at least £25,000 spent on the barracks, 'we do not find there is £500 expended in mere ornament,' soothed his overseers, 'which is mostly occasioned by hewing some stone, and making cornishes round the front eaves, without which so large a pile of building would look miserably.' The copious Irish barracks that followed stirred controversy. For some they were 'often the finest buildings in the town' and a guarantee of security. But others looked upon them as ill-built importations, sited to protect powerful private interests not those of the populace as a whole.[105]

Under the Hanoverians barrack-building spread to Scotland. After the 1715 rebellion, four large garrison-forts with semi-standardized barracks were constructed in remote Highland locations.[106] Though these dour fortresses of pacification symbolized political will, their authors did not deem the clans worth cowing with any costly expression of might. Unlike Vauban's border citadels, here was an architecture for the frontier of civilization, not between civilizations. Likewise, the prolific Highland road- and bridge-building campaign undertaken from 1725 under General Wade and the 'Board of Ordnance in North Britain' indulged in few frills. From the architectural standpoint its main importance was that the enterprising Adam family, masons to the Board, used it to wax rich and come to the fore.[107]

On the other hand, for barracks in the old military stronghold of Berwick upon Tweed, poised between England and Scotland, display was at least contemplated. For the Ravensdowne Barracks 'apud Bervicences', as Nicholas Hawksmoor touchingly termed the border town, that architect in 1717 produced an enriched sketch design, perhaps at the behest of his friend Michael Richards, surveyor-general to the Ordnance. Though it was cut down out of recognition, trace-elements of Hawksmoor's Roman romanticism do peep out from a few other Ordnance buildings just then.[108]

Things seemed to be stirring: and indeed there are glimmerings of a more enlightened attitude to construction at the Ordnance between 1714 and 1722 under the régime of Richards and his political master, the Duke of Marlborough. Something like a true corps of engineers was at last decreed; naval building projects got under way at Chatham, Portsmouth and Plymouth; an Architect of the Ordnance was even created, though the post went to a large, efficient mason-contractor, Andrews Jelfe, and soon degenerated into a sinecure.[109] In the event, it turned out to be just a puff in the wind. Except on the coast, a phobia about flagrant military presence in English cities persisted. Anything like, for instance, the grand metropolitan barracks suggested for Hyde Park in 1718 by Dubois, self-styled 'architect and engineer', could never be proceeded with. Whatever the Army might have hoped, Hanoverian London could not countenance such a scale of military architecture close to City and Parliament.[110]

THE CIVIL SIDE

So much for the military picture. What about the growth of civil infrastructure and its relationship to English architecture and taste? That is puzzling. National endeavour and competitiveness were as sharp as in France; the idea of progress through continuous material development, for mankind in general and England in particular, went back at least as far as Francis Bacon; and the Royal Society was adorned by eminent intellects who relished debating questions of economic, technical and cultural moment, and distributed 'trades and manufactures' among their members to investigate and report on. They were well aware what Colbert tried to do for France. And yet, before the eighteenth century, the call for a concerted programme of national infrastructure and urban development seems to come only from a few lone voices like Sir William Petty and John Evelyn.

The truth was that politics after 1660 seldom allowed even these high talents consistently to convert theory into practice, or specific ideas and advances into general improvements. Take the over-arching career in significant English construction during this period, that of Christopher Wren. Scholars on Wren agree that he cannot be only, or perhaps even primarily, categorized as an architect. The most original study of him in recent years has focussed upon his 'mathematical science'.[111] Though the modern term scientist is easily misconstrued, it is a reminder that Wren tackled a range of technical issues (forces in an arch, truss or dome; the properties of materials; proportions) with a

26. Trinity College Library, Cambridge. Axonometric drawing showing inverted arches in foundations and wide-span floor carrying heavy bookstacks over an open colonnade with central row of columns.

keener sense of principle than any other European architect of the day – saving perhaps his collaborator, Robert Hooke.[112] 'It seems very unaccountable,' wrote Wren, for example,

> that the generality of our late architects dwell so much upon this ornamental, and so slightly pass over the geometrical, which is the most essential part of architecture. For instance, can an arch stand without butment sufficient? If the butment be more than enough, 'tis an idle expense of materials; if too little, it will fall; and so for any vaulting.[113]

As a constructor or 'building scientist' Wren had no peer; he did new things, took risks within limits and sometimes suffered failures, yet always enforced high technical standards. His most ambitious buildings were also his most experimental ones: the Sheldonian Theatre; Trinity College Library (ill. 26); St Paul's Cathedral. He was as absorbed by how things were put together as by how they looked. James Fergusson sums him up thus:

> The truth of the matter appears to be, that, both from the natural bent of his mind and from the circumstances of his education, Wren was more of an engineer than an architect, and, consequently, always preferred the display of his mechanical skill to the expression of his artistic feelings.[114]

Here engineering is defined in terms of skills, not tasks. It is because the tenor of most building tasks that fell to Wren was aesthetic and symbolic that he is naturally remembered as an architect. Palaces and churches were his mainstay as Surveyor-General of the King's Works and a Commissioner for rebuilding the City of London. Yet they do not account for the whole of his constructive range. Before he had built any buildings, Wren was offered the job of fortifying Tangier, shrewdly refusing. In the greater burden that did partly fall to him, the replanning of the City after the Great Fire of 1666, Wren's role turned out to be a compromise between the aesthetic and the technical – an exercise in the art of the possible. Engineering came into that task in more senses than one. As things transpired, it proved easier for him to make headway with the specifics of what we now call 'structural engineering' – the art of making complicated buildings stand up – than in the broader 'civil engineering' of London's infrastructure.

As is well known, Wren, Hooke, Evelyn and others made ideal sketch plans for rebuilding London, all set aside because of politico-commercial imperatives.[115] The impetus behind these plans appears to start with John Evelyn, dilettante, traveller, publicist and scourge of insularity. Ever since the Commonwealth, the royalist Evelyn had been railing against the scruffiness, disorder and pollution of London compared to neat Holland and elegant Paris – deemed so even before Louis XIV's building campaigns.[116] At first this was political point-scoring. But after Charles II's return Evelyn aspired to real influence. He called for action and an enlarged view of national and civic planning. 'We see how greedily the French, and other strangers embrace and cultivate the design', he was still urging in the 1679 edition of his *Sylva*:

> what sumptuous buildings, well furnished observatories, ample appointments, salaries and accommodations they have erected to carry on the work; whilst we live precariously, and spin the web out of our own bowels.[117]

Soon after the Restoration, Evelyn became the driving force behind a commission for reforming the streets of Westminster. To his call for concerted efforts to update and renew London there is reason to suppose that he had already enlisted his young friend Wren, the budding practical architect and technician that Evelyn was not.[118]

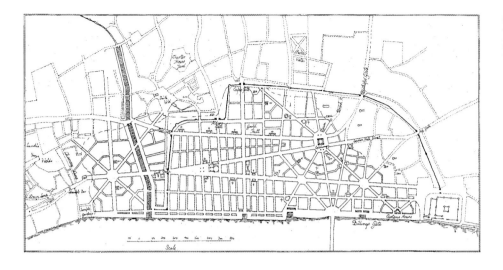

27. Wren's plan for London showing the Fleet Canal running north-south left of centre, and open space along the intended Thames embankment.

When the Fire struck, Wren was not long back from his unique, eight-month visit to Paris of 1665–6. His purposes there were catholic – to confer with astronomers and architects, examine improvements, maybe to check out domes. From an architectural standpoint, his timing was too early.[119] The glories of Louis XIV's reign had yet to come. Though Wren was able en passant to catch Bernini, 'the old reserved Italian', most of the completed architecture he reported on was domestic.[120] Process and infrastructure afforded more practical lessons. Afterwards, Wren claimed 'the opportunity of seeing several structures . . . while they were in rising, conducted by the best artists, French and Italian, and having daily conference with them and observing the engines and methods'.[121] One of these certainly was the Louvre site, then at a temporary standstill. But when asked by an acquaintance

> which he took to be the greatest work about Paris, he said the Quay or Key upon the river side, which he demonstrated to me, to be built with so vast expense and such great quantity of materials, that it exceeded all manner of ways the building of the two great pyramids in Egypt.[122]

That remark could refer to the long-term royal endeavour of embanking the Seine, going back to works under Henri IV on the Ile de la Cité. But what Wren probably meant was the recent masonry embanking in front of the Louvre and opposite, where Louis Le Vau had embarked on the Collège des Quatre-Nations (ill. 262). Those monumental river walls were preliminaries to the 'architecture' to follow on both sides, far from complete when Wren was in Paris. They unified the two banks (there was even talk of a bridge) and gave notice that these sections of the river frontage were out of bounds for any but official moorings and landings.[123]

So it is the more striking that the only major planning features of Wren's post-Fire sketch plan for London which he persisted with, once rebuilding got under way, were the Thames Quay and the related Fleet Canal (ill. 27). Even so, they were a failure. The canal, constructed by Thomas Fitch to a layout and technical brief from Wren and Hooke (1672–4), was too short and narrow to test its designers' ingenuity, too cheap to manifest their artistry. It survived only sixty years. The quay as a whole never got started, despite great efforts. The expense of monumental embanking might make sense as a prelude to a sequence of royal architecture. It did nothing for a great tidal port where, even above London Bridge, every foot of frontage for wharves and lighterage was precious. St Paul's got a special quay for unloading Portland stone: that was about all.[124]

From the 1690s, Evelyn looked back regretfully on the high civic hopes and subsequent compromises of Charles II's reign. London's low-key rebuilding showed that the ideal of a noble infrastructure had not been successfully articulated. In raising fine churches and palaces, Wren had done as much as his compatriots could accept. On the other hand he and Hooke, as two of the six rebuilding commissioners, had helped set technical rules and standards for street-paving, draining and house-construction for the new London that were to elicit applause from eighteenth-century visitors, not least Voltaire. In a modern merchant culture, the monumentality of St Paul's was always going to be an exception. If English urban style was to be framed not by any exalted vision but by the passive authority of building regulations, that might be a price worth paying for a measure of political freedom.

Still, Wren and his colleagues in the circle of the King's Works continued to hope for more, not least on the basis of their experiences in reconstructing London. Asked, late in life, to comment on the proposals of 1711 for building fifty new churches around London, Wren's response was as radical in civic spirit as his plan of 1666, recommending free-standing churches 'among the thicker inhabitants' but cemeteries in the 'out-skirts of the town'.[125] A mere fraction of that vision was realized. For English public architecture to progress, it was necessary to stick to building-types that people respected, and ride on the rare puff in the political wind.

Two Observatories

To the point of parody, the national observatories that France and England built in the 1660s and '70s mirror the contrasting tastes and capacities of the two powers. They also illustrate the shifting of sands between ceremonial and useful architecture, between symbolism and the embryonic agenda of national development, that blurs the borders between architecture and engineering.

28. Louis XIV visiting the Académie des Sciences in 1671, engraving by Sébastien Le Clerc. The construction of the Paris Observatory can be seen in the background.

Grave and commanding, the Paris Observatory (1667–72) is one of just three buildings realized by Claude Perrault, the first high intellectual in the cerebral lineage of French architecture.[126] The Observatory (ills. 28, 29) was to house Colbert's newly founded Académie des Sciences and host astronomical observations and debates. The brief came mainly from the astronomer Adrien Auzout, with whom Wren met in 1665 and doubtless talked observatories. Perrault's austere design had a square outline flanked by three octagonal towers, all raised above a military platform with the imprint of a fort. It was built of ashlar blocks and vaulted throughout, in the finest craft of stereotomy; there was no structural timber. Inside, a grand stair and gallery hinted that ceremony and debate enjoyed the edge over experiment and observation.

During construction, Giandomenico Cassini (the discoverer of Saturn's rings and Jupiter's moons) was appointed astronomer royal and demanded changes; Perrault accused him of talking nonsense but had to submit. The towers were altered, the roofline was raised to accommodate rooms for the staff, and the roof-terrace itself acquired quadrants and telescopes – not, it seems, anticipated at the outset. Soon however the main arena for practical work shifted out of the Observatory itself, into a modest building erected next to it and into the grounds, which never took on the military aura first envisaged. Nevertheless the axis of the Observatory and its environs offered authority and space enough for the Académie des Sciences to meet and dine there. It also served as the co-ordinating centre for mapping the national territory, the programme enjoined upon the astronomers by the Government and inaugurated in 1676, using Jupiter's moons as reference.

England's smaller observatory (1675–6) followed on from France's, just as its army hospital at Chelsea and naval hospital at Greenwich were cut-price imitations of the Invalides. It came about at the instance of Sir Jonas Moore of the Ordnance Office.[127] Following criticism of the Navy's poor performance during the third Anglo-Dutch War, Moore argued that progress depended upon improving navigational aids through more systematic astronomical observations, to be undertaken by his protégé John Flamsteed. He drew the Royal Society into the enterprise, and after some debate a site for the new observatory was conceded on Greenwich Hill. It was to define the national and in due course international meridian: which adds to the ludicrousness of the fact that it lies neither exactly north-south nor on the axis of the naval hospital. The whole project has a touch of farce to it. The Government offered scant support; Moore had to raise money for the construction 'by selling off surplus gunpowder and taking bricks from a surplus store at Tilbury, second-hand wood, iron and lead from demolition work at the Tower'.[128] He even had to equip it at his own expense.

Wren having chosen the observatory site, Robert Hooke took over most of the design in consultation with Moore. Like Perrault, the architects offered a military hint ('a little for Pompe', Wren put it[129]), since they were reusing the foundations of a tiny hilltop fort. But whereas Perrault's nod to defence implied power and a political programme, Hooke turned it into a whimsical, almost satirical eye-catcher (ill. 30).[130] The outcome is a fortlet with a sham frontispiece overlooking the park, a folly in antique Jacobean style with turrets and false windows: 'a very English precursor of arch postmodernism', John Bold has called it.[131] Here, between penny-pinching and improvisation, the national penchant for mocking or faking might and grandeur drew fresh breath. Forty years later on the farther side of Greenwich Hill the ex-soldier Vanbrugh followed suit, domesticating the 'castle style' in a rash of military-style villas for private amusement.

The planning of the Greenwich Observatory followed on Lilliput scale the lines of its precursor in Paris. There was just one grand room, five sides of an octagon. Wren and Hooke may have hoped it might serve as a rendezvous for the Royal Society, but that came to nothing, though feasts and meetings were certainly held there. For the first astronomer, John Flamsteed, it was of little use as a practical workplace, while the basement that he and his family had to inhabit was cramped and mean. Like Cassini, Flamsteed made his main observations from a bare wall behind the Observatory, where a great mural arc and sextant were stationed. In neither Paris nor London was it yet feasible to integrate the instruments of science or civil technology with an official architecture.[132]

To judge the two buildings not by their magnificence but by their results, the Paris Observatory yielded rapid fruit in the shape of the great mapping survey of French territory. In the long term, the cheaper Greenwich Observatory scored the more spectacular success, since Flamsteed and his successors were able to establish longitude at sea – an advance of imperial consequence. That was never the prime goal of the Paris

29. Observatory, Paris, as completed. View by N. Pérelle showing the north side facing towards Faubourg Saint-Jacques.

30. The Royal Observatory, Greenwich, engraved by Francis Place, c.1676.

Observatory. In any case, the success of both institutions stemmed from the quality of their astronomers and instruments, not from their buildings. The latter's main function was to concentrate personnel, morale and thus effort. But that the question of results raises itself at all hints that competition, direct or otherwise, was starting to penetrate the values attributable to civil as well as military construction.

From Surveyor and Mechanic to Engineer

If state-led architecture and engineering did little to direct the enhancement of the English realm after 1660, the same period saw the birth of an unofficial, decentralized culture of civil improvement that gave British engineering a separate slant from the French model. Practical hydraulics and mechanics were pre-eminent in that development; architecture was all but absent.

At the roots of the movement lay surveyors. 'Engineers there were none, except those who used petards for military purposes', wrote Trevelyan of seventeenth-century England.[133] That is because the authors of civil works called themselves surveyors. The term in English is an old one, which could embrace architects, civil engineers, cartographers and other trades or skills now distinct. What surveyors had in common was a relationship with property. In the first instance they were measurers, delineators, adjudicators and improvers of land. They came into their own in the sharper definition and recording of private property that followed the Reformation.[134] Men good at mathematics and mechanics developed a sideline in hydraulics – the crucial skill for infrastructural works. Such, for instance, was the cutting of the New River to supply parts of London with drinking water in 1609–14; or the Dutch-influenced efforts to drain and reclaim large portions of the Fens, starting to show results in the 1650s; or the many inland navigation improvements of the era. Undertaken by landowners or companies with at most the passive approval of the Crown, such works tended to be geared to profit. They offered little incentive for display; indeed the fact that they usually turned out costlier than expected added to the motives for economy of execution.

If there was no call for elaborate architecture to go with these schemes of improvement, where after the Glorious Revolution investment started to pay off was in bigger, better machines or engines: waterwheels, pumps and the like. From about 1690 the skills of the surveyor-technicians who directed English hydraulic works veered in a mechanical direction, towards the craft of millwright, instrument-maker, inventor – or engineer, in the English sense of the term.

Captain Thomas Savery, the pioneer who took the steam engine 'out of the scientist's laboratory and into the workshop',[135] is a case in point. After a youthful spell of soldiering in the Low Countries, perhaps as a military engineer (he later translated Coehoorn's treatise on fortification), around 1694 Savery set himself up in London as an inventor, specializing in pumps and their application, with water-supply and the draining of mines in mind. Though he enjoyed some success with the former, mines proved beyond him. Credit for making the steam engine work commercially passed to his successor and fellow-Devonian, Thomas Newcomen, the inauguration of whose Dudley Castle Engine in 1712 presaged a new technological age.

Less inventive but more prolific than Savery was his contemporary George Sorocold of Derby, known in his lifetime as 'the Great English Engineer' – meaning not a military expert but a civil constructor of engines.[136] He was perhaps the first technician thus labelled. Emerging in the 1690s as a consultant in water-supply for towns, Sorocold carried out schemes with undershot waterwheels for Derby, Leeds, Portsmouth, Bristol, Norwich, Great Yarmouth and King's Lynn, and devised similar projects less fully taken up elsewhere. That side of his work culminated in his renewal and extension of the waterworks at London Bridge from 1701. Sorocold also improved private estates in both England and Scotland and pursued various navigation and dock improvement schemes. It cannot be proved, but he may well have been the engineer for that archetype of large water-powered British mills, Lombe's silk mill at Derby, opened in 1721 after his death (ill. 31).[137]

Much is obscure about the career-patterns of Savery, Newcomen, Sorocold and the pioneering generation of English engineers, duly so called. But far from being horny-handed mechanics, as they used to be pictured, they were figures of fair education and more than local standing. Sorocold, for instance, 'was clearly a man of some substance and would not have relied on his engineering work for all his income,' judges Mike Chrimes.[138] What he was not was an architect-engineer of the type depicted in this chapter. He was a specialist in practical mechanics and hydraulics.

Building professions, we have seen, separate only when there is continuous work enough to feed specialisms. Sorocold's ability to range across England improving the water supply to towns speaks much for his skills and energies. But it says more about the capacity of those towns to fund improvements in the wake of the Glorious Revolution. Relying on that better infrastructure, the Georgian generations commissioned architect-builders to raise market halls, assembly rooms and the other trappings of polite provincial society.[139] Meanwhile the pumps and waterwheels that made all that possible were housed in humble timber huts or under the arches of bridges, thumping away at one remove from the public gaze. The men who erected and maintained them, pioneers of the British engineering profession, were not public officials employed by military or municipal authority but ingenious mechanics working for private hire. Here was a culture of water supply miles apart from the mediaeval conduits and baroque fountains of yore. In an increasingly mechanical and democratic context, it was growing harder to integrate engineering with architecture.

31. Reconstruction by Anthony Calladine of the Italian Works at Lombe's silk mill, Derby, as it was on completion in 1721. The first great mill of the Industrial Revolution in Britain, possibly designed by George Sorocold.

MORALITY AND INFRASTRUCTURE

From Sorocold, Savery and Newcomen the line runs straight to John Smeaton, to the creation of a society for 'professional' civil engineers in 1771, and to the consulting configuration of British engineering which that society confirmed. The slant of the early millwright-engineers and their successors underlines the fact that the national linkages between construction, efficiency, progress, and public magnificence remained as fragile as they had been during the age of Wren – and have since remained.

Not that the low ambition of the public realm went unnoticed, especially among Tories. Evelyn's example has been cited, and he had successors. With the nation's power and wealth increasing, even under the laissez-faire Hanoverians a few Englishmen continued to call for a more inspiring use of resources in construction. Testimony to that effect from an unexpected quarter can close this section.

In 1732 Alexander Pope published the best-known English poem to touch upon architecture: his *Epistle to Lord Burlington*. Written when Palladianism was in the ascendant, the *Epistle* is commonly interpreted, on the basis of its earlier pages, as a diatribe about taste in building and landscaping. A deeper reading reveals it in the terms in which Pope and his early editors framed it, as a 'moral essay' calling for architecture and gardening to be addressed as ethical or political activities, and for a national culture dedicated to public improvements.[140]

Always loath to leave his verses unglossed, Pope himself prefaces them with an 'argument' headed: 'Of the *Use* of Riches. The Vanity of Expence in People of Wealth and

Quality'.[141] Accordingly, much of the poem satirizes frivolity and waste in private architecture – the eternal self-indulgence of the English country house. Not, the reader discovers as the couplets roll by, that Pope would have noblemen refrain from lavishing their
wealth upon bricks and mortar, arbours and grottoes. Indeed it is their patriotic duty to
do so, as a means of spreading employment as well as taste.[142] Only, they must study
restraint and civility, respecting the laws of nature and the practice of the ancients, as
Pope's patrons Lords Bathurst and Burlington are alleged to have done.

There follows the peroration:

> Who then shall grace, or who improve the Soil?
> Who plants like BATHURST, or who builds like BOYLE.
> 'Tis Use alone that sanctifies Expence,
> And Splendor borrows all her rays from Sense.
> His Father's Acres who enjoys in peace,
> Or makes his Neighbours glad, if he encrease;
> Whose chearful Tenants bless their yearly toil,
> Yet to their Lord owe more than to the soil;
> Whose ample Lawns are not asham'd to feed
> The milky heifer and deserving steed;
> Whose rising Forests, not for pride or show,
> But future Buildings, future Navies grow:
> Let his plantations stretch from down to down,
> First shade a Country, and then raise a Town.
>
> You too proceed! make falling Arts your care,
> Erect new wonders, and the old repair,
> Jones and Palladio to themselves restore,
> And be whate'er Vitruvius was before:
> Till Kings call forth th'Idea's of your mind,
> Proud to accomplish what such hands design'd,
> Bid Harbors open, public Ways extend,
> Bid Temples, worthier of the God, ascend;
> Bid the broad Arch the dang'rous Flood contain,
> The Mole projected break the roaring Main;
> Back to his bounds their subject Sea command,
> And roll obedient Rivers thro' the land;
> These Honours, Peace to happy Britain brings,
> These are Imperial Works, and worthy Kings.[143]

The shift of direction in this passage, and its earnestness, are manifest. Abandoning
satire, Pope first checks himself. The proper improvement and adornment of estates, far
from being selfish, serve to supply and enrich the nation. Their herds nourish and clothe
the population; their woods and forests augment commerce and build towns. Warming
to the theme of public weal, there is a yet graver sphere for your reforms, the poet urges
Burlington: harbours, churches, bridges, sea-defences: public works – the tasks of kings
and governments. His patron must chivvy the slothful Hanoverians, neglectful of national infrastructure. That, Pope resoundingly concludes, is the truest 'use of riches', to which
the embellishment of private property can only be a prelude.

In George II's England this is not quite familiar territory. Notes by the author offer the
best guidance:

> The poet after having touched upon the proper objects of Magnificence and Expence,
> in the private works of great men, comes to those great and public works which
> become a Prince. This Poem was published in the year 1732, when some of the new-

built Churches, by the Act of Queen Anne, were ready to fall, being founded in boggy land . . . others were vilely executed, thro' fraudulent cabals between undertakers, officers, etc. Dagenham-breach had done very great mischiefs; many of the Highways throughout England were hardly passable, and most of those repaired by Turnpikes were made jobs for private lucre, and infamously executed, even to the entrances of London itself: The proposal of building a Bridge at Westminster had been petition'd against and rejected . . .

Even after the need for Westminster Bridge had been recognized, Pope goes on to cavil, it was in danger of being entrusted to an incompetent 'place-man' and 'carpenter', Thomas Ripley.[144]

Here is a string of alleged defects about four types of public works – churches, river-defences (the damaging Dagenham breach in the Thames estuary had taken sixteen years to repair), roads and bridges. They prompt Pope to call for an enlarged scope in public construction. 'Temples', such as the London churches built out of taxation revenue in the reign of Queen Anne (including the two troubled by settlement, St John's, Smith Square and St Anne's, Limehouse) imply architecture of high ambition. The 'broad Arch' may also be so construed. Pope's harbours, public ways and moles, on the other hand, are plainly civil engineering. Yet this most fastidious of poets does not scruple to yoke such manful projects to Lord Burlington's revolution in polite taste.

The tone of the passage has much to do with its timing. If in Wren's generation, infrastructural development had been disappointing, at least there were fresh royal works – palaces and much else – to applaud or deplore. But by 1730 the historic struggle between English monarchs and their greater subjects had ended in victory for the latter. Royal patronage had imploded along with royal power. Private patrons and architects were being sucked into the vacuum of central initiative. Figures like Burlington and his circle became the focus of taste. But as Pope could see, they were rattling around in too much space for their quite trivial ideas.

Dead periods are often fallow ones. This one prompted Pope, growing more earnest as he aged, to articulate a thought emerging out of Colbert's and Vauban's France that was to become an enduring theme from the Enlightenment: that the prime object of building should not be the glory of kings nor private citizens, but the advancement and welfare of the whole community. Offered to a self-interested polity, Pope's appeal found no immediate echo. But in time, no thoughtful architect or engineer would be able wholly to escape these ideas.

3.   *America 1660–1900*

COLONIES OFFICIAL AND UNOFFICIAL

When in 1682 the first French explorers under La Salle braved the Mississippi and descended into the region they christened Louisiana, the first building of substance they raised was a timber fort.[145] 'The Spaniards would have built a church; the English, a tavern', scoffed a later commentator. In fact, fort-building was standard practice among all the three powers then vying for American territory. That outpost, run up by La Salle's motley 'coureurs de bois', outlaws really, would have been rough and ready. In the next phase, a single draughtsman accompanied the official expedition sent out from the homeland under Lemoyne d'Iberville in 1698 to secure the lower Mississippi Valley for France. As a result Fort Maurepas, in timber still but tidier and probably tougher, was set up near the present Biloxi.

After a further lapse of years France proceeded to a concerted phase of colonization – the so-called Mississippi Scheme. Now the professionals moved in to develop rivers and harbours. In 1719 a team of four engineers arrived: Le Blond de la Tour, de Pauger, de

32. New Orleans, plan of c.1720, showing the town laid out as a strict grid apart from a central cruciform church next to the waterfront. Water provides the main line of defences, though outlying houses are protected by stockades and a battery of cannon is positioned in front of the church. This unsigned and undated plan was probably prepared by the French military engineers who came to New Orleans in 1719. The canal on the left is inscribed: 'canal projetté au depart de l'autheur'.

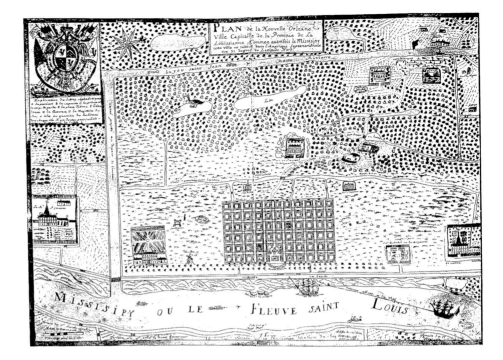

Boispinel and Franquet de Cheville. Seconded from the Génie, they brought experience under Vauban and his successors to the laying out first of Fort Louis (New Biloxi) and then of New Orleans. Dreadful attrition followed among the some 7,000 Europeans and 2,000 slaves imported to Louisiana under the Mississippi Scheme; none of the four survived long. But de la Tour and de Pauger had time to set New Orleans on its feet, devise a simple grid (ill. 32), lay out the Place d'Armes, and build barracks, a hospital, houses for the directors of the Company of the Indies and a hurricane-resistant church. One thing was omitted: the circumvallation that the engineers proposed. Costly though it would have been, its neglect was more so, for a massacre ensued in 1729. Even so, despite intermittent attempts New Orleans acquired no adequate defences until 1791, when the town was under Spanish control.

Here is a neat model for official town-founding in the New World. First to arrive, soldier-adventurers throw up rudimentary security for traders. Later come settlers and with them experts, in modern terms mostly engineers or surveyors, to lay out a town, measure and assign property, ensure water supply and maybe elaborate the defences.[146] They too build the first public buildings, unless an architect is on hand to give a symbolic lift to the church, magistrature or main square.

In reality the pattern of colonial town-founding was seldom so cut and dried, even where the European crowns controlled it. Much depended on national habits. In the Spanish model, soldier and priest had been yoked in uneasy co-operation from the first. For their earliest towns in the Americas, organization on the ground was naturally hand to mouth. For just that reason, royal directives for settlements were sent out to viceroys and their subordinates as early as 1513, to be codified sixty years later in the town-planning sections of the famous Laws of the Indies. That enduring code – which covered only 'pueblos' or civil communities, not 'presidios' (forts) or missions – gave soldiers, priests and settlers data enough to lay out a grid, a main plaza and a good site for the church without much technical help. In that way official town-planning became 'a fairly routine activity governed by a set of increasingly specific regulations', judges John Reps.[147]

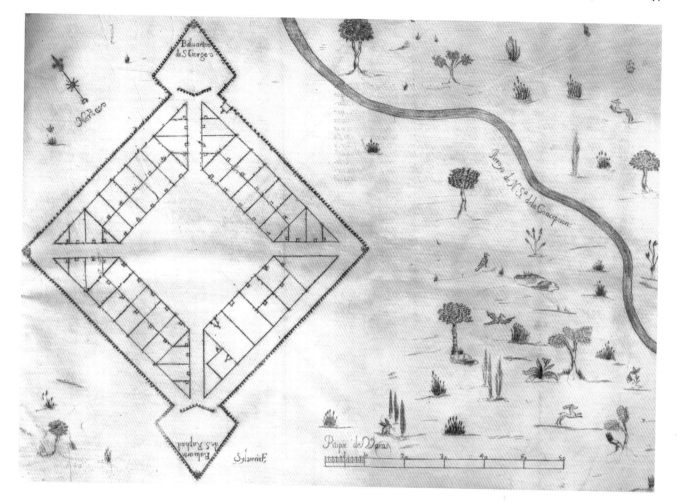

Studying the early Viceroyalty of Peru, Valerie Fraser has accordingly found little trace of professional involvement. 'There were no architects in South America during the sixteenth century', she concludes: 'construction of the colony was not in the hands of theorists or highly-trained specialists, but was the responsibility of all colonists.' In Lima the street layout, property allocation and water supply from the 1530s were sorted out between the magistrate and the 'alarife', a technical Spanish term best construed as surveyor. The Lima alarife may well have been of Indian stock (there being strong stone-building traditions in the indigenous culture), while most of the craftsmen who built the churches and other public buildings were certainly Indian.[148]

Almost two centuries later, when France's Mississippi Scheme galvanized Spain into pressing official colonization northwards into Texas, its trusted and tried town-founding system was still in place. Soldiers and priests took the lead, starting with presidios or missions. The former consisted of a simple square with inward-facing accommodation, the whole protected by a wooden stockade with makeshift corner bastions (ill. 33). The first Texan pueblo at San Antonio (1730) conformed still to the Laws of the Indies, applied by a measuring surveyor adhering to the usual strict instructions. Not before the settling of California in the 1770s onwards do we hear of engineers, so named, working on town-layouts.[149] At much the same time, architects rather than priests start to be credited with design responsibility for the mission churches of Texas and other newly colonized regions of the Spanish-American south-west.

This absence of specialized military engineers from Spanish inland settlements should not surprise. Though defences were needed, properly manned stockades sufficed against native populations inexpert in firearms. As in the Scottish Highlands, so in

33. Plan of the Presidio of Los Dolores de Los Tejas, East Texas, 1722, from a drawing by Juan Antonio de la Pena y Reyes. A standard arrangement with four wedges of building facing a square, wooden stockades and two corner bastions.

pioneering territory there was no call for permanent or symbolic fortifications. More helpful as a means of promoting allegiance to His Most Catholic Majesty were churches of adequate architectural impressiveness.

Not that there were no early colonial fortifications of European grandeur or professional authorship. But those, at least in Africa and the Americas, tended to be ranged against competing powers and therefore to be coastal or riparian.[150] In the sixteenth century they were devised by an assortment of intrepid freelance Italians and others, working for the colonizers in tough conditions and with an underskilled labour-force. The heroic early Spanish examples were at Cartagena and Havana. After a devastating French pirate raid in 1555, for instance, it was decreed that Havana should be guarded by the Castillo de la Real Fuerza (1558–82), followed up by two high outposts flanking the harbour mouth, the Morro and the Punta (c.1593–1610). Their construction was fraught with frustrations over local skills and labour. Though Bartolomé Sánchez, the engineer in charge at the Real Fuerza, imported forty stonemasons from Spain, that did not save him from dismissal over slow progress. 'One can only marvel that a building of such quality and sophistication could have been built with such primitive resources', enthuses Juliet Barclay. The Real Fuerza did its job; Havana prospered, sacked no more. Topped in one corner by the addition of a pencil-like tower and bronze statue, the fort justly became an architectural symbol for its city (ill. 34).[151]

After 1660 the tempo quickened. An upsurge of fortification reflected the systematizing of European trade and settlement policies and the pitching of new predators into the colonial pot. Along the Gold Coast of West Africa, Prussians, Danes, Dutch and English fortified rival trading posts within spitting distance of one another.[152] Across the Atlantic, Spain upgraded its defences around the Gulf of Mexico and the Caribbean, the better to guard its harbours and treasure-fleets against English, Dutch and French raiders. Such was the first permanent military installation on North American soil, the grand bastioned fortress of Castillo de San Marcos, built by the Havana engineer Ignacio Daza in 1672–87 to replace a timber one guarding the settlement of St Augustine, Florida, then a century old.[153]

This sharpening of colonial rivalries sits alongside fortification in Europe as a factor in the intensification of military engineering among the major powers, and the definition of state corps first in France and then beyond. Hence it was a logical step for engineers to take part in the type of town-founding operation represented by the Mississippi expedition of 1719. Wherever state engineering corps were strong, officially founded new towns and major town extensions or reconstructions came in time to be laid out as a matter of course by members of such corps, military or civil. Nor did that apply only to colonies, but to European cities as well. Many examples could be given, from the brilliant emergency rebuilding of the Baixa after the Lisbon earthquake of 1755 by Pombal's military engineers, to the great urban renewals of the nineteenth century – the replanning of Paris, the extension of Amsterdam, and so on. Engineers, scientific, obedient, management-minded, did the plans; architects, more creative, less disciplined, designed the public buildings within them. Descartes' distinction was reaffirmed.[154]

Yet in North America the New Orleans model of military technicians planning towns proved the exception. In part the method failed because France lost all its possessions there to Britain and Spain in 1763. But underlying the weak showing of engineering-led town layouts was the phenomenon of unofficial settler-colonialism. North of Mexico, the officially founded town was never the norm. Even in colonial times, most new communities began not with fresh arrivals from Europe, but with those already present in the New World striking out further. Soon enough deteriorating relationships set in between mother country and offspring, as the settlers started to do things in their own way.

Nor were colonies inspired from Europe always official. The British settlements in particular embraced a pattern of colonization at one remove from authority – of emigrants fleeing the restrictions or clutches of the home country, whether to make their fortunes under licence from a company or to practise their religion in peace. Such colonists built

FACING PAGE

34. The Castillo de la Real Fuerza, Havana, erected under the direction of the engineer Bartolomé Sanchez, 1558–82.

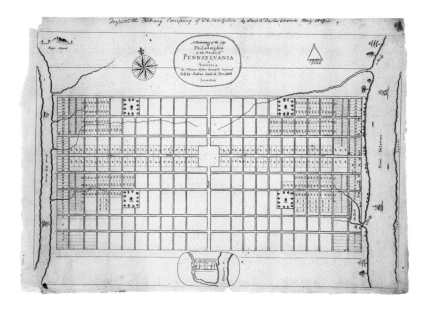

35. Plan for Philadelphia by Captain Thomas Holme, 1682, showing site plan, grid layout of the town, garden squares, and the subdivision of the grid close to the rivers into economical plots.

their towns in a subdued style, with little by way of defences or formal technical direction.[155] A good example is William Penn's Philadelphia, most urbane of the early British foundations, plainly but prettily set out from 1682 to a grid with garden squares by Penn's 'surveyor general', Captain Thomas Holme (ill. 35).[156] Despite its professional authorship and small flourishes, the New Orleans layout was hardly more commodious or effective.

The European nations naturally tackled town-founding in the New World in line with their specific polities and temperaments. But the style of settlement cut across national stereotypes. When a native population had to be held in check or coaxed into loyalty by a minority of outsiders, modes of military government and development tended to converge. In India, for instance, so late as the 1850s following the disaster of the Mutiny, Britain went so far as to drop the indirect company rule upon which much of its empire had been based, and impose a model of military-engineering-led development and architecture through its famous Public Works Department or 'PWD' – a sort of Indian Génie.

That approach could not work in the settler-culture of North America. There, an unceasing influx of colonists gave the white settler the buoyancy to take development, infrastructure and architecture into his own hands. In the process he would sweep aside both the colonizing powers and the indigenous peoples, while proclaiming that providence had chosen the territory for his own: 'the land was ours before we were the land's', as Robert Frost put it with such sinister ambiguity. No Colbertian ideal of centralized technical authority could withstand all that rampant energy. But that it exerted a surprisingly forceful backstage role in American destiny is proven by the strange history of the US Corps of Engineers.

THE CORPS

Without the French alliance, the patriots would probably have lost the War of Independence. At Yorktown, the climactic action of 1781, almost as many Frenchmen fought as Americans, exacting revenge for the loss of their colonies. And in the support France lent its insurgent ally, technicians made as big a difference as fighting men.

Except for a handful of veterans from colonial campaigns, the rebel states had few engineering experts when they declared independence in 1776.[157] Fearing for their towns in a world of warfare focussed on sieges and strongholds, they dispatched an agent to France to recruit help. It was not easy. The Frenchmen asked for salaries the Americans could not afford, and a status they could not guarantee. Their first catch was Philippe du Coudray, a notable artillerist, brevetted in September 1776 with the rank of 'Major-General in the Forces of the United Colonies' and dubbed commander of a non-existent 'Corps of Artillery and Engineers'.[158] After various intrigues, du Coudray and a ragtag of volunteer-officers managed to foregather in Philadelphia, among them a juvenile artist, Pierre Charles L'Enfant (ill. 36).[159] Further French gentlemen-volunteers trickled in, jockeying for place and sometimes claiming skills in engineering as well as combat.

Meanwhile Benjamin Franklin managed to negotiate a formal alliance with France. Hence the prompt dispatch early in 1777 of four Génie engineers, with Louis du Portail in charge. He was the man the Continental Army needed. An officer of standing, vigour and judgement, du Portail was on record as insisting that the focus of military engineering should be neither abstract science nor high-class construction but the arts and requirements of attack and defence.[160] With du Coudray already in post, artillerist and engineer inevitably quarrelled. Then in September 1777, du Coudray drowned in an accident. That left du Portail in control, and things began coming together.

Soon du Portail was writing a memorandum to General Washington, urging that distinct companies of sappers be formed under his overall command. Labour was the priority, so he had most to say about recruiting sappers. About their officers he added:

> If it be important to choose the privates in the companies, it is much more so to choose the officers. The Congress ought, in my opinion, to think of forming engineers in this country when we shall be called home. The companies of sappers now proposed might serve as a school to them. They might there acquire at once the practical part of the construction of works . . .

36. Pierre L'Enfant (1754–1825), portrait medal.

Washington and Congress acquiesced. Three companies were formed up in time for the campaigns of 1778; and in the general orders it was explicitly stated that 'as this Corps will be a school of engineers it opens a prospect to such gentlemen as enter it and will pursue the necessary studies with diligence, of becoming engineers.' Later, Congress ratified the scheme, resolving that 'the engineers in the service of the United States shall be formed into a corps, and styled the "corps of engineers"'.[161]

The first, short-lived American corps of engineers originated therefore in wartime urgency; gentlemen-officers learnt their trade in the field alongside the men. Despite tensions between English-speaking troops and their French superiors, du Portail welded together an impressive team of sappers and officers. The latter ranged from Washington's friend, the bullish, half-literate surveyor-millwright Rufus Putnam, through the Franco-Polish officer and idealist Thaddeus Kosciuszko to L'Enfant, the artist-draughtsman. Burgoyne's supposition from American mistakes at Ticonderoga in 1777 that 'they have no men of military science' came to be words he might have eaten.[162] Neither the temporary capture of du Portail nor his long leave of absence after Yorktown hindered the efficacy of the new corps.

Wartime constructions are seldom meant to be permanent. Apart from a gigantic mole or bridge forming part of the defensive works at Ticonderoga, little the patriot engineers built was impressive. But they did embark on one stronghold that was to be their lasting memorial: West Point, christened the key to the continent because it commanded the Hudson and bottled up the British in New York. West Point became the corps' main base, and the focus of debate over its future when the war ended.

A flurry of memoranda went the rounds in 1783–4. The departing Frenchmen were adamant that the infant republic needed both a permanent corps and a school. But a sapient du Portail could forsee the difficulties:

> A Vauban . . . should be necessary in this moment to the United States and nobody, unless he thinks himself as able a man as that famous marshall, can undertake, without the greatest diffidence, that difficult work. And he who would undertake it, without any fear, proves that he has not the least idea of it.[163]

Putnam too wanted the corps maintained, but his topographical and Anglo-Saxon cast of mind preferred to list the places he wanted to see fortified, with West Point at the centre of the web as 'the Grand Arsenal of America'. Washington summed up army sentiments:

Of so great importance is it to preserve the knowledge which has been acquired through the various stages of a long and arduous service, that I cannot conclude without repeating the necessity of the proposed institution, unless we intend to let the science become extinct, and to depend entirely upon the foreigners for their friendly aid, if ever we should be again involved in hostility. For it must be understood, that a corps of able engineers and expert artillerists cannot be raised in a day.[164]

To one Frenchman the corps' perpetuation was of personal interest. That was the versatile L'Enfant. Though lacking formal grounding in engineering, the young artist had risen to the rank of major and made valuable connections. Back in America after a spell in France and touting for work, L'Enfant rounded off these paper plans in December 1784 with his vision of a corps of peacetime engineers, and the role it could play in developing the country – under his leadership:

The duty of the said Corps shall be to attend to and have the direction of all the fortified places, that of all military and civil building, the maintenance of the roads, bridges and every kind of work at the public charge. Surveys of the several places shall be by them made and properly drawn with a view to make out an atlas of the whole continent from which the supreme power may be able to obtain a more just idea of the situation and form a distinct opinion upon its advantages and defects. To these plans shall be added proper notes and remarks with schemes for taking advantage of good positions or of preventing the defects of some unavoidable inconveniency.[165]

That came far too late. Why, questioned Congress, should the constituent states of a debt-ridden nation, secure at last, pay for a central body of experts to sit around anticipating a hypothetical threat or planning hypothetical developments? Du Portail having gone home, the corps had been wound down a year earlier. The sappers and miners departed West Point, leaving a token garrison behind them.

Yet the issues broached by L'Enfant and the others never left the political agenda. With hindsight, the renewal of a small artillery and engineering corps at West Point in 1794 and, after further stops and starts, the founding of a military academy there in 1802 were inevitable. Those steps came soon enough for the value of French technique, discipline and instruction to be remembered; indeed some French veterans became teachers in the early academy.[166] Bolstered by the fresh war against Britain in 1812–14 (when L'Enfant turned down a West Point professorship[167]), French hegemony over American military engineering remained preponderant until the Civil War. France, mostly a staunch friend, had an explicit model for furthering engineering and construction; Britain, potentially hostile, still had none. 'I do not know how it has happened', puzzled the superintendent of the infant West Point academy: 'I cannot find any full English Idea [for] what the French give to the profession'.[168] Still, a balance had to be struck. Many Americans feared France's revolutionary tendencies, while in language and craft traditions the country's slant still leant towards Britain.

Once again, war and the threat of war had propelled a nation to forge and then maintain a specialized workforce, so sharpening and focussing its expertise in construction. Could those personnel and skills be put to peacetime use, in the manner airily sketched by L'Enfant? If you were used to centralized governments, it did not look hard. Even after its revolution, France had managed to maintain not one state corps of engineers but two. But under a democracy in constant negotiation between states and interests, the politics were different. British experience hinted at the problems; and both the geography and the hard-won constitution of the United States were less centralized than Britain.

Nevertheless a peacetime role for the American corps was a bargaining-chip. If the country had to have a permanent federal officer corps, it might as well do something useful. That was the clinching political deal hammered out which allowed the Corps to gain permanence and its academy to flourish.[169] The compact has lasted ever since. The US Corps of Engineers exists primarily to assist its country's armed forces, but will turn its skills to any civil task that the Federal Government invites it to undertake. In a centrifugal, free-enterprise society, such interventionism is always controversial. So the Corps is often presented as exercising a covert or sinister role. It could never have grown up but for the United States' ravenous need for infrastructure, and the pace of technological change while it was getting into its stride.

The majority of its early practical efforts for peacetime American development leant towards waterways and their management. *Structures in the Stream*, the title of Todd Shallat's study of the ante-bellum Corps, sums up the thrust of the work. Under its aegis coasts were surveyed and made safer, harbours improved, river navigations eased, canals and dams built. In the interior the Corps and its younger affiliate, the Corps of Topographical Engineers – at times subsumed within the senior organization, at times separate[170] – became instruments in the Jeffersonian strategy whereby the Federal Government assisted in the opening up of new territory by guaranteeing major communications. That was the thrust of the Gallatin Report of 1808, formally the 'Report of the Secretary of the Treasury on the subject of Public Roads and Canals'. For federalists hoping to further infrastructure, Treasury support and the science of the Corps had to be harnessed in tandem. For those who feared federalism and monopoly, any accommodation between the two bodies had to be resisted. So the interventionist vision was tarnished by endless manoeuvres and compromises.

Under the General Survey Act of 1824, the Corps acquired a wider official remit for planning public works; 'the military will be incorporated with the civil, and unfounded and injurious distinctions of every kind be done away', dreamed Monroe.[171] But it never took on much above-ground civil building of the kind which concerns this book, federal building projects apart. Though the topographers surveyed and mapped away, they were usually obliged, says Shallat, to 'stop just short of construction, planning without building, promoting commerce and educating the nation through the science in their surveys'.[172] That was the American way. Meanwhile the West Point engineers, snobbish about their practice-bent topographical colleagues ('we see no where among them, not any, not a single individual . . . whom we could see introduced into the Corps of Engineers without mortification or pain'[173]), focussed upon the higher knowledge of their subject and the defensive works they felt to be their real raison d'être.

When the Corps did undertake works, they took too long and cost too much – in short, were done too well in relation to the generally low standards of ante-bellum civil construction. Only in a few specialized fields like fortification and lighthouses (entrusted to the Corps in the 1830s) was its grip on practical building secure. But it could never bury itself so far in science as to lose its political way, as the Génie had done. Since the United States possessed no Corps des Ponts et Chaussées, it was down to the army engineers to honour the compact of 1824 and proffer technical advice on any great improvement meditated by government. Likewise, after initial reluctance the West Point academy was forced into training engineers who shifted into the buoyant civil sector, and so raised the general standard of construction (see pp. 448–50). The want of a hard and fast division between civil and military endeavour proved catalytic for American development.

What the engineers of the Corps did not do was to interfere in the urban development of the constituent states, laying out towns or building public buildings. Those, the constitution had confirmed, were for private and local initiative. Meddling in such matters being out of the question, any ambitions in that direction were suppressed. They came to be countenanced in one city only – Washington, the heart of federalism. With a glance at the tenor of Washington's development, this sketch of the shifting forces behind 'imperial works' in three countries can draw to a close.

Engraved for the Massachusetts Magazine, June 1789.

View of the FEDERAL EDIFICE in NEW YORK.

37. Federal Hall, New York, as improved by L'Enfant, from the *Massachusetts Magazine*, June 1789.

WASHINGTON AND L'ENFANT

This time L'Enfant was ahead of the game.[174] His notion of a peacetime corps of engineers rebuffed, he had metamorphosed into a successful architect and decorator in New York. Public recognition came with his recasting of Federal Hall (ill. 37), undertaken by New Yorkers in the hope of making it 'the permanent residence of the Federal Legislature'.[175] George Washington was inaugurated first president in the spruced-up building in April 1789. But he had bigger ideas which L'Enfant and the commander he idolized then most likely discussed. That September, at any rate, Washington received a begging letter in two parts. In the longer half L'Enfant asked to be made 'Engineer to the United States', so that the nation's neglected coastal defences could be addressed. Here was his old plan, set out now in personal terms but capped with the reminder that

the advantages to be derived from the appointment will appear more striking when it is considered that the sciences of military and civil architecture are so connected as to render an engineer equally serviceable in time of peace as in war.

Prefacing these remarks came some briefer lines on the topic of a federal capital. It would need to be 'on such a scale as to leave room for . . . aggrandisement and embellishment', urged L'Enfant. Would Washington do him the favour of employing him 'in this business'?[176]

All that came long before Congress approved the lower Potomac for the new capital in July 1790, and longer still before L'Enfant received instructions in March 1791 from Washington's Secretary of State, Thomas Jefferson, to proceed to Georgetown, where he would find the surveyor-mathematician Andrew Ellicott making a survey of the designated federal territory. Somewhere in that interval Washington, unimpeachable in influence, personally appointed the voluble Frenchman to a major but undefined role in planning the new city. There was no competition, nor is there evidence that L'Enfant showed the President any indicative sketches. The arrangement lasted just eleven months.

Washington and Jefferson bent over backwards to be generous to L'Enfant, and posterity has done the same. If vision and idealism are enough, that generosity is justified. But the planning of cities demands more. Normally L'Enfant is referred to as an architect or as an engineer, and he has not seldom been called brilliant. In neither role did his thinking go as deep as it did far.

At heart L'Enfant was an artist-designer. All we know of his early years is that his father was a painter good enough to become an academician, that the son spent some of his childhood at Versailles, and that he entered the Académie Royale de Peinture et Sculpture in 1771, aged seventeen. Of the next five years we are ignorant. Then came L'Enfant's American adventures. Quick, bold, gallant and far-sighted, he had a good war, learning on the move about fighting and fortification. Before he felt himself slighted and grew touchy, his charm must have been great. But creditable though his architecture and decoration in New York were, they offered slim justification for the promise that the President certainly gave him of designing Washington's main buildings.[177]

As for laying the city out, it is far from clear that that belonged to his brief. His claim to do so was based on no practical experience. Paris and Versailles he knew well, and a memory of the latter's autocratic geometries appears to pervade his plan. The rest is guesswork, too often informed by sources doubtfully available to L'Enfant. Perhaps the most attractive guess comes from André Corboz, who thinks he may have seen some of the imaginary urban projects devised by Perronet's students at the Ecole des Ponts et Chaussées.[178] That makes sense; the L'Enfant plan for Washington is redolent of the brilliant student project. He is not known to have seen other European cities besides Paris. Nor is he likely to have studied any before getting to Georgetown, for once there he appealed to the superior knowledge and library of Jefferson, who by return sent him maps of no less than twelve cities.[179]

But inexperience seldom hinders ambition; and L'Enfant was intent on 'a new and original way', he told Jefferson.[180] The grand plan he concocted with such speed and brio seems to have been mugged up largely out of his own head. The upshot was no capital city tailor-made for democracy, not even a backdrop for the neo-Roman republic of virtuous farmers that Washington and Jefferson seem to have aspired to, but an absolutist fantasy. The wonder is not that the artist came unstuck, but that the dream survived. Over time, it was to have questionable influence on the way government in the United States came to think of itself.

In surveying the federal territory, we have seen, L'Enfant had the official assistance of Andrew Ellicott, a competent mathematician, astronomer and instrument-maker, later a professor at West Point. The Ellicotts (Andrew was assisted by his brothers Joseph and Benjamin) came from a family of millwright-mechanics who had turned to official land and boundary surveying.[181] Theirs were skills in the English tradition. Had Washington been a regular, English-style town like Philadelphia, Ellicott might have had the planning of it, moving from the survey of overall terrain first needed to the gridding of a smaller area within it, then breaking it down into blocks of carefully selected size, dividing them into lots and working out in what order they should be auctioned. These seem to have been the procedures which Jefferson anticipated when he made his own sketch for a smallish federal city on a grid: only with wide roads, ample space for the President and the Capitol, and potentially pleasant 'public walks' between them, stretching down to the water (ill. 38). Sophisticate and Francophile though he was, Jefferson always remained devoted to the methodology and psychology of the grid. L'Enfant on the other hand despised simple grids, calling them 'tiresome and insipid . . . a mean continuance of some cool imagination wanting the sense of the real grand and truly beautiful'.[182]

The interpellation of a plan on a far vaster scale and headier layout (ills. 39, 40) made a rough ride inevitable, despite Washington's initial support. L'Enfant's formal artistry was evident in the elevation of the Capitol, and the conversion of Jefferson's public walks into a graceful mall flanked by buildings, shops and a canal, leading to a kind of place royale before the president's house. There were dashing touches too, like the trick of taking waterpipes under Capitol Hill and unleashing a cataract below the legislature. But the whole thing wanted practicality. The plan lacked defences (which would have come in handy in 1814, when the British burnt the public buildings), and broke up the inevitable grid with symbolic diagonals named after the states of the union. Pretty on paper, diagonal axes were less telling on the ground and always problematic to develop, as had been known since the radial towns of the sixteenth century. Nor had L'Enfant attended to the valuing or disposing of lots. Perhaps he thought others would sort all that out. His strategy was for the Federal Government to raise a million dollars to pay for the public portions of his art-vision: this in a climate when the states grudged every penny

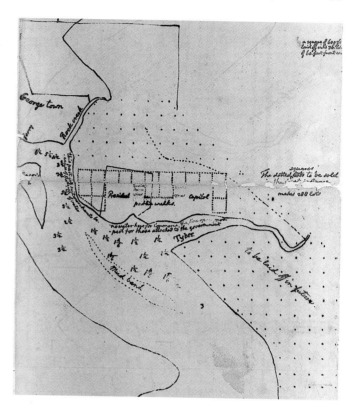

38. Thomas Jefferson's sketch-plan for Washington, 1791. The relation between the White House and the Capitol is anticipated, linked by public walks leading down to the water. This arcadian arrangement is framed by squares for division and sale as lots; further potentially gridded areas are 'to be laid off in future'.

39. L'Enfant's plan for Washington, 1791, showing the transformation of Jefferson's sketch into a much larger city with a canal and grid-busting diagonal axes.

for the capital project. It says much for the trusting hearts of Washington and Jefferson that they embraced L'Enfant's plan. By August 1791 he had set it in aspic.

Things soon soured. 'Three separate acts of procrastination, impetuousness or insubordination', as Reps calls them, terminated L'Enfant's role in planning the capital.[183] Pulling down a house that stood in the way of one of his future avenues could be put down to a silly squabble. But his refusal to release his plan in time for the major auction of lots in October, and the confusion and potential loss of income that engendered, were a real warning. Even the indulgent Washington, at fault for not defining his architect's brief, had his limits. The commissioners whom he had appointed to oversee the city, men of no enlarged outlook,[184] were now given paramount authority over L'Enfant, who would countenance as patron none but his chief. A trial of strength ensued. The outcome of this third offence was predictable. 'He was ultimately dismissed by a government that admired his enthusiasm but could ill afford it.'[185] And so the Frenchman flounced out, protesting his idealism to the last:

Permit me also to assure in the most faithful manner that the same reasons which have driven me from the establishment will prevent any man of capacity, impressed with the same disinterested views by which in every stage of it I have been actuated . . . from engaging in a work that must defeat his sanguine hopes and baffle every exertions. Should this business fall into the hands of one devoid of these impressions,

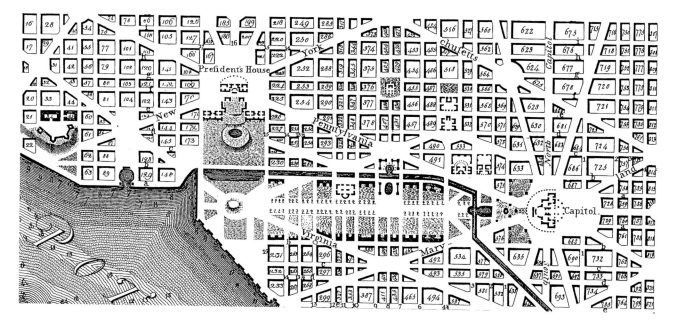

and of course insensible to the real benefit of the public, how great so ever his power may be, self-interest immediately becomes his only view, and deception and dishonour issue.[186]

There spoke neither the flexible architect nor the disciplined engineer but the egotist artist-designer. Still, L'Enfant had done enough. Though a humiliating wrangle about compensation ensued, deception and dishonour were averted. Instead there followed a scramble toward completing Washington's public buildings in time for the politicians to arrive in 1800; delays thereafter as the economy of the new capital stuttered and stalled; and then the litany of alterations that any city-plan is certain to undergo. Through them all ran an uncanny fidelity towards L'Enfant's paper-dream of the city as a work of art on the part of his many successors – not least the corps he had longed to command.

WASHINGTON AND THE ENGINEERS

Once again it was war that drew the engineers into the picture. During the confrontation of 1791–2, L'Enfant's Man Friday had been Isaac Roberdeau.[187] From a Philadelphia family with West Indian interests, Roberdeau had learnt his engineering skills in England. After Andrew Ellicott had brought him into the Washington project, he shifted over to help L'Enfant. Such was his loyalty to the Frenchman that he disobeyed the Commissioners and found himself briefly in prison as a result. Roberdeau tried in vain to help L'Enfant with his next project too, an abortive plan for the town of Paterson, New Jersey, before setting up as a surveyor-engineer in Philadelphia. Later, his hope for a federal career seems to have been blocked by Jefferson.

Twenty years on, the war of 1812–14 led to the formation of a Topographical Bureau within the War Office, later the Corps of Topographical Engineers. Roberdeau enlisted, served in it and taught at West Point. In 1818 he was assigned to Washington to head the young body and steer it towards national planning. Local uses were already being found for the engineers already there. The Capitol having been burnt by the British, they had helped Benjamin Latrobe make good the damage. Under Roberdeau, they laid on spring water to the White House and were drawn into grading newly laid-out streets.[188] He must often have reflected on the irony of his position.

Then came the General Survey Act of 1824, removing legal limits as to how federal engineers might be employed. In respect of national buildings there was one great

40. Central portion of Andrew Ellicott's plan for Washington, 1792, incorporating minor modifications to L'Enfant's plan. This detail of the central area shows the original Mall's restricted length, and the siting of commercial or public properties between the canal and the Mall.

obstacle: the rights, pretensions and competences of architects. The prospect of federal employment naturally attracted architects, both for Washington's own buildings and for those that the government was starting to sprinkle across the country: customs houses and marine hospitals came first, post offices following. Though Latrobe had for a time been called 'Surveyor of Public Buildings', there was no official federal architect, certainly no building organization – no Bâtiments du Roi, not even a limited King's Works. The quixotic politics of the capital, where patronage was a way of life and an all-change threatened every four years, made any private architect's involvement in federal projects insecure. By contrast, army engineers as government employees offered continuity – and the prospect of higher building standards.

Washington's first public buildings had been designed by a ragbag of architects, builders and amateurs, some skilled in construction, some not. Though things gradually improved, there was little reliable connection between quality of design and of construction before 1829. In that year Robert Mills, most professional of Washington's early architects, returned to the capital.[189] Mills had worked in his youth on the White House under Hoban, possibly too on the Capitol under Latrobe. By now he was a thoroughly experienced architect and technician. With the Jackson administration getting into its stride and the capital's population showing signs at last of concerted growth, building was again in the air. On the strength of the 'Jacksonian surplus' Mills amassed a portfolio of customs houses, hospitals and Washington buildings, notably a headquarters for the all-influential Treasury Department, started in 1836. By then he had been dubbed 'Architect of the Public Buildings'. Later that year the administration changed. A rival, Thomas Ustick Walter, promptly unleashed a broadside against Mills's manner of construction. Walter was probably wrong; Mills took great pains with building and fireproofing. He survived, going on to design the General Post Office and supervise the Patent Office, but he had been damaged. In 1842 he lost his job and was bundled out of his office in the incomplete Treasury Building. Procurement of public buildings in the capital reverted to a free-for-all.

With Mills's departure, the campaign of the army engineers to set their mark on federal building construction stepped up. From the mid-1840s the Topographical Engineers began supervising marine hospitals across the country. In Washington, Colonel John J. Abert, Roberdeau's successor as head of the 'topogs', and the Philadelphia architect William Strickland concerted plans for a new War and Navy Building, as well as alterations at the Capitol. Sponsored by a 'Committee of Buildings of the War Department', these plans were avowedly aimed at avoiding architectural mistakes and extravagance.[190] Neither came to anything. But Abert's ambitions were confirmed when he entered the competition of 1850 for completing the Capitol. Likewise Joseph Totten, head of the main Corps of Engineers, was drawn at this time into the preliminary stages for the Smithsonian project. Interventionism was in the air.

Fresh arrangements for federal building made under the Fillmore administration (1850–3) ushered in a brief golden age for the Corps in Washington, lasting up to the Civil War. On the architectural side, the Treasury Department came up with the notion of a full-time 'Supervising Architect' – in the first place, Ammi B. Young – who would design and maintain a high proportion of federal buildings. The arrangement exasperated private-sector architects, but this time it stuck. Above Young, the Department set up a Bureau of Construction, headed by an army engineer, Alexander Bowman. The idea was for Young and his in-house staff to design and plan, while Bowman and his engineers imposed methodology, consistency and management on the far-flung projects, so as to combine 'improved character and construction' with 'great savings to the public': so the Treasury claimed. Or, as Bowman put it in his first annual report:

> The preparation of the plans, specifications, estimates, and contracts in this office, under the immediate direction of the department, where the number of occupants and the precise amount of business to be transacted in each building are known, has

41. Centering for the arch of the Cabin John Aqueduct Bridge, Cabin John, Maryland, 1859. Built by Montgomery Meigs and the French-trained Alfred L. Rives as part of a new aqueduct carrying water to Washington, this low-slung masonry arch at 220 feet was the longest in the world for some forty years.

many advantages. Errors committed in buildings already in use can be avoided, a proper apportionment of office-room made, and such an arrangement of the different offices as will facilitate the transaction of business effected. This can be better done where the conveniences and inconveniences of similar buildings are subjects of frequent discussion with those who occupy them, than if the buildings were designed by some one less acquainted with the uses for which they are required, and who would probably be more likely to make a beautiful than a suitable structure.[191]

Here is the classic argument for the in-house team, public office or bureau d'études in respect of buildings in series. Mutatis mutandis, it might have come from the pen of Vauban, or from an architect in the British public sector of the 1960s. The technical skills of engineers are not mentioned, so probably they were being employed mainly because they had been trained to manage, as architects were not. But it is clear that Bowman's team did more than just supervise construction.

In Washington itself, the Corps of Engineers now burst into open activity. That ebullition coincided with the War Department's campaign under Jefferson Davis (Secretary of War, 1853–7) to seize hold of the capital's public development. It was under Davis that the egregious Captain (later General) Montgomery Meigs reached the apogee of his career. Meigs was assigned in 1851 to survey the District of Columbia's water supply, and came up with a scheme for an aqueduct of Roman magnificence. Railroaded through Congress and constructed with equal aplomb, the Washington Aqueduct is acknowledged today as one of the great American engineering projects (ill. 41). But it was tinged throughout with Meigs's overweening vanity. 'Over the whole fourteen miles of construction', comments Albert Cowdrey, 'he had his name engraved on bridges, gatehouses, staircases in pipe vaults, even on the derricks and hoisting gear'.[192]

The Washington Aqueduct belonged to a field in which the Corps could claim authority – that of water management. Not so Meigs's intervention at the Capitol.[193] In 1853 the War Department at last got its way and took over that interminable project; Meigs the army engineer was thrust upon Walter, the architect in charge, because of the latter's poor handling of contracts. Had Meigs possessed tact, the pairing might have worked. So long as Jefferson Davis was in power, he got his way. Afterwards, the tide of his authority receded and he was eventually replaced, to everyone's relief. The best that can be said about his contribution to the Capitol is that he made the great iron dome,

42. The Capitol as completed, from a lithograph by E. Sachse. The Mall appears to the left of the dome encumbered with gardens and minor buildings, and Pennsylvania Avenue to the right. The dome is surmounted by the figure of Armed Freedom, raised in December 1863.

designed in the main by Walter and his able assistant, August Schoenborn, 'technologically feasible'. Too much of Meigs' time was spent on questionable decisions of taste. His non-partnership with Walter is a specimen of the entirely avoidable bad relationship between architect and engineer. Still, Meigs's contribution to Washington did not end in shame. Late in life he built the fine Pension Office (1882–7), now the National Building Museum. It is testimony to the fact that when acting on his own, like so many army engineers, he could be an excellent architect.

In time of civil war, Thomas Crawford's effigy of 'Armed Freedom' was hoisted into place on top of the Capitol Dome to mark its completion in December 1863 (ill. 42). Unthinkable in the demilitarized days when Washington, Jefferson and L'Enfant had dreamt of an undefended capital replete with virtuous public buildings, bowery carriage avenues and peaceable riverside walks, the figure symbolizes how much the United States had changed, and how fast. It is also a reminder of the fact that federal armed forces had insinuated their way to the heart of government, in peace and war alike, by virtue of their needed techniques and competences.

The rest of the Corps of Engineers' part in the Washington story need not keep us long.[194] After the war, ever fiercer lobbying by architects, backed by a professional institute based in the capital, prevented the Corps from taking over major buildings outright. But it was always there in the background, ready to monitor or be superimposed upon any project mired in political or constructional difficulties, as happened with the Library of Congress (1887–94). Talk thereafter of getting the Corps to supervise all federal construction projects came to nothing.

In broad oversight of the city, its participation actually strengthened. In 1867 an Office of Public Buildings and Grounds of the Army Engineer came into being, with all the fed-

43. First plan for reordering central
Washington and the Mall, by Colonel
Theodore Bingham of the Corps of
Engineers, 1900. Most of the land
freshly reclaimed from the Potomac is
destined for military encampments
and parade grounds.

eral lands within the District of Columbia in its purview. From that developed the Corps
of Engineers' share in shaping Washington as a whole. Until the city acquired its own
elected government, a member of the Corps always sat on the three-man board,
appointed by the President, which governed the District of Columbia. 'As a direct result',
notes Albert Cowdrey, 'the Army Engineers acquired an unprecedented role in the regular, peacetime government of an American city'.[195]

Naturally it was the heart of Washington that became their special stamping
ground, particularly the Mall, confused and chopped up since L'Enfant's time,
though his monumental idealism had never quite been forgotten. Engineers completed
Mills's Washington Monument in 1877–84 and built Bacon's Lincoln Memorial
in 1914–22. More important was the preliminary work, often ignored or reviled, that
set the scene for the McMillan Commission and the full-throated L'Enfant revival of the
early twentieth century. It began with the Corps' strength, water management.
Between 1882 and 1900 the engineers reclaimed more than six hundred acres of
tidal flats and marshes from the Potomac at the west and south-west end of the
Mall. The effect was almost to double an area of already perilously large open
urban space, and to provoke rival claims as to how it might be filled in honour of
the capital's centennial.

Having done the spadework, the engineers were keen to make suggestions for a new
unified layout for the whole area. So it was that Colonel Theodore A. Bingham, head of
the Office of Public Buildings and Grounds, came up in the spring of 1900 with the first
viable plans for a 'centennial avenue' along a restored Mall (ill. 43). On the evidence of
architects, Bingham has been dismissed as 'apparently ambitious, impulsive, and rather
impressed with his own knowledge'.[196] In fact his plans were not half bad, though they

44. Perspective of Union Square at the east end of the Mall, made for the Senate Park Commission, 1902. The monumental language imposed upon Washington by the Commission's architects was far more ponderous than that conceived by the city's early architects and engineers.

took up much of the reclaimed land with military parade grounds and the like. Given the professional politics of the centennial, they stood no chance of adoption, but they offered a guide for much that followed. Having accepted partnership with a landscape architect, whose scheme was rather worse than his own, Bingham finally cooked his goose with some clumsy plans for extending the White House. He was certainly no architect.

All this played into the hands of Glenn Brown, the secretary of the American Institute of Architects, who had been lobbying all the while, looking into L'Enfant's plan and making his own equivalent. It was through Brown's manoeuvring that the McMillan Commission eventually fell into the hands of those veterans of the Chicago World's Fair, Daniel Burnham and Charles McKim, along with Frederick Olmsted junior. In the end it was the civil architects and landscapists of the City Beautiful movement who revived L'Enfant's vision – with an imperial pomposity that the military engineers, bent on tasks before them, had never thought to ape. And so a wearying Beaux-Arts version of what it was supposed the capital ought to have been came to be laid over Washington in L'Enfant's name (ill. 44). Though much revised, the so-called Senate Park Commission plan was essentially carried through. Its frigid, authoritarian classicism has continued to define the city in the minds of the public and its rulers ever since.

A LAST WORD

The story of construction in the hands of the Corps of Engineers could be extended onwards: to the inflation of its numbers and activities in response to the institutionalized violence of the twentieth century; to its peacetime role in building dams and infrastructure, foreshadowing the much-publicized activities of the Tennessee Valley Authority; or, back in the Washington area, to the gigantic project for the Pentagon, hastily drawn up by army engineers in 1941; and to the transfer of Colonel Leslie Groves from building that behemoth to a brand-new section of the Corps, the 'Manhattan District', charged with creating a secret new community to develop atomic energy as a weapon of war.[197]

A few of these topics find a later mention in this book (pp. 358, 395). But even the loose time-limits of this chapter have been exceeded, and its narrative must stop. The one thing left is to remind the reader that its theme has been not so much military construction and constructors as the way in which the policies and ambitions that beget them have engendered techniques, institutions and ideas useful to architecture and the entire public realm. So it seems right to end with France and leave the last emotions to an Englishman, Arthur Young, responding to the sight of the Canal du Midi:

> Leave the road, and crossing the canal, follow it to Béziers; nine sluice-gates let the water down the hill to join the river at the town. A noble work! The port is broad enough for four large vessels to lie abreast; the greatest of them carries from 90 to 100 tons. Many of them were at the quay, some in motion, and every sign of an animated business. This is the best sight I have seen in France. Here Louis XIV thou art truly great! Here, with a generous and benignant hand, thou dispensest ease and wealth to thy people. Si sic omnia, thy name would indeed have been revered. To effect this noble work of uniting the two seas, less money was expended than to besiege Turin, or to seize Strasbourg like a robber. Such an employment of the revenues of a great kingdom is the only laudable way of a monarch's acquiring immortality; all other means make their names survive with those only of the incendiaries, robbers and violators of mankind.[198]

2

IRON

1. Europe to 1850

ARCHITECTS AND FABRICATORS

It would be idle to try and pinpoint the first structural occurrence of iron in building. What can be done is to qualify the uses to which iron has been put and place them in the hierarchy of construction. Cramps, straps, bolts, nails, hinges and ties: such products of the forge were as old as the iron age itself. Plain in comparison with weaponry and armour, farm tools or even domestic ironmongery, they were subordinate features, added to supplement timber joints or masonry openings and so secure the grosser elements that bore the brunt of the struggle against gravity and decay.

The conspicuous elements in construction do the most to define architectural style. So though iron had contributed to structure since antiquity, and it was familiar to the great mediaeval builders, it was always regarded as supplementary.[1] Much of it lay hidden. Yet it could be bold and naked too, as in the open ties across the vaults of Sancta Sophia, or sinuous and ornamental, as in the door hinges of great Gothic churches – features that warn us not to simplify the division between structural and other uses of ironwork.

As with weapons, gearing and utensils, so also with construction, the wider availability of forged iron from around the time of the Renaissance intensified its use at almost all levels of western building.[2] At the loftiest level, there was growing recourse to iron in the peripheral 'chains' or continuous ties hidden within masonry to counter the lateral thrust of wide-span vaults and domes. Perhaps in emulation of the ancients, Brunelleschi's famous dome at Florence Cathedral included a measure of these 'catene di ferro', in the form of rods, clamps and bolts.[3] In the 1520s or 30s Sansovino repaired one of the domes at St Mark's, Venice, with a girdle of seemingly continuous iron,[4] while the dome of St Peter's, Rome, raised at the end of the sixteenth century, contained yet more iron in its chains. By 1682 Filippo Baldinucci, contributing to long-running arguments about cracks in the dome of St Peter's, could claim that 'there was never an architect, feeble though he might be, who did not know it was impossible to raise a structure of such a form and size without reinforcing it with chains in many places'; and the response of the experts who finally tackled its structural problems in the 1740s was to add more chains with more iron.[5] The dome of St Paul's in London (1704–7) likewise conceals several series of iron chains (ill. 46).[6]

From time to time further examples of chains binding domes, steeples and other features come to light.[7] In the same way, Perrault's Louvre colonnade of 1667–74 renewed the Hellenic use of hidden iron ties to support classical entablatures with wide intercolumniations (ill. 47). That technique was to help neoclassical architects who favoured trabeation over the arch when using the orders on a monumental scale.[8] In another usage, Wren employed hidden iron hangers to help truss wide-span floors – in vertical form at Hampton Court (ill. 48) – as diagonal straps at Trinity Library, Cambridge.[9]

From early in his architectural career Wren was manifestly at ease with iron reinforcement. Asked to report on the stability of Salisbury Cathedral in 1668, he noted that the original 'Artist' had braced the great spire with 'many large bandes of Iron within and without, keyed together with much industry and exactnesse'. Though critical of the

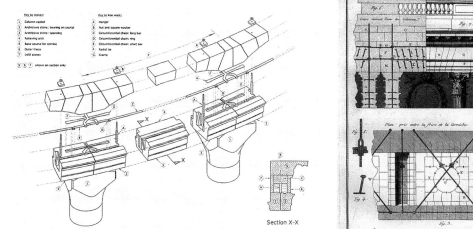

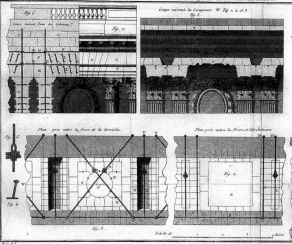

Section X-X

46. Isometric drawing showing part of the iron 'chain' system over the peristyle of St Paul's Cathedral, drawn by Robert Bowles. One of several such chains in the dome of St Paul's. Hidden iron bracing of this kind and complexity was common by this date (1704–7).

47. Iron reinforcement of Perrault's Louvre colonnade, from Pierre Patte, *Mémoire sur les objets les plus importans de l'architecture* (1769). The hidden iron increases the width of span feasible with intercolumniations of limestone ashlar.

48. Section of Wren's south range at Hampton Court drawn by Daphne Ford, showing various uses of hidden iron ties. Note the hanger helping to stiffen the 30-foot span of the top floor by linking it to the roof truss.

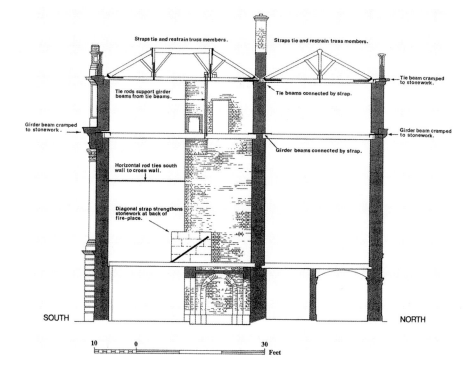

procedure because of the uneven quality of the old iron and its susceptibility to rust, Wren had no compunction in recommending further iron bracing, so long as it was fabricated 'at some port towne where they worke Anchors and other large worke for shipps, for I have found by experience that large worke cannot be wrought sound with litle fires and small bellowes.'[10] Nothing seems to have been done immediately at Salisbury. But when a few years later Wren built the tower of his first major London church, St Mary le Bow (1671–3), an 'anchorsmith' supplied the ring of iron cramps and wedges over the main tower arch.[11] The great chains at St Paul's follow on logically.

Already we have drifted into discussing the use made of iron specifically by architects. At the Louvre, for instance, Rowland Mainstone believes that the decision to use iron rested with Perrault, who had 'a lively interest in the use of metal in architecture'.[12] But in all this, there was no question of showing or expressing the ironwork. The chains, ties and hangers remained extras behind the scenes allowing designers to say better what they wanted to say, in their predetermined language of architecture. As yet this conceal-ment of a structural aid seemed unexceptionable, indeed inevitable.

The spur to these extensions of construction came in civic buildings of special expense and style designed by official architects – able constructors like Perrault and Wren with a scientific bent, yet primarily architects, not engineers. True, professional boundaries were blurred. Even so, the main types of 'engineer', so denominated, who dealt with con-struction – mainly military or civil officials holding appointments under national, colo-nial or urban régimes – appear to have played scant part in these pre-industrial stirrings of an iron architecture.

The relationships involved in this aspect of procurement were therefore chiefly between architect and fabricator – 'ferraio', 'serrurier', or smith. Metallic construction is commonly associated with engineers. Yet in buildings proper it is striking how long it takes before the engineer in the sense given above gets much grip upon iron construction. As we shall see, the pattern of collaboration between architect-designer and smith-founder-fabricator survives into the age of the railways and beyond. Some of this chap-ter pursues the relations between architects and engineers proper in buildings making use of iron. But much of it is about how architects engaged with those who supplied, made and fixed iron – the fabricators and contractors.

Who are these fabricators and contractors, on whom the extension of iron construction will so much depend? The pattern of their employment is fluid. Industrialization revolu-tionizes the iron-working trades. Labour divides and regroups in fresh permutations. The smiths and serruriers of the eighteenth century give way to a multiplicity of founders, fab-ricators, suppliers and fixers. We may call them engineers if we like, but from the profes-sional standpoint they start out as engineering contractors or subcontractors. Up to about 1850, just as architects and builders shade into one another but can broadly be distin-guished by their contractual duties and functions, so a line can be drawn in most European countries between official engineers (supplemented by a handful of what we should now call consulting engineers in private practice) and engineering contractors.[13]

In the second half of the nineteenth century, the wheel turns again. Trained engineers flow out of the new technical schools, taking over the design of iron structures and aspects of contracting alike. Gustave Eiffel, proud to label himself 'ingénieur et con-structeur', is a case in point. It would be correct still to call him a contractor-fabricator; at heart he was more of an entrepreneurial contractor than a designer. But because of his numeracy, his articulacy, and his appetite for innovation, posterity defines Eiffel as an engineer. The transition happens naturally. For though in many European countries nineteenth-century professional protocols (extended and strengthened in the twentieth century) discourage architects from becomng contractors, the fluidity and breadth of engineering prevent the professional bodies for engineers from doing the same. In so far as engineers engaged in construction need safeguards for their status, that is increasingly secured by their education – in the richer French phrase, their 'formation' – not by the criterion of whether they are remote from trade.

As we trace the shifting story of the relationship between architects and engineers in respect of iron, it will be well to keep in mind that as a material which permits, indeed requires, a high degree of preparation and fabrication off site, iron lends itself well to what is today called the 'design and build' approach. At all stages of our story, the iron engineer is often a contractor too. And the closer the link between design, fabrication and assembly, the greater the likelihood that the architect will be cut out from the loop. As industrialization intensifies, it becomes vital for him to impose a controlling idea or vision. This, we shall see, is what architects often try to do when building with iron.

In the development of iron construction were the fabricators leaders or followers? The burden of this chapter is that there is no single pattern. Take founders. In British struc-tural ironwork of the early industrial period, they might be of several types. Some sup-plied castings to order; others were drawn some way into the design and sizing of

columns, trusses and beams; others again progressed to designing or building complete structures and can be classed as engineering contractors or even engineers. Similar developments took place in mechanical engineering, where the smarter makers of iron machine-parts turned themselves into designers and erectors of complete machines.[14] In the following pages sundry types of founder and ironworker will be seen contributing to the building concept and process. Almost always there is a different nuance of relationship between designer and fabricator.

Iron and the Auditorium in England, 1690–1820

The natural moment to start the clock for an 'iron architecture' is when iron becomes a primary and visible element in building structures. That first took place in England around 1700. It owed more to fabricators and inventors than to architects or engineers. In France, by contrast, architects were to the fore in promoting the structural development of ironwork in buildings. Yet despite the national instinct for clarity in construction, until well into the nineteenth century most such iron in French buildings went on being hidden in walls, roofs and floors.

The first occasion we know of when iron confronted the world with its unaided structural potential, not as a supplement but as the primary means of support, was conspicuous. In 1692 the House of Commons needed internal renovation. Wren took on the job.[15] Among the improvements were galleries all round, carried on console brackets except on the short entry side. Here, to save space and safeguard sightlines, they were supported (to quote from a memorandum of Wren's after the work was finished) on 'two iron pillars and capitalls of Tijou's worke'.[16] This means Jean Tijou, the Huguenot smith who wrought rich gates for Wren at St Paul's and Hampton Court. Next year Tijou published *A New Booke of Drawing*, which included a plate showing just such a column, base and capital, all of iron.[17] The memorandum's wording, the change from the conventional timber columns shown in his original drawing for the chamber, and the authorship of the publication all point to these columns being Tijou's invention, though the fact that Wren was at ease with iron can only have helped. Here then was an idea advanced by a craftsman known for his ingenuity, and keen to widen the market for his material and trade. Tijou's prowess was manifest in the exuberant capitals, while the columns were doubtless cast from charcoal-smelted iron, in the then-customary manner of English cannon.

50. The Iron Bridge at Coalbrookdale Bridge, painted by William Williams, 1780.

There was a sequel. In 1707 came the Act of Union. England and Scotland were formally united; Scottish Members of Parliament joined their English counterparts in London. Accordingly, the House of Commons galleries were deepened to accommodate the fresh influx of members, and needed new supports. Four extra 'iron pillars' of the same appearance were now extended along each of the two main sides for the sake of space and sightlines (ill. 49). This second episode is ill-documented, but Wren was seemingly less involved, Tijou not at all. On this occasion the bunchy Corinthian capitals were of wood not iron, lavishly carved by Grinling Gibbons.[18]

Here then by negotiation between architect and fabricator, bare structural ironwork was for the first time slipped into the role of open support in the interior of a building, and a major public building at that. The apparent motive was neither fire-risk nor the strength of iron but space-saving, while its unembellished plainness and slenderness were set off with separately crafted capitals.

Not until the 1760s did open structural ironwork resurface within British buildings. Once it did so, it picked up speed. The fresh rhythm proceeded from improvements in iron manufacture. Coke-smelting spread and sand-casting broke new ground, making cast iron easier and cheaper to make, while fresh puddling and rolling processes from the 1780s eased the production of wrought iron. Within the building industry, cast pipes, railings, stoves and other fittings became common and affordable, allowing the ironfounder to encroach on the preserves of plumber, smith and even carpenter. Individual components of architecture now became the target for ironmakers hungry for markets.

That the earliest extant iron architecture was promoted from the supply side is confirmed by its most famous symbol, the Iron Bridge at Coalbrookdale of 1777–9 (ill. 50).[19] Bridges were the aspect of early iron architecture to which specialist engineers contributed the most (see pp. 315–28). But not yet at Coalbrookdale. The first stimulus and design for the bridge came from T. F. Pritchard, an architect with all that profession's appetite for novelty. Pritchard is not credited with special mathematical or engineering skills. What he did possess was closeness to the Shropshire ironmasters. He had designed fireplaces executed in iron by Abraham Darby's Coalbrookdale Company, and he seems to have built a brick bridge at Stourport on an iron centre, temporary or permanent, in 1774–5. 'Whoever inscribed on Pritchard's portrait "Thomas Farnolls Pritchard, Archt, Inventor of Cast Iron Bridges, 1774" was essentially correct', Barrie Trinder has written.[20]

51. St Chad's Church, Shrewsbury, interior. George Steuart, architect, 1792–4. Early use of superimposed iron columns continuing through the galleries to help support the ceiling and wide-span roof.

Yet, mortally ill before the Coalbrookdale bridge started on site, Pritchard died before its completion and cannot have 'sized' its cast-iron parts. Though the idea may have been his, the bridge was primarily an ironmasters' accomplishment and advertisement. Nor did its mastery have to do with engineering calculations of the type deployed today. A triumph of empirical manufacturing, it is a monument to the virtuosities of the Coalbrookdale Company's foundry and labour force, to the drive of ironmaster John Wilkinson, and to Shropshire's iron-working pride.

Compared to the Iron Bridge, contemporary uses of exposed iron inside buildings were limited. They consisted at first just of cast-iron columns as subsidiary supports within buildings, in place of timber. That repeated what Tijou had done before and needed limited courage or skill. But the whereabouts of these usages is striking. If such columns were inserted in mills or workshops before 1790, we do not know of them. Instead they are first found in auditoria where sightlines mattered, notably churches. The practice may have developed from earlier churches in which timber supports to galleries had hidden cores of iron.[21]

Open iron columns first turn up regularly in the English iron-making districts, and in churches where craftsmen or tradesmen had some autonomy. Again the initiative starts from the supply side, with founders or smiths. The earliest recorded example seems to be that of St John's, Leeds, where a smith, Maurice Tobin, supported galleries on twelve 'Cast Pillers' along the south and west sides of the church, during renovations of 1764 by the architect John Carr.[22] A craftsman of local prominence and ingenuity, Tobin worked elsewhere with Carr and was to build a precursor of the Iron Bridge in the grounds of Kirklees Hall nearby.[23] Next came two minor Liverpool churches, St Anne's (1770–2) and St James's, Toxteth (1774–5). In the surviving example, St James's, the gallery columns are not circular but have a thin, quatrefoil section. That church at least was designed not by an independent architect but under a speculative surveyor-builder, who probably allowed his tradesmen some leeway.[24] Further examples of the kind in iron-producing districts suggest the presence of a local fabricator pushing his product.[25] This pattern of ironfounders' ecclesiology would come to a head in the Liverpool churches of John Cragg.

Thus far the evidence for architects actively promoting the expression of structural iron in churches is thin.[26] But from about 1790 bolder ventures take shape in Shropshire, cradle of the English iron revolution. Here an accomplished neoclassical architect, George Steuart, designed two churches (All Saints, Wellington, 1788–90 and St Chad's, Shrewsbury, 1792–4) with attenuated cast-iron columns adorned with the orders, not only supporting galleries but projected up to relieve the roof spans (ill. 51). Another church in Shrewsbury, John Carline's St Alkmund's (1794–5), this time Gothic, had a full set of windows with cast-iron tracery.

This cluster of examples again implies the influence of the local trade. But it could be and was resisted. Neither of Telford's two Shropshire churches of the 1790s, at Bridgnorth and Madeley, seems to have embodied a trace of ironwork, despite that architect-engineer's nascent mastery of the medium in bridges and aqueducts. Nothing could better highlight the tenacity of ideas about architectural propriety than the contrast between Telford's masonry-and-timber churches and his contemporary iron aqueducts at Pontcysyllte and Longdon upon Tern. Here the boldest of British engineers held out for learned conservatism.

In the event, the eighteenth-century interior where an architect made proudest use of open ironwork was not a provincial church but a sophisticated secular project, Henry Holland's Theatre Royal, Drury Lane, London, of 1791–4.[27] In the buildings mentioned hitherto, sightlines and ironmakers' interests explain why the material is there. Drury Lane owed its lavish iron to a third factor – the burgeoning movement for protection against fire, always a fear in metropolitan building, most of all so in theatres.

The iron at Drury Lane took two separate forms. One was the concealed presence of 'fireplates' – insulating devices of rolled wrought iron, laid as a protective sheeting between the timber joists and the floorboards. Fireplates had been patented in 1773 by David Hartley, libertarian MP and amateur inventor. As the researches of Frank Kelsall have shown, they were employed by many London architects, not least Henry Holland, in the last quarter of the century.[28] Fireplates worked well in multi-floored buildings, notably houses, by stopping fire spreading from one level to the next. But they were of little avail in auditorial buildings. At Drury Lane, Holland also adapted Hartley's fireplates for use as a safety curtain.

More blatant were the structural supports. Round the horseshoe of the auditorium ran tiers of superimposed iron columns at 15-foot centres, carrying three and sometimes four tiers of boxes. Next to the stage they took the form of square-shafted pillars, but around the circumference they turned into slender candelabra equipped with tripod bases and saucer-shaped heads, in keeping with Holland's light, Louis-Seize decoration (ills. 52, 53). Here in a fashionable London building beyond the reach of the ironfounders, a chic architect had no qualms about historicizing iron on the basis of ancient bronzework and ironwork. The fabricators (Lacy and Horsley) seem just to have executed patterns to Holland's designs. Because the columns were only simple supports, the relationship could be simple too.

52. Theatre Royal, Drury Lane, section of auditorium as rebuilt to the designs of Henry Holland, 1793–4, showing the iron columns rising through the tiers.

53. Coriolanus performing at Holland's Theatre Royal, Drury Lane: illustration by Ackermann. The swelling profile and tripod feet of the balcony columns may have been exaggerated.

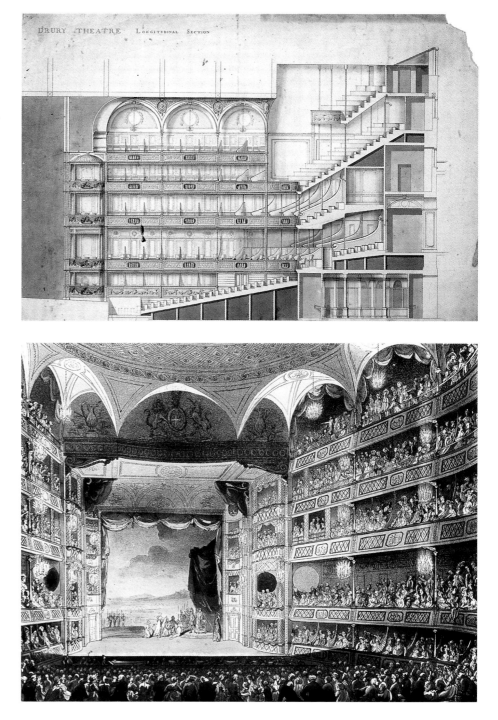

In this respect Drury Lane contrasts with its predecessor by a few years, Victor Louis's Théâtre Français (1786–9), a building surely known to the Francophile Holland. Both theatres used iron prodigally, in the hopes of warding off fire. But their construction differed altogether. The Paris theatre (pp. 81, ill. 61), used ironwork to span the auditorium roof and support the boxes, but hid it from sight. The London theatre flaunted iron columns for its galleries, gracefully ornamented. But the structural principles needed no more study than any other of the early buildings with open iron columns. Above them, the roof was entirely trussed in carpentry.

54, 55. St George's Church, Everton, Liverpool. John Cragg, ironfounder, with Thomas Rickman, designer. Left, all-iron interior: right, masonry exterior.

While Drury Lane was in progress, Holland also refitted Covent Garden Theatre (1792), where he shoehorned plainer tiers of iron columns into the back of the auditorium.[29] But neither cast-iron columns nor fireplates, it was soon discovered, offered a guarantee against fire. Both theatres burnt down within months of one another, in 1808–9. The next English theatre known to have used iron copiously, John Foulston's Theatre Royal, Plymouth (1811–13), reacted to this twin calamity. Foulston designed almost the whole internal structure of iron, encompassing the framing of the roof, the staging for the seats and even the balcony fronts. But in the French manner, nine-tenths of it was hidden away. Only the open gallery columns hinted at arrangements which would not now be known, had Foulston not published his drawings for Plymouth at the end of his career, claiming to have built 'the only fire-proof Theatre in the country'.[30]

During the 1790s the well-known developments in 'mill construction' gathered pace, and hidden iron began seeping into country houses and other English buildings. Postponing these topics, let us jump ahead to the three Liverpool churches of 1813–16 that represent the climax of early auditorial architecture in iron. These churches – the extant St George's, Everton and St Michael's, Toxteth, and the demolished St Philip's, Hardman Street – are not unknown. But they have missed the fame they deserve for their originality, their place in the history of building in series, their beauty even (ills. 54, 55). They fall between stools. They do not belong to the 'progressive' building-types that have hogged the limelight in the study of historic iron structures. Nor do they meet the criteria for revived Gothic architecture applied to churches after their time. In other words, they have pleased neither architects nor engineers. That is unsurprising, for they were initiated by an ironfounder, John Cragg.

By 1812 a revival of English church-building was stirring. Cragg was ahead of the game. A Liverpool ironfounder of sanguine temperament and bookish interests, he had already 'made his fortune and his energies were by that time devoted to church and local affairs'.[31] A patent Cragg applied for in 1813,[32] along with a pamphlet in which he called for the revival of Gothic on the grounds of 'not only taste but economy', reveals that he hoped to build not one church but many.

> In the best days of Gothic Ecclesiastical Architecture, lightness was a most distinguishing characteristic of the style, both here and upon the Continent. It remains for us to consider the practicability of restoring this style of building, in sacred edifices, with its wonted loftiness, airiness and simple elegance, at an expense which may be borne, and in materials that will endure.
>
> The friable nature of English stone is ill calculated for fine carving, or to endure the climate when delicately carved, and the cost of workmanship in these days is another discouragement; let us then only construct plain walls and the simple mouldings of stone, and substitute metal for the lighter and ornamental parts.
>
> Iron-Ore is inexhaustible in our own mountains. The labour is purely British. Cast-Iron, alone, will supply the lofty pointed arch – the column of slender proportions – the rich and flowing tracery of the mullions of the windows – the cusped pannel, and the crocketed pinnacle adorned with foliage, and the finial of equal beauty.

Cragg mentions too 'the rapidity with which a Gothic structure may be erected, the parts which in stone-carving would be so dilatory, being cast from models finely carved'.[33]

St Michael's, Toxteth was the first of his churches. A model was made by April 1812 and a trial arch set up in the foundry.[34] As a fabricator, Cragg focussed on breaking down the architecture into separate elements and then working out how to design, make and put them together. Here in modern terms was a proprietary building system, meant to maximize use of the ironfounder's product. But Cragg had an Achilles heel: his ignorance of construction and Gothic architecture alike. At first he relied upon books. He also procured a design from an architect (J. M. Gandy) which Thomas Rickman judged to be 'shewy but very incorrect'. It was to Rickman, not yet then a professional architect but a clerk in a Liverpool insurance company, studying and lecturing on mediaeval architecture in his spare time,[35] that Cragg next applied for assistance. Rickman's efforts in laying a veneer of archaeological respectability upon the ironmaster's ideas marked the start of his career as an architect proper.

Rickman's diaries give a rare insight into how things can go wrong between designer and fabricator when the latter is dominant. Hired too late and as a draughtsman, not an independent architect, he never controlled the erection of the churches. Largely he seems to have been limited to designing the skin of Gothic masonry enclosing the iron interior. Often he was kept in the dark. When the second church at Everton was proposed, Rickman went to a meeting 'and was much surprised to find J Cragg there and that he produced my drawings'. His frustrations were frequent: 'if he had studied buildings instead of books he would have altered many things himself'. At one point he notes: 'I have now clear'd all the improprieties I complain of which are not attributable to the iron &c.'. Then Cragg proves reluctant to alter 'his obnoxious arch over the great one so he must have his way indeed though it may be in trifles mended his iron work is too stiff in his head to bend to any beauty . . . I think he would change it if he could but does not like to give it up'. On site at Toxteth, Rickman finds 'they have set up canopy iron set offs to buttresses but all the ironwork is crooked and queer and does not look as if it ever would look like any thing but patchwork'. Such was the amateur architect's repudiation of churches today inseparable from his name.

A little later, parliamentary Acts of 1817 and 1818 heralded a proliferation of English church-building through government grants. Something like Cragg's system for iron churches might then have taken off. Now in professional architectural practice, Rickman made a church model for one diocese with some iron elements for replication.[36] But no such general model found favour. Though iron is to be found plentifully in the roofs, arcades, gallery supports and tracery of English churches of the 1820s, nowhere was the Liverpool experiment repeated. Cragg's inflexibility and egotism doubtless contributed to that. But there were other factors. As the movement for a 'correct' Gothic gathered strength, architects and antiquarians grew wary about iron in churches. The Crown architects who vetted the designs for the so-called Commissioners' Churches never smiled upon the idea, even though two of them, John Nash and Robert Smirke, strewed iron about in their own secular buildings. After Pugin's time, expressed iron in Anglican churches lived on only in projects for prefabrication and colonial export,[37] while on home soil the Church of England fell back on architecture to assert continuities with the national past.

There is scant structural logic about the Cragg-Rickman churches. They have their interest as early instances of double-height 'cage construction', that is to say, a full cast-iron frame within a skin of enclosing masonry. As against that, pointed arches support flat ceilings not vaults, and the structures are braced and belted for safety. A mediaevalist would find the decorative ironwork risible. Yet a visit to the Everton church in particular, where upper galleries replace the clerestory at Toxteth and swell the interior volume, is uplifting. Here perhaps for the first time can be savoured the spatial qualities peculiar to iron architecture and cherished by latter-day critics. The main effect is less of space than of the lightness called for in Cragg's pamphlet. Hamfisted detail, inconsistency and improvisation may all be forgiven, as supports, arches and spandrels weave their tenuous web. A curiosity and a dead-end no doubt: but a vision of iron architecture unlike any other.

MILLS AND WAREHOUSES

The 1790s marked a turning point in building with iron. Hitherto its advance into English buildings had been episodic, linked to opportunity more than science or art, with fabricators in the vanguard. Only then did architects get to grips with the appearance and decoration of iron. Only then too did those pioneers of British structural engineering, Telford and Rennie, focus upon iron bridge construction.

The same decade saw the flowering of the 'first iron frames', so called, in English mills. Mills and warehouses dominated the pioneering historiography of early iron construction. Stripped of their machines, goods, noise, stink and drudgery, these ancestors of the skeleton frame looked cool, spacious and logical. They appealed to modernist tastes. Hence enthusiasm for the type in the post-war decades, spurred on by progressivist chroniclers of technology and architecture.

At the time these first iron frames evolved, the building of multi-storey mills and factories enjoyed low status. They seldom exercised the leading designers of the day. British structural engineering had grown in good measure out of the millwright's craft.[38] Yet the generation of consulting engineers that rose to eminence during the 1790s like Jessop, Rennie and Telford spent little of their time on mill carcases, though Rennie had begun as a millwright and designed milling machinery. As for architects, their participation in mill-building between Samuel Wyatt's brick-and-timber Albion Mills (1783–6) and Lockwood and Mawson's giant mill for Sir Titus Salt at Saltaire (1850–3) was subdued at best.[39]

Against that must be set the record of the purpose-built warehouse. Mills and warehouses tend to be lumped together as one and the same building-type. But their pro-

56. Branching cast-iron supports of the 'Skin Floor' at the Tobacco Warehouse, London Docks, before conversion. Daniel Alexander, architect, 1811–14.

curement seldom took the same path. In the early industrial revolution the mill was commonly a one-off, lone building or complex, sited with respect to clean and free-running water, not urban context. But the dock warehouse typically belonged to a broader scheme of infrastructure and civic extension. Civil engineers and architects sometimes therefore took on the design of warehouses as elements in a bigger whole.

In Britain's docks, early iron construction featured most often in single-storey buildings, cranes and other machinery, while multi-storey dock warehouses usually had internal frames and roofs of timber. Such, for instance, was the structure of Telford's warehouses at St Katharine Docks, London (1826–30).[40] The notable exception came from an architect-engineer, Daniel Alexander, whose Tobacco Warehouse at the London Docks (1811–14) boasted branching columns of cast-iron over brick-arch vaults (ill. 56). Built at a time when timber was hard to get during Napoleon's blockade of the Baltic, that magnificent curiosity stood to one side of the main line of development and had no successor. Well into the nineteenth century, dock warehouses persevered with timber beams and even timber supports, whereas mills in the textile districts converted to iron supports quite quickly. These variations probably had much to do with the price of timber and iron in different regions at different times.

Who then were the pioneers of iron mill construction, if not engineers of the budding professional kind? They were factory-owners, inventors and mechanics who counted construction as one among many concerns. The men who devised the first iron frames, William Strutt and Charles Bage, were manufacturers with mathematical and technological aptitude. Inventive businessmen in the enlightenment mould, they were engineers only by courtesy. Having set their mark upon mill-building in the 1790s, they moved on: Strutt into hot-air heating and the chemistry of bleaching, Bage into an elusive quest to improve the power loom.[41]

Their successors were a small band of 'design and build' mill-builders who erected and equipped factories not only for themselves and their partners, as Strutt and Bage had done, but for hire. Around 1800 there were two such firms of note: the great Birmingham enterprise of Boulton and Watt; and Murray, Fenton and Wood of Leeds. Boulton and Watt devoted only a tithe of their energies to mill-building; their scope was far broader than that of the ingenious, working-class Matthew Murray, 'a great scoundrel but a very able mechanic', as James Watt junior called him.[42] Nevertheless both firms can be grouped under the rubric of professional millwrights or mechanical engineers. Their primary task was the design, erection and maintenance of machines. For them the building was an extension of the machine.[43] Just as machine-parts were increasingly made of iron, so too were buildings. That pattern remained true for the third generation of mill-builders, centred upon Manchester.[44] William Fairbairn, for instance, came to mill construction in middle age, after an early career in millwork and other moving machinery.[45]

It follows that the early use of iron in mills was about efficiency, not architecture. As in Holland's Drury Lane Theatre, so in the three famous buildings erected by William Strutt between 1792 and 1795 for his father's firm, the Derby Mill, the extant Milford warehouse and the Belper West Mill, iron was employed to protect against fire, and as a visible means of support. But there the similarities end. Whereas Holland looked for

expression and visual integration, Strutt sought system and simple construction. His combination of bare column, cast and bolted head, iron tie and transverse jack arch was thorough and rational, in a way no previous building employing iron had been. But it was there to save space and therefore money and to avoid combustion, not to be seen. Machines and materials must in any case have obstructed the spaces. Skempton and Johnson in their classic article on the first iron frames claim only that the Strutt mills presented 'a fully integrated structural scheme'.[46] To go further would be to confound visible architecture with efficient construction.

Strutt had to use timber beams, their soffits protected by plaster. Had he been able to procure beams of iron to match his columns, he would probably have done so. That is what Charles Bage, in correspondence with his friend Strutt and collaboration with the Shrewsbury ironfounder William Hazledine, achieved at the Marshall, Benyon and Bage flax mill at Ditherington outside that town (1796–7). The birth of first cage construction and then the skeleton frame may fairly be traced back to the union of cast beam with cast column at Ditherington, and of Bage the mathematical businessman with Hazledine the fabricator (ill. 57).

Ditherington, as Skempton and Johnson say, is 'one of the most remarkable buildings in England'.[47] But visibility or even legibility was beside the point. As in the Strutt mills there was a jack-arched floor system, here born upon anvil-shaped beams with only the undersides exposed. Thereafter, iron beams varied in profile and underwent a regimen of testing and research until they settled down in the 1840s into I-sections.[48] There is no evidence that anyone cared about exposing them. The early profiles possess a fetishistic

57. Floor of the Ditherington flax mill, Shrewsbury. Charles Bage, designer and proprietor, with William Hazledine, ironfounder, 1796–7. Cruciform columns with swelling profiles support transverse jack-arches.

58. Early cast-iron beams in Britain, from Skempton and Johnson's seminal article, 'The First Iron Frames', 1962. These experimental profiles have a fascination akin to that of Gothic mouldings.

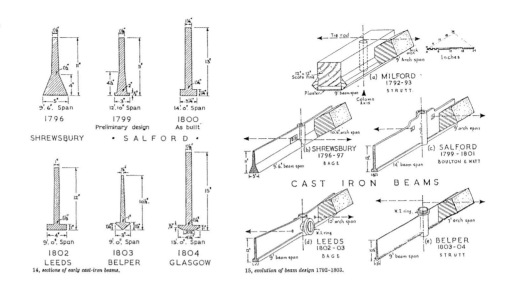

attraction when drawn out together (ill. 58). But they were not originally thought of as expressive: mills are not auditoria.

The case of early mill columns is more enigmatic. The Ditherington columns, for instance, possess a grace absent from those in Strutt's mills (which differ little from the early columns in St James's, Toxteth). But their star section and swelling profile were devised by Bage and Hazledine not for aesthetics but for bending moments, so as to give maximum depth of section at mid-height. The argument that a functional shape is a beautiful shape is always worth reflecting on, but hardly explains how innovation happens. The most that can be said is that the Ditherington columns convey the care lavished on their design. The puzzle is compounded by a set of surviving mills in Scotland and the north of England of 1806–17, some associated with Murray, Fenton and Wood, where the iron columns have elaborately moulded capitals and bases.[49] Were their builders deliberately applying decoration or 'architecture'? Or did they borrow castings from ironfounders' moulds previously designed for some church or public building? We can only speculate.

France before the Revolution

Despite the different structure of its professions, France displayed the same pattern as Britain, whereby architects explored what might be made of iron before engineers took much heed of it. But first, some remarks about the material conditions affecting development in that country.

France exhibited its prowess in early iron construction in face of a charcoal-dependent iron industry, strong on crafts skills but resistant to change. Few native iron ores available in the eighteenth century were of high quality, nor were coal and iron often found close to one another.[50] That inhibited coke-smelting, raised the cost and limited the production of castings, and reinforced native partiality for wrought as against cast iron. With ground to make up, Louis XVI's government invested effort in acquiring English casting skills, mainly through the ironfounding brothers John and William Wilkinson. A first outcome was the naval cannon foundry at Indret on the Loire (1777–9). It was succeeded by continental Europe's first full steam-blown, coke-smelting foundry at Le Creusot in Burgundy. Here too the primary object of the plant was to produce cannon. Hardly had it opened in 1785 before it became snagged in the toils of the revolutionary years.[51]

Meanwhile architects had been using growing quantities of wrought iron in bar form to further their own ends. This initiative centred upon Paris, far from the iron-producing districts, and focussed not upon industry but upon major public buildings and the 'archi-

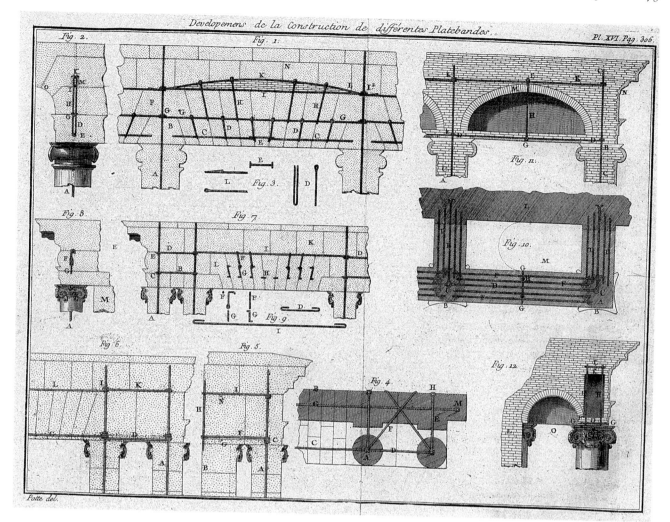

Devcloppemens de la Construction de différentes Platebandes.

Pl. XVI. Pag. 306.

Patte del.

tectes du roi', working with the skilled serruriers who had fashioned iron since the middle ages. In the first place they were bent on extending the old habit of strengthening masonry or carpentry spans by means of hidden reinforcement. The ties used by Perrault to support wide intercolumniations at the Louvre have been mentioned (p.65, ill. 47). The soft French limestone used there, excellent for cutting and carving, was less suitable for flat spans than the tough marble of ancient classical temple-fronts and porticos. The Louvre colonnade signalled the start of a movement for reinforcing limestone entablatures with networks of iron ties. The practice was taken up, for instance, by Servandoni in the frontispiece of St Sulpice (1732 etc.), and by Ange-Jacques Gabriel in his buildings facing the Place de la Concorde (1758 etc.).[52]

These and further examples were depicted and analysed by Pierre Patte in a mémoire of 1769 that inaugurated a French tradition of publishing hidden iron construction (ill. 59).[53] Armed with Patte's data, an architect would not have needed much help to repeat the practice. Such famous neoclassical porticos of the 1770s as Ste Geneviève by Soufflot and the great theatre in Bordeaux by Victor Louis were likewise reinforced, as were trabeated colonnades within a number of church interiors. In all these examples, the hidden ties were supplements used to extend the scope and scale of monumental masonry buildings: specifically, to promote the aesthetic of column and lintel.[54]

The next stage was the wrought-iron truss in roofs or floors. A composite iron floor beam – in effect a shallow truss – and an iron roof truss were invented by the obscure

59. Details of iron reinforcement in four French-designed projects, from Patte's *Mémoire* (1769). 1–3: Saint Sulpice by Servandoni. 4–6: Sainte Anne by Pierre Desmaisons. 7–9: from the Palais-Royal by Constant d'Ivry. 10–12: from a project in St Petersburg by Lamotte, showing reinforced brickwork.

60. Section of the Opéra at the
Palais-Royal by Pierre-Louis Moreau,
1764–9, showing long iron hangers
over the proscenium opening and
copious strapping of the trusses.

60. Section of the Opéra at the Palais-Royal by Pierre-Louis Moreau, 1764–9, showing long iron hangers over the proscenium opening and copious strapping of the trusses.

M. Ango in the 1780s. Both were publicized and soon had successors. Ango may have been predated by Soufflot, who before his death in 1780 had been experimenting on the strength of iron, and designed a small wrought-iron roof with top-light for a staircase in the Louvre. Just before the Revolution, the first part of the Louvre adapted to show pictures acquired a much larger experimental truss-cum-rooflight.[55]

But the most telling extension of the technique came in a theatre. Pierre-Louis Moreau's short-lived Opéra at the Palais-Royal in Paris (1764–9) had already built iron into the boxes and used long hangers dropped from the roof trusses to help suspend the ceiling span (ill. 60).[56] The burning-down of this auditorium in 1781 prompted Victor Louis to employ even more iron in its successor, the Théâtre Français (1786–9). A sectional drawing of its roof, published by Rondelet in 1817, has often been reproduced (ill. 61).[57] It shows a framework of wrought iron spanning the full 25 metres of the auditorium by means of ingeniously linked shallow trusses. The theatre's galleries and floors, along with the rest of the rebuilt Palais-Royal, were also, we are told, framed in iron with light 'hollow pot' filling – the other French advance in fire-protection and floor-technology during the twilight of the Ancien Régime.[58] In her study of French iron architecture Frances Steiner leans towards crediting the technique of all this to Ango.[59] But Charles Eck, writing in 1836, asks us to honour not only 'the artist' (Louis) but 'also the constructor, who has remained anonymous, for putting those who have followed them on the road to innovations of this type'.[60] By this he must mean the unknown serrurier, critical to undertakings of this kind.

As with masonry reinforcement, none of this ironwork was exposed, portions of the Louvre rooflights excepted.[61] Nor was it believed that it should be. Debates among architects of the time prized clear construction, not exposed construction. If iron could help make auditoria bolder, wider and safer, that was ambition enough. In the tangle of trusts over the ceiling of the Théâtre Français, architect and serrurier worked their magic behind the scenes, like stage technicians. When Rondelet published the roof a generation later, he was keen to point out that it ought to have been done differently. But he never suggested that it should have been shown.

Nevertheless there were stirrings of interest in open ironwork for bridges before the Revolution. The Coalbrookdale Iron Bridge had elicited curiosity and some minor experiments from French engineers, despite cast iron's cost and a proper wariness as to its limitations.[62] Had peace prevailed, no doubt things would have moved faster. As it was, France's first important iron bridges – the Pont d'Austerlitz and the Pont des Arts – had to await Napoleon and his plans for embellishing imperial Paris.[63] So too did the country's earliest major example of open iron construction within a building, at the Halle au Blé. That project was thirty years in gestation and consisted only of an addition to an existing structure. But as revealed in 1813, the outcome was worth waiting for. As an early instance of the intricacies of authorship when building with iron, it warrants attention.

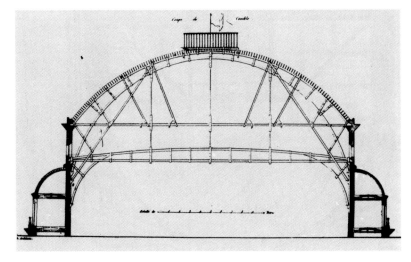

61. Section of the Théâtre Français by Victor Louis, 1786–9, from Rondelet. An ingenious system of hidden wrought-iron trusses spans the auditorium.

THE HALLE AU BLÉ

The original Halle au Blé for Paris's wholesale grain market opened in 1767.[64] As built to the designs of Le Camus de Mézières, it consisted of a ring with an arcade at ground level and smaller rooms above, all faced in ashlar and surrounding an open courtyard 120 feet in diameter (ill. 62). Very soon, proposals were made for roofing the courtyard. In 1782, two fresh suggestions of equal originality came forward. The one which concerns us came from François-Joseph Bélanger. Then a fashionable architect of houses and gardens, Bélanger combined vitality, breadth of interests and a streak of cunning that helped him survive scrapes during the Revolution. He had studied physics in his youth and was curious about technique; on a visit to England he had sketched bridges, wheels, pumps, a pottery kiln and cannon-boring alongside country houses and parks.[65] Now he proposed to cover the space of the Halle au Blé with a dome of iron ribs covered in copper. Though an exceptional idea, it chimed with the fashion for iron structures prevalent during the 1780s. There are few details, but we do know that Bélanger's technician was Deumier, 'serrurier de la ville et des bâtiments du roi', a craftsman experienced 'in the art of fabricating works in iron on a large scale'.[66]

For the moment, Bélanger's brainwave was passed over in favour of a carpentry dome devised by rival architects, Legrand and Molinos, and made up of small timbers according to a system sketched out centuries before by Philibert de l'Orme. Timber indeed was as much a subject of innovation as iron at the time.[67] The Legrand and Molinos dome was hailed for its ingenuity and lightness. But it proved a headache to maintain and burnt down during repairs in 1803. The calamity led to a plethora of ideas from the architects and contractors of Paris for its replacement in every possible material: stone, timber, iron and canvas. Among them Bélanger resubmitted his iron-ribbed dome, with a number of modifications.

In 1805 the most plausible proposals were put before the Commission des Travaux Publics, which included Legrand and Molinos; all were rejected. But commissions of the Napoleonic era were fickle. Soon a fresh one was appointed and recommended a stone dome, this time to be rebuffed by the Government on grounds of expense. After further

62. Plan, elevation and section of the Halle au Blé, showing the relationship between the masonry hall of 1766–7 and the iron dome of 1810–14. From Gailhabaud, *Monuments anciens et modernes* (1850).

manoeuvring behind the scenes, the Minister of the Interior asked Becquey de Beaupré to report afresh on Bélanger's iron design. Becquey, who had served on both commissions, was 'ingénieur en chef' at the Ponts et Chaussées and co-designer of the iron Pont d'Austerlitz, just then nearing completion. He duly advised in 1807 that a dome made (like his bridge) mainly from cast iron with some wrought-iron features 'could fulfil the views of the Government'. A reconvened commission concurred, adding the rider that 'this operation departs from the normal rules and can only be entrusted to experienced hands'.[68] Officially, the aim was 'both to avoid the fire which destroyed the former roof, and to encourage a method of manufacture much more economical than works in wrought iron'.[69]

So far the design was just a concept; technical development had to follow. In the sequel, the execution of the second dome for the Halle au Blé became a state construction project in the formalized French style, with Bélanger one of several leading actors. Two of his competitors were given consolation prizes. Rondelet (who had hoped to build a stone dome) became the unenthusiastic inspector; Brunet, a contractor-engineer about whom one would like to know more, was made 'controleur' of the works.[70] The main fabrication of ironwork was assigned to the Le Creusot foundry, just then undergoing a restructuring under the Paris businessman Jean-François Chagot after years of chequered fortune.[71] Eck adds the name of the ferronnier Roussel père ('one of the most accomplished practitioners of the day'[72]), who was the contractor and perhaps supplied the wrought iron.

In the event Brunet's role was such as to inspire rumours that he had designed the dome. He published a book about it in 1809, when Le Creusot was starting to make the iron but nothing had yet happened on site. At the same time he wrote to the Minister, confirming Bélanger as 'the sole author of the designs which he has presented . . . I have only been charged with some calculations so as to avoid a model which in view of the great diameter of this cupola could only be inaccurate'.[73] Bélanger had conceived the dome as high and hemispherical, as much perhaps because that was the simplest shape to set out as because he believed it to be the most efficient structural profile. Brunet's book shows that what its author did was 'size' the work, in other words determine the dimensions and shape of the timber models used as templates for the castings. Then before the design was finalized, says Steiner, Bélanger made some adjustments to 'increase the aesthetic effect upon the viewer standing on the floor below'.[74]

Sigfried Giedion claimed the second dome of the Halle au Blé (ills. 45, 63) as 'the first time that architect and engineer were no longer combined in one person'.[75] That formulation simplifies the chain and nature of partnerships. Brunet's task was not that of a modern engineer calculating the forces in a structure and designing the elements to suit them. Nor was he the only figure besides Bélanger involved in the dome's evolution. What can be said is that here we have a first detailed picture of the kinds of collaboration that iron construction will make necessary. We also see an architect promoting visible iron, then warping or refining the technology for the sake of appearances. In this respect Bélanger will have many successors.

Assembling the Halle au Blé dome commenced on site in October 1810, recorded by a young assistant of Bélanger's, Jacques-Ignace Hittorff (ill. 64). Bélanger's main role now became that of tireless propagandist for the project, needful as the work dragged on. Here was a new form of art, he claimed. Squiring the painter David to the site, he informed him: 'You have admired the outcome of a new conception which is the first in this mode to offer present-day Europe the idea that with encouragement, artists can sometimes steal a spark from the torch of Prometheus'.[76] The turn of phrase betrays the architect's passion for concept over object. The result, at any rate, was enduring. Much of this earliest of iron domes remains within the structure of the enlarged Bourse du Commerce, built to supplant the old ring of the Halle au Blé in 1888–90.

The Halle au Blé was built in phases: first the external ring, then as an afterthought the dome within. Yet it set a model for much civic architecture in the nineteenth century: that of a masonry perimeter containing small rooms, prefacing and shielding a large, iron-roofed space for public uses. How these should connect became a recurrent theme

64. Workmen on the scaffolding for
the dome of the Halle au Blé, drawn
by the young Jacques-Ignace Hittorff,
c.1810.

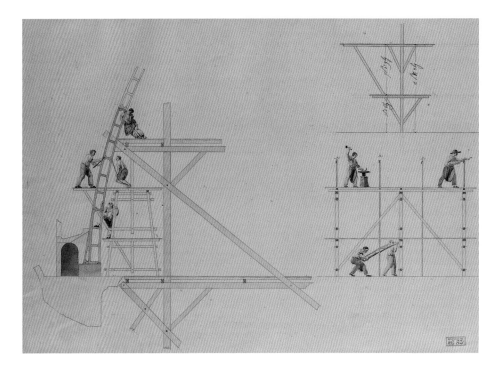

of debate. Should the inside space show on the outside? Should the two relate in style or
technology? How should they join? Such questions have long preoccupied architects and
engineers. At the time of the Halle au Blé, they had still to surface. What is surprising yet
prophetic is that the details of Bélanger's dome should have eschewed historical allusion
or ornament. Perhaps that reflected the low status of a grain market and the need to
minimize costs more than any fixed beliefs about iron and decoration.

Prussia and Russia

In the waning days of his Empire, Napoleon came to admire the completed Halle au Blé,
attended by Bélanger's friend, the imperial architect Fontaine.[77] It is sometimes said that
Napoleon stimulated the use of iron in French construction. He was certainly alive to its
capabilities. In the false dawn that followed the Peace of Amiens in 1802, he reported to
the Institut that for the new monumental Paris then in prospect, a manner of building
was needed suitable to the changes and inclemency of the climate: 'the mode or means
must one day be cast iron, the same iron employed during the war to serve the victory
and in peace to elevate the trophies of it'.[78] Yet he queried the construction of Paris's
second iron bridge, the little Pont des Arts, given France's wealth of building stones in
comparison with England. No doubt Napoleon, the ex-artillery officer, thought of iron
as a tool of modern warfare which might incidentally serve the arts of peace.

All this smacks of raising iron production by the means that enlightened absolutists
had always applied to new arts and manufactures – stimulating demand by example. We
have seen how the early experiments with iron were mostly state projects in Paris controlled by royal architects. Surveying the full course of French iron architecture, Frances
Steiner has noted 'the encouragement given by every political régime, beginning with
that of Louis XVI'.[79] Development in France was top-down, entailing a hierarchy of
relationships between patron, designer and builder alien to the bottom-up, British model.
Other European states seeking to catch up on iron in the early nineteenth century
embraced the French model of state-sponsored development, while drawing what practical expertise they could glean from Britain. Two of these, Prussia and Russia, are sufficient to shed light on the relations between different interests and professions as they
started to exploit iron in the construction of buildings.

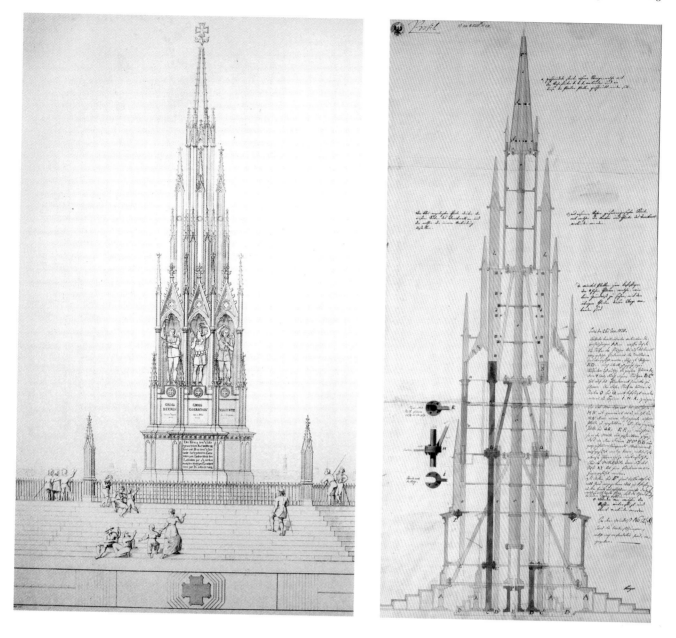

The development of Prussian iron coincides with the modernization campaigns insti-tuted under Friedrich Wilhelm III.[80] In 1796, the year before his accession, a royal iron-works was established at Gleiwitz in the Silesian mining district. After trips of enquiry to Britain, this was followed in 1803 by the Königliche Eisengiesserei or royal foundry in Berlin itself. The structural first-fruits of Prussian iron construction were a pair of small bridges in parks. War then intervened, and the foundry plunged into making military hardware.

The effects of French invasion on the growth of German national consciousness at this time are familiar. Cast iron, its symbolism and its styling furnish a sideshow to that drama, with the great architect Schinkel in the leading role. During his youth Schinkel had eked out a meagre income as an independent architect, designing for the applied arts, in ceramics and metalwork. In 1810 through Wilhelm von Humboldt's mediation he became an employee of the Prussian royal building service, the Oberbaudeputation. His

65. Kreuzberg Monument, Berlin, by Karl Friedrich Schinkel, 1818–20. The line engraving in Schinkel's col-lected works does not convey the cast-iron character of the memorial.

66. Section of the Kreuzberg Monument. A hidden cast iron core fabricated at Prussia's Royal Ironworks supports the decorative cast-iron panels and pinnacles.

first task was a mausoleum for Queen Luise, a symbol of German nationhood, who died that year. Schinkel wanted a passionately Gothic chapel. Instead, Friedrich Wilhelm sketched out a Doric temple, which was duly built at Charlottenburg. But at Gransee, where the Queen's body had lain in state, Schinkel did succeed in erecting a sarcophagus sheltered by a high Gothic canopy in cast iron, made at the Königliche Eisengiesserei. So proud was the foundry of this novelty that it was shown to the Berlin public before its transfer to Gransee.

Then in 1813 the War of Liberation erupted. To mark it, the King announced and Schinkel designed the Iron Cross. The material of this new and exalted decoration for military valour aptly symbolized endurance, modernity and austerity. After the defeat of France, Schinkel created a series of cast-iron Gothic war memorials for battlefield sites, and then the National Memorial to the Fallen at the Kreuzberg (1818–20), again entirely of cast iron and Gothic (ill. 65). All these too were fabricated at the royal foundry, to which Schinkel was now furnishing a stream of exemplary essays in applied design, military demand having dwindled.

Here was something other than French or English essays in iron thus far: a kind of alchemy whereby humble iron ore was refined into works of patriotic commemoration and national policy. Schinkel was the first to supply iron architecture with deeper meanings. Yet all that had little to do with the material's structural potential. At the Kreuzberg Memorial, for instance, beneath the architect's array of pinnacles lay an unexpressed cast-iron core designed separately by Krigar, foreman of the Königliche Eisengiesserei (ill. 66). Though of iron all through, in this respect it prefigured those grosser public monuments of the next age, the Albert Memorial and the Statue of Liberty, which also were to need iron frames to prop them up. Here was a sleeping dragon, not to wake until the time of the Eiffel Tower.

Only in the next phase in his career, as Prussian ironwork switched from a military-symbolic role to an industrial-commercial one, did Schinkel use iron in buildings proper. Once again this grew out of government policy. It was preceded by the famous visit of 1826 to France and Britain by Schinkel and his colleague Peter Beuth, director of the Königliche-Preussische Gewerbeinstitut.[81] Here was no mere artist's tour, but a technical fact-finding mission by a pair of senior civil servants. Beuth having paid a preliminary visit in 1823, they now went to garner systematically what they could about industrial developments, and apply the lessons back home. In Britain they made a beeline for engineering structures and factories, paying scant heed even to the smartest conventional architecture if it lacked technical interest. Iron featured high on their agenda, and took up many of Schinkel's fluent sketches and notes (ill. 67).

68. Albrechtspalais, Berlin, the imperial staircase, 1830–3. A ceremonial
cast-iron composition, improved and
refined on the Nash stairs Schinkel
saw at the Royal Pavilion (ill. 73).

Yet once home and building, his iron construction in Berlin remained that of the artist-impressionist. It burst out in the lavish staircases of two palaces built or altered for sons of the King, the Karlspalais (1827–8) and Albrechtspalais (1830–3). The branching stair of the Albrechtspalais (ill. 68) copied and refined Nash's fanciful bamboo stairs from the Brighton Pavilion (ill. 73), which Schinkel had noted. Here, distinct from the masonry core of the palaces, were experiments in updating classicism with ironwork – about reproportioning features hitherto crafted in timber. The artist's hand showed in the discipline with which streamlined support was picked out from decorative infill.

In pursuit of his late style (his 'radical abstraction', Pevsner called it) Schinkel also drew on the mill construction he had seen in Manchester. Such were the additions to Beuth's Gewerbeinstitut (1827–9) and the more famous Bauakademie (1831–6): tokens both of the commitment to technical education which was part of Prussian Government policy (see pp. 445–6). By a process of borrowing, adaptation and refinement, the Gewerbeinstitut and Bauakademie rid the British mill of its dourness, raising it to educational uses and urban dignity. The Bauakademie in particular ranks high in progressivist history, as a step on the road from mill to modern office building. But iron featured sparsely in either building. The Gewerbeinstitut, equipped with workshops for the crafts and trades, had columns of iron but beams and roofs of wood. The higher-status Bauakademie, where would-be architects and engineers were taught academically, sidestepped the material altogether, save for some chic, ornamental iron heads to the windows, tucked in for visibility's sake (ill. 69). In structural terms it can hardly be deemed exemplary. It was the image of mill construction which Schinkel cribbed, not the reality.

69. Bauakademie, Berlin, by Karl Friedrich Schinkel, 1831–6. Brick and terracotta front with cast-iron window frames.

Werner Lorenz has taken Schinkel to task for dogmatizing about construction while being behindhand in technical knowledge and proficient only in cast not wrought iron.[82] In comparison, his contemporary Georg Moller of Darmstadt, or his assistants and pupils Persius and Stüler, trained at the Bauakademie, displayed greater structural competence.[83] The Prussian best versed in iron during the 1830s and 40s, says Lorenz, was J. C. Borsig, carpenter turned entrepreneur and machine-maker. Borsig's inarticulacy did not hinder him from carrying out brilliant technical feats in iron construction, fathering the German steam locomotive and influencing the whole industrial development of Berlin.[84] Nevertheless Schinkel raised the profile of Prussian iron construction, not once but twice, at both the symbolic and the pragmatic stages of his nation's development. In technique, he had the architect's instinct; he understood less about iron than appeared. But that did not hinder him from designing in it creatively, and using his authority to promulgate iron construction beyond Prussia's borders.[85]

Many of the princely states of Europe saw iron as a key to invigorating their economy and culture alike from about 1830.[86] But the country where autocracy, modernization and iron architecture ran closest in harness was Russia. Here development owed much to an old czarist tradition of buying in foreign experts. Technologists from Scotland and architect-engineers from France united with native talent to raise an impressive array of early iron structures.[87] As in Paris and Berlin, most of the initiatives took place in and around the capital, St Petersburg.

Russia's ancient iron industry in the Urals had been revitalized by Peter the Great. Second in size only to Sweden's in the eighteenth century, it exported a high proportion of its bar-iron. Early iron structures became fashionable. During Catherine the Great's reign (1762–96), much experimentation with iron bridges took place in the grounds of royal palaces; John James claimed to have identified no less than eleven, all small and ornamental, at Tsarskoe Selo alone. Most seem to have been designed by Catherine's foreign architects, Cameron and Quarenghi, and fabricated by the State Arms Works at Sestroretsk.

Few of these ventures involved cast iron, since Russian forge and foundry technology had fallen behind that of Britain. Changes followed the creation or extension of three state foundries (Petrozavodsk, 1786; Kronstadt, 1789; and St Petersburg, 1801). In charge of all three was Charles Gascoigne. Managing director of the Carron Company, Falkirk, Gascoigne was lured to St Petersburg in 1786. To the fury of British industrialists and officials, he took with him a team of skilled Scottish workers. A further private foundry was set up in St Petersburg by Charles Baird, a member of the Gascoigne group. Both Gascoigne and Baird possessed drive and breadth of skills and interests. They were the engineer-counterparts to Charles Cameron, the Scottish architect who had already found favour with Catherine the Great and likewise brought compatriots to Russia to help fashion Tsarskoe Selo and other palaces.[88]

The prime purpose of these foundries was armaments production – to supply cannon equal to or better than those the Russian Government had already been ordering from Carron. But they also made castings for tools, machines and structures. Factory buildings for the Admiralty at Kolpino, erected by Gascoigne in about 1803–7 with help from members of the Cameron team, William Hastie (architect) and James and Alexander Wilson (smiths), became a showcase for cast-iron construction.[89]

In 1810 a fresh ingredient enriched the mix: Russia's Corps of Highway Engineers, founded on the lines of the Corps des Ponts et Chaussées to improve Russian infrastructure. This body recruited architect-engineers from several countries (many with some French training), and boasted excellent numeracy and literacy, as well as a collective purpose. The leading figures in the new Corps were its founding director, the Spanish-born Augustin de Bétancourt, and his deputy and successor, P.-D. Bazaine. What was essentially a subcommittee of this body undertook the co-ordinated reshaping of St Petersburg, centred upon the Admiralty.

So by the end of the Napoleonic Wars, St Petersburg enjoyed multi-national, multi-disciplinary skills in construction. Iron loomed large in the state-commissioned buildings, bridges and monuments that ensued. Many of these structures entailed collaboration between architects and the technologist-ironmakers of the new foundries. Experts from the Corps of Highway Engineers were sometimes drawn in for advice, sometimes into experiments of their own like St Petersburg's suspension bridges (see pp. 320–2). Monuments such as V. P. Stasov's Moscow Gates, a colossal, all-metal Doric propylaeum of 1834–8, could be blatant in their display of iron. In the buildings proper it was largely confined to hidden roof and dome structures, though Loudon, visiting Russia before things had gone far, complained that supports for towers, domes and spires there were 'disfigured by wrought-iron tie rods in every direction'.[90]

The French tradition of trussing wide-span roofs, renewed from the 1820s, offered precedents for this genre of ironwork. In Russia it took a course of its own. The pioneering venture was the outermost dome of St Petersburg's Kazan Cathedral.[91] This wrought-iron structure was designed and calculated by its architect, A. N. Vorokhinin, and fabricated in the state foundries 'under Mr Gascoigne's control' between 1806 and

70. Archive hall in the Main Headquarters building, St Petersburg. Carlo Rossi, architect, with Matthew Clark, ironfounder, 1819–23.

1810, so just preceding the wider Halle au Blé. Later structures saw a better balance between architect and technologist, well represented by the achievement of Matthew or Matvej Clark, as chronicled by Sergei Fedorov.

Son of a Gascoigne emigré, Clark worked his way up within the state foundries to become manager of the St Petersburg ironworks established in 1801. Later, he designed and built a steam engine for one of the capital's mints. By 1817 he seems also to have been acting as an engineering contractor advising on structure: negotiating with architects, notably Rossi and Stasov, about what they wanted to do and how it could be constructed, and then supervising the fabrication. The first structure in St Petersburg to involve Clark was the handsome archives hall within the vast Main Headquarters, designed by Rossi and built in 1819–23. It is framed in cast iron throughout for fireproofing purposes but gracefully plaster-vaulted, the ironwork obtruding only in the columns with their pretty capitals and simple gallery fronts (ill. 70). After that, Clark retreated into devising hidden structures for wide spans: with Stasov, for the dome of Trinity Cathedral (1827–31, blown off in 1834); with Rossi, for the superstructure of the Aleksandrinsky Theatre (1828–32); and finally for a set of roofs over the Winter Palace (1838–42).[92]

Clark's hidden roofs, spanning between 20 and 29 metres and climaxing in forests of trusses over the Winter Palace (ill. 71), surpassed his earlier work. Yet they were also improvisations derived, Fedorov reckons, from his empirical background in the foundries. The school-trained Bazaine had little respect for Clark's structural methods. In response to Bazaine's criticisms of the Aleksandrinsky Theatre roof, it is said that Rossi swore he would hang himself from one of the beams if anyone came to harm.[93] As usual, it was the articulate architect who defended the structure, not the technologist himself.

Clark probably did not care if his trusses were hidden or what they looked like, so long as they worked. Whether work of this genre was based on calculation, testing or rule of thumb, it entailed the deference and subordination of technologist to architect. Since

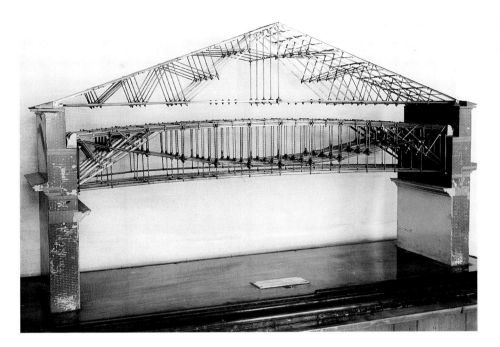

71. Model of the hidden roof trusses over St George's Hall in the Winter Palace, St Petersburg. Carlo Rossi, architect, with Matthew Clark, iron-founder, 1841–2.

time immemorial, roof structures in all classes of building had been tucked invisibly away. In this respect, Clark belonged to the tradition of structural illusionists. But by 1840, illusion in all walks of architecture was starting to lose ground.

Hidden Iron Comes of Age

France, we have seen, had led the way in promoting hidden floors and wide-span roof trusses of wrought iron. After the 1780s, revolution and war imposed a long pause. In the first edition of his *L'art de bâtir* (1802–17), Rondelet could be brief on iron within contemporary buildings, because little was going on. Not until the 1820s did concealed iron roof trusses reappear in Paris. The earliest example of note was the roof of the Bourse in Paris, raised under the supervision of the architect Eloi Labarre by the serrurier Albouy in 1823.[94] The practice then picked up. By 1836 Eck could illustrate a host of examples, mostly from government buildings or theatres and all the fruit of architect-serrurier collaborations.[95]

The span of these trusses often exceeded that of the earliest contemporary markets and arcades with open iron roofs. In some ways they foreshadow the crisply angular French railway-station roof. But the sections produced by the iron industry were still quite primitive. So these roofs were designed, says Bertrand Lemoine,

> very empirically, often with a redundancy of members, such as a doubling of certain pieces, the joists especially, to counter the forces acting on flat bars and the risks of spreading . . . The beautiful drawings of this time, washed in blue, conventionally the colour for metal, attest both a technique fumbling in its infancy as well as the continuities in the old art of the serrurier, much of whose concern was about getting the materials ready for forging.[96]

In France then, as in Russia, the serruriers had developed roofing systems of their own within parameters defined by architects. Not until after 1836, when the railway engineer Polonceau first tried out the light, graceful truss that bears his name in a shed on the Paris-Versailles 'left bank' line,[97] did French iron roofs acquire both visibility and coherence. The two qualities made their presence felt simultaneously.

In Britain, where responsibilities began and ended is, as always, less clear. From the first, architects and founders had used bits and pieces of iron without much definite

72. Corsham Court, the hall. Nash's first blatant essay in domestic iron-work, 1797–1800, using the Coalbrookdale Iron Company.

philosophy. That persisted after 1800. The classic case is John Nash, whose art and technique are obscured by a smokescreen of improvisation. In Summerson's words, Nash was 'quite merciless in his use of iron as a general fac-totum'.[98] That is not to say that he was not original and ingenious. He was among those who, after a short hiatus, took up iron bridge design anew in the 1790s. No matter that his first attempt collapsed; its replacement, designed on a patent voussoir system and cast at Coalbrookdale, was soon up again and lasted a century. Blatant or suppressed, orna-mental or plain, iron threads in and out of Nash's domestic architecture too, from Corsham Court (1797–1800) to the Brighton Pavilion (1815–22) and beyond (ills. 72, 73).[99]

The Brighton Pavilion is a reminder that iron construc-tion need not be smothered in science and solemnity. Before it was linked with dour modern industry, iron's delicacy and pliability were among its allures. Why then should it not join with other materials and be extemporized for effect? Cast iron, Nash's partner Repton had remarked, was 'peculiarly adapted to some light parts of the Indian style'.[100] Three elements in the comedy of the Pavilion can be singled out: the iron columns with crowning copper palm-leaves in the great kitchen, tallying with the iron and copper batterie de cuisine; the 'bamboo' staircases which, we have seen, Schinkel sketched and imitated; and the ad hoc corsetting of the roofs, climaxing in a 'birdcage' within the great onion dome. Principle is not to be sought here; one can only gape at Nash's fertility and instinct for occasion.[101]

By the time of Buckingham Palace (1825–30), Nash was riddling floors and roofs with iron, employing castings up to 35 feet long, as well as fake columns for external colonnades.[102] Haste and changefulness underlay these dispositions. But it would be incautious to credit the ironwork design beyond Nash or his office. Its suppliers, the Welsh ironfounding giants R. & W. Crawshay, had grown fat upon armaments and were unfa-miliar with structures. They passed much of the order on to a hapless Staffordshire foundry which, like other subcontractors for the palace, went bankrupt in the frenzy of parliamentary retaliation that followed George IV's death in 1830. Nash was sacked, and every detail of the unchecked expenditure on Buckingham Palace put to scrutiny.

The safety and the principle of iron beams came into question as well as their cost. Three experts, the consulting engineer George Rennie and the founder-engineers Bramah and Rastrick, pored over the work. They found much to criticize. But the old architect put up a vigorous defence. In so doing he made two strong averrals:

> No founder ever furnished me with a design for any casting I ever used . . . the Architect ought to be the most competent judge of the form of the castings he requires and their applications and strength.[103]

These statements cannot mean that no beam ever went into a Nash building that he had not personally 'sized' or 'proofed'. What they signify is that he had a broad understand-ing of iron construction, knew what uses could be made of it, and asserted overall con-trol and competence. And indeed the floors and roofs of Buckingham Palace turned out to be sound – unscientific but sound. No doubt the beams had been overspecified.

Nash also claimed to have been 'the principal user, and perhaps I may add, the intro-ducer of cast-iron in the construction of the floors of buildings'.[104] If by that he asserted priority in stitching hidden cast-iron girders into the upper floors and roofs of structures

73. Royal Pavilion, Brighton, detail of the cast-iron 'bamboo' staircases.

74. Composite truss of cruciform cast-iron girders with wrought-iron ties, for a span of about eight metres. From an addition to Worthy Park, Hampshire, by Robert Smirke, c.1816. English adaptation of the French Ango truss of the 1780s, possibly by Thomas Tredgold. Drawing by James Sutherland based on a sketch by John Reeves.

other than mills, he must share the honours with Robert Smirke – one of those who sat in judgement upon Buckingham Palace. That specific practice started in both architects' country houses around the first decade of the nineteenth century (ill. 74), before they tried it in their public buildings.[105]

Smirke was as committed to structural innovation as Nash, but lacked any grain of the quixotic or intuitive. He saw new materials (concrete as well as iron – see p. 209) as aids to solidity and employed them with a semblance of science. As in the Brighton Pavilion, so in Smirke's British Museum, iron ran a gamut of uses, but with the sobriety befitting a permanent public building.[106] The one place where it saw the light of day was the roof of the strictly temporary Elgin Room (1816). Throughout the public parts of the museum as it rose by stages, almost all the roofs were spanned by hidden iron. The front colonnade, not built until the 1840s, relied on iron to reinforce the spans of its colossal entablature in the French neoclassical manner, though it was constructed by British methods.

The King's Library, the first part of the British Museum to be built (1824–5), incorporates the best-documented early iron girders

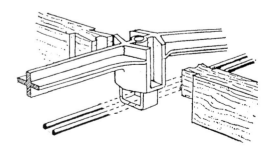

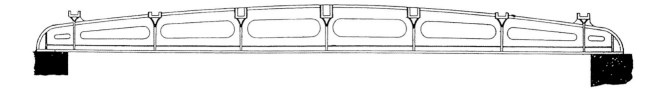

75. The King's Library, British Museum, by Robert Smirke, 1824–5, with detail (above) of the cast-iron girders supplied by John Rastrick to span the neoclassical ceiling and carry the upper floors.

in wide-span floors and ceilings. Smirke's collaborator here was John Rastrick, afterwards the most hostile of the experts who examined Buckingham Palace. Rastrick was a typically versatile figure of the period: part fabricator, part builder, part mechanical engineer. He had started out in Shropshire with a branch of the Hazledine dynasty, and built a fine cast-iron bridge at Chepstow (1814–16) before shifting to Stourbridge and a partnership in the foundry Foster, Rastrick & Co.; later he laid out a few early railway lines and built a few locomotives.[107] In the early phases of the British Museum, Rastrick was plainly Smirke's adviser – in today's terms his consultant engineer – as well as his supplier. Previously Smirke had managed to procure iron beams 30 feet in length for his country-house floors, but for wider spans had employed shallow, pieced trusses in the French mode, familiar in Britain by this date. For the King's Library, Rastrick advised Smirke against a truss and supplied him with single castings up to 39 feet long, with convex top flanges and voids in the web (ill. 75). Rastrick claimed he could have cast to almost any length. But the low tensile strength of cast iron imposed a limit on loading and had kept down spans in mills and warehouses. The skill lay in designing the beams and getting first-rate iron.

The dissemination of the hidden girder owed much to technical publications. Foremost among these was Thomas Tredgold's treatise on cast iron (1822), which gave rules not just for calculating sizes but also for the proof-loading that responsible constructors performed on all beams, with the maximum deflection permissible. James Sutherland has gone into the question of who designed or sized such beams and what were their qualifications.[108] There is no simple answer, he concludes. Tredgold, the era's leading British writer on building construction, spent ten years in the office of his uncle, the architect William Atkinson, latterly as his chief assistant, and published several technical books during that time.[109] At the end of his life he wrote a classic definition of the profession for the infant Institution of Civil Engineers (p. 416). On those grounds he is revered today as a founding father of British engineering. Yet Sutherland's judgment of Tredgold is that 'on cast iron work he was not a structural engineer. He could more correctly be said to be a manufacturer's designer detailing pre-fabricated parts'.[110] That sounds like Brunet's role at the Halle au Blé. William Turnbull, another author of early textbooks, also worked for an ironfounder.

Between architect, engineer, founder and technical writer, the iron girder gradually settled into a regular place in construction. At first it was still deep and variable in profile. As calculation and testing improved, and casting gave way from the 1840s to wrought-iron I-beams, rolled on rollers of increasing scale and adjustability and better able to deal with tensile forces, girders became shallower and standardized. A little familiarity with the textbooks allowed architects and builders to specify them without specialist help. Soon, architects were inserting them wherever it was awkward for upper walls to coincide with lower ones. Smirke's pupil, the country-house architect William Burn, was among the first in Britain to do so as a matter of course. It helped free up the plans of houses and other multi-storey buildings where small and large spaces had to co-exist. It was especially handy over shops, which acquired better display windows and opened out internally in consequence. A carpenters' strike of 1845 in Paris is the usual point of departure given for the widespread adoption of rolled girders and joists in French floor construction – a practice in which France raced ahead. A decade later Francis Fowke could claim that making whole floors that way was 'almost unknown in England'.[111] But soon afterwards flooring systems combining iron girders or joists with some form of infill proliferated across Europe and America as a means of fireproofing buildings (pp. 185–6).[112]

Small girders might be buried in the floor, but deeper ones for wider spans had to be dropped below ceiling height. As with iron so later with concrete, dropped beams might 'express' the construction but could be ugly and obstructive. That caused architects problems of appearance. Usually they were cased in timber or plaster. In the King's Library, the flat neoclassical aesthetic inhibited Smirke from underlining the presence of deep beams; in later parts of the British Museum, he was bolder. But such beams were seldom exposed in smart interiors before the 1860s. Propriety apart, the fire regulations in London and many other cities discouraged open structural iron within inhabited buildings after some nasty experiences with buckling cast-iron columns.

Hidden iron reached its British apogee in the Houses of Parliament – a monument that can be read with equal validity as modern or old-fashioned.[113] On the one hand the vast building marks high tide for the old cultural tradition that construction and services should defer to the symbolic claims of style under the charge of a plenipotentiary architect. At the same time the Houses of Parliament has a strong claim to be the 'first modern building' in terms of process. No less than the Crystal Palace, it represented the transfer to architecture of techniques learnt in constructing the canals and railways. As the most intricate building yet assembled, its procurement entailed subdivision and delegated management. Experts of every kind contributed to the Houses of Parliament

76. Roof space above the House of Lords, showing the cast ironwork supplied by Henry Grissell in about 1845–6.

76. Roof space above the House of Lords, showing the cast ironwork supplied by Henry Grissell in about 1845–6.

under Charles Barry's aegis; and more than one British building profession came to maturity over the span of its construction (1839–70).

One figure was missing throughout the process: that of the independent consulting engineer. Today this seems unimaginable for so ambitious an undertaking. Such structural design as occurred before tender took place in Barry's office. At first it may have been assigned to a 'surveyor', F. H. Groves. After 1844 it fell mainly to Alfred Meeson, 'a trained engineer on whose skills Barry leaned heavily', Michael Port tells us.[114]

What did this mean for the ironwork? Like Smirke, Barry had been an enthusiast for iron construction since the 1820s and specified rigorously for quality and proof-testing. As the previous Houses of Parliament had burnt down and the new one was to house public records, the incidence of iron in the new building was higher than ever, ramifying throughout the floors and roofs to the internal framing of the two great towers and the very outside plates of the roof itself. Few of those who revel in Pugin's passionate craftsmanship in the House of Lords appreciate that its benches rest upon iron staging and its roof is spanned by all-iron trusses (ills. 76, 77).

We need not credit all the structural details to Barry or even Meeson. The Houses of Parliament is permeated by team effort. Just as in embellishing it Pugin enjoyed a tolerably free hand, so also its structure represents the contractor and engineering fabricator collaborating within a framework set by the architect. The complexities of railway construction had by now taught firms to organize work in that way. Much of the building was contracted to Grissell and Peto, a firm that had previously undertaken Barry's Reform Club but by now was entangled in vast railway works. A high proportion of the parliamentary roofs and floors were subcontracted to Thomas Grissell's nephew, Henry Grissell of the Regent's Canal Ironworks, a fabricator deeply involved in prefabricating iron buildings and in railway construction.[115] Such working drawings for the roofs as survive appear to comprise not Barry's independent thoughts but the fruit of prior negotiation and agreement. We can be confident that not only their fabrication but their design can be laid at the door of this enthusiast for iron.

IRON EXPOSED: HOTHOUSES AND BEYOND

It was Henry Grissell who went on in 1857–8 to supply all the ironwork for the Royal Opera House and Floral Hall.[116] Their architect, Charles Barry's son E. M. Barry, had been closely connected with the Parliament works; that no doubt secured Grissell's employment. This pair of London landmarks, juxtaposed and built as an ensemble, shows how status and function guided the way iron could weave in and out of Victorian structures: here hidden away, there on flagrant display.

Iron was fundamental to both buildings, yet in looks they could hardly have differed more (ill. 78). In the opera house, the ceiling hung from massive wrought-iron girders, the balconies too were canted out on girders, and the stalls sat on iron staging, as in the Houses of Parliament. But it would have been against all social and dramatic proprieties to expose the material. Illusion was fundamental to historic theatre architecture: the suspension of disbelief depended, so to speak, on the suspension of ceilings. Everywhere the iron was concealed by a skim of plaster. Yet the Floral Hall next door, destined as a market by day and a concert venue by night,[117] resembled a festive conservatory. Here the whole structure, built by the same architect and fabricator using the same framing materials, candidly acknowledged the era of iron-and-glass architecture ushered in by the Crystal Palace.

To set the Floral Hall in its technical context, one must go back to the first flowering in Britain of open-span structures in exposed iron. The suppression or sheathing of iron within buildings favoured architectural control. So did the relations of procurement change when iron became conspicuous, and the mere exposure of a few columns gave way to entire iron environments? What happened when the dwarves were let out of Nibelheim? The short answer is that a muffled struggle took place.

Between 1815 and the eruption of railway technology in the 1840s, open-roofed structures in iron fall into two broad categories. There were the building-types in which architects tended to have a hand: hothouses or conservatories; arcades; and markets. But there were also various separate, less public and therefore less 'architectural' types of open roofs. These included the single-storey iron workshop or factory, and the wide-span covered ship-building slip or shed, which shifted from timber to iron construction during the 1840s.[118] What follows focusses on a subset of the first group, the iron hothouse or conservatory, because it highlights tensions between architecture and technology and boasts their most precocious progeny.

It is tempting to think of the hothouse, like the bridge, as an autonomous structure discharging a single function. It can but need not be so. Many conservatories were attached to houses, while even the independent ones deferred to their context and their proprietors' tastes as well as the welfare of plants. The looseness of English terminology is illuminating: 'hothouse', 'greenhouse' or 'glasshouse' favours the plantsman; 'conservatory' embraces a broader view.

From the start, iron in hothouses responded both to the gardener's desire for health and growth and to the architect's obsession with form.[119] Let the gardener come first. European horticulturalists in the eighteenth century wanted bigger glazed surfaces in their hothouses. To manage that, glazing bars of wrought iron and other metals were sometimes proposed. In Britain, little concerted development took place until 1805, when the newly founded Horticultural Society raised the issue of scientific hothouses. Among participants in the ensuing debate was the young J. C. Loudon – at that stage of his life mainly an agricultural improver, and never wholeheartedly an architect, let alone an engineer.[120] Over the next decade – coinciding with the Continental System, when Baltic timber prices were high – iron began to seep into commercial hothouses by way of glazing bars and trusses, protected against rust by paint or sometimes copper.

Then in 1815 George Mackenzie indicated to the Horticultural Society, with illustrations, how the glazed surfaces of hothouses might be curved to get better light and

FACING PAGE
78. Royal Opera House and Floral Hall, Covent Garden. E. M. Barry, architect, with Henry Grissell of the Regent's Canal Ironworks, 1857–8. Iron, critical to the construction of both buildings, was exposed in the hall but hidden in the theatre.
79. Palm House at Bicton Park, c.1825. Conjectured to have been made by W. & D. Bailey using the Loudon bar.

80. Cast-iron conservatory added to Carlton House by
Thomas Hopper, architect, 1807. The style of Henry VII's
Chapel, Westminster Abbey, is adopted.

secured by cast-iron glazing bars. Loudon was among those who took this up, producing specimens of a curved bar in wrought (not cast) iron for the Society's examination and 'acuminating' (otherwise, pointing) the resultant structure's profile. This simple-seeming component required research and development, and the erection of experimental hothouses in Loudon's suburban garden at Bayswater. In 1818 improvements on the original bar were patented by the ironmaster William Bailey of the firm of W. & D. Bailey, to whom Loudon made over his rights.

The Loudon and Bailey bar was the key to a revolution. It led the way to the curvilinear school of hothouse design and hence to a specifically British school of curved iron-and-glass roof in other building-types. By 1824 Loudon could point to twenty-four such structures by the Baileys, of which the great domed conservatory at Bretton Hall with a base diameter of a hundred feet was the most spectacular. A survivor is at Dallam Tower, while a similar structure at Bicton is attributable to the Baileys (ill. 79).[121] Neither Loudon nor any other architect is known to have designed them. They were ordered direct from the Baileys, who understood the ideas well enough to vary their shape and design to the site, usually with the backing of a wall. To the modern eye they are, quite simply, beautiful – perhaps the first independent iron structures apart from bridges for which that can be said. Here elegance came by working from the component outwards.

Architects began by tackling hothouses from the other end of the spectrum, the whole building and its setting. The first stab at an all-iron conservatory is a drawing in the royal archives for a pitch-roofed and square-planned structure, high and open, now thought to be a speculative design made in 1798 by Nash for the Prince of Wales.[122] In time the Prince did build a cast-iron conservatory at Carlton House to the designs of Thomas Hopper (1807).[123] A tour de force of Gothic archaeology (ill. 80), it may have made for atmospheric promenading, but as a place to grow things in it was an absurdity. Most horticultural architecture was less dogmatic. An example of creative compromise between architect and gardener was the camellia house at Wollaton Hall (1822), designed by Jeffry Wyatville and made by the Baileys' main rivals, Jones and Clark of Birmingham (ill. 81). Without masonry backing of any kind, it may be the earliest surviving all-iron building. But it is not curvilinear; it resembles earlier orangeries, only with thinned-down piers and stylized sash bars in between.

The architectural conservatory soon came to an accommodation with the curvilinear, horticultural type.[124] The main arena for this compromise was the aristocratic garden. The Earl of Shrewsbury's Alton Towers, the Duke of Northumberland's Syon House, and the Elector William II's Wilhelmshöhe, Kassel, all spawned grand conservatories of balanced, symmetrical design, in which masonry piers vied with metal-and-glass or timber infill, roofs and domes.[125] Among the finest was the Syon House conservatory by Charles Fowler (1827–30). Its Bath-stone casing chimes with the great house nearby, while its cast-iron columns and high dome of gunmetal ribs defer to the hothouse revolution with a nod (ill. 82). But only a nod: 'So far as plant culture alone is concerned, no

81. Wollaton Hall, interior of the camellia house by Jeffry Wyatville, architect, 1822. An early example of a free-standing iron and glass structure.

82. Conservatory at Syon House by Charles Fowler, architect, 1827–30. Classical interior encasing interior with cast-iron columns and a dome with gunmetal ribs.

83. Shop in construction next to the Hungerford Market, London, showing iron stanchions between the windows: drawing by George Scharf, 1834. Charles Fowler, architect, Grissell and Peto contractors.

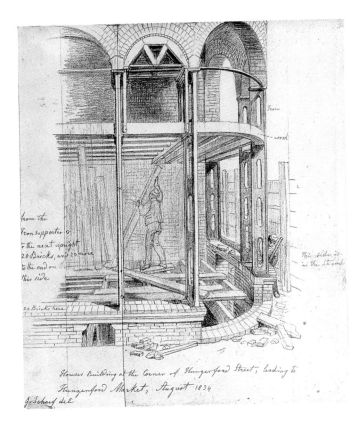

arrangement can be more unsuccessful', judged a horticulturalist.[126]

Architectural enthusiasm for open iron in the run-up to the railway age should not be exaggerated. Fowler is a case in point. Howard Colvin calls this able constructor 'one of the few nineteenth-century architects who, like the engineers Telford and Rennie, were able to handle structure and form with equal assurance'.[127] But he was never obsessed with displaying ironwork. Most of the market buildings that won him renown were sturdy classical buildings in stone, with open colonnades and timber roofs. In his great London markets, Covent Garden (1828–30) and Hungerford (1831–3), the original portions were of granite, brick and timber, with some hidden iron girders. Only in afterthoughts to the two projects did the iron emerge, notably at the Hungerford Market, where shopfronts towards the Strand were constructed in 1834 with ingenious stanchions (ill. 83), and an independent shelter with cantilevered roof added next year for the fish stalls.[128] That latter piece of brio has won Fowler a place in histories of architectural ironwork, but it was not typical. The iron roofs at Covent Garden date from the 1870s and '80s, when all such markets came to be covered. At that juncture, Fowler's son stated that he and his father both believed 'that iron is not the best material for the roof of a market'.[129]

In the 1830s and 40s developments in hothouses and conservatories altered the balance between architecture and technique, and paved the way for the railway-spans and other open iron structures to come. Instead of the grounds of great houses, the typical settings for this next generation of progressive structures tended to be botanical gardens. In these locations, programmes of science and instruction sidelined older, vaguer instincts of recreation. Here technology allied itself with plantsmanship to triumph over architectural preconception.

84. The Great Stove at Chatsworth, 1836–40, by Joseph Paxton, assisted in obscure measure by Decimus Burton.

Such for instance were the original iron 'serres' in the Jardin des Plantes in Paris, designed by the polytechnician-turned-architect Charles Rohault de Fleury (1834–6).[130] Devoid of masonry or style above the basement, this composition would have stretched a full 170 metres of curvilinear wings and rectilinear central pavilions, had there been money to build its full length. Though unfinished, it was enough to whet Parisian appetites for an open iron architecture. But in the gathering struggle between the champions of architecture and of technique, the most absorbing episode was to be the building of the Palm House at Kew (1846–8).

KEW

Like all such institutions, the Royal Botanic Gardens at Kew has always had to strike a balance between research, education and pleasure. Its transition from royal retreat to government-controlled institution in 1841 reflected the sea-change then taking place in attitudes towards natural history. In fact the gardens already boasted a harbinger of these changes in the Aroid House, first designed as a conservatory for Buckingham Palace by Nash, then taken down and re-erected at Kew in 1836 by Wyatville.[131] The masonry casing and iron interior of the Aroid House read like a compromise between the pleasure-buildings designed for the earlier, private and aristocratic Kew of the eighteenth century and what was prospectively in store.

The protagonists over the Palm House were Decimus Burton, architect, and Richard Turner, a Dublin ironfounder. Thanks to the researches of Edward Diestelkamp, we know more about Turner than other of the enigmatic fabricators – 'engineers' by courtesy – who did so much for developing iron architecture.[132] His forbears were ironmongers, smiths and minor building speculators in the Dublin of the Protestant Ascendancy. From about 1830 Turner ran the family firm, diversifying into hothouses and their accompaniment, hot-water heating. He was soon supplying Loudon-esque hothouses to the Irish gentry and nobility. His first work for a botanical garden was the Belfast Palm House (1839–40). There he came under an architect, Charles Lanyon. Only two wings could at first be afforded, built without Lanyon's architectural masonry. Their width and height seem to have prompted Turner to go beyond Loudon and the Baileys and find a fresh way of supporting the roofs at mid-span, to avoid trusses and maximize light. The few decorative touches at Belfast do not suggest that Turner thought much about the topic of ornament for iron structures.

85. Kew Palm House, plan of the penultimate design, by Richard Turner and Decimus Burton, March 1844.

Hearing in 1844 that Sir William Hooker, the director of the reformed Kew, intended to build a palm house, Turner sought permission to send in a design. The final plan of a longitudinal glasshouse with a higher, wider centre was present in his first project. Its submission triggered the appointment of Decimus Burton by the department that would have to foot the bill. Burton had connections with government and botanical experts and had already built plenty of conservatories. In most, attached to houses, 'an elegant and chaste masonry structure framed large sash openings', says Diestelkamp.[133] But a few were independent conservatories on the curvilinear principle. In both types he had sometimes used iron columns at mid-span. So Burton could move between simple neoclassicism and the stylelessness of the new hothouses.

Burton had also been consulted on the Great Stove (1836–40), the largest of Joseph Paxton's celebrated timber-and-glass conservatories for the Duke of Devonshire at Chatsworth (ill. 84). The Chatsworth sequence is an exception to the rule that advanced structures during these years were in botanical gardens. Though adept with timber and glass, Paxton at that point was not yet versed in iron. Burton's role at Chatsworth was and is disputed, but may have had to do with inserting cast-iron columns to help carry the span. Mrs Paxton, at any rate, suspected his intervention: 'how goes the Great Stove on,' she wrote to her husband in 1837, 'you must really cut the London architects . . . they are gaining experience at your expense'.[134] Once it was finished, Burton complained to the Duke of Devonshire that Paxton had received all the credit in a travel book. The Duke responded diplomatically, by recording in his guidebook to Chatsworth that Paxton's executive role had been 'cordially met and assisted by Mr Decimus Burton'.[135] The architect, then, was not averse to collaboration, but he expected recognition.

The process for the Kew Palm House starts with a skirmish in which the protagonists review one another's designs and dissect them politely. The stances they take are not as might be predicted. Burton objects to Turner's first design on grounds of style: far from being too styleless, it is the reverse – too Gothic, too ornamented. He then produces his own version with 'the classic and ecclesiastical avoided – appropriateness, with as pleasing an outline as the case will admit of, are the objects aimed at, and all extraneous ornaments are dispensed with'.[136] This design the ironfounder in turn condemns as 'wildly extravagant' and having a 'much encumbered', ill-lit interior.[137] But Turner's lower status and lack of English experience prompt him to seek an accommodation with Burton. Soon they are working together and have come up with a new design fifty feet in span at the centre, rationalized to reduce the number of components and ease their production.

At this point Turner faced a challenge recurrent in relations between architects and fabricator-engineers. The agreed design for the Palm House ended up almost entirely of iron and glass, apart from foundations and stub walls, and included innovations on which Turner had spent time and money. He therefore expected to build it himself. But the authorities had no guarantee about its cost. They therefore insisted upon tenders.

The tender, a standard device of the market economy, had much to do with the emergence of consultant engineers for buildings in the the later nineteenth century. When iron technology was young, experts were few. In any advanced use of iron for construction, the tasks of design, manufacture and site-assembly were apt to coalesce. This put the fabricator in a strong position. The architect might strive to co-ordinate the whole, but as in all situations where subcontractors offer complete 'packages' for the design and execution of elements in a building, his powers were limited. That is why Barry's management of the Houses of Parliament was remarkable. The alternative, commoner as iron structures proliferated, was to hire an engineering expert to work out the structure under the architects before a contractor was chosen. The working drawings could then be put out to tender among the growing number of firms able to tackle iron construction.

At Kew, things turned out well enough for Turner. The one other firm that competed, Jones and Clark of Birmingham, sent in a design of their own that was soon rejected.

86. Kew Palm House erecting, 1847.

87. Kew Palm House today.

The Turner-Burton design was then put out to tender between the two firms. So as to be impartial, Burton employed an outside engineer to prepare the working drawings – a procedure which exasperated Turner, as his details were altered and, so he felt, misconstrued. But he won the tender and was able to substitute his own intentions as the work went along. In the primary agreement of August 1844, Grissell and Peto were appointed main contractors, no doubt on the strength of their management experience; Turner, by now overextended financially, became the iron subcontractor under them.

This is not the place to rehearse the innovations in wrought-iron construction embodied in the Palm House (ills. 85, 86, 87). Suffice it to say that the changes in structural detail and planning which took place during construction sprang from Turner, working with Hooker. Burton's role, as it had been from the start, was to act as the front man and keep the confidence of the government paymasters. Persuasion is what an architect is partly there to do. Presumably the sparing Greek ornament of the Palm House ironwork is his too. Perhaps his biggest contribution to the project was the high Italian campanile built as a chimney for the flues, a full 500 feet away in distance, light years away in style. And yet Diestelkamp concludes:

> It is evident that Decimus Burton was responsible for the styleless character of the Palm House and it is very likely that the controlled restraint of the elevations was due to his Neo-classical taste and predilection for what was in this case a simplified form of repeating elements.[138]

As for Turner, the construction of the Palm House was an engineer's passion; reason and emotion were bound together in it. When all was over in 1848, Burton could report that 'Mr Turner is a loser to a very serious extent on the contract'. Turner explained this as a consequence of the new I-shaped deck beam he had used in order to achieve 'beauty, utility and stability'.[139] It was the first of these words that cost the most.

2. *A Railway Interlude*

The Early Railway Station

So vital are the railways to our story that they merit space for themselves. In English-speaking lands they represent a step-change in elevating those masters of the iron road, the engineers, to heroic status. Surely in such relationships as they afforded between engineers and architects, the former enjoyed absolute sway? Taking examples from the annals of iron railway architecture, this section argues that architects took their share of the spoils and honours.

Locomotion aside, the primary job in building a railway may be defined as civil engineering – that of determining, designing and constructing the line and level of the track. In that sphere engineers held undisputed sway. But in the British context a railway engineer could mean several things.[140] We are easily mesmerized by the giants of early railway construction – the Stephensons, the younger Brunel and Locke – into supposing them typical railway engineers. They, along with their many rivals, were the companies' engineers-in-chief. Hundreds served below them. First came their sizeable office staffs and their site representatives, the resident engineers. On the other side of the fence were the 'engineers' directing the contracting firms, their own office staffs and their own site supervisors. The boom years of the 1840s led to a parlous shortage of them all. 'Railways were the cry of the hour, and Engineers were the want of the day', wrote Francis Conder in a vivid memoir:

> So came to the front, – military men, accustomed, perhaps, to rapid and skilful sketching of the country, able with the theodolite and with the sextant, but unacquainted with the other requisites of their improvised profession . . . mining surveyors, accus-

VIEW OF THE INTERSECTION BRIDGE ON THE LINE OF THE S^T HELENS & RUNCORN GAP RAILWAY,
CROSSING THE LIVERPOOL AND MANCHESTER RAILWAY NEAR THE FOOT OF THE SUTTON INCLINED PLANE.
Drawn by Charles Vignoles, Esqr C.E., F.R.A.S., M.I.C.E.
DEDICATED by Permission to EDWARD GREENALL, ESQ^{re} of Wilderspool, Cheshire
London Published 1833 by R. Ackermann N° 96, Strand

tomed to use the dial, and to the intricacies of subterranean work; mathematical engi-neers, good, no doubt, to direct the smithy and the lathe, but unaccustomed to works of magnitude; – architects who generally limited their claims to the construction of stations, and who seemed to be regarded as interlopers by all who, on any of the above grounds, called themselves engineers.[141]

In the earliest days railway structures, even bridges and tunnels, came second to the projection of the lines, and stations were scarcely thought about at all. Nor, though the iron revolution gave the railways birth and breath, were many of the primitive structures of iron. Bridges were the obvious candidates. On the Stockton and Darlington Railway (1823–5), George Stephenson built just one small bridge in a mixture of cast and wrought iron. When he proposed to cross the line's widest river in iron, the county surveyor (Ignatius Bonomi, an architect) intervened and substituted stone instead.[142] On the Liverpool and Manchester Railway (1828–30), the iron bridge over Water Street, Manchester, built on a slight skew, was important in the history of bridge design, but other bridges and viaducts were of brickwork.[143] Occasionally there were bridges of ash-lar masonry like the skew bridge at Rainhill (1829), or the 'architectural' bridge where-with Charles Vignoles took the St Helens and Runcorn Gap line over the Liverpool and Manchester in 1831–2: trabeated not arched, with coupled Doric columns carrying a frieze and triglyphs (ill. 88).[144] Nearly all the above were designed by junior engineers in the ambit of George and Robert Stephenson. Young men most of whom had served their apprenticeships with machinery, the Stephenson disciples found themselves cata-pulted outwards to grapple with line lay-outs and structures.

88. The Intersection Bridge at Runcorn Gap carrying the St Helens and Runcorn Gap line over the Liverpool and Manchester Railway, aquatint by S. G. Hughes. Charles Vignoles, engineer, 1831–2. The grandest of early railway bridges, in stone with coupled Doric columns perhaps derived from bridges by the Rennies.

What about the stations? Wayside halts were minimal and of timber, while the earliest termini were plain. 'It cannot be said that the question of the design of the station bulked large in the Company's deliberations', says Ron Fitzgerald about the original Manchester terminus of the Liverpool and Manchester Railway.[145] Called at first a 'coach office', it affected the simplest stuccoed Greek-ness. Next to it stood a warehouse which Fitzgerald thinks may have been designed by T. L. Gooch, draughtsman to the older Stephenson. The first Liverpool terminus at Crown Street was similar, but had a verandah with iron columns facing the railway. At either end of the line wide wooden roofs were added over the tracks for shelter a year after the opening (ill. 89).

Nevertheless in Liverpool a note of fancy was also struck from the start. The early railway companies had to sooth civic sensibilities. Accordingly John Foster junior, architect and surveyor to Liverpool Corporation, was invited to be engineer for the original tunnel into the town in 1826. Though he declined, Foster went on to design at least one if not both of the company's initial indulgences, the Edge Hill boiler house chimneys and engine house. The chimneys took the guise of high Doric columns, while the engine house became a castellated gateway to the tunnel spanned, absurdly, by a 'moorish' arch (ill. 90). Here was born the romance, the compensatory irrationalism, of railway architecture. That this caprice was built in brick rather than the stone Foster wanted adds to the touch of the ludicrous that tinges many early English railway structures.[146]

The line's extension into central Liverpool (1835–6) begat a second precedent. Here Foster devised a full classical front for the new station at Lime Street. At his prompting no doubt, the Liverpool and Manchester directors took this design to the Corporation with the message that 'the great public entrance into Liverpool should not be inferior in style and design to the most approved Public Buildings of the Town'.[147] They sued for a subsidy to make up the cost between a plain brick elevation and one in ashlarwork with

89. Crown Street Station, Liverpool, one of the two original termini of the Liverpool and Manchester Railway. Drawing by Talbot Bury, 1831. The timber roof is an early addition.

90. The 'Moorish arch', at Edge Hill, Liverpool and Manchester Railway. John Foster junior, architect, 1830. Drawing by Talbot Bury, 1831.

columns – and received most of what they asked for.
Liverpool Lime Street as built therefore had a fine front
by Foster. Behind it were tucked lowly offices and a timber-and-glass station roof, erected by another architect
working with a local builder.

That station lasted a decade. Everything behind the
front was then remodelled to the enlarged scale of the
1840s under Joseph Locke's supervision. The architect
William Tite tacked on side offices, after which came a
new bridge at the back by the resident engineer, Edward
Woods, and an iron roof supplied (in the teeth of
Locke's advice) by Richard Turner of Palm House
fame.[148] The completion in 1850 of this first of the

91. The Euston Arch, soot-blackened in 1934. Philip Hardwick,
architect, 1836.

92. Railway stations from A. W. N. Pugin, *An Apology for the Revival of
Christian Architecture* (1843): an afterthought in the manner of *Contrasts*.
Left, robustly Gothic over- and under-bridges with small stations at
low and high levels. Right, satirical sketches of the London and
Birmingham Railway termini: Euston, with the pitch-roofed sheds
behind the arch (top); Curzon Street, Birmingham (bottom). The
author proposes a defter counterpoint between railway technology
and ancient architecture.

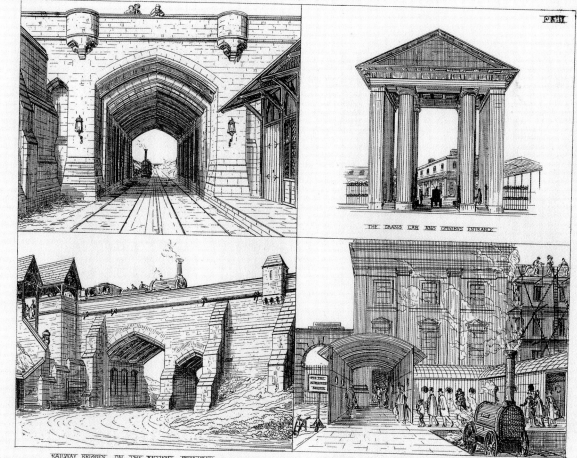

RAILWAY BRIDGES ON THE ANTIENT PRINCIPLES

THE GRAND CAB AND OMNIBUS ENTRANCE

A SHOW FRONT WITH CONVENIENT ADDITIONS &

93. Euston Station, the original interior. Charles Fox, engineering assistant to Robert Stephenson. Drawing by Talbot Bury, 1837. Fox's light wrought-iron trusses foreshadow the development of shed roofs in French railway stations.

great arched railway roofs, without intermediate supports, marked the moment when technology derived from the curvilinear hothouse tradition got its grip on the railway terminus. But the second Lime Street Station confirmed the pattern of disconnection set by its predecessor. An architectural front making some figure presented itself to the town; engineers and builders tackled the functional parts behind. That split between the public-aesthetic and commercial-technical sides of railway architecture widened, as the technologies of station front and station interior diverged.

London's first major railway terminus, at Euston, offers the classic instance.[149] Once the profitability of passenger railways was clear, the capital became the great prize. By 1835 the course of the London and Birmingham Railway had been set, and a site selected at the edge of the metropolis. Robert Stephenson, engineer for the line, delegated the detailed design for the last few miles into Euston, including the original station sheds and platforms, to Charles Fox, an engineering assistant then in his mid-twenties. But the company wanted a fine frontispiece towards London and unlike the Liverpool and Manchester directors was willing to pay for it. Hence the Euston Arch, commissioned from the architect Philip Hardwick in 1836 and built forward and apart from the station (ill. 91). Two flanking hotels were appended as afterthoughts, and then in 1846 a hall and board room to one side of the station itself, designed by Hardwick's son.

The incoherence of this ensemble became as notorious as the arch's splendour. There were extenuating circumstances. When the lay-out for Euston first emerged, a second station for the Great Western Railway was anticipated on the site to the west of the arch and roadway, answering the London and Birmingham station to their east. The GWR then pulled out, choosing Paddington instead. That is why the arch was not aligned on the axis of the one station built. Nor was the physical disconnection between frontispiece and station imprudent. Railway stations underwent rapid evolution in their early years. It made sense to preface the working parts with a detached screen which need not be affected by alterations.[150]

Still, the shift between them was always bizarre. It soon attracted criticism – and Pugin's nose for humbug (ill. 92).[151] Arriving, you swept like a hero through a version of the triumphal arch, democratized to hail the railway traveller: a propylaeum in baseless Doric, fashioned from eternal masonry and symbolizing a 'national moment'.[152] Alighting, you were cut down to size, as you penetrated a pair of mere sheds (ill. 93). No matter that their roofs were carried by the first wrought-iron trusses in any station, to contemporaries they looked makeshift and spindly. Only in this bald setting did you encounter the magic machine that would hurtle you away. Hardwick's arch and Fox's roofs alike represented a modern miracle. But they seemed to be in mutual opposition. The pride and the substance – the 'architecture' – were in the wrong place.

Could stations fronts be reconciled with station interiors? Critics have never quite got over the schizophrenia of Euston. Railway termini have been judged ever since by whether the form and purpose of the interior are manifest on the outside. There is certainly pleasure to be had from 'reading' the form of a station from its façade, so that its inner arrangement may be deduced from its public front. In termini of the generation after Euston, like King's Cross in London or the Gare de l'Est and first Gare du Nord in Paris, the offices were still mostly along the sides, so that the sheds could show through at the ends. Such 'head stations', in the French manner of rational construction, have routinely been praised for their clarity.[153]

But as time went on it made less sense to plan in that way. After 1850 British railway directors wanted hotels in front of their termini, which made end-on visibility of the sheds impracticable. Soon broad passenger concourses at the heads of stations introduced frontal cross-axes, while the sheds over the platforms themselves abandoned the scale and span attained in the costly iron roofs of the 1860s and 70s. Later again, with electrification, came the chance of banishing the tracks underground, leaving front, offices and concourse masters of the world of appearances (pp. 126–31).

The display of interior space could only ever be a side issue in planning a railway terminus. A better guide to excellence in each age and type of station-building is how parts of separate character relate – in plain words, how they join. Already at Lime Street and Euston we have seen these parts assigned to different professions. How well they cohered in later stations would depend on permutations of the partnership between architect and engineer. In Britain, where these relations tended to be loosely defined, the results might be exhilarating but the points of junction were too often botched. That accusation can be fairly levelled at St Pancras, for all Sir Gilbert Scott's attempts to adjust his hotel to W. H. Barlow's noble trainshed. Elsewhere, more formalized relationships sometimes led to a crisper station architecture that prefigured collaborations over other modern building-types.

There has been much engineering rhetoric about station-building. 'Stations are designed by engineers', one critic has written: 'the architect only comes along later to decorate them.'[154] That may be true of a few termini. Yet in the heady days of British railway-building most rural and town stations were allotted to architects, whether independent practitioners or employees of railway companies. The core of the job was usually a simple office building, made from traditional materials, as often as not tinged with a dash of style. It might be linked to a station canopy of timber, iron or a mixture of both. Only when the canopies became complex roofs did engineers enter the picture.[155]

The independent architects responsible for the scores of smaller British stations built in the 1840s and 50s, before standardization crept in, worked intuitively. Christian Barman once classified the results into the functional, the 'social' (tuned to their context) and the 'hieratic' (glorifying the railways and towns they served).[156] The picturesque thread inaugurated at Edge Hill was the commonest, but people could be irritated by its restlessness. The traveller finds, wrote *Punch* in 1848, that

almost every place he arrives at presents some deviation from the style of architecture he has just left behind him . . . he is very likely to encounter an old English ticket office, a Turkish water-trunk, a Swiss engine-house, a Grecian goods depot, and an Italian terminus, all within the limits of fifteen miles of railway.[157]

Sometimes it is supposed that behind the romance of these structures lay an urge to distract a public fearful of train-travel or even iron architecture. But context is at least the usual conscious motive. Thus Sir William Tite:

> I have done my best to mould the forms and modes of mediaeval architecture to the unusual requirements of railways . . . I believe I was the first man who attempted to make a Gothic railway station, because it suited what I may call the genius loci. Whether I succeeded or not you will be able to say when you visit Carlisle. It was very troublesome, and did not go very well with the platforms and sheds and roofs but I had no difficulty with the refreshment rooms and halls.[158]

94. Newcastle-upon-Tyne Central Station, exterior. First unexecuted design by John Dobson, architect, 1846. All Dobson's initial efforts went into the design of this monumental station front.

95. Newcastle-upon-Tyne Central Station, perspective of trainshed by J. W. Carmichael and John Dobson. Designed by Dobson in 1849–50 with help from Thomas Charlton of Hawks, Crawshay and Co., who fabricated the elliptical wrought-iron ribs.

Tite exemplifies the types of uses competent architects could be put to in railway-building.[159] Established in the City of London by the time of the 1840s railway mania, he amassed a fortune as a surveyor valuing the vast tracts of land involved in its transactions. He was on equal terms with the great railway engineers, who found him 'an admirable man of business'.[160] Architecture was never a sideline for Tite. But it had to jostle for its place amidst his many directorships and duties. His railway career began in the late 1830s with the London (Nine Elms) and Southampton terminuses of the South Western Railway. Though never wedded to a single railway company, Tite built up a relationship with the demanding Joseph Locke, who retained him, says Victoria Haworth, 'primarily for the purpose of designing impressive station buildings'.[161] Sometimes his role went further. When Locke laid out the Paris–Le Havre line with William Mackenzie (1841–7), Tite not only designed some of the original stations, including two at Rouen, but was brought in to draw out the second Barentin viaduct after the first collapsed.[162]

The spread of iron roofs for the bigger stations contributed to a shift in power from architect to engineer after 1850. Before then there was a precarious balance. Though Fox at Euston and Turner at Lime Street offered technical pointers for the future, both were subservient to the powerful engineers who laid out the line. Occasionally, architects retained overall control of station-building, even where there was extensive ironwork. That was what happened at Newcastle upon Tyne.[163]

The eminent local classicist John Dobson had been involved in projects for a central station at Newcastle since 1836. After many chops and changes he was able to build most of an enlarged scheme in 1847–50. His civic front towards the town, erected in a grand Roman manner, hogged the limelight when the station was published (ill. 94). But what of the interior with its unfolding roof, which follows the curve of the platforms and consists of three 60-foot spans carried by elliptical wrought-iron ribs (ill. 95)?

The curve was the making of the station. Arising from difficulties attendant on slashing a railway high over the Tyne into the city's heart, it had been set out by T. E. Harrison, engineer for the Newcastle and Berwick Railway, after a sketch by Robert Stephenson. But the trainshed itself was certainly designed by Dobson, pronounce John Addyman and Bill Fawcett. The aesthetics of the curve may have led him to insist on elliptical ribs rather than settling for a pitched-roof-and-truss system. Dobson was sixty when Newcastle Station was erected, with years of building experience behind him. But his familiarity with iron construction was slight. Anything but straight runs of wrought iron was still hard to roll in the 1840s. For the shape he wanted, he had to rely on the ingenuity of his ironwork contractors, Hawks, Crawshay & Sons of Gateshead, also then building Stephenson's contiguous High Level Bridge. 'The success of this splendid architectural adventure', Tom Rolt has written, '. . . was largely due to Dobson's own invention of special rolls for producing the curved wrought-iron ribs'.[164] For invention, read inspiration. It was Hawks, Crawshay's foreman, Thomas Charlton, who manipulated the rolling mills to produce the troublesome sections, with Dobson doubtless enthusing in the background. Here we see an architect pressing a fabricator to do something new.

Newcastle's triple iron sheds had their influence, notably and immediately on Paddington. But at Paddington the architect came close to elimination and the engineer took the credit. It is one of four case-studies of terminus stations built after 1850 that follow. They form an intermezzo to this chapter, after which the chronological thread will resume.

PADDINGTON

I am going to design, in a great hurry, and I believe to build, a Station after my own fancy; that is, with engineering roofs, etc. etc. It is at Paddington, in a cutting, and admitting of no exterior, all interior and all roofed in . . . Now such a thing will be entirely *metal* as to all the general forms, arrangements and design; it almost of

96. Temple Meads Station, Bristol, interior view by J. C. Bourne. I. K. Brunel, engineer, 1840–2. A wide timber span expressed in hammer-beamed Tudor-Gothic but covertly supported by wrought ironwork.

necessity becomes an Engineering Work, but, to be honest, even if it were not, it is a branch of architecture of which I am fond, and, *of course*, believe myself fully competent for; but for *detail* of ornamentation I neither have time nor knowledge, and with all my confidence in my own ability I have never any objection to advice and assistance even in the department which I keep to myself, namely the general design.

Thus Isambard Kingdom Brunel to his coadjutor, Matthew Digby Wyatt, in a celebrated letter of January 1851.[165]

Brunel, that prodigy of relentless energy, incarnates our image of the Victorian railway engineer. But he was untypical of the breed. His upbringing, if precarious, was educated and metropolitan. His father's and his own early employment leaned towards the civil as much as the mechanical side of engineering. Unlike the Stephensons, Locke and other British railway engineers of humbler background, Brunel reckoned to do architecture. But he was always 'in a great hurry'; and this, as the excitability of his letter betrays, made him hungry for help yet hard to work with. In any Brunel project, much went on beneath the surface of his absolute control.[166]

Like other railway engineers, Brunel had been in no haste to make use of iron in the infrastructure of the Great Western Railway. Locomotives and rails had to be of iron, but brick or timber could serve for most early stations and bridges. He was justly suspicious about the strength of cast-iron beams. If not of brick, then bridges and viaducts might be of laminated or trussed timber, strong enough to carry the light trains on the broad gauge of the GWR's original west-country affiliates, and quick to build and alter if costly to maintain. In his roofed structures too Brunel stayed loyal to timber, trussing the wider spans with wrought-iron plates or ties. Such were his early engine house at Swindon and goods shed at Bristol.

When it came to stations and 'architecture', the results were uneven. In the west of England, Brunel tended to mediaevalize. But his Gothic was old-fashioned and he would not delegate to good architects. Temple Meads Station, Bristol (1840–2), the best surviving instance, has been studied by John Binding.[167] There were three loosely linked elements at Temple Meads: an office building for the Bristol directors of the GWR, the depot (with Brunel's drawing office) and the train shed. All were stone-faced, with timber construction and some iron columns inside. The first two involved some help from independent local architects: William Westmacott at the design stage, perhaps S. C. Fripp in the working-out of the offices. Otherwise Brunel sketched things out and his assistants worked them up.

As in engineering, so in architecture, Brunel was extravagant. He had to cut back the Temple Meads office building after the directors put it out for costing to another local architect.[168] Even so, this must have been Pugin's chief target when he inveighed that in the larger GWR stations 'mock castellated work, huge tracery, shields without bearings, ugly mouldings, no-meaning projections and all sorts of unaccountable breaks . . . make up a design at once costly, and offensive, and full of pretension'.[169] Here the 'preterpluperfect Goth' out-rationalized the ardent engineer. As for the famous train shed, its 72-foot timber span underwent investigation and testing on the part of Brunel and his office. It is a superb technical achievement, but is not constructed as it looks (ill. 96). 'A composite construction of wrought iron and timber',[170] it has extra beams hidden above what is seen. And the proportions of the hammerbeams to the span might make a mediaevalist wince.

Had the GWR's original Paddington terminus been built along the lines sketched out by Brunel in 1836, it too would have had wooden roofs. But by the time the company resolved to rebuild the temporary Paddington in 1850, things had moved on. The wrought-iron station roof

97. Alternatives for the Paddington trainshed by I. K. Brunel, 10 December 1850. Dating from before Fox and Henderson's involvement, the sketches show Brunel's aesthetic interest in the tripartite treatment of the 'country' end of the station. To the right, sketch for division of the transverse bays, also tripartite.

had established itself, first in trussed and lately (at Newcastle and Liverpool Lime Street) in arched form. Meanwhile its pioneer, Charles Fox, was busy building the Crystal Palace. As a member of the Great Exhibition's building committee, Brunel had been to the fore in the controversies that led up to the commissioning of that watershed in iron architecture. He had even produced an official design, before he realized that Paxton's last-ditch alternative, as worked up by Fox, was better. That revelation affected Paddington. 'While Brunel was planning the new Paddington Station in late 1850 and early 1851, he was also attending weekly meetings for the Exhibition and visiting the site in Hyde Park', Steven Brindle has written.[171]

At the same time as they gave Brunel the go-ahead for the station, the GWR directors agreed to build a hotel in front of it to the designs of the architect P. C. Hardwick. Paddington then was a suburban village, and the directors saw no reason for endowing their station with a civic frontispiece. Their decision had an enduring effect on British railway architecture. The positioning of the hotel against the end of the sheds, without an axial way through, turned Paddington into an enclosed experience. In consequence, the station's dignity had to be expressed by a structure 'in a cutting, and admitting of no exterior, all interior and all roofed in', as Brunel put it to Digby Wyatt. That more or less had been the aesthetic of the Crystal Palace – thrilling internally, but a disappointment to many on the outside. Paddington was the first station whose interior received priority

98. Paddington Station, view looking inward before 1892, showing the Dietterlin-like column capitals, the 'drops' attached below the main longitudinal girders, and the applied decoration and 'stars and planet' piercings to the principal ribs. I. K. Brunel, engineer, with Matthew Digby Wyatt, architect.

98. Paddington Station, view looking inward before 1892, showing the Dietterlin-like column capitals, the 'drops' attached below the main longitudinal girders, and the applied decoration and 'stars and planet' piercings to the principal ribs. I. K. Brunel, engineer, with Matthew Digby Wyatt, architect.

from the start. Co-ordination between Brunel and Hardwick seems to have been minimal. The tasks of planning a hotel and a station are, after all, poles apart.

Brunel chose his team for Paddington from his contacts over the Great Exhibition. Charles Fox's firm of Fox, Henderson and Co. became ironwork suppliers and contractors without even having to tender. Fox had developed his iron roofs since Euston, building wide spans for the royal dockyards and the railways, including work for the GWR at Swindon. Knighted for his part in the Crystal Palace, he was at the height of his career and about to embark on a great roof at Birmingham New Street Station. But Paddington was not Fox's finest hour. He was overstretched, and it was due to his firm's delays that building stuttered on until 1854. Matthew Digby Wyatt, the secretary of the Great Exhibition's Executive Committee, was employed according to Brunel's invitation as his 'architectural assistant', but by the engineer, not the GWR. Owen Jones, the architect-designer who had coloured up the ironwork of the Crystal Palace, became colour-consultant on the same basis, seemingly on Wyatt's recommendation to Brunel. Wyatt also tried to draw in the sculptor Alfred Stevens to decorate the royal waiting room, but the idea never got beyond sketches.

Who did what at Paddington? Brunel's sketchbook (ill. 97) shows the lines of almost everything fixed in principle: the disposition and profile of the three original sheds with their eccentric 'transepts' for turning locomotives; the plans and façades of the flanking departure building; details too, from the columns and capitals down even to the drops or pendants under the purlins – fake features added in timber to give a sense of termination to those intermediate ribs that do not rest upon columns. The bulk of the expressive opportunities lay with the ironwork. Everyone – Brunel, Fox, Wyatt and Jones – had their say in this. Fox seems to have altered Brunel's segmental arches to a five-centred profile, perhaps to avoid tie-rods, while Wyatt and another young architect, Charles Fowler junior, developed his sketches in the usual way of assistants.

In 1851 Wyatt had built little but written a good deal about decoration and its principles. In one article he had gone out of his way to praise Brunel's 'independence of meretricious and adventitious ornament'[172] (he cannot have been thinking of Temple Meads). Yet extra ornament, with the moorish flavour traceable back to Foster's arch at Edge Hill, was just what Wyatt added to Paddington, to its great advantage. It took the form of the famous flowing patterns that embellish the ends of the sheds, and of sundry ornamental castings and timber pieces tacked on to the ribs and capitals themselves (ill. 98, 99). Perhaps Wyatt felt that the grandeur of the occasion and the need to mend the damage of dualistic Euston saved this sleight of hand from the charge of adventitiousness.

FACING PAGE
99. Paddington Station, transverse view in 1985, looking towards the stationmaster's rooms.

Theorizing about structure and ornament is one thing; carrying theory through into reality is another. By this date, the enrichment of structural ironwork had been episodically addressed by different figures – Holland, Cragg, Nash, Schinkel, Labrouste and others. Wyatt's efforts at Paddington are no less wayward than some that had gone before. But they usher in a campaign during the 1850s to make concerted progress in this sphere, usually on the basis of a contrast between plain wrought frame and decorated cast infill.

In Britain the strongest inhibiting factor against iron ornament, as against iron architecture, was not principle but parsimony. At Paddington, the GWR directors had never asked for more than basic sheds. They blotted out the one trial area of roof-colour painted by Owen Jones in 1852, and denied any responsibility for paying Wyatt. Brunel shot back:

> if I had requested the Directors to call in an architect, as all other Companies have done for their stations, their charges would have been doubled and tripled, and I may safely refer to the Great Northern and the Euston Square and Camden Town stations as proofs that the building of *railway stations* by architects instead of engineers involves an excess . . . in works which in our case would not have been less than £20 or £30,000.[173]

That was disingenuous in view of Brunel's extravagant instincts. But he knew the terms in which the average company director thought. Gradually after 1850, things would go the engineer's way.

The Gare du Nord

Since in France the top ranks of architects and engineers defined themselves by their distinct education, the distribution of tasks between them on that country's railway stations tended to look more orderly. That did not always make the process straightforward, as the case of the Gare du Nord in Paris illustrates.[174]

Early railway-building in France had been hampered by a question: should it be controlled by the State or by private enterprise? Following debate in the Chamber, the Corps des Ponts et Chaussées had drawn up a national plan for railway routes in 1833–5, only to be confounded by an Orleanist law that put the onus on private money and initiative. The results were inevitably piecemeal. In 1842 the Government took the matter up again, resuscitating the idea of a co-ordinated network of strategic lines fanning out from Paris. The State would plan and build four central termini in the capital, then lease them back to the railway companies. Designing the four termini – the Gares de l'Est, du Nord, de Lyon and Montparnasse – would be managed or overseen by the Corps. Thus did the national instinct to control major construction projects in Paris and treat them as exemplary assert itself. What happened elsewhere could be left to the companies.[175]

That arrangement put the ideals of France's construction élite before the working needs of the railway companies and their own engineers, who vehemently disputed it. But at least in a few cases, the Ponts et Chaussées was able to plan the termini on its own terms. The first Gare du Nord (1843–6), predecessor of the present station, was a case in point. It was allotted to Léonce Reynaud. One of a small band of French architect-engineers of this period who fully merit the dual title, Reynaud had switched across from the Ecole Polytechnique to the Beaux-Arts.[176] Professor of architecture in succession to Durand, first at the Polytechnique, then at the Ponts et Chaussées, he was among the group of architect-constructor-educators in the circle around Labrouste who strove to keep architecture and engineering together in the 1840s (p. 444). As a designer, Reynaud is best remembered for his lighthouses, but he could work in the terms of either profession. The first Gare du Nord is a rare example of a single figure straddling the whole design and construction of a major terminus.

The layout of Reynaud's station was lucid (ills. 100–2).[177] In front, a court opened out with arcades to the side for access to the suburban platforms in separate wings. The building itself was single-storey, yet high and well-windowed in the plain, round-arched style favoured for functional buildings by Durand's disciples. Across the façade ran a

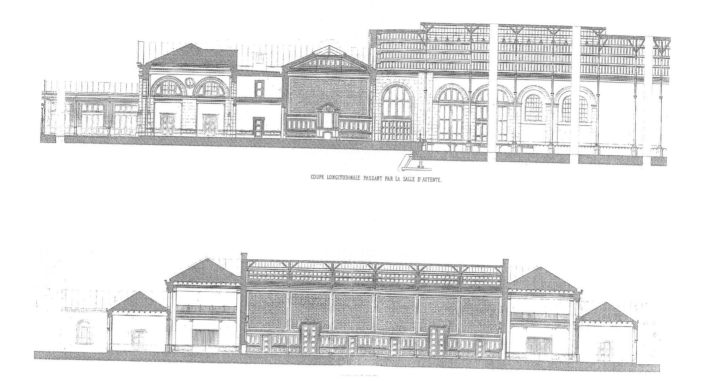

COUPE LONGITUDINALE PASSANT PAR LA SALLE D'ATTENTE.

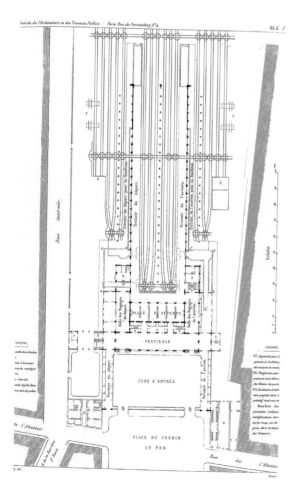

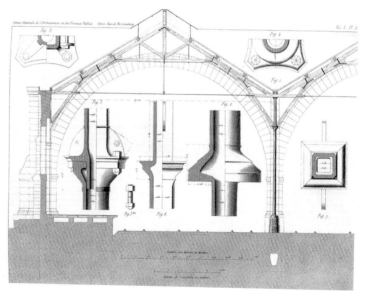

100, 101, 102. The first Gare du Nord, Léonce Reynaud, architect-engineer, 1843–6. Illustrations from *Revue générale d'architecture*, 1845. Top, long section through station from courtyard (left) through vestibule and waiting room to trainshed (right, abridged), and cross-section through waiting room and offices. Left, plan of station, showing tracks outside as well as inside the trainshed. Below, details of the shed roof, showing cast-iron columns and short-section timber trusses with iron ties.

103. Gare de l'Est, front before it was doubled in width. François Duquesney, architect, 1847–9. Expression of the shed is gracefully reconciled with the need for railway offices, housed in the projecting wings.

broad 'vestibule' for circulation – in effect, the earliest passenger concourse – with a lofty waiting room behind.[178] Then came twin arched sheds roofed by iron-tied timber trusses, 'using the short scantlings obtainable only in Paris'[179] and resting on a central row of iron columns. At first Reynaud had wanted a single iron span of a daring 110 feet; that had proved infeasible. The shed had its own scale and materials and did not appear on the façade. What gave harmony to the station were its semicircular arches and cornice lines, which ran right through at a single level. Sober geometry plus Reynaud's capacity to handle architecture and roof-structure alike drew the parts together.

Once in possession, the Compagnie du Nord was underwhelmed by all this clarity. It found the station hard to operate and needing many changes. The *Journal des chemins de fer* portrayed Reynaud as an architect who had gone into the project blind:

> One must have had experience of the movements of passengers and goods in a big station, not on paper but on site and for years, in order to be able to come up with the right ideas on so intricate a matter. Once the general arrangements are in place, then everyone can assume what is within his own competence; the architect can be put in charge of the buildings, and give them the grave taste fitting for such monuments.[180]

That may reflect sour grapes over the law of 1842. By the time the first Gare du Nord was finished, yet another railway law had been passed. In respect of station-building, its provisions were influenced by the Société Centrale des Architectes, a pressure-group bent on improving the prospects of Paris-based architects. Compared to the openings available for budding engineers within government service, official employment for architects was restricted. So when the new law was debated in 1844, the society proposed that the Minister of Public Works should oblige the railway companies to employ architects for stations, shelters, administrative buildings, warehouses and workshops, while engineers saw to the cutting, embanking, tracks, bridges and tunnels. The lobby, led by Jacques-Ignace Hittorff, prevailed.[181]

This ordinance, offering a fresh framework for railway architecture, came into force from the time of the Paris–Lyons railway (1847). But problems of demarcation remained. A new generation of engineers was developing the systems of wrought-iron roofs, based on the light Polonceau truss, which made French railway roofs graceful, affordable and exportable.[182] It would have been ludicrous to exclude them from stations. Nor did all railway companies desire to employ independent architects. Easier to manage and co-ordinate were in-house architects and engineers, whom the biggest companies now began to foster.

One such was the Compagnie du Nord, operators of the Gare du Nord. Though Reynaud's terminus had been planned for expansion, the changes made to it and the growth of traffic soon numbered its days. A memorandum of 1858 from within the company raised a question about its replacement: should the sheds be prefaced and masked by a 'monument', or should they show through?[183] That must refer to the nearby Gare de l'Est – last of the early Paris terminals to be built (François Duquesney, architect; P.-S.-A. Cabanel de Sermet, engineer; Jacquemart, serrurier, 1847–9). Here too the front building included a cross-vestibule, behind which rose the end gable of the single original trainshed, flanked by offices (ill. 103). The link between front and shed has earned the Gare de l'Est the title of 'the finest station in the world'.[184] But such a parti sacrificed a valuable hunk of prime space. The expansion of railway traffic entailed also an expansion of railway offices. In front of the station was the natural place to put them.

In the event, the Compagnie du Nord resolved to erect a separate headquarters to the west, and to give their rebuilt station (sited south-west of the old one) a proud new frontage fully 180 metres in length. There would be ample space in the centre for the train shed to peep through, along with offices at the ends and sides. It used to be supposed that these dispositions and the plan of the station itself were due to the great neoclassical architect Hittorff. But the researches of Karen Bowie have shown that not just the plans but the lines of the front were first worked out by the company's in-house architects, Lejeune and Ohnet.[185]

104. The front of the second Gare du Nord. Jacques-Ignace Hittorff, architect, improving designs by Ohnet and Lejeune, 1861–3.

105. Design for the Crystal Palace, by Hector Horeau, 1850. A likely source of ideas for the roof of the second Gare du Nord.

106. Gare du Nord, looking back towards the façade, showing the bold wide-span roof and the high branching columns detailed by Hittorff.

Then early in 1861, not long before construction started, the 68-year-old Hittorff was imposed upon the project. Haussmann's replanning of Paris was in full swing around the station; and the wisdom of combining a civic gesture with advertisement commended itself to the company's canny president, James de Rothschild. At first Hittorff seems just to have reworked the Lejeune-Ohnet elevation. His instinct was to articulate the front by distinguishing the centre from the wings and lowering the blocks in between. That meant a loss of offices and cannot have been countenanced. Instead, Hittorff had to coax the orders so as to squeeze upper storeys of offices into an enriched façade, without losing the oversailing line of the train shed behind. The upshot was the noblest of all station fronts in the monumental classic idiom (ill. 104).

All this sumptuousness conjured up the luxury of second-generation railway travel – 'the factory transformed into a palace', as Béatrice de Andia has put it.[186] As a result the frontispiece of the second Gare du Nord collided with progressive views of architectural propriety, which set more store by unity and consistency than ebullience. Before Hittorff's front was even finished, Viollet-le-Duc's young disciple Anatole de Baudot attacked it, complaining that while its broad central gable might correspond to the line of the shed behind, its three arched windows were arbitrary. The idea of a frontispiece was on its way to becoming a moral failing – frivolous 'confectionery'.[187]

How responsibilities for design worked inside the station is obscure. Having perfected the front, Hittorff seems to have worked his way inside and designed or redesigned the

107. Castings for the trainshed of the Gare du Nord at the works of Alston and Gourlay, Glasgow, 1862. In this carefully composed photograph it seems likely that the seated figures on the left are the ironfounder-partners, while the young man with the rolled umbrella may be Charles Hittorff.

shed with wider spans than the in-house architects had envisaged. He was always interested in structure, ancient and modern. His engagement with iron went back to the Halle au Blé (pp. 81–4), and we shall later examine his Panorama des Champs Elysées. It ought to have been too late for him to have much say. The platform layouts at least must have been fixed by the company's engineer, Couche (who goes almost unnoticed in modern accounts of the Gare du Nord),[188] along with Lejeune and Ohnet. But the arrangement of the great pitched roof, 238 feet wide, supported at mid-slope by rows of slender iron columns, resembles Hector Horeau's rejected design for the Crystal Palace (ills. 105, 106). Since everything designed by Horeau reflected that architect's cavalier approach to iron structures, surely only architects – whether Hittorff or Ohnet and Lejeune – would have adapted it for the Gare du Nord? If so, in the Paris of the early 1860s, the engineers had not yet monopolized the iron train shed.[189] What we do know for certain is that Hittorff (like Digby Wyatt at Paddington) enriched the ironwork by styling and colouring the iron trusses, columns and capitals. These beautiful, isolated castings were worked out in conjunction with his son Charles Hittorff, who liaised with the fabricators, the Glasgow firm of Alston and Gourlay (ill. 107).[190]

Despite the rationalists' objections to Hittorff's front, the procurement pattern of the second Gare du Nord was repeated in several later French termini. In-house engineers and architects working together laid out the plan and blocked out the bulk of a station and even its elevations, leaving smart architects to come in late and liven up the main

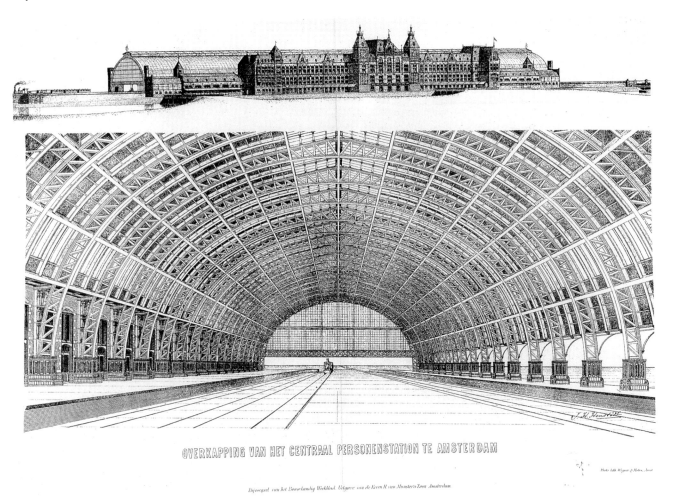

OVERKAPPING VAN HET CENTRAAL PERSONENSTATION TE AMSTERDAM

Bijvoegsel van het Bouwkundig Weekblad. Uitgave van de Erven H van Munster & Zoon Amsterdam

108. Amsterdam Central Station. Perspective of the trainshed, with detail showing its relation to the station building, from *Bouwkundig Weekblad*. Van Prehn, Leyds and Eymer, engineers, P. J. H. Cuypers, architect, 1882–9.

front and spaces. That happened, for instance, at Tours (1896–8) and again at the Gare d'Orsay (1898–1900), where Victor Laloux twice imposed 'confectionery' on a predetermined arrangement. In the case of the Gare d'Orsay, there was even a competition, which was explicitly devoted to civic show. Even though Laloux was able to impose important modifications, his essential function was to give face and style to programmes agreed in advance.[191]

AMSTERDAM CENTRAL STATION

Amsterdam's Central Station (so Aart Oxenaar has written) was a 'monumental keypoint' in the city's structure – the finale of one of the greatest water management operations undertaken in the Netherlands.[192] The drainage of a decrepit city; the revitalization of its waterways, harbours and docks; a shorter connection with the sea; overhaul of the interchanges between road, river and rail traffic: all these culminated in the bringing of the railway into the heart of Amsterdam and the construction of a terminus in front of the Open Havenfront in 1882–9. The project symbolized the renewal of Holland's and Amsterdam's prosperity. Here was a national as well as a local undertaking, involving the mighty Waterstaat – effectively the country's planning and transport agency. The railway companies, subject as elsewhere in continental Europe to firm state control, came low in the pecking order.

Hitherto the design of all stations on the Dutch railways had fallen either to Waterstaat engineers or to architect-engineers in the service of the companies. They fit-

ted their elevations to standard plan-types, classified from first to fifth class. The outcome was an orderly, often handsome architecture, like the early classical terminus at Amsterdam Willemspoort (1842). In essence the plan of the Central Station as built consisted of the standard first-class station design, expanded and endowed with an exceptional architectural treatment.

In 1875 Van Prehn, engineer to the chief of the three lines destined to meet at the station, suggested to the interior minister that P. J. H. Cuypers should be chosen as architect in association with A. L. van Gendt, the in-house architect for the line, who had already been working on platform arrangements. This proposal was readily accepted. Just a few months later, Cuypers was also confirmed as architect of the Rijksmuseum. To the fury of many, Amsterdam's two most prestigious commissions for years were thus simultaneously awarded to a Catholic and a Goth. The disputed status of the 'old Dutch' style, which Cuypers brought to maturity in the two buildings, meant that the exemplary display of style loomed large in the working-out of the station.

Cuypers was noted for his efficiency. But besides these two great commissions, he was much taken up with church-building. He had never dealt with railways. Van Gendt was meant to manage all that; only he died early in the process. The project was not for a 'head station' but for a lengthy side-station flanking the lines. In such stations the links between building and platforms are vital. The architects' first sketch plan omitted a proper system of access to the farther platforms. That alarmed the engineers of the companies destined to operate the station. Van Prehn along with Van Hasselt, on behalf

109. Amsterdam Central Station, detail of front with historicizing reliefs over entrance to the royal waiting room.

110. Amsterdam Central Station, the buffet. Decorated interior with open iron girders and jack-arch construction enriched by painting.

of the other companies, therefore went on a tour of modern German stations (the German and Dutch railway systems were closely allied). Van Hasselt then tried to introduce another architect, but was rebuffed. In due course a fresh plan with a modern circulation system for passengers and baggage was sorted out.

Cuypers' chief preoccupation meanwhile was with the main station building facing towards the city. Despite the dogged historicism and decorativeness of his major buildings, he is often defined as a 'rationalist' – a term generally justified at the Central Station by his use of iron. As in most heavily used buildings of its date and size, structural iron abounds in the body of the station, but it is largely hidden. For those in the know, it is referred to in the ceilings of the halls and the lusciously ornate waiting rooms, where the iron joists are clad in terracotta and have jack arches hopping between them. Only in the columns of the plainest waiting room, for the third class, is the ironwork exposed (ill. 110). The message is clear: structural iron was admissible, but as an industrial product it ranked low in the hierarchy of expression. On the station front, electricity, industry and steam were portrayed in old-world allegorical reliefs – symptoms of Cuypers' public caution. And yet he was also toying with a bolder attitude to iron during the construction of the Central Station. A few minutes' walk away, exposed girders span the broad nave roof of his Dominicuskerk (1884–6). The train shed was bound to feature iron, but Cuypers neglected it until it was too late. Despite some early talk of a single iron span, he believed he had settled with Van Prehn for modest twin roofs with columns on the platforms, subordinate to the front building. Then Van Prehn died in his turn. Responsibility for the roof passed in about 1881 to his successor, Leyds, and his colleague, the bridge specialist Eymer. Since the Central Station was a monument, the engineers felt the shed should partake of that monumentality. Eymer therefore designed a roof 45 metres in clear breadth, 23 metres high, and at over 300 metres the longest arched shed in the world at that time. When Cuypers found this out, he mustered what objections he could, apprehensive that the shed would oversail his own building, especially at the ends (ill. 108). Fellow-engineers also grilled Leyds about this extravagance, but he stuck to his guns:

> Would not the departing traveller immediately lose the impression which he has received in the beautiful great hall and in the waiting rooms when passing from the forecourt through the building to the platforms, if he suddenly appears under the (low) roof overhead? And would the arriving traveller who has already heard about the the beauty of Amsterdam and its new Central Station, not get a more pleasing impression under the great roof, which has not a single tie and not a single junction between trusses to obstruct the eye, than under the twin roofs?[193]

That touched a sore point. Cuypers had endowed the front building with a richly wrought iconography of painted friezes and carved ornament (ill. 109). Even if modern travellers had the time and disposition to notice it, they could hardly be expected to decipher it all. Had the money and artistry been spent in the right place? Or might the immediate, less demanding emotion of spatial experience count for something? Might it even be more democratic? And if such spaces as the Amsterdam Station roof could touch people more readily than could architects with their obscure styles and intricate detail, were the engineers not the better artists, indeed the real ones? That question, foreshadowed back at Euston, was starting by the 1880s to be posed in earnest. For all his 'rationalism' Cuypers can hardly have seen things that way, as he reluctantly tidied up the end gables of the engineers' shed with flanking turrets and a flurry of iron ornament.

Grand Central Terminal

Grand Central Terminal in New York (1903–13) stands for the moment when the railway station is handed back to the architect on a plate. Developments in a related field of his discipline cause the engineer to efface himself from the image of the terminus. In reality his structural, civil and mechanical capacities are omnipresent. Yet he now retires to the wings. Electrification delivers the glory back to the architect.

Like Paddington and the Gare du Nord, the present Grand Central replaced an existing station. But here the previous station stood not nearby but on the self-same site. So complex had it become, that replacing it in phases while trains still ran absorbed much of the skill and expense of this giant civil engineering project.[194]

Sited at what was then the urban edge of Manhattan, the former Grand Central Depot of 1870–1 had arisen following Commodore Vanderbilt's fusion of several lines under his control. Typically of big terminuses in its day, its front building (architect, John Snook; engineer, Isaac Buckhout) and its wide-span shed (architect-engineer, Robert G. Hatfield) cohabited but failed to harmonize (ill. 111). Both were added to in dribs and drabs, keeping pace with increases in traffic among the three companies that used the station. A concourse across its front and unified facilities for the companies were spliced in as late as 1898. By then building development in Manhattan had pushed northwards around the station's fringes, causing nuisances and quarrels between the companies and their neighbours.

Electrification, with the opportunity it brought of sinking tracks and platforms below grade, was the key to the simultaneous construction of Manhattan's two great terminuses. Pennsylvania Station, on a virgin site, was planned a little before Grand Central, started later but finished earlier (1905–10).[195] It was the brainchild of Samuel Rea, 'a quintessential railroad man',[196] chief engineer and later vice-president of the Pennsylvania Railroad, who had been trying to get his line across from Jersey City to Manhattan for years, and Alexander Cassatt, the company's new president. They went on a European tour of inspection in 1900–1, visiting the Gare d'Orsay in Paris, the first major station built for electric traction, also Frankfurt and Dresden. On return they chose Charles McKim as their architect.

McKim's first achievement was to dissuade Cassatt from putting a high hotel atop the station. Instead a romantic vision of giant Roman thermal baths asserted itself for the exterior and main public spaces. Open steelwork was conceded in the concourse, but at the architects' request it was distorted into arch form by the engineers, Purdy and Henderson, to make it blend with the rest of the composition (ill. 112).[197] Hence passengers

112. Concourse of Pennsylvania Station approaching completion, September 1909. The transition from the Roman approaches to the underground tracks, with steelwork manipulated at McKim, Mead and White's wishes for appearance's sake by the engineers Purdy and Henderson. Guastavino tile vaults to left.

descended to the electrified tracks and trains, tucked below in the nether world of technology. It speaks volumes about the size and diversification of big American architectural firms by this date that McKim, Mead and White had personnel enough to set out a clear division of labour for Penn Station. One senior assistant, W. S. Richardson, looked after the creative side of the project and worked up its 'Romanitas', with what help an ailing McKim could supply. Another, the Delft-trained technician Teunis van de Bent, co-ordinated with the main contractors, the George A. Fuller Company, and the various engineering interests, civil, structural and mechanical-electrical.

At Grand Central, things only began to move after a fatal collision in the smoke of the Park Avenue tunnel just outside the station early in 1902. This induced the city authorities to set the New York Central company a deadline for switching to electric traction in Manhattan. To tackle the obligation, the company promoted William Wilgus from the ranks as chief engineer, and planner for the recasting of Grand Central Terminal. Wilgus, though an excellent organizer, lacked the clout enjoyed by Rea in the Pennsylvania company. His background was in mechanical engineering, and his first job was to sort out the host of issues associated with electric traction – capital outlay, weight of locomotives, type of live rail and current, safety, and the range of the zone to be electrified. Not all these problems could be solved definitively. Teething troubles occurred when the system was introduced. In the fall-out after a further accident, Wilgus lost the company's confidence and resigned in 1907. When the great rebuilding plan he devised for Grand Central was completed, he was all but written out of the publicity. He remained resentful for the rest of his life.

Wilgus's hold on posterity has been tenuous, because his name was attached to a plan and a process, not to definite physical features. The superimposition of offices over the

station building, with main-line and suburban platforms suppressed to different base-ment levels; the separate circulation systems for trains, vehicles, passengers and baggage; the innovation of ramps instead of stairs; the phasing of the works into eight enormous 'bites', with excavation schedules modelled on those used for the Panama Canal; the planning of 'Terminal City' above the buried tracks along Park Avenue once the railway engineering was complete: all these were worked out by Wilgus and his team. But fickle humanity remembers only what touches the eye or heart. Hence a fixation upon the main architect of Grand Central, Whitney Warren.

Nepotism was rife in the selection of architects for Grand Central. A limited compe-tition took place for the above-ground buildings in 1903. The four entrants could suggest aesthetic and commercial dispositions of their own, but were steered clear of the plat-forms and tracks. The choice fell upon the railroad specialists Reed and Stem, who pro-posed a court of honour towards Park Avenue, access by elevated roadways, and a revenue-raising block of twelve storeys over the main concourse. It did not escape notice that Wilgus was married to Charles Reed's sister. Afterwards an architect not invited to compete, Whitney Warren, lured his cousin W. K. Vanderbilt, most influential of the company's directors, into adding in his own firm of Warren and Wetmore – to Wilgus's disgust. A contract between the two firms, the so-called Associated Architects, was signed early in 1904. In time, acrimony erupted. After Reed's death in 1911, it culminated in a lawsuit over fees and Warren's expulsion from the American Institute of Architects. But it was he who ended up ruling the public architecture of the station.

Whitney Warren had one big idea: to replace the Reed and Stem office block with a lower but grander concourse building, an idea obviously cribbed from Pennsylvania Station. If Alexander Cassatt could be seduced by McKim into sacrificing profit and sky-

113. Exterior of Grand Central Station during construction, 1912. Fronts designed by Warren and Wetmore, architects. Their partners, Reed and Stem, had little to do with this concourse building – the icing on the cake of a decade-long project.

114. Grand Central Terminal, sketch
for portion of underground platforms
and tracks by Warren and Wetmore.
The residual 'architecture' conveyed
by this drawing was lost in everyday
reality.

line to thermal grandeur, should not the New
York Central directors do the same? There are
moments when businessmen can be coaxed into
spending money on high architecture: this was
one of them. To public opinion just then, a 'City
Beautiful' gesture seemed both virtuous and wise.
'The railroad terminal' (wrote one journalist)

is the city gate. Without, it rises in the superior
arrogance of white marble from an open
square as an architectural something . . .
Within, this city gate is a thing of stupendous
apartments and monumental dimensions – a
thing not to be grasped in a moment . . . Civic
pride demands a fitting gateway and it is in the
best interests of the rail road to do its share.[198]

Warren's training at the Ecole des Beaux-Arts had directed him to the goal of organizing hierarchies of symmetrical circulation and public space. He had learnt the lesson well. The Associated Architects' design of 1904, remarked a magazine, was 'so much like a Beaux-Arts projet that it has rather caught the public fancy'.[199] The success of Grand Central confirmed the Beaux-Arts thesis that efficiency could be reconciled with a ruthless priority for aesthetics. The great concourse, the fine front towards Park Avenue and the warren of subsidiary public spaces and corridors integrate discipline with a bland beauty (ills. 113, 114).

The cost was an utter separation of the worlds of architecture and engineering. The tracks at the Gare d'Orsay had been sunk and semi-separate, but at least they could be seen. In Penn Station, parts of the platforms were visible; descending from the con-

115. Grand Central, the sectional
perspective.

course, the steelwork abruptly bared itself and opened up like a trainshed, making Montgomery Schuyler worry about the abrupt transitions which 'always occur when the modern architect and the modern engineer work in conjunction'.[200] But at Grand Central the cut-off between the aesthetic and technical realms was total. The passenger dropped from marble halls to spaces of claustrophobic banality. Yet behind the masonry itself lay steel frames of a complexity equalling that of the tracks and trains below (ill. 115). The names of the engineers and contractors who worked them out in subservience to the architects are even obscurer now than William Wilgus's. Whitney Warren had supplied a fantasy world which exhilarated – until you plumbed the nether regions of reality.

The rift between worlds at Grand Central redistributed a break always there in station architecture, going right back to Liverpool Lime Street, Euston and almost all the stations discussed. Even at self-contained Paddington, it betrayed itself in the laying-on of architectural decoration upon engineering structure. The Gare du Nord and Amsterdam Central Station are two among hundreds of variations on the vexed theme of juxtaposition in nineteenth-century urbanism. Each time that civics and efficiency met, a stand-off took place between the old architecture, rooted in half-shared memories and hard to decipher, and the new engineering, spatial, technically baffling, yet rapidly read and to that extent accessible.

Must we look on these fissures as shortcomings, incompatibilities, symptoms of cultural confusion? Human needs and human moods are various; there are bound to be breaks between worlds. Sometimes differences between neighbours need to be ironed out, sometimes not. But policing borders and managing disputes call for qualities not only of technique and art but also of respect and tact. To manage that well in buildings, the mutual understanding and sympathy of the senior building professions will always be crucial.

116. The Crystal Palace: the first 24-foot compartment erecting in Hyde Park, 1851. Components designed and fabricated by Fox and Henderson.

3. *Britain and France, 1850–1900*

Having trespassed forward, we now pick up the main lines of our story again from the mid-nineteenth century. During the 1850s and 60s iron structures on all scales proliferated and extended to many new tasks, amid broader public awareness of what they could do. That leap forward portended shifts in relations between architects and engineers. This section resumes themes of national development, looking briefly at Britain, then in greater length at France.

The Crystal Palace and The Issue of Permanence

The revelation of the Crystal Palace in 1851 marks a familiar watershed.[201] That vast, uniquely publicized structure opened the world's eyes to what an iron architecture might look like, and what the railway approach could do for building construction. Its story needs no repetition. But in relation to the present theme, a few points must be made about its procurement and its purpose.

The Crystal Palace myth lays its design at the door of a single individual of genius. After an abortive competition, the balance of eminent engineers and architects on the

117. Transept of the Crystal Palace from Kensington Road, 1851. This view shows the architectural styling applied to the elevations and transept, whose authorship was disputed between Joseph Paxton and Charles Barry. The ridge-and-furrow roofing was entirely Paxton's.

building committee (I. K. Brunel, William Cubitt and Robert Stephenson on one side; Charles Barry, C. R. Cockerell and T. L. Donaldson on the other) had floundered and failed to come up with a viable design. At the last moment Joseph Paxton, a horticulturalist, saved the Great Exhibition from disaster by sketching out a design which was logical, simple and beautiful and could be built very fast. So goes the story.

Paxton deserves his pedestal; the vital moment of inspiration was his. He refined his ideas and inventions as the structure evolved, and helped detail and supervise it as it went along, while over the rebuilt version at Sydenham (1853–4) he exercised a more personal control than over the original in Hyde Park. Paxton was tenacious of his claims to authorship of the Crystal Palace. But once his design was hailed as the answer, he was never alone. The project then reverted to being profoundly collaborative, as it had been before his intervention.

Revisionists make much of the role of Charles Fox.[202] Once the young Stephenson assistant who had designed the Euston roofs, by the time of the Crystal Palace he presided over Fox and Henderson of Birmingham, pre-eminent among the engineer-fabricator railway-contractors of the 1840s and 50s. It was the technical address of Fox, with

118. The Crystal Palace in its enlarged and enriched form at Sydenham, view looking southward from one of Brunel's watertowers in the 1930s.

fifteen hectic years of bridge-building and roof-building behind him, that saved the schedule by translating Paxton's idea into a fast-track working methodology (ill. 116). That meant teamwork in design as well as construction. Just as Fox came to the fore under Stephenson on the London and Birmingham Railway, so now the Crystal Palace, in both its Hyde Park and its enlarged Sydenham configurations, saw new talents cut their teeth as lieutenants in the Fox and Henderson team, among them C. H. Wild, R. M. Ordish and F. W. Sheilds. We shall meet them again.

Nor was that all. On the architectural side, the roles of Matthew Digby Wyatt and Owen Jones in embellishing the Crystal Palace's interior are well known. Less so is the degree to which the Paxton structure itself was 'architecturalized' by the Building Committee, even at the peril of delay. Authorship of the great barrel vault, the after-thought that permitted Hyde Park's full-grown elms to survive within the height of the building, was hotly disputed between Paxton and Charles Barry (ill. 117). What is not disputed is that Barry altered the proportions of the basic system – the 'form and distribu-tion of the arches and filling in frames, as well as of the columns'. As a result, says Jan Piggott, the proportions and mouldings of the columns took on a palpably Italianate

119. The 'Brompton Boilers', first portion of the South Kensington Museum: a come-down from the Crystal Palace. William Dredge, engineer, with C. D. Young and Company, contractors, 1855–6.

tinge.[203] Owen Jones also participated in this inflection and enrichment of detail. His most famous contribution, the interior colour scheme, was partly intended to make up for the fact that minor architectural details failed to tell within the great volume. It too was meant to help turn construction into architecture.[204]

The Crystal Palace then was a collaboration: a nursery for engineers, and briefly too an academy for debate and discovery between architects, engineers and fabricators of many ages and experiences. There being no such true school in Britain, some of the most creative constructors of the 1850s and 60s had their true training there.

Barry's tinkering with the Paxton design is apt to raise hostility among purists and modernists, but deserves understanding. Such changes were far from symptoms of root-and-branch opposition to iron or to novelty. Instead, the architects sought to temper raw technology to the dignity of what everyone acknowledged as an exemplary structure – to inch their way towards what Cockerell called 'an iron order of architecture'.[205] Owen Jones, for instance, in a run of bourgeois projects of the 1850s, probably undertaken with structural help from his engineer brother-in-law, C. H. Wild, tried to enrich iron architecture with opulence and colour.[206] Alas, all those that came to fruition proved ephemeral.

Ought an entire new architecture to be based on a style of building seemingly destined only for a short life? There lay the dilemma. Like most of the famous exhibition structures in iron that followed it, the Hyde Park Crystal Palace was expected to be temporary. Conventionally the values of architecture had been set upon permanence, and thus upon monuments. A temporary building is the opposite of a monument. There are, of course, degrees of temporariness. Much ordinary housing does not last so long. Several of the building-types mentioned in this chapter, mills for instance, were thought of as short-life structures. Therefore neither housing nor mills much attracted architects. As for exhibition buildings, they might have to be built fast, but they soon came down as well. The early examples enjoyed only makeshift subdivision, negligible heating and lighting, and little enrichment. For architects intent upon monumentality, such structures could be left to engineers or railway contractors. But civilization was bowling along too fast now to hang on to its old respect for permanence. Railway architecture was bedevilled because no one knew how permanent it would be, or how to marry the monumental with the ephemeral – the Euston problem.

The Crystal Palace was just as confusing. As it unfolded, it transcended the trade show or entertainment venue to take on the function and prestige of a museum. That purpose was half-present in Hyde Park, more fully so at Sydenham (ill. 118). So popular and so educational did the Crystal Palace prove that it ushered in a spate of English museum-building, in which its example was sheathed in a cloak of architectural permanence. The upshot was schizophrenic, as the stories of the South Kensington and Oxford Museums attest.

TWO MUSEUMS AND ONE ARCHITECT

Henry Cole, the eminence behind South Kensington, had been the Prince Consort's closest ally in the planning of the Great Exhibition. On the strength and income of its

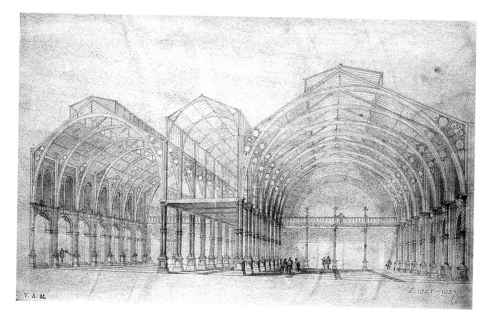

120. Perspective of iron construction for the South Court, South Kensington Museum. Francis Fowke and Godfrey Sykes, designers, 1860–1.

121. South Court, South Kensington Museum, under construction, 1861–2.

triumph, he was appointed head of what became the Science and Art Department. Out of that grew in 1855–6 London's South Kensington Museum, and then the various cultural institutions around it. Cole therefore stood high among official Victorian guardians of the arts and sciences. But he was also a modernizer. His views about building having been coloured by the exhilaration of the Crystal Palace, he was wary of architects (see p. 460) and indeed of permanence.

When a museum was first mooted for the Science and Art Department, Prince Albert asked the German architect Gottfried Semper, then in exile, to design it a permanent home. Semper's ideas turned out too costly, so John Physick relates. According to Cole, it was Albert who then suggested that they start with an 'iron house'.[207] An iron building with tiers of galleries was duly procured in 1855–6 from C. D. Young and Company, one of a band of engineering contractors who by now could supply 'iron structures for Home and Abroad'. Its author was the obscure William Dredge.[208]

The 'Brompton Boilers', as this structure came disparagingly to be called, was not unambitious, since it rose to two full storeys without bracing walls of masonry (ill. 119). The re-erected Crystal Palace having just opened at Sydenham, it was thus recommended to the Government:

> Irrespective of its simplicity and cheapness, and the remarkable facility with which it can be constructed, it enjoys the great advantage from a pecuniary point of view, of being designed of a material which possesses a permanent pecuniary value, to which the cost of the labour employed in its construction bears only a small proportion. While, therefore, it could on the one hand be at any time taken down and re-erected, if necessary, on another site, or in another form, at a very trifling expense, it could, on the other, be re-sold . . .[209]

Here permanence is construed in terms of the component, not the building, which is taken as temporary. Such an argument was bound to appeal to a British government department.

If the architectural press was predictably hostile to the Brompton Boilers, Cole too was disappointed:

> The public laugh at its outside ugliness and us . . . The light is so bad below the wide galleries, that nothing can be exhibited well there. Above the galleries, the angle of light is quite wrong for pictures. The iron produces excess heat in summer, and cold in winter. It offers no protection against fire, which will burn the contents and prevent ready succour from the outside.[210]

Yet far from turning to an architect for the next stage, Cole resorted to a Royal Engineer. It was under Captain Francis Fowke that the South Kensington Museum took permanent shape in stages from 1856. The building for the International Exhibition of 1862 nearby was Fowke's too, as would have been the Natural History Museum which replaced it, had Fowke not died. His successor Major-General Henry Scott then built the Science Schools and the Royal Albert Hall. The original core of the South Kensington complex, London's showpiece of mid-Victorian cultural values, came thus to be designed by military engineers.

To employ military engineers on prestigious civil projects was not new, as the previous chapter explored. But by this time it was rare for them to design major urban buildings in European countries, if common in their colonies. In Britain the move was wholly novel and irregular. In part it reflected the discredit that had fallen upon systems of choosing architects for public buildings. Then as now, important commissions were mostly decided by competition. Every episode since the Houses of Parliament (the Crystal Palace included) had shown how cumbrous competitions can be.[211] Albert and Cole wanted results. Probably they envied the French system of public patronage, whereby officially trained architects took on one major commission at a time, became civil servants and could count on generous back-up.[212]

Remote from the normal tenor of British architectural practice, the Royal Engineers had been running a small school of architecture and engineering since 1812 (pp. 458–60). Some of its graduates were at the forefront of structural design. It was under the Royal Engineer Godfrey Greene, for instance, that the multi-storey iron frame completed its development with the Boat Store, Sheerness (1858–60). Francis Fowke came out of the same nursery. But it would be wrong to categorize him as a pure engineer. He was an all-rounder: an architect, engineer and inventor, who had seen service with the War Department in the West Indies and built a barracks before coming to the fore as

122. Oxford Museum, the iron court. Francis Skidmore, designer and fabricator, with Deane and Woodward, architects, 1857–9.

secretary to the British Commission for the Paris Exhibition of 1855. What commended him to Cole was his commitment to efficiency and progress. Fowke's choice as architect and engineer to the Science and Art Department was only personal. It did not commit the Royal Engineers as a body, though he got his superiors to lend him some staff on the grounds that it would help with their training. More helpful were two civilian assistants in his South Kensington office, John Liddell and J. W. Grover – Liddell for architecture, Grover for structure.[213]

Fowke was employed at South Kensington not as an 'art-architect' (to use the phrase of the time), but as a team-player who could get things done to time and budget. A parliamentary committee was told that he had more than 'the ordinary knowledge of a general architect' and had rescued Edinburgh from 'a very uneconomical' design for a museum there.[214] Yet economy was only half what was wanted. Henry Cole worked to two timescales, the temporary and the permanent. Fowke's brief at the South Kensington Museum was to provide a backdrop, a carcase of plain iron construction upon which art could be added as money allowed.

The method was tried out in the North and South Courts of the South Kensington Museum, the first parts of the building destined as permanent. In the spirit of the day, the fabric was meant to instruct through an exemplary display of ornament. To that end, Fowke worked up a scheme with Godfrey Sykes, an artist-modeller brought in by Cole. The iron structures once built (1861–2), the rooms were then opened and the painted decoration followed on in phases, notably in the South Court (1864–86).

Because the ornamental cast ironwork had to be built along with the construction, that came first and set the style (ills. 120, 121). It was Sykes, in the role Digby Wyatt had taken at Paddington, who designed these details.[215] The outcome was a second-rate array of barleysugar columns (Fowke himself thought them 'like bed posts'[216]), capitals, pierced arches and spandrels. They raised the question whether repetitive castings would do for the interior of a major museum. A lick of paint having cheered things up, the gradual enrichment of the walls and roof allowed the ironwork to blend into the whole. But even with the supplement of decoration, it proved no clinching advertisement for iron museums. 'How far is such a structure as this to be considered permanent?', asked the architect-loving *Building News* when the South Court reopened with its gaudy paintwork:

> Will it last, if not destroyed by violence, as long as an average thirteenth century cathedral, or as long as the more massive monumental buildings of the Romans? It can hardly be supposed so. Yet an important national edifice on which large sums of money are to be spent, ought to be of a durable character . . . very little margin has been allowed for the weakening influence of time.

For this drawback the magazine could see that 'modern notions of economy are more to blame than any individual architect'.[217] Paint alone could never transform engineering background into architectural foreground.

For a moment in the 1860s the South Kensington procurement methods sponsored by Cole threatened to cut architects out of prestigious British public buildings. But the tide receded after the premature deaths of Prince Albert, Fowke and Sykes. It was Alfred Waterhouse, an art-architect outside Cole's circle, who inherited Fowke's design for the Natural History Museum.[218] To an extent Waterhouse stole the royal engineer's clothes. The great building which he raised relied throughout upon structural ironwork, hidden except in the roofs by a veneer of terracotta. The Natural History Museum combines delicious detail with the swagger that eluded Fowke. But it was managed in a way that would have appalled him. It came in very late and over budget, and led to 'the total and absolute ruin of the contractors'.[219] By the time it was finished in 1880, the iron revolution on which it depended had lost its drive in Britain. Among major architects in the last third of the nineteenth century, none was more lavish than Waterhouse in the use of iron and steel in the sections of his buildings.[220] Yet he treated them almost always as a means to other ends and showed scant interest in their expression. In the older architectural tradition, they were there to do the job he wanted – to assist behind the scenes in conveying scale and permanence.

At South Kensington the iron had come first; the art was added as an afterthought. In Oxford it was the other way round. 'Cast iron columns substituted for wrought iron. Diagonal roof of glass. Quite a muddle.'[221] So Henry Cole noted when he inspected the Oxford Museum (1855–60) on its completion. Technical mishaps notwithstanding, the court of this museum thrills the heart in a way that Cole's earnest projects fail to do. It shares the fresh naïveté of the Cragg-Rickman churches (ill. 122).

The 'University Museum' at Oxford was intended to further the study of natural history, and to bind the wound then opening between religion and science in English universities. That a covered court of iron and glass should have been part of a building programme in the Tory Oxford of those years, bastion of the Gothic Revival, is curious. It is said to have been suggested by Philip Pusey, brother of the eminent Oxford divine, Dr Pusey. Such roofs were still rare, though Paddington had just been finished.[222] Nevertheless the technical challenge of covering the court was not stressed in the competition brief, nor did the winning architects, the Irish Gothic revivalists Deane and Woodward, pay it much heed. What they cared about was perfecting the masonry buildings round the court, in an exemplary array of building stones and carved ornament.

'The evolution of the roof design is shrouded in mystery', says Freddie O'Dwyer in the fullest study of the Oxford Museum.[223] We do know that when it came to the contract, the notion of an alternative roof in stone, brick and timber was floated. That idea could, O'Dwyer believes, have come from Ruskin, who had the ears of Benjamin Woodward and of Henry Acland, the museum's chief promoter. Ruskin was engaged at this time both with the museum project and with the topic of iron and its applications, on which he delivered a lecture of glittering complexity in 1858.[224] Certainly Acland and Woodward were influenced by Ruskin's vehemence against industrial production and railway-style structures like the Crystal Palace, and his preference for wrought and beaten metalwork over castings. But though Ruskin designed some wrought-iron brackets for the court (none seemingly executed), he had little say in the big decisions that determined its fate. Mainly he goaded in the background.

Given the court's size and its use for display, an alternative to iron and glass cannot have been taken far. So Deane and Woodward must have been relieved when instead of their sketchy idea of plate arches on cast-iron columns an allegedly cheaper proposal was advanced by Francis Skidmore. Rashly, 'considerations of economy and confidence in Mr Skidmore's ability' induced them to advise that his offer be accepted.[225] A jeweller and silversmith by training, Skidmore was a gifted artist-craftsman and an enthusiast: 'very clever, but very wild', the Queen was later told.[226] His specialism was the Gothic enrichment of church-metalwork; gas-brackets, altar rails and minor church screens were the nearest he had thus far got to structures. But he was eager to explore every aspect of what metals might do for a Christian architecture, he told ecclesiological lecture-audiences.[227] His ideas for Oxford seem to have followed from the model iron church for export he and another architect, William Slater, had worked up in 1856.[228]

Skidmore's first roof for the Oxford Museum seems to have started with the details of spandrels and capitals and worked back from there to the structure. This, arches included, was of thin wrought-iron tubes, and assumed the lightest of coverings. Anxiety about leaks prompted Deane and Woodward to add to the weight of glass it had to carry. As it neared completion, a small portion of the roof collapsed in February 1858.

123. Detail of the Oxford Museum roof from *The Builder*, 23 June 1860, showing Skidmore's second design as amended on William Fairbairn's advice.

Cheerfully bearing the blame, Skidmore called in William Fairbairn of Manchester. That eminent engineer's report spared the blushes of all concerned. But he advised taking down the whole roof, substituting arches made of rivetted plate and angle iron, and replacing the wrought-iron tubes with double columns.[229] That was duly followed. Just how much else was changed we do not know, but Skidmore's capitals and wrought-iron outer foliage probably remained the same (ill. 123). The result looks breathtaking, but it is an exotic hybrid. The *Building News* summed up the fiasco:

Of all the roofs that ever were built, the most unfortunate was that which only a few months ago covered in the great Quadrangle, and delighted every beholder with its graceful beauty . . . It was supported, or, rather, was intended to be supported, by rows of iron shafts arranged in groups, each shaft scarcely thicker than an ordinary gaspipe . . . The spandrels of the arches were filled with exquisite foliage, made of wrought-iron, and in true and natural forms. Nothing more light and elegant could have been devised, and it was the admiration of everybody. But, alas! 'all that's bright must fade;' and this bright roof faded very prematurely, and with it a very considerable sum of money, which (as the old saying goes) 'might just as well have been thrown into the sea'. The shafts were light and graceful, and the umbrageous spandrels were imposing; but the latter were heavy, and the former were weak; and, almost before the roof was quite up, it was obliged to be taken down . . . It was left, we presume, to Mr Skidmore, which

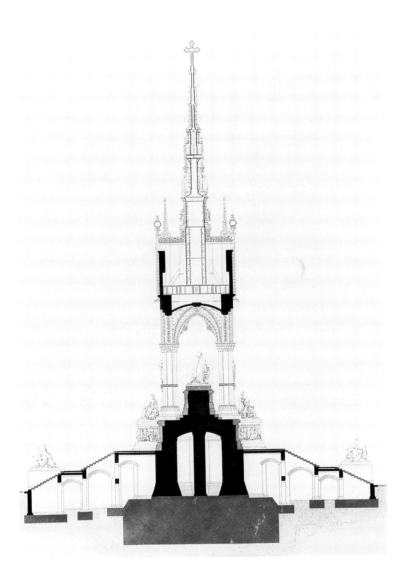

124. Section of the Albert Memorial, showing the foundations and the outlines of framework for the fleche, designed by F. W. Sheilds to carry Sir Gilbert Scott's architecture in a more scientific version of the arrangement at the Kreuzberg Memorial (ill. 66).

was a mistake. He is great in small constructions, and in detail; but a vast iron construction of this description ought to have been committed to the care of an experienced engineer, whose rude general plans Mr Skidmore could have elaborated.[230]

The contrast between South Kensington and Oxford suggests that in the mid-Victorian building world a permanent, symbolic museum architecture was incompatible with efficiency. You could follow the railway manner of construction, run something up in iron and hope to add in the art by degrees; or you could cling to the architectural route, strive for effect and hope that technique would not let you down. The dilemma is familiar still, not least in the making of museums and galleries. Such alternatives for prestigious buildings were only now becoming clear, as professions and techniques diverged more manifestly.

Within the current of mid-Victorian Gothic, wary of technology, only one leading architect made much attempt to reconcile these trends. That was Sir Gilbert Scott. Unlike most of his fellow-Goths, Scott was always curious about 'modern metallic construction'. In his *Remarks on Secular and Domestic Architecture* (1857), Scott noted that iron roofs and arched bridges were usually beautiful, while iron beams and girder or tubular bridges were 'always ugly, unless artistically treated'. Therefore

It will become a question whether the iron beams can, even in a building of a finished character, be allowed to be exposed. If they are so, they must certainly be more highly finished than any I have yet seen; but as brass has been most successfully introduced as a constructive material, or for decorating construction . . . there can be no reason why iron should not be so. This will add another to the class of ceilings, i.e. ceilings with ornamental wrought-iron beams, (perhaps decorated with brass,) and the interstices filled in with diaper, or other ornamental work in plaster, perhaps with stone or terracotta cornices. I would also suggest, whether porcelain may not be brought in to heighten the effect.[231]

These musings anticipate Scott's iron girders of the 1860s in the setpiece interiors of his two foremost secular buildings, the Foreign Office and Midland Grand Hotel. There the bottom flanges are exposed but the sides are cased in ornamental plaster or ceramics and flanked by ventilation extracts (ill. 125).[232] As a way of reconciling Pugin's insistence upon truthful structure, the resources of modern building technique, and the demand for rich interiors, these elaborated beams convey Scott's genius for compromise. Though by now ubiquitous in the floors of major buildings, previous iron beams had almost always been concealed within any interior above the warehouse class. That bothered the Goths: as William Morris was to put it, 'If you *will* have railway architecture, why don't you show it?'[233] By showing iron construction half-naked, Scott offered architects a middle way through the ideological snare of the beam-problem.

Having created worries about truth for themselves, architects had to ask for help to solve them. What makes Scott further pertinent to our story is that he seems also to have been the first English architect habitually to have called in consulting engineers to help him. That despite or perhaps because of his own competence as a constructor, for Scott had profited in his youth from employment under the contractors Grissell and Peto on that advanced building, Charles Fowler's Hungerford Market (p. 102).[234] Though fragmentary, the evidence merits reporting.

On the first such recorded occasion, in about 1863, Scott asked F. W. Sheilds, formerly resident engineer for the re-erection of the Crystal Palace at Sydenham and author of a textbook on iron, to design the hidden structural frame of the Albert Memorial (ill. 124). At that juncture Scott was already working closely with Skidmore on the memorial. Despite the mishap of the Oxford Museum, he went on trusting Skidmore as an art-metalworker of consummate skill and indeed allowed him to take the contract for supplying all the ironwork for the Memorial. But Scott was too shrewd to let him design the frame. Employing Sheilds meant that Skidmore could concentrate his craftsmanship on the jewel-like metalwork exposed to view.[235]

In 1865 Sheilds went on to report for Scott on the tower of Salisbury Cathedral. Normally in his vast restoration practice Scott felt able to deal with even the most massive and precarious church crossing towers on his own; Salisbury was an exception. Following Sheilds' report, yet more iron was added in to the much-strapped lantern stage of the tower. This may have been the first time an engineer was formally consulted over an English cathedral restoration.[236]

During the 1860s Scott developed a relationship with a second engineer from the Fox and Henderson stable. This was R. M. Ordish, a remarkable if shadowy figure in Victorian structural engineering. After Ordish set up his consulting practice in 1858, he amassed a varied portfolio of work. Bridges came first, but he also took in washing from

125. Midland Grand Hotel, St Pancras Station. The Dining Room (formerly Coffee Room) in 1911, showing iron brackets and beams and pierced panels for the ventilation system. Sir Gilbert Scott, architect, 1871–3. The original colour scheme, suppressed by the date of this picture, would have made the ironwork stand out.

architects and on one occasion personally designed and procured an all-iron, multi-storey hotel sent out to Bombay. Scott 'consulted Mr Ordish on all important questions of structure', according to the latter's obituary. He was certainly involved in Scott's restoration of the Westminster Abbey Chapter House, and in the roof for the winter garden at the Leeds Infirmary.[237]

These jobs overlapped with the building of the Midland Grand Hotel at St Pancras. Since Ordish is known to have helped W. H. Barlow design the great station roof there, it is tempting to ascribe a role to him in Scott's hotel, where open and hidden iron construction proliferates.[238] But there is no evidence of it. What we do know is that in designing the structure of the hotel, Scott relied upon an iron-floor system supplied by the firm of Moreland, one of many such patent fireproof packages in vogue at that time. Such systems seem to have helped the bolder Victorian architect-constructors to erect complex buildings without consulting independent engineers in any but exceptional circumstances. Thus Alfred Waterhouse, Scott's closest successor in this respect among major London architects, relied upon 'Lindsay's steel decking' for the floors in many of his later jobs.[239]

Scott's employment of versatile figures like Sheilds and Ordish should not be construed as the regular modern relationship between architect and consulting engineer. For the architect, it was a matter of drawing occasionally on expertise; while for the engineers these would have been minor distractions from bridge-building and kindred tasks. Nevertheless, by the 1860s Scott had discovered that if he wanted to do anything unusual with structural iron, it was helpful to look beyond the contractor-fabricator. The pattern of employing an independent engineer to further an architect's aims was starting dimly to make itself felt.

Paris in Context, 1839–1889

The French architects are far before us . . . we are forced to confess, although our manufactures produce more iron than the rest of the world and our engineers were the first to initiate its daring application to the construction of ships and bridges. In this part of building progress English architects and builders have lagged behind without any excuse or extenuation.[240]

Britain was falling behind its competitors. The British parliamentary commission's reaction to the Paris Exposition of 1855, here cited, intones a dirge to be repeated ad nauseam. Typically, it follows from an encounter with one of the world exhibitions – events that generated as much anxiety and rivalry between nations as stimulus and delight. Less typically, here architecture and engineering are linked. In the ensuing debate, engineering and engineering education were to be fixed items on the agenda, but architecture soon disappeared. The reason is not far to seek. In all its forms, engineering is a progressive subject whose products and services can be sold. Architecture is only superficially progressive. Its shifting values do not lend themselves to reliable appropriation by markets or governments. This topic is further pursued in Chapter 6 (pp. 456–8).

Yet the two disciplines were often linked in the public mind, and nowhere more effectively so than in France. The leadership in iron architecture noted in 1855 manifested itself further in the succeeding Paris exhibitions of 1867, 1878 and 1889 (and to a lesser extent that of 1900) as well as in a string of permanent buildings. Despite France's political instability, its architects and engineers produced iron structures of consistent confidence and artistry. How did they do so?

We can discard one hypothesis: that the two professions worked together on iron buildings in equality and harmony. Architects and engineers in nineteenth-century France (most of all in Paris, the eternal proving-ground of French architecture and therefore the focus of the following pages) were divided by their formal training. There had long been a stand-off between the two groups over their education and their spheres of operation

126. Les Halles, with Saint-Eustache in the background.

and influence. This provoked reams of comment, then and since. Sigfried Giedion saw it as a 'battle' ending in the triumph of the engineers. But the researches of Helene Lipstadt and others suggest that the war was a phoney one, fought largely in the columns of the architectural press.[241] Architects, of whom there were generally too many with too few jobs, always feared that engineers, with their focussed education and obvious public utility, would take buildings away from them entirely. If that sometimes happened in the provinces and colonies, the wholesale threat that the architects apprehended never transpired. As for the engineers, they took little heed of the controversy.

The furore was essentially a common-or-garden demarcation dispute, but it did not make for easy relations. Apart from special exhibition structures like the Galerie des Machines of 1889, in few of the buildings to be described was there an equal or even a consulting relationship between professional architect and professional engineer. The commoner pattern in permanent buildings was that of collaboration between architect and engineering contractor. In that relationship, the architect easily maintained the status he felt was his due.[242]

A better clue to how France got ahead in iron architecture lies in the articulacy and resilience of the progressive strain in French culture throughout the century. Itself a fruit of the Revolution, the Saint Simonianism of the 1820s, passing into Comtean positivism, linked art, intellectual life and technology in a forward-looking alliance. Though often anti-establishmentarian, its adherents succeeded in endowing the idea of progress with an artistic allure that the British legatees of the Crystal Palace failed to generate. The eagerness of the so-called rationalist architects, men like Labrouste, Reynaud, Baltard, Viollet-le-Duc or Jourdain, to integrate iron with architecture on their own terms is well known. They were never isolated. That industrialists, politicians (early on at least, Napoleon III was among the enthusiasts[243]) and vulgar materialists should have championed iron is natural. But alongside them, many intellectuals glorified iron as the representative material of their times. They needed no persuading that iron structures meant art as well as power.

Emile Zola is the most famous case. Son and grandson of engineers, Zola promoted iron as a symbol of spiritual as well as material progress in his novels. Back in the 1830s, Victor Hugo had sounded the note of reaction in *Notre Dame de Paris*. Its famous epigram, 'ceci tuera cela', taught that printing had replaced architecture as the true bearer of culture at the dawn of the Renaissance; artistic renewal could only come from going backwards. Zola's *Le ventre de Paris* of 1873 challenged that head on. Noticing the old church of St Eustache framed within the central arcade of Les Halles (ill. 126), the novelist purloined Hugo's catchphrase and reversed it into a prophecy that iron would replace stone:

> 'A strange encounter', he said, 'that stump of church framed under this iron avenue . . . The one will do for the other [ceci tuera cela], soon enough now . . . Do you believe in chance, Florent? I suppose it can't just have been the needs of alignment that led to a rose window of St Eustache facing straight down the middle of Les Halles like that. You see, it's a complete manifesto: modern art, realism, naturalism, you might call it, has grown up facing the old art . . . Have you seen the kinds of churches they build nowadays? They look like everything you can imagine, like libraries, like pigeon hutches, like barracks; but surely nobody can believe that the Good Lord lives in them. The Good Lord's masons are dead, and it would be wise not to build these ugly stone carcases with no one to house inside them . . . Since the start of the century, only one original monument has been built, a monument which hasn't copied from anywhere but has grown naturally in the soil of its time, and that's Les Halles . . .'[244]

Here France's foremost iron structure built in the wake of the Crystal Palace is unequivocally hailed as a monument. Later, when Zola visited London, he chose to live within sight of the Sydenham Crystal Palace.

Iron, then, could be monumental. Under the norms of French art, monuments were made by architects. And indeed Les Halles was designed not by an engineer, nor by a team of engineers and architects, but by an architect of established reputation, Victor Baltard, working with a skilled iron contractor, Pierre-François Joly. For all major urban buildings involving iron, the architects continued in the driving seat, as they had done since the Ancien Régime. Infrastructure of all sorts might be left to the engineers; culturally, the right of architects to bear the torch was never in real dispute.[245]

How then could French architects integrate into their projects a material which needed ever more skilled techniques to calculate and fabricate if they were to be built? The truth is that little knowledge of technique is needed to produce adventurous designs in a modern material. Writing of the architect Hector Horeau, famous for a visionary series of wide-span projects in iron dating from the 1840s to the 1860s, François Loyer tackles the point:

> The problem is often raised, quite wrongly, we think, about Horeau's engineering knowledge. The use of metal is certainly so linked to technology that it would seem as if an architect could not make use of it without having some knowledge of construction. But ignorance of the problems of stone construction did not prevent others from realizing structures of great audacity. The problem of nineteenth-century technology is no more compelling than that of the adequacy of the structural system conceived in the Gothic era: it hardly matters whether or not the principles of the skeleton worked as those who designed it thought it did; what matters is their attitude in respect of the visual representation of a technical problem.[246]

In other words, unbuildable though Horeau's designs for exhibitions and market halls might be, they could still stimulate iron architecture's development. With luck, they might have been picked up, translated and built by an able technician, just as an architect today may design something which can be built only after the intensive intervention of a consultant engineer. What mattered was an architect's ability to open eyes and raise the game by articulating a technical challenge in the language of art. Just that happened; Horeau's scheme for Les Halles made its impact on what Baltard built there, while the shed of the Gare du Nord derives from his rejected submission for the Crystal Palace (p. 122, ill. 105).

Still, Horeau died a disappointed man who would have done better, had his technique meshed with his vision. Why did it not? As an architect trained at the Ecole des Beaux-Arts back in the 1820s, his upbringing had given him some preparation for the adventurous road he chose. Though the Beaux-Arts is often castigated still for its backward-looking agenda, competitions in 'serrurerie' had been set as a regular part of the construction syllabus there in Horeau's time (p. 437).[247] He could hardly have imagined his venturesome conceptions without the stimulus of such projects. More likely, Horeau had failed since then to build up the relations with collaborators which could have helped him make the link between the art and the technique of iron structures. He persisted in seeing design as an ideal and isolated process: powerful for propaganda, essential maybe to the conception of a monument, but of less use in furthering a career and getting buildings built.

ARCHITECTS IN CONTROL OF IRON: THREE PROJECTS

What happened then when architects did actually have to build with iron? Let us take three conspicuous Parisian projects dating from between 1838 and 1856, the period when French architects were deemed to be getting ahead, and see what falls out.

Least familiar of these projects now is the earliest: the short-lived Panorama des Champs Elysées, built in 1838–9, taken down just seventeen years later. Circular panoramas, devoted to the display of large-scale illusionistic paintings round their circumference, were a special nineteenth-century building-type, confined to large cities and resorts. They absorbed as much architectural care as did theatres, though their life was often even shorter. The Champs Elysées panorama, promoted by the painter-entrepreneur Charles Langlois, was the biggest yet. The brief required an unbroken roof-span of fifty metres – wider than the Pantheon in Rome – with a suspended hood over the viewers in the centre to hide the roof and the flood of top-lighting.

The neoclassicist Hittorff was the architect chosen. Ever since he had worked as Bélanger's assistant on the Halle au Blé, Hittorff had been fascinated by iron technique. He even looked out for it when he visited the Greek temples of Agrigento. Two Parisian theatres of the 1820s in which he participated had acquired iron roofs and safety curtains. Faced now with the challenge of a circular building, Hittorff's first instinct was to crib from the Coliseum for the elevations,

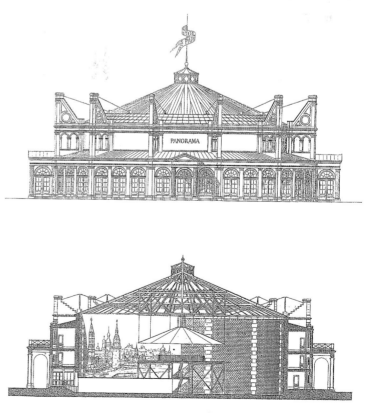

127, 128. Panorama des Champs Elysées. A. Elevation of the original cable-hung scheme with six cables designed by Jacques-Ignace Hittorff, 1837. B. Section of revised scheme with twelve cables and intermediate supports introduced close to the anchorages, showing also the internal viewing platform and suspended hood. From *Revue générale d'architecture*, 1841.

129. Bibliothèque Ste Geneviève,
exterior. Henri Labrouste, architect,
1844–50. Opulent, controlled, austere, in immaculate masonry: the
image of French public architecture.

130. Bibliothèque Ste Geneviève, the
reading room. Henri Labrouste with
François Calla, serrurier, 1847–50.

while resorting to suspension-bridge technique for the roof-span, below which the hood
would be suspended from an internal iron framework.

Suspension bridges were cheap and fashionable in France during the 1830s (pp. 323–4).
They were however engineering structures whose methods had yet to be adapted to
buildings. Hittorff had confidence enough in his own abilities to risk raising the world's
first cable-hung roof without professional help. But he soon realized that the iron frame
would cost too much. So in his preferred project, never published, the internal roof framing over the hood was of light timber hanging from six intersecting cables, which
emerged from within to be slung from external 'contreforts' – half pylons, half buttresses
– rising high from the top of the perimeter wall. Since the roof was meant to be hidden,

the only visible part of the structure was the external junction of cable and contrefort: in this Hittorff took pride.

There was natural apprehension about so novel a structure. The number of cables and contreforts was therefore doubled to twelve (ill. 127), not (according to Hittorff) because his first project was unsafe but 'because of the fears which are always the first impression created by an application of any new system of construction having an appearance of audacity, and which almost always degenerate into a disheartening opposition'.[248] That was not enough to reassure Langlois. His reservations, 'supported by a distinguished engineer', says Hittorff, prevailed with the Conseil des Bâtiments Civils and forced him to alter the cable system once again.[249] In the Panorama as built, the point at which the cables were fixed to the contreforts was lowered; they then rose and passed over a compression member – a kind of mini-pylon – before entering the roof (ill. 128).

Hittorff looked upon the finished structure as a costly compromise, and had his pound of flesh by publishing his side of the story – the only one we have. Well enough equipped to make the necessary calculations, derisive of his opponents' fears, he regarded himself as in absolute charge throughout. It was the proprietor, not he, who consulted an engineer. Nevertheless despite his protests about cowardice and waste, Hittorff proved flexible in practice in order to maintain his control. Nor do we hear more than the mere names of the fabricators: Chavier, who made the cables, and for the other ironwork Roussel, a firm involved back at the Halle au Blé and in Hittorff's theatre projects.

Despite the classical dress of the perimeter, Hittorff makes the telling point that he deliberately sought 'industrial' aid for the Panorama. The lowering of the cables, he asserts, prevented him 'from the chance of applying my suspension system in such a way as to obtain architectural forms with an industrial rather than a monumental means of construction'.[250] This must be one of the first occasions on which a high-class architect spoke about a major urban project in such terms.

Our second example is the Bibliothèque Ste Geneviève (1844–50), best-known of early French 'rationalist' projects and one of the most often analysed of buildings.[251] In reaction to the positivist enthusiasms of Giedion's generation, some recent interpretations have underplayed the famous library's ironwork. Aspects of how its forceful internal skeleton of iron came into being therefore merit remark.

By the time that the Bibliothèque Ste Geneviève was awarded to Henri Labrouste in 1838, open ironwork was accepted in arcades, markets and conservatories. But in prestigious French public buildings – those commissioned by the State – iron roofs up to this time had remained hidden. So Labrouste's resolve to combine open iron with masonry had more to do with culture than structure. It set out a thesis about what a modern 'monument' might be like, to be debated with his peers.

As David Van Zanten has set out, French official architecture at that time was a closed shop.[252] Major government buildings were designed by those few architects who, having excelled at the Beaux-Arts and gone on to Rome at public expense, became to all intents and purposes civil servants on their return. You served your time and perfected your practical knowledge as an inspector on projects built by other government architects. If like Labrouste you were diligent, that was the stage at which you learnt to be a practical constructor as well as a designer. Your hope was to be appointed 'architecte en chef' in due course for a major project by the Conseil des Bâtiments Civils, the board that commissioned and maintained major government buildings. Since senior government architects were in a majority on the Conseil, they rather than the occupiers of the building were the primary clients. Your concept would be bandied back and forth between you and the Conseil until agreement was reached. At that point you could form your agency for detailed design and construction, which might go on for years.

In this case, the Conseil des Bâtiments queried Labrouste's notion of roofing the main library space in open iron. It persisted for years in preferring a stone vault, but never actually vetoed an iron one. Authorized in 1843, the building took seven years to complete. A serrurier for the roof was appointed only in 1846. That was Roussel, whom we have met before. Roussel then brought in the experienced ironfounder Christophe-François Calla, the final design then being studied and worked out between Labrouste and Calla. Only at a late stage did Labrouste decide to hide the main roof structure itself above plaster vaults. Robin Middleton believes that the idea to cover was probably Calla's, taken up to obviate the need for falsework.[253] In the final solution, the twin naves were crowned with openwork iron arches supporting tunnel vaults of mesh and plaster (ill. 130). A complete arch was laid out flat for inspection in July 1847, after which assembly and erection of the metallic parts proceeded.

Open ironwork below the vaults, hidden ironwork (as in previous buildings) above: if that seems mendacious, it only follows the tradition of great Gothic churches which hide their roofs above their vaults. In both cases the arches and columns confront the eye: and that is the point. As the Bibliothèque Ste Geneviève design unfolded at its leisurely pace, iron roofs were changing fast. Even in 1838 Labrouste could have covered the library's meagre nineteen-metre breadth with a single iron span, by means of Polonceau trusses, for example. He toyed briefly with an open hall. Though that might have been cheaper, it would also have been less modulated, less 'architectural', while the iron, even if exposed, would have been skied in darkness above eye-level. That no doubt is why he divided the library into two equal naves, based on the mediaeval Parisian refectory of St-Martin-des-Champs.

The decision made a clash of proportions between stone and iron unavoidable. Labrouste relished this. In the main library he gave the central row of iron columns eccentrically high stone pedestals, rising to the level of the galleries along the walls, as if to hint at two storeys instead of one; in the vestibule below, stumpy piers of masonry bore mincing little iron trusses. Exploitation of the abrupt proportional transitions between iron and masonry seems to start at the Bibliothèque Ste Geneviève. But the strategy was architectural, not structural. Not that the slenderness of the iron members was new or illogical. The delicacy of French iron sections compared to the robust sections prevalent in Britain can be traced back to the cost of the material, the suspicion of casting and a sharper culture of calculation. The novelty was the challenging of sensibilities by setting fatness against thinness. Iron and stone were instruments in Labrouste's architectural orchestra, to be sized and enriched and conjoined experimentally so as to explore their contrasting tones.

Eventually the thinness of iron members came to be enjoyed.[254] That took time. At first Labrouste's library disconcerted those who felt that monumental buildings should be just that: monumental. One critic opined that in the vestibule the architect had forgotten 'that everything true is not always beautiful and that what is solid and sufficient materially might not satisfy the eyes. One sees and feels before one reasons and more quickly than one reasons'.[255] Gottfried Semper, likewise missing a sense of comfort in the reading room, blamed the spindliness of its ironwork.[256] Charles Garnier, an enthusiast for iron construction so long as it was thoroughly tamed, generalized the point in 1857: 'it is . . . the inability of iron to provide masses and supports satisfying to the eye which necessitates its rejection for any artistic construction'.[257]

It was no part of Labrouste's programme for his library to surrender the idea that an architectural monument must display its permanence and meaning. A thick external skin for a building in which people were to sit for a long time, not move about as in stations or markets, was needful in the 1840s, and long remained so. That meant masonry (ill. 129). Combining a thick skin with an open, airy, internal frame of iron, and reconciling the consequent proportions and details, were to be items of the rationalist programme for the rest of the century. As for 'meaning', that too was still mostly conveyed by stonework – by the lists of names cut into the front and by the busts against the pedestals carrying the reading-room columns.[258] The conventionalized iron ornament of the Bibliothèque Ste Geneviève adds little to that. Robin Middleton interprets the enrichment of these castings as an attempt by Labrouste to carry into

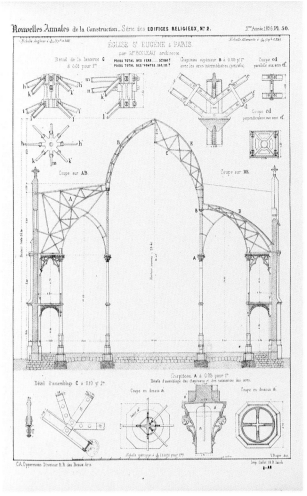

public interiors the kind of 'civic decor' which had filtered into the iron street furniture of Paris since the Restoration. If that is right, it represents a generalized endeavour, not anything specific to the library.

As with the Panorama, the Bibliothèque Ste Geneviève was a design over which the architect claimed total control. But though Labrouste was a modestly brave constructor, that control is about 'architectural effect above all rather than technical prowess', as Bertrand Lemoine puts it.[259] The late change which gave the roof the form it does has an almost casual air. And when writing up the building after its completion, Labrouste named the serrurier who made the splendid locks, but not the fabricators of the roofs.

Our last example in this series of architect-led structures in Paris is the church of St Eugène: a building that since its construction in 1855–6 has been as routinely disparaged or patronized as the Bibliothèque Ste Geneviève has been praised.[260] That despite the fact that it boasts not the least august of nineteenth-century iron church interiors. It also stands for a resumption of ideas about producing major buildings in series which we saw set out by John Cragg in Liverpool forty years before.

Much of the hostility to St Eugène stemmed from the marginal status of its author. Louis-Auguste Boileau was self-educated and proud of it – a rarity in Parisian architecture. 'Through working in the joinery trade at an age when you can lose time at school,' he wrote, 'I had the advantage of familiarizing myself with the resources of practice and of escaping the influence of the prejudices which reign in the teaching of the fine arts.'[261] Boileau started off by making Gothic fittings for church restorations. Hence he

131. Ecclesiology in iron: the interior of St Eugène, Paris, by Louis-Auguste Boileau, 1855, from *The Builder*, 21 February 1857. Columns, ribs and tracery are all of iron, as is the hidden superstructure above the vault. The proportions are based on the mediaeval refectory of St-Martin-des-Champs.

132. Section and details of the ironwork of St Eugène, from *Nouvelles Annales de la Construction*, 1856, showing the hidden framework over the vaulting.

was drawn into progressive Catholic circles, and into restoring and extending churches himself. His only wholly new one at this stage of his career, at Mattaincourt in the Vosges (1846–59), was an orthodox enough essay in Gothic revivalism, apart from its rather curious cast-iron pews.

Boileau returned to Paris from the Vosges in 1849 bent on making a mark in architecture. With church commissions in mind, he pondered how Gothic might be reconciled with modern styles and ways of building. The idea was in the air, but Boileau pursued it with an idiosyncratic vigour. The first stage was his so-called 'synthetic cathedral', a vast, vaulted fantasy on a cruciform plan. It came in two versions. The first scheme, meant for masonry, drew notice when exhibited as a wild-looking model in 1850. Then under the probable impact of the Crystal Palace, Boileau reconfigured its interior in iron and glass, reducing the Gothic elements but keeping the vaults.

The revised design was published by Boileau in his *Nouvelle Forme Architecturale* (1853), a blatant bid for work. 'Starting from the most advanced architecture ever known in terms of construction and art, in other words the Gothic, retaining its special symbolic and emotional expression, and developing the principle of the skeleton ['ossature'] on which it depends, suppressing the flying buttresses which impede it, eliminating every dishonest aspect of its form and sparing all the materials which play a double role in the construction: that is [the author's] programme.'[262] The key to it all was iron, though little was said about it, perhaps because Boileau lacked experience in iron construction. He claimed endorsement from a number of progressive architects and figures connected with the Edifices Diocésains. His warmest supporter seems to have been the architect-archaeologist Albert Lenoir.

If a cathedral was never going to come Boileau's way, parish churches might just do. So he started to develop and publicize a design – a system almost – for iron churches which he was to pursue for the next fifteen years. St Eugène was the prototype. It formed part of a programme of new churches planned by the Hierarchy for crowded or developing quarters of Paris. The speed and economy which iron construction promised would be a boon. Boileau was brought in after the priest appointed to St Eugène had read an article about a church he had proposed elsewhere in Paris, with an endorsement from Lenoir.[263] Cheaply and rapidly erected, the church was consecrated in December 1855.

After the radicalism of the synthetic cathedral, St Eugène is cautious to the point of correctness. A broad hall church with vaulted nave, aisles and passage aisles, richly painted and dark now with stained glass, it possesses a spaciousness belying its constricted site. But it gets its dignity from the thirteenth-century aesthetic to which Boileau clung, not from the all-iron structure (ill. 131). Only the slim supports, ribs and tracery hint at the material; even then the columns are cribbed from the refectory of St-Martin-des-Champs – the same model that Labrouste had reinterpreted at the Bibliothèque Ste Geneviève. The potential liveliness of the iron roofing is, in the customary way, hidden above sheet-metal vaults (ill. 132). As for the exterior of St Eugène, it is thin, plastery and insubstantial: a shell pared down to permit maximum floor space, lacking the depth that Gothic cusps and mouldings cry out for. There Boileau fails.

The naïveté of St Eugène in so prominent a location prompted instant and cutting criticism.[264] Soon Boileau found himself drawn into a polemic with Viollet-le-Duc. The essence of Viollet-le-Duc's objections, oft repeated since, was that Boileau had taken a modern material and warped it for historicizing purposes, instead of grasping its properties and potential. It was a strong case, if set out in an unpleasantly disdainful tone. But the controversy did lead Viollet-le-Duc to reflect for the first time on the character and development of modern iron construction – on which in time he would have much more to say.[265]

Today we can perhaps relax and look at Boileau's venture less prescriptively. His real trouble was his technical isolation. The most challenging previous attempt to devise a system for iron church-building had come from an ironfounder. Boileau was no such thing. He was a joiner who claimed to put practicality above fine-art prejudices. Yet he seems to have entered into no true technical partnership in developing his iron designs, even when they came to construction. The booklet on St Eugène mentions in passing that he was helped by his two young sons, one of whom, Louis-Charles Boileau, we shall meet again, and lists the fabricators; that is all.[266] In construing iron architecture as a matter of design alone, Boileau hardly differed from the school-trained architects who presumed to condemn him.

By putting style before materials, no doubt he sought to give the clergy what he thought they expected from a church. If so, the stratagem failed to bring him the commissions he hoped for. Boileau gained no general support from the Hierarchy for his system. During the 1860s he did acquire a smattering of provincial churches in vaulted ironwork, as well as the main French church in London.[267] They showed, says Robin Middleton, 'little advance on St Eugène, if indeed they attained its standard of consistency'.[268] If some may have shared dimensions or components, there was no progress based on the industrial cycle of design, production and improvement. As time went on, Boileau revealed his colours by devising further iron fantasies on the lines of his synthetic cathedral. These could be as imaginative as he liked, since they stood no chance of being built. They confirmed the lure of architectural control – specially strong maybe for someone who had worked up to it from trade.

To summarize this section, our three disparate adventures in iron building by architects around the mid-century – panorama, library and church – are linked by the instinct for control. Hittorff is confident that he can calculate and guarantee the safety of a major 'industrial' structure, to use his own term, but submits to belt and braces because he knows that only thus will he stay in charge. Labrouste plans to bring the iron roof within the ambit of monumental architecture, with its arcane rules and agenda, there to tame and aestheticize it by dint of example. Boileau, least equipped yet most ambitious of the three, aims explicitly to alter the direction of nineteenth-century construction. To that end he embarks on a series of parallel projects ideal and real, the technological implications of which he pretends to hold in his own hands.

One lesson of these projects is that the better French architects of this period – even those trained at the supposedly impractical Ecole des Beaux-Arts – possessed knowledge enough to design safely and even experimentally in iron without engineering assistance. That confirms the observations of the British delegation in 1855. So long as spans were modest, iron structures could be empirically managed by architects. Hence the energy and variety of iron architecture on a craft scale in the second half of the century down to the domestic experimentalism of Art Nouveau, triumphantly exemplified by Victor Horta's virtuoso houses in Brussels.[269]

Where then in all this were the fabricators – the serruriers who, we saw before, had helped French iron construction along since the eighteenth century? Hittorff mentioned his collaborators at the Panorama in passing; Labrouste at the Bibliothèque Ste Geneviève did not bother to do so, although Calla appears to have played some part in the final roof design; Boileau too minimized his iron fabricators and contractors. So long as the serruriers were subordinate tradesmen, that was possible. But things would change, as products of the schools started to percolate into the fabricating firms, and lowly serruriers reinvented themselves as 'ingénieurs et constructeurs'. This stage we next address.

SHOWS AND SHOPS

We have seen how British museum-building reacted to the Crystal Palace. In France, the 'expositions' of the second half of the nineteenth century set an even firmer mark upon the architecture of shopping. The modern department store was a stepchild of the Paris exhibitions. Over time, stimulating consumption became a goal of these mammoth dis-

133. Palais de l'Exposition, Paris
Exhibition of 1867. This thoughtful
if unloved building was the outcome
of collaboration between Léopold
Hardy, architect, Jean-Baptiste
Krantz, engineer, and Ernest Gouin,
fabricator and contractor.

plays, along with education and beating the national drum. Once exhibits were destined
for sale, their setting had not just to classify and instruct but also to allure. That meant
co-ordination between the engineer, expert in running up impressive spaces, and the
architect, pedlar of art and ornament.

For most of the great Paris shows, architects and engineers were yoked together in
some semblance of partnership, not usually quite explicit. In 1855 the main building, the
Palais de l'Industrie, was expected to be permanent; engineers were given the big iron
hall while architects got on separately with the masonry frontage and offices. The exhibition of 1867 was the first where a high proportion of the displays were on direct sale.
The structure that housed them did its best to respond. The core of the event took place
in a vast, elliptical enclosure of iron, the Palais de l'Exposition (ill. 133). It was an engineers' conception, guided by the intellectual polytechnicians Frédéric Le Play and
Michel Chevalier, and worked out by an experienced railway engineer, Jean-Baptiste
Krantz. To draw up some of the girders and verify calculations, Krantz employed a protégé, young Gustave Eiffel, then in the throes of setting up his own contracting firm, with
railway bridges in his sights. The fabrication and construction itself were likewise
entrusted to a railway contractor, though one with a polytechnical background, Ernest
Gouin.[270] Laid out with an excess of logic and subdivision, the 1867 building was too
compartmentalized to be memorable. One critic inveighed against its 'cold, dry, alge-
braic, industrial formula', typifying 'the spirit of the engineer who calculates, specifies,
and pares down the cost again and again without troubling about the aesthetic and moral
effect of his work'.[271] It vanished unregretted.

As the later shows became consumer-oriented, the flair of an architectural touch grew
in value. An architect, Léopold Hardy, had been marginally involved in 1867. Playing a
bigger role, he came back again with Krantz in 1878. From then on, in 1878, 1889 and
1900 the main exhibition buildings and their subsidiary pavilions waxed ever grander

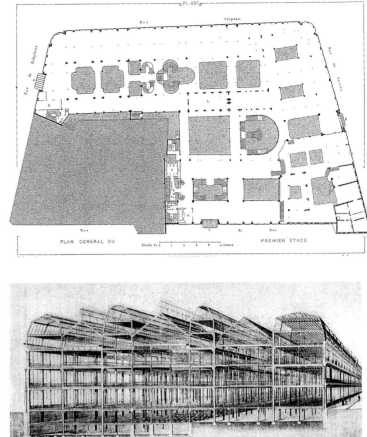

134. Main staircase at the Bon Marché, Paris, *c.*1875, from *Le Monde Illustré*. Louis-Charles Boileau, architect, Armand Moisant, engineer-fabricator.

135. Ground floor of the Bon Marché, Paris, showing framed construction, thin walls and light wells, from *Encyclopédie d'Architecture*, 1880.

136. Iron skeleton of the Bon Marché, from Julien Turgan, *Les constructions métalliques et entreprises générales, Moisant-Laurent-Savey et Cie*, 1890.

and richer. To some, lavishing monumentality on buildings doomed to destruction seemed wasteful. These concerns, which we have already witnessed at the Crystal Palace, underlay the controversies attending later exhibition structures, notably in 1889.

No such doubts inhibited the rampant new department store. By 1867 the Paris shop on several storeys round galleried courts or light-wells was on its way to becoming a fixture. Markets, arcades, 'magasins de nouveautés' and bazaars with iron-framed toplit halls or passages had already convinced retailers that iron and glass could maximize well-lit selling space. But the organization and display techniques of the great shows helped give an enhanced image and coherence to the new shopping-type. Enterprising shop-keepers, noting the extra trade generated by the exhibitions, soon mimicked their scale and manner of display. Unlike them, the stores were there to stay. In flux they might be, ephemeral they were not. City-centre buildings in fashionable streets, they needed style to counteract their perpetuum mobile. Change had to be reconciled with permanence, bustle with dignity, clutter with space, ornament with structure. In the process, iron architecture acquired fresh brio and popularity – with the architect to the fore.

The earliest store conspicuously to strike out on these lines is Aristide Boucicaut's pioneering Bon Marché, rebuilt in phases between 1869 and 1887.[272] In the first phase (1869–72), the architect J.-A. Laplanche furnishes the ashlar front typical of Second-Empire Paris, but with ample shop-windows, behind which lie ceremonial stairs and multi-storey light-wells. 'The construction remained traditional and the iron did not show', says Bertrand Lemoine.[273]

After the hiatus of the Franco-Prussian War and the Commune, Laplanche is replaced for the three later phases of the main Bon Marché building (1872–4, 1875–6 and 1879–80) by the Boileaus: that is to say, Louis-Charles Boileau, with Louis-Auguste hovering in the wings. Every whit as verbal and as devoted to iron as his father, the younger Boileau is nevertheless Beaux-Arts-trained and pragmatic. Reflecting on the Bon Marché,[274] he pinpoints the challenge of the department store: that of creating a structure that has minimal mass and surface, yet is stable, safe from fire, heavily serviced and richly ornamental. This, Boileau fils forecasts, calls for a fresh approach, whereby the building is conceived in terms not of its masses ('les pleins') but of the space ('le vide') it envelops. As for ornament,

> The architect's usual studies are little help here; what is the use of having learnt how to set out and proportion mouldings or ornaments on the abundant surfaces of stone . . . when there are no more surfaces available to receive them? [275]

Throughout Boileau's work, the front of the Bon Marché runs on in the style set by Laplanche. But from 1872–4 the iron frames of the central skylights, along with the spacious galleries and branching staircases, open up beneath a double skin of glass (ills. 134, 135, 136). In the last phase of 1879–80 the intermediate spaces between the great lightwells open up too; the last obstructive stone piers vanish, all the services being threaded through ornamented iron columns and floors. Boileau calls the result 'one single network of iron'.[276] Here, to use the American term, is multi-storey 'cage construction' – familiar from warehouses, factories and offices, now taken up in a chic and therefore architectural building-type.

This liberation by stages of the internal frame within the Bon Marché needed the good will of the regulatory authorities, always crucial to the progress of applied building technology. According to Boileau, the Paris voirie had to be persuaded that a five-storey building could be stable without substantial internal stone piers. But thanks to the progress of iron building and 'new studies on the part of our architectural administration, the objections of 1869 no longer existed in 1878'.[277]

How was this all procured? Responsible for the ironwork in 1872–4 and 1875–6 was an engineer-fabricator-contractor of the new, educated type, Armand Moisant.[278] A 'centralien' – a graduate-engineer of the Ecole Centrale des Arts et Manufactures (p. 436) – who had won his spurs working for Joly, ironwork contractor for Les Halles, Moisant had set up his own firm in 1866. He is best known for having made the frame of the famous Meunier chocolate factory at Noisiel. Boileau fils, who became his close friend and the architect of his country house, states explicitly that the principles for building the later phases of the Bon Marché were worked out together between Moisant, 'ingénieur et constructeur', and himself. For the final section of 1879–80 a second centralien, Gustave Eiffel, took over as fabricator-contractor.

Eiffel is sometimes described as the 'engineer' for the Bon Marché. That description applies to the last phase alone, and only within limits. The opening-up of the store had come from Boileau fils and Moisant. The design of the roofs in the 1879–80 section looks more bridge-like than Moisant's; and we know that Eiffel solved a minor problem about the differential expansion of iron beams and plaster ceilings. Probably he supplanted Moisant on the last phase because he offered a better price. By the scale of his railway-bridge enterprises, then in full flood, this was a small if smart job.

In normal building procurement, the contractor is appointed only when the design has gone far enough for rough pricing or formal tender, even if attached to that price is a means of construction that affects the design. So an engineer-contractor working with an architect will generally be at a disadvantage, because he comes in when the design process has already gone some way. The architect has a start in the race. Thus Boileau fils remained in charge throughout the later phases of the Bon Marché, even though Moisant and to a lesser extent Eiffel contributed to its revolutionary construction and appearance.

Despite Boileau's faith in iron as a way of solving technical and spatial problems, the tenor of his comments on the Bon Marché shows that he saw the prime challenge of the department store as artistic – meaning decoration. Though tritely classical, its iron details vied in richness with anything previously done. In the next generation of shops, architects were to elaborate decorative ironwork of ever more intensity on the back of frames devised by engineer-contractors. The emergence of Art Nouveau, the bourgeois style par excellence, is indissociable from the take-off of iron construction for shops. Yet the structural partnerships on which it relied were only half-acknowledged. So much emerges from the stories of the Printemps and the Samaritaine, best-documented of the Paris stores that followed upon the heels of the Bon Marché.

Following a fire in 1881 the Printemps was rebuilt, in the usual phases, to a unified design by Paul Sédille.[279] An art-architect and man-about-town, Sédille mainly built bourgeois houses.[280] Possibly his knack for fashion and decoration commended him to Jules Jaluzot, owner of the Printemps. At any rate, he had built a small portion of the premises consumed by the fire. He had also styled the pavilion showing the products of the great Le Creusot iron foundry at the 1878 Exhibition, which may have helped familiarize him with iron structures.[281]

Framed construction after the Bon Marché model was selected for the Printemps on grounds of speed and space. That entailed early decisions about technique. Point-loaded frame construction bearing a heavy superstructure requires meticulous foundations when the subsoil is poor, as both Paris and Chicago were discovering at this time: to design them, an engineer expert in the great Parisian 'travaux publics', Zschokke, was brought in. As for the superstructure, this was the thinking:

137. Magasins du Printemps. Paul Sédille, architect, 1882–5.

> Cast iron requires models which take a very long time to set up, and time also to pour at foundries in the départements . . . Wrought iron lent itself to relatively quick execution in the well-equipped workshops of a large Parisian constructeur within reach of the Printemps site. By virtue of the basic elements (plates and angle-irons) available to the architect it also imposed the obligation of very sober forms, deriving from the materials themselves and therefore characteristic. Wrought iron was therefore chosen as the exclusive element, as it were, of construction.[282]

That avowal of sobriety sits strangely with the florid store manifest in old illustrations. But it shows that the basic grammar for the Printemps was worked out by Sédille in collaboration with the ironwork contractor. Here the choice fell on Baudet, Donon and Company, located in suburban Argenteuil in the premises of the now-defunct serruriers Roussel and directed by Emile Baudet, a 'centralien' of the same stamp as Moisant and Eiffel.[283] Once again architectural progress gets a leg-up from an engineer not of the official Ponts et Chaussées breed, but one equipped by the Ecole Centrale to tackle large industrial contracts. The size of recent railway bridges, pursues the article quoted above, has allowed firms like Baudet, Donon to equip themselves for fabricating pieces

138. Cross-section through Magasins du Printemps, showing iron construction of floors and main staircase. Paul Sédille, architect, Baudet, Donon and Company, fabricator-erector of the ironwork. From *Encyclopédie d'Architecture*, 1884.

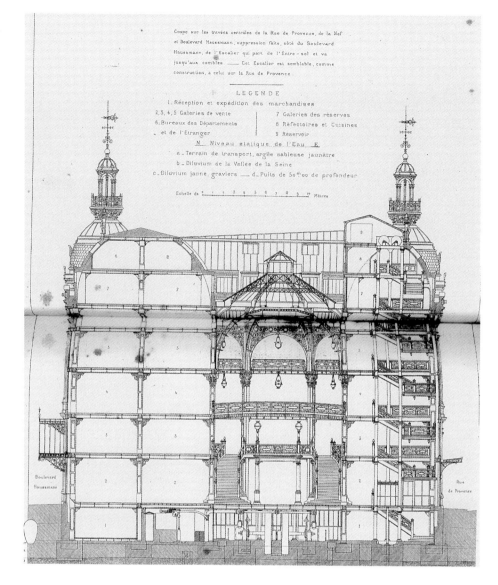

on a larger scale; and these have affected the design of the 'ossature' – the frame. Standard plates and angles have been taken in from foundries like Le Creusot as part of the process, while Baudet, Donon have also carried out most of the 'serrurerie d'art' at the Printemps.[284] The engineer-contractor's tasks therefore range from basic manufactured components to the applied arts.

Since the store's success was deemed to hang as much on ornament and palette as on the extra space furnished by the iron frame, in publicity for the Printemps the décor took pride of place. Sédille exerted himself to outdo Jaluzot's rivals by orchestrating artists, materials and colours. Outside, metal and glass obtruded between the piers and above the cornice, but there was still masonry enough for the leading lines of the decoration to mimic monumentality (ills. 137, 139). Enriched by an array of sculptors including the celebrated Chapu, the Printemps façade earned itself a full article in the *Gazette des Beaux-Arts*.[285] The all-metal interior (ill. 138) was trickier because of the insubstantial frame, as Boileau fils had found. More vibrancy than had obtained at the Bon Marché was wanted to detain the eye. Sédille argued that wrought-iron ornament would look too slender, fail to stand out from the backdrop of the frame and need regular repainting. Instead he chose a contrast of tones and materials: bronze (mostly gilded), marble, cast-iron, and

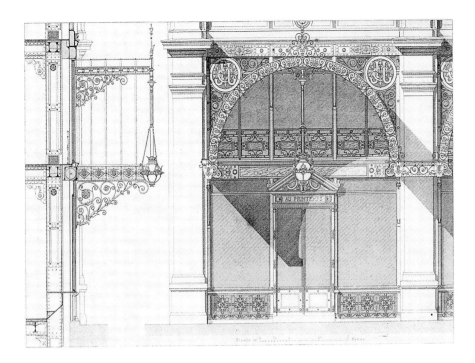

139. Ironwork of an entrance bay of the Magasins du Printemps, with initials of the owner, Jules Jaluzot, in the spandrels. From *Encyclopédie d'Architecture*, 1885.

140. La Samaritaine. Interior of Building no. 2 with completed cupola behind, July 1907.

enamel, along with stained glass and the gaudy paint hues splashed over ironwork ever since the Crystal Palace.

In terms of rationalist architectural theory, much debated in France, it is a moot point whether such superfluity of ornament followed from the nature of construction, or was an imposition and disguise. The case can be argued either way. If non-structural features diverge in character or position from the elements and means of construction, or if they

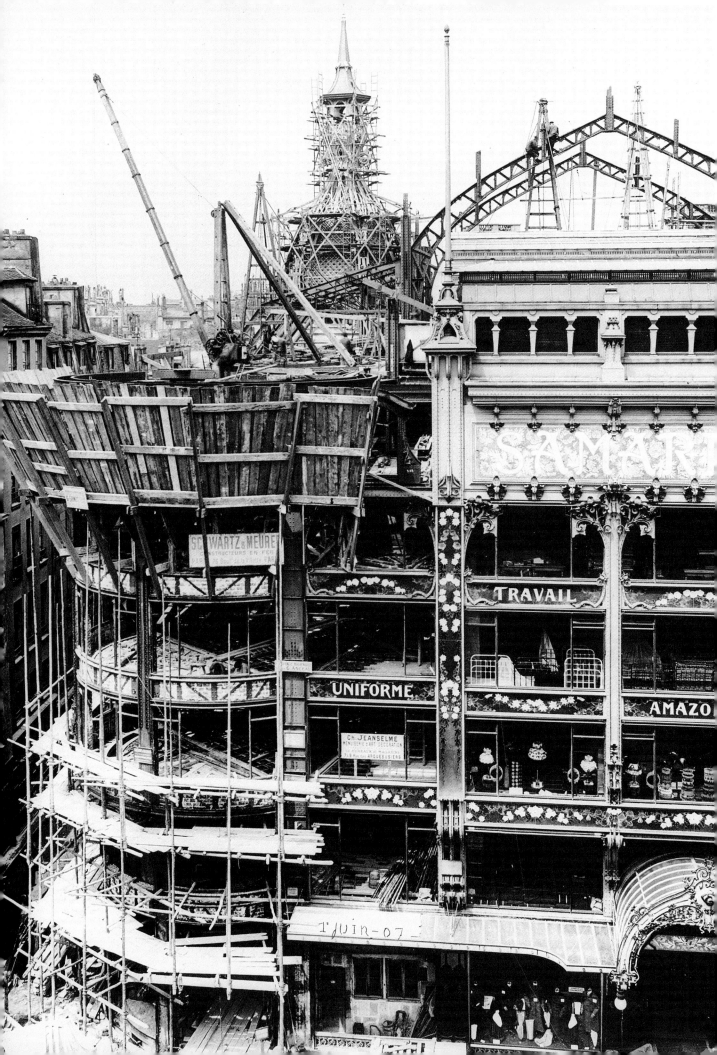

envelop them, they can be proscribed as merely ornamental; if they draw attention to those elements, the same features can be legitimized by the same theory. Yet the doctrine of structural primacy will always allure, because structure alone holds out the promise of unity. It is the one thing a building cannot do without. That is why the idea that everything else in architectural expression should derive from structure is both logical and attractive.

Since more buildings were now of iron, it seemed to follow that their decoration should grow more directly from the structure than in stores like the Bon Marché and the Printemps, which clung still to classical conventions of ornament. If the frame was of iron, should not the keynote of decoration also be of iron, and integrated with it? That way lay a chance to reconcile the richness of the monumental tradition, crucial to French architecture's self-esteem, with iron structures. Such a programme had been hinted at by Viollet-le-Duc in his lectures of the 1860s, as he started to move beyond his beloved Gothic. In those ideas, it has long been recognized, lay the seeds of the 'organic' style we know as Art Nouveau. Our concern here is not the theory of the matter, but how an all-iron decorative architecture claiming to arise out of structure was procured in practice.

The store that set out these ideas most blatantly, the Samaritaine, comes late in the short trajectory of Art Nouveau. By the time its rebuilding commenced in 1905, reinforced concrete structures had penetrated the market for modern framed buildings. A shift towards simplicity even in public buildings was taking hold, with Hermann Muthesius calling from Berlin for a sober new 'building-art', and execrating the 'whiplash curve' of Paris and Brussels. Frantz Jourdain, the main author of the Samaritaine, had acquired the great job of his career too late.

142. La Samaritaine in the 1930s, showing the junction between the Art Nouveau store and the addition of 1925–7 by Sauvage and Jourdain: the cupolas have been removed.

FACING PAGE
141. La Samaritaine. Building no. 2 in construction, with one of the cupolas behind, June 1907. Frantz Jourdain, architect, with Schwartz et Meurer, ingénieurs-constructeurs.

In Meredith Clausen's study of his career, Jourdain cuts a figure of brilliance, wit and paradox as a propagandist, but of less definition as a practical architect.[286] In theory, he had the answers. A torrent of articles from his pen lambasted reaction, artistic and political, and warned architects that they had to get the hang of iron technology if they were to ward off the engineers, supposedly at the gates. After 1889, for instance, in the wake of the Eiffel Tower, a fresh frenzy of 'ingénieurophobie' does the rounds, prompting impassioned articles in *L'Architecture*. Jourdain rounds off the series.[287] For him, architects are the heroes of 1889. His epigraph comes from Maupassant: 'When you see an engineer, take a gun and shoot him.' The two professions are like dogs and cats, Jourdain says. Engineers possess intelligence and even good taste, but can't be creative. 'The engineer instinctively goes for the ugly, as the duck makes for water.' But the tactic is not to ignore him, as reactionaries would like; it is to steal and recut his clothes. A fair part of Parisian Art Nouveau architecture comes out of a campaign to appropriate and inflect the iron building-style of the 1889 exhibition – an event that fixed but also ossified Jourdain's thinking.[288]

Jourdain had been engaged with the architecture of shopping as far back as 1882, when he made up a graphic brief for a modern, iron-framed department store to help his friend Zola in the planning of *Au Bonheur des Dames*. Fittingly, this first store of Jourdain's was composed in the words that were his strength. But Zola was writing about the 1860s. Alive to anachronism, he toned down the iron content, though in the newly rebuilt store of his novel he did evoke 'la vie sonore des hautes nefs métalliques'.[289]

Au Bonheur des Dames was published in 1883. About then Jourdain also encountered Ernest Cognacq, chief proprietor of the Samaritaine, soon becoming his architect. Among the minor works to the existing premises that ensued was an entrance canopy of blue-painted iron, fabricated in 1892 to Jourdain's designs by Bergerot, Schwartz and Meurer, a firm then just expanding from horticultural greenhouses to winter gardens and station roofs.[290]

A decade later, Cognacq was ready at last to reconstruct his premises. The sites and phases of the Samaritaine are involved. Mention need be made only of the famous 'No. 2' building, started in 1905, completed in 1910 (ills. 140, 141).[291] The whole of its steel frame was exposed right out to the front, there to be decked in prickly wrought-ironwork, floral panels in ceramics and enamels, and inscriptions. If baring the frame to the weather and the street still defied convention in some measure, it completed a process started in small shops of the 1860s, and enlarged since by stages in the stores of Paris, Brussels and Berlin.[292] The point about Samaritaine No. 2 was its extent. That allowed the rhythm of column, beam and infill ('ossature' and 'remplissage'), to unfold in a manner more reminiscent of an American than a European city. But besides the tautness of, say, Sullivan's Carson Pirie Scott Store, the Samaritaine's decorative tissue looked wayward and untidy. To flaunt a theory of iron structure and ornament was one thing; to integrate them needed the touch of the born designer. Lacking that gift, Jourdain lavished on his street fronts gaiety of vivid colour: friezes in daffodil yellow, and metalwork in peacock blue and green.

The flamboyance of Samaritaine No. 2 belied some conservatism of procurement. Iron or steel construction was hardly an avant-garde choice in 1905. By opting for framed and decorated construction, Jourdain was copying the structural methodology of previous stores like the Printemps, while updating their ornament and palette. He teamed up with the ingénieurs-constructeurs Schwartz and Meurer, as that firm had now become, for both the structural and the decorative ironwork, then brought in a team of fine and applied artists for the other materials that embellish the frame. 'To what extent Jourdain himself was responsible for the design of decorative features is not clear', says Clausen.[293] The interior of the Samaritaine proffered fewer contrasts than the Printemps. Luxurious bronze was discarded in favour of enamel and of Art Nouveau wrought-iron ornament such as Schwartz and Meurer supplied in their great greenhouses.

The quirkiness of the Samaritaine highlighted the limits of translating engineering construction into the conventions of monumental architecture. Like many decorative styles, Art Nouveau lost allure when the scale got too large. The cupolas on the southern angles of building No. 2, for instance, were clumsy versions of suaver features at the Bon Marché and Printemps. No one liked them. In 1922 a conspicuous southern extension to No. 2 was proposed, bringing the Samaritaine down to the Seine. In his first scheme, the veteran Jourdain clung to the idea of an all-metal façade, if without the ornamental baggage of yore. But when it came forward, Paris's Comité Technique d'Esthétique insisted that the building be clothed with stone and that the offending cupolas be taken down – and won on both points. The extension was built in 1925–7, under the tutelage of Jourdain's friend Henri Sauvage. Casing its steel frame came a front of impeccably respectable masonry (ill. 142). Engineering was one thing after all, it now seemed, street architecture another.

In this section, we have seen how in one special urban building-type, the department store, the Parisian architect was able to exert his hold over an iron-framed architecture and play with its expression and enrichment. Crucial to that process was a set of suburban engineer-contractor firms (Moisant; Eiffel; Baudet, Donon; Schwartz and Meurer) who shared with the self-effacing serruriers of earlier years the combining of fabrication with erection, but were now becoming half-partners in design. Many such firms owed their access of authority to a directing mind in the shape of an engineer or engineers trained at the Ecole Centrale des Arts et Manufactures, and to a scale of operations that encompassed large bridges and railway contracts.

It needed only a different type of job with other requirements and conventions for these roles to be reversed, and for the ingénieur-constructeur to change from the architect's handmaid into an aesthetic force in his own right, who would inspire his professional counterpart, yet frighten and confuse him. That role-reversal was to be undertaken by Eiffel. Its defining monument would be the 300-metre tower of 1889 that bears his name – erected not in some unfrequented valley where the railway engineer held sole sway, but in the heart of Paris where design hierarchies mattered more than anywhere in the world. It is with the broad professional context of that famous commission and of the 1889 exhibition that this account of architect-engineer relationships and French structural ironwork can end.

Eiffel and 1889

Gustave Eiffel made his name as a fabricator-contractor of railway bridges.[294] Railways, we have seen, were largely the province of engineers, stations apart. But it is as well to be clear what kind of engineer is meant at any given moment. On the early continental railways, most major bridges were designed by independent engineers or consultants to the companies, not engineer-contractors of fabricators. In France that tended to mean graduates of the Ponts et Chaussées who, after a spell of state-service, shifted into lucrative private practice as railway engineers. But as contracting firms attracted their own trained engineers and became better capitalized and equipped, railway companies began under the eye of their chief engineer to seek competitive bids in the form of design-and-build packages, even for ambitious bridges. In expectation of such work, some 'centraliens' ventured into the perils of contracting, relying upon networks they had built up within the tight-knit world of railway construction.

Eiffel is the one of whom by far the most is known. After leaving the Ecole Centrale, he found his first job with the Paris engineer-fabricator-contractor Charles Nepveu. That 'ingénieur-constructeur inventif et prolixe', as Bertrand Lemoine calls him,[295] made engines, rolling stock and all kinds of iron equipment, so his assistant's tasks were as much mechanical as structural. Nepveu failed in 1856, but he was able to pass Eiffel on to his successors, the Belgian firm Pauwels & cie, as supervising engineer for the great railway bridge over the Garonne into Bordeaux (1858–60). Such were the young man's

COMPAGNIE ROYALE DES CHEMINS DE FER PORTUGAIS

PONT SUR LE DOURO
A PORTO

M. D'ESPREGUEIRA, Directeur de la Cie.   M. LE FRANÇOIS, Ingénieur-Conseil.

G. EIFFEL & Cie
INGÉNIEURS-CONSTRUCTEURS A LEVALLOIS-PERRET — PARIS
ENTREPRENEURS GÉNÉRAUX

Longueur totale du pont entre culées . . 330ᵐ00	Ouverture de l'arche centrale 160ᵐ00
Hauteur du rail au-dessus de l'étiage . . 61ᵐ00	Flèche de l'arc métallique. 40ᵐ00

143. Maria-Pia Bridge over the Douro, Oporto, with train crossing. Gustave Eiffel, ingénieur-constructeur, with Théophile Seyrig as chief designer. Probably an engraving of the view made by Stephen Sauvestre in 1877.

prowess at site organization and technique that he acquired further bridge contracts for Pauwels in the south-west. Most of Eiffel's railway work followed from the name and contacts he made at Bordeaux. Finding Pauwels too in decline on his return to Paris, he resolved to set up his own contracting firm in 1866.

Before Eiffel could win the all-metal bridge contracts that made him famous, he had to equip capacious workshops at Levallois-Perret outside Paris – a laborious and hazardous task managed on family loans. The other requirement was cutting-edge bridge design. Eiffel was not primarily a designer. That skill came in the main between 1868 and 1879 from another talented 'centralien', Théophile Seyrig, who also added to the firm's capitalization. The outline of the first designs, starting with viaducts in the Allier département, came in fact from the Orléans railway company's engineer, Wilhelm Nordling; Eiffel contributed manufacturing precision, innovations in assembly, and site efficiency.

From the start, contracting for buildings was a small, supplementary part of his business.[296] It spread the load in case the railway market fell off, while Paris jobs like the Bon Marché brought publicity. More precious still were the exhibitions. Having been marginally involved with the 1867 Exhibition, Eiffel secured prominent representation in 1878. There, working with different architects, he erected the front part of the gigantic main palace on the Champs de Mars and two supplementary pavilions. In an open ideas-competition for the show, Eiffel also proposed a bold footbridge to link the two halves of the site, crossing the Seine in a single span over the top of the Pont d'Iéna.[297] It is in connection with this unbuilt 'passerelle' and with the small iron-and-brick pavilion for the Compagnie Parisienne du Gaz that we first meet the architect closest to Eiffel's retinue, Stephen Sauvestre.

In Autumn 1877 Sauvestre had accompanied Eiffel, his young family and his new lieutenant for technique, Emile Nouguier, to the opening of the great Maria Pia bridge conceived by Seyrig to carry the railway into the town of Oporto. That commission took the Eiffel enterprise to a new pitch of prestige, innovation and grace. With its deck sweeping across the gorge of the Douro, perched upon a single segmental arch that swelled to its crown and tapered to its feet, the design had triumphed over strong opposition in the preliminaries. Now it was complete, Sauvestre was there to make a watercolour of the bridge for presentation to the King and Queen of Portugal (ill. 143).

Sometimes described as 'architecte attitré' or consultant architect to Eiffel, Sauvestre was not a full-time employee but an independent practitioner with good connections. Thirty-one at the time of the Douro opening, he had grown up in a progressive family (his father was a Fourierist) and been among the earliest successful graduates of the small, autonomous Ecole Spéciale d'Architecture.[298] Sauvestre had a number of villas to his personal credit, but was also a ready collaborator. Though no original designer, he had the Parisian knack for making metallic construction exuberant. In the passerelle competition entry the footway itself, slung between pairs of single-span, Douro-style arches, bore ample enrichment. This we may assume to have been Sauvestre's contribution, on the strength of its successor in 1889.

The genesis of the Eiffel Tower goes like this. It starts long before the 1889 Exhibition has been officially announced, as a spark in the brains of Eiffel's two great deputies: Emile Nouguier, 'chef de technique', and Maurice Koechlin. An ETH-Zurich-trained engineer, Koechlin has joined the firm in 1879 after Seyrig drops out, and as 'chef du bureau d'études' has steered through the greatest of Eiffel's bridges, the Garabit viaduct. Towers are now in the air, the firm having lately completed the internal iron framework for Bartholdi's Statue of Liberty. After conferring with Nouguier, Koechlin goes home one day in June 1884, sketches out and roughly calculates a 300-metre tower on a base with sides of 125 metres (ill. 144). A stiff-looking thing with the wiry force of the Garabit, it is in its essential structure what will be built. Mildly dismissive when shown the idea, Eiffel nevertheless does not object to Koechlin and Nouguier pursuing it.

They then take it to Sauvestre. In Bertrand Lemoine's words:

The latter gives shape to the design, until then just a diagram; he links the four ribs and the first level with monumental arches, destined to increase the impression of stability which the whole must give, and to figure in due course as an entrance to the Exposition; on the first level he sites a great hall of glass, rather like the one on the bridge project of 1878; he embellishes each face of the second level with sculptures of the famous etc. Thus 'decorated', the project is shown to Bartholdi and features in the decorative arts exhibition held that autumn at the Palais de l'Industrie in the Champs-Elysées.[299]

Seeing the tower thus styled (ill. 145), Eiffel now takes it up. In September 1884 he applies for a patent in his own name and those of Koechlin and Nouguier. He also negotiates a contract with Sauvestre whereby the latter will pursue the decorative side of the project and supervise the tower on site, should it come off, for 3 per cent of the total sum.

Subsequent lobbying pays off when the exhibition gets its final go-ahead in April 1886. A competition is held, open to all architects and engineers in France, with the proviso

144, 145, 146. Preparatory designs for
the Eiffel Tower. A. Sketch by
Maurice Koechlin, Eiffel's chef
d'études, showing the projected
height of the 300-metre tower against
famous monuments. B. Design as
embellished by Stephen Sauvestre.
C. Executed project.

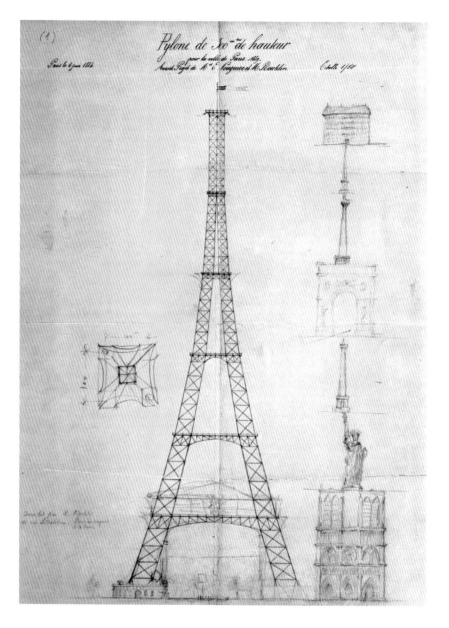

that the main buildings are to be of iron or steel. A 300-metre tower on a 125-metre base
is embedded in the brief. So the cannier competitors simply tuck the Eiffel design somewhere into their site plan, whether embellished with their own 'architecture' or otherwise. The winners are Ferdinand Dutert (architect), who has been working on his own
ideas for the Champs de Mars since 1884; Eiffel and Sauvestre (engineer and architect);
and Jean-Camille Formigé (architect). In this competition phase, Sauvestre has to come
up with exhibition pavilions to accompany the tower. In the event Adolphe Alphand, the
veteran engineer in overall charge of the exhibition buildings, parcels out the main site
between the victors. To Formigé he allots the long side pavilions, and to Dutert the great
Galerie des Machines at the back; Eiffel is given the chance to raise his tower in the position he and his team have always wanted, facing the Seine. A design long conceived,
patented and publicized can now be developed and built (ills. 146, 147).

   In the old view, the 1889 Exhibition represents the triumph of iron and therefore of
the engineer: the Eiffel Tower is offered in ultimate proof of that.[300] That is poor history.
Two of the three victors in the 1886 competition were architects, confirming their profession's tightening grip on the great exhibitions since the 1850s. Moreover, although the

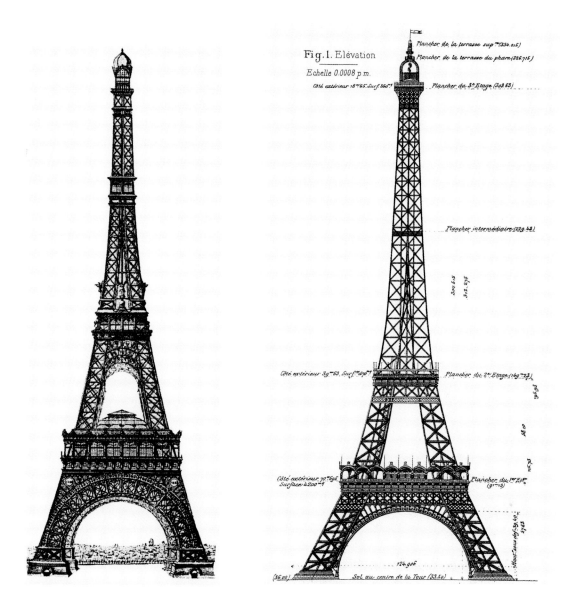

Fig. 1. Elévation

tower was unquestionably an engineering conception, drawing upon the technique and panache accumulated by the Eiffel firm in bridge design and construction, its fate too depended at a critical juncture on an architect.

Many commentators have judged Sauvestre's contributions an embarrassment – fiddling and dishonest impediments to the pure structure sketched out by Koechlin. The accusing finger points to the false swagger of the segmental iron arches at the base, only hung from above once the structure of the first floor was in place. Yet Eiffel himself only took note of the scheme once Sauvestre had intervened. Perhaps the shrewd engineer did so because he knew that concessions to bourgeois taste would be needed to get the tower built, and cared little for the 'architecture'. When confronted with the famous 'protest of the artists' against his tower, he certainly did not cite Sauvestre's embellishments. Instead he retired behind the old engineering smokescreen to the effect that a beautiful structure follows natural law:

> Do not the laws of natural forces always conform to the secret laws of harmony? The first principle of the aesthetic of architecture is that the essential lines of a monument should be determined by their perfect appropriateness to their end.[301]

As if there were neither choice nor ambiguity in the matter! Whatever Eiffel's motives, it remains the case that architectural embellishments were vital to the Eiffel Tower's success in 1889, just as it is also the case that those same embellishments were vandalously destroyed in the name of modernist simplification forty-eight years later, at the time of the 1937 Paris Exhibition. Autres temps, autres moeurs.

Looking back from that later date, it seemed necessary to ridicule the indignation of the protesting artists. The 'triumph of engineering' in 1889 appeared to have opened the door to a more rational architecture. The disconcerting ebb in the tide of great structures with externally expressed ironwork in the two next world exhibitions, Chicago 1893 and Paris 1900, had therefore to be presented as a betrayal of that high ideal. Professionally in fact, the tide went on flowing in the same direction, with architects gaining a tighter hold upon these great shows. The upshot was an ever-richer style of exhibition architecture down to about 1905, when simplification started to seep in.

One last aspect of the Eiffel Tower must be mentioned. That is the uncertainty about its life-cycle. Eiffel had dubbed it a monument, with all that implied about permanence and architectural values. But though the author of a great exhibition structure might hope and fight for its survival, that could not be guaranteed. The ornamental extravagance of the pavilions in the later Paris exhibitions in part represented a pitch for permanence on the part of their architect-designers. In 1889, Formigé's colourful Beaux-Arts and Liberal Arts pavilions, and Bouvard's central dome, yet more riotous, were of that nature. But neither survived the planning for the next exhibition on the Champs de Mars in 1900. Few expected the Eiffel Tower, obstructive and raw despite Sauvestre's trappings, to last. It boasted the scale but not the propriety of the monument. For all its size, it looked temporary.

The tower survived its first thirty years not because it was popular, nor because it yet represented exemplary values or symbolized France, but through a succession of near-shaves. Following its first triumph, Eiffel hoped that the tower would go on bringing in publicity and money. But receipts dwindled throughout the twenty years after construction, apart from a predictable jump during the 1900 exhibition. It was lucky to survive the planning for that event (in connection with which Sauvestre suggested mini-towers to the sides to take visitors by lift to the second stage). After close debate in 1903, the architect Pascal adjudicated in favour of retention. The tower then went on receiving short reprieves until the 1920s, when reports in favour of its usefulness for meteorology, astronomy, telegraphy and radio secured it for posterity. Thereafter it was to turn into the pre-eminent symbol of the engineering aesthetic, with all the paradoxical qualities of that aesthetic – beyond the technical grasp of most people, yet photogenic, and therefore readily presented, grasped and remembered, even by those who have never seen it. No one dares any more to deny its beauty.

That French architects were as eager in 1889 as they had ever been to tackle and tame iron construction, and not acquiesce in the ancillary role of a Sauvestre, appears from that other celebrated icon from the exhibition, the Galerie des Machines.[302]

In terms of his 'formation', Ferdinand Dutert, its principal author, conforms to the stereotypes about the Ecole des Beaux-Arts. Grand Prix de Rome 1869, he returned with a portfolio of archaeological drawings, only to be subsumed for years into what is now called arts administration. What may have given Dutert his industrial frame of reference were his liaison duties with textile and engineering manufacturers in the Département du Nord, where in tandem with the Galerie des Machines he built the design school at Roubaix (1886–9).

Like Nouguier and Koechlin, Dutert was nurturing a proposal for the 1889 exhibition almost five years before the event. His plans, including a gigantic machine hall just where it was built across the back of the Champs de Mars, received a preliminary approval in March 1885. Like the team of Eiffel and Sauvestre, Dutert was obliged to enter the competition held next year – something of a charade. As has been seen, both were declared

winners along with Formigé, though Adolphe Alphand was not able definitely to assign buildings between them. As yet Dutert had no collaborator, nor had his Galerie des Machines taken on the finality which the Eiffel Tower had long assumed. Its overall breadth of some 150 metres was distributed between five aisles.

In August 1886 Alphand parachuted in Victor Contamin as engineer in charge of metallic construction for the whole exhibition. A 'centralien', Contamin united the post of chief structural engineer to the Compagnie des Chemins de Fer du Nord with teaching and writing on the strength of metals. If, as seems likely, he was already acquainted with Dutert, the appointment confirms the intention to allot the machine hall to the latter. Before that could be agreed came negotiations and changes. The ground at the back of the Champs de Mars was known to be poor; and that along with the need to maximize exhibition space inside the hall suggested reducing the number of point-supports and foundations. The first idea was to have three aisles of 50-metre width. Then Dutert (Alphand is definite that it was he[303]) came up with the idea of an enormous nave covered by single trusses of 110 metres, with minor flanking aisles and side-walls on light foundations. Eiffel rose to this challenge with a counter-proposal, but could not beat the researches of Contamin and Dutert during the autumn of 1886 on truss-form and price. Though Alphand was at first disinclined to believe in the practicability of a truss one-

third as wide again in span as St Pancras (at 73 metres still the widest roofed space at that time), Contamin's calculations persuaded him.

The Galerie des Machines (ill. 148) prefigures one of the classic patterns of procurement for major modern building projects, whereby architect and engineer meet on equal terms and the contractor is not centrally involved in the design. It can be broken schematically into sequential phases. First the architect has the idea, which he must set out in persuasive words and sketches, before he knows how feasible it is. Then the engineer works out if and how it can be done, coming back to the architect if visible changes are implied. Lastly the architect repossesses the project, refines, details and enriches it, sees it through construction (where the engineer may again be crucial) and – not least important – presents and interprets it when all is done.

But in even a well-documented case like this, the partition of credit between the collaborators is fraught with uncertainties. We know the idea of the 110-metre span came from Dutert's but not, for want of early drawings, how far he defined it in the first instance. Its translation into the specifics of the 'determinate' three-pin truss without costly and obstructive ties must surely have started with Contamin the railway engineer, since its previous history derives from Austrian and German railway-building.[304] We know also that Contamin, upheld by Alphand, fought and won an argument against

148. Galerie des Machines, Paris exhibition of 1889, view of end gable showing scale. Ferdinand Dutert, architect, Victor Contamin, engineer.

149. Galerie des Machines, interior. Photo by Glucq from L'Album de l'Exposition, 1889, showing the roof and ground-floor hinges as they appeared when the hall was full of exhibits.

Dutert on the subject of the ends of the hall, where the architect had hoped for deep hips – no doubt to give better definition to the exterior. These would have entailed variations in the design and impeded the famous fast-track construction method adopted, whereby two contracting firms using different methods of erection, Fives-Lille and Cail, each started with a truss at the centre and raced each other outwards in opposite directions until they reached the end walls.[305]

So much tells in favour of the engineer. What then did Dutert do, besides attend to the incidentals? The answer is that he accentuated and refined the structure with the controlling instincts of a Prix de Rome. According to his deputy, Victor Blavette, once Dutert had given in on the hipped roof,

> he became inflexible about the rest, and everything remained as he drew it, from the whole frame to the smallest details. Technological assistance [sa collaboration scientifique] was confined to determining mathematically the thickness attributable to the metal.[306]

That may be an exaggeration. The point one would most like to elicit is whether the virile, angular profile of the truss was Dutert's. The German three-pin trusses preceding the Galerie des Machines took an elliptical or segmental form, but that was neither in the mainstream of French shed roofs nor consonant with the logic of half-trusses meeting at a high, central hinge.

What we do know is that, late in the design, Dutert relished taking off the cast-iron guards at the feet of the trusses and exposing the great hinges in their nudity (ill. 149). This was to become the memorable, fetishized detail of the empty structure, though it cannot have been so clear to view when the hall was packed with machinery. His committee had asked for a base and he prepared designs for the guards, Dutert said,

> in case he found that this small resting-point for the great trusses caused a feeling of insecurity for the public; but, since he found that the public rather liked, than otherwise, the non-concealment of the construction, he had determined to let it remain exposed.[307]

The alternating of panel-sizes in the great arches, and the discreetly curvaceous styling of purlins and minor trusses are also known to have been Dutert's. Details of this kind remind us that a prefabricated language such as wrought iron or steel still allows some licence for art. Given a large, repetitive building and a designer alert to the tolerances of manufacture, there will usually be room and money for 'specials'.

Such structural nuances were as much embellishments as the stained-glass tableaux which enriched the ends and centre, the painted coats of arms along the side panels of the roof, the ornate electroliers at the base of the stairs, the pomp of the main entrance and vestibule, or the all-over dark-yellow tone of the ironwork (in which Dutert prevailed over Alphand's preference for blue). Yet because this styling of the ironwork was integral with the structure, it had professional significance. Since the wide-span trusses of the Galerie des Machines could be plausibly attributed to Dutert, that gave them superiority over the seeming afterthought of applied art. If an architect could be presented as controlling or at least inflecting the actual design of trusses, not just enriching them (the priority of Art Nouveau) that marked a professional step forward. A distinction could henceforward be drawn between what architects would foster as 'detail' and what they would condemn as 'ornament'.

For that reason, amidst a range of views about the buildings of 1889, several architects of progressive mentality praised the Galerie des Machines while condemning the Eiffel Tower. For the modernist, enamoured of structure, that is a puzzle. Banister Fletcher, who studied the exhibition with care and interviewed both Dutert and Contamin, articulates the ostensible answer. In the Eiffel Tower, he says, the decoration 'has been applied "après coup", or as an afterthought', whereas in the Galerie des Machines it is 'worked in the composition itself of the structure . . . by the juxtaposition of voids and solids'.[308]

Beneath that, a shrill repetitiveness in Fletcher's language betrays the deeper reason – the leitmotiv of professional insecurity:

> It is a great thing that an architect should have conceived a building of this sort. It is not generally so . . . that such a bold step should be taken by an architect is a matter of congratulation to architects all over the world. Iron has been the architect's watchword . . . The architectural world has been waiting for years for such an application of iron as this building shows; a building which will have infinitely more power in influencing architects and the treatment of construction than all the lectures or papers on the subject that were ever written.[309]

4. *America*

LATROBE: IDEALS AND REALITIES

'You forget', Benjamin Latrobe reminded a French friend concerned that he had not yet made his fortune,

> that I am an Engineer in *America*, that I am neither a mechanic nor a merchant, nor a planter of cotton, rice or tobacco: you forget, for you know it as well as I do, that with us the labor of the hand has precedence over that of the mind; that an engineer is considered only as an overseer of men that dig, and an architect as one that watches over others that hew stone and wood . . . The service of a republic is always a slavery of the most inexorable kind, under a mistress who does not even give to her hirelings civil language. This kind of treatment extends from her first political characters to her menials, her public Architects.[310]

In the year of that letter, 1811, the versatile Latrobe had been some time 'Surveyor of the Public Buildings of the United States', was adding to the Capitol at Washington and in the throes of a Catholic cathedral for Baltimore. Since arriving in Virginia from England fifteen years before, he had installed steam-powered waterworks and a rolling mill at Philadelphia, helped set up a power-driven forge and sawmill in the Washington Navy Yard, improved the navigation of the Susquehanna River and planned a Chesapeake and Delaware Canal, while creating a handful of America's noblest neoclassical buildings into the bargain. He had also designed a lighthouse for the mouth of Mississippi, and was just restarting a waterworks scheme for New Orleans, an ill-starred project that led eventually to the death of this most illustrious if hapless of early American architect-engineers.[311]

Latrobe's weary words remind us how much had still to be done thirty-five years after the American Revolution to construct the new country, even in its longer-settled communities. In the herculean efforts demanded, doers reaped more credit than designers. Needy, thrusting enterprises do not much bother with demarcation. With a whole continent to tame, a watertight distinction between architect, engineer and builder will be sought in vain until American cities and territories have been well established.

Not that Latrobe himself drew any boundary between the building professions. As few have done, he united technical proficiency in civil (and to an extent mechanical) engineering with high aesthetic talent. His English training had comprised some three years on canals and harbour projects, possibly under Smeaton, certainly under Jessop, leading British engineers of the 1780s, followed by a spell with the reputable architect S. P. Cockerell.[312] Those men had distinct skills, in token of how architecture and engineering in Europe were drawing apart. Though some few still moved fairly seamlessly between buildings and infrastructural works, no European contemporary stretched to Latrobe's range of endeavours. In part this was because America possessed too few experts for specialization yet to be possible. Latrobe urged the need for more: 'we import blankets, scissars, and wine, why should we not import knowledge? A good civil engineer

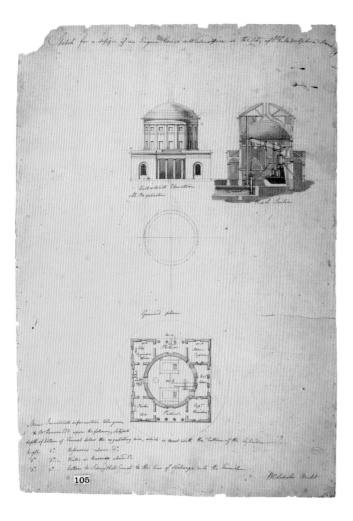

150. 'Sketch for a design of an Engine house & Water office in the City of Philadelphia', signed 'B. H. Latrobe Archt'. Neoclassicism and the steam engine reconciled. Notes on the bottom of the drawing refer to information required by Nicholas Roosevelt, supplier of the engine.

is an acquisition peculiarly desireable', he wrote in 1817.[313] Yet he relished having his fingers in many pies.

Running like a litany through Latrobe's correspondence is a caustic, semi-paranoid resentment at the want of American understanding and respect for independent designer-technicians. 'Here I am the only successful Architect and Engineer', he wrote from Philadelphia in 1804. 'I have had to break the ice for my successors, and what is more difficult to destroy the prejudices the villainous Quacks in whose hands the public works have hitherto been, had issued against me. There, in fact lay my greatest difficulty.'[314] He harboured a special fury towards the Carpenters' Company of Philadelphia, whose members collared all the house-building and stopped him getting architectural work, so he believed: 'It is not their fault that I have maintained my professional character . . . there is not a single instance in which I have been consulted in which some carpenter has not counteracted me'.[315]

Whatever the task in hand, Latrobe tried to champion the design-giving, fee-charging character of the independent professional. Yet such were his circumstances and enthusiasms, that often he got drawn into investing in commercial enterprises, even into making things himself, though that ran counter to his principles and dignity. In 1812, for instance, Latrobe impulsively set up a shop in Washington to build the steam engine for his New Orleans waterworks himself. Work soon fell away because of war with Britain and he found himself taking on odd jobs. He tried to keep cheerful but the bitterness showed: 'my neighbors the printers found me a much cleverer fellow at mending their presses, than I was considered in building the Capitol. Even the tavern keeper near me . . . has found out that I can be very useful at mending his pokers and gridirons. Thus by keeping a Blacksmith shop, I have actually maintained my family for some Months.'[316]

The Washington 'Blacksmith shop' brings us to iron and its place in Latrobe's philosophy and practice of construction. Like all his contemporaries, Latrobe built fundamentally with brick, stone and timber, though the last he held in some contempt, associating it with impermanence and his foes, the carpenters. Forged iron was in fair supply when he started building on the East Coast, but its quality varied; cast iron was less developed. In the European way, Latrobe used the former to supplement his major masonry structures. Wrought iron in bar form served to reinforce the dome of his first major American building (1798–1800), the Bank of Pennsylvania in Philadelphia, and was likewise deployed in his later vaults and domes.[317]

What really excited Latrobe about iron was its promise of wider technical progress through the steam engine. When he first proposed steam-powered waterworks for Philadelphia in 1799 (ill. 150), there were only three steam engines in the whole of the United States, says Darwin Stapleton, and no established engine-builders.[318] Much of the instability in Latrobe's career arose from tangles consequent upon his infatuation with steam engines for powering waterworks, mills or boats. For all his talents he could never control their procurement. In the nightmare case of the New Orleans waterworks, getting the right engine, boilers and other equipment made and shipped to Louisiana plagued the last nine years of his life. A veritable odyssey took him to Washington,

Pittsburgh, New York and ultimately Baltimore, whence a used engine and a mill for boring wooden pipes were sent south in 1818–19. For this enthusiasm he paid the ultimate price. 'With the suction pipe laid, the engine in operation, pipes in the streets, and (as his wife later claimed), water actually flowing through the streets, Benjamin Henry Latrobe died on 3 September 1820.'[319]

One by-product of the steam engines touched Latrobe's architecture. While he was planning the two engines for the Philadelphia waterworks with Nicholas Roosevelt and his engine-builder, James Smallman, they hatched the idea of building a rolling mill next to the Schuylkill engine and using some of its spare capacity to supply a rising demand for sheet iron (ill. 151). It was modelled on steam-powered English rolling mills built by John Wilkinson with Boulton and Watt in the 1790s, which Smallman may have known.

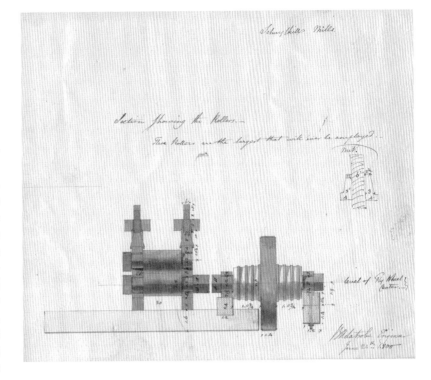

151. Section of the rollers for the Schuylkill Mills, Philadelphia Waterworks, signed 'B. H. Latrobe Engineer June 24th 1800'. One of a set of drawings manifesting Latrobe's skills in mechanical design.

Latrobe went ahead and designed a set of rollers. The mill was duly constructed, starting up production in Autumn 1802. Plagued by debt and technical troubles, it soon failed and had to be taken over by the city of Philadelphia. While it lasted, though, Latrobe promulgated roofs of Schuylkill iron. Jefferson agreed to buy some for Monticello ('the President's iron') in 1803; while Latrobe specified sheet-iron roofs for two minor works of his own, Nassau Hall, Princeton, and the Bank of Columbia, Georgetown.[320]

Where iron was concerned, therefore, this foremost of early American architect-engineers ended up behaving more like an entrepreneur than the professional he was proudest to be. Latrobe placed high hopes in iron for the technical development of his adopted country. But what possessed him was its potential for machinery, not for structures. In his buildings he made only pragmatic use of the material, as ingenuity and occasion allowed.

CAST IRON IN NEW YORK

In so far as Americans experimented with iron for fixed structures during Latrobe's lifetime, Pennsylvania, the richest of the founding states in sources of ore, lay at the heart of things. Here it was that James Finley, the inventor-farmer, devised his cheap suspension bridges for rural communities from about 1800 (pp. 315–7). Matters had gone little further. Only in the 1820s did the Philadelphia architects William Strickland and James Haviland start slipping cast-iron columns and floors into their buildings.[321]

The changes of that decade can be traced back in part to the suspension of British imports during the War of 1812–14. That prompted urgent new investment in industrial plant, not least the iron trades, and the updating of foundry techniques. There followed the making of the Erie Canal (1817–25), a turning point in American construction. Its scale and range, along with the mistakes that were made in laying it out, rendered the canal, so it has been said, a practical school of civil engineering. Once completed, it transformed communication between the coast and the iron-ore districts of up-state New York and, indirectly, Pennsylvania. Now the twin cities of New York and Brooklyn leapt ahead in shipbuilding and in commerce. Foundries burgeoned along either bank of the East River. Soon after the Erie Canal opened, gas and water mains in Manhattan switched from timber to iron, and the first iron shop-fronts appeared.[322] Hence over time

emerged the culture of cast-iron fronts which today gives its name to a district of New York. Consonant with American circumstance, it was driven by patents and competition, and only marginally connected with the professions.

Cast-iron building-construction began with the fabricator-founders of the waterfront and matured in the hands of an inventor-mechanic, against a backdrop of growing anxiety about urban fires. A seventeen-block fire on Manhattan's Lower East Side in December 1835 confronted New Yorkers with the issue. That same year Jordan L. Mott built a foundry for fabricating storefronts. A patent for casting columns followed. Soon Mott was displaying his 'complete iron storefront', that is to say a multi-storey façade, to

the American Institute, one of those public-spirited businessmen's societies common to English-speaking cities in those days.[323] But the take-up was slow. Not until 1846 did the Boston 'ironsmith', Daniel Badger, most prolific of the New York cast-iron builder-fabricators, open his Manhattan foundry, later pretentiously termed the 'Architectural Iron Works of the City of New York' (ill. 152).[324] At first Badger made rolling iron shutters and only started creating complete iron fronts and buildings from 1853.[325] That puts him behind and in debt to the most absorbing figure in the story, James Bogardus.

Thanks to Turpin C. Bannister and Margot Gayle, we know a fair amount about Bogardus. By origin a watchmaker and die-cutter from the Catskills, he was employed in

153. The Sun Iron Building, Baltimore, around 1900. Robert G. Hatfield, architect, with James Bogardus, inventor-engineer, 1850–1.

New York by 1828 and frequenting the American Institute. During the late 1830s he was in London trying to market an engraving machine; he also went to Italy where 'the architectural designs of antiquity' impressed him. If, as was later claimed, Bogardus thought there and then of 'emulating them in modern times by the aid of cast-iron',[326] maybe he knew about Mott's presentation to the American Institute, made while he was away. By 1841, at any rate, he was back in New York peddling a fresh invention, a grist mill.

Bogardus began then as an inventor of mechanical bent, expert in gearing. His shift into building with iron stemmed, say the Gayles, from a model he made for housing his grist-mill machinery. His first known building work consisted of a shop-front and extra storeys in 1848 for a Broadway pharmacy. Larger jobs followed, until by 1851 Bogardus and his investor-partner Hamilton Hoppin could advertise themselves as 'eccentric mill makers, architects in Iron', with premises in the Manhattan ironworks district. Though he undertook some pre-assembly at his works, Bogardus was never a primary fabricator. His buildings were put together from components cast by different founders, notably James L. Jackson but sometimes also Badger.[327] The inventor's contribution was to work out the 'system' of four basic cast-iron elements: column, trough-beam, spandrel panel and cornice. These were the core of his patent of 1850. The resulting façades were by no means minimal. In order for Bogardus to compete with New York's latest commercial palazzi in stone or stucco, they boasted copious Italianate decoration: starbursts, 'rinceaux', Medusa's heads and the like. We do not know who designed them, but any carver able to make a mould for stucco could do the same for cast iron.[328]

If all the above has been about fronts, that is because Bogardus and the founder-fabricators started from façades. There a show of smartness, impermeability and, so it was imagined, fireproof-ness could be put on at a lower cost than masonry. But soon they worked inwards, from 'architecture' to 'engineering'.

The first Bogardus building supposed to have had floors and internal walls of iron was his own works in Duane Street (1849–50) – no doubt a showcase for his system, but demolished after a decade. Better recorded is the Sun Iron Building, erected for a leading Baltimore newspaper in 1850–1 (ill. 153). There the constraints of the system collided with the needs of a nineteenth-century printing building, housing heavy presses in the basement, rental premises on the middle floors, and journalists and compositors on the top storeys. Hence collaboration and compromise: according to David Wright's study of the Sun Iron Building, 'no one individual was in total control'.[329] Since Baltimore was a city with an iron-founding culture of its own, the proprietors of *The Sun* insisted that the procurement of cast iron be shared between local firms. Bogardus's main contribution was the front. For this he teamed up with Robert G. Hatfield, a young New York architect-engineer. Like other technicians of the transition between cast-iron fronts and skyscraper frames, Hatfield is hard to classify. He went on to specialize in iron buildings, wrote about fireproof flooring-systems and aspects of structural technique, and designed the first trainshed roof for Grand Central Station (1870–1).[330] The façade of the Sun building, a showy confection dotted with life-size iron statuary, must have stretched and tested the Bogardus language to the limits. Inside, the plan was sorted out by a Baltimore architect, William Reason, while the structure mixed cast-iron columns from the Bogardus range with timber floors.

In a better-known printers' and publishers' building in Manhattan, different hands again combined. The Harper Building (1854–5) is prophetic of the distributed relations that would attend the making of skyscrapers (ills. 154, 155, 156).[331] The burning of Harper and Brothers' premises in 1853 convinced John Harper that their replacement should be iron-framed, with heavy, fireproof floors. Structural technique just then was on the move. Since the Sun building, the Crystal Palace had been raised in London and copied on a smaller scale in New York (1853). British cast-iron technology, on which Bogardus, Badger and the New York founders based their construction practices, was starting to be supplemented or supplanted by wrought iron.[332] That had much to do with

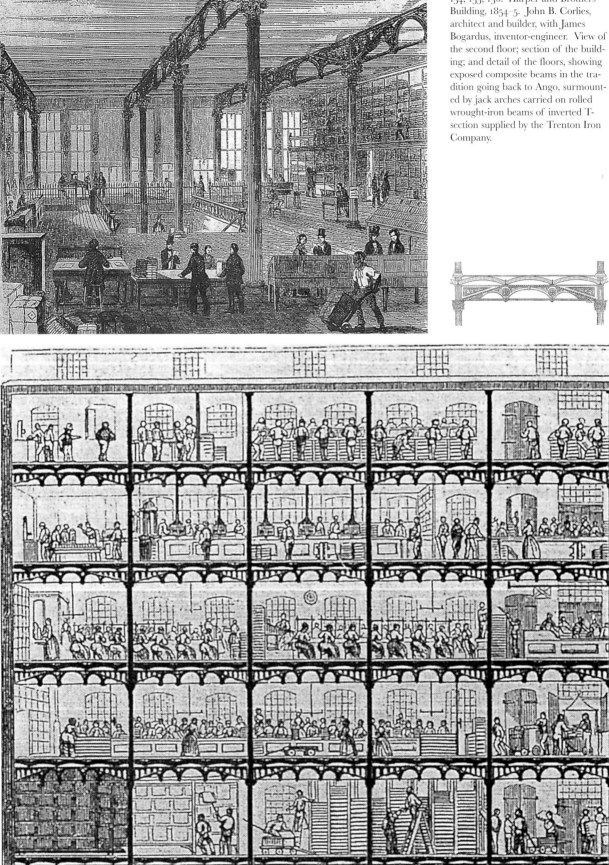

154, 155, 156. Harper and Brothers'
Building, 1854–5. John B. Corlies,
architect and builder, with James
Bogardus, inventor-engineer. View of
the second floor; section of the building; and detail of the floors, showing
exposed composite beams in the tradition going back to Ango, surmounted by jack arches carried on rolled
wrought-iron beams of inverted T-
section supplied by the Trenton Iron
Company.

the awakening of American railroads in the 1850s. Railroad engineers were starting to devise their own truss-forms, while better-capitalized fabricators invested in large premises and new rolling mills for rails and other wrought-iron sections along railway routes outside the cities, just as the 'ingénieurs-constructeurs' shifted to the outskirts of Paris.

The Harper building (in fact two buildings linked by a common staircase and court) had side walls of masonry but a complete internal iron frame and one fancy iron front for show. As in Baltimore, we cannot be sure that Bogardus controlled more than this main façade, which repeated the arcaded language of the Sun Iron Building. The interior layout was due to the otherwise obscure John B. Corlies, 'architect and builder'. Its cast-iron elements, meaning the ornamental columns and cast portions of the composite main girders, were designed (so it was reported) as well as manufactured by Jackson, Bogardus's most regular founder. On top of these came fireproof jack-arch floors, supported by a novelty: inverted 'deck beams' of wrought iron in T-sections, manufactured by Peter Cooper and Abram Hewitt of the go-ahead Trenton Iron Company, using newly invented rollers able to cope with deeper flanged beams.

If rehearsing the various contributors' names for the Sun Iron and Harper Buildings is tedious, it helps to scout the idea of a unifying technical or aesthetic mind behind their construction. Though for the fronts Bogardus suggests himself as a candidate, for their appearance he relied much upon Hatfield. Inside the buildings, a medley of collaborators present themselves. Perhaps the closest to a controlling intelligence at Harpers was that of the proprietor himself. Corlies, the architect-builder, may have been as much project-manager as designer. If that is right, no architect or engineer in the modern conception of those professions is at work here. Instead we witness patentees, designers, fabricators and subcontractors banding together to erect new-fangled but wholly pragmatic buildings.

As an innovator, it has well been argued, Bogardus's best claim to fame depends not upon his commercial fronts but upon his original series of iron-framed shot towers and fire towers of the 1850s.[333] But the present enquiry is less concerned with such pure 'engineering' projects than with commercial buildings, which better convey connections and tensions between the professions and more typical patterns of procurement. In that respect, Manhattan's early cast-iron fronts represented just a start. The New York school of iron fabricator-contractors flourished and expanded up until the Civil War and well beyond.[334] In these years the vogue for cast-iron fronts all but dominated commercial premises in the city, and spread all over America.

Even after demolitions, New York boasted a staggering 2.75 miles of iron frontage in 1943, claims Bannister. Façades were the main focus, often of 'excellent quality' and architect-designed. Badger's Architectural Iron Works set the trend. His extant Haughwout Building (1856–7), famous for its Sansovinian cast-iron front by J. L. Gaynor and for including a passenger elevator from the start, is a fine early example (ill. 157). Badger worked with many architects, since clients often began with their own designer. But the package could equally well begin with the contractor or fabricator, the architect trailing along behind.[335] If the client had no designer in mind at all, Badger could furnish one in the shape of George H. Johnson, a versatile young architect-technician whom we shall meet again in the context of fireproofing. Johnson could as well design fancy architectural fronts or the fairly complete iron frames now becoming feasible, among them the Singer Factory (1857) and US Warehousing Grain Elevator, Brooklyn (1860).[336]

As commercial buildings soared higher and grew in prestige, New York's architects were able to fight back, contesting the ground they had lost to the cast-iron lobby. Not that they had been innocent of iron hitherto. In many architect-controlled buildings of the late 1840s and 1850s it had started to thread its way into internal structures.[337] But for those who took their art seriously, the cheap, showy cast-iron fronts were an intolerable vulgarity. Luckily for them, iron – cast iron in particular – turned out not to possess the fireproof virtues it had promised. That, coupled with the stirrings of improved education for construction professionals, allowed architects to get a grip on

the market for commercial buildings after the Civil War, and to profit from the iron revolution without dropping the dignity of masonry fronts, status-giving for their owners and creators alike.

The key figure in this recovery was George B. Post, confirmed by Sarah Landau's researches as the leading architect of early New York skyscrapers.[338] Post had been school-trained at New York University, where a course in civil engineering with a modicum of architecture had started in 1854, just before he arrived.[339] On graduating he joined the small 'atelier' of Richard Morris Hunt, the first New York architect to trade exclusively on art as the means to eminence in that city, following his own 'formation' at the Ecole des Beaux-Arts. A balance between parallel school-based approaches – the engineering technique learnt at NYU and the Beaux-Arts pride in art fostered in Hunt's atelier – marked Post's whole subsequent career. It is reflected in Manhattan's best commercial architecture of the Gilded Age.

To gauge that balance, just two of Post's achievements need be mentioned. The first, the Equitable Building of 1868–70, figures in the histories as a step in the much-studied

158. The Equitable Building.
Gilman and Kendall, architects, with
George B. Post, architect-engineer,
1868–70 etc. This view shows the
building after 1876, when the roof
had already been raised, but before
Post made a series of large additions
from 1885.

evolution of the iron frame (ill. 158). That frame was due to Post; but the richly modelled masonry elevations, towering for their day at seven-and-a-half storeys, were not. There had been a limited design competition, itself indicative of post-bellum progress for architects in the commercial field. After havering between two entries, the building committee awarded the façades to Gilman and Kendall but the ironwork, elevators and vaults to Post, who also superintended construction.[340] In 1868, therefore, the notion that a major office building should be a single concept emanating from a single intelligence had still to make headway. The division of labour seems not to have bothered Post. Effectively he acted as the project's engineer, busying himself with other specialists in perfecting the building's wrought-iron frame and advanced services. In later additions to the Equitable, Post featured once more; and in a gesture anticipating Chicago practice, he rented the suite atop the building for his own office.

159. New York Produce Exchange, exterior. George B. Post, architect, 1881–4.

160. New York Produce Exchange, the internal 'cage' of the exchange room in construction, 1883.

If the Equitable made Post's reputation, his crowning achievement was the New York Produce Exchange of 1881–4 (ills. 159, 160).[341] Again there was a limited competition, this time won fair and square by Post. But if designing the inside and the outside was now united, their character continued to be divided. That is the reproach which modernist chroniclers of the skeleton frame used to cast in the teeth of New York's early skyscrapers. Within, a complete iron frame or 'cage' and spectacular toplit court; without, an overwhelming brick front towards Bowling Green of thirteen broad bays' width and ten storeys' height, broken into bands of changing scale. In this august front are premonitions of Richardson's and Adler and Sullivan's round-arched masterpieces.

Once again, the internal structure of the Produce Exchange forwarded the metal-framing of buildings. Post was proud of that. But he remained adamant about differentiating exterior and interior. His reasons were partly technical: 'I have never enclosed a cage in solid mason work. I have never dared to. I have always built the cage detached inside, anchoring the walls to it, so that in the cage in case of corrosion, it could be painted and repaired', he averred in 1894.[342] Aesthetics were also relevant. Outside, monumentality and public uplift – the ideal of the Beaux-Arts architect; inside, efficient construction and maximum rentable space – the province of the proficient engineer. Because of Post's equal capacities as artist and technician, he could see both sides of the coin and flip it. His stance is a reminder that an insistence on revealing structure is an aesthetic not a technical position.

CHICAGO: THE BACKGROUND

We know when Chicago felt ready for 'architecture', because John Van Osdel has pinpointed the moment. Forty years after Fort Dearborn was founded, in the winter of 1844, wrote Van Osdel (then a young carpenter-architect),

> when builders were their own architects, some leading builders proposed to me that I open an architect's office pledging themselves not to make any drawing or construct any building of importance without a plan. With this promise I undertook to do so . . . . No one had ever used an architect and it was difficult to convince the owners of the necessity of such a branch of the building business.[343]

Things had moved on from the days when the Carpenters' Company of Philadelphia, so Latrobe believed, blocked him from practising as an independent designer.

Shift the picture on again a generation. Chicago boasts a rich cultural and religious life to balance its booming trade, and abounds in villas with ample gardens. Parks to rival London, Paris and New York are being planned; the boosterish name 'Garden City' is going the rounds; Van Osdel is prospering, as are others in his wake. Here migrating architects from Europe or the East Coast may hope to make their mark.

Among the hopefuls is Major W. L. B. Jenney. Thirty-six when he arrives in 1867, he is ready to marry and settle down after a roving youth. Jenney goes into partnership with Sanford E. Loring, a former assistant of Van Osdel's. The pair soon get work and by way of advertisement issue a book of designs. Its preface offers a little homily on architecture for the local bourgeoisie. 'The first step towards building should be to send for an architect . . . It pays to pay an architect for spending time in developing one's wants on paper before the work is commenced':[344] that is the core message. Soon after the book's publication Loring drops out, to develop a terracotta business.[345] Now on his own, Jenney becomes chief engineer to the West Chicago Park Commissioners, and architect to the affluent suburb of Riverside. In 1870 it is announced that three new parks for West Chicago will be created by the consortium of 'Jenney, Schermerhorn and Bogart, Architects and Engineers': Jenney will design the layout and the park buildings, while the Chicago engineer L. Y. Schermerhorn (in partnership with John Bogart of New York) is to sort out their infrastructure.[346]

161. Reconstruction in Chicago after the Great Fire, 1872. Prominent is the front of Burling & Adler's Marine Building at the corner of LaSalle and Lake Streets, Chicago, with iron construction visible behind.

Faint mystery attaches to Jenney. Honoured in his working lifetime for pioneering the skeleton frame (a steamship was even named after him on that account[347]), he has not escaped the odour of plausible adventurer. In a snide swipe at his first Chicago employer, Louis Sullivan judged Jenney 'a free-and-easy cultured gentleman, but not an architect except by courtesy of terms. His true profession was that of engineer'; only to add a few lines later that 'the Major was not, really, in his heart, an engineer at all, but by nature, and in toto, a bon vivant, a gourmet'. That might be dismissed, had P. B. Wight not likewise remarked that Jenney 'could talk building better than anyone else I knew, but he knew very little how to construct and design then [in the 1880s] and depended on others'.[348]

Be that as it may, Jenney had enjoyed as robust a technical and practical training as any American architect of his generation. Born into a Massachusetts whaling family, he got his start in engineering at the Lawrence Scientific School at Harvard before transferring to the Ecole Centrale des Arts et Manufactures in 1853 – a rare move then for an American. He took the entire construction course, emerged a fully fledged 'centralien', and passed some two more years in Paris, on and off. Just before the Civil War, Jenney was acting as a consultant to a company channelling European investment into American railroad construction; William Sherman was its president. Hostilities then opened, and on Sherman's advice he joined up.

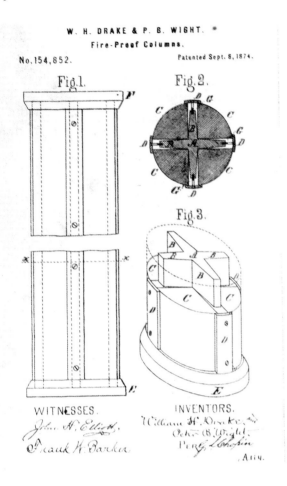

Fig.1.   Fig.2.

Fig.3.

WITNESSES.   INVENTORS.
John H. Elliott,   *William H. Drake,*
Frank H. Barker.   *Peter B. Wight,*
                             Attg.

162. Fireproof column patented by
W. H. Drake and P. B. Wight, 1874.
A cruciform iron core is encased in
hardwood.

Like the War of 1812, the Civil War gripped the United States in a technical crisis, but of larger proportions and longer-term consequences for construction. Armaments had to be made, troops equipped, transported and fed, roads and railways repaired or laid down, forts and depots thrown up, bridges built and rebuilt: all with a terrible urgency. Hence a swift expansion of the US Corps of Engineers and its merging in 1863 with the Corps of Topographical Engineers. The military-industrial effort mobilized in the North to stave off defeat and then crush rebellion transformed sensibilities. The lessons, like the traumas, were not to be forgotten. 'The engineer corps of the army during the war', remembered Jenney,

> was an excellent school to learn expedients. The problems as usually presented were to rebuild a bridge burned by the enemy; to construct a pontoon across a wide, rapid stream; to fortify a mile or so of front. Usually only a few hours were allowed for this work. The army furnished any number of men and army wagons, a limited number of tools and sometimes boats. All materials must be found in the neighborhood . . . The means were always found and the work was done.[349]

On the strength of his training and connections, Jenney rose from lieutenant to captain and finally major, learning in the process to lead and to delegate. At the end of the war he was active in Sherman's march to Atlanta and claimed to have been head of engineering in that devastated city.[350]

On one of his tours of inspection for the Sanitary Commission, F. L. Olmsted bumped into Captain Jenney superintending a military canal near Vicksburg. He has left us a glimpse of him. 'He has had a half artist education in Paris', the great landscapist wrote home, 'and was warm on parks, pictures, architects, engineers and artists.'[351] Peace once re-established, Jenney wrote to Olmsted touting for work, bemoaning the 'little knowledge and little desire for Art' in the Mid-West, and seeking a situation 'in which Architecture, Gardening and Engineering were associated'.[352] At that juncture Olmsted could or would do little for him. But that Jenney, once set up in Chicago, acquired a role in the Olmsted-designed suburb of Riverside and the layout of the West Chicago parks must surely have to do with their Civil War acquaintance.[353]

The whiff of art-dilettantism about Jenney's early years in Chicago is belied by his later career. If his change of direction was gradual, its ultimate cause was abrupt. The great fire of October 1871 shifted the centre of gravity for architectural practice in the city. The business district, worst hit, had to be urgently rebuilt. That meant iron, and recourse to the fast-track 'expedients' of war. A famous photograph shows a building (by Burling and Adler) well advanced just months after the fire. Its masonry front asserts civic pride regenerate; behind lurks the ghost of an iron frame (ill. 161). Then after eighteen months' frenetic building, confidence collapsed. The ensuing slump lasted until the end of the 1870s. But the focus of endeavour remained steady. When recovery came at last, central Chicago thrust outwards and upwards. Famously, the continuities of the 1880s fostered a frame of mind and a style of mercantile architecture in which the skin of a building was increasingly determined by its bones. Those like Jenney whose skills straddled architecture and engineering came to the fore.

To retell the story of the 'Chicago school of architecture' lies beyond present purposes. What belongs to them is to review some of those who promoted certain advances or deployments of technique which underpinned that 'school', looking at their backgrounds, motivations and interactions. To do so, we must begin with the aftermath of 1871, when there was little originality to boast about in Chicago's commercial architecture.

THREE AREAS OF TECHNIQUE

A starting point is the catastrophe itself. How were such events (another bad fire smote Boston in 1872) to be guarded against? Not just by building in iron, experts knew. In the dawn of cast-iron fronts the material had been advertised as indestructible. By the Civil War, losses and disasters (like the burning of the New York Crystal Palace in 1858) had scuppered that claim. Iron stayed the material of choice for fast and cheap building with maximum rentable floor space. But it needed covering.

A spate of new methods for fireproofing buildings were patented after the Chicago Fire. Scores of such patents had been issued in Europe and America since the 1830s, mostly for flooring systems.[354] They emanated from inventors, amateurs, fabricators and builders, plus a smattering of independent architects and engineers. If these last were in a minority, that may be because a patentee had to be able to fabricate and market the system if he were to further his invention. The professional's best recourse was to associate a fabricator in the application, or hope one would take it up once the patent was issued.

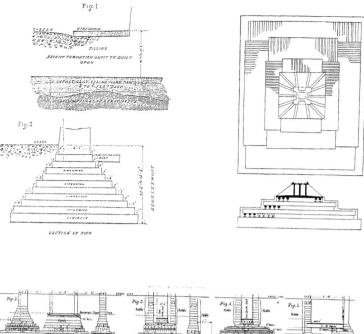

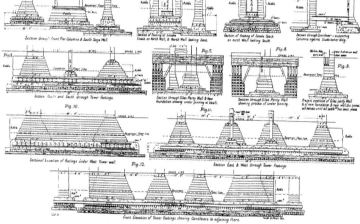

This veneer between commerce and the professions, always thinner in the United States than Europe, proved specially permeable in the wake of the Chicago Fire. Three architect-technicians now patented and, harder, successfully marketed types of flooring and cladding, mostly in light brick or terracotta. Two of them came from New York, where increasing thought had been devoted to fireproofing iron. In 1871 Daniel Badger's former designer George H. Johnson, working with Balthasar Kreischer, a firebrick manufacturer from Staten Island, registered a tiling system for floors, perhaps on the basis of French precedents. After the fire, Johnson travelled to Chicago to market fireproofing materials, soon extending the Johnson-Kreischer range to hollow-tile floor voussoirs and partitions as well as to the cladding of columns.[355] Then in 1874 Peter B. Wight, formerly also of New York, and his partner William Drake patented a fireproof column clad in hard oak (ill. 162) In the same year Jenney's old partner Sanford Loring, of the Chicago Terra Cotta Company, which had been active in reconstruction work, patented a porous terracotta, lighter and less brittle. This Wight promptly adopted instead of oak for his columns.[356]

The most intriguing of these patentees was Wight. His career contradicts the rule that chic architects do not forsake the high claims of their calling for business. Far from a technician, during his early New York career Wight had been an art-architect of Ruskinian persuasion. Among the motives that drew him into fireproofing seems to have been distaste for that city's sham cast-iron fronts. Lecturing in 1869 about iron, he sounded a

Skyscraper foundations in Chicago:

TOP LEFT
163. Jenney's Home Insurance Building, showing isolated pier foundations with pyramidal footings consisting of alternating cut masonry and rubble courses on a concrete base.

TOP RIGHT
164. Burnham and Root's Rookery Building. Plan and section of footings, showing a cast-iron shoe atop grilles of steel rails encased in concrete.

BOTTOM
165. Adler and Sullivan's Auditorium Building. Here the two systems are combined and developed. The steel grillage extends in a continuous raft under the footings of the towers, and everything sits on heavy baulks of timber. Drawings courteously lent by Joseph Siry.

166. Thomas Tallmadge venerates the frame of the Home Insurance Building during its demolition, 1931.

167. Home Insurance Building. William L. Jenney, architect, 1884–5. Colour lithograph by Louis L. Prang.

defensive note: 'We are . . . forced to use a material which, though not combustible of itself, will do little work if exposed to heat . . . it must be conceded that with the best we can do with this material, there is danger.'[357]

When the Chicago fire struck, Wight was going through a lean patch in New York. Within a month he had found a berth in the Chicago architectural partnership of Carter and Drake, inspected the ruins and addressed the American Institute of Architects' convention on the reasons and remedies for the destruction. Hard on his heels came John Root, summoned from New York as Wight's office foreman.[358] So intensive was the initial reconstruction that Carter, Drake and Wight designed over fifty buildings in central Chicago between 1871 and 1873.

The slump allowed Wight to secure and advance his fire-proof patent. Slowly he withdrew from designing complete buildings into the role of fireproofing contractor and structural consultant. The setting-up of the Wight Fireproofing Company in 1881 completed the process. All through the ensuing boom in which the skyscraper took shape, Wight fulfilled fireproofing subcontracts for iron-framed commercial architecture in cities of the Mid West. In Chicago he took on contracts for Burnham and Root (previously both his assistants), Adler and Sullivan, and others; and he did the fireproofing for Jenney's Home Insurance Building. Eventually cheaper competitors emerged, and concrete flooring systems made inroads upon the lighter hollow tile which Wight championed. So in the 1890s he changed tack once more, relinquishing his business for architectural journalism.

To act as a subcontractor for a building designed by your former assistants was perhaps unique. But though Wight tired of being an architect, he never tired of architecture. His chequered career is best seen as a symptom of the subdivision and rescrambling of skills and tasks attending the birth of the skyscraper and its affiliates. Bowing to economic and technical circumstances, Wight acquiesced in and maybe relished that fragmentation.

In a second area of commercial construction technique, less tied to patents, Chicago architects were equally active in innovation until about 1885. That was foundation technology, crucial if buildings were to be higher and bigger (ills. 163–5).[359] This sphere of expertise developed in tandem with the increasing point-loads exerted by framed buildings upon footings. The Chicago soil consisted of a shallow layer of good ground over untrustworthy sand, clay, gravel and water, with bedrock far beneath at levels then more or less unreachable.[360] If you built anything heavy in the city, you needed fair structural know-how to prevent your building from subsiding unevenly. That was among the reasons why architects with engineering capacity rose to the top there. But before 1890 instinct and canniness came before exact calculation in designing foundations, while building regulations allowed a degree of latitude. The successful Chicago architect, wrote one journalist, had to be 'permeated by the peculiar surroundings of our locality . . . in sympathy with the soil, so to speak, . . . adscriptus glebae . . . as much a product of the soil as one of our scrub oaks, or one of our cotton woods.'[361]

The sharing of local science on how to tackle this terrain goes back to *The Art of Preparing Foundations for All Kinds of Buildings, with Particular Illustration of the 'Method of Isolated Piers,' as followed in Chicago*, a tract issued in 1873 by Frederick Baumann, 'architect'. Detached, caustic, 'educated in Germany to the point of cynicism' (so Louis Sullivan put it),[362] Baumann is an éminence grise in the Chicago story, and among the clues to the manifold impact of German habits in technique and art upon the city's cul-

THE CHICAGO BUILDING OF THE HOME INSURANCE CO.
OF NEW YORK

ture.[363] Having trained in 'architecture and building' at the prestigious technical schools of Berlin, he was among those who decamped after the revolutions of 1848–9 and made their way to Chicago. At first Baumann worked mainly as an architect (he was briefly in partnership with Van Osdel), but he also spent a spell as a building contractor. Later, on the eve of Jenney's Home Insurance Building, he published his short *Improvements in the Construction of Tall Buildings* (1884) which, it has been argued, first precisely defined the concept of the independent iron-framed tall building.[364] If Baumann was not notably a creative figure, he was no mere technician, for he could expound the ideas of Schinkel and Semper and was close friends with Sullivan.[365]

Baumann argued that foundations should consist of 'isolated' footings under each pier or wall, and not form part of a rigid overall raft which might split or tilt in case of uneven settlement. The base area for each footing was then to be calculated from the load it had to carry and the bearing capacity of the soil.[366] Baumann's method seems to have been taken up in fits and starts.[367] Adler and Sullivan adopted elements of it during the 1880s, as did Jenney (ill. 163). But as buildings grew taller and heavier, the pyramidal stone footings enlarged and started to invade basements, which were needed for mechanical and other plant. Without piling, they could not be dropped below the level of the compact upper layer of ground, which stopped at an average of fifteen feet below grade.

Enter therefore a second architect, P. B. Wight's former protégé, John Root. In the sequence of high and heavy buildings he designed during the Burnham and Root practice's heyday, starting with the ten-storey Montauk Block of 1881–2, Root devised and extended a system of shallower footings, each consisting of a grille of steel rails in tiers encased in cement upon a pad of concrete. In early versions the pad was unreinforced (ill. 164), but for the massive Monadnock Building, among the last things Root designed before his death in 1891, the bridge engineer and foundations expert William Sooy Smith was called in and advised reinforcing the pad itself with steel. Besides this system for footings, Root was among the first to cantilever out the flanks of foundations in tall structures to avoid disturbing neighbouring party walls, as was done in the Rand McNally Building of 1888–90.[368]

Root saw himself as much as an artist as an engineer; we shall return to the balance he struck between art and technique. To complete the foundation story, during the 1880s it grew clearer that footings in the upper soil would never by themselves withstand the rampant height and concentrated pier-weight of commercial buildings in the Chicago Loop. Thereafter permanent piling came back into play, as mere architectural ingenuity withdrew before the technology of deep caissons, taken from bridge engineering. By the mid-1890s foundations, like other aspects of skyscraper technique, were becoming subject to strict building codes, and therefore proof against casual empiricism.

The change was gradual.[369] It owed the most to William Sooy Smith, the consultant brought in on the Monadnock Building. Smith was a West Point-educated engineer of railroad bridges who had reached the rank of brigadier-general during the Civil War interlude. One of his early bridges across the Savannah, started in 1859, is supposed to have made the first major American use of pneumatic caissons.[370] From about 1885 he was much consulted on foundations for Chicago buildings, in the first instance notably by Dankmar Adler. The pattern was for Smith to test the soil and offer advice, on which basis Adler himself then designed the foundations. At the Auditorium (1886–9), Smith deep-piled the entire ground as a testing device, to ascertain the pressure it would stand; Adler then set out the footings, mixing Baumann's and Root's methods with notions of his own (ill. 165).[371] In later Adler and Sullivan buildings for which Smith's advice was sought, like the Schiller Theater and the Stock Exchange, wooden piles driven to about fifty feet underlay the footings; at the latter (1892–3), to avoid damage to a next-door building during piling, one of the lines of piles was replaced with Chicago's first concrete-filled caissons – the method which gradually prevailed in skyscraper buildings thereafter.[372] The regularity with which

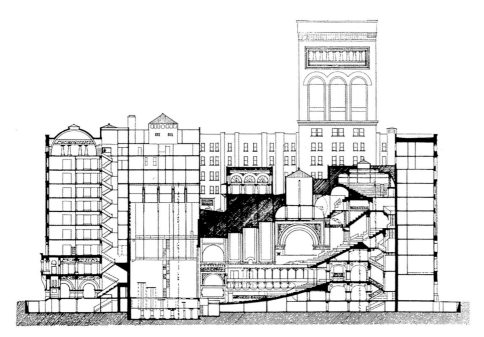

168. The Auditorium Building.
Adler and Sullivan, architects,
1886–9. An early view of the completed building.

169. The Auditorium Building.
Section showing the integration of a
vast theatre with stage and flytower
into a hotel and office building.

Smith was consulted on foundations in Chicago up to his retirement in 1910 confirms the shift in this area of building technique from architectural empiricism to the accumulated experience of bridge engineering.

From fireproofing and foundations it is time to turn to the frame itself, and judge what balance of professional skills drew it to completion in the 1880s. The convention has been to ascribe the great Chicago advances to architects, with a measure of technical assistance from engineers after steel replaced iron. But recent scholarship has shifted the spotlight on to that third party who has run through this chapter like a dark horse, the fabricator.

Louis Sullivan summarized the position like this:

> the Chicago activity in erecting high buildings finally attracted the attention of the local sales managers of Eastern rolling mills; and their engineers were set at work . . . It was a matter of vision in salesmanship based upon engineering imagination and technique.[373]

Sullivan's remark has now been filled out by a chapter of Thomas J. Misa's *A Nation of Steel*, which offers a gripping account of the early Chicago skyscraper from the supply side.[374] Misa sets the steel frame's emergence against the backdrop of a predatory American iron industry, diversifying successively into rails, armour and automobiles – besides construction. By the mid-1880s the demand for rails that had buoyed ironmakers since the Civil War was near its peak. The switch from rails of wrought iron to ones of steel was all but complete. Go-ahead companies like the Carnegie-owned concerns around Pittsburgh therefore sought out fresh markets to justify their investment in steel-making plant and the special labour skills it entailed. Parallel with their volume business in rails had gone the fabrication of iron and steel railway structures, notably bridges and elevated tracks. That drew the companies into employing in-house or consultant structural engineers of their own. A move into building construction followed logically. A steel building-frame, it is occasionally said, was not really so different from a bridge or elevated track set on end; early framing technique was sometimes known indeed as bridge construction.

Not that the birth of the modern skeleton frame in the 1880s can be equated with the switch from iron to steel. Though the processes overlapped and connected, they were distinct in principle. The frame reached its logical conclusion in Chicago, after a century of evolution, because entrepreneurs there pressed hard for rapidly built buildings offering maximum rentable floor area. The main components of such a frame did not have to be of steel. During the 1880s and 1890s many building structures mixed steel, wrought iron and cast iron. Nevertheless steel with its thinner, lighter and stronger sections took up less space than iron, resisted fire better and imposed a smaller load on footings. It was the future. But it took time, steady demand and fresh know-how to respond to its properties. Whether and how steel was used in the early skyscraper era depended on its cost (generally above iron in the 1880s but falling) and its supply at a given moment. Not least, it hung on the availability of one of the few experts, preferably bridge engineers, who could as yet translate a building's needs into the language of the steel fabricator.

Misa looks at two famous Chicago buildings of the mid-1880s in the light of overtures from the Carnegie companies. The earlier is Jenney's Home Insurance Building (1884–5), that recurrent orientation-point in histories of the Chicago skyscraper (ills. 166, 167). For Misa, the Home Insurance Building matters not because it may or may not mark the true commencement of the new type (the fact is that it was so regarded[375]), but because it represents 'the first salvo by Pittsburgh steelmakers in taking over the Chicago structural trade'.[376] The ironmakers of North Chicago had been slow in diversifying from rails into major structural components for the building industry, even in wrought iron let alone steel. So for some years, Jenney had been cultivating relations with iron fabricators from all over the Mid West, inviting them to his office to 'figure' when tendering for major buildings.[377] Part of the brief for the Home Insurance Building was that its piers should be exceptionally narrow. This may have been the condition that generated contact between Carnegie Steel and Jenney's office. As a result, Pittsburgh steel was substituted for wrought iron on the original top three floors – only. A tentative début: and the Home Insurance management insisted that a separate consultant engineer recheck all the calculations. But once the building was up, there was no holding Carnegie or steel. 'No fewer than seventeen of the twenty-nine classic Chicago School buildings constructed from 1885 to 1895 used the Carnegie company's steel columns and beams', remarks Misa.

We can take on trust Jenney's account that he personally set the conception for carrying most of the outer walls of the Home Insurance Building upon its frame. Yet Sullivan and Wight have told us that he relied heavily upon assistants. Who then actually 'designed' the frame (and in such a pared-down building there was little more than the frame and its consequences to design)? The likely answer is that it was worked out by office assistants in collaboration with the fabricators. We have the name of Jenney's principal technician for the building: George B. Whitney.[378] A graduate of the University of Michigan and former bridge engineer, Whitney was the type of expert who could talk the Carnegie language. Once steel creeps in, some such engineer with outside experience can usually be found in the penumbra of the framed building, in or out of the architectural office, mediating between architect and fabricator. Soon, in Joseph Freitag's words, the architect will start to 'turn the details, if not the entire constructional scheme, into the hands of the engineer, either as an employé or co-partner'.[379]

Such a figure or figures can be detected in the second structure examined by Misa. A mightier work of architecture altogether than the niggardly Home Insurance Building, the Chicago Auditorium by Adler and Sullivan (1886–9) was no minimal commercial frame but something altogether more taxing: a complex of half-block size, incorporating a hotel, offices, a tower, and a vast, wide-span theatre.[380] Metal-framed in part, it also boasted load-bearing external walls of meditated depth and richness (ills. 168, 169).

It was once believed that 'there were no consulting engineers on the Auditorium. Except for minor details, Adler did the whole job.'[381] To the contrary, reveals a memorandum from Adler's 'Man Friday', Paul Mueller, a slew of consultants poured in.[382] Two, Charles L. Strobel and Edgar Marburg, acted for Carnegie. As Carnegie Steel and its subsidiary, the Keystone Bridge Company, supplied the steel for the Auditorium, it might be imagined that Strobel and Marburg were mere subcontractors' representatives. In fact both were high-class engineers with a roll-call of bridge designs to their credit, and experience in managing as well as inventing structures.

Yet another of those German-trained technicians who did so much for American construction, Charles Strobel was perhaps the crucial figure in the transfer of steel technology from bridges to buildings.[383] During his years with Carnegie he rewrote the group's handbook for structural steel and devised an influential form of column, the Z-bar. Acting semi-independently, he lived in Chicago from about the time of the Auditorium, making it his business to familiarize the architectural and engineering community with steel, before reverting to bridges after the slump in 1893. That Adler was able to span his later theatres with great trusses, and at the Schiller Theater even superimpose extra office storeys over the theatre roof, was due to Strobel and Marburg.

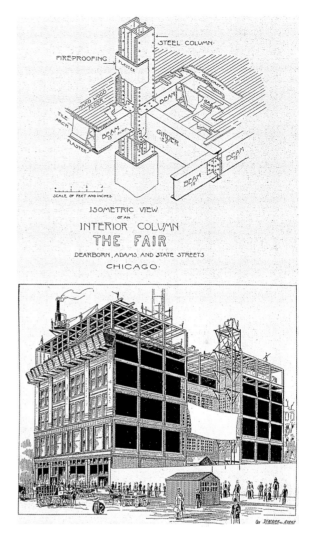

170. Perspective and column detail of Jenney's Fair Store during construction, from the *Inland Architect*, 1891. Already the steel columns and connections have become standard.

Whether large architectural offices went to consultants like Strobel for their steelwork, like Adler and Sullivan, or employed engineers in-house, as did Jenney, was at first rather hit and miss. Take Burnham and Root. According to the reminiscences of Paul Starrett, a technically minded assistant with the firm between 1888 and 1894,

> the entrance of iron and steel into building construction brought the engineering side of the office work into great importance and made me feel that I really had a role in the play. I found that calculating the strength of columns and beams in a steel frame was fascinating. As the steel frame came into use, it was the custom of architects to have their draftsmen figure the size of the different steel members from the handbooks put out by the various steel companies. Inevitably, this resulted in a lot of amateurish engineering.[384]

Following a scare when a half-built building partly collapsed in 1888, pursues Starrett, Daniel Burnham decided to avail himself of in-house expertise and hired the 'brilliant young bridge engineer' E. C. Shankland, afterwards a partner in the firm.[385] But outside engineers were also drawn in. Corydon T. Purdy, of whom more shortly, helped with the frame for the Rand McNally Building; while George B. Whitney, Jenney's right-hand man for the Home Insurance Building, was consulted on the Masonic Temple (1890–2).[386] Farming out structures was well under way at Burnham and Root before John Root died in 1891.

In the 1890s outside consultancy becomes a fixed pattern. Typically, a graduate from one of the proliferating engineering schools garners industrial experience by designing, making and managing steel structures for a fabricator. He enters building construction as that fabricator's representative, then sets up as an independent consulting engineer, and offers help to commercial architects on framing and design. Firstly therefore, steel fabricators barge into the Chicago market for metal frames; then in response, consultant engineers are hired, so as to shield architects and their clients from structures designed and detailed for the convenience and profit of the fabricators.[387] The poacher has turned gamekeeper.

That shift in apportioning preparatory design, standard for the steel-framed buildings of the twentieth century, takes place with seemingly lightning speed. It is settled enough by 1896 for a Chicago skyscraper architect to tell Montgomery Schuyler that he starts his task by getting from 'my engineer' the data about the thickness of steel required for 'my design' – a token of divided domains.[388] At this juncture a fresh term enters the lexicon of construction: 'architectural engineering'. The first manual of high building construction, published by Joseph Freitag in 1895, goes by that name. Its introduction adumbrates Freitag's hope for the 'perfect union' of the two building arts held out by the steel frame.[389] But the fracture in the design process points in another direction.

CONTRACTOR AND ENGINEER

To grasp the growing demarcation between architect and engineer in respect of steel-framed buildings, one last change in Chicago procurement practices must be added. That is the rise of the general contractor for large commercial buildings – a far-reaching development that scholarship has hardly explored.[390]

Before the 1880s most big commercial buildings in Chicago arose by means of multiple independent contracts managed by developers or their representatives, usually their architects. That hands-on tradition sheds light on the business-like attitude of Chicago architects after the Fire. When, in the rescrambling of procurement relationships that attended the triumph of the frame, well-capitalized general contractors began to take over large building projects, several who took the plunge and succeeded in that line had started out as architects.

Among them were Paul Starrett, whom we have already met, and his older brother Theodore; both commenced with Burnham and Root before making separate leaps into contracting.[391] But the outstanding pioneer in big-scale commercial building contracts was not a Chicagoan by origin or training. This was George A. Fuller, a former assistant with the Boston architects Peabody and Stearns.[392] Fuller set up as an entrepreneur-builder in 1881 and moved his operations to Chicago soon afterwards. With a 'disrespect for traditional methods',[393] he preferred to procure the whole package for large commercial developments on a cost-plus basis, after negotiating with architect and client. His breakthrough came with Holabird and Roche's Tacoma Building of 1886–9. Between then and his death in 1900, Fuller had a hand in 'virtually every large office building in Chicago', claims Robert Bruegmann.[394] In connection with the Tacoma Building, we know that he made various suggestions including the substitution of rivets for bolts in the frame. Fuller therefore was intervening actively in matters of design and structure, with cost, efficiency and above all control in mind.

How did that affect the shifting relations between design professionals? The question is best addressed by tracing the career of Corydon Tyler Purdy, the engineer who worked most closely with Fuller's firm. Purdy is a copybook example of the early style of consulting structural engineer expert in steel frames.[395] After graduating from the University of Wisconsin, Purdy began as a humble municipal engineer. Things opened up for him after he moved to Pittsburgh in 1887. He worked on shop floors and rolling mills before becoming a materials inspector and draughtsman to Carnegie's Keystone Bridge Company, awash then with Chicago building contracts. In 1889 Purdy opened a 'bridge engineering' bureau in that city. Commercial contracts promptly flooded in. His first big project was Burnham and Root's Rand McNally Building (1889–90), said to be Chicago's first all-steel frame. But his main connection was with Holabird and Roche. For that firm he sorted out the frames in a sequence of major office buildings, and through them he linked up with George A. Fuller.[396]

The sequel of Purdy's career took him to New York, whither it will be illuminating to follow him. During the slump of 1893–6, worse in the Mid West than on the East Coast, he opened a Manhattan office of his firm, Purdy and Henderson. His aim, he said in retrospect, was to insert a consulting step between New York's architects and the steel fabricators, on whom he felt the former had become dependent. The New York office flourished. By 1903 Purdy and Henderson were being recommended to McKim, Mead and White by Paul Starrett as 'the best steel designers I know'.[397] Purdy was not alone. His progress is entwined with the fortunes of the George A. Fuller Company, also making inroads on the East Coast in the late 1890s under Fuller's son-in-law, Harry Black. Together they built up an alliance that won Chicago procurement methods a foothold in New York. Their success is symbolized by the famous Flatiron Building (1901–3), the Fuller company's Manhattan headquarters, designed by D. H. Burnham and Company of Chicago with robust, rusticated stone facings upon a steel structure engineered by Purdy and Henderson (ill. 171).

It has long been objected that the neoclassical skyscrapers like the Flatiron Building which sprang up after the World's Fair of 1893 ignored or negated the steel skeletons that carried their façades. If they did so, that was not just because of changes in culture or fashion. By 1900 procurement operated in a more compartmentalized manner than during Chicago's rumbustious commercial heyday. To be sure, such divisions had existed ever since iron shaped the making of New York's Harper and Equitable Buildings. But back in those days there had been a fluidity and an unpredictability about the various roles and responsibilities. Now things were settling; changes in technique had elicited changes in construction management. Once steel became familiar, large and well-capitalized general contractors like the Fuller and Thompson-Starrett companies, bent on industrial efficiencies, had taken hold of the market for commercial construction and broken the brief dominance of the steel suppliers. In part, the formalized separation between architects and engineers was a side-effect of this revolution in organizing and sequencing building. With frames and fronts designed by the two professions within a

171. The Flatiron Building erecting,
1901. D. H. Burnham and Co.,
architects, Purdy and Henderson,
structural engineers, George A. Fuller
Company, contractors. The mature
skyscraper represents a partnership
between architect, engineer and
contractor, with the last by no means
just a passive partner.

system of overall building management, the opportunity and perhaps the inclination for exchange, intimacy and improvisation had declined. Architects remained honoured for their imagination. But in large commercial buildings they had shed some of their responsibilities for design and some of their supervisory power alike.

In an editorial on 'engineering-architectural problems' that appeared as the Flatiron Building neared completion, the *Building News* set out the state of affairs with regard to skyscrapers:

That construction has outstripped architectural expression is a statement that will scarcely be disputed when we contemplate the buildings for huge offices and tenement houses in New York . . . Specialism in design cannot go further . . . The most elaborate mechanical plants for obtaining power, warming and ventilation, and electric light, regardless of cost, have been introduced; the structures have been riddled with ducts and vents, fans and exhausts, boilers and motors, electric conductors and wires, till we wonder whether these appliances have not been the main motive of the designers, who have planned structures to suit them – whether, in short, the engineer has not had the greatest share in the work. In the steel-skeleton building, at least, the architect's design is subsidiary, and confined to a mere casing of terracotta. He has little to do with the steel framing of piers, uprights and girders, which is generally entrusted to a firm of structural experts.[398]

But the *Building News* had forgotten about one factor: art.

Art, Size and Leadership: Adler and Sullivan

Chicago boasted no architect in 1844, and the city's subsequent fame for construction derives largely from technique. How then did its architects come to be accorded such respect? Given that the trend in procuring complex buildings was towards specialization and fragmentation in design, coupled with tighter financial management, why were they seen as determinants in the process?

Our review of the main conditions affecting the tall building – fireproofing, foundations and framing – has hinted at some answers. Until about 1890 Chicago's architects brought their ingenuity to bear upon a fresh and fairly open field of technique. Nor, any more than Latrobe, need we classify figures like Jenney, Root or Adler as either architects or engineers. They were both, or rather half-and-half. The division between the design professions, adumbrated during the age of steam and iron, was incomplete still when that of electricity and steel dawned. Moreover, Chicago during the period under review as yet boasted no full-time school of its own for either profession. Instead, it sucked in talent trained elsewhere and often rich in experience. That fostered an exchange and informality palpable in the open-mindedness of the clubs, associations and publications dedicated to construction and in their zestful debates.

Over and above that there is art, and the sense of leadership it can afford. All architects must strike the delicate balance between art and technique. One who brilliantly did so was John Root, whose improvised foundations have been mentioned. In many ways Root resembled George B. Post of New York. Their formal education was identical; both attended the same engineering course at New York University a decade apart, and were stylistically and technically versatile. But the exhilaration of 1880s Chicago turned Root into the more charismatic figure. Like all his contemporaries there, he professed belief in an architecture that unflinchingly answered to material conditions. Yet he also saw himself as an idealist and artist, and was hailed as such.

Root was unsystematic. 'He grasped at novelties like a child with new toys', remembered Louis Sullivan. 'He thought them efficacious and lovely – then one by one he threw them away.'[399] A vivid tribute from his brother and one-time assistant, Walter, explains his approach to technique:

In a technical and narrow sense, John's mathematical and engineering abilities were deficient. He had not time to learn and keep up the many branches of constructional detail. He was rusty in his calculus and trigonometry, ditto much of his applied mathematics. I do not believe he would have cared to trust himself to calculate an important truss; it was not necessary nor desirable that he should. With the increasing press of work in the office, the custom developed rapidly of having specialists work out the various problems; for example, after an idea like the steel-rail footing had been developed, engineers were consulted as to the best methods of execution. But John had such a quick perception that he could suggest to a specialist an idea which would illuminate him, and enable him to work out a solution of a hard problem in a new and brilliant manner.

172. The architect-engineer Dankmar Adler in about 1895.

173. Louis Sullivan in 1891.

172. The architect-engineer Dankmar Adler in about 1895.

173. Louis Sullivan in 1891.

Captain Eads, I think, was not technically informed in the details of his own work, yet he is considered by engineers one of the greatest lights in the profession in his day and generation. He could not calculate a complex truss, but he could develop the general scheme of an enormous and difficult bridge and see its practicability when others felt discouraged.[400]

The mention of Eads is telling. By invoking him, Walter Root hints at a distinction not between architects and engineers, or between artists and technicians, but between leaders and followers. In any complex building, it is implied, creativity lies with him who can imagine, inspire, and pass the torch to collaborators. That line of thought sheds light likewise on Jenney's reputation for a broadbrush approach to design and technique. It has since found favour in large design firms of all kinds, in which the notion of a controlling mind and a controlling idea goes with that of delegation.

The issue of size was pervasive in Chicago. Large buildings meant large firms. Burnham and Root and its successor-firm were among architecture's first big commercial practices. To operate efficiently, such firms had to establish functional hierarchies: for instance (as we have seen), what range of structural skills to keep in-house or farm out. Such matters were the province of Daniel Burnham, American architecture's first great businessman. Burnham saw architecture as essentially a 'human' problem, that of managing people.[401] Outstanding designers seldom make natural managers. It was Burnham's achievement to channel and control Root's abilities, and those of others like and unlike him. No artist himself, he was shrewd enough to see that the key to architecture's appeal and to his firm's success and growth lay in the magic of art. Efficient engineering and good management had to be overlaid and indeed led by the irrational imagery, the imagination, of art.

The tension between these factors lies at the heart of the relation between architect and engineer. It will always be personal and elusive. In the Chicago context, it is usefully illuminated by Burnham and Root's rivals Adler and Sullivan, the pre-eminent architects of the city's halcyon years, with whom this chapter can close.

Technology and art are hardly as one in the works of Adler and Sullivan, as romantic modernists once argued. In their Auditorium Theater, the applied arches and ornament of the acoustic ceiling conceal the Carnegie trusses above; the passionate naturalism of the Wainwright Building's façade counteracts the frame it conceals. Another way of thinking about the connection is that of compensation. Technology is taken as necessary and beneficent. But far from being sufficient, it needs the supplement and relief of art. The partners' skill lies not in obliterating the difference between structure and ornament, but in juxtaposing the two in such a way as to make them stand for difficult, interlocking paths in modern culture.

Dualism ran deep through the Adler and Sullivan practice. In their small office atop the Auditorium, Adler and his assistant Paul Mueller sat at one end, Sullivan and Frank Lloyd Wright at the other: the solidity of German technique seemingly balancing mercurial Celtic artistry. But the relation should not be caricatured.

Ever since Wright reminisced about his time with them, Dankmar Adler (ill. 172) has been patronized as a noble engineering dogsbody to Louis Sullivan's art. He came to think that way himself. 'Of late years', wrote Adler in an autobiographical sketch, 'owing to the preeminence in the artistic field of my partner Mr. Sullivan, I have devoted my efforts to the study and solution of the engineering problems which are so important an incident in the design of modern buildings.'[402] After the partnership broke up during the slump of the 1890s, Adler went so far as to find work briefly with an elevator company as its 'consulting engineer and sales manager'.[403]

Adler was twelve years older than Sullivan. His experience of building in Chicago went back to before the Civil War. During that conflict he spent a short but valuable spell with the Corps of Engineers, learning intrepidity and making up for his lack of technical education.[404] By 1880 he was becoming an adventurous constructor and planner. He could never detail his buildings to make them sing, as those few he designed after parting with Sullivan betray.[405] Yet Adler was no banal technician, riding the crest of constructive experiment. He felt strenuously about the environmental quality of his buildings, and devoted his utmost powers to that end.[406]

The best proof lies in Adler's theatres. Since the 1860s he had created auditorial spaces of all kinds: churches, synagogues, college halls, above all, music halls and theatres. Since the days of the iron auditoria mentioned early in this chapter, theatres had grown almost schizophrenic in nature, as we saw with the Royal Opera House in London (p. 99). For fire-protection as much as for delight, fast-built iron frames allowing wider spans and deeper balconies were plastered over with a veneer of decorative make-believe. Adler understood the goal of engineering technique in theatres to be not honest construction but comfort and illusion. The intelligence with which he channelled his talents to those ends entitles Adler – not Sullivan – to be counted among the great theatre planners and designers.

What little is known about Adler's culture hints at familiarity with the German stage tradition. From his researches for the Auditorium he certainly accrued international expertise in stage machinery. But his outstanding technical achievement, forwarded in each major project from the Central Music Hall of 1878–80 onwards, was neither structural nor mechanical but acoustical: that of adapting the ever-larger volumes demanded by impresarios, politicians and preachers to the spoken voice, in an age before amplification. The method was based upon Scott Russell's diagram of the 'isacoustic curve', and then coaxed out of Adler's canniness with materials, surfaces and spaces. Here was an initiative on a par with the other empirical inventions of the 1870s and 80s by Chicago architects. Though Adler never thought of exposing the structures of his theatres, the isacoustic approach found its visible counterpart in the shape of the soaring plaster ribs overarching the finest of the Adler and Sullivan theatres, the Auditorium and the Schiller. It chimes with the falling-away of architectural inventiveness in technique we have noted at the end of the century that Adler's approach to sound receded once the

first of the modern acoustical engineers, Wallace Sabine, embarked on his experiments at Harvard in 1896.[407]

To summarize, Adler was an architect-engineer with talents for progressive technique and man-management. But he also grasped that in major projects all that should defer to a level of art which he could not supply. Art implied abundance and escape: whether the Gesamtkunstwerk of theatrical illusion, or enriching the building's carcase to mollify the ruthlessness of construction. In that higher sphere Adler was shrewd enough to sustain and draw out a partner with the flair for disciplined superfluity which so driven a culture as Chicago needed and, intermittently, valued.

Louis Sullivan (ill. 173) wrote his autobiography in isolated old age. Its picture of a consistently unfolding mission may well be discounted, along with some of its peevish details. But most of it is about his youth, and on the ambience of 1870s Chicago its voice is persuasive. It depicts an immature young man, wavering between the ideals of art-architec-

175. Guaranty Building, looking up the façade in 1952, showing the rich, repetitive terracotta ornament.

ture and of engineering. In his youthful romanticization of engineering, the progressive strain in American literature made up for what the post-bellum spirit of city and nation could not yet supply. A wider reader than most architects, in the fallow period after his return from the Ecole des Beaux-Arts in 1875 Sullivan devoured books. A cross-over between technology and art was nascent in the broad strain of speculative thought that linked Emerson and Whitman with Darwin; it was tempting to imagine engineers as creative visionaries and modern Michelangelos. Sullivan's brother Albert, then close to him, was starting to carve out a railroad career; and Louis mentions a moment when he contemplated abandoning architecture in favour of bridge engineering. Then the impulse passed. The idea of the bridge was what enthralled him, he remembered, not the process or the reality; he was a platonist, therefore an architect. And so he retreated to general scientific reading. 'Yet science, he foresaw, could not go either fast or far were it not for Imagination's glowing light and warmth.'[408]

In the event Sullivan turned out a fair technician. Seven short months with Jenney followed by time in various Chicago offices gave him the basis he needed. When he joined Adler he still had much to assimilate. 'I was developing a little technical knowledge myself', he remembered of the period when he became a partner in 1883. But the one innovation he claimed for himself, setting the bare, low-wattage light bulbs of the early Edison era in sparkling tiers, first done at McVickers Theater (1883–5), was really decorative.[409] His main role with Adler was to put flesh upon bones. It was from the starting point of appearances, as Sullivan found the space to grow, that he developed his own redemptive interpretation of the link between technology and art. The Auditorium, on whose plutocratic surfaces decoration was sanctioned and encouraged, gave him facility and helped sow the seed. But on Sullivan's own account it did not germinate until 1890, when a different kind of commission, the Wainwright Building, hove into view.

The Wainwright Building in St Louis, the Guaranty Building in Buffalo and other Adler and Sullivan office projects of 1890–5 have commonly been attributed to Sullivan alone. That may not be misleading. Adler had tired of travelling, so his younger partner fronted the out-of-town projects, proliferous in the firm's final years, and took more initiative. In addition, the structures of office buildings were settling down by now to the conventions of the frame. Compared to the complex Auditorium and Schiller Buildings, such office-only jobs did not challenge Adler. He certainly helped with 'the solution of the engineering problems', to use his phrase; we know he was vexed when Sullivan dropped his name from the credits for the Guaranty Building, completed after the partnership had ended. But the problems that had absorbed Chicago architects since the Fire were all but solved; responsibilities for sorting out the design of structures had been redistributed. It was time to return to the surfaces. That meant Sullivan.

In the *Kindergarten Chats*, most pregnant of Sullivan's writings, the interlocutors come at last to the question: 'What is an Architect?'. The dialogue proceeds by elimination. The architect is not a plumber, a bricklayer or a mortar-mixer. '"Neither is he a surveyor, a mechanical engineer, an electrical engineer, a civil engineer, is he? Otherwise why are these men so called? Why are they not called architects?" "That is well put."' In the end, the definition has to be arrived at by getting at the architect's function. But Sullivan does not speak of design. Instead, we are told that 'the architect *causes the building* by acting on the body social . . . the architect is a *product* of the body social, a product of our civilization . . . we approach him from two sides – as a product and as an agency; and so of course I come at once to his true function, namely the double one: TO INTERPRET AND TO INITIATE!'[410]

It is helpful to turn back from that passage, dating from 1902, to 'The Tall Office Building Artistically Considered', a more accessible essay long popular with modernists. Written in 1896 before the downturn of Sullivan's career had become definite, it sums up half his experience of the late Adler and Sullivan office buildings – that of the positivistic initiator. From the practical and mechanical conditions of the modern office, he extrapolates to the fronts that are his real interest. Here we are on the familiar ground of an external architecture that reflects or expresses the use to which the building is put and, in this case, the engineering conditions set by the steel frame. If a building fails to do that, asks Sullivan, how can it be true to its own nature? 'Form' (as he says elsewhere) 'ever follows function.'

As every student of the phrase knows, that is only a beginning. 'Interpretation' must follow if an architect's creations are to rise to the level of significance. In Sullivan's view, it must start from his origin in and consciousness of the 'body social'. The frantic exuberance of his designs after 1890 (ill. 176) is about giving back to the American people through ornament the delight and the contact with nature he believes they have lost through technology, routine and densification – the demise of 'garden city' and the rise of conurbation. Those forces are inevitable and progressive. Their rigour must be accepted and even celebrated, and their means of production availed of in ornament

FACING PAGE
176. Crown of the Bayard Building, New York, from *American Architect and Building News*, 1900. Louis Sullivan with Lyndon P. Smith, architects, 1898–9.

177. The completed Schlesinger and Mayer Store (later Carson Pirie Scott). Watercolour by Albert Fleury. Louis Sullivan, architect, 1899–1903.

and structure alike. There is no looking or going backward with Bellamy or Morris. But homage once paid to a business-led democracy and the science of modern engineering, they must be transcended. Only the architect can do that, thinks Sullivan, by means of his 'double' function, 'to interpret and initiate'.

That high philosophy doomed the architect to splendid isolation. It set him aloof from the mere technician and builder – something Chicago culture had avoided up until Sullivan's time. The separation tallies not only with Sullivan's disengagement from Adler but with the trend towards professional fragmentation in the building art of the 1890s. Likewise, the drift of classical skyscraper architecture in the years after the World's Fair ran away from structural issues towards appearances. What turned Sullivan resentful was the superficiality of the classical reaction, its irrelevance to modern American circumstance and, worst of all, its victory. He who had thought hardest about raising high the democratic ideal in the street and appealed over the heads of the building trades to the people was to be shunned by the people themselves – or so he felt. It would have been a miracle if so intellectualized an architecture had been widely understood.

It cost money too. Barr Ferree pointed in 1895 to the Guaranty Building (ills. 174, 175) as 'possibly the most richly decorated commercial building in America'.[411] Tall office projects of such lavishness would be few. One factor in the decline of Sullivan's career was that his self-imposed programme for responding to 'democracy' made him an expensive architect. Though modern offices might be a good place to reach and touch the urban American, the nature of their financing meant that few speculators could be expected to stretch their budgets to his vision. Sullivan, like Wright, idealized the entrepreneurial businessman of the Mid West because he needed him. He was inevitably disappointed by the response.

In the end another building-type offered Sullivan a more natural alliance between ornament and democracy in his understanding of the term, and therefore the best occasion for the unique marriage he sought between art and building. Happily he acquired one chance to explore it.

I believe that there is an impression abroad that I deal only with the more ornate forms of architectural construction. This is an error for I have done a great deal of work in heavy construction, plain factory buildings, machine shops, etc. . . . As to my general ability etc. I will gladly refer you to any officer of Schlesinger and Mayer whose new retail store I am erecting. I believe the time record made thus speaks for itself; and you know well enough that no contractor however capable would make such a record were not the architect's drawings and specifications well-nigh perfect, and every contingency discounted by him.[412]

So wrote Louis Sullivan to the directors of the Carson Pirie Scott Company in May 1903. Their firm was contemplating a new wholesale store in Chicago, and had made enquiries of him. Naturally they knew about the Schlesinger and Mayer building nearby, just then finishing, which Carson Pirie Scott indeed were to buy only a few months later.

Their overture came to nothing; and when Carson Pirie Scott enlarged the Schlesinger and Mayer store in 1906–7 they asked D. H. Burnham and Co. to continue the lines of Sullivan's earlier front. Nor did Sullivan build on the scale of Schlesinger and Mayer again. It is hard to read the defensiveness of Sullivan's letter without that hindsight.

The Schlesinger and Mayer store (1899–1903) was the one big Chicago building erected to Sullivan's designs after he parted from Adler. A department store, we saw in Paris, warranted a richness that an office block did not need. That, hazards Joseph Siry in his study of the store, may be why Sullivan was chosen. When they announced the project, Schlesinger and Mayer went out of their way to conjure up a picture of opulence in marble, bronze and mahogany. But they did not omit to mention fireproof construction and an all-steel frame.

Like the Parisian shops, the new Schlesinger and Mayer (ills. 177, 178) occupied the site of the existing premises, so had to be built in phases, imperilling the implacable unity promised by Sullivan's perspectives. Here the phases were confined to two. For the first portion, just three bays raised on Madison Street in 1899, Adler was retained by the company to design the power plant and much of the equipment, though not, it appears, the steel frame. David Mayer, the client, had known the former partners for years and must

have wished to avail himself of both men's skills; evidently, they could still work in tandem. The second and larger phase of building followed in 1902–3. By then Adler was dead and Sullivan took sole charge. This time the ubiquitous George A. Fuller Company came in as contractors. The caisson foundations going down to bedrock were of special complexity, because they had to be inserted underneath while business carried on in the old shop. That job was no doubt overseen by Fullers, in co-ordination with the architect's office staff. But Sullivan was proud enough of the achievement to write it up for an engineering journal.

He never did the same for the flamboyant shop fronts of Schlesinger and Mayer (ills. 179, 180). Perhaps he preferred them to speak for themselves. They are of iron, but not the structural iron whose progress has been the burden of this chapter. By 1900 the frame can be so far taken for granted as to mean just a backdrop for the reflective architect – something to raise efficiently and clad appropriately.[413] Instead the double-storey shop fronts furnish a kind of commentary on the frame, from which they emerge, boldly set forward. They are far from just decorative, for Sullivan, so one commentator on the shop remarked, 'accepts every exigency prescribed by modern commercialism . . . He accepts the modern machine, and demonstrates its capacity to assist him in evolving a work of art. He does not despise the task of designing a commercial building, but rejoices in it.'[414] Ample and efficient display matters; the glass is the most up-to-date technology can offer, so disposed as to shed the best quality of light and to draw in fresh air, as Adler would have wished.

But that is only the start. The enrichment of the shop fronts, remarks Siry, takes up the longest section of the building specifications for the building:

the use of ornamentation is to be very general. On the exterior store front work is to be exceedingly rich and delicate, and is to cover not less than, say 95% of the surface. This work will consist of geometrical and foliated designs. The scheme of ornamentation is to be very elaborate, with very fine and delicate detail.[415]

'Exceedingly rich and delicate' are rare and emotional words for a specification. They are borne out by the result. Tours de force of the modern casting

179, 180. Schlesinger and Mayer Store (later Carson Pirie Scott), details of the double-storey shop fronts from the *Inland Architect*, 1903. A. The corner. B. Windows

art, uniting the imagination of Sullivan's draughtsman George Elmslie, the finesse of his modeller Kristian Schneider, and the industrial virtuosity of the makers, the Winslow Brothers, they mark a climax for two centuries of interchange between the ferrous industries and art. Compared to these fronts, the rigidity of earlier ornamentalists in cast iron appears merely ham-fisted; even Art Nouveau ironwork looks wayward and half-hearted. The intimation of 'motifs coming forth from the surface of iron, as if an inner vitality had promoted their emergence'[416] transforms dead and obdurate matter into a source of life and hope. The inorganic, to use Sullivan's terms, has been subverted into the organic. In the Schlesinger and Mayer shopfronts, the energy of iron and the power of art are deployed in protest against spiritual impoverishment. That at a time when architecture is haplessly starting to rid itself of ornament, and the miracles of technology threaten to divorce man from nature.

3

CONCRETE

1. *Styles of Concrete 1800–1914*

BEFORE REINFORCEMENT

Who developed modern concrete for buildings? In epitome, the champions of the concrete renaissance – not too strong a word – that took place in nineteenth-century Europe and America were a farrago of engineers, architects, building tradesmen, manufacturers, chemists, ingenious amateurs and entrepreneurs. Progress lurched unevenly for most of the century between experiment, patents, publication and wider exploitation. Then after about 1885, in the second stage of the renaissance, the crucial mechanism for the diffusion of reinforced concrete became a series of patent systems manipulated by resourceful engineer-contractors, leaving architects at a disadvantage from which it took them a generation to recover.

'The history of concrete', a sapient science-journalist has written, 'is almost but not entirely without interest.'[1] The saga of modern cement and concrete starts with efforts to improve upon the 'puddle' used for mortars, linings, rafts or adjuncts to pier-foundations in river and marine projects.[2] Here was engineering territory – the province of water-engineers. To improve the setting and strength of limes, it was advisable to add a special binder. Pozzolana, the volcanic binder mentioned by Vitruvius which allowed Roman concrete to set under water, was the favourite, still available from Italy in the eighteenth century, but costly to import. A northern equivalent, 'tarras' or 'trass', could be obtained from German or Netherlandish sources. Mixed with hydraulic lime, this was used for instance by the Dutch for sea-defences. Broken tile or brick might be substituted, but was less good.

By the last quarter of the eighteenth century, varying compacted mixtures of lime with sand and gravel or pebbles had become quite common in France for underwater walls and linings (though not for piers), and occasional in England. They were customarily known in both countries as 'béton', an old French word of uncertain etymology; the English term 'concrete' came into currency only in the 1830s, and early forms are now normally called lime-concrete to distinguish them from later cement-concrete. Bélidor described having seen such works carried out in the navy-port at Toulon in 1748, connecting them with ancient practice; and further government-supported trials were going on there in 1777 using a French equivalent to pozzolana.[3] George Semple, the author of the Essex Bridge, Dublin, recommended building pier foundations in that manner in his *Treatise on Building in Water* of 1776, but never actually did so. By the 1780s the chemistry and mixing of ingredients for cements in maritime works were eliciting experiment and debate in the ambit of the Corps des Ponts et Chaussées.

The next stage follows the paradigmatic model of modern technology whereby research leads to publication, then to commercial take-up. That happened first and loosely in England, then more methodically in France. Both steps focussed on cements – the key to concrete's development. The English story goes back to the country's first consulting engineer of eminence, John Smeaton, who in the 1750s had tried out various cements, not as mass-foundations but as mortars, for his Eddystone Lighthouse. These

FACING PAGE

181. 25 bis Rue Franklin, A. & G. Perret, architects, 1903–4. Detail of the bays, showing the decorative faience cladding over the concrete by Alexandre Bigot. The building in the background is the Trocadero.

trials were sparked off by his visit of 1755 to Holland, where he observed a selection of trass-rich cements in use for the Den Helder canal. Smeaton's Eddystone investigations were not published until 1791, just before his death. They concluded that the best limes for 'water building' included high natural proportions of clay. That 'overset the prejudices of probably more than 2000 years' in favour of hard limestones, and stimulated a search for new natural sources of cement, soon identified among so-called 'septarian' deposits, common enough on the English coast, with good proportions of alumina and silica present in the limestone.[4]

At first the exploitation of these cements was mainly directed to the competitive English market for mortars, external stucco renders and mouldings. Commercial patents for burning and grinding a naturally occurring 'cement-stone' into powder were awarded to James Parker in the 1790s. Parker (allegedly 'Rev Dr Parker', but his background is mysterious) soon disappeared. But his 'Roman cement', requiring no very elevated temperature to burn, proved effective both as a strong, quick-setting mortar in engineering works and as an external render. It continued in wide international use and repute for half a century. After the patent lapsed in 1810, rival processes proliferated. Most were controlled by building tradesmen like James Frost, Joseph Aspdin, William Lockwood, James Pulham and Isaac Johnson, chronicled in Major Francis's history of the English cement industry. At first they produced alternative natural cements to Parker's. But after Vicat's experiments in France had been published, many shifted into making 'artificial cements', derived from a mixture of sources and requiring more care in manufacture. The most successful was Aspdin's 'Portland cement', made from burning chalk and clay at high temperatures to the point of vitrification. After an obscure early history, the process became more reliable. By the late 1840s Portland cement was establishing itself as the basis of most modern strong cement and concrete mixes.[5]

In France, the paradigm model of research begins with Louis Vicat, an individualist among Ponts et Chaussées engineers. Having quarrelled with his superior, Vicat was sent off in 1812 to construct the Pont de Souillac over the Dordogne. Work ground to a halt, as the Empire imploded. He therefore embarked on systematic trials of the different balances and handling of ingredients needed to make a consistent hydraulic cement. The upshot was, in A. W. Skempton's words, 'the first major work with lime-concrete foundations in which no pozzolana was used'.[6] Vicat's initial findings were published in 1818, with underwater setting still chiefly in mind. His masters, notably Louis Bruyère, a senior official at the Ponts et Chaussées, welcomed, debated and added to them. With their support, an experimental plant was set up in 1821 close to the recently completed Pont d'Iéna in Paris by Maurice de Saint-Léger, a captain in the Génie. Its artificial cements were adopted in concrete foundations or linings for the approaches to the bridge and for the Canal Saint-Martin, allowing Saint-Léger to set up a permanent factory in suburban Meudon. Over and above their strength and reliability, they proved cheaper than natural alternatives, at least in Paris. Cement factories soon sprang up elsewhere. Meanwhile Vicat pursued the subject on, declining promotion in favour of research. He crisscrossed France obsessively in search of good deposits, publishing a running list and analysis of hundreds of quarries. His book on cements and mortars of 1828 was translated into English in 1837, and he was still updating his knowledge in the 1850s.[7]

Vicat's researches, says Antoine Picon, came 'in the nick of time to allow the Corps des Ponts to assume a dominant position in a sphere of which it had hardly made mention before then'.[8] There was a sense abroad of the need to energize French entrepreneurship, and the Ponts et Chaussées was reshaping itself to meet that aim. The promulgation of Vicat's unpatented findings, his championing of artificial cements from a mixture of sources, and his purchase on costs boosted the growth of a factory-based cement industry. In the French way, that started with public works and the backing of the engineering élite. But by the time of the July Monarchy a number of commercial pro-

ducers were manufacturing high-class cements. The British grip upon the market for cements in France, pronounced in the decade after Waterloo, had been countered.

The experiments of Smeaton and Vicat, and the consequent growth in commercial range and chemical understanding of cements, were preconditions for the spread of concrete construction into general building and architecture after 1830. Since both were engineers, their seminal researches gave their profession an authority over the material's development that it had yet to assume in relation to iron. But a further strand remains to be mentioned – the earliest in which an architect as such is identified with a practical advance in concrete technology. That is the extension of concrete foundations from bridges, harbours, quays and canals into buildings proper.

The point of departure was a hint from the engineer John Rennie. He himself seems to have used concrete foundations only in the loose sense of a mixture of rubble and lime. But while building Waterloo Bridge, Rennie stumbled upon an impressive deposit of concrete formed when a cargo of hydraulic lime had sunk and coagulated with Thames gravel. That led him in 1813 to suggest a specified balance of these materials for the repair of troubled foundations at London's Millbank Penitentiary (ill. 182). The idea was taken up by Robert Smirke, the architect-constructor in charge of the work, and his builder brother-in-law, Samuel Baker. Smirke and Baker installed large rafts of lime-and-gravel concrete in specified proportions first at Millbank (1816–17), then at Lancaster Place in the Savoy Precinct (1820–3) and the Custom House (1825–7). As all these jobs were on the banks of the Thames, they were river works, yet they were also buildings proper. By the 1830s Smirke and others were using concrete raft foundations for major projects as a matter of course, well away from water. In the first full English monograph on cements and concretes, his friend the Royal Engineer Charles Pasley could assert with confidence that 'the merit of introducing this immense improvement systematically and generally into the practice of modern architecture is undoubtedly due to Sir Robert Smirke'.[9]

182. Millbank Penitentiary, from an Ackermann aquatint of 1817. Built on a raft of lime-concrete devised by Robert Smirke, architect, Henry Baker, contractor, and Durant Hidson.

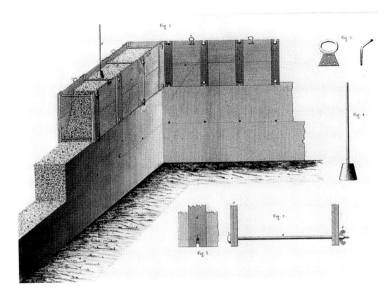

183. Making of a concrete wall from François-Martin Lebrun, *Traité pratique de l'art de bâtir en béton*, 1843.

Having thus been drawn into the base of buildings, concrete construction was ready by the 1830s to rise into above-ground walling, filling, floors and roofs. Here it would have fresh forces to withstand. Mixtures devised to resist salt-water or subsidence might not lend themselves to spanning openings by arch or lintel or to the cycles of the weather. In this sphere the pattern of engineering research promoting development broke down. Concrete above ground did not much interest Vicat, and it was viewed with hostility by the influential Pasley. The first such above-ground experiments were undertaken therefore not by engineers but by private architects or builders, without much technical help. Early developments took two main directions. The more familiar is that of in-situ walling poured within formwork, in the manner of pisé – the earth construction cased and rammed within shuttering, standard in many parts of Europe. The leap was almost self-evident. As Peter Collins pointed out, a former 'building labourer', François Cointeraux, had since the 1780s been championing a primitive concrete soup made simply by adding mortar to the normal concoction of pisé. In a memorandum of 1829 Clément-Louis Treussart, a military engineer of much practical experience who had long been interested in béton and disagreed with Vicat's conclusions, revived and respecified the Cointeraux idea.[10]

The first concerted French attempts to build entire structures in béton were unsurprisingly undertaken in a pisé-building area (ill. 183). Their instigator was a regional architect in private practice, François-Martin Lebrun of Montauban. Treussart seems the immediate source of his inspiration; though Lebrun is said to have relied on Vicat's researches and dedicated one of his books to him, his formula of lime, gravel, coal-ash and earth was unsophisticated. His bold proposal of 1831 for a three-arched bridge in concrete was turned down; but he did go on to build a small canal bridge, a house for his brother, and a small Protestant church, extant, at Corbarieu (1836). Strikingly, Lebrun vaulted all his mass-concrete spans – floors, cellars, roadway and church roof. But the roof soon gave trouble and had to be replaced in timber. The church is plain and has always had, along with brick dressings, an external render to shield its coarse, permeable core. Concrete could not yet be compacted sufficiently for weather-protection with one of the modern cement-renders to be omitted; nor was anyone calling out for it to be exposed.[11]

There was an alternative to in-situ concrete: pre-cast or even factory-made concrete construction. This was first actively exploited by the Ponts et Chaussées from the 1830s for harbour works, not buildings. In 1833 the engineer Poirel set up a plant manned by 'condamnés militaires' to make huge concrete cubes for the breakwater at the port of Algiers. The success of this operation and of the blocks led to its repetition in later French ports.[12] But under Pasley's influence the English held back. When concrete blocks were proposed for a breakwater at Dover in the 1840s, there was suspicion. 'Since those works at Algiers began, I have talked with French engineers about them; the masses are dissolving, many of them', apprehended George Rennie.[13]

More pertinent to architecture, smaller factory-made blocks light enough to be transported for general building use were meanwhile being tried out in England. The pioneer was William Ranger, a versatile figure who ventured in turn into manufacture, design, railway contracting, public health inspection and technical teaching. In 1833 Ranger, then basically an engineering contractor, took out a patent for manufacturing concrete

184. Westley Church, Suffolk.
William Ranger, architect, 1835–6.
One of the earliest buildings to
survive from the nineteenth-century
concrete renaissance.

blocks or 'artificial stone'. His interest in concrete appears to have begun in Brighton, where he undertook his early works.[14] There another contractor, Thomas Cooper, had built a high sea wall of concrete in the 1820s. Ranger moved to London to exploit his blocks, which were factory-produced at a plant on the south bank of the Thames. They were made out of lime-concrete, to which a smooth sand-lime render was added for the surfaces.[15]

During the 1830s Ranger was employed as a contractor by the Admiralty, building dock foundations and walls in mass concrete as well as blocks. But his blocks were also taken up experimentally for ordinary wall-construction. They survive in a church exactly contemporary with Lebrun's Corbarieu, erected in a plain Gothic to Ranger's own designs at Westley, Suffolk (1835–6). Now at least, the blocks are covered by an overall render (ill. 184). They may have been used with more verve in a little Greek-style school of 1836 built at Lee in the London suburbs to George Ledwell Taylor's designs (ill. 185). Taylor was architect to the Admiralty and worked with Ranger on the dockyard projects. But he had also travelled to Greece and Italy and (with Edward Cresy) published on archaeological aspects of architecture. The school was demolished long ago and only a perspective remains to record it. So we cannot be sure whether its paraphernalia of mini-porticoes, pediments and columns were moulded from lime-concrete blocks along with

185. Blackheath New Proprietary School, Lee, in the South London suburbs. Built of concrete blocks to the designs of George Ledwell Taylor, 1836.

the walls – whether, in other words, it explored how far concrete, like stucco, could furnish a cheap expressive equivalent to masonry in the historic styles.

Having moved into railway contracting in 1836, Ranger soon got into difficulties. The setback may have accelerated the collapse of block as a means of early concrete construction in Britain. But there were other factors, notably Pasley's blanket condemnation of above-ground concrete construction in his *Observations on Limes, Calcareous Cements etc.* (1838). Few entrepreneurs or experimenters were spared in this exacting work, and Ranger came off as badly as anyone. After condemning the inadequacies of Taylor and Ranger's underpinning work at Chatham Dockyard, Pasley reported test-results on the latter's lime-concrete, which he found weaker than the best stone and brick. Some tendency to exfoliate under frost was also reported from the dockyards. Advising that concrete ought to be confined to substructures, Pasley rose to a climax of negativity in a passage on Treussart:

> He seems to have been led away by the same sort of enthusiasm, which prevailed in this country a few years ago, and which caused concrete and artificial stone to be valued so far above their real merit, that many persons seemed to think that they would supersede stone altogether: but the failure of this substance in the Docks and Wharfs in Her Majesty's Dock-yards at Woolwich and Chatham, which were at first to have been built with concrete exclusively, have brought it to its proper level, by reducing it from the rival to the humble companion and assistant of stone and of brickwork laid in cement.[16]

In scientific terms this prudence may have been fair. As the founder-initiator of the Royal Engineers' little technical school at Chatham (pp. 458–9), Pasley wanted to establish what is now called 'best practice' for his pupils. John Weiler, who has examined his career, points out that his views were shared by other engineers at the time as well as by his friend, the architect Smirke.[17] Yet Pasley addressed the properties of concrete in engineering works alone, not its commercial or expressive potential. Herein lay the drawbacks of the research perspective: the terms of its rationality could tend to caution. Engineers could hold back innovation in architecture as well as foster it.

Later, engineers recognized that the loss of British momentum in concrete construction began with the *Observations*. 'The writer who has most discredited its general employment was our own Sir Charles Pasley', remarked Henry Scott in the 1860s.[18] In consequence the British market for factory-made cement and concrete products for

Pair of Semi-detached Houses
Erected in Sydenham Road Croydon.

Ground-Plan:

buildings narrowed itself down to specific features or components – cills, copings, ornaments and the like. With the Victorian building and transport boom, this became a large and profitable industry; by 1900, many external features in ordinary houses that had once been of stone were of factory-made concrete. Here was a niche market offering substitutes for existing components and details, not a fundamental means of construction. The prefabricated cement and concrete industry confirmed trends set at a higher level of artistry in the Georgian period by the famous Coade stone. A factory-based approach was always likely to follow fashion, not lead it.

The role of factory production in the pre-history of modern concrete construction tends to be underestimated. Soon enough, cement plants were making objects for the street, the workshop, the farm, the garden or the park: pipes, pots, cisterns, artificial rocks.[19] Houses or large items of equipment like reservoirs and water-towers were only cast in situ because they were too heavy and bulky to be transported. Concrete construction in blocks and slabs too enjoyed spasmodic revivals. In Britain, prefabricated (and lightly reinforced) slabs nailed to a timber frame featured in the system patented by the Croydon builder W. H. Lascelles in the 1870s, and worked up into fashionable styles by leading architects (ill. 186). Lascelles' system was berated by progressive commentators for failing to show the material's character.[20] But it enjoyed such success as it had because the architects were working round the builder's brief, not the other way round. Commerce's interest in a material is limited to properties affecting its economic potential.

187. François Coignet's house and chemical works, St Denis, bird's eye views of 1864.

So though ideas for prefabricated systems often started with architects, they persisted only in the hands of builders or builder-engineers with a nose for the market. Concrete block needed the housing circumstances and crises of the twentieth century before it could make continuous headway.

The factory also contributed to the in-situ tradition. That is easy to miss. In-situ concrete differs from masonry, timber or (usually) brick construction in that the material constituting the building's core is created as part of the act of construction. It seems to restore primacy to the site. Nevertheless in the next stage of a 'concrete architecture', neither the carpenter who made the formwork nor the site-builder who supervised and poured the mixture nor even the architect or engineer who conceived the design was the crucial figure. That was left to the intermediary who controlled the formula for the ingredients. Such was the figure most responsible for giving France a lead in promoting mass concrete for above-ground structures after 1850, François Coignet.

Coignet was not primarily a builder. He was a resourceful Fourierist manufacturer who turned out a plethora of products from his works at Saint-Denis (ill. 187). His development of monolithic concrete construction began from 1852 as a means of using the slag by-products from his other enterprises; he first applied the process in his own factory and an experimental house. Coignet's move into contracting came later, in 1861. It was prompted by an instinct for enterprise and by the fact that his 'béton aggloméré', as he latterly called it, required good compaction of the concrete, entailing special mixing machinery and control of the site. Most of his concrete contracts had to do with the great infrastructural renewal of Paris under the Second Empire; his involvement in buildings was restricted.[21]

Coignet attracted Peter Collins, in modernistic quest during the 1950s for self-revealing materials, because his compacted concrete could be exposed in elevations without a render. But in monumental architecture its inventor assumed that it would need a

188. Concrete aqueduct carrying waters of the Vanne over the valley of the Yonne. Belgrand, Humblot and Huet, engineers, François Coignet, contractor, 1870–3.

covering of 'plaques of marble and varied stucco' or, at the least, paint.[22] Since the main marketing point of his concrete was to save money, it was unlikely to be employed in that way. Coignet cared little for the expressive or the structural potential of monolithic construction. His was a product to sell: an ersatz substitute for masonry foundations or walls, tough and fireproof. As a means of cheap construction, it appealed to social rather than aesthetic progressives. Its enduring advertisement, and the most spectacular above-ground use of mass concrete before reinforcement, was the bold viaduct of 1870–3 over the valley of the Yonne, designed as part of an aqueduct taking the waters of the Vanne to Paris (ill. 188). The fact that it was built of Coignet's concrete had little effect upon its form. Nevertheless its use added to the Roman connotations of the Haussmann-Belgrand plan for renovating the capital's water supply.[23]

Coignet's most promising collaboration with an architect, the small church in the Paris suburb of Le Vésinet (1864–6), was a disappointment: no meeting of minds took place. In principle Louis-Auguste Boileau, its designer, was a structural progressive, having already devised the iron-framed 'system' for Gothic churches described in the previous chapter (pp. 149–51). St Eugène, Paris (1855–6) was the much-criticized prototype. Le Vésinet had already been projected on similar lines, with a frame of iron and an external casing of stuccoed rubble, when for the latter element Coignet and his material were imposed on him by the client, 'homme de progrès, porté à favoriser les inventeurs'. The result (ill. 189) disconcerted Boileau. He found the walls damp, prone to cracks, wanting in the crisp line and surface attainable with stone, and not all that cheap. The true economy and future of 'béton aggloméré', he judged, lay in 'the additional elements of decoration which have to be reproduced in great numbers'. At that stage, an enthusiast for iron architecture could believe that concrete's future lay in prefabrication.[24]

Le Vésinet was evidently not well made. Likewise, technical difficulties meant that the viaduct over the Yonne had to be built three times, reported Paul Séjourné. For such

189. Church at Le Vésinet from *The Builder*, 1 November 1865. Louis-Auguste Boileau, architect, François Coignet, contractor. Concrete walling with an internal iron frame, evident from the profile of the aisles.

reasons, after a flurry of popularity in France, Britain and America between about 1865 and 1875, concrete construction for engineering works and cheap buildings alike lost some of its momentum. Pasley's prognostications seemed to have been right. Then came reinforcement. Reinforced concrete was to liberate the structural potential of concrete and in the long run to strengthen the hand of professional engineers. But ideas about its expression lagged further behind, becoming after 1900 a fresh source of anxiety for those ever-insecure professionals, the architects.

THE ADVENT OF REINFORCEMENT

The possibilities of reinforcing cement or concrete to improve their performance in tension had been glimpsed as soon as above-ground concrete structures were contemplated in the 1830s. It was in 1832 that Marc Brunel erected his famous experimental semi-arch in brick at the Thames Tunnel yard, achieving a cantilever of sixty feet by reinforcing with hoop iron the same Roman cement as had been used for setting the bricks in the tunnel. Brunel's stated motives were to find out 'how far cheaper substances might be used for the construction of arches and also to show how far the cumbersome apparatus of centering might be dispensed with'. If that looks restricted in retrospect, it is a reminder that savings were the leitmotiv of early concrete construction.[25]

Many combinations of iron and concrete were devised and marketed for fireproof floors over the ensuing decades. But they were thought of as filling round iron joists, not as calculated iron strengthening of concrete. After 1863 the two materials were known to have similar coefficients of expansion. Papers, patents and experiments on iron and concrete litter the second half of the nineteenth century, the Anglo-American amateur Thaddeus Hyatt's of the 1870s being the most consistent and impressive.[26] Why was their outcome so fitful? One theory is that patents retarded development: 'infallibly, the competitive game passes from the site to the law-courts', argues Gwenael Delhumeau. Perhaps so in France. There a former gardener, Joseph Monier, took out sundry patents for farm-equipment, small structures and elements of buildings in cement with round iron bars. For lack of technique, resources and professional credibility he failed to exploit them to the full. Yet once German licensees got their hands on Monier's system, they raced ahead.

A surer explanation for the take-off of reinforced concrete after years of stalling lies in the improvement and control of the Vicat-derived Portland cements, and the substitution of steel for iron in reinforcing bars. The crucial development of better cements first took place in the highly regulated commercial culture of Germany, and largely accounts for the triumph of Monier's German licensees. Reinforced concrete, says Skempton, 'could not be put into effective practice until the strong Portland cements became available from the German factories with their scientific management'.[27] That involved closer chemical analysis, the finer grinding of raw materials, and improved kiln design and firing. The pioneers were the chemically trained manufacturer Rudolf Dyckerhoff, the private consultant Wilhelm Michaelis, and the testing engineer Johann Bauschinger. National specifications for cements were agreed in Germany in 1878, followed by France in 1885.

It is in Germany therefore that the model of engineering research stimulating commercial take-up best fits the case of reinforced concrete.[28] Even there, the experts responded in the first place to what entrepreneurs proposed. Germans were well aware of concrete's potential. At least one housing development of the 1870s, Victoriastadt in Berlin, availed itself of a British system of mass concrete, while there had been international interest in the Monier patents since his products were touted at the Paris 1867 Exhibition. In 1884 two contractors, Conrad Freytag and Philippe Josseaux, agreed to purchase the Monier rights for different areas of western Germany, with an option for extending them. But it was a third contractor-engineer, Gustav-Adolf Wayss, who in the next year made the decisive leap by taking on northern Germany, including Berlin. He went on to buy up Freytag and Josseaux, extend the Monier rights to the rest of Germany and Austria, and promote 'extensive experiments with structural members of reinforced concrete'. In 1886 Wayss asked Matthias Koenen, an architect-engineer with mathematical abilities then running the construction of Paul Wallot's Reichstag in Berlin, to test some experimental vaults. Having published the encouraging results, Koenen was drawn into Wayss and Freytag, the great contracting firm which survives to this day.

The ten thousand copies of the promotional booklet, *Das System Monier* (1887), which Wayss and Koenen issued simultaneously in Berlin and Vienna, mark the commencement

Das System Monier

___

Eisengerippe mit Cementumhüllung.

☗

Herausgegeben

von

G. A. Wayss

Wien, I. Bez., Elisabeth-Strasse 3.

———◈———

Wien 1887.

Druck von Becker & Horuberg, Berlin S.W., Hollmann-Strasse 22.

190. *Das System Monier.* Title page of the Vienna edition of the 1887 booklet distributed by G. A. Wayss. The earliest influential publication on reinforced concrete.

of concerted reinforced construction in Europe (ill. 190). The subtitle, 'Eisengerippe mit Cementumhüllung' – iron skeleton with cement infill – intimates that the system was first thought of as yet another fireproof flooring method, only with the iron or steel in bar form accurately positioned to counter tension forces. Quantities of Monier-system flooring were indeed laid in the latter stages of the Reichstag. As one element among many, it could have no effect on Wallot's architecture. Yet the Monier licence was broad, allowing the exploration of many technical possibilities and forms, notably arches, slabs and, later, shells. Wayss and Freytag exploited their monopoly with energy, relying on Koenen's data and the testing back-up of Bauschinger and others. Most early applications were to engineering and manufacturing equipment – things like reservoirs, tanks and vats that had been at the heart of Monier's own enterprise. But portions of buildings were present from the start (ill. 191). So too were bridges, fast increasing in scale from little footbridges to the 37-metre span of the Wildegg Bridge in Switzerland (1890). It has been claimed that 320 Monier-system bridges were built in Germany and Austria between 1887 and 1891.[29] It soon became clear that calculated reinforcement could give arched bridges an economy of materials and a visible leanness infeasible with masonry or mass concrete. They gave notice that reinforced concrete would challenge the look of structures. In time, architects would have to take notice.

Other early applicants of reinforced concrete diverge from this German pattern of alliance between state-supported technology and private entrepreneurship. Ernest Ransome must come first, since he built one of the earliest concerted series of permanent structures in the technique.[30] Ransome belongs to the strain of inventor-constructor: his father was an English manufacturing engineer who back in 1844 patented a concrete based on heating flints in an alkaline solution. He is thought to have sent out his son to represent him in the United States. Starting with concrete blocks, Ransome junior moved on to take out a patent for twisted square bars for monolithic construction in 1884. A trickle of reinforced structures followed on around his base in the Bay Area of California, mostly styled by a reputable San Francisco architect, George W. Percy. Highlights included a small bridge in the Golden Gate Park (1889) and the chic Leland Stanford Junior Museum at Stanford University (1890–2), in both of which the concrete was exposed and sedulously mimicked classical masonry (ill. 192). Doubtless neither Ransome nor Percy saw anything backward about that. But more interest has naturally been shown in Ransome's early Bay Area industrial interiors, built with floor slabs carried by reinforced beams and shallow jack-arches, all constructed monolithically.

For whatever reason, Ransome after his head-start proved slow to develop the promise of his system, while the Europeans raced ahead. At some point he left San Francisco for Chicago. Only in 1897 did he extend his network by licence to the East Coast; and only from 1903, when rivals were challenging his methods, did he come up with the sophisticated Ransome 'unit construction' system, whereby portions of the structures were pre-cast and integrated with in-situ elements. Thereafter the exteriors of his buildings shed their load-bearing cavity walls and converted to full frame-construction with wide window-openings (ill. 192).

In *A Concrete Atlantis*, Reyner Banham addresses Ransome's industrial architecture with quixotic gusto. Engaging with him as if he were a designer rather than an inventor, he is

TOP

191. The Alberthalle, part of the Krystallpalast at the Leipzig Exhibition of 1887. Arwed Rossbach, architect. The building combined circus, diorama and auditorium on different levels. *Leipzig und seiner Bauten* (1892) reported that the whole construction used the Monier system but may have exaggerated. In its early architectural uses, reinforced concrete was typically used only for portions of structures.

BOTTOM

192. Leland Stanford Junior Museum, Stanford University. George W. Percy, architect, Ernest Ransome, concrete contractor, 1890–2. The construction of the original museum was entirely in concrete. Later additions in masonry suffered more severely from the San Francisco earthquake of 1906.

193. Topping out on a block of the United Shoe Machinery Company's headquarters, Beverly, Massachusetts, c.1904. The concrete frame adopts Ernest Ransome's unit construction system.

half-amused, half-annoyed by Ransome's 'cultural timidity' in permitting his early structures to be styled by conservative architects, or garnished with limp surface rustication and 'Eastlake/Queen Anne' details. How, he asks, could so profound an innovator have failed to carry architects and engineers along with the implications of his ingenuity? The answer is twofold. A creator of components or even of a building system may care little what the overall result looks like, or about the cultural limitations of his time, so long as they guarantee him work and opportunity. There is a deeper point. The delayed spread of Ransome's methods within a free-market economy contrasts with the swift take-up of the Monier system in more highly regulated Germany. Ransome and Wayss both enjoyed something approaching monopolies on reinforced concrete for about a decade. An ingenious individualist, Ransome paddled his own canoe. Wayss, on the other hand, grasped the value of linking patents to prompt technical publication and official support, and so achieved greater rapport with the professions. Organization, not originality of design, was the key.

THE FRENCH SYSTEMS

Creating an organization was the key to the triumph of François Hennebique, pre-eminent in the worldwide diffusion of reinforced concrete after 1890. By 1905 this self-educated builder was carrying all before him, with a fifth of the world market in this new means of construction, fifty design offices under his auspices, and 10,000 workmen on site building in Hennebique concrete every day. Analysis of Hennebique means analysis of his organization: not that of a brain in a body but that of a tree or a nervous system, argues Jacques Gubler; not an empire but a diocese, prefers Gwenaël Delhumeau in his study of the firm.[31]

Unlike many pioneers of reinforced concrete, Hennebique started off as a practical builder: in Delhumeau's phrase 'un homme de chantier',[32] someone who knew about materials, men and how to fit them together. A master-carpenter by training, he had run his own building firm for over twenty years before he began exploiting concrete systematically from about 1889. Arras was his original base of operations; and France's northern borders and Walloon southern Belgium, teeming with industrial growth, were the launchpad for his system. Hennebique took out his first concrete patent of 1892 in Brussels, and kept his headquarters there till 1897, the laxer building regulations of Belgium's enterprise-culture making structural innovation easier there.

Nevertheless France was the bigger market and better centre of operations.[33] It was from Paris with its aptitude for communication that Hennebique came to preside over the international market for reinforced concrete. Like Ransome, Hennebique himself thoroughly grasped the principles and practice of reinforcement. Once his system took off he abandoned building for design. To that extent he was an engineer. But not having a school training or direct access to academic means of disseminating engineering research, he needed and secured articulate publicity – the touch that eluded Ransome. A brochure of 1892, *Plus d'incendies désastreux*, his equivalent to *Das System Monier*, was smartly picked up by the technical press. Henceforward Hennebique bombarded building professionals and clients with publicity, culminating in annual congresses of his representatives and a magazine, *Le béton armé*.

Hennebique's engagement with architects and engineers followed from the way in which he licensed his patents and the types of structures to which they were relevant. Rights for using the system were delegated from headquarters to regional licensees – building firms who competed for jobs, carried out the reinforced concrete element in contracts and remitted ten per cent to the centre. Between headquarters and the licensees, as the business expanded, there grew up another tier, the regional or national agents who were there to represent Hennebique and ensure the jobs were properly designed and built. In practice the division of labour between the centre and the agents was fluid. More than in other early reinforced concrete methods, the core of the technical system consisted of standard but regularly updated building elements – beams, columns, heads of columns, roof-trusses, floors etc., together with details of the reinforcement's design and positioning, including the trademark Hennebique 'étriers' or stirrups. The skill was how to apply all that to the variety of commissions. Some of this design work was done in-house by the central office or bureau d'études, some of it by the agent, who ran a similar bureau. Both were essentially specialist providers of engineering design.

The survival of the Hennebique archive has allowed Delhumeau to glean much about his own staffing. In 1896 there were some eleven in the Brussels office: his original partner from the building industry, Septon, in charge; three members of the Hennebique family (his son Edouard as manager and two daughters helping with the accounts); a 'chef d'études', Henry Foort; and under him two engineers, two architects and two draughtsmen. A middling-sized family firm, in other words. Next year the Paris bureau d'études was being run by Hennebique's son-in-law, Georges Flament, with rather more engineers; and by 1905 the complement had risen to 65 'engineers', draughtsmen and typists. The best, including Flament, had enjoyed formal training at the Ecole Centrale des Arts et Manufactures, a rich source for engineer-constructors under the French educational system (p. 436). After an induction, sometimes quite brief, into the Hennebique methodology, a chosen few were sent out as agents.

Of certain agents we know more. Not all were alumni of the Hennebique nursery. Samuel de Mollins in the Suisse Romande was a school-trained engineer from Lausanne with local connections and a name of his own. At his initiative the firm in 1894 put in for but failed to win one of its first major bridge contracts, in Geneva. In the aftermath, Hennebique rebuked de Mollins for too much independence, but the venture brought him cachet in a country where professionalism was respected. De Mollins did his best to woo the great Zurich academic engineer Wilhelm Ritter in an attempt to challenge the Swiss dominance of Wayss and Freytag. In vain: the German firm's chief engineer, Emil Mörsch, transferred to become a professor at the Zurich ETH, and wrote the first major textbook on reinforced concrete.[34] After 1900, three of Hennebique's most successful agents, Eduard Züblin in Strasbourg, Giovanni Antonio Porcheddu in Italy and Louis Gustave Mouchel in Britain all enjoyed measures of freedom in developing the patents. According to one of his subordinates, Mouchel was not a trained engineer; but then neither was Hennebique. To our knowledge only one agent called himself an architect: René Martin in Rouen.

Hennebique had to deal also with the independent professionals charged with the structures for which the licensees tendered. This was often a delicate relationship, as use of the system might involve reconfiguring some of what had already been designed. Hennebique's earliest experiments had been with reinforcing floors in houses designed by architects. But the first complete buildings erected in the 1890s under his patents were industrial, mostly around Lille and Tourcoing – mills, refineries and warehouses or single-storey sheds with north-light roof trusses. Industrial clients seldom wanted 'architecture', only cheap, sturdy, fireproof construction, along with the unimpeded spans feasible with a reinforced concrete frame (ills. 194, 195). Of modest status, their architects or engineers usually went along with the system, giving Hennebique and his agents leeway in

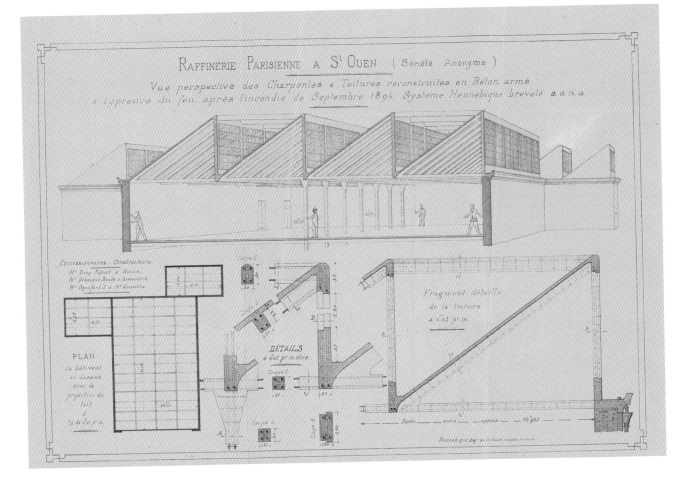

194. Raffinerie Parisienne, St Ouen. Plan, details and perspective of shed with north-light roof truss. François Hennebique's first major job in the Paris region, 1894.

195. Load-testing in warehouse of the Chambre de Commerce, Nantes, built to the Hennebique system, showing his standard columns and two-way grid of dropped floor beams. Clériceau and Tessiers, architects, Ducos builder, 1902–3.

the design. In this way its elements gradually asserted their own character, starting internally before working their way outwards, as in the iron-framed mills of a hundred years previously. By 1900 the ubiquity of Hennebique structures meant that a reinforced concrete 'look' – a framed look when it came to buildings – was growing familiar in France and beyond. It had not been self-consciously designed. But it set its mark on the next generation.

When the Hennebique system was applied to the average industrial job, it had no need to compromise; it could look like and be itself. Structurally demanding or smart work was different. Like previous purveyors of concrete, Hennebique did well out of infrastructure – docks and harbours, railways, retaining walls and so forth. Big bridges were rarer. His system had developed to cope with short trabeated spans, while the Ponts et Chaussées of the day was largely preoccupied with steel bridges. The Pont de Châtellerault over the Vienne (1899–1900), a major three-arched bridge with spans of 40–50–40 metres, was the exception. It followed from a rare decision by the Ponts et Chaussées to try out concrete and trust Hennebique to make it work. The outcome was successful and useful to his publicity machine, but involved no specially creative partnership between engineer and entrepreneur.

As for architecture, any new building technology usually starts off by abasing itself to

the discipline's traditions and fashions. That was the pattern with Ransome's Bay Area buildings. Hennebique did the same. His wooing of architects dwelt upon showing that his concrete could build anything they dreamt up, from the swish staircase of the Grand Palais in Paris (1899–1900) to a complete traditional Armenian church in Tiflis (1903) or the Indo-Cambodian villa of the Baron Empain at Heliopolis, Egypt (1907–10) – surely the queerest of the early monolithic structures (ill. 196). With reinforced concrete you can replicate almost any kind of structure; it must have been fun, if costly in formwork, to show that. Commoner were the jobs where a money-saving concrete frame and floors were prefaced by a frontispiece of brick or stone for neighbourliness and dignity. Here the early preoccupation was not honest appearance, but organization and order: archi-tect and agent had so to design the building and arrange the contracts, that the licensee could erect the concrete-work and leave the site clear for the masons to arrive and com-plete the elevations and enrichments. The same applied to the fixing of stonework to a steel frame, also then in vogue.

The thought that architects should temper their styles to concrete construction took time to penetrate and win acceptance, if far less long than had been the case with iron-work. The speed of that shift can be put down to three factors: the general quickening of cultural and technological change; concrete's capacity to provide solid walls, which iron and steel could not; and the persistence of rationalist rhetoric in French architec-tural discourse, with which went lip-service at the least to the nature of materials.

One early champion of cautious adaptation was Louis-Charles Boileau, son and suc-cessor to the architect of St Eugène and the church at Le Vésinet, prolific journalist, sec-retary-general of the Société Centrale des Architectes, and main designer of the Bon Marché store. In that last context, Boileau fils had been among the first to articulate what iron entailed for architectural composition (pp. 153–5). Some twenty years later, contem-plating reinforced concrete for a block of stables and workshops for the Bon Marché in 1895, he encountered Hennebique. He toured a selection of Hennebique structures and was impressed; so was his friend and builder, Dumesnil. So Dumesnil undertook small trial works and became Hennebique's main licensee for Paris.

In a first flush of enthusiasm, Boileau wrote some articles evangelizing on behalf of the method. They led up to the architect's eternal question: what form should concrete take, in other words, what ought concrete buildings to look like on the outside? Unlike iron, he said, concrete offered surfaces which you could stucco, marble, paint, or even fresco – and Boileau lavished a full article on the future of fresco. All this went little further than Coignet years before; in the city street, even for workshops and stables, it was not proposed to show concrete. An expert might have deduced that the Bon Marché stables had a framed structure, but a passer-by would not have noticed, since it was pitch-roofed and brick-clad (ill. 197). Nevertheless now the debate had been taken up by an architect.[35]

There was a sequel. Late on in the preparations for the 1900 Paris Exhibition, Dumesnil and other licensees asked Boileau to design them a Hennebique-system terrace and pavilion, wherein their efforts could be illustrated. The project came to nothing. But the liberty of a free-standing exhibition structure prompted Boileau to experiment. Though the overall style of his proposal was conventional, he used surface dec-oration in such a way as to allude to the reinforce-ment within the beams (ill. 198).[36] So naïve a response to rationalist propaganda betrays only how marginal was architects' engagement in devel-oping the concrete systems. They mattered little to Hennebique, except as a conduit for publicity and business.

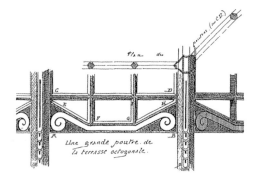

198. Proposed terrace and pavilion for exhibiting Hennebique's achievements in reinforced concrete. Louis-Charles Boileau, architect. The decorative detail of this unexecuted project for the Paris 1900 Exposition draws attention to the reinforcement within the beams. From *L'Architecture*, 13 January 1906.

By contrast, the one concerted French attempt to subject the early reinforced concrete systems to so-called architectural rationalism ended ignominiously. That was the programme of Viollet-le-Duc's spiritual heir, Anatole de Baudot, to champion Cottancin's 'reinforced cement' against Hennebique's system.[37] The history of that rivalry runs pari passu with a jockeying between two competing architectural clubs in 1890s Paris: Boileau's Société Centrale and de Baudot's Union Syndicale des Architectes. Both groups were keen to turn the new material to advantage – on their own terms.

Paul Cottancin was a 'centralien' – an engineer trained at the Ecole Centrale. In 1889 he took out a first patent for a concrete with a high cement content, little aggregate, and reinforcement in mesh form. Unlike Hennebique, who took specific components as his starting point and then assembled them, Cottancin's was a 'continuous' system. Extended by stages to take in the reinforcement of brickwork, it promised well for thin shells, vaults, arches and floors and enjoyed successes for infrastructural works as well as buildings (ill. 199). But Cottancin was vulnerable to the assaults of his competitors: he lacked the resolute independence of Hennebique or the ability to justify his ideas by calculation and so win over the engineering establishment. 'The constructors in reinforced cement have got to understand that the era of prophets is over', growled Napoléon de Tedesco, one of Cottancin's rivals. 'The engineering world is quite prepared to apply [the system], but it will not agree to do so unless the means for checking the proposed constructions is available.'[38]

Early on, Cottancin tried to interest architects in his method and approached de Baudot, who was puzzling out how Viollet-le-Duc's doctrines might be better applied. In his *Entretiens*, the latter had bestowed his blessing upon ironwork combined with masonry walling as heir to the Gothic tradition. That lacked unity, felt de Baudot. He also found iron and steel structures inflexible. For the architect, the advent of Cottancin's reinforced cement was an epiphany, like finding a philosopher's stone, says Marie-Jeanne Dumont. It could do everything Viollet-le-Duc had wanted, yet remain a single building system and keep faith with the Gothic spirit of wall, arch and vault.

199. Hall and roof in the Cottancin system near Paris, *c.*1900. From *Concrete and Constructional Engineering* July 1906.

De Baudot first tried out Cottancin on some floors in his own house and then on portions of a school. The system's real test came at Saint-Jean de Montmartre (ills. 200, 201). That famous Parisian church blends the integrity and awkwardness that mark much French rationalist architecture. So litigious was the progress of its construction that an English expert in concrete dubbed the enterprise 'the folly of the century'.[39] The project started out against a background of antagonism between the reluctant civil authorities, who had to give permission for any new church, and its promoter, the zealous Abbé Sobaux. These difficulties were compounded by a clash between architects and concrete patentees. Edouard Bérard, an architect attached to the Edifices Diocésains organization that was de Baudot's power-base, believed he had been promised the commission and asked Hennebique to sort out a structural solution for the precipitously sloping site. But by the end of 1894 Bérard and Hennebique had been displaced by de Baudot and Cottancin. Only once that dispute had been sorted out could building get going in 1897 – against the backdrop of a separate suit brought against Hennebique by Cottancin for infringement of patents. With the deep basement complete and the superstructure well advanced, the collapse of a concrete footbridge at the 1900 Exhibition, with fatalities, led to a municipal stop on the works at the church, which had not been fully notified to the authorities, and to scrutiny of their novelty and alleged frailty (as well as 'ciment armé', the superstructure relied on the even less familiar reinforced brickwork). That and a second stop delayed the project by two full years. Before Saint-Jean de Montmartre could

200. Saint-Jean de Montmartre, builders on site erecting ribs of the vaults in Cottancin's reinforced brickwork, *c.* 1902.

be completed in 1904, Cottancin had lost control of his company and system and parted bitterly from de Baudot. Hence fresh recourse to the law courts. Soon afterwards he went bankrupt.

How much of this sorry tale was due to bad relations between French Church and State, and how much to Cottancin's inability to justify his system to the appointed committee of experts, is hard to divine. Peter Collins presents it as one of heroic innovation obstructed. But it was not unreasonable for the authorities to insist on examining the principles of 'continuous construction'. The Cottancin system was never prohibited, and seems to have been in growing use at the time of his failure.[40] All along, its inventor's difficulty was his inability to come up with the proofs everyone wanted. He was an old-style, intuitive engineer-designer in an age of creeping mathematization and regulation. In his case, the exactitude of modernity inhibited creativity. As for de Baudot, he looked upon Cottancin's system as a means to furthering preconceived ideas about architecture. That those ideas had implications for structure did not make them less arbitrary than the conventionalities of style. The deterioration of his alliance with Cottancin hardly suggests an equal or comfortable relationship with his technologist. Certainly, Saint-Jean de Montmartre shows deeper study of how concrete might be made expressive than the shallow response of Boileau fils. Yet it also reads like an intellectual essay, from which the pretty ceramics of Alexandre Bigot on the front and the stained glass within furnish welcome relief.[41]

Just two years after Saint-Jean de Montmartre opened, a lengthy official evaluation of France's concrete systems culminated in a ministerial circular of October 1906. The outcome endorsed proofs by testing and calculation against mere empirical ingenuity, and strengthened the hands of the pure, professional engineers against their horny-handed rivals, the ingénieurs-constructeurs.[42] At first the Ponts et Chaussées had been slow on the uptake over reinforced concrete, just as they had been years earlier over iron. Now they had woken up to the challenge; Charles Rabut had for some time been teaching and updating a course on the technique at their Ecole. Typically for the top-down thinking of the Ecole, Rabut's ideal was that independent engineers would come up with struc-

tural solutions which the systems-owners – Coignet, Cottancin, Hennebique and the rest – would then adapt to the case in hand. With the patent-holders still dominant, things seldom yet worked that way round. But the Ponts et Chaussées contrived that the only one of its alumni to have invented a promising concrete system, Armand Considère, became secretary of the evaluating committee. The membership consisted of five Ponts et Chaussées engineers including Rabut, two Génie officers, two architects, two systems-men (Edmond Coignet and Hennebique), and a cement manufacturer.

Cottancin's system got a mauling by the committee. To the extent that it survived afterwards, it was used mainly for restoration work, as in the post-1918 reconstruction of the Rheims Cathedral roofs. Nonetheless until his death de Baudot and a dwindling band of followers in the Union Syndicale des Architectes went on advocating 'reinforced cement', as opposed to 'reinforced concrete'. The latter had only triumphed, de Baudot argued down to his death, because architects left construction to engineers heedless of composition. Once the two were put back together, the 'puissance créatrice' of the new material could be released.[43] He was not wrong. But when that took place in the hands of the Perret brothers, the results turned out quite different from what de Baudot had pressed for. In part that was because they were based on a better grasp of practical building.

DEMISE OF THE SYSTEMS

The Perrets could only have emerged when they did. All over the western world, recommendations by expert commissions like the French one led to new national or urban codes for reinforced concrete round about 1905–8. Suddenly it became simpler to build major buildings in the material. Acknowledged by now as a cheaper technique than load-bearing masonry for large buildings of up to about six storeys, there was every incentive to use it. The short-term effect was to boost new patent concrete systems. Among the main beneficiaries were Considère, whose matured system naturally embodied his own committee's recommendations, and the American-based Kahn system (pp. 243–5). These and others now rose to challenge Hennebique, Wayss and Freytag and other early patentees. But in the longer term, as old patents expired and a choice between methods became realistic, Hennebique-style modes of control broke down. Under the tutelage of teachers like Rabut, official and private consulting engineers adept in concrete emerged. In parallel, architects started thinking harder about how to design in the material and what a concrete architecture might look like.

In summary, the early years of reinforced concrete present a chequered picture from a professional standpoint. In those countries where the technique made rapid inroads between 1885 and 1910, pure engineers contributed little to its discovery – 'one of those paradoxical inventions which disconcert science', noted Arthur Vierendeel.[44] The prime movers tended to be entrepreneurial contractors, some from the building trades like Wayss and Hennebique, others largely inventors like Ransome. Though these men are sometimes called engineers, few had enjoyed much formal education, while the ingenious Cottancin, despite his training at the Ecole Centrale, failed to conform to the rigorous science expected of his profession by 1900. Yet the proprietors of the concrete systems relied upon trained engineers – both those of the second rank to act as their agents, fill their offices and work out their schemes, or those of higher status to guarantee their materials and to test and extend their methods, as happened most promptly and conspicuously in Germany. The formulation of rules and codes for cements, the disposition of reinforcement and the sizing of elements, together with the advent of courses on reinforced concrete in the schools, strengthened the position of official and independent engineers after about 1905. But they never controlled the technique or the trade. The subordination of the pure engineer to the engineer-contractor survived the concrete systems, extending throughout the inter-war period and beyond, as the trajectories of Freyssinet, Arup, Nervi and many others betray.

FACING PAGE
201. Saint-Jean de Montmartre, the completed front. Anatole de Baudot, architect, 1897–1904.

Architects enjoyed only a tangential relationship with the pioneer engineer-contractors in reinforced concrete. They certainly did not control them. The architect who did the most to further reinforced concrete before 1900 was Matthias Koenen in Berlin, essentially a technician, who extended and published the Monier system as licensed by Wayss. The commoner role of architects was to act as propagandists by styling concrete buildings (whether the material was hidden or shown) and by enthusing on the technique to a wider public. George W. Percy championed Ernest Ransome's early efforts in that way, as Boileau fils did for Hennebique, while Anatole de Baudot tried to do the same for Cottancin. Only in that last case did the relationship imply much about architectural style – and only in that case did a setback occur. Though the proprietors of the early concrete systems sometimes needed architects and were willing to ape established styles, they saw no reason to surrender to some architectural agenda supposedly based on the technique they peddled. How relations worked out as such agenda matured forms the topic of the second half of this chapter.

In one major country alone did architects do more for the advancement of early reinforced concrete. That was Britain, a latecomer in adopting the technique. Why Britain should have lagged as it did is a puzzle, given the abundance of buildings and patents for concrete there up to about 1880. The first full building to exploit the technique, Weavers Mill, Swansea, dated from as late as 1897. It was constructed by Mouchel, Hennebique's British licensee; and for a few years Mouchel enjoyed a near-monopoly, until the Kahn system from America and some others made inroads into a growing market.[45] Not until 1907 did a British engineer, E. P. Wells, introduce a homegrown system.

No less tardy in advancing the steel frame, British construction suffered from technological inertia at this time. At the root of the problem lay the hidebound Institution of Civil Engineers, which for years treated reinforced concrete with alleged 'indifference or hostility'.[46] Though early British engineering specialists and writers of textbooks on the subject there certainly were, they seem to have operated in an institutional vacuum. It was therefore left to three very different architects not so much to advance the technique of reinforced concrete as to fight for its take-up.

The first was an autodidact: William Dunn, a Scot from a poor family, who taught himself higher mathematics, took engineering qualifications to add to his architectural skills, and strove to mitigate the ignorance of his architect-colleagues about modern construction.[47] Dunn was the prime mover behind the British equivalent to the French investigating committee of 1901–6, the Royal Institute of British Architects' committee on reinforced concrete of 1906–7. Its chairman was Henry Tanner, the Chief Architect to the Office of Works, and therefore in effect (though in the British way, informally) the representative of government. Having noted the savings big private companies such as the Great Western Railway had made by using the Hennebique–Mouchel system, Tanner took it upon himself to explore the material in a series of large postal sorting offices and to promote a series of model regulations through the RIBA's committee. Advanced technical input into this committee was lacking, because the 'Civils' refused to join in. But the upshot was a workmanlike set of recommendations that soon passed into urban building regulations, easing the path for British constructors to build in reinforced concrete.

Perhaps the most impressive contribution came from a private architect in a small way of practice, Edwin Sachs.[48] Born in London to a Hamburg merchant family, Sachs trained and first worked in Berlin, returning with a range of international and technical enthusiasms then rare in British architecture. His passions were theatre design and fire prevention. Sachs came to see building regulation and concrete construction as means to these ends. In 1897 he set up the British Fire Prevention Committee, which tested materials including concrete in a small permanent fire-testing station (Europe's first) built at his own expense. After reinforced construction took hold, he became the moving force behind the Concrete Institute. From its founding in 1908 this organization blossomed into the rendezvous for experts in concrete design, evolving in time into the Institution of Structural Engineers. Britain's second major membership organization for construction engineers was thus inaugurated by an architect.[49]

Though Sachs ceased to guide the Concrete Institute's later activities, so long as he lived he did control and edit *Concrete and Constructional Engineering*. That monthly periodical, started in 1906, sought to promote debate and information with an objectivity absent from the sundry in-house journals promoting the systems. *Concrete and Constructional Engineering* was perhaps the liveliest of the early specialist periodicals on concrete and cement. The range and literacy of its articles compared with the best architectural magazines, and helped raise awareness of the subject's range across professions and nations.

Though Dunn, Tanner and Sachs were all architects, none of them spent much time bothering about the external expression of concrete. The priority was to get the material rightly specified, built and regulated, and thus understood, accepted and diffused. Yet as they went about their work the climate was changing. Architects around the world, even in backward Britain, were starting to echo the questions posed by Boileau fils and de Baudot: what should a concrete building look like?[50] To answer that question, no one could afford to work alone. Before there could be effective styles of concrete architecture, there had to be effective styles of partnership.

2.   *Styles of Partnership*

THE PERRETS

The Perrets were builders before they were architects.[51] What they never were, at least in name, was engineers. Yet like any major building enterprise of their day they employed in-house engineers and consigned the detailed working-out of their designs to a bureau d'études akin to those of Eiffel and Hennebique. Just as Eiffel and his contracting contemporaries were known as 'ingénieurs-constructeurs', so the older Perret brothers described themselves as 'architectes-constructeurs'. That label offers the most insight into their skills and procedures.

The Perret story begins with the building firm set up by Claude-Marie Perret, stonemason and Communard, during his exile in Brussels in the 1870s. It was already prospering before the amnesty of 1880 and Claude-Marie's return to Paris.[52] Along with him came a large young family: all three of the sons who were to make the Perret name illustrious, Auguste, Gustave and Claude, were Brussels-born. Back in Paris, a period of speculative enterprise ensued. Auguste and two years later Gustave entered the Ecole des Beaux-Arts and the atelier of Julien Guadet, champion of the classicizing faction within French architecture's disputatious rationalist tendency. Thus they followed the time-honoured custom whereby successful builders' sons step up a social rank and join the design professions. Auguste (ill. 202) soon stood out for his talents and sturdy temperament.

But the mid-1890s saw a downturn in the father's fortunes. The older sons therefore curtailed their time at the Beaux-Arts to help support the family, while young Claude Perret may have lost his chance of higher technical training. Eventually the ailing father started the hand-over to his able sons; in 1902 the firm became 'Perret et ses fils',[53] and in 1905 Claude-Marie died, aged 58. That early immersion in building after their time under Guadet gave the brothers a rounded 'formation' that most Beaux-Arts architects failed to acquire. But they never doubted the primacy of design. Auguste above all was not going to relinquish the pleasures and powers of architecture.

Our sense of the first practical design activities of the older brothers is sketchy. Very likely it was intermittent and tied to their father's fortunes. As early as 1894 Auguste is found helping on a series of buildings for the family firm in the Ménilmontant district of Paris. On his return from military service he is joined by Gustave; they start styling themselves A. & G. Perret. An urban industrial building dates from 1897. There is then a blank in Paris until they embark from 1902 on a sequence of four apartment-houses in the

smart 16th and 17th arrondissements.[54] That chimes with the sons' revival of the building firm. A single big commission further afield fills the gap: the Casino at Saint-Malo (1898–9), an exuberant essay in the 'style Normande' (ill. 203). There were independent contractors for the casino, but a photograph showing Claude-Marie Perret on site hints at family involvement. Concealed behind the florid finishes, one portion of the upper floors was in reinforced concrete; the architects may have wanted more. We are told that the experiment set them off investigating sites where concrete was used.[55]

Of the four Paris apartment blocks that Auguste and Gustave Perret designed between 1902 and 1906, all but the famous 25 bis rue Franklin display conventional stone-clad fronts, construction and planning.[56] Rue Franklin, second in the series (1903–4), seems to have been the only one of the four wholly controlled by the Perret family, without an adjunct developer-client.[57] Destined to house both the office of the rejuvenated family firm and rooftop flats for the brothers, it was also the trickiest site to develop. Here was a chance to experiment and advertise – to show how space and grace could be squeezed out of a shallow plot by cutting structure to the bone and employing an attenuated concrete frame. If it did not invent the notion of frame and infill in architecture, this brilliant building revolutionized it. It also offered a sop to humanity by refreshing the surfaces between frame and fenes-

tration with Alexandre Bigot's flamboyant leaf–patterns in faience (ills. 181, 204). Later, Auguste Perret felt he needed to excuse the all-over ceramic covering of frame and infill (laboriously distinguished): they had believed then, he said, that the reinforcement needed that protection. In view of what has happened to many early buildings in which reinforced concrete was exposed, including those by the Perrets, the decision was surely wise. In any case the Bigot panels lent allure to what would otherwise have been an austere, unmarketable building. The time is surely past for it to be insisted that the concrete should have been shown and the decoration suppressed.

25 bis rue Franklin fulfilled the hopes of the rationalist camp in French architecture by marrying clarity with suavity in a way unseen since the Bibliothèque Sainte-Geneviève. That is why it became one of the most pored-over buildings of its century, and why narratives of the architectural response to concrete often start with Auguste Perret. Much of his later work hits the same felicitous note. The Perrets were to stamp artistic authority and personality on concrete in a way Boileau junior, de Baudot and others never managed. But the virtues of their architecture concern this chapter less than its means of procurement and the collaborations that entailed.

Having, so it seems, tried out concrete before only in a fragment of the Saint-Malo casino, the Perrets were still novices in its use both as architects and as builders at the time of Rue Franklin. They needed to learn, and to hire specialist subcontractors for their concrete frame in order to do so. In 1903 the choices were limited, though growing. Hennebique and Coignet were the obvious people to go to. Maybe fearing that they would be constrained by those established systems, the Perrets opted instead for the little firm of Latron et Vincent, ex-employees of Edmond Coignet who had just set up on their own. Like many of the senior staff under Hennebique and Coignet, Latron was a 'centralien', in other words a trained engineer. He and his partner had built no previously framed buildings on their own account. As they had to avoid infringing existing patents, realizing the frame at Rue Franklin must have been an education for both sides. The key to the method was a type of 'béton fretté' – concrete with a style of spiralled reinforcement just then coming into vogue. Auguste was pleased with the result: 'work completed in ninety days and in absolutely perfect condition', he told the commission looking into concrete systems.[58] We do not know what became of Latron et Vincent. But for the Perrets the experience was decisive. Only on the basis of the lessons from Rue Franklin could the bond between architects and building firm be tightened and Auguste come up with his trademark boast: 'Je fais du béton'.

The second well-known building in the Perret oeuvre is the garage in Rue de Ponthieu, Paris (1906–7).[59] Famous for baring its concrete frame towards the street front, this was also the first job in which the brothers' building arm – reconstituted as Perret Frères after Claude-Marie's death – tackled the concrete work themselves.[60] That had become easier after the official commission published its guidelines for concrete construction in 1906. All the same they had difficulty, Auguste recalled, in creating gallery floors able to carry both the cars and the rolling iron bridge between them (ill. 205). There seems no trace of an engineer's involvement with this semi-industrial structure. The transition to competence and, soon, fluency in concrete construction raises the question of which brothers did what. Oscar Nitzchké, who was in the Perret circle later on, alleged that Auguste was the architect, Gustave (the shadowiest brother) the engineer and Claude the money man. Experts have dismissed this as 'reductive'. Other than that Auguste and Gustave together did the architecture while Claude ran the building side, the fact of the matter is that we do not know.[61]

If a garage was a banal start, the Perrets were always bent on extending their integrated style of procurement to high-class commissions. So long as a design-and-build firm limits itself to infrastructure and modest buildings, it will usually escape architects' hostility. Not so when it tackles a church or a theatre. Once the brothers took that path, Auguste found himself in a curious position that persisted until the creation of the Ordre

des Architectes in 1940. On the one hand he came to be revered as France's leading architect, and one whose pronouncements and teaching invariably upheld the dignity of his art. On the other hand his stance as an 'architecte-constructeur' linked with a building firm stoked conflicts of interest. That contravened the code for an independent profession written by his old master Julien Guadet and respected by most architects who had attended the Beaux-Arts; and it led to him more than once worming his way via his chosen medium of concrete construction into jobs begun by other architects. There were always two faces to Auguste Perret: the champion of a noble, indissoluble link between art-architecture and construction, and the thrusting entrepreneur on the Hennebique model, deploying his skills and network in the interests of his close-knit clan.

The notorious occasion when the Perrets displaced another architect was the Théâtre des Champs-Elysées. It was foreshadowed by the cathedral at Oran in Algeria (ill. 206) – the first of many jobs carried out by the Perrets in that country.[62] Oran Cathedral had been designed in 1900 by the architect-cum-archaeologist Albert Ballu. Claude-Marie Perret, an old friend of Ballu's, was expected to build it, but Paul Cottancin nipped in with a lower bid using his ribbed system of 'ciment armé'. With the crypt complete, work

204. 25 bis Rue Franklin, Paris. A. & G. Perret, architects, with Latron et Vincent, engineers for the concrete system, 1903–4.

205. Garage, Rue de Ponthieu, Paris, interior with rolling steel bridge. A. & G. Perret, architects, with Perret Frères, contractors, 1906–7.

206. Oran Cathedral nearing completion in February 1910. Albert Ballu supplemented by A. & G. Perret, architects, Perret Frères, contractors, 1908–12.

ground to a halt after problems with subsidence and the demise of Cottancin's fortunes. In 1908 the Perret sons, adept now in reinforced concrete, were brought in to finish the job. It was their trickiest assignment yet, entailing pendentive domes and a measure of compromise between Cottancin's system and their own framing methods. In the sequel Ballu's design was simplified and streamlined by Auguste and Gustave Perret, without demur on the older architect's part. Oran Cathedral emerged from its scaffolding in 1912 as much the Perrets' building as Ballu's, and with the first appearance of their trademark concrete 'claustra' in the windows to prove it.

What happened at the Théâtre des Champs-Elysées was not dissimilar, only with a grander cast of characters and the added spice of scheming, backbiting and publicity that art-projects in Paris attract. The skein of events surrounding that august 'temple of music' is summarized here only in so far as it illuminates professional relations and the modus operandi of the Perrets.[63] The theatre already had an architect, Roger Bouvard,

207. Théâtre des Champs-Elysées, Paris, the grand stair. A. & G. Perret, architects, Perret Frères, contractors, 1911–13.

and a consultant engineer, Eugène Milon, by the time the present site on the Avenue Montaigne was bought in January 1910. The architect-decorator Henry van de Velde, head of the Weimar Kunstgewerbeschule, was passing through Paris just then. Claiming expertise in modern music-drama, Van de Velde was drawn into the job and paired up with Bouvard by Gabriel Thomas, the theatre's main fund-raiser. His appointment added a cosmopolitan touch to the enterprise.

Many changes of plan ensued, dictated by differences of taste, shortage of funds, and the dangers (pointed out by Milon) of inundation if the orchestra and stage foundations were dug too deep. In the last such revision before the Perrets came into the picture, further advice from the engineer on constructional issues and their effect upon sightlines led to a circular reconfiguration of the auditorium by Van de Velde early in 1911. On the latter's advice, concrete construction was now considered and Perret Frères submitted a tender. Their bid, confirmed after investigation by Milon, offered radical savings. But it went

208. Ateliers Voisin-Marinoni, Montataire, addition of 1926, showing the north-light semi-vault system patented by Perret Frères with Louis Gellusseau.

beyond the tender drawings by proposing modifications to the auditorium's structure. Once aboard so prestigious a vessel, Auguste Perret was not to be deflected, inexorably imposing his own course and his immaculate concrete classicism, here first made manifest as the topmost layer in this palimpsest of a project (ill. 207). The infrequency of Van de Velde's appearances in Paris and his dilettantism in matters of construction were a gift to the Perrets. But the main reason they prevailed was their ability to save the management money without depriving the theatre of artistic authority. By the summer of 1911 Van de Velde was off the job in all but name and Bouvard reduced to an administrator.

Structurally, the Théâtre des Champs-Elysées may have been no trickier than Oran Cathedral. It had undergone much engineering scrutiny before the Perrets submitted it to the disciplines of concrete framing. Nevertheless the job's arrival in their portfolio coincides with our first inkling of an interdisciplinary technical staff and the formalization of a bureau d'études, linking the two Perret firms and freeing the architect-brothers for creative endeavours. To help build the theatre they in April 1911 hired as 'chef des études et des travaux' the engineer Louis Gellusseau, a 'centralien' and an ex-employee of Edmond Coignet, like their first coadjutors in concrete, Latron and Vincent.[64] For many years Gellusseau became a fixture in the office. Though a technician, he was no dogsbody. He was soon entrusted with, or volunteered for, the job of publishing the structural solution for the Théâtre des Champs-Elysées in *Le Génie Civil*. The task was repeated for the Ateliers Esders and, with outside help, for the church at Le Raincy.

Gellusseau was also bound up with the industrial projects that became the staple of the Perrets' work during the First World War and its aftermath, the toilsome years of 'la première reconstruction'.[65] Then it was that the brothers developed their system of shallow-vaulted floors and parabolic roofs in shell concrete. It enjoyed its first outing in the Wallut warehouses at Casablanca (1916–17), reaching maturity in a bevy of post-war workshops, factories and warehouses – all now demolished or radically changed (ill. 208). In a competitive climate between rival concrete flooring and roofing techniques for housing and industry, the Perrets were far from the first to propose such vaults, though their striking combination of north-lights with concrete semi-vaults in place of saw-tooth trusses in steel may have been original. So it was prudent for them to patent a specific method. At the second stage of that process in 1920–1, the patent application was made in the names of Perret Frères and Gellusseau. Though Gellusseau's name dropped out in the sequel, it is safe to assume that the engineer had something to do with the subsistence of vaults in Auguste Perret's architecture, otherwise so partial to trabeated classicism.

Though Auguste and Gustave called themselves 'architectes-constructeurs', their office as a whole was not a bureau d'études on engineering lines.[66] The small architectural section seems to have been quite separate. Here the crucial figures were Georges Brochard and Marc Conchon, faithful subordinates who over the course of a lifetime's work learnt the knack of transforming thumbnail sketches by Auguste into finished drawings embodying what he meant, as architectural underlings often do. Other engineer-'centraliens', Francis Jarnoux and Jacques Minet, overlapped with or continued after Gellusseau's departure, but seem to have had no special stature. 'In most cases', Guy Lambert has written, 'the bureau d'études techniques was subject to the pre-eminence of the architectural design or to the production needs of the building firm'.

During the 1920s the Perrets perfected their design-and-build methodology. Sometimes they repeated their old trick of supplanting another designer by showing that their integrated approach would be both cheaper and better. That occurred, for instance, at Notre Dame du Raincy (1922–3), where a hapless architect-engineer found his church-plan purloined and dazzlingly transformed by Auguste Perret.[67] One can only rejoice. The Le Raincy church (ills. 209, 210) combines profound historical reflection with a genuinely rational approach to cheap construction – something the century never managed

209. Notre Dame du Raincy, an early view of the exterior by the Düsseldorf photographer Franz Stoedtner. A. & G. Perret, architects, Perret Frères, contractors, 1922–3.

again. Harmony between A. & G. Perret and Perret Frères was never more palpable than at that high moment. At the same time, as masterpieces can do, Le Raincy presages the end of a road. Thereafter Auguste turns into a cultural authority. His architecture mutates into a fixed style, always resourceful, inventive, fiercely upholding the cause of reinforced concrete. But now tried and tested technical solutions can take up and carry through the design. The challenges of concrete construction have been met; the architect retires to his shell to focus on his art.

From 1930 Auguste starts to be drawn into great public commissions. The rules of that game mean there must be tenders, and the family building firm will have only an outside chance of carrying out the job. If in the halcyon days Auguste is able to practise architecture in the way he wants because Perret Frères are making good money, by the end of his career the tables have turned. 'I really don't know what's to become of us', laments Claude Perret in 1954, with the building business in a bad way. 'It's all because the firm has been suffocated by Auguste's architecture and his reputation – it's tragic and terribly hard to get over: how can we do it?'[68]

To repeat, none of the Perrets described himself as an engineer. Auguste and Gustave doubtless inherited some of their profession's phobia about its alleged rivals. Rationalist rhetoric, bandied between French architects from Rondelet to Le Corbusier, reiterated that they would have to get a hold on construction if they were not to be overrun by engineers. Many had tried to do just that. But no architects had managed so thoroughly to tame a new building technology without losing either artistry or practicality as the Perrets did with concrete. That success in beating others at their game engendered in Auguste Perret a certain hauteur towards engineers. It was in strongest evidence after the First World War, when industrial jobs were the core of the brothers' work and they were really competing with engineering firms.

In fertility of resource and imagination they had at that time just one rival: Eugène Freyssinet, then employed by the firm of Limousin to create big factories and sheds, not to mention the bridges that are his first claim to fame (pp. 351–5). Together, Auguste Perret and Freyssinet occupy the summit of the twentieth-century French achievement in reinforced concrete. But they get there from different points on the compass. The architect harnesses known technique to perfectionist ends, while the engineer forwards technology, often leaving the details of his work rough. There is a difference too in their scale of work: Limousin and Freyssinet exert themselves on grand infrastructural contracts, while the Perrets build nothing seriously large until late in their careers. Scale and its artistic consequences prompt Auguste – not without a touch of the green eye, for the Perrets had tried their hands at the genre without luck – to damn Freyssinet's stupendous airship hangars at Orly (1921–3) (ill. 211) with faint praise:

Given their purpose and context, no attempt has been made to turn it into Art. The arch, in the form of a catenary or a parabola, has not been destroyed; you see their purpose right away. So they have character, maybe they even have style. But is this Architecture? No! Not yet! Here is the work of a great engineer, not of an architect. When you look at these hangars from afar, you ask what these half-buried tunnels can be. When you see Chartres Cathedral from the same distance, you ask what is this great building, yet you could easily fit Rheims, Paris and Chartres into a single one of Orly hangars, and on the floor surface you could fit five cathedrals. What the Orly hangars need to become architecture is Scale, Proportion, Harmony and Humanity.[69]

To which, years later, Freyssinet supplied an indirect rejoinder:

One of the greatest architects, Auguste Perret, has supplied this definition for architecture: the art of organizing space. We can use that for the art of construction, perhaps the more justly so because space for architects is often a simplified space in which the properties of the material over and above those that have to do with appearances

210. Notre-Dame, Le Raincy, the interior. Stained glass by Maurice Denis with Marguerite Huré.

211. Aerial view of the airship hangars designed by Eugène Freyssinet and built by the Limousin Company, 1921–3.

more often than not play an unobtrusive role. Our own space is imbued with material that resists, material that weighs: this is true space.[70]

The Kahns

The Perret brothers were not the only high-class constructors in concrete to find success en famille. They have a parallel in the story of the Kahn dynasty in Detroit. Here again an architect-brother has hogged the limelight, leaving the technical back-up afforded by his family in shadow.

Albert is the famous Kahn, if apt nowadays to be confused with his later, unrelated namesake Louis.[71] Both at least were architects to the marrow. That must be said at the outset, because Albert Kahn made his reputation as a designer of car factories. It was as their architect, not their engineer, that he defined himself. Nor were factories the only things he designed. He belonged to a generation of American architects keen to prove their civility and artistry. As a young man he excelled in sketching and composition; and he became a discriminating designer of office and university buildings, along with houses for Detroit's plutocracy. Had he not been deflected by the automobile revolution, he might have found another route to fame.

Albert Kahn's background resembled that of Dankmar Adler, the sturdy Chicago architect-engineer of the previous generation (p. 197). Both were children of poor German rabbis, and both had emigrated to the United States as boys – the Kahns arriving in 1880. There is a hint of contact; it is rumoured that when Frank Lloyd Wright left Adler and Sullivan in 1893, Albert was offered the position of right-hand-man to Sullivan but turned it down.[72] By then he had served years in a Detroit architect's office, plus time out on a travelling scholarship in Europe. So his tastes were formed. 'How I do wish . . . that I were a Richardson', he wrote to his fiancée;[73] at the Chicago World's Fair his

STYLES OF PARTNERSHIP 243

favourite pavilions were by McKim, Mead and White. In time a firm of his own gathered momentum, going through various names and partners. A first factory arrived in 1900. Soon came a whole run of buildings for the Packard Motor Company, just establishing itself in Detroit. Nine had gone up before in 1905 Albert Kahn embarked on Packard Plant Number Ten. Here commenced the Kahn revolution in factory architecture. It would have taken a different turn, had he not possessed an engineer-brother.

Albert was the oldest of a large, close-knit family. After him came Julius, and then Moritz, who successively managed the business interests of his older brothers; further down were Louis, who became Albert's second in command, and Felix, eventually head of a big West Coast construction firm.[74] Julius Kahn is of abiding interest in the saga of reinforced concrete.[75] Unlike Albert, who was sent to work young, Julius studied engineering at the University of Michigan, supported by his elder brother. On graduating he held various positions before spending almost a year in Japan (a lifelong enthusiasm), developing a sulphur mine. He returned to join Albert as his chief engineering assistant early in 1901.

In 1902 Albert Kahn acquired the Engineering Building at the University of Michigan – the first of many commissions there. 'Not a particularly outstanding or innovative work', judges his biographer Grant Hildebrand:[76] but it was the first the firm designed using reinforced concrete. As brother Julius was a graduate of the department, he surely had much to do with the job and may have suggested concrete. The experience set him off on a series of experiments. Here is Albert's memory of events:

> He quickly saw the weak spots in the empirical system of reenforcement being used and promptly designed a form of reenforcement along scientific principles. We made tests which were conclusive, confirming his theories. Up to then, concrete beams when tested to destruction failed invariably in shear. In other words, the concrete failed – never the steel. Since concrete was a less dependable material than steel, it was difficult to make accurate calculations. If the reenforcing steel could be caused to fail in testing, a more definite method of calculation would be possible. This very point was called to the attention of the profession at the time in articles published by Capt. John S. Sewall [sic], then in charge of construction for the District of Columbia, who had made innumerable tests on different methods of reenforcement. When, therefore, after tests on my brother's design had caused the steel rather than the concrete to fail, we immediately got in touch with the Captain, with the result that he became interested, invited my brother to Washington, saw the reasonableness of the design, and before even a company was formed to manufacture the reenforcing steel, induced my brother to accept a contract for supplying his form of reenforcement for the entire group of buildings called the War College . . . the so-called 'Kahn' system quickly became established and popular throughout the country and while heartaches during the first years were many, the system won out finally.[77]

The Kahn system of reinforced concrete spread faster than any of its predecessors. The setting-up of the Trussed Concrete Steel Company (later Truscon) to make the 'Kahn bar' took place in October 1903.[78] It seems first to have been tried out in the Engineering Building,[79] before undergoing trials under Captain Sewell at the Army War College, Fort McNair, Washington.[80] Soon it was everywhere in America, and Moritz Kahn had been dispatched to promote it in Europe.[81] By 1908 Julius Kahn could be puffed in a Detroit paper as the 'young engineer who . . . has revolutionized building, and has made it possible to erect a 20 story skyscraper with a poured concrete skeleton', while Albert was described as 'primarily the artist, the man whose interest is in the hitherto ugly lines of concrete'.[82] The implication is of a tacit partnership.

Why was the Kahn system taken up so fast? It claimed to offer marked improvement on what up to 1903 Ransome, Hennebique and their licensees could manage in respect of the stresses to be counteracted in the web of concrete beams. The main novelty was

212. 'Packard Plant Number Ten', interior. Albert Kahn, architect, 1905. The Kahn system achieved extra width for the factory bays.

213. Packard Motor Works, exterior in 1907 before two additional storeys were added. Albert Kahn, architect, 1903 etc.   Some enrichment was admitted along the front but soon fell away behind.

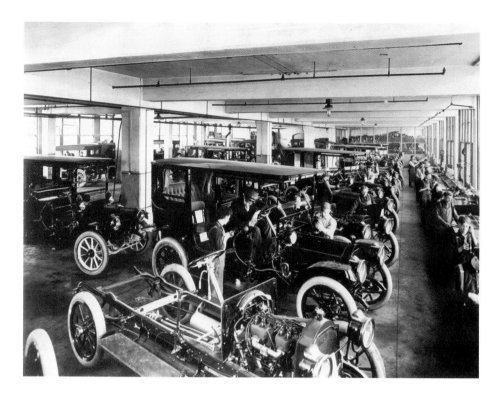

the introduction, at points in beams where shear forces were strong, of steel stirrups angled at 45°, integral with the horizontal reinforcement. The 'Kahn bars' (ill. 214) were designed and positioned to ensure maximum strength and grip on the concrete, and manufactured of high-quality steel, for which Julius developed close relations with steel fabricators. Some experts have judged the Kahn system harshly, Carl Condit calling it 'expensive, redundant, and awkward to handle', while acknowledging its success.[83] Conversely, Albert Kahn claimed that the Hennebique system 'was complicated . . . and with the excessive labor costs of our country, proved rather impractical here'.[84] If that is right, the Kahn system can be added to the rollcall of American inventions stimulated by labour shortages.[85] Just as crucial, however, were the approbation of Sewell and thus of the US Corps of Engineers and the Government, though Sewell regretted that the Kahn stirrup had been patented and so could not be freely diffused.[86] Julius Kahn was both a practical man and a graduate who could talk the language of academic engineers and appreciate their criteria of testing and proof. With a level of official support never courted by Ransome, his system swept the board.

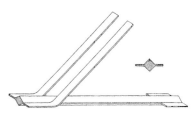

214. The Kahn bar, from *Railroad Gazette*, 16 October 1903.

Packard Plant Number Ten was erected in 1905, just as the new system was being pushed. The Packard complex consisted of a square surrounded by two-storey blocks, later raised to four (ill. 213). The previous units, of conventional 'mill construction', boasted decent brick infill between the piers but descended behind into what Reyner Banham calls 'a kind of architectural null-value condition, stripped of everything but the bare necessities of support and shelter'.[87] The division of their interiors by iron columns permitted a maximum bay-size of 25 × 16 feet. Though the new extension was no nicer to look at, its concrete frame needed no fireproofing for its central columns and beams, and it offered better lighting. In addition its bays achieved an extra seven feet in width (ill. 212). Other 'daylight factories' in concrete certainly existed by then, equal if not broader in bay-size. But Packard Plant Number Ten aligned two forces at work in Detroit: on the one hand, architect- and engineer-brothers of resource, eager to explore their new concrete patents to the full; on the other hand, an automotive industry avid for freer factory layouts, as it fumbled towards the almighty production line.

That alignment lasted just long enough to crown Albert Kahn as the prince of car-factory design. Factory orders from many companies flowed in after Packard Plant Number Ten.[88] But the job that clinched his status came when in 1908 Henry Ford appointed him to design his factory for the Model T. Highland Park was the first of two brand-new plants, equally unprecedented in immensity, that the architect designed for the greatest of the motor moguls. It was the commission of a lifetime. Ford always knew what he wanted, and his construction engineer Edward Gray was involved in at least the later stages of planning, so just what Kahn did and did not contribute at Highland Park is uncertain. But the four-storey model of daylight factory with an exposed concrete frame must have come from him. Since Packard Number Ten he had twice designed Detroit automobile factories on those lines – for the Chalmers Motor Company and the Grabowsky Power Wagon Company. It was wheeled out again for Ford at Highland Park, only extended to a vast 800 feet with a show façade and lower administration block in front. Albert Kahn exhibited his architectural calibre by the layout's authority and by touching long runs of concrete frame and wide steel window with a hint of swagger (ills. 215, 216). He claimed the windows as 'probably the first steel sash . . . at least sash of standard factory type. They were shipped direct from England.'[89] Soon however Julius Kahn's company began to market its own United Steel sashes – a clue that the brothers were still then working together.[90]

For a short time, the Highland Park façade influenced factory architecture the world over. If modernists have preferred its bare backs to its minimally enriched front, that is not how Kahn or Ford would have felt at the time. As for its workings, however, Highland Park was out of date almost before the first block started production in 1910. The method of gravity feed, whereby components worked their way down from the top storey to final

215. Ford Motor Works, Highland Park in about 1913, showing the unrelenting stretch of the four-storey concrete frame and steel windows. Before the building, the massed ranks of the Model T chassis, allegedly a day's output.

216. Highland Park, part of the front. Albert Kahn, architect, 1908–10. The corners and base receive their meed of 'architecture'.

assembly on the ground floor, proved inefficient. It soon dawned upon Ford and his lieutenants that only horizontal production made sense. For that much more land was needed, together with long, open top-lit shops and an overall plant layout dictated by function, not dignity. Hence the development from 1917 onwards of the second, even huger and quite different plant that Albert Kahn built for Ford at Rouge River; and the eclipse of the multi-storey factory in reinforced concrete after its moment in the sun. By the time Turin opened the iconic Fiat-Lingotto Factory in 1923, its heaped-up storeys on a concrete frame were thoroughly outmoded.[91]

For single-storey construction, steel possessed unarguable advantages. So Albert Kahn built most of his factories after Highland Park in that manner and material.[92] The prototype, the Packard Forge Shop of 1911, was also the first of his buildings to look modern in a slick, curtain-walled sort of way, and so to have made the aesthetes salivate. Kahn styled his later industrial buildings with unquestionable flair (ill. 217). In respect of his work in steel, he increasingly cast himself as a moderate modernist, paying lip-service to European factories by Behrens, Gropius and the rest. But that was never the whole picture. He was robustly opposed to the 'degenerate', exposed-concrete architecture espoused by the 'Ultra Modern' Europeans and scornful of their amateurish entries in the Chicago Tribune competition.[93] A formalistic, Beaux-Arts note persisted in his repertoire. Concrete continued to offer a backbone to structures on that side of his work, like the Ford Engineering Laboratory at Dearborn (1922) with its 'dry, neoclassic exterior',[94] or the administration building at the Glenn Martin aeroplane plant (1938). But if in Albert Kahn's hands reinforced concrete, the revolutionary building material of his era, started out showing itself baldly, it ended up hardly expressed at all.

That change in the layout and structure of Kahn's industrial buildings corresponds with what is known or can be deduced about the pattern of relations between the brothers. For a while they worked in something like partnership; the rise of Julius's concrete system chimed with the take-off of Albert's industrial practice. In 1907 Albert designed for Julius the Trussed Concrete Office Building in central Detroit, an eight-storey block on a concrete frame with white-brick cladding (ill. 218). The establishment of Albert's office on the top floor symbolized their closeness at the moment when demand for the concrete-framed daylight factory was at its height. A parting of ways seems to have taken

217. Ford Glass Plant, Rouge River. Albert Kahn, architect, 1922. An example of Kahn's fresher industrial architecture after he switched to single-storey steel plants.

218. The Trussed Concrete Building
(later Owen Building), Detroit. Albert
Kahn, architect, 1907. Albert had
his office on the top floor for four
years; the rest of the building was
Julius Kahn's headquarters until 1914.

place about the time that Albert shifted into steel for factories. By then Julius was busy
setting up a subsidiary in Japan, while in 1914 he moved the Truscon headquarters to
Youngstown, Ohio, closer to steel suppliers.

In 1918 the firm of Albert Kahn, 'architects and engineers', expanded to a larger
office. Besides the associates (Albert's brother Louis, Ernest Wilby and J. F. Hirschman),
its letterhead of that time carries the names of a chief structural engineer (J. T. N. Hoyt)
and chief mechanical engineer (F. K. Boomhower). Busy on Rouge River, the firm was
then eighty strong and highly organized, like the industrial clients Albert served.[95] In the
inter-war period its socio-industrial scope widened as a result of commissions to build
factories in the Soviet Union and to carry out planning tasks for the Tennessee Valley
Authority. By 1942 war-work had swollen the firm to 450 architects and engineers, a formidable industrial concentration in its own right.[96] Kahn long declined to employ graduate architects because he believed they had been too far encouraged in self-expression
to accept the discipline he asked of his workforce. That may have been a defence of his
own artistic prerogative, which he maintained to the last. But it also recognized that the
modern building was a multi-disciplinary activity requiring the architect to act from the
start like a true team-player. As he told his fellow-professionals,

The average architect, without the assistance of men who can deal with the structural
problem, the sanitary, power, sprinkler, heating and ventilating, and cooling problems,

is apt to fail. It is imperative that groups of men conversant with these fields join in the handling of an industrial plant. Nor is it sufficient that the architect tell the owner that he expects to call in specialists to help at the proper time. The main subjects must have consideration from the outset, the very first conference. All must work in close touch with each other to gain the desired result expeditiously; wherefore the combination must exist at the outset.[97]

Wright and Mueller

It is exciting to him to rescue ideas, to participate in creation. And together we over-came difficulty after difficulty in the field where an architect's education is never fin-ished.[98]

Thus did Frank Lloyd Wright sum up both a personal debt to Paul Mueller and a con-nivance – a Faustian pact almost – that grew up between avant-garde architects and engi-neers in the making of modern buildings. Time and again, engineers have 'rescued ideas' in exchange for a share in the adrenalin-rush of creativity.

As technical training expanded in nineteenth-century Europe and America, young men with a grounding in structures often went into contracting instead of becoming con-sultant or government engineers. Paul Mueller was one such. It is as Frank Lloyd Wright's builder of choice for twenty-five years from 1904 that he is now remembered. He had done much before then. Mueller was one of those proficient Germans who were the making of the Mid West. He arrived in Chicago in 1881 aged seventeen, having attended what he called 'the government school of mining and civil engineering in the Saar Basin' and passed a polytechnic entry examination.[99] He worked at first with a suc-cession of architects, engineers and iron and steel companies, getting expertise in the new style of big building. Then he lifted into smarter circles. In 1886 he was hired by Adler and Sullivan as an engineering assistant on the Chicago Auditorium, acting as Dankmar Adler's right-hand man. He quickly became 'foreman of the whole office' and stayed till 1892.[100] Wright's autobiography draws an affectionate sketch of Mueller, all beard and guttural accent, exercising an earnest, youthful authority. Mueller was the first person Wright met when he applied to join Adler and Sullivan late in 1887. The two hit it off. In a published plan of their office atop the newly completed Auditorium, just four pri-vate rooms are indicated: those of Adler, Sullivan and their respective personal assistants, Mueller and Wright (ill. 219).

Both men preferred to learn by doing. At the start of his career, Wright had spent months working on and off for the architect-engineer Allan Conover in Madison, and

219. Adler and Sullivan office on top of the Auditorium, 1890. Adler and Mueller at one end of the office, Sullivan and Wright at the other.

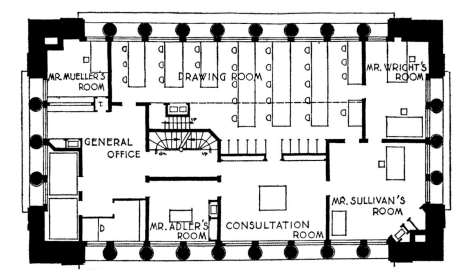

220. Larkin Administration Building,
Buffalo. Frank Lloyd Wright, architect, Paul Mueller, engineer-builder,
1904–5.

221. Larkin Administration Building,
interior. The piers and beams are of
steelwork encased in plaster.

then enrolled for two semesters in Conover's engineering course at the University of Wisconsin; but the formal side, he said, 'meant nothing so much to him as a vague sort of emotional distress, a sickening sense of fear'.[101] No doubt he picked up a lot about iron and steel structures and about foundations chez Adler and Sullivan, in part via Mueller.

Mueller's weightiest job there was on the Auditorium Building itself. A lengthy witness statement he wrote years later, in connection with a lawsuit over settlement, lays out in rare and rich detail the manifold interests and responsibilities that Adler and Mueller had to co-ordinate. It offers an antidote to any view of architecture as an individualistic art. And yet the pair certainly understood their structural-managerial role to comprise the making of conditions wherein Sullivan (and perhaps his assistant Wright) could be creative. Looking back, Mueller felt there had been too much consultation over the Auditorium. 'I am sorry to say that Adler & Sullivan were so solicitous of the opinions of others', he wrote, having in mind Sullivan's old teacher William Ware, who insisted that two extra storeys be added to the tower after its foundations were already in.[102] The testimony says little about Mueller's exact contribution. But his grasp of detail some thirty-five years after the construction contrasts with Wright's loose, romanticized memory of past projects and struggles. That complement of temperaments was to be needful.

After Adler and Sullivan, Mueller joined a big construction firm and supervised many of the structures built for the Chicago World's Fair of 1893, before turning independent contractor-builder. He erected large buildings of all sorts, including churches, but seemingly not houses. So there was no reason for Wright and Mueller to work together until the former's practice grew beyond its early suburban scope. Then, the picking-up of old threads amounted to a shift from the in-house fellowship of Adler and Sullivan to a version of the same thing 'in the field', and from the challenge of big steel buildings to middling-sized ones in reinforced concrete.

The renewal of the link is bound up with projects for Wright's early clients, the Martin brothers. Probably Wright and Mueller's first collaboration was the celebrated Larkin Administration Building, Buffalo, started in 1904 (ills. 220, 221). Though littered with novelties in planning and servicing, the Larkin Building was conservative in structure; what looked like concrete was in fact a steel frame, obscured in and out behind thicknesses of brickwork. There was plenty of money for Larkin, so perhaps the economy of concrete did not need to be invoked. Furthermore, architect and builder alike had been brought up with steel and as yet knew little about reinforced concrete. Well-embedded steel was their natural response to Darwin D. Martin's brief for 'absolutely fire-proof construction'.[103]

Like others, Wright had been toying with ideas for designing in concrete since at least 1901.[104] But the first of his buildings constructed with a reinforced-concrete frame was the E-Z Polish Factory between Chicago and Oak Park, built by Mueller for W. E. Martin (ill. 222). Its lower storeys were put up in 1905 while Larkin was under construction, the rest being added later. Delays on the original portion caused a blistering row between Wright and his client and obliged Mueller to show the saintliness required of Wright's collaborators: 'Mr Mueller is ground between the obstreperous millstones

222. E-Z Polish Factory. Frank Lloyd Wright, architect, Paul Mueller, builder. Lower storeys, 1905, upper ones added later. Wright's first executed building in reinforced concrete.

223. Unity Temple, Oak Park. Frank Lloyd Wright, architect, Paul Mueller, builder, 1906–8. The front has doubtless been creeper-clad to mitigate the unsightliness of the bare concrete.

224. Unity Temple, plan (left) and section (right) of the walls, from *Concrete Engineering*, September 1907. The walls are clearly shown as lightly reinforced both horizontally and vertically. To an extent this prefigures Wright's later practice.

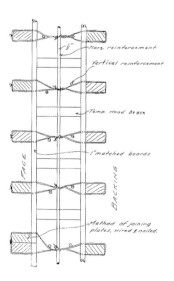

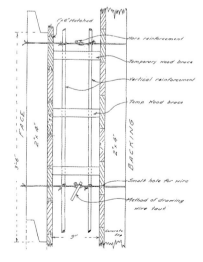

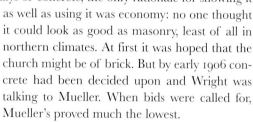

and smiles and smiles, attributing no preponderance of blame to either party, amiable and well-poised gentleman that he is'.[105] E-Z Polish was a decent, disciplined 'daylight factory', to use Reyner Banham's term, but in no way 'precocious', as Frampton has labelled it.[106] It had a brick-faced front over a concrete frame of which we know nothing. Probably Wright and Mueller had to work it out in tandem with one of the specialist concrete firms. Here at any rate was something to learn from.

Concrete came into full play only with their next collaboration: Unity Temple, Oak Park (1906–8), a project that has been fully studied by Joseph Siry.[107] That a mail-order building in a dim neighbourhood of Buffalo could look smarter than a church in the chic suburb of Oak Park shows how far old priorities and proprieties were dying by this date. Just as Larkin was extravagant, so Unity Temple had to be cheap, the congregation having little to spend. As ever in the early days of concrete, the only rationale for showing it as well as using it was economy: no one thought it could look as good as masonry, least of all in northern climates. At first it was hoped that the church might be of brick. But by early 1906 concrete had been decided upon and Wright was talking to Mueller. When bids were called for, Mueller's proved much the lowest.

It is a measure both of Wright's apprenticeship with concrete and of his personality that he never proposed a frame for Unity Temple. The enclosure and weight of walls always mattered to him. In his eyes the key to saving money was standardizing the formwork and simplifying the profiles. These Mueller and he could manage; the scientific design of concrete they could not. So at first not only the foundation walls but also those of the superstructure were to be of mass concrete. Only the spanning members were reinforced, the calculations and estimates being put out by Mueller to a specialist engineer. Wright long stuck to this partiality for horizontalism in reinforced-concrete forms. 'First among them is the slab – next the cantilever – then the splay', he pronounced in 1927.[108]

Just before pouring it was decided lightly to reinforce the church's wall-cores, at considerable extra cost (ill. 224). This puzzling variation may have stemmed from Wright and Mueller's lucubrations over the concrete surfaces. Knowing the unsightliness of normal concrete mixes, Wright at first wanted to line the inside of the formwork with a mortar of cement and red granite aggregate before the concrete for the wall-cores was poured. Duly polished once the forms were removed, this smarter outer lining would have become the surface of the building. After alternative sample panels were made up, he changed his mind. Instead the walls were all cast in one, with light-coloured gravel aggregate all through, and reinforcement added to what may have been a thinned-down core.

Though unrendered concrete can be found in buildings and engineering structures back to Coignet, Unity Temple was the first radical work of architecture to make a show of it (ill. 223). By 1908 smart European architects like Perret and Wagner were using concrete liberally, but still fought shy of exposing it. So flagrant a display of raw walling was less a manifesto for truthfulness than a substitute for what could not be afforded – 'the finished result in texture and effect being not unlike a coarse granite', wrote Wright, echoing a plea then common to excuse exposed concrete in lowlier buildings.[109] Soon enough the surfaces became shabby. Unity Temple was a courageous feat of design, but Wright never repeated its raw texture. Much though he revelled in concrete's versatility, Wright did not share the European infatuation with exposing it in the mass, preferring to tame and manipulate it into moulded blocks. 'Aesthetically', he wrote in 1928, 'it has neither song nor story . . . it is supine, and sets as the fool, whose matrix receives it, wills.'[110]

In Wright's memory of events, Paul Mueller 'comes to the rescue' of Unity Temple. 'Doesn't lose much on it in the end.'[111] The truth is less happy. Mueller's building company had to leave the job before it ended and declared bankruptcy soon afterwards. But the failure left him free to supervise the two great experimental commissions of Wright's early-middle period, Midway Gardens (1914) and the Imperial Hotel in Tokyo (1919–22). Concrete was fundamental to both these lost masterpieces. But it was not deployed in ways that chroniclers of twentieth-century structures might lead us to expect.

Looking back once more with his romantic eye, Wright painted Mueller at Midway Gardens (ill. 225) as a noble slave-driver, pushing a complex project on to completion in just four months.[112] On the technical side Mueller relied upon Clarence Seipp, a Cornell-trained engineering graduate who specialized in reinforced concrete and now came in as his junior partner to supply the in-house assistance wanting at Unity, and calculate the floors and columns. But when Cement Era hailed Midway Gardens for restoring 'the good name of concrete as an architectural material',[113] it alluded not to the structure but to the array of precast and incised concrete blocks and sculpture gracing the walls and balconies throughout the pleasure garden. This novel collaboration involved Wright, the sculptors Alfonso Iannelli and Richard Bock, their mould-maker Ezio Orlandi, and 'a special core of workmen who had previous experience of concrete casting' for Mueller's firm, reports Anthony Alofsin.[114] Midway Gardens lay at the artistic and technical origins of Wright's concrete-block system, normally referred to his Californian houses of the 1920s.

Not clever with money, Mueller became implicated in the finances of the loss-making Midway Gardens.[115] He seems to have been willing to build anything for Wright, to 'rescue ideas' regardless of profit: to quote the inbred language of Taliesin, he was 'obedient to cause'.[116] Accordingly at the age of 55 he betook himself to Japan in 1919 to take on the hazards and delays of supervising the Imperial Hotel. As at Midway Gardens, the

225. Midway Gardens. Frank Lloyd
Wright, architect, Paul Mueller,
builder, 1914. Brick-clad with concrete dressings and statuary.

226. Imperial Hotel, Tokyo. Frank
Lloyd Wright, architect, Paul Mueller,
builder, 1919–21. Concrete structure,
brick cladding, oya-stone dressings.

hotel's core was of concrete. Here it was even less visibly alluded to, since carvings in lava were substituted for the cast-concrete blocks first proposed to set off the brickwork (ill. 226). Yet reinforced concrete was notoriously the Imperial's salvation, its performance in the Tokyo earthquake of 1923 becoming a key publicity incident in Wright's career.

By then the choice of reinforced concrete for major structures in earthquake zones was more or less axiomatic. But there were many theories about how to apply it. The Imperial Hotel legend rests on two beliefs: that Wright designed its foundations on a new intuitive principle which he worked out himself; and that the hotel survived the earthquake outstandingly well. The first was true in part; the seismic specialist R. K. Reitherman has shown that the second was untrue.[117] The idea of a concrete mat 'floating' on the underlying mud lake and secured not with deep piles but with a forest of short reinforced ones at two-foot centres (ill. 227) was Wright's. So too was the notion of a superstructure with a low centre of gravity, broken up into separable units. In Reitherman's analysis, the 'good but not outstanding' performance of the Imperial in 1923 – no better than many Tokyo buildings with the deep piles shunned by Wright – had more to do with the superstructure than with the foundations. Indeed the foundations were among the factors cited in the arguments that led to the hotel's demolition in the 1960s.

What about the authorship of the structure? In a quarrel with Wright, Rudolph Schindler alleged that 'the structural features which held the Imperial Hotel together were incorporated only after overcoming your strenuous resistance'.[118] That may refer to Julius Floto, who identified himself as the structural engineer for the hotel in an article written after the earthquake, following communication with Wright. There the obscure Floto reveals that he was intimately involved in designing the foundations and superstructure. Instead of taking Schindler's line, he attributes the hotel's success in resisting the earthquake to Wright's personal choice of foundation technique. On the superstructure, Floto is fuller. Its reinforced mushroom floors, posts and slabs, he says, 'were originally designed by the writer in accord with the building code of the city of Chicago'. But Wright had chafed under the ponderous safety factors:

> He tells me now that, in building, my computations were disregarded and that much lighter sections were everywhere substituted, making in effect a design which eliminated all the strength usually provided for the live loads.

Far from being alarmed, Floto pronounces the change 'entirely logical', since the superintendence was excellent and the workmen careful.[119] Here Mueller's importance is implied if unmentioned. In a letter to Sullivan after the earthquake, Wright was more explicit. After ascribing the hotel's survival to his 'principle of flexibility', he went on to acknowledge that

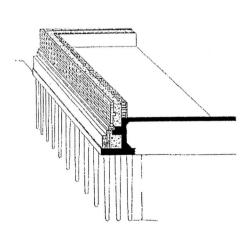

227. Imperial Hotel, foundations. Frank Lloyd Wright, architect, with Julius Floto, engineer. 'Floor slabs balanced over central supports as a tray rests on a waiter's fingers to prevent failure under earthquake strain', reads part of the caption, echoing an image famously employed by Wright. The novelty was the frequency and shallowness of the piling – at two-foot centres to a depth of eight feet.

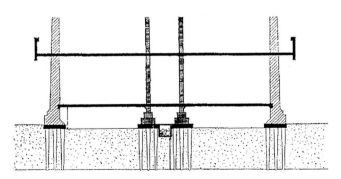

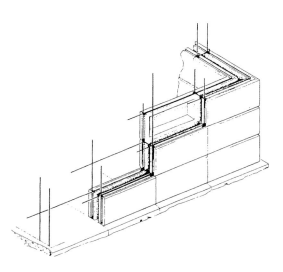

228. Double-skin wall of the Johnson Wax Building, showing the inner cavity and light reinforcement. The technique goes back at least to Midway Gardens.

229. Westhope, Tulsa. Frank Lloyd Wright, architect, Paul Mueller, builder, 1930–1.

'Mueller's untiring attention to the execution of the details of the programme counted too in the final result. Nothing of any importance was put into place without his superintendence.'[120]

The Imperial Hotel's walls consisted of two leaves of brickwork. Steel bars bound the leaves together, while light reinforcement occupied the intervening cavity, into which concrete was poured. Wright must have been happy with the technique, as he adopted it with a layer of insulation for his later Johnson Wax Building (ill. 228). A similar system underlay the walling of Wright's concrete-block and textile-block houses of the 1920s. It derived from amalgamating the Midway Gardens experiments in concrete block with the double skin of the Imperial Hotel. Wright, not Mueller, invented these procedures. But through patience, endurance and commitment to practicalities, the latter proved the midwife to their birth.

The Imperial Hotel marked the climax though not the end of Paul Mueller's collaborations with Wright. He was to have been the contractor-manager of the ambitious textile-block resort at San Marcos in the Desert, Arizona (1928–9), a project that collapsed with the Crash. Perhaps in compensation, Wright induced his cousin Richard Lloyd Jones to employ Mueller on a large concrete-block house, Westhope at Tulsa, Oklahoma (1930–1), one of the few jobs he was able to salvage and build during these years (ill. 229). But the house leaked, looked forbidding and was disliked by its clients. Worse, Mueller diverted a large part of his advance to pay debts and had to give up the contract in tears and disgrace. It was an ignominious finale: he died three years later.[121]

What is to be made of Mueller's role in Frank Lloyd Wright's career? One might dismiss him as a humble builder, carrying out innovations dictated by a higher intelligence. His technical knowledge was evidently too slight, or too restricted to iron and steel, for him to be a modern consulting engineer in concrete – not that Wright would have

brooked one so close to him. A better model is that of the engineer-contractor: not of the usual entrepreneurial kind, but of a far rarer breed ready to take on the financial risks of avant-garde architecture. Perhaps it is best to see Mueller as an instance of the engineer *qua* construction manager. In the making of exceptional buildings with exceptional architects, such a figure sometimes matters more than the engineer *qua* structural consultant. Whatever the perspective, several of Wright's most exciting aesthetico-technical advances first developed in projects under Mueller's management.

From that standpoint, what happened after Mueller's disappearance is illuminating. A propos of the Lloyd Jones house, Vincent Scully has written that Wright seemed to have 'reached an impasse . . . and to be searching for something new'.[122] That tallies with the foundation of the Taliesin Fellowship in 1932, an event usually interpreted in social or pedagogic terms. Another way of looking at it is as Wright's regrouping of his technical comradeship in-house, as in the days of Adler and Sullivan. Now he drew his support in the main not from Mueller or Seipp or Floto but from 'apprentices'. Those with engineering skills like Mendel Glickman and Wesley Peters were specially vital to the Taliesin set-up. Peters, who had an engineering degree from MIT, emerged as Wright's son-in-law and right-hand man.

Ideally, the small domestic contracts of Wright's later practice were run by Taliesin apprentices.[123] But on a larger scale, sympathetic outside partnerships continued to matter. The success of the original Johnson Wax Building, for instance, owed much to Ben Wiltscheck, a university-trained architect turned builder who was a friend of the client. Wiltscheck became a strategic ally of Taliesin in negotiating between the poles of Wright's organic intuitions about structure and the rigidities of the local building code. The mushroom columns of Johnson Wax, conceived by Wright, calculated by Peters and Glickman and tested to destruction before the camera, marked a well-publicized triumph of imaginative genius over scepticism. It was then down to Wiltscheck to get the building beautifully built.[124] Where such a partnership with the contractor was wanting, things could go wrong. Wright's structural intuition was not always trustworthy. In the case of Fallingwater, subcontractors without the tried and trusted status of Mueller or Wiltscheck fought and lost a raging battle with him to put in the extra reinforcement they insisted was needed for the great cantilever beams of the living room – with the apprentices caught in the crossfire. The house long suffered from the ensuing difficulties.[125]

OWEN WILLIAMS

Owen Williams was one of several specialists in concrete engineering to profit from the waning of the systems and emerge with a public profile of their own. In his case, the technically backward context in which he found himself propelled him into an independence from architects that verged upon truculence.[126]

Williams' background was modest. His first job was for an electric tramway company. Like many British engineers of his generation, he supplemented it with technical education at a London night school. Graduating in 1911, he shifted to firms specializing in reinforced concrete systems. Soon he joined the Trussed Concrete Steel Company – Truscon for short, British licensees of the Kahn system. Here Williams went through various posts: senior designer for a six-storey cabinet-making factory (for which there was also an architect),[127] chief estimating engineer for a host of projects, and resident engineer for a fuel works. Neither that eminently practical start, nor his subsequent war work on flying boats and experimental concrete barges and lighters, foreshadowed a taste for architecture. Setting up for himself after the war, Williams reverted to the industrial buildings that were his abiding speciality, concocting a system for framed factory construction with precast elements only a little different from one he had devised for Truscon. Probably he expected to continue as an engineer-contractor in reinforced concrete. All that changed when at the age of 31 he acquired the job of consulting engineer to the British Empire Exhibition at Wembley, in partnership with the architects Simpson and Ayrton.

230. Wembley Stadium. Simpson and Ayrton, architects, Owen Williams, engineer, 1922–3: perspective by Maxwell Ayrton.

The genesis of the designs for Wembley is obscure.[128] The choice of architects, and the resolve to give the 'most modern construction and permanent character' to the complex, eschewing 'the sugar-icing character that has characterized past efforts at exhibition buildings',[129] probably preceded Williams' appointment. At the project's heart lay a permanent national sports stadium. In the event that came first (1922–3), while the exhibition had to be postponed to 1924. The Wembley Stadium (ill. 230) relied on recent American precedents built in concrete, which was probably how Williams came in. For the exhibition he inherited a formal layout incorporating all sorts of activities, besides the main buildings and stadium created in collaboration with Maxwell Ayrton. The leitmotiv of the whole, it was resolved, would be exposed concrete. The Wembley exhibition proved to be a jamboree for concrete in a country that seemed to have shunned it. Even the rails of the 'Never-Stop Railway' were in the material. The selection as main contractor of the concrete enthusiast Sir Robert McAlpine – 'Concrete Bob', he had been called in his Scottish youth[130] – confirmed the policy. In conjunction with McAlpine, Williams became responsible for making it all happen. For his dispatch in handling the onerous brief he earned a knighthood when the exhibition opened. The only structures to elude his grasp were the ephemeral pavilions of the empire.

As was seen in the previous chapter (pp. 134, 167), exhibition buildings tend to get caught in a muddle between permanence and impermanence. Since the Chicago exhibition of 1893, many of them had been made of timber, iron or steel with a coating of 'staff' – quick to erect and quick to take down afterwards. Though there were often minor experimental concrete pavilions or bridges on display, it was not the material of choice for the major buildings, because poured concrete belonged to the language of permanence. But at Wembley the organizers had plumped in advance for just that. Besides the stadium, the two big 'palaces' of industry and art were earmarked for later use as trade displays. A show of psychological endurance was also wanted, to counter the sapping effect of the late world war. When the exhibition opened, one visitor was pleased to note in the main buildings 'a monumental dignity that suggests very forcefully the

231. Palace of Engineering, British
Empire Exhibition, Wembley, with
'Railodok' car passing in front.
Simpson and Ayrton, architects,
Owen Williams, engineer, 1923.

strength and solidity of the British Empire'.[131] That was the effect wanted, and that was
where Ayrton came in.

Ayrton's architectural tastes were those of a progressive conservative. While content to
feed the appetite for monumentality, he appreciated that in Britain's reduced circumstances it had to be cheap and modern – in other words of concrete. He had had some
experience of the material,[132] and no doubt saw the substitution and show of concrete
instead of stone as claim enough to being up to date. Yet to mimic dressed masonry in
concrete did no more than Ransome had done thirty years before at Stanford. It was an
outmoded thing to attempt, showing how behind Britain had fallen in making architecture out of reinforced concrete. The Ayrton–Williams collaboration at Wembley is a classic case of an architect pre-empting decisions about materials and expression and
obliging his engineer to live with the consequences.

Much of Ayrton's energy went into tinkering with the relief and grain of the concrete
surfaces – wall-texture always being a priority in English architecture, and one dear to a
man who had worked with Lutyens back in Arts and Crafts days. The upshot was not a
resounding success. Of the big palaces (ill. 231), C. H. Reilly asked:

> Are they great garages or aeroplane sheds? . . . Naked concrete seen near at hand
> always reminds me of the war and its effects . . . I admit it has character and strength,
> but to my mind it suggests that the British Empire is very far down the scale of
> civilization . . . Has the war really made us cave men again?[133]

Not that Williams was necessarily unhappy to go along with the massive walls and fake-
stone details imposed by his architect-colleague. Simply sorting out the right balance of
block, mass and reinforced concrete for these exteriors was a challenge he must have relished. Besides, he was learning for the first time about expression. Wembley offered him a
crash course in architecture. In the process he acquired a taste for monumentality. Behind
those ponderous fronts, he seems to have been left much to his own devices, coming up for
the immense bare halls with a clever mixture of spanning strategies in steel and concrete.

232. Wansford Bridge. Sir Owen Williams, engineer, with Simpson and Ayrton, architects, 1925–8. Perhaps the handsomest bridge of Williams' partnership with Ayrton, built in mass concrete.

Ayrton and Williams got on well at Wembley, Williams claiming that 'whatever success concrete has here achieved is beyond doubt primarily due to that happy co-operation'.[134] But to speak of 'a marriage of true minds', as one commentator did,[135] is to exaggerate. It was not far removed from the old arrangement in railway termini, whereby the engineer did the spacious interior while the architect styled the façades. At any rate, the publicity attached to the British Empire Exhibition allowed its self-taught engineer to find his own voice. Responding to a compliment on the common sense he had shown at Wembley, Williams asked:

> After all, what was engineering but common sense? Theory and calculation were only the servants of engineering, not its masters. Telford, who was a shepherd's boy till he was 19, and who built the Menai Suspension Bridge, was not handicapped because he had been a shepherd's boy, but rather he built it because he had been one.[136]

As to the lessons of the partnership, they led him towards conclusions that would eventually lead to a parting of the ways with Ayrton and Ayrton's profession.

> A considerable amount of time must elapse before the 'concrete sense' can be acquired; that is to say, before any individual can achieve singly a complete and easy mastery of both the architectural and the engineering technique. The engineer and the architect have a long road to travel before their separate roles can be played by one man. Till that end is achieved the fullest expression of concrete can not be attained. But the goal may be reached more quickly by sympathetic collaboration on both sides.[137]

Increasingly Williams began to cast himself as that 'one man'. But as David Yeomans and David Cottam point out in their incisive study of his career, he was reluctant as yet to press his views about architects' limitations beyond the circle of professional engineering colleagues.

In the next phase the tables were turned: Williams acquired a number of road bridges to design and asked Ayrton to help him style them. The Owen Williams bridges of the 1920s fall outside the scope of a chapter devoted to concrete buildings. Indeed on account of their ingenious quirks of structure and look they stand apart from the run of contemporary bridge development. At that time architectural input in bridges was receding. So it is striking that Williams should have felt the need to hold hands with Ayrton, particularly for a run of remote bridges in the Scottish Highlands where propriety was hardly at issue. Perhaps Williams as yet lacked the confidence to style things for himself. Ayrton was still absorbed with concrete surface and mass. By modernist criteria his contribution to the bridges was retrograde. But it is difficult to regret it in so imposing a structure as the Wansford Bridge taking the Great North Road over the Ouse. At Wansford (ill. 232), virtuosity in unreinforced concrete combines with a stripped-Gothic angularity and gritty texture to create one of the most memorable of English bridges.

By 1931 Williams had come to the end of the road with Ayrton, witness some barbed remarks he made after a talk by the latter on 'modern bridges'.[138] He was now winning chances to show his paces in buildings more intricate and public than framed factories. Engineers, or at least he himself, he felt, could by dint of their grasp of structure, function and efficiency make a better fist of modern buildings than architects, with all their shilly-shallying lip-service to context and style. The emergence of Owen Williams the architect raised the spectre that profession had always feared – that engineers would displace them. Williams did little to quiet that apprehension. He had had few cultural advantages, but he had rapid powers of assimilation – he seems to have picked up his modern architecture on the run – combined with sharp business sense and an increasing bluntness.

It is tempting to caricature Williams as a brilliant architect-hating barbarian and set him against the respectful patience of Ove Arup and the next generation of British concrete engineers. The truth is subtler. If Williams had disliked architects, he would never

233. Dorchester Hotel, design by Sir Owen Williams, engineer-architect. Perspective by Keith Murray, 1930.

234. Boots Wets Building, Beeston, corner. Sir Owen Williams, engineer-architect, 1932–3. Flat-slab construction with mushroom heads brilliantly aestheticized without loss of economy.

have put up with Ayrton for so long. Nor did he ever totally cease to work with them. If bad blood developed between Williams and the architects, that was largely because on the occasion he drove deepest into their territory, he was rebuffed in his hour of triumph. That was the notorious Dorchester Hotel episode of 1929–30.[139]

The appointment of an engineer to design a major hotel in London's Park Lane came about because the syndicate financing it needed their building in a hurry. Their leading shareholder was Williams' old colleague and admirer from Wembley, the builder Sir Robert McAlpine. Sure he could do the job, Williams hired extra architectural staff and came up with adequate hotel plans as well as a predictably clever concrete structure, to be faced in artificial stone (ill. 233). All went well until with the builders on site, the hotel company's consultant architect, Morley Horder, publicly resigned in protest at not having been involved, offering outspoken objections to a concrete hotel in Park Lane.

As the concrete was to be clad, it looked as if Horder had got the wrong end of the stick. And yet within a month Williams was off the job. The pretext for his resignation was his refusal to compromise over the ceremonial interiors, where his talk of a 'great whitewashed barn' came up against the syndicate's insistence on the 'period opulence' that film stars and minor royalty were thought to desire at the Dorchester.[140] If that was Williams' real sticking point, it was a foolish one; he knew nothing of such work, nor would any architect have declined to work with a decorator on a hotel interior. Probably deeper jealousies were afoot, prompting Morley Horder's late letter. Afterwards, Williams talked of two kinds of racketeers, 'those who get out their guns and those who get out their elevations'.[141] That was a swipe at Curtis Green, the honourable architect with whom he had declined to be yoked on the Dorchester interiors and who now inherited the whole job. Curtis Green changed the fronts much more than the plan, as the building had gone beyond radical revision. In a gesture of defiant pride, Williams sent all the drawings round to his supplanter in a taxi.

Williams' later buildings divide into those on which he worked with architects and those he designed on his own. With one exception, only the latter have kept their grip upon the architectural imagination. Most are industrial buildings, the masterpiece among them being the Boots Wets Building (1931–2). There Williams read the world a lesson on how to make thrilling architecture out of flat-slab concrete construction on mushroom columns with full-height glazing (ill. 234). Technically it had all been done before, but never so dynamically, while in type and quality of space Boots Wets stands comparison with Frank Lloyd Wright's Johnson Wax Building or Perret's lost Atelier Esders. Williams was in his element at Boots; adding an architect would have made things worse. That is less certain in buildings like the Dollis Hill Synagogue (1937–8), where the advanced structural concept of the folded plate dominates walls and roof alike but cannot excuse poor internal finishings; or even the bizarre BOAC Maintenance Headquarters at Heathrow Airport (1950–4), in which four hangars, their 336-foot spans sporting virtuoso cantilevers and counterweights (ill. 235), are planned cheek by jowl with offices running alongside and crosswise between them. Together with the surprises and excitements Williams could stage, there was always a touch of bloody-mindedness to his architecture. Before he died, the Brutalists had claimed him as one of their own.

All Williams' pairing with architects after he parted from Ayrton seem to have involved urban buildings for the newspaper trade. Papers in those days were typeset by natural light on the top floor, written and managed on the middle floors, and printed on massive press-lines in the basement. In that unique building-type, the skill lay in reconciling the structure of the three elements while separating their planning.[142] It was the former task that put a premium in London's Fleet Street on Williams' versatility with concrete, while specialist architects took care of the latter. His first contribution was hidden behind the architects' heavy front at the Daily Telegraph (1929–30). From there he moved on to make a dazzling job of the complex reconstruction and extension of the Daily Express

235. BOAC Maintenance Headquarters, Heathrow Airport. Sir Owen Williams, engineer-architect, 1950–5. Spans of 336 feet are made feasible by the characteristic Williams counterweights. This is the only one of the four great hangar spans still visible.

236. Daily Express, Fleet Street, London. Sir Owen Williams, engineer, with Ellis and Clarke, architects, 1931–2. The curtain wall was largely the creation of Bertram Gallannaugh, junior partner in Ellis and Clarke.

(1931–2) nearby. Not content with that, he wrested control of its façade – the first large curtain wall in England and a snook cocked at the extravagantly pompous Telegraph (ill. 236). Not everything at the Express was due to its expansive engineer. Bertram Gallannaugh, junior partner of the architects, Ellis and Clarke, is said to have suggested and certainly detailed the cladding of black vitrolite and glass. But Williams was sufficiently in the confidence of Lord Beaverbrook, the owner, to take on the later Daily Express buildings at Glasgow and Manchester on his own. In his last London newspaper job, however, the Daily Mirror (1955–61), his firm reverted to working out the structure within a modernist matchbox-on-end supplied by architects.

A creative individualist like Owen Williams with a sense of self-worth and a disinclination to court popularity will always create friction. Two points are worth adding. Firstly, though he was frequently paired with architects, there is always a sense of him working alongside rather than with them. That applies even to his relationship with Ayrton. No doubt personalities were involved. But perhaps too at this stage of the architect-engineer interface, intimacy was not thought desirable or just not imagined. Secondly, if Williams posed a more genuine threat to British architects than any previous engineer, as a whole they were remarkably open towards him. That is partly due to the liberal ethos of English architectural criticism at the time, partly also to the fact that Le Corbusier and others had bidden the younger generation revere the achievement and aura of great engineers. Beyond all that, there was no escaping the truth that the boldest, most charismatic modernist buildings built in Britain between the world wars had proceeded from the pencil not of a school-trained architect but of a rude engineer.

DUIKER AND WIEBENGA

No career sums up the contradictions in the modernist philosophy of building construction like Duiker's. Not even Le Corbusier's architecture so vividly conveys the zeal of the 1920s avant-garde. A few years ago his masterpiece, the Zonnestraal Sanatorium, Hilversum (1926–8), was all but a total wreck. Concrete architecture lends itself poorly to the charm of ruins. And yet out of rusting reinforcement, splintered glass and shabbiness shone the poignancy of ideals tempered by experience. That has vanished now the building is restored.

Why should a sanatorium for consumptives have become a write-off in two generations? The reasons are many. Tuberculosis was a scourge then, not after antibiotics. Focussing on the present, the building's authors made no claim on posterity. Like a machine, it served a specific purpose; that purpose having ended, it was ready to be scrapped. The sanatorium (ill. 237) had to be built cheaply and quickly. The structure was light, open and uninsulated, because the doctors desired to force patients into air and sunlight ('Zonnestraal' means sunray). Too little was known about concrete technology in the 1920s; the construction was experimental; the builders made mistakes. And yet the

passion of Duiker's design set up expectations. Posterity has been drawn into reversing the results of those priorities and errors. Architectural sentiment outlasts the logic of functional construction.

Jan Duiker was an architect to his marrow. Yet in the spirit of the 'Nieuwe Bouwen' he presented himself as poised between professions. His title, in the formal Dutch style, was 'ir. J. Duiker, b.i.' – 'bouwkundig ingenieur', a common formulation in his day, usually translated architectural or constructional engineer.[143] On his letterhead he called himself 'architect-ingenieur'; and when he died in 1935, aged just 45, he was hailed as 'a man of genius who had fulfilled Berlage's prediction that the professions of architect and engineer would shortly be united.'[144]

He had been well trained in construction at the Technische Hogeschool Delft, where Duiker and his partner Bernard Bijvoet met. Not only has Delft always been the most prestigious Dutch school of architecture; until 1900 it was just about the only one. It had started out in the German polytechnical tradition, stressing the structural and practical sides of the subject. Changes took place just before Duiker commenced his studies in 1908. A movement had gathered pace in Amsterdam for a more artistic education for architects.[145] Delft now responded. Civil engineering and architecture divided, the latter falling under the guidance and inspiration of Henri Evers. *Bouwkundig Weekblad* wrote:

The Architecture Department of the Technical University in Delft was originally, as a child of its time, based on science and technique . . . Thanks to the insights of men like Professor Evers and Klinkhamer, who understand the need of this age, [it has] gradually departed from this exclusively scientific standpoint – which is fatal to true architecture according to the newer and better insights – in order thus to introduce

237. Zonnestraal Sanatorium, Hilversum. Bijvoet and Duiker, architects, J. G. Wiebenga, engineer, 1926–8. An early image showing the sanatorium in its pristine state, to which it has now been restored after near-total dereliction.

238. Groningen Middle Technical
School. J. G. Wiebenga, architect-
engineer, 1922.

239. Van Nelle Factory, Rotterdam,
general view. J. A. Brinkman and
L. C. van der Vlugt, architects,
J. G. Wiebenga, engineer for the main
block, 1926–30.

education in the art disciplines on a more and more extensive scale, whereby they rightly hope to restore the honour of the title 'Bouwkundig Ingenieur', which it has missed for too long.[146]

Evers has been described as 'a fervent anti-rationalist'.[147] That does not mean that he was impractical or ignorant of construction, but that he inspired his students to believe in architecture as an art-form. When he embarked on the great work of his career, Rotterdam Town Hall (1913–20), he hired Bijvoet and Duiker as assistants. Their professional aspirations had been set, but not yet their style. They began their career in the early 1920s by designing good brick houses in suburbs of The Hague for an enlightened speculator-builder.[148]

On the engineering side at Delft along with Duiker was J. G. Wiebenga.[149] For a while their careers interlocked. They worked together on Zonnestraal, the Nirwana Flats in

The Hague and some other projects, including an entry for the League of Nations Competition. To label Wiebenga the technician and Duiker the creative force would be to oversimplify. Their relationship was fluid and unstable. Wiebenga too was an architect-engineer, but belonged to a generation that often repressed artistry under a cloak of objectivity and efficiency. A good musician (Duiker was musical too), a photographer, someone who could quote Robert Frost at the opening of one of his buildings, he had wanted to be an architect but his father steered him towards engineering. Restlessness was his Achilles heel.

In 1912, straight after Delft, Wiebenga took a job in one of the construction firms specializing in reinforced concrete, Stulemeijer and Co. The licensing systems for the material were breaking down just then in Holland, as elsewhere. Few architects yet knew enough about concrete technique to take much advantage. For those building engineers who did, it was a moment of opportunity. Wiebenga designed factories, warehouses and a sizeable shell dome for a church.

Then teaching drew him into the mainstream of progressive architecture. In 1922 he was appointed director of the Middle Technical School at Groningen, housed in temporary quarters. Soon Wiebenga was designing and building the new school. The job should have gone to the city architect, Mulock Houwer, who insisted that the design, set at the back of a formal open space in a developing quarter of Groningen, needed to be matured. He objected vocally to the disregard of context displayed by the engineer's rapid sketch. But the municipality was in a hurry. Wiebenga undertook to prepare the drawings in six weeks and have the building ready in thirteen months: that did the trick.[150] As completed, the Groningen MTS was a very early manifestation of the so-called Nieuwe Bouwen (ill. 238). Though Wiebenga asked the young architect L. C. van der Vlugt to help him, he claimed to have done almost all the job himself. The school turned out a consistent but rather graceless building, almost devoid of 'architecture'; even the concrete construction is played down. In the first modernism of the 1920s, efficiency without rhetoric or emotion really did sometimes reign supreme.

Wiebenga soon left Groningen for a spell in the United States, working for East Coast engineering firms. The drift of the articles he wrote on his return to Holland in 1925 concentrated on process – the organization of American work more than American building science. Nevertheless the core of his skills was concrete construction. It was in

240. Van Nelle Factory, interior of one of the main floors, showing Wiebenga's flat-slab structure.

241. Nirwana Flats, The Hague. Bijvoet and Duiker, architects, J. G. Wiebenga, engineer, 1928–30.

refining concrete technique that he proved his collaborative worth, foreshadowing those later engineers whose expertise released or reacted to architects' creativity. Three famous projects of the late 1920s demonstrate this: Zonnestraal and the Nirwana Flats, already mentioned, and the Van Nelle Factory at Rotterdam. In each Wiebenga played a vital but varying role. None was quite a harmonious partnership.

At the Van Nelle Factory (ill. 239),[151] Wiebenga acted as a consultant engineer helping young, inexperienced architects. J. A. Brinkman, then still a student in 'civiele techniek', inherited this prestigious job in 1926 when his father, the architect Michiel Brinkman, suddenly died. He asked Van der Vlugt, a former assistant of his father's, to help him. He it doubtless was who brought in Wiebenga as engineer for the multi-storey part of the complex. The storey-heights and cladding of the factory depended upon the concrete frame, whose design in turn hinged on the incidence of natural light on the manufacturing floors. Brinkman and Wiebenga clashed over the proposed direction of the main beams, in danger of obstructing the light. For some reason the architects went on opposing Wiebenga even after he had redesigned the frame with flat-slab floors and mushroom columns (ill. 240), but he won the battle.

At Zonnestraal,[152] Bijvoet and Duiker (specifically, the latter) were unambiguously in charge; heavy engineering was irrelevant. Duiker first toyed with a timber scheme, but presumably called in Wiebenga when he switched to concrete. It is not clear where his work began and ended, but he had the skills to stretch what little money there was as far as possible and ensure the sanatorium could be built fast. No doubt he tightened the discipline of the design. Short spans of three metres were chosen, for instance, so that the shuttering could be removed faster. Wiebenga was probably not involved in site supervision, which was inadequate; the concrete mix turned out to be poor in quality. Where he may be to blame for later problems was in failing to insist on expansion joints.

The Nirwana Flats started as Wiebenga's project.[153] His American visit had given him the idea of waking up The Hague's sedate housing with a cluster of high flats. He put much energy into promoting this radical speculation. Sleekness of image was essential to the process. Duiker made the first sketches late in 1925, showing ten tall blocks in a Corbusian parkland setting. The scheme went through many simplifications before Wiebenga could secure finance for the only block to be built (ill. 241), started in May 1928. So the collaboration was more drawn out and intimate than at the sanatorium. Doubtless the challenge of an eight-storey concrete frame was the engineer's main input on the design side. But relations broke down during construction. Wiebenga was frustrated that Duiker used the many changes demanded by his backers to protract the design process in the name of perfection. Confronted by a new proposal for collaboration, he declined:

> As Heer Duiker falls entirely short of the mark in organization and construction, it means for my part that in collaboration with him I cannot trust him with that side of the work, and so have to relinquish just the most pleasant side of the business. Added to which, on the building side I become the victim of his mania for change, so that I have to redo the work four or five times.[154]

For his part Duiker felt that Wiebenga, eager to move on, was neglecting the job. Efficient delivery sat uneasily with an architectural philosophy of perfectionism.

Wiebenga's credentials suited the ideologues of the Nieuwe Bouwen nicely. On the occasions when he took part in their debates, he mouthed the modernist engineering clichés: efficiency begets beauty, look to motorcars as a model for architecture. Yet in the buildings he built between 1928 and 1934, when he was in municipal service, he developed a marked aestheticism. At Aalsmeer, working without an architect-collaborator, he built a school with a suavity absent from the Groningen MTS. The same holds of a hospital wing at Zwolle, though there his first design was toned down by another hand. For a moment it looked as though Wiebenga had turned architect. But the troubles of the decade and of his temperament worked against him: Zwolle having dismissed him in 1934 for unreliability, he was thrown back on mundane engineering tasks. Although his career ran on into the post-war years, Wiebenga never recovered the balance that set him briefly among the memorable modern architect-engineers. That he acted in turn as architect, consultant, entrepreneur and municipal employee indicates the instability – or one might say flexibility – of those few inter-war building engineers able to address concrete creatively. And his relation with Duiker, if short and fraught, foreshadowed later ways of working.

Wiebenga lived too long, dying in 1974. Today it is only the short-lived Duiker who is really remembered, even in Holland. In the outpouring of grief that accompanied his death there was natural exaggeration: when all is said and done, the record of Duiker's buildings reveals an ingenious designer, not an all-round technician. But the intuition and humanity of the few designs he realized in his maturity gave him the power to engage and move hearts. In that he showed himself a pure, untrammelled architect. On a larger stage, the same holds true of this chapter's last case-study.

LE CORBUSIER

The grip that Le Corbusier still maintains over the imaginative realm of architecture makes his record as a constructor hard to judge. What matters about his art, those who revel in it will feel, is not how it was made but its fertility of idea, its power to excite and confront. In order to woo his audience, he reaches ruthlessly beyond the materials he uses and the men and women who help him transform them.

That must be borne in mind as one examines the ambiguous relations between Le Corbusier, the concrete that was his chief medium of building, and the world of engineering. Many of his practical dealings with engineers were fuzzy, unlike the sharpness with which during his propagandizing heyday he flung down the challenge they represented. That challenge included the famous encomia in *Vers une architecture* on modern engineering (p. 5), and the reproach to architects he saw embodied in artefacts like the Eiffel Tower or Gabriel Voisin's aeroplanes. In a typically dramatized anecdote, Le Corbusier relates how in his Paris student days the Métro's chief engineer came to address a class at the Ecole des Beaux-Arts on reinforced concrete, only to be howled down with the retort: 'Tu nous prends pour des entrepreneurs?' – 'Do you take us for a load of contractors?'[155] The engineer is the man of honour, it is implied; the architects are arrogant ignoramuses. Yet the engineer in the story remains unnamed, an abstraction. It was with artists who could react as unreasonably as those students that Le Corbusier felt at home, not with technicians.

Le Corbusier's basic training was in a Swiss industrial community.[156] The art school in his home town of La Chaux de Fonds was there to turn out skilled engravers and designers for the watchmaking industry. Had his eyesight not given him trouble, had his devoted, art-minded teacher, Charles L'Eplattenier, not therefore pushed him towards architecture for his fourth year, in 1905, nothing might have become of him. From watchmaking he learnt the deepest lesson an architect can take from handicraft – a feeling for the nature and limits of materials. The boy made fast progress. But L'Eplattenier's grounding in architecture was narrow; he could not help much with construction or structures, a point his pupil eventually cast back in his face. Le Corbusier was among the many architects who manage to make up for a lopsided training. Yet something also told him that by keeping his distance from technique he could also keep his freedom. He never sought to be pigeonholed as a professional architect – indeed as anything.

For the very first house Le Corbusier built, the Villa Fallet, another local architect, René Chapallaz, just six years his senior, helped the nineteen-year-old 'put his plans in order'.[157] That arrangement turned into a formal partnership for his next two houses, mostly built while he was away on his travels. For one of them, the Villa Stotzer (1908), Chapallaz introduced Hennebique-system concrete floors absent from the original plans.[158] Already a division between art-designer and executive technician had arisen. In these early craft-based villas at La Chaux de Fonds, Le Corbusier hardly had to confront the border between design and construction. But he knew how much he had still to learn. A ferocious worker and assimilator, he left Switzerland with the ambition to do so: first in Vienna, then Paris and Berlin.

The Paris months, between March 1908 and December 1909, were decisive. Allen Brooks's chronicle of his progress there shows him flailing around at first, then falling in with the rationalist agenda in architecture. For a moment he toyed with studying technique seriously. After listing various technical classes at the Ecole des Beaux-Arts, he seems to have sat in on the construction class alone. There he witnessed the teacher rapping out 'formulae in advanced mathematics by the kilometre . . . no one could grasp any of it and we felt that this degradation would weigh on the whole of our lives'.[159] Like many hapless architectural students before and since, Le Corbusier was allergic to rote learning. He bought no technical books in Paris, and when lent the French translation of Emil Mörsch's great textbook on reinforced concrete, just flicked through it.[160] It was the rationalist ideal, not any deep curiosity about how buildings are made, that drew Le Corbusier towards concrete. That ideal also elicited his first manifesto, a breathless, rambling sermon sent home to L'Eplattenier:

> The architect must be a man with a logical mind, he must mistrust and oppose the passion for plastic, sculptural effects, a man of science yet with a heart, an artist and a scholar . . . They talk about an art of tomorrow. It is going to happen, because mankind has changed its means of living and way of thinking. The programme is new and the context is new. You could talk about an art that is coming because this context is one of iron and iron is a new medium. The dawn of this art is becoming dazzling,

because from iron which is subject to decay they have made reinforced concrete, an invention with extraordinary consequences which will become a proud landmark in the story of peoples seen through their monuments.[161]

By then he had been some weeks part-time with Auguste and Gustave Perret in the Rue Franklin office: 'the most amazing, original place I've ever seen, all in new materials', he told his parents.[162] He stayed on until he left Paris. Not yet too grand, not yet too busy, Auguste treated him with avuncular generosity. During his seventeen months with the Perrets, Le Corbusier had the chance to draw deep at the wells of design, construction and culture. One day he would be told to read Viollet-le-Duc, another to visit Versailles. All around him the gospel of concrete was being preached. He will have grasped that with technical backing most design-ideas could be accommodated to a concrete frame, from the Rue Franklin apartment block to Oran Cathedral – on the Perrets' books at the time. Most of Le Corbusier's later structural thinking is implied by works they had then built or were building. But while Auguste Perret believed that a concrete architecture should also express an established public culture, Le Corbusier came to repudiate that shared culture in favour of one directed and manipulated by avant-garde art-individualists.

What Le Corbusier chose not to learn, he did not learn. He declined – for the moment – to go to Versailles; and it is doubtful how much concrete technique he actually mastered under the Perrets. During his time there he dabbled in maths lessons with an engineer, but worried that going far down that path would mean loss of freedom.[163] Afterwards, he spoke of his need to sit at the feet of an engineer, but did nothing about it.[164] When he built his next two houses back in Switzerland, he had their few concrete details specified and calculated from Paris.[165] And when in 1916 he finally ventured into

242. Villa Schwob, La Chaux de Fonds. Swiss and, perhaps, French flags fly on the topped-out villa in about December 1916. Charles-Edouard Jeanneret (Le Corbusier), architect, Terner and Chopard, engineers for the concrete frame.

a concrete frame for the first of his published houses, the Villa Schwob (ill. 242), an obscure engineering firm from Zurich helped him deploy it with an un-Perret-like timidity and extravagance. Its use added to heavy cost overruns that culminated in litigation and the architect's final escape from La Chaux de Fonds.[166]

If a design diagram were the key to building in concrete, Le Corbusier would have been at ease with the Villa Schwob. For by then he had devised his famous 'Dom-ino' concept, purporting to show how walls could be freed from bearing loads by the conjunction of minimal concrete piers and slabs. Dom-ino has been gently demythologized by Eleanor Gregh and Allen Brooks. Its origins belong to Le Corbusier's scrappy activities just before and during the First World War. These involved his closest engagement with an engineer, Max Du Bois. How their relationship worked is hard to say, since Le Corbusier later erased it from the record: scholars have had to rely on thin documentation, supplemented by reminiscences from Du Bois in his old age, by which time he had developed his own fixed view of the past. Rightly or wrongly, Brooks at least regards Du Bois as 'the real unsung hero' in the architect's career.[167]

A fellow-Swiss, Du Bois had been known to Le Corbusier since childhood. Already settled in Paris when the latter first visited in 1908, he it was who had translated Mörsch's *Eisenbetonbau* into French and lent it to his inexperienced compatriot, so he was well versed in reinforced concrete. It was he too who proffered advice to Le Corbusier on concrete detailing. Quite soon Du Bois turned entrepreneur, running a string of engineering interests and delegating technicalities to a colleague, Juste Schneider. In the engineer's ambit Le Corbusier too drifted into entrepreneurship, even claiming later to have abandoned architecture 'and immersed myself in industrial and economic problems'.[168] That is an exaggeration: he always faced several ways. But around the time of the First World War, the influence of Du Bois was profound.

Bored and restless back in La Chaux de Fonds, Le Corbusier angled for the job of designing an electrical factory prospectively to be built on land belonging to the Du Bois family. Early in 1913 he changed tack. Now he badgered Du Bois to join in on something which he called 'ma proposition Monolythe', thought to be the precursor of the Dom-ino idea. Engineers, he blithely informed his mentor,

don't study proportion, which is innate and belongs to the architect's domain. It's both outside and inside construction. And I feel I am absolutely ready for that . . . as a man can only survive in this filthy society by the support that he can find around him (alas not on his own), support me on this occasion and I undertake to be ready and reciprocate with all the energy and devotion and good will I can possibly offer.[169]

Du Bois seems to have been disposed to help out his keen young friend, but for the moment this was put on one side.

Their real collaboration began only when they met again after war was declared, first in Switzerland and then in Paris. It took the form of two projects. One was an orthodox competition entry for a multi-arched railway bridge in masonry over the Rhone near Geneva – later published by Le Corbusier without mention of Du Bois and Schneider. The other was the developed Dom-ino system. Dom-ino started out as among the earliest of the countless schemes and techniques for low-cost housing elicited by wartime destruction. In 1914 its proponents had Belgium in mind. But so long as Belgium was under German occupation, it was impracticable. Du Bois

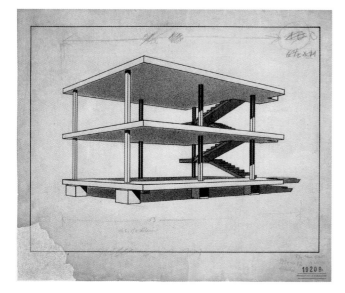

243. The Dom-ino system, perspective of 1915.

perhaps saw it as little more than a desk exercise. 'Probably, as an engineer,' says Gregh, 'he considered the idea from a constructional point of view and saw nothing extraordinary; hundreds of firms and individuals were having similar ideas.'[170] Le Corbusier on the other hand persisted, hoping to patent it and anticipating royalties. When that did not work, he recycled it as a piece of theory.

The lucid separation of frame and walls represented by the neat Dom-ino diagram (ill. 243) has become world-famous. Yet there was nothing new in 1914 about such an arrangement. It was, Colin Rowe has remarked, 'no more than an abstraction of the structure of all those Chicago buildings in the Loop'.[171] The one novelty perhaps was its application to small houses. The challenge at that stage and scale of concrete construction was to join thin piers and thin slabs at right-angles without wasting space and money by adding bulky heads to the piers or dropped beams beneath the floors. In effect, Le Corbusier was asking for the benefits of flat-slab construction without its thickness or its mushroom heads. Positioning and calculating the reinforcement were crucial; for that, its inventor needed help. He was still pushing the Dom-ino idea in 1915, when Perret gave it vague encouragement and asked him about financing, and Schneider helped sort out the floor slab. Eventually Du Bois put in for a patent, which for some reason was never granted. So when Dom-ino was published after the war in *L'Esprit Nouveau*, it was still entirely untested. It made its impact as a diagram not a system, an imaginative vision of pure floors and pure support for architects who had not yet woken up to such things. But that was hardly how it had started out.

When Le Corbusier finally decamped to Paris in January 1917, Du Bois found him a niche in the structure of his companies, part-time so as to allow him to carry on with his painting. Chiefly he provided architectural services to the Société d'applications du béton armé, acronymically SABA, touring provinces remote from the battlefields to check on concrete constructions at power stations and hydro-electric works. During wars, architects are eclipsed and engineers come to the fore. So in most of this work he played second fiddle to Du Bois and Schneider. Here at last was a chance for Le Corbusier to get to grips with construction. Some of that he may have drawn upon in a set of schemes he made for large-scale abattoirs.[172] Yet in laying them out what preoccupied him was not engineering as construction but engineering as idea – in this case the fad of American scientific management: 'by day I am an American . . . read Taylor and practice Taylorism'.[173] These unbuilt abattoirs are the only Corbusian designs to foreshadow the bald industrial structures imminently to be glorified in the pages of *L'Esprit Nouveau*. As if all that were not enough, late in 1917 Le Corbusier also took on from Du Bois the management of a works for making clinker bricks, a project that landed him in severe financial scrapes after the war. He was also advancing sundry schemes for wartime workers' housing. Already he impresses as having boundless vitality and resource, but too many strings to his bow, too little time to follow things through.

It is of a piece with that changefulness that when after the war Le Corbusier started *L'Esprit Nouveau* with Amédée Ozenfant, he subscribed himself 'industriel', not architect. Though the alliance with Du Bois was fading, the clinker brickworks was still in production (it closed in 1921 after a few years of ramshackle operation). Le Corbusier also had other hare-brained schemes on the go just then involving scrap metal and coal-dust briquettes,[174] so the title was half-plausible. Stanislaus von Moos has argued that the choice has a double significance.[175] Firstly, Le Corbusier's frame of mind was set by the war. The adrenalin-rush of his wartime activities under Du Bois drew him to romanticize the destiny of modern industrial organization and the aesthetics of engineering. Many architects were jolted by the world wars into embracing the same forces – beneficent if their productivity could be harnessed, perilous if their brutality was not tempered. Le Corbusier was never at real risk in 1914–18. On the whole, the closer that architects stood during the great twentieth-century wars to construction and the further from destruction, the more shrilly they called for architecture's make-over through engineering.

244. Scrawny cousins in mock fight: Pierre Jeanneret and Le Corbusier on the beach.

But Le Corbusier had another reason for posing as an industrialist. By presenting *L'Esprit Nouveau* as a vehicle for drawing architecture and industry together, he hoped for advertising, endorsement and hence personal commissions. In the original magazine, Von Moos points out, the pictures of cars and planes often take the character of promotional material – an aura they lose once reprinted in *Vers une architecture*. Aspiring to grand collaborations in housing, Le Corbusier aimed for the ear and the favour of great manufacturing engineers, Michelin, Citroen, Voisin and the rest – hence the 'Maison Citrohan' and the 'Plan Voisin'. Those moguls proving deaf to his blandishments, he had to be content with Henry Frugès of Pessac. A resentment towards engineers accordingly creeps into the latter numbers of *L'Esprit Nouveau*. When a manufacturer of lighthouse reflectors declines to advertise on the grounds that 'our light is not decorative', the insulted editors shoot back: 'Common mentality of the engineer, enveloped in the high pride of his figures. Total incomprehension of anything not in the narrow field of his investigations. The old, discouraging, clichéd story, one example among so many others. No mutual understanding.'[176]

As the hope of latching on to industry faded, Le Corbusier reverted to self-promotion by means of an art-led, ideas-led, publication-led architecture, bolstered by episodic fantasies about mass-production. Nevertheless his buildings needed to be built. Concrete, he knew from his time with the Perrets and Du Bois, offered most leeway for the radical design freedoms he sought. For that Le Corbusier needed an office organization, along with technical and engineering help.

The mainstay of the office between 1922 and 1940 was his cousin, Pierre Jeanneret (ill. 244), another of the tribe of able, compliant technicians whom art-architects contrive to attract.[177] Having studied architecture, painting and sculpture at Geneva's Ecole des Beaux-Arts, Jeanneret had enjoyed a higher-grade formal art-training than Le Corbusier himself. For want of a monograph, the depth of his technical skills is uncertain. But after Le Corbusier asked him to join him in Paris, Jeanneret spent part of his first two years with the Perrets; and he seems to have set most mark on the larger projects, especially those with elements of prefabrication. Between the wars it was he who kept office hours and did the calculations, drawings and correspondence, while the master was travelling or making art and propaganda. After a spell on his own, Jeanneret became Le Corbusier's man in Chandigarh for long years from 1951, contributing many buildings of his own there as well as helping to ensure that his cousin's designs were faithfully built.

On the construction side, the main builder of the classic Corbusian villas has remained in the shade. This was Georges Summer. As if to underline the craftsmanliness of these one-off, suburban houses, Summer is often described as a mason, though formally he called himself 'ingénieur-constructeur'.[178] Reconciling what was intended with what could be built must have involved good instincts on his part and an ease with small-scale concrete construction. Probably, thinks Tim Benton, he relied on generous structural tol-

245. A corner of Pessac in November 1993, showing Le Corbusier's housing in varied states of repair.

erances.[179] Summer was solely responsible for the core of the Villas La Roche-Jeanneret, Cook, Stein and Church, as well as the Pavillon de l'Esprit Nouveau. His relationship with Le Corbusier and Jeanneret seems to have been trusting, articulate and to some extent creative; though compliant, the builder would fight his corner if he thought an idea was wrong. It was naturally to the dependable Summer that Le Corbusier turned when an industrialist and idealist, Henry Frugès of Bordeaux, at last entrusted him with a sizeable housing project that had to be not just imagined but built.

The first idea at Pessac (1924–6) and the preceding pilot scheme at Lège was for Frugès to create a bureau d'études in his sugar refinery and carry out the housing under a local engineer.[180] Well before he got down actually to designing the houses, Le Corbusier had the brainwave of spraying on the concrete for the development with an American cement gun, to which he later added the idea of using a compressed straw block as the base material. Frugès was persuaded to spend a sum amounting to the cost of eight or ten dwellings on the cement gun, which required a skilled crew. The local man, Poncet, did his best with the call for industrialized methods, managing to install a few simple pre-cast vaulting slabs, but was out of his depth; progress was slow and clumsy. When Le Corbusier finally arrived to inspect things in April 1925 there was a blistering row. An independent engineer's report that gave some support to Poncet was brushed aside as 'utterly spineless'.[181] The only remedy was to bring down Summer and an élite of workers from Paris. After experimenting with mixes and further wasted time and money, the 'gun-ite' idea had to be abandoned in favour of conventionally rendered concrete blocks.[182] Nor was that all. Le Corbusier had lost interest in Pessac by 1926, as fresh jobs flowed in. Somehow, between Frugès and his architects, the authorities were never consulted about the road network or the drainage. So the Pessac houses (ill. 245) were left unconnected and uninhabitable for two years. At a period of low inflation they had cost between three or four times the original estimate, and

246. The roof of the Unité
d'Habitation, Marseilles, in 1949 with
(left to right) Shadrach Woods,
Vladimir Bodiansky and Georges
Candilis of ATBAT.

double the average cost of working-class dwellings at the time. Severely compromised,
Frugès went to live abroad. He had paid dearly for his housing 'laboratory'. His reward
was to be remembered; and the world enjoys the experiment he had endured and
defrayed.

The story of Pessac can stand for others. From the Villa Schwob downwards, in a succession of Corbusian schemes the construction verges on débâcle and the result comes in
hair-raisingly over budget, hard to maintain and repair, impractical in use: the Villa Savoie,
the Salvation Army hostel in Paris, the Marseilles Unité d'Habitation and the Carpenter
Center at Harvard are just four fully documented examples.[183] Foreknowledge and even
intention must be assigned to the pattern. The shambles at the Salvation Army hostel followed, says Brian Brace Taylor, from the architects' insistence on treating each project 'as
an opportunity for them, as members of an avant-garde, to challenge the traditional modes
of operation of the building industry'.[184] In Le Corbusier's view, you could not push the
boundaries of art without also pushing those of technique; anyone who refused the consequences was timid and bourgeois, blind to the challenges of the times. But while the art was
scrupulously controlled, meditated and revised, the technique often ran after whims and
fashions that he failed to investigate or take prior advice upon, like the cement gun. Le
Corbusier was far from the first architect to be bold with technologies that he had not mastered, and to exploit others to help him do so: in England, John Nash comes to mind (pp.
92–3). The difference was that Le Corbusier glorified the means as well as the ends, and so
led clients and the public into supposing that he was in control of a process relentlessly
biassed towards effect. While he supplied modern architecture with evergreen images and
concepts, he also bequeathed to it the habit of justifying sleight of hand.

Though the limits of this chapter are otherwise set by the Second World War, no account of Le Corbusier's dealings with the engineers can close without some mention of Vladimir Bodiansky, the Ukrainian-born technician who masterminded the making of the first Unité d'Habitation at Marseilles (ills. 246, 247).[185] Bodiansky was possessed of a resource and dynamism equal to Le Corbusier's, so they were bound to clash. His picaresque career is a parable of his age's rootlessness. Trained as a bridge engineer in Moscow, he had before he was thirty helped construct a railway from Bokhara to Kabul, fought in the Tsarist army, transferred to be a pilot, escaped Russia, qualified as a French aircraft designer, served in the Foreign Legion, and spent three years in the Belgian Congo building another railway and masterplanning the town of Albertville. Between 1923 and 1931 he was back in France designing automobiles and aeroplanes, latterly for his own company. Almost the only engineering sphere Bodiansky had not dabbled in by then was that of radical building structures. That followed when he became chief engineer to the building firm of Mopin and devised its pioneering system of prefabricated concrete panels on a steel frame for apartment-building, first used on the Drancy estate outside Paris by Beaudouin and Lods (1931–4) and taken up in England at the Quarry Hill Flats, Leeds (1937–9). He also helped Jean Prouvé work out the structure for the all-metal Maison du Peuple at Clichy (1935–9).

Bodiansky was just the avant-garde technician whom Le Corbusier needed for the monumental buildings he aspired to after the Second World War. They had met on the wartime interdisciplinary committee convened by the latter to promote his personal agenda for reconstruction. When peace came, they went together on a delegation to the United States looking at housing and planning, then crossed the Atlantic again in 1947 to

247. The Unité d'Habitation, Marseilles, aerial view in 1951 shortly before completion, with the Boulevard Michelet to the right driving towards the Mediterranean. The flats no longer stand in splendid isolation; other apartment blocks have sprung up all around.

serve as the French delegates for planning the United Nations project in New York, Le Corbusier on the board of design, Bodiansky on the panel of specialist consultants. Meanwhile the Unité d'Habitation, long meditated in the architect's mind, was about to bear fruit in Marseilles. Here was the man to get it built.

The means chosen to do so was ATBAT, the Atelier des Bâtisseurs. Essentially ATBAT was a glorified, personalized, but 'largely unpaid' bureau d'études.[186] The core of the work in a normal bureau d'études on the French model is the interpretation of design intentions and their transfer into drawings and calculations for construction. In art-architecture, less rational, there is fluidity about how the transfer is resolved and who does what. In Le Corbusier's mind, his personal atelier under André Wogenscky was to communicate the ebb and flow of creative ideas for the Unité which Bodiansky's team would unswervingly execute. ATBAT on the other hand had been set up with much fanfare not just as a functional building operation but as the consummation of all the debate on professional partnership in construction. It had four separate sections and anticipated doing far more than executing Corbusian masterpieces. Since the Marseilles Unité was the first of the ATBAT projects, everyone was eager to make the experiment work. But it depended on communication and equity between the participants.

In the event the vaunted claims to an exemplary process fell away.[187] Despite boundless devotion on the part of the thirty-plus talented young architects and engineers in the two offices at the centre of the Unité design, there was overlap, confusion and delay. The heart of the problem lay in Le Corbusier's refusal to visit Marseilles regularly or, given that fact, to delegate decisions to his personal site-representatives. So Bodiansky and the ATBAT side were forced to take decisions that raised the architect's hackles. They had, for instance, repeatedly to compromise the proportional purity of the fabled Modulor. So far from a device delivering a cheap and universal means of rational construction, as Le Corbusier claimed, the Modulor turned into a weapon of remote control that hampered the construction process. Inside the building, though less outside, the dimensions were repeatedly fudged.

No doubt it was Bodiansky's passion for technical innovation and risk-taking that commended him to Le Corbusier. Yet the structure of the Unité testifies to his sobriety and caution. Because of the post-war shortages of materials and labour, everything in this high-cost project had to be simplified. 'The essential aim of all our efforts and of the constant modification and rethinking of all our studies', wrote Bodiansky afterwards, 'was to make a work which anyone could execute, for good architecture should be constructed just like that, simply and without acrobatics'.[188] Much of the simplification followed from shortages of steel. To Bodiansky was largely due the early change from the steel frame Le Corbusier hankered for to the more economical concrete, while during the construction he was able to impose hollow floors instead of steel-rich slabs. To the ATBAT team also fell the working out of the transfer structure in the shape of hidden longitudinal beams that sustain the upper floors, reducing the load-bearing function of the robust pilotis to near-symbolism. Many major decisions had to be made late during construction because of the ad hoc Corbusian way of working. Predictably, friction between the offices grew. With the core complete in 1949, there was some sort of definite break between Le Corbusier and Bodiansky. When the Unité finally opened in October 1952, the architect was so graceless as not to thank or even mention the engineer who was the second most important figure in its history – one who worked unstintingly to see a memorable conception painstakingly realized.

Le Corbusier lost as much as he gained by his flaws as a collaborator. With all respect to Pierre Jeanneret, he had never had a partner of Bodiansky's technical calibre, and was not to do so again. For his other late works in France he had to sort out new arrangements without the drive and commitment of the Marseilles team. As for ATBAT, founded with high hopes as a means not only of supporting but of extending Le Corbusier's creativity, it carried on without the protagonist. It was its affiliate, ATBAT-

Afrique, which registered most successes in the 1950s in Algeria and Morocco, perhaps because colonies and developing countries offered fewer institutional obstacles to architect-engineer co-operation than European ones.[189] Bodiansky continued to direct the technical side of ATBAT, but it died with him in 1966. Nearly sixty when the Unité was finished, he devoted his later years to consultancy and teaching, constantly travelling. The internationalism of his career, and indeed the growing internationalism of Le Corbusier's own practice, with his willingness to build in countries of which he knew little and on sites he had never seen, are a reminder that procurement takes fresh patterns when architects and engineers work in many countries or at long distance. In such circumstances the gap between design and construction is bound to widen. The phenomenon of long-distance procurement, common from the later twentieth century, is separately addressed in Chapter 5 (pp. 392–4, 409).

4

THE BRIDGE

1. The Masonry Bridge

THE BRIDGE ON THE DRINA

Conflict and oppression are the theme of *The Bridge on the Drina* by Ivo Andric, a novel about Bosnia's history.[1] It chronicles four hundred years of religious battles between the communities co-existing in the border town of Visegrad. The thread that holds it together is Visegrad's eleven-arched bridge over the Drina, built by the Turks – a bridge that stood steadfast over four centuries, until it was mined by Austrians retreating before the Serbian advance in 1914.

At the outset a great vizir in Istanbul, kidnapped into slavery from Bosnia as a boy but now rich and powerful, resolves to build a bridge at Visegrad. His purpose is to abolish the ferry and expiate a childhood memory of his initiation into captivity by boat across the Drina. He appoints a harsh overseer and an Italian chief mason, who presides over Christian masons from Dalmatia. Years of laborious chaos follow, with little to show for them. There are accidents, setbacks, stoppages for lack of money. An act of sabotage exacts excruciating punishment. Then the overseer is dismissed and more enlightened management obtains. Piers are founded, great timber centres set one by one and arches of limestone turned upon them. The townsfolk begin to believe. At last the broad bridge is complete. It brings prosperity to Visegrad's people, Moslem and Christian alike. A stone 'han' or caravanserai, sturdy as the bridge, is built at its foot to house the concourse of traders. Over centuries this han declines and disappears, a victim to changes in régime, consumption and accommodation. The bridge meanwhile holds fast: shot at, bashed about, ill-maintained, and in due course demoted by an upstart railway. Finally the Austrians drill their holes and blow their charges: there, Andric's saga ends.

Here is an engineer's story: about courage, effort and technique; about the benefits a magnificent and useful monument can confer across generations; about amazement at its construction and pride in its endurance. But there is another side to Andric's presentation of the Drina bridge. At its crown, the width is corbelled out over the central pier to make a place called the 'kapia' – an open terrace, with steps, benches and a commemorative inscription in Turkish. Resting points of this kind, sometimes marked by monuments, were not rare at the midpoint of old bridges. On the kapia take place major and minor events in the course of the novel, as of Visegrad's history: courtships and executions, eating and drinking, deals, arguments, conspiracies, rituals of welcome and farewell, ogling and idling. In a community riven between mosque and church, the kapia and by extension the bridge itself become the symbolic centre of the town, the focus of activity, memory and cohesion.

And so the bridge on the Drina becomes imbued with extra human value over and above its sturdy usefulness: which is one way of defining the point at which an idea of architecture enters into construction. Lest it be supposed that such things are the construct of Andric's imagination, or that all such symbolic references are positive, it must be added that in 1992 during the Bosnian war, some fifty years after the book was writ-

FACING PAGE
248. Waterloo Bridge, arches.
John Rennie, engineer, 1811–17.

ten, Moslems were flung from the reconstructed Visegrad bridge and shot at by Serbs for sport as they fell.[2] Visegrad now belongs to the pariah Serbian Republic of Bosnia.

Here then is a hint of what an engineering attitude towards a bridge might be, and what an architectural attitude might add. Can we read from that where architecture comes into a bridge, or where engineering ends? Or say how architects and engineers co-exist in bridge-building, and to what effect? These are the matter of this chapter: which takes as its starting point urban bridges even older and more august than Visegrad's.

ROME

Between the invention of printing and the Industrial Revolution, ancient Rome's array of masonry bridges informed and prompted the memory of ambitious bridge-builders. Out of a former sequence of ten, five remain: going downstream and using modern names, they are the Ponte Milvio (once to the north of the city); the Ponte Sant'Angelo; the Ponte Cestio; the Ponte Rotto; and the venerable Ponte Fabricio. All have been grievously altered. But enough is left for us to know that their principles of construction conform to the pattern set throughout the Roman empire in the many stone bridges which their makers built to last.

Scholars of ancient engineering have taught us to put those principles in context. Bridges, they insist, were particular incidents along the Roman road or aqueduct system. Primarily they were exercises in efficient construction. They may be beautiful, says Piero Gazzola, but it is a mistake to judge them as classicizing objects:

> It is quite misconceived to judge the various categories of ancient art by the standards of classicism alone and to deny aesthetic value to those that diverge from classical exemplars. Economy was a fundamental element in Roman aesthetics, implicit in the very nature of the job . . . The vast majority of bridges are connected . . . to road construction rather than to architecture.[3]

That is borne out by what little we know of those who built Roman bridges. In eleven cases the builder is named. Most seem to be 'curatores viarum' – officials charged with making and maintaining roads. 'Aquileges' – conduit or aqueduct masters – are also mentioned. Only one, Apollodorus of Damascus who designed Trajan's famous timber bridge over the Danube, is called an 'architecton', and that only in retrospect, from the sixth century.[4] The semi-circular or segmental profile of arches; the tendency to make piers massive, entailing a high proportion of pier to span; a complacency over arches of unequal height and, sometimes, irregular width: all that bespeaks pragmatism. Roman bridge-builders had few geometrical or proportional or aesthetic obsessions. Instead, they settled for forms easy to set out and practical to construct, to the long-sighted standard of their masonry tradition.

For any bridge, after all, firmness is the ultimate criterion. Ingenious or graceful it may be, but it must stay up. As wind, water and temperature do not relent, the difficulties of safe construction in bridge-building are vast, incidences of eventual failure high, and their consequences sore. So the bridge has always been first and last the province of the constructor. He must ensure that what is designed can be built, and that what has been built can be maintained. Others may inspire, add to or refine upon what he does; his is the basic contribution. The humanity of the kapia depends upon the undergirding arch which his crew of toilers have risked their lives and he his reputation to establish. Therefore the unshakeable ideal for the bridge is magnificent construction. In terms of the masonry massing of the Romans, that means the Pont du Gard or Segovia aqueducts and their long posterity. In the sleeker language of modern technologies, it means the crossings of Menai, Brooklyn, the Pont Garabit, Salginatobel or Sydney Harbour.

Yet many bridges go further. Where construction ends and architecture begins, no one can definitely say. It has long been part of architectural rhetoric that the two ought to be integrated – a clear incentive to blur the boundaries. Nevertheless some constructions are

clearly more 'architectural' than others. That is easiest to see in classical buildings. Over the centuries classicism developed a hierarchy of expression, reflecting cost and status. Proportions and excellence of construction enter into every good classical bridge. Coursed, dressed and minutely jointed masonry, and costlier materials like travertine or marble, do not. Nor do items like entablatures, corbelled or balustraded parapets, columns or pilasters or niches against the piers, rostral cutwaters, or that ultimate give-away of the monument, commemorative sculpture. All that is decoration, maybe: a word which architects, engineers and critics have come to despise.

Even the practical Romans built a few architectural bridges. Gazzola regards them as peculiarities – isolated monuments that could only be built once the art of bridge-con-struction had been perfected.[5] Commoner were those that integrated architecture into the components of the bridge's structure, in the guise of some of the features listed above. Occasionally there was also architecture on top. Houses seem not to have been permitted on Roman bridges, let alone the fanciful style of superstructure imagined in the Renaissance. But there are bridges like the Pont Flavien at St Chamas with arches at either end, prefiguring the pylons of suspension bridges (ill. 249); or like the Puente Alcantara over the Tagus, which has an arch in the middle. Though such arches may have been gated to allow the screening or taxing of travellers, they were probably com-memorative in chief.

Where are such architectural bridges to be found? Mainly where there was extra money to be spent and it seemed proper to spend it: that is to say, on the approaches to or within towns and cities. The notion that such a display was only proper to the city (one writer calls this kind of bridge 'urbane'[6]) persisted. Only in the eighteenth century, when classical hierarchies were crumbling, did French engineers endow rural river-crossings on their improving road-system with 'architecture', and English landowners lace little spans across brooks on their estates with grandiloquence. Vanbrugh and Hawksmoor's bridge at Blenheim (1708–11) may be the first to signal this outbreak of pretension (ill. 275).[7]

Ancient Rome therefore is the natural place to look for the origins of the urban, or architectural, bridge.[8] Two in the surviving sequence there are specially pertinent to later bridge-building. The remains of the Ponte Rotto (probably of the Augustan period, but drastically recast in the sixteenth century) are ornate, with cutwaters shaped into prows or rostra, and pilasters against the piers – features later much copied. More striking,

250. Fischer von Erlach's reconstruction of the Pons Aelius and Hadrian's
Mausoleum, now Ponte Sant'Angelo
and Castel Sant'Angelo, 1721, with
inset of bronze Hadrianic medal.

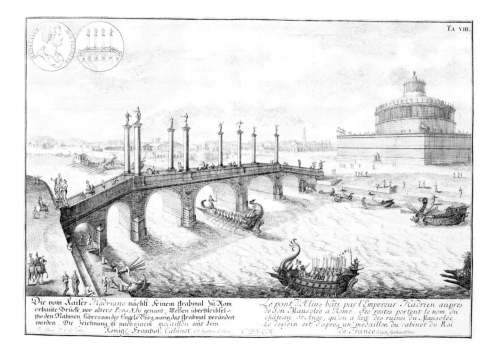

though, is the Ponte Sant'Angelo, now the most decorated of ancient bridges: also much
altered, but better documented.

The Ponte Sant'Angelo that we see today was reconfigured in 1892–4; the statues on
it date from 1667–72.[9] It started out as the Pons Aelius, a bridge leading to the
Mausoleum of Hadrian. To imagine the mausoleum you must reclad the Castel
Sant'Angelo with fine-jointed marble, invest it with an extra storey, and crown it with
statues and a quadriga. To imagine the bridge, you must take the outermost spans of the
present five on either side and replace them with ramps over pairs of shorter arches, giving seven arches and a rhythm of ABCCCBA, where the arches CCC alone support a
level roadway. Along the parapet of this central section, a Hadrianic bronze medal tells
us, stood eight high and hefty pillars topped with terms (ill. 250). The medal no doubt

251. Ponte Sant'Angelo after straight-
ening of end arches in 1894.

exaggerates their proportions, and we cannot be sure that they were permanent. But they accord with other bridges in Rome which bore 'hermae' or statues of the emperors.

The line and figure of the original Ponte Sant'Angelo were not unusual for older bridges. Cities like Rome first prospered because traders had ready, low-lying access to a navigable river. Before the practice of embanking became common, the river margins in such cities tended to be lower than today, the stream wider and slower. To reconcile the low level of the bridge foot and abutments with headroom in the middle of the river and the need to keep the superstructure safe and dry in times of flood or high water, urban bridges were ramped up to the crown. As a practical problem for masonry bridge-builders, that hardly competed with the safe founding of piers or construction of arches. In single-arched bridges the gradient had to be candid, and could be hard work for carts and carriages, especially once four wheels became as common as two. In longer, multi-arched bridges the rise and fall could be lower and subtler. But it caused headaches for aesthetic ambition, especially for those keen to deploy the language of classical architecture.

The incline obliged bridge-builders to build arches of unequal span, narrowest near the bank, widest in the centre: usually ABA for a bridge of three arches, ABCBA for five arches, and ABBCBBA for seven. That gave a tolerable rhythm, though it needed more than one size of centering, and disrupted the balancing of equal, contiguous arches that formed the basis of later bridge-building theory. Trickier was the rise in height of arches, piers and parapet. Classicism tends to find gradual changes of level problematic. It is awkward enough when there are slight changes at the base, but most people do not look down. Higher up and in full view, it is more palpable.

Classicizing bridge-builders therefore resorted to various devices for the profile, trusting to the fact that the sight of a bridge is seldom the perfectly perpendicular one shown in elevation. The solution adopted by the builders of the Pons Aelius was to ramp up the two ends and then set out a level centre for architectural display. That happened too at another in the select class of Roman 'architectural' bridges, the Ponte Augusto at Rimini – drawn by Palladio and therefore much copied (ill. 252).

The recasting of the Ponte Sant'Angelo as a level bridge in the 1890s belongs to an era when wharves were being banished from central-city locations and replaced with embankment walls, roads and promenades (ill. 251). Such older bridges were largely then either replaced or doubled in breadth to the detriment of their proportions. Often they acquired high abutments on made-up ground, carriageways suitable for tram and motor

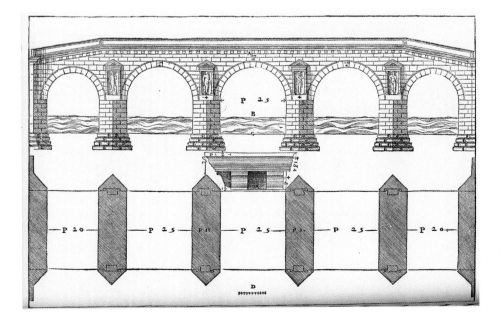

252. Ponte Augusto, Rimini, as drawn by Palladio for his *Quattro Libri dell'Architettura* (1570). A fundamental source for Palladio's Rialto designs and other bridges with statuary against the piers.

traffic, and long approaches reaching into the deeper structure of the city. In that way, bridges became locked into town-planning; and architect and engineer had to meet the challenge of their design on enlarged terms. Though the Ponte Sant'Angelo escaped demolition or 'doubling', its level roadway and approach integrated it into the circulation pattern of modern Rome. As that comprehensive vision of urban efficiency gained ground, the old image vanished of the Pons Aelius as a gesture of magnificence and benefaction at the edge of the city.

THE BRIDGE OF MAGNIFICENCE

In the event, the influence of the Ponte Sant'Angelo outlasted the stilted reality of the original bridge. Through a series of transitions in the imaginations of Alberti, Palladio and Piranesi it became the starting point for the so-called 'bridge of magnificence': in other words, not the engineer's vision of an artery that people passed safely over and under, but an architectural monument, as the Pons Aelius had indeed been intended.[10]

The first step was taken by Alberti. By his time Hadrian's mausoleum had long been the citadel of the popes, and the Ponte Sant' Angelo the chief approach to the Vatican from Rome, festooned with gates and chapels. Maybe it was this that led the father of Renaissance architectural theory to declare, on what grounds no one is sure:

> Some bridges even have a roof, like that of Hadrian in Rome, the most splendid of all bridges – a memorable work, by heaven: even the sight of what I might call its carcass would fill me with admiration. The beams of its roof were supported by forty-two marble columns; it was covered in bronze, and marvellously decorated.[11]

That cue was to be seized upon by Palladio, as he contemplated how to tackle the Rialto bridge in Venice.

The old Rialto bridge was a timber construction, lined by booths and rising to a pair of drawbridges in the centre.[12] It needed major repair or reconstruction once a genera-

253. Fortified bridge at Cahors.

tion. An intention in principle to replace it in stone had emerged before a fire destroyed the market quarter next to it in 1514, though not the bridge itself. Even then, it was to be over seventy years before the rebuilding came to fruition. Because of the bridge's unique position and proximity to major trading spaces, it was viewed from the first as a prestigious planning project. Venice was then at its full tide of faith that architecture could further civic and imperial dignity.

But the Grand Canal is not the Tiber. If its width of some ninety feet meant that a new bridge might just get away without a pier in the water, by the same token the conflict between street-level and headroom for navigation was acute. The nature of the ground beneath any potential weight of abutments was also treacherous. To add to their difficulties, designers for the Rialto faced pressure to renew the shops on the bridge. At that time, the belief that urban bridges ought to be clear of encumbrances on the superstructure had yet to take shape. Roman bridges, we have seen, had not been so impeded. But many mediaeval bridges had had superstructures of one type or another – fortified towers, gates, chapels or just houses. One might divide them into 'war bridges' (as Walter Shaw Sparrow calls them);[13] sacred bridges, of the kind that the Ponte Sant'Angelo had become; and ordinary secular bridges, their surfaces colonized by overspill from replete towns and cities. Seldom did bridge and superstructure form a conscious architectural unit, as with the bridge at Cahors (ill. 253). Most were accretive. Later, urban authorities and technicians relished sweeping away such obstructions. Then in the romantic period painters and travellers began to savour them for their picturesqueness. Only recently, in reaction against tyrannous motor traffic, have such inhabited or 'living bridges' been lauded for human energy and intercourse of the kind chronicled by *The Bridge on the Drina*.[14]

We can watch the shift in attitudes towards bridge superstructures played out in six-teenth-century Paris.[15] When the Pont Notre Dame was rebuilt in 1500–12 it was lined with houses and shops, orderly towards the street but turning their back upon bridge and river. Yet the Pont Neuf, constructed at the other end of the century under royal patron-age, was left uncovered, despite proposals to build on it. The dates of these two bridges are bookends for the drawn-out bickering in Venice over the Rialto. In the interval, archi-tecture came up with an alternative both to the inhabited bridge-street of mediaeval tra-dition, focussed on city and carriageway, and to the engineering style of bridge practised by the Romans, preoccupied with construction and crossing. The bridge of magnificence was a noble idea. But it was destined to remain, with rare exceptions, on paper.

Fra Giocondo, the Veronese mathematician and designer, had acted as a consultant over the Pont Notre Dame. On his return from Paris, he went to Venice and made a plan-ning scheme for the burnt-out Rialto, though not perhaps for the bridge itself. Rebuilding in the market area (not to Giocondo's design) soon began. But the bridge question dragged on for decades between committees. In 1554 came a fresh start: the authorities, we are told, 'sent to different parts to get designs from the first architects of Italy, to cre-ate something excellent and unusual.'[16] In other words, 'architecture' was solicited. Among those who sent in designs were Vignola and Sansovino; both are lost.

The scheme that is not lost is Palladio's, proffered at the same time – and architectural in spades. A drawing survives of his original project, while a revised version, most likely an afterthought, became famous through publication in his *Quattro Libri*. At first Palladio proposed compact forums at either end of a five-arched bridge modelled on Rimini and/or the Ponte Sant'Angelo, with a superstructure along the level central sec-tion bearing a brief temple front and shops. In the revision the flanking forums were abandoned and, in a gesture of imaginative monumentality, thrust up on to the bridge itself. In Palladio's words:

> Wherefore, as well to preserve the grandeur and dignity of the said city as very considerably to encrease the revenues of the same, I designed the bridge so broad as to make three streets upon it . . . Besides this, there were to have been made galleries at each head of the bridge, and in the middle over the great arch, wherein the merchants should keep their exchange, and which should have occasioned no less ornament than convenience.[17]

To carry this hopefully lucrative platform of streets and shops, the end arches had to be suppressed and the bridge made half as wide as it was long. And in memory of Alberti's imaginary 'forty-two marble columns', the whole superstructure was lavished with extra temple fronts bearing a free-standing order. If the original project was extrav-agant, the revision was surely never meant as more than an ideal – indeed the site was not identified, to imply universality. It was the first published bridge fantasy. Expense apart, to get the level centre for his exchange, Palladio needed flights of steps at the ends encroaching deep on to valuable land. Such breadth of superstructure would also have aggravated the problems with abutments and piers.

Modern engineers have sometimes puzzled at the discrepancy between the pomp and superfluity of this design, and the 'scientific', 'thoroughly satisfying' trussed carpentry bridges to which Palladio also gave space in the *Quattro Libri* (ill. 255).[18] The distance is indeed striking. The technical deftness manifest in these timber mountain bridges has deserted the Venetian project. Context and decorum answer the conundrum. Palladio knew that the ancients employed carpentry for bridges as often as masonry, and respected it according to its lights. Its sphere was the rural bridge, or the temporary military structure

254. Canaletto, 'Caprice with Palladian Bridge at the Rialto', 1743–4, showing Palladio's published design as if it had been built.

255. One of several designs for trussed carpentry bridges in Palladio's *Quattro Libri dell'Architettura* (1570).

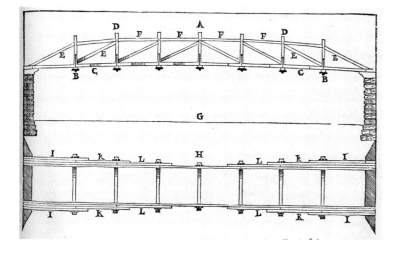

like Caesar's bridge over the Rhine. It amounted, in modern terms, to light engineering. But at the heart of its urban territory the Serenissima, so he and his backers believed, merited permanence, learning, and corroboration of its claim to Roman dignity.

Here then was an image not of practicality but of propaganda – something that would long be remembered as magnificent, not because it was built but because it was published. As architectural ideas can do, Palladio's revised conception took on a life of its own. Designs by Scamozzi and Guglielmo di Grande during the final round of projects for the Rialto bridge in the 1580s already betray its influence.

The high tide of reverence for Palladio's vision had to await the climax of masonry bridge-building in the eighteenth century. By then living bridges were out of fashion. So the commercial imperative which gave Palladio his pretext was forgotten; only the disembodied idea of the design remained. In Venice, both Canaletto and Guardi painted capricci of the Rialto showing Palladio's bridge in place, as though it had been built (ill. 254). The connoisseur and art-theorist Algarotti still pined for it.[19] In England it inspired a bevy of toy bridges in Georgian parks and paper bridges of magnificence for urban river crossings. So powerful is architectural publication – and so devoid of context. Let us deviate along this path a little further, to shed light on the way that architects like to imagine bridges when free from the constraints of site or construction.

For the superstructure-rich bridges of magnificence popular among English and French architects after 1750, there was a second catalyst after Palladio's Rialto design: the draughtsmanship of Giambattista Piranesi.[20] An obsessive delineator of Roman masonry construction, Piranesi read antiquity in the light of his Venetian upbringing and sensibility. That included apprenticeship under his uncle, the hydraulics engineer and architect Matteo Lucchesi. Nevertheless he attracted suspicion of technical ignorance. William Chambers said of him that he 'knew little of construction or calculation, yet less of the contrivance of habitable structures, or the modes of carrying real works into execution, though styling himself an architect.'[21] Piranesi's feeling for structure, in other words, was rhetorical. He could imbue architects with a vision of engineering without going into its substance.

Piranesi's passionate bridge drawings are of several kinds. There are the engravings of Roman bridges – mainly in Rome itself, but including Rimini. They are faithful enough, though a cross-section of the Ponte and Castel Sant'Angelo shows a crazy depth of masonry foundation. Then there is the perspective of Robert Mylne's Blackfriars Bridge erecting in London, done from Mylne's drawings in Rome and calculated to exaggerate its scale. And there are the glimpses of baleful arches in the *Carceri*: stage-sets seen in perspective, leading nowhere, resting on nothing (ill. 256).

Most to the present point is the original 'ponte magnifico': an imaginary design for a grand level bridge with round arches and a liberal complement of superstructure, which Piranesi first published in 1743 (ill. 257). He recast it – in a guise closer to Palladio's second design for the Rialto – as the dedicatory plate for Volume 4 of his *Antichità Romane* (1756). These drawings became the departure point for a surfeit of bridge exercises and competitions, splendid yet vacant, which infested drawing boards in London and Paris for half a century. Though some were for specific river-crossings, never were the superstructures devoted to practical ends, let alone so menial a use as the Rialto shops.

How far did Piranesi influence bridges that were built? He certainly lifted their ambition and styling. The queer coupling of columns in front of piers, which appeared in the 1743 drawing, was much imitated in British bridge design, while grandeur of detail enjoyed wider vogue in his wake. Nor can it be coincidental that the level, multi-arched bridge became a priority soon after he published. Even so, it seldom clung to the half-circular arches Palladio and he had taken from the Romans and assumed as the norm for masonry bridges. In short, the bridge of magnificence highlighted conflicts possible when architecture was superimposed upon construction. That had already been foreshadowed in the last gasp of the Rialto controversy, to which we must belatedly return.

If Palladio's project was improbable in the Venice of the 1550s, it was impossible a generation later, when the Rialto bridge saga reached its climax. By then the economy and the architectural or Romanizing party were both weaker, though the partisanship endemic to Venetian patronage was unabated. In 1587–8 the Senate once more called for schemes and plunged into committee quarrels over alignment, architecture and number of arches – one

256. Piranesi, *Carceri d'invenzione*, Plate IV.

257. Ponte magnifico, from Piranesi, *Antichità Romane*, volume 4 (1756). Prototype of the grand level bridge with coupled columns against the piers. The inscription reads: 'Bridge of magnificence with loggias and arches, erected by a Roman emperor whose equestrian statue appears in the centre. This bridge is viewed from beyond an arch attached to the said bridge on one side, and a similar arch appears in the background attached to the main bridge.'

258. Rialto Bridge, Venice, as reconstructed by Antonio da Ponte, 1588–91.

Ponte magnifico con Logge, ed Archi eretto da un Imperatore Romano, nel mezzo si vede la Statua Equestre del medesimo. Questo ponte viene veduto fuori di un arco d'un lato del Ponte che si unisce al sudetto, come si vede pure nel fondo un medesimo arco attaccato al principal Ponte.

or three? After many manoeuvrings the bridge was entrusted to Antonio da Ponte, an official connected to the salt administration. In contemporary terminology da Ponte was a 'proto', defined by Manfredo Tafuri as 'masterbuilder, foreman, or superintendent of buildings'.[22] In other words, he belonged to the technical or engineering end of the professional spectrum. Though perhaps only semi-literate, he was good at accounting. Da Ponte's was not a learned bridge. For such reasons some people resented his accession to the job. He had to endure bookish criticism from Marcantonio Barbaro, who championed fresh 'architectural' designs sent in by Scamozzi in homage to the late Palladio.

In times of retrenchment an empirical attitude to construction reasserts itself. At one level, da Ponte seems to have won out because of his grasp of arch thrust and foundation technology, acquired maybe from military experience. His abutments, which have withstood the test of time well enough, were submitted to intense scrutiny. Making the spandrel courses of masonry radiate from the common centre from which the arch was struck was the chief technical novelty in the superstructure, and one that had a long progeny. Da Ponte also handled the intractable change of levels well.

What then of architecture (ill. 258)? Compared to Palladio and Scamozzi's dogmatism, da Ponte was able to flavour technical competence with flexibility of expression. Scamozzi's projects, argues Tafuri, 'contained too much language: those who opposed this semantic excess were rejecting an architecture that was the slave of its own autonomy'.[23] In those terms, the Ponte di Rialto represents more than the triumph of technology over art. The architectural element in da Ponte's treatment is more relaxed and extroverted: it radiates an artisan freshness absent from Palladio's connoisseurship. Here is a case of an engineer arbitrarily beautifying his construction, sometimes by cribbing details from rivals or books. The solecisms da Ponte committed have long attracted scholarly wrath.[24] But the popular visitor sometimes knows best. So too does Ruskin. His antennae ever alert to spontaneity, he found the Rialto bridge the best Venetian building of his 'Grotesque Renaissance' – 'safe, palpably both to the sense and eye'.[25]

RATIONAL BRIDGE-BUILDING BEFORE PERRONET

In any Renaissance city proud of its worth, a bridge over a wide waterway was likely in some degree to pit construction against prestige. But an accommodation was possible. So much is shown by the most prophetic bridge of the sixteenth century, the Ponte Santa Trinità in Florence, designed around the time Palladio was drawing the red herring of monumentality across the trail.

The Ponte Santa Trinità (ill. 259) raises the question of the place of pure design in masonry bridge construction.[26] Bartolomeo Ammannati was an architect, sculptor and draughtsman. No doubt he believed in 'disegno' in the Florentine, conceptual sense as much as his friend Vasari, popularizer of the term. But as explained in Chapter 1, that did not make him aloof from technique. Ammannati had learnt his craft in Venice with Jacopo Sansovino, an architect with a technical bent to his practice. On returning to Florence he became one of several experts employed on Duke Cosimo's campaign to augment and embellish the city's water supply, working across the boundaries of technique, architecture and sculpture. That is how he came to be given the Santa Trinità bridge to reconstruct.

Ammannati does not present a strong personal profile, in part perhaps because he was a natural collaborator. Starting wth the Pitti Palace, he worked regularly in the 1560s with his nephew Alfonso Parigi and the extended Parigi family, masonry-contractors of sophistication, as their surviving notebook shows. It was to the Parigi that the Ponte Santa Trinità contract fell. A rigid legal line was seldom drawn in those times between designer and contractor. Bringing in the family was commonplace. At the Rialto, for instance, after the first contractor for the abutments was dismissed, the da Ponte clan took over. The Florence bridge was therefore a 'design and build' job, built by a family team. If Ammannati had breadth of skills and connections, it would be wrong to ignore the input of the Parigi.

The need for reconstruction followed from one of the River Arno's destructive flings in 1557, which damaged three bridges. The preliminary stages are unclear, but Ammannati was involved early on in repairing one of the other bridges. There is scant sign of a competition for the new Ponte Santa Trinità. Because of its cost it had to wait until 1567–9; it is usually assumed that Ammannati's design dates from then.

Florence at the time felt different from Venice. With the fear of flood recurrent, safety came first. The bridge therefore contains no heavy hint of magnificence. Decoration is

259. Ponte Santa Trinità, Florence.
Bartolomeo Ammannati with the
Parigi family, 1567–9.

not omitted. There are statues of the seasons at the ends and cartouches at the crown of the arches, while their spandrels are here first framed into panels – a feature with a long later history. But what has engrossed students of the Santa Trinità are its three revolutionary arches. Their profile is far flatter than in previous bridges, their span is unequal (87, 96 and 86 feet), yet they spring from a common height. Are they elliptical, parabolic, catenarian, struck from two centres, or drawn freehand from aesthetic instinct? Theorists of the arch have argued that their line is mathematical, art historians have derived it from Michelangelo's tombs in the Medici chapel.[27] To each his or her own preoccupation.

Everyone agrees at least why Ammannati and his team departed from the semicircular or segmental profile ordained by Roman precedent. In order to escape the violence of the river, they wanted to combine a minimum of piers with abnormally high-springing arches. The configuration of the site meant that the rise from bank to crown could be reduced to a gentle curve amenable to traffic. But with only two freestanding piers, the consequence was arches with a ratio of rise to span of an unprecedented 1:7 rather than 1:3 or 1:4.

Ammannati could hardly have settled on that proportion without trial with the Parigi as to how arches of such breadth and flatness might safely be built. Yet the Parigi notebook ignores the line but not the feat of construction. What it and archaeology confirm instead is how important in bridges are features either hidden or soon removed – the design of centering, machinery and foundations, and the Roman-style core of cement-concrete in the piers, as revolutionary as the line of the arches. The unsung skills of on-site vigilance and decision-making (we know that Ammannati was on site 'quasi del continuo' in 1567–8) add an extra invisible factor to the creativity of process.[28] Any doubt about the honour in which Ammannati held the ardours of process is dispelled by his fetching relief in front of the Pitti Palace commemorating the labours of a mule, men and machinery (ill. 260). Something similar might usefully have been hung on his great bridge, as a reminder for design-historians.

The Ponte Santa Trinità belongs to the class of bridges that engineers most admire, where invention and detail are integrated with – or subordinated to – the purpose and manner of construction. That is one meaning attached by architects to the term 'rationalism'. It stands at the head of a rational tradition in masonry bridge-building, which after the gap of a century was to be carried forward most consistently in France.[29]

There, administration always explains much. French bridges sat within a national framework for communications whose origins, set out in Chapter 1, go back to Colbert (p. 27). With its lines of command and delegation from the centre to the regions, and its efforts to oversee highways and waterways as a whole, that system at its high point came close to the old Roman model of managing communications. But it took a century to achieve. The hiatus in the last, cash-strapped years of Louis XIV was grave, and marked by deteriorating infrastructure. In 1714 was published the first modern technical book on bridges: the *Traité des ponts* by the savant Hubert Gautier,

260. 'Ammannati's mule' near the entrance to the Pitti Palace betrays his concern for the construction process. The inscription above the relief reads: 'She fetched, carried and drew the carts, the stones, the marbles, the timbers, the columns; she even bore them herself.'

who among many jobs had worked as an official engineer in the Languedoc. It coincided with the recasting of the Corps des Ponts et Chaussées along the lines of Vauban's Génie. The reforms of Daniel-Charles Trudaine, the great intendant appointed to administer the Corps in 1734, and the start of in-house training for its architect-engineers in the 1740s were further advances. All these were preliminaries essential to the school of bridge-building that flowered after 1765 in the work of Perronet and his disciples.

Even at that peak, the coherence and power of the Ponts et Chaussées should not be exaggerated. French masonry bridges under the Ancien Régime never emanated from a uniformly trained corps with unwavering purpose and authority. That image derives from the military aura of revolutionary and Napoleonic France, when a sharper discipline asserted itself. Nevertheless, a line of development had established itself in French bridge design well before Perronet and indeed before the formalizing of the Corps des Ponts et Chaussées. That is the more remarkable, given that up to 1750 the senior figures in French bridge-building were not engineers who had risen through the Ponts et Chaussées or the Génie but architects identified with royal and aristocratic service.

Not that a crude dividing line can be drawn. Under the Ancien Régime, architects of the Bâtiments du Roi were distinguished from engineers of the Génie or the Ponts et Chaussées by their office more than by their technical skills. All had a grounding in the same masonry technology, and all would have claimed some competence in hydraulics. The slant of their employment then prompted them to specialize, and we sum up their professional character from that. Architects such as Jules-Hardouin Mansart, the fifth Jacques Gabriel and Germain Boffrand were used to supplying a rich architecture when the circumstances were right. Yet none of the bridges they designed was conspicuously magnificent or ornate. In that genre they sought elegance through precision of construction, not superfluity. As designers of masonry bridges they took their cue from Ammannati, not Palladio.

The degree of enrichment on these bridges would hardly have been left to the architects themselves. That can only have followed from high-level decision-making, perhaps going back to Colbert. However it was arrived at, the evidence of the eye is clear: French bridges were to be sober. Take the Pont Royal (1685–8), commissioned from J.-H. Mansart after the previous timber bridge collapsed.[30] Here was an urban bridge built at the high tide of the grand siècle in the heart of Paris, next to the Louvre and Tuileries and not far down-river from the gaiety of the Pont Neuf. It was even paid for from the royal purse. Yet it is a plain, quietly proportioned five-arched bridge with a gentle rise to the centre, thick vaults and almost no decoration (ill. 261). 'If its ornament has been neglected', remarked Félibien with a hint at reproach, 'the solidity given to it promises that it will last indefinitely'.[31]

Expert in masonry arching and vaulting though the architects of the Bâtiments du Roi were, when it came to crossing major rivers they needed help. For the Pont Royal the royal administration therefore brought in François Romain or Romeyne, a Flemish monk from Ghent who had just successfully completed the eight-arch stone bridge at

261. The Pont Royal, Paris. Jules-Hardouin Mansart, architect for the Bâtiments du Roi, François Romain, engineer, Jacques Gabriel IV, contractor, 1686–8.

262. Coffer-dam in construction for the Pont Royal, drawing by Lievin Cruyl, 1687. In the background the Louvre is to the left, the Pont Neuf in the centre, and to the right the embankment in front of the Collège des Quatre-Nations. The foundations were certainly the responsibility of François Romain; the employment of Cruyl, a fellow-Netherlander, to draw them is of interest.

Maastricht for the States of Holland. A technician of the Netherlandish type versed in surveying and waterworks,[32] Romain was no doubt hired for the foundations. As it turned out, he probably supervised the whole bridge. The contract for the Pont Royal had gone to Mansart's close collaborator Jacques Gabriel the fourth, who had built a good deal of Versailles. Gabriel died after a year, but that did not impede the speedy execution of the work. Afterwards, Romain stayed on with the embryonic Ponts et Chaussées and was regarded as a valued technician for the rest of a long career.[33]

There were novelties in the Pont Royal's construction which it is tempting to attribute to Romain. The engravings we have of it erecting (etched by another Netherlander in holy orders, Lievin Cruyl) belong to a tradition of illustrating building process that France was to make its own (ill. 262). Still, the experts agree that the sober line and super-structure of the Pont Royal emanated from Mansart, perhaps with some input from Gabriel. In one key respect the bridge heralded the future: its arches were slightly bas-ket-handled or elliptical. Since the Ponte Santa Trinità, arches that broke with the time-honoured semi-circle or segment had turned up once or twice in France. But the Pont Royal presaged a concerted shift towards the ellipse. The change came just as mathe-maticians (Hooke in England and de La Hire in France) were starting to analyse the line of forces in an arch or vault – a development crucial to the emergence of engineering theory and design. Yet all the important early bridges with such arches were styled by men now thought of primarily as architects, not engineers.

The troubles of the French economy from the 1690s meant that bridge-building stag-nated for some time. One project already planned that did go ahead in 1704–10 was a bridge designed by Mansart for Moulins.[34] Its thick piers, roadway rising to a pro-nounced crown and three wide elliptical or 'basket' arches varied the theme of the Pont Royal. No sooner than completed, it was swept away when the Allier flooded in 1710. There followed the breaking of the old Loire bridge at Blois in 1716. How to deal with wide inland rivers prone to spate now stimulated fresh thinking. Obstructions to flood-water – hefty piers and vulnerably low arches – came under suspicion. From this practi-cal condition was to evolve the level bridge, high-springing shallow arch and attenuated pier perfected by the school of Perronet. When the Allier at Moulins was tackled again in 1756–64, its replacement was a level bridge of the reformed type, which had thirteen arches yet left twice as much room for the river.

One last audacious bridge sums up the Bâtiments du Roi's approach to the genre before that reformation began. That was the replacement bridge at Blois, of 1717–24 (ills. 21, 263).[35] Its main author was Jacques Gabriel the fifth (son of Mansart's assistant for the Pont Royal), appointed premier ingénieur when the Corps des Ponts et Chaussées was strengthened in 1713–16, but like Mansart primarily an architect. The last of France's great hump-backed bridges, it repeated over its eleven arches the now-familiar 'baskets' or ellipses: between them, two of the intermediate piers were thickened up for extra stability, offering an articulation of 4+3+4 arches. Since crossing the Loire was a major feat and the bridge made a fine line of approach into the town, this time embellishment was allowed in the form of a cartouche by a chic sculptor, Guillaume Coustou, topped by an obelisk of 14 metres, set perilously over the crown of the central arch. But the bridge as a whole remained

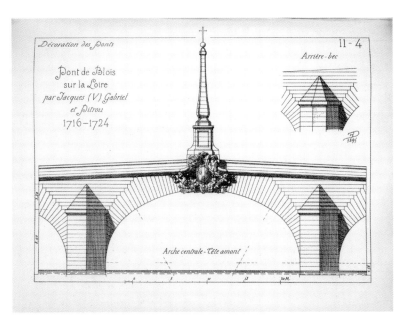

263. Bridge at Blois. Drawing of central cartouche and obelisk by Guillaume Coustou from Dartein, *Études sur les ponts en Pierre*, vol. 2.

264. Pont Cessart over the Loire at Saumur, among the first of France's great level bridges. Jean-Baptiste de Voglie and Louis-Alexandre de Cessart, engineers for the Corps des Ponts et Chaussées, 1756–70. Elevation from L.-A. de Cessart, *Description des travaux hydrauliques* (1806).

plain. There were certainly problems with money at Blois. As often, the division of responsibilities is hard to interpret. While Dartein is sure that the design was wholly Gabriel's, technique and supervision seem to have been shared, not quite amicably, between his lieutenant from the Ponts et Chaussées, Robert Pitrou, and Jean-Baptiste de Régemorte, another engineer with long military and civil experience.[36] Behind the designers of all these masonry bridges stood men with superior skills in technique and process. If no full distinction was made between architects and engineers, the specialisms that separate them today always lurked in the background.

Triumphantly completed, Blois became a model all over Europe. Then, after a long hiatus in making major French bridges, comes a rash in the 1750s: Mantes, Moulins (again), Orléans, Saumur, Toul, Tours and Trilport. With their sleeker profiles this group stands for a leap forward in rationalized bridge-building. All boast elliptical arches rising from high-water mark; several have slimmed-down piers. Mantes and Orléans venture a cautious change in proportion of pier-width to arch-width, from the traditional 1:5 to roughly 1:6. The longest (Moulins has thirteen arches, Saumur twelve and Tours fifteen) now boast equidistant piers and arches and carry entirely level roadways, turning them into causeways across the river. The larger bridges also usher in new procedures. Saumur, for instance, becomes a laboratory for foundation techniques carried to maturity at Moulins.[37]

Now, most – not quite all – of the principals in these bridges are engineer-officers who have risen within the clarifying career structure of the Ponts et Chaussées. Saumur (ill. 264) was designed by Jean-Baptiste de Voglie and Louis-Alexandre de Cessart, among the first to come up through the Corps' drawing school and acquire its in-house training. Mantes and Orléans were begun by J. H. Hupeau, who had trained as an architect but ended up as premier ingénieur at the Ponts et Chaussées. Tours fell to Mathieu Bayeux, a senior engineer at the end of his career. Moulins is the odd one out. It is attributable to and was published by Louis de Régemorte, a regional figure who had inherited family interests along the Loire but enjoyed respect in Paris.

Embarked upon in the 1750s while political confidence ran high, these bridges are the fruits of Trudaine's success in earmarking and guaranteeing resources for the French road system. At last the Corps des Ponts et Chaussées could enjoy a continuous building programme of the kind that had been the making of Vauban's Génie. Bridges of prestige, if not of magnificence, were the tip of this iceberg. They attracted glory and made the banalities of an engineer's job rewarding.

PERRONET AND GAUTHEY

Founding-father of the École des Ponts et Chaussées and premier ingénieur of the Corps from 1763, Jean-Rodolphe Perronet inherited the promise of the new situation and exploited it to the full.[38] Perronet's great quality was many-sidedness. Other chapters touch on his roles as administrator and educator (pp. 438–40); here he appears as a bridge-builder. Without challenging its masonry traditions, his eminence in that genre tipped its balance towards the engineers. Technically, Perronet combined audacity with conservatism. His innovations were not in calculation or technical theory; he relied on simple formulae for calculating forces in an arch. Instead he supplied leadership and courage, plus the cultural authority to lay claim to the rational tradition in construction for his Corps and its school.

'Perronet's interest in architecture', Robin Middleton has written, 'was dominated by intellectual preoccupations'.[39] That intellectualism was never confined to design. Engineers of the Ponts et Chaussées knew that design and process had to be kept in balance. From his early road-building experiences Perronet developed a passion for the sequencing of operations, vital to communications contracting. That led on to studies in productivity and economy. But in due course, as a result of teaching in his school and his contacts with the smart end of architecture, he became equally absorbed by cultural theories of design.

Perronet's bridge-building career starts with a design made for Trilport in 1754 but rejected as too ornamental. Refusing to rid his cutwaters of their accessories, he lost the job.[40] So architectural expression was on his mind from the start. When he became premier ingénieur, finishing Hupeau's bridge at Mantes was the first major task. The collapse during construction of one of its arches led him to ponder anew what expression in bridges might mean. Those reflections were bound up with the empirical engineering lesson he deduced from that collapse – that individual, balanced arches were obsolete, and that bridges with piers slimmer even than those of the 1750s and arches far wider and flatter would be practicable, safe, possibly cheaper and certainly more beautiful.

The first – and profoundly controversial – upshot of these meditations was Perronet's celebrated Pont de Neuilly, on the prolonged axis of the Champs-Elysées (1768–74). Before the centres had been struck at Neuilly, in an intensively creative phase round about 1770–2 he had made the preliminary designs for his two most famous later bridges: that at Pont-Sainte-Maxence (built 1774–85) and the Pont de la Concorde in Paris (built 1787–92), as well as a further one erected at Nemours after his death. Throughout the sequence, Perronet linked the goal of constructing lighter, faster-built bridges better able to withstand floodwater, with a set of esoteric arguments then going the rounds in Paris about how structure and architecture should relate. That 'rationalist' theory is familiar now, the way in which Perronet applied it to his own work less so.

While the Neuilly bridge was erecting, the controversies in Paris over the structure of Sainte Geneviève (the Panthéon) came to one of their intermittent climaxes. Soufflot, its architect, aiming in response to the Abbé Laugier and other theorists at an expression of lucidity and 'point supports', proposed to rest his dome to all appearances on columns rather than heavy piers (ill. 266). Pierre Patte attacked the security of this arrangement. Hitherto friendly with both men, Perronet became drawn into the argument. In two letters published in 1770, he supported Soufflot. The second makes an analogy between Soufflot's proposal, Gothic construction and the skeleton of animals – not quite the first nor certainly the final appearance of that specious idea. There follows Perronet's confession of his theoretical motivation as a bridge-designer:

> We have in the Academy the advantage of possessing architects knowledgeable in these principles who like you, sir [Soufflot], will be able to give us models of solid construction which without throwing away the elegant proportions presented by the monuments of antiquity will also approach the boldness and lightness of Gothic work, without revealing as they do that kind of framework or skeleton which I have tried to attribute to them when I compare them to the structure of vessels or animals. It is after reflections of this kind that I have dared to attempt the making of bridges more boldly constructed and with less materials than has been done hitherto. I do not expect them to be favoured by everyone, and in comparison with other more massive bridges they may be supposed less durable. But I shall have no more concern about their solidity than you need have, sir, about that of your beautiful and splendid monument.[41]

In the throes of personal creativity, therefore, Perronet saw the Sainte Geneviève experiment as a parallel to his own programme for uniting grace with economy of means.

265. Pont de Neuilly, the inauguration. J.-R. Perronet, engineer, 1768–74.

266. Sainte Geneviève, now the Panthéon. J.-G. Soufflot's early design for the interior, with the dome partly supported by point-supports in the form of columns. Engraving after P. A. de Machy, 1761.

Vue Intérieure de la nouvelle Eglise de S.te Genevieve à Paris.

On avertit que Cette Vue est le veritable projet
dessinée d'apres le Tableau peint par M.r Machy au Louvre 1761.

Paris chés Mondhare M.d d'Estampes rue S.t Jacques à l'Hotel Saumur et à S.t Jacques proche la Font.ne S.t Severin.

267. Thomas Telford pays homage to Perronet: 'cornes de vache' – the chamfering away of the inner arch to allow free flow for flood waters – at Over Bridge, Gloucestershire, by Telford, 1827–9.

How was this worked out in the bridges? At Mantes, Perronet noted that the mass of a pier offered no guarantee of stability, and that some of the forces in arches could be transferred laterally across piers to the abutments. So at Neuilly, his lost masterpiece (ill. 266), Perronet raised the proportion in width of pier to arch from 1:6 to a radical 1:9. The rise of the arches accordingly diminished. They were still elliptical, but streamlined with 'cornes de vache': in other words, the facing voussoirs of the arches were chamfered away towards the springing. The pretext for this feature, much copied over the next half-century (ill. 267), was to contrive a wider roadway above while flattening the portion of the arch most exposed to the river in times of flood. But the intention was also aesthetic: to restore the facing plane of the elliptical arch to the line of a segment – but a shallow, bounding segment now, not the static self-sufficiency of the old semi-circle.

In his series of three-arched bridges designed in 1770–2, Perronet dropped ellipses altogether in favour of these 'stretched' segmental arches butted against the piers. These he then pared down into separate shafts with voids in between beneath the roadway, presented as massive classical columns, single or clustered. Here was his personal response to the neoclassical ideology of post and beam – so-called trabeated construction. The effect was all but to eliminate the arch, that time-honoured principle of the permanent masonry bridge. Indeed in a bridge at St Dié designed in 1785, Lecreulx, one of Perronet's close disciples, made his trio of arches so flat that with the help of cornes de vache they gave the illusion of post and beam. Because it used less material, trabeation should also have pointed to cheaper bridges, as Perronet had told Soufflot.

Perronet himself took these ideas furthest at Pont-Sainte-Maxence, a small-town crossing on the Oise where streamlined plainnness was appropriate. A famous perspective (ill. 268) offers the complete image of the modern bridge: shallow arches bound with grace and economy from point-support to point-support, answering the level line of the carriageway. Photographs taken before the Pont-Sainte-Maxence bridge was destroyed are less electrifying (ill. 269). The piers look ponderous, the junctions between shaft and arch not without awkwardness. Neuilly and Pont-Sainte-Maxence were brave structural

268. Perronet's bridge at Pont-Sainte-Maxence, 1772–86. Perspective by the Ponts et Chaussées alumnus J.F. Eustache de St-Far, from J. R. Perronet, *Description des projets et de la construction des ponts*, 1783. The ideal image of the level masonry bridge.

269. Detail of Perronet's bridge at Pont-Sainte-Maxence, from Paul Séjourné, *Grandes Voûtes*. The separation of the piers into double drums is clearly shown. The juxtaposition of columns and flattened arches appears less elegant than in the famous perspective. The bridge was destroyed in 1914.

experiments, meticulously monitored. But they turned out neither as cheap nor quite as secure as Perronet had hoped. Though the fears of collapse put about by his enemies were unjustified, early French bridges in this style did suffer from problems. 'Want of sufficient power of resistance in the abutments has occasioned partial failure in most, and total failure in some, of the bridges that have been attempted in masonry with the very flat segmental arch', noted William Hosking in 1843.[42] Independent column-piers and flat segmental arches did not long continue after Perronet.

Accursed lightness! Must thy cult and thy altars fix their abode in my country? Why not take thy strange proliferation, thy delusions, thy poisons, to the rest of the world? Once tainted by the same disease, other nations will no longer be able to point their fingers at us in scorn.[43]

So mocked a member of the Ponts et Chaussées council when confronted by a Perronet-style bridge design. The outburst foreshadows later aversion to the thinness of iron construction. It also shows that though the great engineer's experimentalism was upheld by the Trudaines, it never enjoyed unchallenged support. This was more than conservatism. After Neuilly, Perronet's efforts to transmogrify masonry arched bridges into point-supports and beams had become too much of an end in itself, as he strove to reconcile an arbitrary theory of architecture with safe, economic bridge-building.

It is the less surprising then that the last of the bridges built during Perronet's lifetime, the Pont de la Concorde in Paris (ill. 270), turned out a disappointment.[44] Here after years of waiting, he was able in old age to build what might have been a modern bridge of magnificence. All along, he had sought a fresh architectural language of bridge design – not an add-on, as in the school of Piranesi, but something arising out of the develop-

270. Pont de la Concorde (Pont Louis XVI), as built. J. R. Perronet, engineer, 1786–94. This view, taken before the width was doubled in the 1930s, shows the continuous masonry attached to the columns under the bridge against Perronet's wishes.

271. Pont de la Concorde (Pont Louis XVI). Elevation, plan and section of Perronet's matured scheme, from the Collection des ponts de France, showing giant columns curiously mixed with a rising bridge.

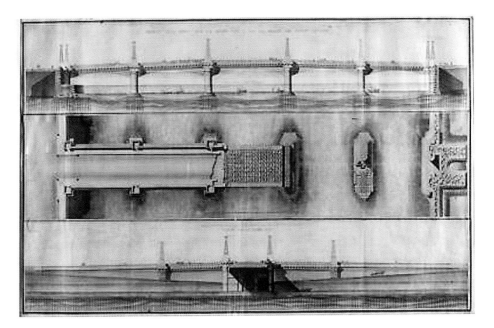

ment the Corps had overseen since 1750 which, he hoped, might reconcile the monumental instincts of architects with the paring-down of mass in the interests of economy. The Pont de la Concorde was Perronet's last great opportunity. That the bridge was altered against his will; that the closing stages of its construction got caught up in the French Revolution; that it was doubled in width in 1929–31 – all these were misfortunes. Yet it always betrayed an unease between the line of development championed by the Ponts et Chaussées and the inevitable irrationalism of magnificence.

Talk of a bridge aligned with the Place de la Concorde (then the Place Louis XV) was as old as the square itself. Perronet's scheme of 1771 was first postponed, then revived and revised in 1786–7 as the Pont Louis XVI. A level bridge was impossible because its approaches would have encroached on to the Place. At a late stage the bridge was raised to allow more headroom for the towpath, accentuating the rise of the five arches to the centre. All that boded architectural difficulties.

272. Scheme by Etienne-Louis Boullée for improving the superstructure of the Pont Louis XVI, c.1790.

Perronet was bent on hanging on to his stretched arches and his hefty, individual column-piers, meant here to be massive and monolithic. But despite his age and honours he was interfered with. For fear of instability, he was forced to integrate the columns into solid piers stretching beneath the full width of the bridge. The administration also insisted upon deeper arches springing from a lower point on the piers, which entailed more masonry in the spandrels. Yet even in the original design, the stretched arches sprang from below the capitals of the column-piers, not from their heads as at Pont-Sainte-Maxence. The scale of the columns and their Doric capitals imposed a modillion cornice at odds with the graceful balustrade above, the first such balustrade on a French bridge, which responded instead to the roadway and the Place. On top of that came the headache of adjusting the heights of the capitals to the varying gradient of the super-structure (ill. 271). Here one can feel the force of John Soane's petulant remark that 'Perronet and Labelye, although excellent engineers, were very bad architects.'[45]

On the column-piers, Perronet was forced into painful pleading:

People used to the normal proportions customary for the orders in architecture may find those of the external columns . . . too short. But if it is remarked that the columns are piers whose strength must be proportionate to the weight they have to carry, it may be felt that for that reason bridge architecture requires the supports of arches to be short and strong; besides, the feet of the columns rest in water, so they can easily be imagined to be of any height desired. To be considered also is that the reflection from the water will make the columns appear double in height, and end up giving them the proportion of those in the temple at Paestum. It will be the more necessary to lend oneself to this way of thinking because when the waters are higher, the columns will become even shorter in proportion, but one can hardly reproach this type of column for that; and indeed one ought not to compare them strictly with those which are employed in the orders of Greek and Roman civil buildings when they were con-structing out of water.[46]

Among those who could see that the Pont de la Concorde had gone astray was the architect Boullée:

> Civil engineers charged with this type of architecture have done miracles in the scientific sphere, but the artistic sphere has eluded them. In general, the decoration presented by their bridges is not handsome . . . I have made it my duty to subject myself to all the data of the engineer, whose talents I respect . . . I think I have better satisfied the engineer's idea of presenting what may be called a flat-arched bridge by alluding in the decoration to means which far from being repugnant to our senses become pleasant instead, and by presenting in an ingenious manner the arms of the City of Paris.[47]

Boullée was one of several who had a stab at interfering (ill. 272). Alert to associations, he wanted to replace Perronet's clumsy piers with cutwaters in the shape of ships' prows, in a reference to the Parisian coat of arms and the local tradition of 'ponts de bateaux'. Above them was to run an extravagant sculptural frieze. Such adjustments naturally went unheeded. They were attuned to a purely imaginative vision of architectural value – yet one better suited than Perronet's to those turbulent times. Like Piranesi, Boullée was to become famous for what he drew, not what he built.

Perronet's superstructure would have looked better, festive even, had his piers been surmounted as he wished with tall obelisks of iron holding globes or lanterns for lighting. Instead of an empty parade of arches or columns, enrichment was to complement utility. In the event revolutionary fervour curtailed the work. Ponderously antiquarian, the new age clamoured for heroic and military sculpture on the piers. Some was set up in due course, but it did not last. So the bridge's present austerity is accidental. Certainly the style of structure latterly championed by Perronet sat ill with conventional enrichment. The development of the masonry arch was far from over in 1790. But as a statement in a peculiar programme of architecture, the Pont de la Concorde spelt 'fin de ligne'.

273. Pont des Echavannes, Chalon-sur-Saône. Emiland-Marie Gauthey, engineer, 1781–90.

In the twilight of the Ancien Régime, other bridge-builders schooled under Perronet achieved a less dogmatic equation between structure, architecture, appropriate decoration and economy. The outstanding example is Emiland-Marie Gauthey.

Gauthey presents the Enlightenment ideal of the unified architect-engineer.[48] Mathematically gifted, he studies architecture at Blondel's Ecole des Arts, then attends the Ecole des Ponts et Chaussées for a single year, financing himself by teaching maths there. He returns to his native Chalon-sur-Saône in 1758 as a 'sous-ingénieur' in Burgundy's small, autonomous public works service, advancing to chief engineer in 1782. Over thirty-three years he builds bridges all over southern Burgundy (he has twelve on site in 1786), and plans the Canal du Charolais. Yet the calls of provincial officialdom leave this workaholic – vigneron and philologist as well as constructor – energy enough to take on major buildings, including two town halls and a theatre. Perronet, despite his toying with architectural theory, never builds buildings; Gauthey often does so.

Just one of his buildings, the intricate, double-domed church at Givry, ties in with his role in the structural debates of the time. In 1771, suborned perhaps by Perronet, he writes a defence of the dome proposed by Soufflot for Sainte-Geneviève which demonstrates an exceptional grasp of vaulting theory. Simultaneously Gauthey embarks on the Givry church, in an attempt to combine the point-supports of Sainte-Geneviève with economical construction. Later, he is one of the experts called back to advise on the strengthening of Sainte-Geneviève in the 1790s. After his death, his *Traité de la Construction des Ponts*, the first comprehensive treatise on bridge-building, is published by his heir and nephew Navier.

Here is a career poised between practice and theory – the theory of masonry vaults and arches, as understood in Gauthey's day, as much as architectural theory. For all the experimentalism of Givry, a conservative note is sounded in the setting-out of his bridges. There is no structural gymnastics, no straining after wide spans, flat arches or thinned-down piers; an arch remains robustly an arch. Yet by building his foundations and abutments as simply as possible, Gauthey achieved the economies that Perronet had chased in vain by paring down his structures. Dartein is able to show that his bridges cost on average little more than half what the Ponts et Chaussées engineers managed, and sometimes a third of the lavish sums spent in the Languedoc.[49] Gauthey proved a most scrupulous servant of the Estates of Burgundy.

What about architecture? Surely cheap bridges meant plain bridges? To the contrary, the Burgundians liked to see major river crossings dignified.[50] In any such bridge by Gauthey, structure is balanced by expression; only, instead of being forced into one inexorable mould, the architecture comments on the structure. To that familiar manner of embellishing masonry bridges he then adds the spice of ingenuity and discreet experiment. Paul Séjourné found his bridges more imaginative than tasteful but admired him for never repeating himself: 'every one of his fifteen bridges was treated differently: something worth copying.'[51]

Two surviving examples can make the point. The Pont des Echavannes (1781–90), in a backwater of Gauthey's beloved Chalon-sur-Saône is a stripped neoclassical causeway, flat in both roadway and elevation (ill. 273). Every stone and voussoir underlines the economy of means while reinforcing the architecture. Structural and non-structural

elements – 'frame and infill', a later era will
call them – are differentiated. The spandrels
set back neatly, while over the piers where
solidity is needless yawn oval voids or vomitories for floodwater. By contrast, the Pont de
Navilly over the Doubs (1786–90) is playful
(ill. 274). Its arches offer the hint of a point
and therefore of that Gothic which intrigued
Soufflot's circle. Between them come piers
with asymmetrical cutwaters, prow-like
upstream and stern-like downstream, to tally
with the current, while from the spandrels
emerge foliage and urns in rough relief, the
latter delightfully devised to spill rainwater
from the road surface back into the river. In
bridges like these – there are many others in
that golden age of French masonry construction – the struggle between structures that
stand firm and please the architectural eye resolves itself to near-perfection.

274. Pont de Navilly. Emiland-Marie
Gauthey, engineer, 1786–90. Detail of
the downstream side, showing the
blunter piers and the urns in relief
with mouths for draining rainwater
back into the Doubs.

LONDON

Compared with France, the administration of bridge-building in Britain during this same
period looks confused. It was not wholly so, as Christopher Chalklin's study of English
county building has shown.[52] Beyond the towns, there were the rudiments of a regional
system for Georgian roads and bridges. Counties arranged and paid for most maintenance and new work through justices meeting at quarter sessions. As traffic increased and
bridges proliferated after 1765, they employed salaried bridge surveyors or general
'county surveyors', of varying technical competence and background. Consistency grew,
as the counties reformed their practices in the light of new or improved toll-roads
decreed by Parliament – roads which in due course were to employ engineers of Telford's
calibre.

But for the kinds of encounters possible between architecture and engineering in British
bridge-building, it is more fruitful to turn to London with its unruly Georgian energy, its
quixotic management of major building projects, its anti-monumental culture. Six great
bridges were constructed over the Thames between 1735 and 1835. To London's shame,
not one in the sequence still stands. Westminster, Blackfriars, Southwark, Waterloo, and
Vauxhall Bridges: all have gone (ill. 284). To get an inkling of the last of them, the
Rennies' reconstruction of London Bridge, one must visit the arches rebuilt at Lake
Havasu City, Arizona. The growth of traffic has been the main reason for this destruction.
Besides that, the scour of the tidal Thames took a sore toll of their foundations.

Let us look closely at three of these bridges, starting with the well-chronicled
Westminster Bridge.[53] During the long sequence of arguments and operations there,
architecture darts in and out of bridge design episodically, while engineering offers a continuum.

When Westminster Bridge was proposed, British Parliaments still had difficulty with
the notion of direct central-government investment in civil infrastructure, while for the
City of London – that independent fiefdom within a widening metropolis – the bridge
was a suburban irrelevance. In the first instance therefore it had to be funded by lottery.
Once its proximity to Parliament and Whitehall made its advantages plain, it became
almost by stealth a prestige project funded by Government.

The first concrete proposals for a bridge at Westminster came in the early 1720s and
from architects. They consisted of two monumental, all-masonry designs asserting the
bridge's national and civic pretensions from Colen Campbell (who picked up Palladio's

275. Blenheim Palace, the bridge over the lake. John Vanbrugh and Nicholas Hawksmoor, architects, 1708–24.

Rialto scheme) and John Vanbrugh, and a cheaper one of stone piers with a timber superstructure from Thomas Ripley. As yet an exact site had not been chosen, nor was the manner of construction spelt out. Often the point of an architect's first design is to create a vision, something upon which decision-makers and technicians can fix in order to advance. Realistic or realizable it may not be, but it helps lead to the next step.

In 1734–6 the process was repeated on the ground thus tilled. This time the contenders were Nicholas Hawksmoor, John Price, Batty Langley and John James. All were architects, more or less. The veteran Hawksmoor had much the longest experience, and had superintended the hundred-foot arch at Blenheim – though never quite finished, the closest England ever got to a built bridge of magnificence (ill. 275).[54] Three of the four contenders published pamphlets to explain their projects. Hawksmoor's, issued in the year of his death, 1736, is the maturest.[55] It shows acquaintance with Gautier's *Traité des Ponts* and reviews many bridges, ancient and modern, including eleven-arched Blois which, Hawksmoor says, 'will come nearest to our Affair'. He is clear that a Palladian bridge of magnificence will not do for the Thames, with its tides and bustling wheeled traffic. If stone all through proves too costly for British stinginess, stone piers with brick arches will do, 'provided the Bricks are made on purpose'. He therefore offers two variants of a plain, nine-arched bridge with a central, hundred-foot span. Yet Hawksmoor hankered for more. Earlier he had tinkered with a seven-arched bridge and a central span of a full 120 feet; and in the month he died he was wistfully trimming his design with balustrades, urns and statues in niches.

Hawksmoor had done one other thing: he had called in Charles Labelye to calculate the scour of the Thames at Westminster and the effect of the piers on the river's fall. With Hawksmoor's death, Labelye's advent and the fixing of a site, we pass to the engineering phase of Westminster Bridge. Labelye's career offers an obscure point of contact between the world of French official engineering and a culture in which an engineer still meant a military technician or a mechanic.[56] Swiss-born, the son of a French Protestant refugee, he came to England around 1725 at about the age of twenty, so probably without experience of construction. But he was friendly with a famous Franco-British teacher of technology, J. T. Desaguliers, and seems to have had good contacts in France, whither he eventually retired. By the time of the Westminster Bridge project, he was expert enough in harbours and waterways to call himself an engineer, in the modern French fashion. Because of Labelye's persistence, control of London's first monumental Thames bridge fell into the hands not of an architect or entrepreneur but of someone who must be called Britain's first bridge engineer. Though his origins often worked against him, a few understood the value of French technique. With or without Labelye's connivance, the Francophone naval commander and technician Charles Knowles was in touch with Paris in 1736, asking about Blois and seeking a French designer (Robert Pitrou was canvassed) and contractor for Westminster in view of 'the ignorance and vanity of our architects'.[57]

The chain of events whereby Labelye took charge of the Westminster Bridge project shows how the sequence of decisions in major bridge projects could still at this date be separate from one another, to the detriment of unified design and process. At the time of his work for Hawksmoor, mathematical and surveying experts like Labelye were being asked just to predict the fall of water level which the piers might cause at the bridge and therefore the potential dangers for navigation. Technicalities of construction itself – how to found piers in a tidal river and how to construct good centres and safe arches upon them – lay in the future.

When further designs for the bridge were solicited in 1737, Labelye put one in to keep his foot in the door. But the real steps in his promotion were his graduation to designing first a sample stone pier for the prospective timber superstructure, all that the lottery funding looked likely to permit; and then, his original caisson method for founding the piers. The principle of the caisson – constructing a large box on the shore line, floating it out, sinking it, pumping it out and constructing the pier within it, as opposed to the time-honoured, fixed, in situ coffer-dam – was not quite new. The novelty was the size of the caissons, and the fact that they could be reused for each pier in turn. Along with a method of pile-driving invented by another Huguenot in the Labelye circle, James

276. Westminster Bridge. Charles Labelye, engineer, 1739–50. This engraving of 1751 by Thomas Willson is the most reliable image of the completed bridge.

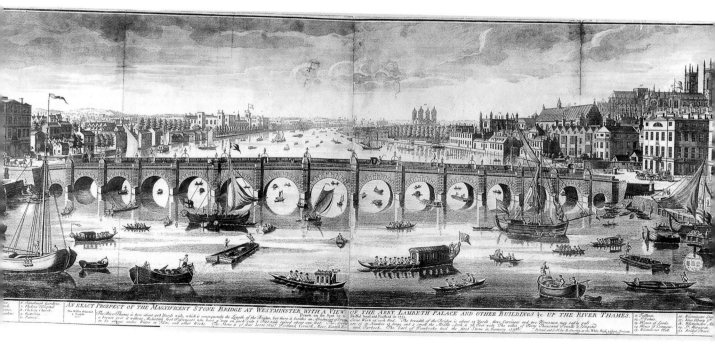

AN EXACT PROSPECT OF THE MAGNIFICENT STONE BRIDGE AT WESTMINSTER, WITH A VIEW OF THE ABBY, LAMBETH PALACE AND OTHER BUILDINGS &c. UP THE RIVER THAMES.

277. Details for one of the first caissons for Westminster Bridge, dated
1739, from Bélidor, *Architecture
hydraulique* (1753). Bélidor got his
information from Labelye's technical
account of the bridge, which was
swiftly translated into French.

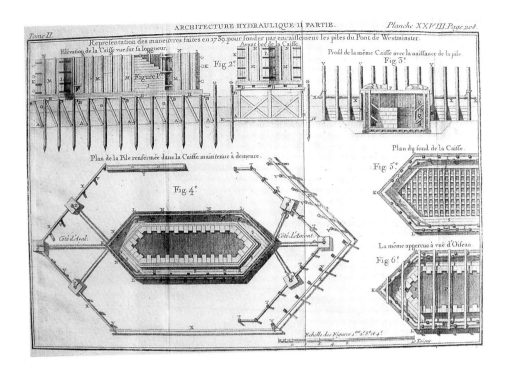

277. Details for one of the first caissons for Westminster Bridge, dated
1739, from Bélidor, *Architecture
hydraulique* (1753). Bélidor got his
information from Labelye's technical
account of the bridge, which was
swiftly translated into French.

Valoué, they got into one of Bélidor's later volumes (ill. 277), signalling that fresh techniques of bridge-building were afoot in Britain.

The innovations in design which earned Labelye the confidence of the Bridge
Commissioners had therefore to do with the apparatus wherewith the permanent structure was built, not with that structure itself. On that basis he received the go-ahead in
1740 to build a stone superstructure, once the threat of a timber bridge receded. Strictly
responsibilities were shared between Labelye, the carpentry contractors for the centering, and the masons. The most active of the commissioners, the Earl of Pembroke (who
may himself have had some help), then added finishings to Labelye's plain arches and
spandrels. The one fresh architectural note was the balustrade, the first to replace a plain
parapet on a major bridge. So 'architecture' came back in the final phase, and from the
hand of a dilettante, as styling or decoration, conspicuous but not intrinsic to the process.
At one stage there was meant to be a central cartouche flanked by the figures of Thames
and Isis. Nothing came of it, although Canaletto with a Venetian's instinct for what was
proper to so great an undertaking elaborated the idea in one of his paintings of the
bridge (ill. 278).[58]

By the time it opened in 1750 (ill. 276), Westminster Bridge was old-fashioned. Its fifteen segmental arches of moderate span (Hawksmoor had hoped for seven or nine) rose
to the crown with a 'lubberly strength',[59] carped a critic, both piers and arches increasing in width along with a rise which some found too steep. Despite major problems with
one pier and consequent delays, it was in the thorough technique of the bridge's constructor that its unity resided. Labelye's claim, in a sober description of his completed
monument, that it contained twice as much stone as St Paul's Cathedral sufficiently conveys the scale and endeavour of an engineer's enterprise.

At Blackfriars Bridge, built in 1760–9, architecture proved more tenacious.[60] As the site
fell within the purview of the wealthy City of London, the funding for a stone bridge was
guaranteed. The City's architect George Dance hoped to build it, but was told to resub-
mit his scheme in what became a quarrelsome competition. Fifty submissions were eventually whittled down to eleven plans from eight authors, of whom two can be classified
as carpenters (Bernard, Phillips), one as an engineer (Smeaton) and five as architects
(Chambers, Dance, Gwynn, Mylne and Ware).

278. Canaletto, detail of Westminster Bridge idealized for Lord Mayor's Day, 1746, with figures of Thames and Isis. The bridge was then years from completion. The figures were never there, nor were the recesses as shown.

Historians of engineering tend to smile on John Smeaton's design, of nine plain segmental arches, radically thin. But the plainness was a tactical error on the part of the non-metropolitan Smeaton: some flourish is needed in competitions. His effort was damned in a mettlesome pamphlet, *Observations on Bridge Building*, as 'the meanest and poorest of all the designs'.[61] Its writer implicitly endorsed Robert Mylne's design as the middle way between William Chambers' triumphalism or John Gwynn's 'turkey carpet' and the blandness of Dance, Smeaton and the rest. Chambers had caught the Piranesi bug in Rome. He offered the full bridge of magnificence, in honour of national heroes. Though bereft of a triumphal arch which was bound to block the roadway, his design boasted a colonnade in the centre, plus urns a-plenty for the pamphleteer to dash to pieces:

> although they [the Romans] invented the very proper form of an urn, as a repository for the burnt ashes of the dead; yet why should an English hero, who has fallen and been stretched out cold in his country's service, be crammed neck and heels into the same form; for no other reason but that 2000 years ago a people superior to every other nation made use of it . . . The fine shaped smoking urns on the top of the bridge, would raise noble ideas in the minds of a Cato or a Brutus, when standing on the ashes of some eminent dead in the Appian Way; but what would these ideas be, what would an Englishman think, when told that they contain a living watchman or toll-gatherer, who sits there hovering over a cinder fire.[62]

The twenty-six-year-old Robert Mylne was the surprise winner. Also lately back from Rome and the aura of Piranesi, he too embraced commemoration. Mylne's proposed Blackfriars Bridge was dedicated to 'the actions of our admirals, since England began first its present dominion of the sea'. In one version there were to be inscriptions along the top, and statues of seadogs against the piers, 'viewed from an element whereon these heroes have raised the character of this nation above all others'.[63]

All that soon disappeared in the usual round of reductions. What remained was a bridge design new in England for its elliptical arches (against which Samuel Johnson conducted an ill-informed polemic on his friend Gwynn's behalf), and with easy approaches and a gentler gradient to the centre than Westminster. These basics said more about Mylne's upbringing in an Edinburgh mason's family than about the cultural baggage he had acquired in Rome or the 'fine shell and lodging' he had lashed out on in London. No doubt all three ingredients contributed to his victory.

Yet Mylne had landed himself in architectural trouble. Once his admirals and inscriptions were whittled away, he was reduced to the bare language of the orders. For that he needed a level bridge. The reconfiguring of property on the northern approach that implied would have dwarfed the expense of the most lavish imaginable ornament. The gradient up to a high arch for navigation was a given, and should have ruled out the orders. Nevertheless Mylne clung to his Piranesian pairs of columns on the piers, meant to flank the missing statuary (ill. 279). The upshot was an order with columns of varying height, a straight entablature at shifting levels above each pair, and an inclining ornamental parapet on top (ill. 280). Pointing out this solecism gave pleasure to critics.[64] One thought Mylne might have employed the coarser Tuscan order, but to the eighteenth-century mind that would have put paid to the last trace of celebration. Facing the same issues at the Pont de la Concorde a little later, Perronet managed the transitions more originally but hardly better. In fairness to Mylne, on a bridge of this scale what stands

279. 'A view of part of the intended bridge at Blackfriars … by Robert Mylne architect engraved by Piranesi at Rome', 1764. To the left, St Martin Ludgate Hill amidst an italianized London.

280. Blackfriars Bridge, half plan and elevation, showing a rise to the centre coupled awkwardly with Piranesian double columns on the piers.

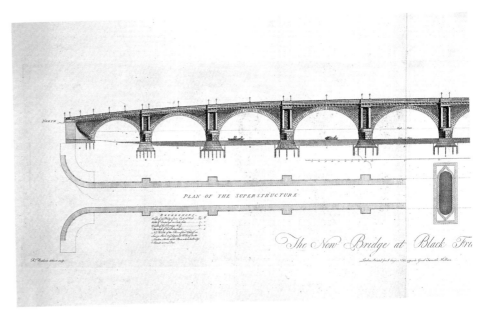

281. Waterloo Bridge in 1895. John Rennie, engineer, 1811–17.

out on a drawing may not do so on site. The eye adjusts to changes in levels, and the prospect of a river bridge like Blackfriars is seldom strictly elevational.

Despite these embarrassments, Blackfriars Bridge was a technical success. Mylne showed himself meticulous throughout, and au courant in the founding and construction of piers. For the rest of his long career, he played the civil engineer as much as the architect. His buildings seem often to betray an engineer's preference for good construction and plainness, at odds with the exuberance of his youthful design for Blackfriars. Yet the vestiges of an architect's attitude were always there. In his several later bridges Mylne dropped the elliptical arches of Blackfriars and reverted to the segmental profile hallowed by architectural tradition. Compared to Smeaton, who set his face against expressive bridges, Mylne was guided, says Ted Ruddock, 'more by convention and less by science'.[65]

As it turned out, the climax to London's sequence of masonry bridges, and the closest that city ever got to heroic monumentality in its river crossings, was supplied not by an architect but by an engineer. John Rennie's Waterloo Bridge (built in 1811–17) responded to Blackfriars, just as Blackfriars responded to Westminster.[66] Building half a century after Mylne, Rennie was far better placed. He had access to the whole developmental cycle of the French masonry bridge and the books of Perronet and Gauthey. He also had solid experience in constructing canals, harbours and bridges – a second career on top of his mechanical prowess as an erector of steam engines for Boulton and Watt and manufacturer of milling machinery.

A Perspective View of the Centering of Waterloo Bridge

282. Waterloo Bridge, steps to Victoria Embankment in November 1904, before enlargement for tramways.

283. Waterloo Bridge, rib of central arch with centering, engraving after a drawing by Edward Blore.

So Rennie, unlike Labelye or Mylne, was an established engineer with an organization. In that organization, along with budding engineers like his two sons, were there experts in architecture? It is no detraction from a great man to suggest that the exalted detailing of Waterloo Bridge may have proceeded from his assistants.[67] Not that Rennie senior – like Mylne, a product of the Scottish Enlightenment – was without culture. From the Lune Aqueduct onwards his grander works have a consciously Roman scale, while he was sharp in noting architectural lapses in others' bridges. Nevertheless his personal genius was mechanical and constructive.

Waterloo Bridge was not intended to be commemorative. It was planned on the British private-sector model as a toll-bridge, to be funded by a joint-stock company without state or municipal subsidy. Long predating Napoleon's defeat, it was denominated the Strand Bridge right up until its opening in 1817 by the Prince Regent, when Waterloo seemed a popular name in default of official plans to commemorate the final victory. Monumentality began with the design; the occasion stumbled in after.

The bridge came to Rennie in succession to the hapless George Dodd, who is alleged to have cribbed Perronet's design for the Pont de Neuilly and then added 'petty' features, 'like gipsies who disfigure the children they steal', quipped James Elmes.[68] Rennie and William Jessop supplied the promoters with a report in 1809 that criticized both Dodd's design and, respectfully, Perronet's technique. Soon Rennie, able to inspire confidence in investors, inherited the project. The proximity of Somerset House and the decision, already made, to take the approaches off the Strand, high above the water, gave him the opportunity for the monumental and level bridge with equal arches that London had not yet seen (ills. 248, 281). The roadway was to draw traffic off the Strand straight across the bridge until it dropped down the southern abutment on to Lambeth Marsh.

By then the Ponts et Chaussées had spent decades perfecting the level multi-arched bridge over broad rivers. A level had always been obligatory for aqueducts, and would be so for railway viaducts. But for road bridges it was not always possible; everything depended on the terrain and the navigation. In England, the first such feat had been Thomas Harrison's Skerton Bridge at Lancaster (1784–8), an architect's bridge on the Rimini model, but with elliptical arches. Rennie admired it, and it was close by that he built his great Lune aqueduct for the Lancaster Canal (1793–7).

Was the level multi-arched road bridge worth it? Not all bridge-builders thought so. Rennie's rival Telford inveighed against the type:

284. Charles Edward Cundall, 'Demolition of Waterloo Bridge', c.1934.

The affectation of preserving the entablature upon a perfect level, has led to making the roadway along the bridge also level, which is nothing less than constructing, at vast expense, a piece of road more imperfect than what is formed by the common labourer in the open country; and besides, this mode of construction gives an appearance of feebleness to the outlines of the bridge. This false taste was introduced by some of the French engineers, and has of late been, in some instances, copied in Britain, and cannot be too early reprobated.[69]

But it afforded a chance for 'architecture'. At Waterloo, the roadway was supported by regular 120-foot arches and 20-foot piers, before which stood yet again the double columns promoted by Piranesi. They had a practical pretext, for Rennie believed (against French practice) that it was important to load the cutwaters. The Doric order in granite, severer than Mylne's Ionic at Blackfriars, was taken from a bridge Rennie had built at Kelso in his native Scotland, here refined (the order is from Segesta) and set off at the ends by superb access staircases (ill. 282). If that was engineer's architecture, this engineer had little to learn from the architects.

The exchange between process and design at Waterloo amounted to a critical summing-up of a century's bridge-building in Britain and France. Instead of the caissons of Westminster and Blackfriars, Rennie reverted to cofferdams; improved the hidden inverted arches now customary over piers to help transmit thrusts horizontally; rethought the design of the centering (ill. 283) for speed and security; adhered to elliptical arches, but modified them and gave a suaver profile to the voussoirs; and substituted granite, here used for the first time as a facing material in London, instead of frost-prone Portland stone. Rennie was proud of the fact that his arches hardly deflected when the centering was removed, unlike Perronet's at Neuilly. The construction of all this was undertaken not by separate trades but by the great canal contractors Jolliffe and Banks, the Rennies' partners for three Thames bridges – Waterloo, Southwark and the rebuilding of London Bridge.[70] That was the modern style of public-works contracting, which had grown up with the canals and was to dominate the railway age. A consulting engineer held undisputed command, with a specialist contractor in civil works under him.

Canova thought it worth coming to England just to see Waterloo Bridge, while Alec Skempton has called it 'perhaps the finest large masonry bridge ever built in this or any other country'.[71] Its magnificence depended on a balance – not an identity – between architecture and construction. Images of the bridge make that balance look imperturbable. But it was not to last. Already, iron structures proliferating in the hands of Rennie himself and others were undermining the hard-won harmony of the masonry bridge.

2.   *The Suspension Bridge*

INTERLUDE

Arched masonry bridges do not cease to be built after Waterloo. Spans stretch ever further; there are thousands of good brick railway bridges; and around 1900 masonry bridge-structures enjoy a sparkling revival and champion in the person of the French engineer Paul Séjourné. Until concrete construction matures, piers and abutments continue to be of masonry; while dressed stone remains a ready medium for dignifying a bridge. But brick and stone have now to work around materials with other properties and, in the case of iron and steel, appearances.

Public works meanwhile are in revolution. Road-building, to which bridge-building has been an adjunct, loses status. Canals proliferate, then pass their constructive skills on to the grosser scale and practices of the railways. In bridges, mathematics and mechanics progress and relegate rule of thumb. New types and spans cannot be designed without calculation or built without testing. Advancing technique and sharper management bring professional contractors and consultants to the fore. All these changes conspire to turn the engineer into the sole author of the bridge-building book; the architect is reduced to limning the occasional page.

The nineteenth-century bridge engineer swells the heart. He tackles challenges, sports skills, takes risks and proffers visions of an order the architect cannot match. A mythical figure, he stands for resolution of humanity's struggle between matter and spirit. For the full heroic effect he must be isolated from his collaborators and subordinates: in our imagination Telford, Brunel, Roebling and Eiffel stand alone. Since most of us cannot comprehend their structures, we allow their size and sublimity to wash over us. 'Everybody experiences far more than he understands', remarked Marshall McLuhan.

Bridges have inspired countless pictures, eulogies and verses. The engineers themselves have turned poets in awe of their tasks. Telford wrote poetry and was a friend of Southey, who travelled with him in Scotland and versified on his achievements there.[72] Thomas Pope, the first to imagine bridging New York's rivers with a single-arched 'rainbow', concluded his *Treatise on Bridge Design* (1811) with a long poem. D. B. Steinman, eminent American bridge engineer and biographer of the Roeblings, published *Songs of a Bridgebuilder* (1960).[73] In prose, the professional literature about the 'romance' of engineering is vast. John Roebling himself was given to utterances of Hegelian abstraction, at the opposite pole from the exactitude of his structures. Freyssinet's late pronouncements on bridge-building have a likewise mystical strain. Treatise-writers on bridges seldom fail to preface their technicalities with aesthetics, aware that they participate in a lofty public art.

Where in all this does architecture come? Architecture is surely closer than poetry to the art of engineering. Should it not be the expression, or the result, of excellent engi-

285. James Finley's patent chain bridge, elevation. From the *Port Folio*, June 1810.

neering? Or can rhapsody express how people feel about modern bridges in a way that architecture can't? Did architecture wither away altogether from nineteenth-century bridge design along with architects, except as an occasional add-on? And is what is left to us now a separate art of engineering?

That is what one engineer-historian, David Billington, believes: that there is something called 'engineering art', not to be confused with architecture at all.[74] Another, Tom Peters, disagrees. Engineering art, he argues, is just a variation upon the 'aesthetics of product' – just another way of categorizing an object as art.[75] For Peters the appeal of the nineteenth-century engineer lies in the 'aesthetics of process': the contemplation of extraordinary and ordinary men joining hands to fashion progress and advancement in the shape of railways, tunnels, bridges and machines.

Heroic and humbling their achievements certainly are. But not everyone looks upon the sublime monuments of engineering in so informed or reflective a spirit. We must surely believe that the effect of their art comes in large part from direct experience, over and above our grasp of the skills which went into their conception or of the ardours involved in their making. As we distil that experience, our feelings are not of one kind in front of engineering and of another kind in front of architecture. We react to a structure designed wholly by an engineer just as we do when an architect is involved. Different skills certainly go into their making; if that is what is meant, we can readily assent to the idea of a special art of engineering. But they do not affect the core of our experience. Though the exercise of a particular set of skills in design may lead to characteristic forms, it entails no unique aesthetic response.

Let us delve further into these issues by way of the suspension bridge. Surely suspension bridges are the especial province of the engineer? With their elongated spans, their lightness, their reliance on precisely calculated, fabricated and tested materials, their sensitivity to load, wind and vibration, they are a type of structure symbolic of engineering authority. But further factors have always touched their design. Beauty, for one; location, for another; tradition, for a third.

THE EARLY SPANS

No scientist, engineer or architect but a Pennsylvania farmer, James Finley, invented the modern suspension bridge round about 1800.[76] Two of the engineers who later developed it, Charles Ellet and John Roebling, also spent seminal parts of their lives on Pennsylvanian farms. That is hardly a coincidence. Early American farmers had to be problem-solvers. At their best they combined pragmatism with enlightenment: Jefferson is the great model. Before the railways, Pennsylvania still offered a blend of pioneering conditions with intellectual stimulus through the clubs, societies and printing houses of Philadelphia.

In these circumstances amateurs like Finley, a county judge and state senator as well as a farmer, developed their ingenuity. Lack of skilled labour was an incentive. A creek

286. Essex-Merrimack Chain Bridge, Newburyport. James Finley, inventor, John Templeman, contractor-engineer, with Samuel Carr, carpenter, 1810. Masonry piers, timber trestles covered in shingles, and short-link chains modified by Templeman from Finley's patent. Watercolour reproduced in *American Architect*, 20 October 1904.

287. Gothic detail for the cast-iron span of 600 feet proposed by Thomas Telford and James Douglass for the venerable London Bridge crossing, 1800.

needed crossing? There would be few good carpenters on hand to build a timber bridge, let alone erect the centering for a masonry one: and both would take time. So Finley, having an empirical knowledge of mechanics and an ironmaster friend, Isaac Meason, built trestles on either bank, draped iron chains over them – in tension and therefore firmly secured to anchorages at either end – and hung a timber deck from the chains. Rope

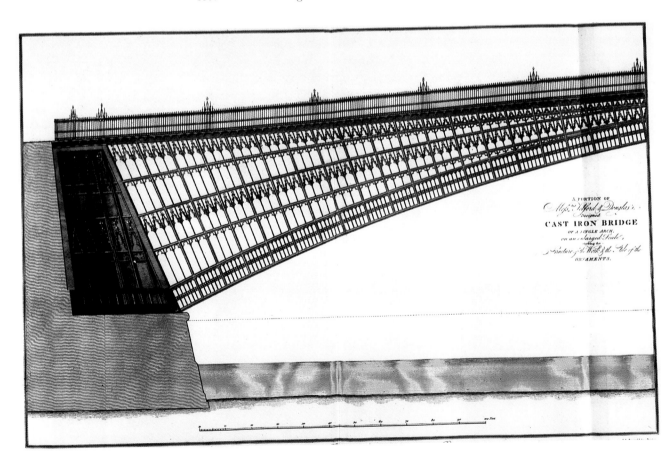

A PORTION OF

Messrs. Telford & Douglas's

CAST IRON BRIDGE

OF A SINGLE ARCH,

on an enlarged Scale,

shewing the

structure of the Work & the Style of the

ORNAMENTS.

bridges were well known; the novelty was the suspended deck, tolerably stable. The tricky part was working out the forces and therefore the best line of suspension for the catenary curve of the chains, the way they met the deck and the fastenings between deck, chains and hangers. But Finley had managed double spans of over 150 feet and was starting to go much wider with his single-span Essex-Merrimack bridge of 244 feet, when his ideas and experiences got into print in 1810–11 (ills. 285, 286).

Though keen to promote his system, Finley thought of his patent 'chain bridges' as temporary or at least needing frequent maintenance or replacement, like most rural timber bridges. 'Happily for me', he wrote, 'utility, economy and despatch are the ruling passions of the day, and will always take preference of expense, idle elegance and show.'[77] Once his ideas travelled, fresh considerations prevailed. First to pick them up were Samuel Brown and Thomas Telford in Britain. Brown was a naval officer who had just started manufacturing wrought-iron chains for anchor cables.[78] A classic example of the inventor-manufacturer diversifying into construction, he saw in suspension bridges a new outlet for his product. Telford was an engineer – or rather a mason-architect turned engineer. He had already given notice of intent over long-span structures when he proposed in 1800 to replace London Bridge with a single arch of iron. He thought of the suspension principle in the first place as a means of erecting temporary centering for the arched iron bridge he contemplated at Menai.[79]

The idea took rapid wing. In 1814 Telford proposed a permanent thousand-foot suspension span for the Mersey at Runcorn. Then in 1818 he returned to the Menai Straits and embarked upon the revolutionary 579-foot-span suspension bridge there. By the time Menai and its shorter companion bridge at Conway were opened in 1826, Europe was awash with suspension projects: all shorter than Menai, many of them just footbridges. Among notable British examples were several chain bridges and piers completed by Samuel Brown, of which the Union Bridge linking England and Scotland near Berwick was the first to open in 1820. London's first suspension bridge, at Hammersmith, was in progress under William Tierney Clark, while Marc Isambard Brunel had had suspension spans for Réunion prefabricated in Sheffield and shipped out to the Indian Ocean.[80]

French reaction had been a little slower. But another entrepreneur eager to diversify, Marc Seguin of the Annonay wool-merchants Seguin et Cie, had by 1826 built a major suspension bridge over the Rhône at Tain-Tournon – the first of about a hundred such bridges by the Seguins in France, Italy and Spain.[81] On behalf of the Ponts et Chaussées, Gauthey's nephew and editor Navier had made two trips to Britain to see what was brewing. After reporting on the development and principle of the suspension bridge, he had embarked on one of his own. Navier's Pont des Invalides, in Paris, with a span not much short of Menai, was expected far to exceed it in science and magnificence. Other countries were not behind. In Russia, for instance, the architect-engineer Wilhelm von Traitteur had started on a bevy of short suspension spans over the waterways of St Petersburg.

A makeshift device had thus turned into something spectacular and seemingly permanent. The question then arose: what should the suspension bridge look like? That could hardly be uppermost in the minds of its pioneers. They had to focus on the forces in a suspension structure, strength of materials, manufacture and proving of components, means of erection and assembly. In particular there was a fierce struggle between the rival claims of chain, bar or wire rope for the suspension cable. These too are the issues that have most engaged historians of the early suspension bridge.

Not that the pioneers lacked the ability or attention to engage with expression. But it took them time to grasp the consequences of permanence. The first attraction of the suspension bridge was that it was cheaper than alternatives. Embellishing such a structure ran counter to the reasons for its invention. On the other hand they cannot have thought that the bare means of suspension were handsome enough in themselves and needed no help. The 1820s did not think that way.

288. Craigellachie Bridge, plan and elevation. Thomas Telford, engineer, 1812–14. Iron span with castellated tollhouses: the aesthetic of contrast is born.

289. Union Bridge near Berwick-upon-Tweed. Samuel Brown, engineer, perhaps with John Rennie for the masonry, 1819–20. The first-completed European suspension bridge for vehicles. Masonry arch at the Scottish end, tollhouse at the English end, where the road turns along the river bank.

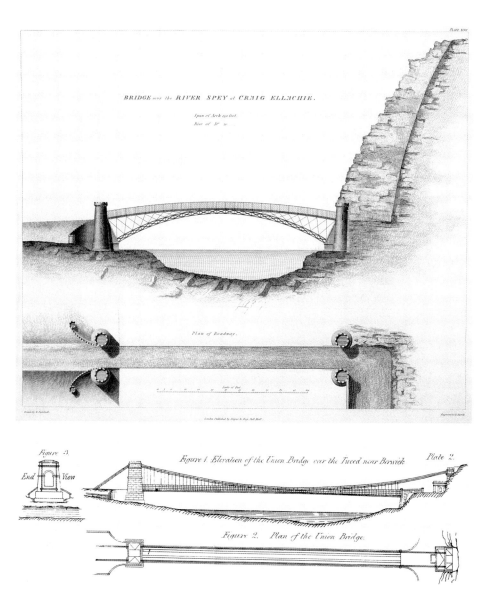

Expression had rumbled in the background since the first appearance of iron bridges. Should the iron itself be stylized? The original Iron Bridge at Coalbrookdale (ill. 50) had toyed timidly with that. Could something be made of the construction itself, or should detailing for dignity's sake be relegated to piers and abutments? To put it another way, should the piers and abutments partake of the same character as the bridge or look different, just as masonry differs from iron? Telford, an exponent of trial and error, tried everything. In his aqueducts of the 1790s the iron panels mimic classical voussoirs. One of the drawings for the London Bridge project of 1800 shows Gothic ornament to the ironwork (ill. 287). At his pretty Craigellachie Bridge of 1812–14 (ill. 288), an aesthetic of contrasts emerges: the plain spandrel patterns of the iron arch play counterpoint to castellated tollhouses in stone on the abutments at the ends. The same uncertainties appear in other British iron bridges, and even French ones. Engineers able to build masonry bridges of much suavity were at sea when it came to styling iron ones.

Suspension bridges soon took up the kinds of contrast deployed at Craigellachie, but with changes that followed from the inversion and greater delicacy of the catenary arch.

290. Széchenyi or Adam Clark Bridge, Budapest. William Tierney Clark, engineer, 1840–9. The sturdy Roman arches are typical of Clark's bridges.

The interplay in arched bridges between masonry abutment and pier, as against iron arch and spandrel, now turned into one between anchorage, abutment and tower on the one hand, deck and cable on the other. Let us trace this through, looking at the towers which were to become the new type's focus of architectural expression.

The towers of suspension bridges (also called piers, portals, pylons and pyramids in the experimental period) were there to raise the cables to their highest point and distribute compressive forces down to piers or abutments. Most of the early towers were built on banks, though some sat on piers in the water (raising issues of proportions when those piers had to be high, as Telford found at Menai). At first they seldom loomed high above the deck. But as spans widened and the base of the catenary curve began to be kept above the level of the deck, they grew taller. Finley's first bridges had low-slung cables raised between plain timber trestles standing on masonry piers and abutments. Although the two trestles on each pier were quite separate from one another, they were cross-braced to one another for stability. The bracing foreshadowed the idea of a gate or arch. As the decks and rails of Finley's bridges were also of timber, their superstructures had some unity. Sometimes the trestles were covered with weather-shingles for protection, as in the Essex-Merrimack bridge, whose gateway-towers took on the guise of a covered timber bridge.

The early batch of British suspension structures prompted several ideas for the towers. First off the mark was the classical arch or portal, for which we have seen there was some Roman precedent in bridges. In a patent application of 1817–18, Brown suggested a miniscule pedimented gateway of stone in the Tuscan order atop the abutments to raise his cables.[82] In his Union Bridge, he carried his chains through the head of an arch in rustic battered masonry (ill. 289).[83] This arch – there was only one because of the peculiar topography of road and site – was the first executed essay in architectural monumentality attached to a suspension bridge. Since Brown had no experience of architecture or construction, John Rennie was brought in to design the masonry elements of the Union Bridge. Already the aesthetics of the suspension bridge pointed to collaboration.

Thereafter the classical arch became a standard device for articulating medium-sized towers in suspension bridges – and not just in Britain. From Tain-Tournon onwards, the Seguins most often prefaced their spans with the plain masonry arch. It could be trite, but in the hands of an engineer with a sympathy for architecture like Tierney Clark it imparted real dignity. Clark, a former assistant of Rennie, used it in all his four suspen-

291. Brighton Chain Pier, *c*.1900, from *67 Views of Brighton, Hove and Neighbourhood*. Samuel Brown, engineer, 1822–3. Cast-iron Egyptianizing panels cover iron A-frames. The pier was not just for amusement but for mooring packet boats to and from Dieppe.

sion bridges: notably the Norfolk Bridge, New Shoreham (1830–3), where the budget permitted semi-triumphal arches with heraldic beasts on top, and the noblest and last of the series, the Széchenyi Híd in Budapest (1840–9) (ill. 290).[84]

An alternative to the masonry arch was the cast-iron pyramid or A-frame. That was used by Brunel senior for the spans he sent out to Réunion and by Samuel Brown for his Brighton Chain Pier, both of 1822–3. As ornament was irrelevant on Réunion, the frames were left bare.[85] But at Brighton, Brown – or whoever did his architecture for him – clad them with cast-iron panels and turned his four suspension towers into Egyptianizing gateways. Why Egypt? Partly because the tapering frames suggested battered walls, partly perhaps because Brighton traded in the exotic. 1822 was the year when Champollion's deciphering of hieroglyphics was announced; things Egyptian were in vogue. The Brighton Chain Pier (ills. 67, 291) was not a bridge but a landing stage for ferries to and from Dieppe. But from then on, Egyptian construction in bridges picked up the resonance formerly reserved for Roman engineering.[86] Its heftiness offered an antidote to the nerve-racking movement of the suspended span, often pronounced on the weakly stiffened early bridges. ('No shuddering, which might have been detrimental', wrote Schinkel on crossing the Menai.[87]) The Egyptian theme was to play powerfully in the architecture of suspension bridges. With the shift in technologies came a shift also in cultural reference.

The shining exception to the plodding treatments of towers for smaller early suspension spans came from Wilhelm von Traitteur. In a batch of five small bridges – two road-bridges and three footbridges – thrown up between 1823 and 1826 as part of the remaking of imperial St Petersburg, Traitteur brought glitter and grace to the genre. Only the footbridges survive, but the series has been rescued from oblivion by the researches of Sergey Fedorov.[88] A state engineer from Baden with some French training, Traitteur had been among the foreign experts drawn into Russia's Corps of Highway Engineers under

292. Panteleimon Bridge,
St Petersburg. Wilhelm von Traitteur,
engineer-architect, Charles Baird,
ironfounder, 1824–5. Drawing for
Traitteur's *Plans … des ponts en chaines
exécutés à Saint-Pétersbourg* (1825).

Bétancourt and Bazaine (p. 89). Enthused by the British experiments with suspension bridges, he begged his masters to have a go. His experiments seem in part to have been preliminaries for a great, thousand-foot span over the Neva, projected by Bazaine but never built.

At the time, iron amounted to an official obsession in Russia. The nature and quality of the iron chains was the critical issue. But Traitteur had no hesitation in plumping for iron also for his portals or towers:

Chain bridges are of a lightness and elegance unobtainable with other systems in use. It would be desirable that this character should be adhered to in their portals, which sometimes present masses little similar to or as bold as the slender style of the span. Portals in solid masonry can only confer an unfavourable impression on passengers. These great masses, thus placed in the presence of so thin and apparently fragile a span, make so much contrast with it that they seem readier to evoke unease than to give assurance. It would appear then that masonry portals should only be applied to very great openings requiring that the span itself should be reinforced throughout, or when the strictest economy and fast construction make them preferable to cast-iron portals.[89]

Here Traitteur came close to admitting that the reassurance of solidity was needed on big bridges. For smaller spans he sought unity – but a unity of materials, not style. The fetchingly ornamented cast-iron portals of the Panteleimon Bridge (ill. 292), the first in his series, contrasted with the suspension chains almost as much as a masonry arch would have done. Likewise in the second, a modish Egyptian Bridge, exuberantly archaeological portals of iron merely thinned down what in masonry would have been a larger mass. The footbridges, notably the extant Lion and Gryphon Bridges over the Catherine Canal, were lighter, witty creations, their chains threaded through the mouths of iron

293. Gryphon Bridge, St Petersburg.
Wilhelm von Traitteur, engineer-
architect, P. P. Sokolov, sculptor,
1825–6. One of two ornamental
footbridges designed by Traitteur for
crossing the Catherine Canal.

294. The Menai Bridge. Thomas
Telford, engineer, 1819–25, as painted
by G. Arnold.

beasts (ill. 293). All Traitteur's bridges depended on juxtaposing festive and workaday elements. They were the nearest the infant suspension bridge came to a pure work of art.

What of the pioneering builders of the long spans? It can be hard to follow Telford's
train of thought about his bridges.[90] For unrealized Runcorn in 1814 he had provision-
ally sketched cast-iron pyramids, consistent in design above and below the deck, like an
iron translation of Finlay's trestles. At Menai he fixed a general plan and then left the
details 'to be devised until the time arrived for their execution'.[91] Since the deck had to
run high over the strait for navigational reasons, the structures above it were small in
scale compared to the massive masonry piers, which reared up from deep water in con-
tinuation of arched side-spans. In the first instance therefore Telford seems to have
shunned 'architecture' above the deck, and thought still of raising the cables on iron trestles, stiffened with cross-bracing and distinct from the piers beneath.

Wiser counsels prevailed. In the end he continued the masonry of his 'pyramids' upwards, tapering their flanks as they rose (ill. 294). These rugged towers, with their quadruple openings and saddles on top, have often won praise for their stylelessness. Some have seen Egyptianism in them, but that is not endorsed by the sources. Without a context that offered architectural hints, under pressure to get Menai finished cheaply, Telford may have been content to omit style as far as possible. Yet at Conway, where his companion bridge to Menai forged a new entrance to a historic town and castle, he readily heeded the dictates of locality (ill. 295):

> In the elevation of the supporting towers, of the toll-house, gateway, breast walls and parapets, attention has been paid to the castellated style; so that the bridge, which is right opposite to the water entrance of the castle, has the appearance of a huge drawbridge, with an embanked approach or causeway.[92]

Conway sets off the sequence of nineteenth-century bridge-towers in romantic Gothic, often acting as railway gateways to ancient towns. It is a reminder that Telford never objected to the 'beautification' of bridges.

In the month that Conway opened in 1826, an accident with lasting repercussions befell the Pont des Invalides, the new suspension bridge just being finished on behalf of the Ponts et Chaussées by Claude-Louis-Marie-Henri Navier in Paris.[93] On his visits to Britain in 1821 and 1823, Navier found the advances in suspension bridge-building exciting but chaotic. After his second visit he drew the lessons together in an authoritative report which soon became the standard work of reference for the new type. Yet the bridge he was then allotted proved a catastrophe and a dead end. That is the more regrettable, since Navier like his uncle Gauthey saw himself as an architect-engineer, and put thought into the aesthetics as well as the technique of his design.

The Pont des Invalides (ill. 296) ought to have been a leap forward in the suspension bridge's development. Its explicit programme was to make this cheap and useful new device august and architectural. On a site in front of the Invalides the goal, Navier accepted, had to be monumentality:

> There exists no urgent necessity to construct a bridge to the Champs-Elysées; there is no obligation to build a suspension bridge in Paris. But if it is desired that one be built, let it be made into a monument; let the character of grandeur be given to this work that the style of construction admits of; let its disposition be calculated with the idea of forming an edifice approved by artists, agreeable to the public, honourable to the administration.[94]

What did that mean in practice? Unlike Traitteur, Navier welcomed the contrast of deck and cable with piers and anchorages. But like Gauthey, he sought also to be

295. Conway Bridge, with Conway Castle behind. Thomas Telford, engineer, 1821–6.

sparing. The masonry had therefore to be splendid but concentrated. Over the anchorages came pedestals surmounted by couched lions in a symbol of strength, while the towers carrying the cable saddles aped monumental columns or pillars instead of arches – two at each end, braced by a plain cross-beam of iron between them. Their character was Egyptian, but incomparably more robust and refined than on the Brighton Chain Pier. In homage to the Invalides, Navier reclaimed Egyptian archaeology for French arms, arts and learning.

There were many novelties in the Pont des Invalides. One affecting appearances was the 'sag-to-span ratio', in other words the depth or shallowness of the cables' catenary curve. If the towers were not to rise ever higher as spans grew wider, the line of the cables would have to become shallower. This Navier justified on the grounds of wind-damage observed on British suspension bridges. Whereas the typical sag-to-span ratio on a Finley bridge was 1:7, on the early British examples it averaged 1:12, while on the Pont des Invalides it became an elegant 1:17.[95] Where Navier got no further than the British was in joining his suspension cables felicitously with the towers. Whether they threaded through loopholes (Union; Conway), or just sat on top (Essex-Merrimack; Menai), the junction between fixed masonry and flexible chain or cable was apt to look crude. Navier avoided defacing his fine columns by drawing the cables through frieze-type blocks on top of them – blocks which also bore the ends of the naked cross-beams (ill. 297). That was hardly better. The difficulty recurred down to the era of all-steel suspension-bridge towers.

The downfall of the Pont des Invalides in 1826 was unmerited. With the bridge all but complete, a modest accident to one of the anchorages turned into a contractual dispute and hence, in the hothouse politics of Restoration Paris, into a vendetta – a tool for attacking the privileges of the Ponts et Chaussées and the Government's decision to build a monument. Navier's bridge was almost frivolously dismantled. The débâcle opened the way for the Seguins, entrepreneurs on the British model, to take over the French suspension bridge. That meant a profit-making type of toll bridge in which cheapness came first and architecture second. Bridges of that kind were plentifully built during the July Monarchy, sometimes with charming towers, arches or pylons. But the State – and the struggling Ponts et Chaussées – were never behind them. When in 1850 a suspension bridge at Angers (ill. 298) collapsed with the loss of over 200 lives, the type all but vanished from France for the best part of a century.[96]

By then suspension spans have fallen from favour throughout Europe. The reason is the railway. Deemed too frail even for the light trains and locomotives of its early years, they slip into a long European purdah. Not so in America, where progress is continuous and Roebling completes his Niagara rail suspension bridge in 1855. Even there, this most graceful of bridge-types rekindles the public imagination only after the triumphant opening of Brooklyn Bridge in 1883.

In railway bridge-building, as it matures, invention is bountiful, art incidental. Before Eiffel the new iron spans for heavy loadings, whether in arch, box-

296. Pont des Invalides. Claude-Louis-Marie-Henri Navier, engineer, 1824–6. Imaginary drawing by Navier of the completed bridge, probably made after it had been taken down.

297. Pont des Invalides, cross-section showing plain blocks for the suspension saddles atop the Egyptianizing towers, iron cross-beams between them, and roadway reinforced with iron trusses.

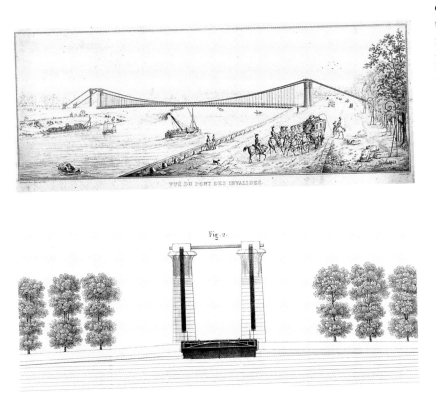

VUE DU PONT DES INVALIDES.

Fig. 2.

298. Basse-Chaîne Bridge, Angers. Joseph Chaley, engineer, with the brothers Seguin, 1839. The bridge's collapse with high loss of life in 1850 put suspension bridges long out of favour in France.

girder or truss form, are mostly graceless. Happier are the sweeping railway viaducts, with their substructures of brick, masonry, iron or timber trestle (the laminated timber viaducts, now gone, must have been a particular delight).

In all this architecture, and indeed architects, take at best a walk-on role. If industrialization is the main cause, it is compounded by the propensity of architects to write themselves out of the script. Just when the technicality of design in iron is tipping bridge-building towards the engineers, the diffusion of a rationalist rhetoric leads architects to repudiate the civic and ornamental role which has been their strength. Thus William Hosking in 1843:

> The usual *materiae architectonicae* are entirely out of place, and out of character, in bridge composition. Columns and approximations to columnar forms and proportions, pilasters, entablatures, niches, battlements, balustrades, towers and turrets, pinnacles and pediments, are gauds and devices, in the application of which to bridge composition the most eminent engineer-architects have failed to produce any thing but meanness, or absurdity, or a combination of both.

In respect of suspension bridges, Hosking echoes Traitteur in pleading for masonry to be abandoned, but goes further:

> the superstructure – all above the floor of the bridge, or the roadway – would be much better, both constructively and as a matter of taste, of cast iron. Not, let it be understood, iron columns and entablatures, or imitations of arches in iron, or mock piers, as at Brighton, but simple, plain, and boldly composed standards, well braced and tied together, and yielding as much space to the roadway as the chains themselves yield. The standards and frames of steam engines often take the proposed character, when the machinist has aimed at the end to be answered in the directest manner, without mystifying himself and making his work ridiculous by the imitation of what are considered architectural forms of decoration.[97]

In due course such attitudes helped to reform bridge design. In the short term they abetted its impoverishment.

The poignant story of the two suspension bridges by the younger Brunel gives an inkling of the downward spiral. Clifton, the better-known of them, should have been the masterpiece in the aesthetic of contrast.[98] A protracted competition took place in

299. Clifton Suspension Bridge. Lithograph of I. K. Brunel's original design. The Egyptianizing pylons were to be clad in cast-iron reliefs to the designs of Brunel's brother-in-law J. C. Horsley.

300. Pylon from unexecuted design for Neva Bridge, St Petersburg, by P.-D. Bazaine, engineer, 1825. The design was published in France in the *Annales des Mines* and is evidently the source for Brunel's Clifton pylons.

1829–31, the committee specifically calling for suspension bridges. The moment was promising: after the completion of Menai, but before the railways became the consuming obsession of British engineers. Architecture still felt it had some claim upon the type. Common ground for many of the entries was Gothic, in picturesque response to a spectacular site. Brunel's initial efforts all included Gothic towers or abutments. So did Telford's intervention, which proposed a shorter span and gigantic towers down to the base of the gorge. But in the second stage of the competition, after toying with several styles, Brunel went all out for an archaeological Egyptian. The switch helped win him the commission. Shown the preliminary sketches, William Beckford exclaimed: 'A truly grand and noble design – how superior the style of the gateways to florid barbarisms!'[99]

Brunel's winning entry cribbed brazenly from Bazaine's unbuilt suspension project for the Neva (ills. 299, 300).[100] Just like that design too, it proposed a lofty synthesis between iron technology and historicizing art and architecture. The intended gateway-towers were surmounted by sphinxes and clad in cast-iron panels adorned with reliefs of bridge-making, to be worked up by the painter J. C. Horsley, while upon the anchorages rested hefty sarcophagi. All this high art was contrived by a self-professed engineer, working so far as we know only with his technically minded father, Marc Isambard Brunel. Even the figure-subjects for the reliefs were sketched out quite fully by Brunel himself: 'very clever' they were, Horsley said, 'excellent in character'.[101] In 1831, Brunel still had the time to balance technical imperatives with the pleasures of art.

Hand-to-mouth funding – that chronic British disease – postponed and stripped what might have been the most beautiful of suspension bridges. In 1835 Brunel had to give up his ornament and recast Clifton's pylons. By then immersed in the multiple tasks of the Great Western Railway, he did so in the bare brick language of the railway age, giving them a stilted shape better suited to the line of forces than to the grandeur of the site (ill. 301). Whether the towers as they exist today represent Brunel's final intentions we cannot be sure, as the bridge was finished only after his death (1862–4) by railway engineers of the non-aesthetic generation, Barlow and Hawkshaw. But the 'simplification and refinement' of the Clifton towers, as some have seen that process,[102] appear to have been due to Brunel. It denotes a falling-off not in his architectural skills but in the British railway culture's capacity to find time and resources for enrichment. Brunel was always intensely alive to appearances, and to the value of detail in design. But his later whirl of practice left little room for a coherent aesthetic.

301. Clifton Suspension Bridge as completed by Barlow and Hawkshaw in 1864.

So much is revealed by his other suspension bridge, the short-lived Hungerford foot-bridge (1841–5).[103] Its London site (ill. 302) was as prominent in its way as Clifton. Any earlier architect-engineer would have felt it merited monumental homage. The project's genesis goes back to 1835, when Brunel had his hands full of the Great Western (not least its countless bridges) and noted to himself that he had 'condescended' to engineer little Hungerford. There are sketches in Brunel's notebooks for the towers, little different from what was built – a variation on the arch theme with a foray into the Italianate, square and tall. Eventually they were subcontracted to a competent architect, J. B. Bunning. Fetching though it was, Hungerford footbridge hardly forwarded the integration of architecture by contrast favoured by Telford, Navier and indeed Brunel himself. So mercenary were the times in which it was built, that it gave way after twenty years to the ugliest, most obstructive of London's river crossings, Hawkshaw's Hungerford Railway Bridge: 'an enormous act of desecration', Jacob Burckhardt called it.[104] It is still ripe for removal.

At Hungerford footbridge the engineer came foremost, the architect styling or mollifying what had in principle been decided. That was the normal pattern on the infrequent occasions when architects were brought in on Victorian railway bridges – when company directors could see that a public gesture was unavoidable. Since architects and engineers were often already working together on station-building (pp. 108–13), it was easy to extend the collaboration to bridges. The partnerships varied according to human nature and circumstances. Architects were not so much brought in late; as we saw at Westminster, they tended to be most useful at the beginning, in the conception and

302. Hungerford Bridge. I. K.
Brunel, engineer, with towers detailed
by J. B. Bunning, architect, 1841–5.
This anonymous painting shows the
short-lived footbridge acting as a
landing stage for steamboats. In the
background, the two shot towers of
the South Bank, and Waterloo
Bridge.

303. Britannia Bridge, a view not
long before the fire of 1970. Robert
Stephenson and William Fairbairn,
engineers, Francis Thompson, architect for the masonry detailing,
1846–50.

presentation of the bridge, and at the end, in its finishing. But their integration in the
core of the process was always in doubt. That was the basis on which, for instance,
Francis Thompson participated in Robert Stephenson and William Fairbairn's Britannia
and Conway Bridges (1846–50).

Hailed for its ingenuity, as often reviled for its ugliness, the Britannia Bridge over the
Menai Straits ranks along with its companion at Conway as the world's first major box-
girder bridges.[105] Yet it was first thought of as a special kind of suspension bridge with
an extra-deep deck for trains: hence the loopholes in its towers, for cables that never were
(ill. 303). The project was a ferment of invention and controversy. It embodied the findings of a definite programme of research on the properties of wrought iron, undertaken
for Fairbairn by the engineering analyst Eaton Hodgkinson. 'The team members,' Tom
Peters has written, 'had different priorities; Hodgkinson was interested in theory,
Fairbairn in its applicability, and Stephenson in the results'.[106]

How then did Thompson, a competent architect of small railway stations, fit in? His status was that of a minor consultant, neither on a par with Fairbairn and Stephenson nor with the authority of Hodgkinson; he often goes unmentioned in accounts of the Britannia Bridge. He made his mark at the beginning, as the empty loopholes in the towers remind us. Thompson's task was to try and redeem, in Montgomery Schuyler's words, 'the repellent baldness of this brute mass of a plate-girder' through which the trains had to pass.[107] He did so by deploying the well-tried method of contrast, adding bold strokes of now-familiar Egyptianism and 'Nubian' lions to the masonry abutments and piers. Here is architectural contrast as palliative, not enhancement. The brutality of the box girder proved a hard nut to crack. Designers have wrestled with its appearance ever since – proof that a good engineering solution may not be a beautiful one.

NEW YORK

Brooklyn Bridge is the moment when the city and the monumental suspension bridge discover one another (ills. 304, 309–11). No bridge has been so intensely experienced and memorialized.[108] Hardly a modern account of it but invokes the poetry of Walt Whitman (hymning the Brooklyn crossing *before* the bridge was built) and Hart Crane, the paintings of John Marin and Frank Stella, the photos of Walker Evans.[109] Brooklyn Bridge has become 'architecture' not because of any precise feature in its design, but because like the bridge over the Drina it is a constant in the New Yorker's mind and memory. Britannia Bridge was a monument to technical courage in a remote setting, destined to be glanced at, if at all, by people passing not even over it but through it in closed compartments at speed. Was it worth lavishing art on such a bridge? Ruskin would have said not. Architecture, he believed, was not for people in a hurry, and railway architecture therefore a contradiction in terms. But a great urban bridge, the first ever landline sprung into and out of teeming Manhattan, confronting thousands each day, kindled the collective imagination. An art-judgement was unavoidable.

The dominant influence on American bridge-building has been Germanic. All three of New York's great bridge engineers, John Roebling, Gustav Lindenthal and Othmar Ammann, enjoyed measures of technical education in German-language institutions before they emigrated. That grounding gave them bottom. How much architecture each absorbed in his youth is less clear. Roebling at least cannot have escaped it. We know that in the 1820s he attended what American sources call 'the Royal Polytechnic Institute in Berlin' – in other words, the Bauakademie, then split between architectural and engineering sections (pp. 445–6). Though Roebling was on the engineering side, an acquaintance with the high architectural ideals surrounding Schinkel was unavoidable; besides, he was friendly with one of Schinkel's star pupils, Friedrich Stüler.[110] A tradition handed

304. Brooklyn Bridge, image by Sempé.

down from a family friend claims further that in Berlin he attended the lectures of another idealist, Hegel, and was even a 'favourite pupil'. A speculative austerity certainly ran deep in Roebling: 'metaphysics was his dissipation', adds the friend.[111] The ardour of his great bridge can well be dubbed Hegelian, so long as the term is not pressed.

Suspension bridges, we have seen, were a craze during the 1820s and Roebling had been drawn to their study – 'for the last years of my residence in Europe my favorite occupation', he told his future rival, Charles Ellet.[112] Once he migrated to America in 1831, harsher influences supervened. He started off farming among fellow-Germans in western Pennsylvania, but proved unsuited to it. After a time he drifted back into surveying and engineering. Ellet remarked that the American engineer was 'reared on a canal bank, nursed in a lock pit, fed in a puddle ditch and oft-time bedded on a couch as hard as the wall he builds'.[113] That was Roebling's experience too. From 1841 he turned manufacturer of a new type of wire rope, picking up a German invention for the mining industry.[114] It was first used for hauling boats and carriages up the inclined planes that punctuated canals and early railways in the Alleghenies. But cables for bridges seem always to have been in his mind.

Roebling resembled Samuel Brown in starting with cables and moving on to bridges. But his product was far superior. Under his promotion, bunched cables of spun steel wire, meticulously specified and produced, won out against chains and bar-links. They were the key to his success as a bridge-builder. Roebling came to see manufacturing as a shrewd response to American conditions. 'My father', wrote Washington Roebling,

305. Delaware Aqueduct, Lackawaxen. Suspension aqueduct in four short spans, later converted to road use. J. A. Roebling, engineer, 1847–8.

306. Smithfield Street Bridge, Pittsburgh. Roebling's first road bridge, 1845–6, of nine spans, with sidewalks passing through the base of the 'towers'.

> always held it as a necessity that a civil engineer (one of the poorest professions in regard to pay) should always, when possible, interest himself in a manufacturing proposition – the rope business being established, his ambition promoted him to greater efforts.[115]

After vainly soliciting employment with Ellet, Roebling began building suspension spans on his own account in 1844. His original aims were severely pragmatic: to prove that with his patent spun cable suspension bridges made up of short, repeated spans could carry heavy loads – dead loads imposed by water in aqueducts, and live ones like six-horse haulage carts. In the earliest surviving example, the Delaware River Aqueduct at Lackawaxen, Pennsylvania, of 1847–8 (now converted to road use), he provided only the 'iron, timber and wire work'; the canal company did the plain, low-key masonry (ill. 305).[116] As yet appearances did not count; in these gawky structures Finley's pioneering spirit lived on.

Pittsburgh, close to Roebling's home at Saxonburg, was his early testing ground. At the junction of great rivers, it can boast as fine a record of bridges as any city. In those days it was an unruly, iron-working town uncontaminated by the civic codes of Europe. Here he built the first of his aqueducts and two road bridges, all suspension structures, all long ago destroyed. His first road bridge (1845–6), crossing the Monongahela at Smithfield Street, consisted of no less than eight short suspended spans, entailing a proliferation of stone piers in the river topped by queer iron pyramids (ill. 306).

> Each span being separated by two separate cables, there are therefore, 18 cables suspended to 18 towers. The towers are composed of four columns moulded in the form of a two-sided or cornered pilaster; they are connected by lattice panels, secured by screw bolts. The panels up and down stream close the whole side of a tower, but those in the direction of the bridge form an open doorway, which serves for the continuation of sidewalks from one span to the other. On top of the pilasters or columns, a massive casting rests, which supports the pendulum to which the cables are attached. The upper pin of the pendulum lies in a seat which is formed by the sides and ribs of a square box occupying the center of the casting. For the purpose of throwing the whole pressure upon the four columns underneath, 12 segments of arches butt against the centre box, and rest with the other end upon the four corners.[117]

Here is experimental engineer's architecture. Invention precedes; bits of style are then commandeered and thrown at the job. Hints for this bridge's appearance probably came from Ellet, who (following Navier) had separated the towers in his first major suspension bridge over the Schuylkill in Philadelphia (1841–2). But using iron for their expression was a change, no doubt tailored to Pittsburgh's susceptibilities. Roebling's later Sixth Street Bridge across the Allegheny River (1857–9) reworked these ideas (ill. 307). This time the all-iron superstructure had fewer but longer spans and was decked with 'gilded domes on the towers and well shaped spires on the tollhouses', he wrote with pride.[118] Like Telford, Roebling was not against conventional embellishment; it had just taken him time to attend to it.

307. Sixth Street Bridge, Pittsburgh. J. A. Roebling, engineer, 1857–9. In Roebling's last Pittsburgh bridge, the radiating stays typical of his later work appear. The main spans are fewer and wider and the towers consequently higher. Behind, Gustav Lindenthal's Seventh Street Bridge of 1884.

Though Roebling's early multi-span bridges could take heavy loads, they bucked and swayed. Few were destined to last long. They also ran counter to one of the suspension bridge's advantages – its reduction or elimination of obstructions in the water. It was not long before he had developed his wire cables to the point where he was ready to proceed to heavier live loads and the kind of wide-span bridge which since Menai and Clifton has hogged imagination of the type. His chance came when he took over from the peevish Ellet the Niagara Gorge railway bridge (1851–5).[119] On a site as fine as Clifton's, a deep deck-girder and the addition of stays allowed Roebling to prove, *pace* Robert Stephenson, that long-span suspension bridges could be adapted to the dynamics of railway loading (ill. 308).

It was a pyrrhic demonstration; as trains soon got faster and heavier, the bridge did not last very long. Its depth allowed room for two levels: trains above, vehicles and pedestrians on the covered timber deck beneath. In visual as well as physical terms, that gave extra weight to the span and foreshadowed altered relations between deck and towers. Roebling's Niagara Gorge towers were again divided. They took the form of tapered twin obelisks of brickwork without cross-bracing, not unlike Ellet's Schuylkill River towers. But their rudimentary Egyptian detailing points to Roebling, by whom sketches for bridge towers in an Egyptian or even Moorish taste survive. As his bridges waxed grander and bolder, their appurtenances grew more conventional.

In his last big suspension bridge before Brooklyn, the Cincinnati-Covington Bridge (1856–67), Roebling for the first time built masonry arches for his towers.[120] Though they veer between the styles, the sturdiness of the masonry saves them from naiveté. It is the same with the Brooklyn Bridge towers (ills. 309, 310). Not even the most moonstruck commentator can pretend that they are quite right. But they are among those structures so memorable that their improvement is unthinkable.[121] To the modern eye, the art of this bridge lies not in the towers' detailing or in the botched and altered abutments, but in the lift of the deck, the cables' grace, and the fanning-out of the stays. That applies to Brooklyn more than earlier bridges because it is so much longer, 1,595 feet in the central

308. Niagara Gorge Bridge. J. A. Roebling, engineer, 1851–5. The first railway suspension bridge, with a separate deck for pedestrians and light vehicles below.

span (half as long again as at Cincinnati), while the towers stand aloof at the river-edge, isolated monuments between spans. The balance of the elements has changed along with the scale. The aesthetic of contrast still operates, but is in retreat.

309. Brooklyn Bridge in context, 1894. J. A. and W. Roebling, engineers, 1869–83.

Nevertheless dissatisfaction with the architecture of Brooklyn Bridge did register, and played its part in future bridge development. During its gestation the New York architect Leopold Eidlitz, Berlin-trained, was so distressed by the design of the towers that he made an oblique offer of free help with them, never relayed to Roebling. What Eidlitz, an architect of rationalist convictions, might have made of the job can only be guessed.[122] But the great critic Montgomery Schuyler remembered the gesture. He seems to have taken it as a cue for the series of essays he wrote about the bridges of New York.[123] In part they reflect the attitudes his friend Eidlitz took to the relation between architecture and construction.

The first and freshest, on the Brooklyn Bridge itself, belongs to 1883, the year of its opening. It appeared not in a technical paper but in *Harpers Weekly*. Schuyler claimed it as the first aesthetic analysis of an American work of engineering.[124] As such, he argues, the bridge is magnificent. But the details of its masonry architecture, so massive that they may survive American civilization and one day survey the 'solitude of a mastless river and a depeopled land', are lamentable. Such a 'scientific constructor' as Roebling, says Schuyler, is apt to look upon 'current architectural devices as frivolous and irrelevant to the work upon which he is engaged, and consoles himself for his ignorance of them by contempt'. There follows an indictment of the falsehood of the arch form, and of the

310. Brooklyn Bridge featured in advertising well before its completion. This image dates from 1877.

311. Brooklyn Bridge: tower and joggers on the pedestrian walkway.

junctions between cables, saddles and masonry. Such charges might have been levelled against most previous suspension bridges. At the end, the precision of these criticisms is tempered by vaguer homage to engineering sublimity – to this 'gossamer architecture', this 'aerial bow', this 'organism of nature'. The work of John and Washington Roebling is beautiful, Schuyler rhapsodizes,

> as the work of a ship builder is unfailingly beautiful, in the forms and outlines in which he is only studying 'what the water likes', without a thought of beauty, and as it is almost unfailingly ugly when he does what he likes for the sake of beauty.

Even the towers may be forgiven, as their faults are 'the defects of being rudimentary, of not being completely developed'.

How torn this eloquence now seems.[125] Schuyler endorses the old belief that only masonry will last, that iron bridges are temporary and cannot be monuments; yet he relishes the freshness and freedom of the ironwork the most. He condemns the Roeblings for architectural insouciance while admiring their unselfconsciousness. He enjoys the archaic robustness of the Brooklyn Bridge towers, yet would wish them more sophisticated. Amidst this equivocation one idea is definite. The art of bridge-building needs debate, in which architecture must stake its claim. When that debate finally took place, Schuyler's voice would again be heard. But America in the 1880s was not ready for it, nor were its engineers yet to be fettered in their control of great bridges.

So much seemed confirmed by the advent to New York of the forceful Gustav Lindenthal.[126] A German-speaking Moravian, Lindenthal claimed to have had some technical education in Brno.[127] For a while he laboured on Austrian and Swiss railways. Emigrating in 1873, he worked up the American way, from mason to railway technician. His spell with the Keystone Bridge Company, a Carnegie subsidiary located in Pittsburgh and crucial to the development of steel bridges and the steel frame (pp. 191–3), is likely to have been seminal.

In 1880 Lindenthal turned consulting engineer. His resounding début was the Smithfield Street bridge, Pittsburgh (1882–3), a replacement for Roebling's earlier experiment (ill. 312).[128] For the former multiple and shaky spans Lindenthal substituted just two, using a German-style 'lenticular' truss that combined suspension chains for the lower members with tubular top members. The lightness of this pretty bowstring form owed much to Lindenthal's aptitude with the new steel. But the spans were set off by two-storey portals of iron in the quirkiest ornamental idiom. They hid slender steel posts that rocked along with the spans and, Lindenthal argued, would have looked out of proportion to the piers and trusses if exposed to the eye. Such bridge-follies, then fashionable in Europe, had their charm but they did nothing to make the architecture of bridges more coherent. Here they exposed a certain fake-sophistication in the young self-taught engineer. The portals were replaced by quieter ones when the bridge was widened in 1915.

The Smithfield Street Bridge and Brooklyn Bridge opened in the same year. Though John Roebling was long dead and his son and successor Washington Roebling an invalid, New York awoke to fresh river crossings in their wake. In 1887–8 Lindenthal followed up

312. Smithfield Street Bridge, Pittsburgh. Gustav Lindenthal, engineer, 1882–3. This picture, from *Transactions of the American Society of Civil Engineers*, September 1883, shows the curious original portals, presumably designed by Lindenthal himself.

313. Hudson River Bridge. First design by Gustav Lindenthal, 1887–8.

314. Pont Alexandre III. Jean Résal and Amédée Alby, engineers, with Cassien Bernard and Gaston Cousin, architects, 1897–1900.

his success in Pittsburgh by proposing to the Pennsylvania Railroad that the Hudson River be spanned with a gigantic suspension structure over twice as long and high as Brooklyn Bridge. It nearly came off, but was blighted by financial slump and counter-proposals. Undaunted, Lindenthal continued to press his idea in various forms and for sundry sites until 1923, when his assistant Othmar Ammann snatched it from under the old engineer's nose and turned it into the George Washington Bridge.

The contrast between Lindenthal's original Hudson River Bridge design of 1887–8 and the version he proposed in 1921 reveals how far the urban bridge-building culture changed over those years. Common to both was scale. The early model had a multi-level deck spanning 2,850 feet clear between towers rising 625 feet above water (ill. 313). In their superstructure masonry gave way to steel – something only feasible for a large-scale suspension bridge once steel construction spread in the 1880s. Smashing into midtown Manhattan, this all-metal behemoth sounds like an unregenerate railroad engineer's project. Not that Lindenthal neglected appearances. Scouting the idea that 'correctly designed structures have an innate architectural beauty, requiring no adornment', he excoriated American engineers for their neglect of aesthetics, and commended various European bridges to them, notably Budapest's Széchenyi Híd.[129] Here was a first response to the debate Schuyler had hoped for. But Lindenthal hardly implied that an

architect should be parachuted in, Eiffel-Tower style, to smooth the great venture's path.

Exactly that had happened by the time of his final Hudson River proposal of 1921. It had been tamed to look conservative and civic. There was now an associated architect, Whitney Warren; the decks were confined to two, automobiles on top, trains underneath; and the towers and anchorages were masonry-clad, with the Manhattan-side anchorage acting as a gross substructure to multi-storey flats.

What had happened to quicken this reversion to 'architecture'? Among suspects the Chicago World's Fair of 1893 stands clearest in the frame, for spreading the gospel that it was time for American cities to temper their vigour with dignity. There were parallel developments in Europe, to whose public works cultured Americans now looked with renewed respect. In broad terms, this outbreak in civic aesthetics can be put down to the long span of prosperity enjoyed by cities at the end of the nineteenth century. Architecture is only well placed to prefer its claims over utility when there is surplus urban wealth: 'culture is the reward of affluence'.[130]

Bridges played their part in that movement. At the centre of the 1900 World's Fair, the Pont Alexandre III in Paris (ill. 314) ushered in a wave of enriched bridge-building, while in London the Institution of Civil Engineers hosted a long discussion in 1901 on the aesthetics of bridges.[131] Kindred developments were afoot in German-speaking countries. In New York the debate took on especial intensity. 'At this point', notes Carl Condit's history of the Port of New York, 'the building of bridges and tunnels began to assume the dimensions of a passion'.[132] Brooklyn Bridge and Lindenthal's first Hudson River project paved the way. Population and trade were uncontainable. Once the cities of New York and Brooklyn united in 1898, horizons expanded. Sights and expectations were raised for the sequence of great bridges built between 1900 and 1965, as arteries flung out in all directions and bound Manhattan to the intricate riparian geography of the boroughs and New Jersey.

By the 1890s, everyone admitted that bridges – big bridges at least – were not right. American engineers lined up to apologize. Some, like Lindenthal, blamed their training, from which architecture and aesthetics were absent, along with a culture of commercialism that promoted undue competition, haste and cheapness. Others pointed the finger at the railroad companies. The first railroad bridges, claimed the architect Thomas Hastings, had usually been designed in the shops of the companies: bad enough. But then they were farmed out as 'design and build' jobs to construction firms: worse still. Only recently had independent engineers started to design this class of work.[133]

Did the architects have the answers? Hardly so, to judge from one response. Invited in 1898 by a bridge-engineer to indict his fellow professionals, Henry Van Brunt chose to vilify the iron or steel truss:

> These compound lintels or trusses are in themselves triumphs of mind over matter . . . They are structures not dedicated to the immortal gods like the post and lintel in the Greek temples, the decorative character of which was largely inspired by religious emotions, but devised to meet secular and practical conditions of an exceedingly unpoetic and unimaginative character. The mind of the architect appreciates the fine economy of these sensitive and complicated organisms, but it also recognizes that they are still in active process of development; that they are on trial, and will not reach final results until they shall have assumed those conditions of grace and beauty which are essential to completion . . . If therefore the ugly character of the present steel trussed bridge is in itself a proof of the immaturity of the science which has produced it, the remedy, of course, must reside in the perfecting of the science, and this process of perfecting will be quickened, if beauty is recognized in engineering as it is in architecture, as an aim and not as an accident of growth.[134]

In brief: engineers, come back when you have done your aesthetic homework. That kind of aloofness was bound to leave architects on the margins of bridge-building. 'That one man should devise a construction and another make it presentable was a proposition

never heard in the world till within a generation', claimed Schuyler.[135] In fact it had happened often before, and would now happen again.

The partisans of architecture were too late to shape the first suspension bridge built in New York after Brooklyn.[136] That was the Williamsburg Bridge (1896–1903), by Leffert L. Buck, one of the Roeblings' assistants on Brooklyn Bridge and a competent engineer in his own right.[137] As head of New York's Bridge Department, Buck had been authorized to start on no less than three bridges, the future Williamsburg, Manhattan and Queensboro crossings. Williamsburg was the first out of the water. Slightly outdoing Brooklyn in span, it had some technical originalities and deployed the new style of steel tower, but it looked crude and cheap. Before it could be completed, Buck lost his job and with it the chance to take forward the other two bridges. This was due to the lobbying of the Municipal Art Society, a pressure-group of architects, artists and their political allies founded 'to provide adequate sculptural and pictorial decorations for the public buildings and parks of the City of New York'.[138] One of their supporters, Seth Low, became Mayor in 1901. Low was able to supplant Buck, parachute Lindenthal into city government for a three-year term as Commissioner to the Department of Bridges, and insist that all major bridge designs should be referred to the Art Commission that had been spatchcocked on to the charter of the enlarged city in 1898.

That entailed architects. The results of the policy were patchy. Its initial beneficiary was Henry Hornbostel of Pittsburgh, brought in to assist Lindenthal.[139] To the credit of both men (Lindenthal was notoriously difficult), they got on well. Next in the pipeline after Williamsburg was Queensboro Bridge, begun by Buck using cantilever trusses with suspension-style piers.[140] Lindenthal and Hornbostel set about revising it (ills. 315–17). Their efforts drew praise from Schuyler:

We are bound to assume that the original cross section of the Blackwell's Island [Queensboro] Bridge was mechanically sufficient. But we have only to look at it to see, from an architectural point of view, what a helpless and hopeless muddle of makeshifts

315, 316, 317, 318. Details of New
York bridge designs by Gustav
Lindenthal and Henry Hornbostel.
Queensboro Bridge (A) before and (B)
after its redesign by Lindenthal and
Hornbostel; (C), model for the revised
Queensboro piers and towers; (D)
model for the Manhattan Bridge piers
and towers. From *Architectural Record*,
October 1905.

it is. Evidently the thought of an expressive or a dignified arrangement had never so
much entered the mind of its designer. The most that academic architecture could do
to it was to cloak and dissemble its chaotic ugliness by a mass of 'features' which would
have been nothing to its purpose. It needed scientific as well as artistic training to take
it apart and put it together again in such wise as to meet architectural as well as
mechanical requirements. And this is precisely what has been done by the redesigner
who combined the scientific and the artistic points of view . . . It is the highest and also
the rarest result of a training in the forms of a historical architecture that the student
shall attain through it the power of invention, or rather of assisting at the evolution,
of new forms, which have nothing superficially in common with those which have
been his academic models and yet which have in common with them the possession of
unmistakable 'style'. This high and rare triumph, in these two designs, Mr Hornbostel
and his engineering collaborators have clearly achieved, and in achieving them have
given fair promise of a 'zukunftsbaukunst'.[141]

That was at model stage. Looking at the completed Queensboro and Manhattan
Bridges, it is hard to feel so ebullient. Why do these modernizing steel bridges not thrill
like Brooklyn? There are excuses. Lindenthal had hardly got his feet under the municipal table before he was ousted by a new mayor in 1904. Queensboro had extra steel
added after that and lost coherence. And though both bridges always had two decks, both
have since been mauled to take more traffic. Nevertheless their infelicities were always
latent. At Queensboro, Hornbostel had started out with misgivings: 'My God, it's a
blacksmith shop!', he supposedly exclaimed on seeing the original design.[142] Despite all
the steel, the piers and the lumbering approaches are partly of masonry. The contrary
properties of the materials, glaring at junctions, never really set one another off. What
Hornbostel attempted and Schuyler applauded was not contrast but compromise – a
stripping-down of classical ornament and its application here in masonry, there in iron
or steel. In the process, Roebling's robustness was dissipated.

319. Manhattan Bridge. Originally designed by Gustav Lindenthal, engineer, with Palmer and Hornbostel, architects, revised by Leon Moisseiff, engineer, with Carrere and Hastings, architects, 1905–9.

The irony is the greater at the Manhattan Bridge, whose politics, writes Gregory Gilmartin, amounted to 'a byzantine tale of personal revenge and corruption.'[143] Lindenthal had designed this suspension bridge with a double catenary made up of chains and eyebars, in a daring bid to dethrone the Roebling Company's woven steel cable. After his sacking it was entirely redesigned by Leon Moisseiff, a brilliant young engineer of mathematical bent in the Bridge Department. By applying deflection theory for the first time to suspension bridges, Moisseiff could offer a leaner, lighter, potentially cheaper structure than Lindenthal's. Though the savings came mainly from the deck, steel had also to be taken out of the towers. Handling the aesthetic consequences was plucked by the authorities from Hornbostel and awarded to the Beaux-Arts architects Carrere and Hastings. The response of the classicizing Thomas Hastings was further to conventionalize the towers, by creating sham arch openings in place of Hornbostel's striking superstructure of lattice work (ills. 318, 319). Once again, as with Roebling, bolder engineering begat greater architectural orthodoxy.

On dry land the architects' touch asserted itself to happier effect. As often with these features, the masonry anchorages of the Manhattan Bridge are structures so remote from the water that they read not as part of the bridge but as monumental incidents rooted deep in the city (ill. 320). Lindenthal had dreamt their volumes might contain 'fine and imposing assembly halls, each larger than Carnegie Hall'.[144] That did not happen. Instead, Carrere and Hastings romanized Hornbostel's designs for the anchorages and appended an august court on the Manhattan approach towards Canal

320. Manhattan Bridge, flank of approaches, Brooklyn side, showing the urban incursion and scale of major bridges. Architectural detailing by Carrere and Hastings.

321. Hell Gate Bridge, the first design. Gustav Lindenthal, engineer, with Palmer and Hornbostel, architects, 1906. Brick pylons with terracotta ornament were planned for the towers. In the event plain granite was used throughout.

Street. In urban context, the urge for civic nobility could be indulged and enjoyed. Yet how to handle a bridge tower alone in a broad river remained a puzzle for aficionados of the City Beautiful.

Given these tensions, it is no surprise that Lindenthal and Hornbostel were at their happiest at one remove from the city and the aesthetic impasse of the urban suspension bridge. Their masterpiece – or rather Lindenthal's masterpiece, with Hornbostel in support – was the Pennsylvania Railroad's Hell Gate Bridge, built in 1912–17 on the high-level New York Connecting Line.[145] Since it fell just within the city limits, employing an architect was obligatory. Suspension and cantilever types were ruled out by the nature of the railway approaches. Instead, Lindenthal came up with an 850-foot trussed steel arch, of a type that was to be a future reference point for inter-war bridges, not least Sydney Harbour. That meant massive masonry abutments, but did not imply towers. Nevertheless in the interests of monumentality, Lindenthal and Hornbostel came up in their first design of 1906 with brick pylons atop the abutments, ebulliently tricked out in terracotta (ill. 321). The upper chords of the arches ended fifteen feet short of these pylons, to denote their freedom from support. Remarkably it was the Art Commission, not the railway company, that urged restraint.

It has been attempted to give decoration to the towers and to the bases of the towers. This attempt has not, in the opinion of your committee, been satisfactory from an artistic point of view . . . Your committee cannot approve of these decorative features, as they would have been ready to approve a strictly utilitarian construction.[146]

Hell Gate not being downtown Manhattan, arches in plain granite were deemed a better counterpoint to Lindenthal's noble truss; and these a chastened Hornbostel duly supplied. For better or worse, in the revision the clear detachment of upper chords and towers was also suppressed.

Lindenthal's chief assistant while Hell Gate was erecting was Othmar Ammann, the Swiss-born engineer who was to preside over major bridge-building around New York City for over forty years.[147] In time, Ammann was to enter clear water with a fair wind of support, unhindered by architects or politicians. Such developments were worldwide; from about 1930, engineers were once more trusted and admired. Urban history and art history can offer their own reasons for the change: the collapse of civic coalitions, or the triumph of modernism. But in New York the moment of transition stemmed from a road not taken, as a by-product of the Depression.

Cool, cerebral and taciturn, Ammann differed in temperament from the fiery Lindenthal. Equipped with top-class education in structural engineering from the Zurich ETH, he arrived in New York in 1904 and soon found bridge-building work. As an assistant engineer with the Pennsylvania Steel Company, the firm responsible for erecting the Queensboro Bridge, Ammann had the chance to mull over Lindenthal's skills at close hand before joining him in 1912. Once Hell Gate was finished, there was little work because of the war. So Lindenthal sent Ammann to manage a New Jersey clay mine owned jointly by himself and George Silzer of that state. He did the job as thoroughly as he did everything, cementing a relationship with Silzer that set him off on his own career as a bridge-builder.

To co-ordinate planning between New York and New Jersey, the Port of Authority of New York was created in 1921.[148] The move coincided with fresh schemes for bridging or tunnelling under the Hudson. Lindenthal dusted down his great bridge and revised it for a site landing in Manhattan at 57th Street; Ammann rejoined his old master. But once it transpired that it could never be built, he privately began devising something more feasible on his own, using Silzer, newly elected Governor of New Jersey, as an ally. He told his mother:

> In vain, I as well as others, have been fighting against the unlimited ambition of a genius that is obsessed with illusions of grandeur. He has the power in his hands and refuses to bring moderation into his gigantic plan. Instead, his illusions lead him to enlarge his plans more and more, until he has reached the unheard of sum of half a billion dollars – an impossibility even in America.[149]

Ammann's counter-proposal was for a single-deck, manageable suspension bridge for road and pedestrian traffic only, to run from Washington Heights in lightly inhabited northern Manhattan to a practicable sector of New Jersey. These were the factors that in 1925 earned political approval for the future George Washington Bridge, and for Ammann the job of bridge-engineer within the fledgling Port Authority. He was hired not as a consulting engineer but as a public servant, with the frustrations attending such a post at worst, the continuities accompanying it at best.

What of the bridge's design? Before a second deck was added in 1962, the George Washington was one of the slimmest as well as longest of suspension spans, relying on Moisseiff's application of deflection theory – and indeed Moisseiff himself contributed to the conception. Ammann's design was certainly experimental, unduly so in some engineers' eyes. As a disciple of Lindenthal, if not a wholly loyal one, he again questioned the preference for wire cables as against linked eyebar chains, but again the lobbying of the Roebling Company prevailed on behalf of wire.

As for the towers, only steel-framing made sense; the days of solid masonry were past, while concrete towers had yet to come. In the other urban suspension bridge of contemporary note, the Delaware River (now Benjamin Franklin) Bridge between Philadelphia and Camden (1922–6), a smart architect, Paul Cret, had been brought in to dignify the approaches and anchorages with masonry, but the steel towers stayed unmolested.[150]

FACING PAGE
322. George Washington Bridge, final design. Othmar Ammann, engineer, with Cass Gilbert, architect, 1927.
323. George Washington Bridge, sketch-design for sculpture over anchorage by John T. Cronin, September 1929.

Yet Ammann proved keen to clad his towers. Having won his spurs on Hell Gate with Lindenthal and Hornbostel, he was committed to art. 'A great bridge in a great city', he pronounced after Hell Gate's completion,

> although primarily utilitarian in its purpose, should nevertheless be a work of art to which Science lends its aid. An elaborate stress sheet, worked out on a purely economic and scientific basis, does not make a great bridge.[151]

Slender towers, he said later, harking back to an old notion, give 'an impression of weakness'.[152] He seems at first to have entertained concrete cladding over the steel. Then Cass Gilbert, designer of the Woolworth Building and much else, was appointed consulting architect by the Port Authority. 'It is not my function to design the bridge but to take the engineer's design and develop the architectural forms in harmony with it', he announced.[153] For the styling of the granite facings agreed upon, Gilbert referred Ammann to the monumental rustication of the Pitti Palace – an odd precedent for deference.

The Gilbert drawings for the George Washington Bridge at the New York Historical Society range through the gamut of fantasy for suspension-bridge towers. Gothic, Sanmicheli-style and Baroque are all there, along with the proposition accepted, a sleek Stripped Classic verging on Art Deco. Most deploy the arch (ill. 322); yet there is also a scheme for naked steel towers with cross-bracing of lattice girders and a few faint Tudor touches. There was more to the project's architecture than the towers, of course. As one of the first major bridges fullheartedly dedicated to the automobile, its ample approaches embraced underpasses, plazas, toll booths, maintenance buildings and floodlights as well as the anchorages. All were prospectively monumental, in a streamlined sort of way; and Gilbert's assistant John Cronin had fun sketching out sculpture for the various points of emphasis (ill. 323).

History then played its tricks. The George Washington Bridge was halfway built when the Crash came in 1929. It could not be left unfinished, and indeed much of the approach work survived the cuts that followed – done in too cheap a granite, to

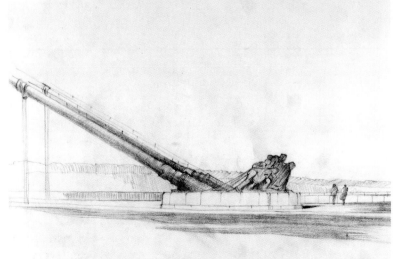

324. George Washington Bridge, aerial view before the decks were doubled. Othmar Ammann, engineer, 1928–31.

FACING PAGE
325. Bronx-Whitestone Bridge. Othmar Ammann, engineer, 1937–9. Photograph of 1941 showing diagonal stays recently added in a first reaction to the Tacoma Narrows collapse.

Gilbert's disgust. The steelwork for the towers was completed and the cladding simply postponed, formally so in 1931. Later, puritan modernists, blind to the coarseness of the undressed towers (ill. 324), persuaded themselves that dispensing with this 'mask' was due to 'popular protest'. Here is Le Corbusier:

> The George Washington Bridge over the Hudson River is the most beautiful bridge in the world . . . When your car moves up the ramp the two towers rise so high that it brings you happiness; their structure is so pure, so resolute, so regular, that here, finally, steel architecture seems to laugh.[154]

Ammann was not one to laugh or cry. But despite an article alleging the contrary in the *New Yorker*, the evidence is clear that he lamented the disfiguring of this, his first great project. While admitting that the naked towers looked 'much more satisfactory' than he had supposed, he regretted the loss of the cladding, and long hoped that it could be reinstated.[155] Yet the setback proved to his advantage. For at the George Washington Bridge the aesthetic of contrast finally died. Thereafter, no architect dared monumentalize the suspension bridge. Even in the Third Reich, when Hitler sketched out triumphal towers of granite for a suspension bridge at Hamburg, the engineers felt confident enough to face him down.

The rest of Ammann's career forms a coda to our particular story. Freed from masonry trappings, he was able under Robert Moses's egregious patronage to build four great modern suspension bridges: the Triborough, Bronx-Whitestone, Throgs Neck and

Verrazano Narrows bridges. Not that he and his in-house team could do just what they liked in these projects. At Triborough, Ammann inherited a previous design which cramped the appearance of the towers. Here too, as at Bronx-Whitestone and Throgs Neck, he was assisted by Aymar Embury II, a socialite friend of Moses and prolific architect who had styled some of the bridges on the early automobile parkways around New York.[156] But Embury, though a traditionalist, was content with a light touch. His role was cosmetic and episodic: a profile here, some streamlining there. The concrete anchorages of the Triborough Bridge, for instance, became sleeker after his intervention. Simplicity had become a virtue shared, unhampered by civic baggage.

Like all good bridge engineers Ammann had deep aesthetic instincts, which prevailed in the new era of style-free bridge-building. 'Architectural embellishments, so admirably exemplified in many smaller, older bridges, have no place in modern large bridges,' he wrote in 1954, as might have anyone then. 'Pleasing appearance must be produced by a clear expression of the natural function of the structure; and by simple, pleasing lines and proportions.'[157] The Bronx-Whitestone Bridge of 1937–9 (ill. 325) is usually regarded as his masterpiece. A 2,300-foot ribbon slices the air between the towers, simple steel uprights braced by not-quite-'truthful' arches close to the summit, beyond which the cables drop sheer to compact concrete anchorages styled by Embury. It is so austerely perfect as almost to be boring. It has certainly never earned the encomia heaped on its less delicate contemporary, the Golden Gate Bridge. Detail in a structure often counts for less than beauty of setting.

Once again history had a trick in store. Leon Moisseiff of the Manhattan Bridge had been a consultant for the Bronx-Whitestone project. His demonstrations of deflection theory had induced Ammann to thin down his suspension decks even further – for elegance as well as economy. Then came 'Galloping Gertie', the notorious episode of 1940 when the Tacoma Narrows Bridge in Washington State collapsed after oscillating uncontrollably in a gale. Moisseiff had designed that bridge, and his reputation fell with it. As a result, almost all recent American suspension bridge decks were retrofitted with deep secondary stiffening trusses. The policy damaged the lines of Bronx-Whitestone and detracted from many subsequent bridges – one of the victims being Ammann's last great bridge at Verrazano Narrows. It has even been argued that the fall-out from Galloping Gertie ended American supremacy in suspension bridge design and handed the initiative back to Europe.[158]

3. *Propriety and after*

MAILLART

A quest for the spirit of Robert Maillart should start in Zurich. Two rivers, the tamed Limmat and the freer-flowing Sihl, run through that city, converging in parallel on the Zürichsee. The town opens its heart towards the Limmat but turns its back upon the Sihl – the river that first calls Maillart to our attention. Swiss urbanity once digested, it is time to strike out for the mountains. Perhaps for Zuoz in the touristic Engadin (ill. 326), where Maillart built the little bridge in bare concrete that is esteemed his true first child, stayed at the village inn and fell head over heels in love with the Italian governess who became his wife.

Cattle are herded peaceably over the bridge at Zuoz of an evening still. But next to it there runs now a by-pass. It is a modern motor road of the type for which Maillart latterly made some bridges, albeit humbler ones than those his German peers built along the first autobahns, let alone later motorway bridges and viaducts. Maillart designed such things with grace; if you leave the car and slip down to the river bed, you will be rewarded. But it is not there that you catch what is most moving about his bridges. To do that you have to leave main roads and take to lanes, tracks almost sometimes, that bank and weave tenuously through Alpine hamlet, forest or open meadow. There suddenly it is, perilously slim: a trough leaping the gap from precipice to precipice. A view of the bridge as a whole can be hard to get; at Salginatobel, the most famous, the old photos (ill. 328) are impossible to replicate. What no image, old or new, reveals is the miracle that these masterpieces of art-structure should have been bestowed on the by-ways of tiny rural communities.

'The war against all that is natural begins already before birth', wrote Maillart to his future wife in 1901, during their courtship. 'One tries to hide as long as possible that of which one should be most proud; one is afraid of appearing ridiculous and yet only imbeciles can make fun of that.'[159] The letter is about pregnancy, but might have been a metaphor for the renaissance of the twentieth-century bridge. Till now the bashful engineer has concealed his inventiveness beneath the skirts of convention. It is time to flaunt his fertility. That is what Maillart did at Zuoz, and those are the terms in which his work has been expounded, above all in a set of devoted studies by David Billington — surely the most compelling analyses of any bridge engineer ever written.[160]

The Stauffacher Bridge of 1899 over the Sihl in Zurich, with which Maillart's career opens, belongs squarely in date and type to the City Beautiful movement.[161] A single-span bridge with granite facings over an unreinforced concrete arch, it boasts high pillars at its ends bearing light-fittings and lions (ill. 327). It is usual to refer the type to Paris and the Beaux-Arts, and indeed there is a little of the Pont Alexandre III about the Stauffacher pillars. But such bridges, road over river, were then erecting throughout the cities of Europe.[162] As in America, architects were commonly brought in, sometimes first-rate ones. Theodor Fischer styled bridges of some pretension in Munich; Otto Wagner did the same in Vienna.

Most of the genre were arched bridges. The choice of materials wherewith to turn arches was wider now than ever. A masonry revival was in train; in Switzerland, the bridge engineer Robert Moser argued that stone structures were permanent, cheap to maintain and national in character. Or there was steel: faced with which, architects reached for what we have called the aesthetic of contrast by enriching the masonry abutments and piers. That worked best on an ample budget. Coming up on the rails was reinforced concrete, taking off rapidly for bridges after 1887. The Monier-patent systems for reinforced concrete were first off the mark and better diffused in German-speaking regions; the Hennebique-patent systems did better in Francophone areas. With access to

326. Zuoz: bridge, village and mountains. Robert Maillart, engineer for the bridge, 1900–1. The substantial building on the left-hand edge of the village is the hotel where Maillart met his wife.

327. Stauffacher Bridge, Zurich. Gustav Gull, city architect, Robert Maillart, job-engineer for the city engineering office, 1899.

both, Switzerland embraced concrete construction and applied to it its traditions of craftsmanship and precision. Hennebique licensees were soon building Swiss railway bridges, which Maillart saw and reported on.

Concrete was taken up because it was cheap and, in its reinforced versions, wonderfully versatile. But exposed concrete was coarse and ugly, and no one knew how it would wear. For thick retaining walls, wharves, temporary buildings or factories it might serve, but for smarter uses a render at least was needed. In any urban structure aiming at permanence or chic, concrete postulated a cladding of stone or tile. No question then but that the Stauffacher Bridge was bound to be clad, if not constructed, in stone. Variant proposals (a single steel arch or two spans in steel or masonry) were before the authorities when Maillart, a bright young graduate from the Zurich ETH, entered the City Engineering Office in 1897. His gambit was to add to the ideas in play that of a single concrete arch, and have it analytically vindicated by his old professor, Wilhelm Ritter. It proved cheaper than steel. So the worked-up design (with the addition of hinges) went to the city architect, Gustave Gull, for cladding and enrichment. As in all such structures where concrete and stone co-existed, Gull's granite only covered what polite society was likely to see. Underneath the arch, the soffits are of bare concrete. Down there besides the pebbly, fast-flowing Sihl, you feel remote from urbanity.

The process whereby Maillart frees himself from this straitjacket of styles and materials runs parallel with a personal development: with his ability to master his destiny, escape the city and construct small road bridges for rural communities. Soon he leaves municipal employment and joins a building firm, holder of one of the Hennebique licences. The bridge at Zuoz (1900–1), where he first ventures on personal expression, is undertaken on its behalf. By the time of his technical breakthrough at Tavanasa (1904–5), Maillart is designing and building for himself.[163] The importance of these single-span structures for the art of bridge engineering is inestimable. Though other handsome light bridges in concrete are being built at this early date, no definite line of development emerges from them – save perhaps in the work of Eugène Freyssinet. Both in his early years (1902–14) as an engineer-contractor and latterly as a consulting engineer, Maillart always undertakes other work, often of originality. But outside his period in Russia (1914–19), his first concern is with bridges.

What Maillart did in his early concrete bridges owed as much to imagination as to technique. He ceased to look upon the arch as supporting the deck of the bridge, and thought of the whole as a hinged, hollow box in which all the elements acted together and redundant parts were pared away, to the gain of economy and aesthetics alike. That was close to Gauthey's attitude. But Maillart was working not with masonry but with an experimental, monolithic material in which calculation ran on all fours with intuition. Here Wilhelm Ritter's analytical skills were vital; on that rock Maillart's success was founded. Ritter however recognized the limits of calculation and had strong aesthetic and pragmatic instincts, which he imparted to his brilliant pupil. The design of the early bridges at Zuoz and Billwil was done almost in partnership. By the time of Tavanasa, when the spandrel panels between arch and deck were eliminated, Ritter's health was failing and Maillart was on his own. An aesthetic of twentieth-century concrete bridge design had sprung to life, with the architects miles away.

There are Maillart bridges of many types and qualities; he was not an engineer who cared to repeat himself. Nevertheless the lean, mountainous spans of the 1920s, Flienglibach, Valtschielbach and Salginatobel, as well as some of the later major road bridges like Felsegg (ill. 329), belong in principle to the line started at Zuoz, raised as it were to a higher pitch along with their altitude. Perhaps the sufferings stemming from Maillart's Russian setbacks can be read into their intensity. By the 1930s architects were taking notice; and Maillart himself had become conscious of his destiny as a torch-bearer for the art of engineering. But he wanted to move on. The bridges of his last years are deliberately less sleek and artful.

FACING PAGE
328. Salginatobel Bridge. Robert Maillart, engineer, 1929–30. The build-up of trees has made this view impossible today.

329. Felsegg Bridge over the River Thur. Robert Maillart, engineer, 1932–3. A broader road bridge with parallel three-hinged arches, situated in a less ardent landscape.

330. Laufenburg Bridge over the Rhine. Robert Maillart, engineer, Joss and Klauser, architects, 1911–12. In the heart of communities, Maillart's bridges express themselves cautiously.

Maillart's story can be read as a parable about the twentieth century's efforts to redefine propriety. Take his first exposed concrete arch at Zuoz. It could never have been done in Zurich, where the citizens expected more than the bare minimum. Nor was it easy in a mountain region, where the farmers were used to covered timber bridges or small masonry spans. The first art needed was the art of persuasion. The canton had been planning a steel-truss bridge. As in Zurich, the counter-attraction of concrete was its price, which Maillart impressed upon the local council. But he was able to add that the concrete would come from local suppliers, whereas the steel truss would have to be shipped in. And:

> because concrete outwardly resembles stone, it will have a simple but elegant appearance and will unquestionably bring honor and embellishment to the community.[164]

Concrete, in other words, was to be looked upon as the natural extension of localized masonry building. The craftsmanly juxtaposition of abutments in rubble with board-

marked concrete arches on many Maillart bridges confirms that. The more bridges like Zuoz that were successfully load-tested and completed, and the less trouble they gave in performance, the more acceptable the argument became. Naturally there was some trouble. Zuoz Bridge cracked close to the abutments; the Tavanasa Bridge was swept away by an avalanche after twenty-two years. Some Swiss communities or officials could not be persuaded. At Wattwil, a Tavanasa-style arch of 1909 had to be given a facing by architects, Joss and Klauser.[165]

All these were in rural territory. In multi-arched and municipal commissions, Maillart seemed less certain what note to strike. In his earliest town bridge, at St Gallen (1902–3), he plumped for concrete blocks; the aesthetic outcome was unexceptional. A pair of Rhine bridges for the towns of Rheinfelden and Laufenburg (1911–12), won in competitions, were essays in streamlined historicism, again in blockwork and designed in collaboration with Joss and Klauser (ill. 330). When in 1911 Maillart entered a third competition with them for a bridge in Bern, he failed.[166] Billington blames all this on the conservatism of juries and the influential Robert Moser's obsession with masonry. But it may also be suggested that Maillart had no expressive equivalent on offer to the puritanical paring-down so suited to Alpine scenery. Some other bridge engineers felt that. 'These Maillart-type arch bridges only look good in special situations, as . . . over a gorge and against a mountainous background', wrote Fritz Leonhardt.[167] Society was all the time becoming more urbanized. Yet a wilderness backdrop was to prove pertinent to the tasks confronting future generations of bridge designers, and so keep the spirit and example of Maillart alive.

FREYSSINET AND SÉJOURNÉ

The early career of Eugène Freyssinet runs in parallel with Maillart's. Here again was a bridge engineer who flexed his muscles blithely in the countryside but showed restraint in towns. Seven years younger than Maillart, Freyssinet belongs to the French pattern of the Ponts et Chaussées engineer trained in Paris (which he disliked) and then dispatched to hoe his row in the provinces. That was in 1905. Proud of his peasant origins, he revelled in the task of persuading rustic mayors in the Bourbonnais to let him build cheap and original concrete bridges.

Then and later, Freyssinet did not pretend to culture. His friends were craftsmen and builders. Self-defined as a technician, he was wary of architecture and averse to contemporary art: 'apart from his work and prestressing there was no modern world for him,' says his biographer.[168] When comrades and critics called him an artist, he denied it:

> Matters of art are foreign to me . . . I am of peasant stock, and come from a district where the harshness of life leaves no room for art. My only training, as a polytechnician, made a physicist and an engineer of me, passionate about professional matters but pretty well ignorant in every other domain, especially that of architecture. I have applied my mind to nothing except research into the properties of materials and into the forms they may be given, and into improving the conditions and ways in which we use them.[169]

And yet like every ardent bridge engineer Freyssinet always wanted to build beautiful bridges. As he aged, he reflected almost mystically on what makes a structure beautiful. He found his answer in the perfection of process – the exhaustion of all other possible solutions until the designer discovers the one that best fits all the conditions and hence becomes logical, satisfying and therefore lovely: a Platonic engineering ideal.

Freyssinet's earliest single-arched concrete bridges are well made but clumsy in a way that Maillart's never were.[170] The arches are efficient but ungraceful, the parapets canted out to broaden the roadway with railings that can be quaintly decorative. These concrete railings of his persist into the 1920s, in defiance of the vogue for streamlining. Unlike Maillart, Freyssinet gains in aesthetic confidence when the scale of the bridge increases.

331. Pont de Boutiron over the Allier near Vichy. Eugène Freyssinet, engineer, 1911–12.

His breakthrough occurs in 1907 when his first patron, François Mercier, backs him to build a trio of three-arched bridges across the Allier in replacement of old suspension spans: at Le Veurdre, Boutiron (ill. 331) and Châtel-de-Neuvre. The problems are the same as those faced by the eighteenth-century Ponts et Chaussées engineers: how to build cheap but sturdy bridges with high-springing arches, so as to withstand a river in spate. But the solutions, at this stage of concrete technology, are entirely experimental and require much technical care. Hinges are a big preoccupation for the early Freyssinet.

His career divides into definite phases. The economy and beauty of the Allier bridges made his reputation and propelled Freyssinet from 1914 into a national career as consulting engineer to the Limousin company. Then after 1930 came the experiments in prestressing (prefigured at Le Veurdre) which were to transform concrete technology. From the middle phase, architects remember his mammoth parabolic hangars for airships at Orly (ill. 211), engineers the great bridge at Plougastel, marking a new high point in processes of technique and assembly. Both were magnificent, independent monuments. But there were also structures in which, departing from this pattern, Freyssinet reined in his native boldness. Two of these are to our purposes, both town bridges in Lot-et-Garonne: one in the centre of Villeneuve-sur-Lot, the other not far away on the edge of Tonneins. Both bridges were touched by the creative conservatism of Paul Séjourné – an engineer whose half-forgotten career merits brief digression.

The achievements of Séjourné serve as a reminder that no fair-minded history of bridge-building can follow simplistically progressive lines. Just when it looked as though steel and concrete had won the day, Séjourné gave masonry bridge design a new lease of life. To appreciate the revival he championed, we must backtrack to the institutional rivalries which beset bridge-building in nineteenth-century France. After Navier's setback at the Invalides, the Corps des Ponts et Chaussées lost the initiative in metal bridges. The famous French iron railway bridges were mostly designed by entrepreneurial 'ingénieurs-constructeurs' like Eiffel. Nevertheless the Ponts et Chaussées continued to hold its own in masonry bridges, both on state roads and on railway lines. By the end of the century it was at last building fine steel bridges into the bargain, conspicuously Jean Résal's Pont Mirabeau and richer Pont Alexandre III in Paris (ill. 314).

The Ponts et Chaussées had always been proud of its masonry heritage. Renewed national enthusiasm for the 'siècle des lumières' under the Third Republic made sections of the corps more so. One of its leading lights, Fernand de Dartein, embarked on the documentation and analysis of French bridges finally published in four magnificent volumes as *Etudes sur les ponts en pierre remarquables pour leur décoration antérieurs au XIXe siècle*

332. Pont Adolphe, Luxembourg.
Paul Séjourné, engineer, 1899–1903.

333. Pont Catalan (formerly Pont des
Amidonniers), Toulouse. Paul
Séjourné, engineer, 1904–10. A masterpiece of the masonry revival in
bridges.

(1907–12). Dartein was more of a teacher than an original engineer; Paul Séjourné was the reverse. He came to notice in the 1880s with a set of wide-span masonry railway bridges in the Tarn, one of which (at Lavaur) owed its form to the proximity of a 'very fine' road bridge of the 1770s.[171] But Séjourné's masterpieces are two urban road bridges, the Pont Adolphe, Luxembourg (ill. 332), a single arch spanning a park in a gorge (1899–1903), and the multi-arched Pont Catalan (formerly Pont des Amidonniers), over the Garonne in the centre of Toulouse (1904–10). The piers and arches in both are wholly of masonry; the latter bridge (ill. 333) has an infill of colourful brickwork and ample vomitories in homage to its neighbour, Toulouse's venerable Pont Neuf.[172]

All this sounds like antiquarianism. Not so, for in both bridges Séjourné stretched masonry construction by the size of his spans and by dividing each into two separate,

334. Villeneuve-sur-Lot, new bridge,
centering for the single arch, 1914.
Eugène Freyssinet, engineer,
Limousin & Cie, contractors.

335. Villeneuve-sur-Lot, the
completed bridge.

parallel ribs carrying a corbelled-out deck for breadth of modern roadway. Here surely was something learnt from iron bridges, though Séjourné referred it back to Perronet at Pont-Sainte-Maxence (ill. 268).[173] The decks were of reinforced concrete. Concrete he was not against – only metal, in whose susceptibility to rust he saw a will to revert to the original ore. At Toulouse he argued that concrete piers would have been more vulnerable to scour than granite. Nor was concrete cheaper than stone, Séjourné maintained, if his rationalized rib construction was adopted. Naturally, stone was more handsome. In the competition for Toulouse, judged by Résal, a stingy steel bridge had been chosen, then set aside; the cheapest concrete bridge cost too much. So they came to Séjourné and masonry, selected 'so as to offer a minimum of satisfaction from the architectural point of view', he dryly remarked.[174]

Appearance was paramount for Séjourné. Most of all, he urged engineers to be alert to context. A bridge should do more than respect its setting, the buildings around it, and local light and colour, it should smack of the soil, grow out of the earth, avoid looking like something dropped down: 'il faut à Toulouse un pont toulousain'. Propriety must be specific, in other words. In towns, discreet decoration might be to the point. At Luxembourg, Séjourné derived his bridge cornice from profiles on local fortifications; at Toulouse, the Pont Catalan boasted wrought railings and carved cartouches. But an architect's help was not wanted. For a Ponts et Chaussées engineer, architecture had never entailed architects: he did it himself. 'Engineers must know their architecture,' insisted Séjourné. 'At the Ecole des Ponts et Chaussées, it too is taught, and very well . . . Intellectual culture must not be narrowed down to utility alone. It has been a crime to sacrifice the old humanities to it – temporarily so, I hope.'[175]

These precepts come from Séjourné's *Grandes voûtes* (1913–16). Its six volumes on bridges spanning more than forty metres expand and update Dartein's chronicle. Together, the series constitute a joint memorial to the heritage of French masonry bridges, as that superb tradition neared its end.[176] But it was not quite over. Séjourné's bridges and book had an international impact, not least upon those who felt his techniques and ideas about contextual building could as well be explored in concrete as in masonry. Among them was Freyssinet, upon whom the Pont Catalan made a decisive impact as he prepared to build in towns of the same region.

The new bridge at Villeneuve-sur-Lot (ills. 334–5), started just before the First World War and finished after it, was Freyssinet's first major work for the Limousin company, and the first in which he had to tackle context with delicacy.[177] The line of the new bridge ran along the edge of the old town, within sight of its ancient predecessor. If the juxtaposition was similar to Toulouse, the scale and contour differed. A single span made sense. Freyssinet borrowed from Séjourné his parallel rib-system but built the twin arches in mass concrete – reaching a 96-metre span, unprecedented without reinforcement. The naked ribs, de-centred by a new method, spanned the Lot throughout the war, and were so illustrated as the single new bridge of importance in the final volume of *Grandes Voûtes*, published in 1916.[178] When peace came, Freyssinet added a polite superstructure of red brick, with arcaded spandrels and a corbelled parapet.

336. Tonneins, bridge over the Garonne. Eugène Freyssinet, engineer, Limousin & Cie, contractors, 1919–22. A narrower version of Séjourné's bridge in Toulouse, arched in concrete.

Séjourné's imprint was even stronger at Tonneins, a bridge of five arches Freyssinet built over the Garonne (1919–22) to replace an old suspension structure. Here the site was less constraining, outside the town though overlooked by it. In appearances the bridge (ill. 336) was close to a crib from the Pont Catalan, with concrete substituted for masonry, but a similar profile of arch, the same handsome flush brickwork and open vomitories for flood water over the piers. Only the twinning of the ribs is missing, for Tonneins did not need a wide bridge.

Given his instinct for courage and experiment, it is telling that Freysinnet looked back on Tonneins, the most imitative bridge he ever built, as 'perhaps the most successful of all my works'.[179] Maybe he was thinking of process, but maybe of aesthetics. If these bridges pick up via Séjourné on the traditions of the Ponts et Chaussées, the resonance is more than just historic or contextual: it is about how old and new ideas, old and new materials, unite to make beauty. The juxtaposition of fair-faced concrete with pink brickwork at Villeneuve-sur-Lot even reminded Freyssinet's biographer of Louis Kahn. Working on his own along lines indicated by Séjourné, Freyssinet was able to come up with a happier compromise between invention and propriety in his town bridges than Maillart ever managed when he teamed up with architects.

THE NEW ROADS

Conscious of his place in history, in 1936 Maillart delivered his reflections on propriety, the nut he had never quite cracked:

> The engineer decides only with difficulty to depart from the forms that conform to tradition and even if he might wish to, owner and public follow him unwillingly. Small wonder then, if new types of structural forms, coming honestly and relentlessly (without looking to the past) from the nature of reinforced concrete, are more readily brought to completion in rather out-of-the-way places; the cities remain 'immune', because there weight will be given to a certain 'monumental' appearance. Even with rural reinforced concrete bridges an 'architectonic' minimum are the abutments required to frame the structure and isolate it from its surroundings. To do away with this seems as revolutionary as to build a house without a base.[180]

These remarks formed part of a lecture on the new bridges of the Reichsautobahn, pioneers of the new motor roads across Europe. They were to transform the context of bridge design as utterly as the railways had done a century before. The ornamental bridges of the City Beautiful, even the suspension bridges of New York up until the Crash, were monuments whose urban context did not lessen their independence. But with the motorway, the bridge was restored to the road and rooted anew in the country-

337. North Avenue overpass, Westport, Connecticut, on the Merrit Parkway, *c.* 1940.

338. Waschmühltalbrücke near Kaiserslautern on the Saarbrücken to Mannheim autobahn, styled by Paul Bonatz, architect, 1935–7. Separate concrete arches with sandstone cladding for the two carriageways.   From *Das Bauen in Neuen Reich* (1938).

side. During the Roman Empire or the golden years of the Ponts et Chaussées, bridges had belonged within road-building programmes. Now the new line and style of the motor roads, broad, open and inviting, drew bridge and carriageway into one and set them together against the landscape. Hence the continuing relevance of Maillart's example. Hence also architectural interest in the type.

As motorway or freeway architecture was first attempted on a national scale in Nazi Germany, its assessment remains controversial. Much has still to be understood. What debts, for instance, did the autobahn system owe to the parks and early parkways of East Coast America?[181] A link can certainly be made between the driveways of nineteenth-century urban parks by Olmsted (Fenway Park in Boston is the notable example) and the early motor parkways around New York (the Bronx River Parkway of 1914–23 being the first). All these had roads set apart for riding or driving, tastefully aligned and landscaped, and featuring single-span bridges in romantic, rusticated masonry. Borrowed from the tradition of the small bridge on the country estate, here were tasks that architects could handle. H. H. Richardson styled two of the bridges on the Fenway; Carrere and Hastings were involved in the Bronx River Parkway; and so forth. In the inter-war parkways, carriageways were doubled and concrete became the norm for bridge construction. But rustic stone facings hung on – along with the architects (ill. 337). They perpetuated the rural illusion, even if you were only motoring home to your suburb.[182]

Some of the American ideal of privilege, pleasure and virtual ownership of the countryside carried over to the early autobahns.[183] It was in no way at odds with the political and regenerative ambitions of German road-building, derived from Italian programmes that had stalled.[184] Since the Nazis drew much of their support from rural areas, the dynamic Fritz Todt, engineer, party member and general inspector of roads from 1933, was keen to make the connection. Todt's policy of associating architects with engineers in this prestige programme was calculated to please Hitler. Besides the bridges and road layouts there were petrol stations and rest areas to design, along with 'Autobahnmeistereien' every sixty kilometres – service areas with little motels where you could put up for the night. For all this, Todt brought in architects like Werner March, Friedrich Tamms and Bruno Wehner to assist his teams of engineers. A better-known name, Paul Bonatz of Stuttgart Railway Station fame, soon joined the list of consultants, with the role of setting out a broad design philosophy or 'Arbeitsstil' for the bridges. There were competitions for important structures, while in different parts of the Reich regional architects came in for particular bridges and buildings.

Not all structures of the Nazi era were hectoring stone edifices or essays in simpering rural revivalism. Along the early Autobahn they ran the gamut of expression according to their purpose, size and location. There were flat-roofed and pitch-roofed petrol stations, 'heimatlich' rest places and plain administration buildings. So also the bridges, of which some 9,000 were built between 1933 and 1941. Many were standardized. The overbridges usually updated the parkway look, having one broad and solid arch spanning both carriageways, faced in coursed rubble. The big bridges and viaducts amounted to just three per cent of the total. Allowing for differences to do with the terrain to be crossed, there was a tendency in the larger structures for steel trusses to give way over time to concrete or even masonry arches, as military spending grew and steel was diverted into armaments. In that way militarization nudged architecture into a reaction which ideology had not insisted upon. But propriety and monumentality were certainly issues, around cities in particular. They were present also in border regions, where the Reich felt a special mission towards display.

339. Adolf-Hitler-Brücke (later Rhein-Brücke), Krefeld. Friedrich Voss, engineer, 1933–6. Third Reich bridges were not all reactionary or historicizing. From *Das Bauen in Neuen Reich* (1938).

Many of the early big bridges off the autobahn engineers' drawing boards consisted of deep, solid plate-girders resting on piers of brick or stone-faced concrete. The clarion call for all-masonry viaducts was given by the Lahn crossing near Limburg, styled by Bonatz himself, which made up for the conventionality of its twelve railway-type arches with exquisite proportion and detailing. Far from being specifically Nazi, the Lahn viaduct and its successors (ill. 338) pursued stripped-classic directions explored by Bonatz during the Weimar Republic, in the Stuttgart Station and his streamlining of tidal barriers and weirs along the Neckar. Bonatz was as happy to style for speed, as in the low-lying concrete bridge built over the Danube at Leipheim with the engineering contractors Wayss and Freytag. More persistently 'reactionary' was Friedrich

340. Millennium Bridge: the opening. Foster and Partners, architects, Ove Arup and Partners, engineers, Anthony Caro, sculptor, 1998–2000.

Tamms, a champion of Roman-aqueduct style for his bridges and viaducts. The Tacoma Narrows collapse gave German war-time propagandists a chance to justify such heavy structures, arguing that 'abolishment of mass leads not only to formlessness but to failure'.[185]

The styling of the autobahn bridges raises questions about the role of architects in engineering works after the First World War. Were they there to offer a veneer of civilization and pleasure to the educated classes, or to further hegemonic political programmes? And was there much difference between the symbolic agenda of the civic coalitions and democracies and those of the totalitarian régimes precipitated from the war? It was not just for Fascist governments or in the old styles that architects were drummed up to play with the edges and details of vast structures. In the United States, where they streamlined the dams and power plants of the Tennessee Valley Authority, the motive was no less propagandistic – to offer a smart façade for reconstruction, and make heroic engineering projects accessible to the masses. As one employee put it:

> They hear the hum of the generators and they are impressed by the size and importance of it but they don't understand it. It is completely foreign to their knowledge of things in the past, but the things that they do understand are handrailings in stairs, doorknobs, lighting fixtures and floors.[186]

The onslaught of a dogmatic modernism during the 1920s and 30s added to the uncertainty. Modernist rhetoric insisted that architecture update its thinking about techniques and materials, and that the engineers had been the true creative constructors of the recent past. If that were so, architects had best withdraw from areas like bridge-building. Maybe there was nothing for them to add – a conclusion which some champions of architect-free engineering, like Owen Williams in Britain, had the gall to underline. The position of the 1840s–60s, when the architectural grip on bridge-building weakened and architects connived in their own redundancy (p. 325), was repeating itself. This time concrete not iron was the technology that had upset the applecart, and roads the backdrop instead of railways.

Architects could still add touches of styling to modernist bridges, on the lines of what Embury did for Ammann in the 1930s and 40s. However subtle, that was marginal involvement, less visible than the old towers and abutments of suspension bridges. Where and how might they be better employed? They had long ceased to have much input in rural bridges; and their hope of setting an architectural mark upon the freeways collapsed with fascism. Only cities, those strongholds of the profession, remained. When Friedrich Tamms resurfaced after the war, it was in an urban setting, as art-adviser for Düsseldorf's post-war bridges.[187] In cities and towns, people had cause and leisure not just to ride across bridges, but to walk, stop, think and look. There a thin line of architectural input was maintained. Even under the levelling hand of modernism, urban propriety was not dead. To the question of where, the answer was the same as in the days of the Romans. But the question of how had become a puzzle.

If, in modernist terms, architecture was defined as the excellent expression of how a structure was used and how it was made to stand up, bridges might indeed be left to the engineers. Compared to buildings, their purpose is simple; most of their complexities arise out of solving structural issues. Nevertheless even in the most positivistic of twentieth-century periods there were always some who shied away from the idea that modern architecture was just about function, in either the purposive or the structural sense of that word, and clung to older ideas about meanings, ideas and emotions. How now to accommodate those vaguer, looser feelings? No longer could they be met from the arcane and hard-learnt culture of past architectural styles. Propagandistic programmes were suspect, and a shared social understanding could not be taken for granted. Only two ideals remained unexhausted: the proprieties of context, resilient still; and an appeal to the magic of art, located now not in common values but in the charisma of creative individualism.

MILLENNIAL

'Bridge Wobble Not My Fault – Architect', howled the London newspaper placards on 12 June 2000. 'Architect Lord Foster says surprise sway is an "engineering issue".'[188] The headlines marked bathos for a bridge whose image had been as tailored as the structure itself, all the way from conception to completion (ill. 340). Not only the Millennium Bridge had wobbled, a month after its triumphal opening. So too had the ideal of seamless professional partnerships in the making of modern public monuments.

It would be churlish to exaggerate either deviation. The physical wobble was no disaster, but a minor engineering setback, blown out of scale by the hype attracted to the project. If it took a year and a lot of money to put right, in hindsight it adds to the richness of the Millennium Bridge saga.[189] If too it was Norman Foster's immediate instinct to run for cover, he was soon back with his collaborators, sitting subdued alongside the four engineers from Ove Arup and Partners as they told the press why the bridge had been shut precipitately and what they thought had gone wrong.[190] Later, Fosters contributed generously to the remedial works. For all that, the two types of wobble point to a single truth. In bridges, the buck stops with the engineer. Other issues, other people may come and go, lead or follow, but in any modern bridge it is he – or she[191] – who bears the brunt of the work and provides the continuity. If engineers take the blame when things go wrong, it follows that they ought also to have the lion's share of the credit.

What became the winning design for London's Millennium Bridge competition of 1996 originated not with an engineer nor even with an architect but with a sculptor, Anthony Caro. There were two reasons for his presence, global and local. Europe was in the grip of a revulsion against inartistic bridges, just like the reaction a century before. It is often stated that the Millennium Bridge was London's first new bridge since Tower Bridge, itself among the quirkier products of that earlier reaction. That is true if brand-new crossings are meant. But there had been replacements. The reconstruction of the time-hallowed river-crossing at London Bridge (1967–72), for instance, was the kind of traffic-engineers' bridge which the next generation reprehended – an impoverished replacement for the Rennies' masonry bridge, devoid of historic or human appeal.

Engineers and architects concurred in that feeling: and in the trickle of better urban bridges, footbridges notably, that began to appear from the 1980s, the pairing of architects with engineers at the conceptual stage became common. In figures like Santiago Calatrava and Marc Mimram there also emerged new versions of the old-style architect-engineer specializing in smart bridges. Calatrava became the first bridge-designer beatified by architectural stardom since Maillart. In 1996, the year of the Millennium Bridge competition, an international exhibition was held on Living Bridges – one token of renewed passion for the genre.[192] For it, Calatrava produced a design linking St Paul's with Bankside by means of an 'inhabited bridge'. Footbridge fantasies for central London had proliferated over the years, and an inhabited bridge was never plausible. What drew Calatrava's site into the realms of realism was the Tate Gallery's decision to install its modern collections in the majesty of the redundant Bankside Power Station across the river from St Paul's. Here was the local factor which made the art-idea so appropriate as to be unavoidable, and beckoned Caro along with other artists and art-architects on to the stage.

The early stages of the Millennium Bridge repeated those haphazard British arrangements for bridge-building which so easily end in inadequacy or débâcle. An open competition is held with much fanfare: then on the strength of the winning entry, planning permission and money are sought. One thinks of Brunel's trials at Clifton. At first the project was just a pipe-dream in the minds of a few enthusiasts. Formal government support was wanting, though the local authorities on either bank looked benignly on the venture. The *Financial Times*, a paper with a record of architectural patronage, sponsored the competition. The brief was for a pedestrian bridge, with some latitude as to location and no fixed budget. Involving architects was mandatory, involving artists welcomed. The art-factor became an incentive for fund-raising purposes, since once a design was in hand, a trust needed to be formed and public and private contributions solicited.

Of the six competition entries shortlisted, two included a famous artist who worked with steel and leant towards a kind of engineering aesthetic. Richard Serra teamed up with Frank Gehry, Anthony Caro with Foster and Partners and Ove Arup and Partners. Caro had started an engineering course at Cambridge before leaving to work with Henry Moore. It was he who rang up Norman Foster suggesting a collaboration, and Foster and his right-hand man of the time, Ken Shuttleworth, who then contacted Chris Wise at Arups. Such were Arups' size and fame that different teams within the firm often worked on different entries for any major competition, independently and without conferring. This particular submission was clearly prestigious. It won.

'Is this bridge architecture, engineering or sculpture?', asked the collaborators in their original submission document. 'It has been largely created by three individuals, *from* these three individuals – a sculptor, engineer and architect. As authors of the project, we see it as a shared creative act'.[193] Thanks to Deyan Sudjic's informative booklet about the Millennium Bridge, we can get behind that. From start to finish, the partnership was genuine and sustained: everyone continuously reviewed, rationalized and refined what everyone else had done. Yet some plain truths remain. The engineer-individual (Wise) devised the conception of the bridge. The architect-individual (Foster) resolved its alignment. The sculptor-individual (Caro) came up with a thrilling idea for the southern abutment which probably got the team into the last six, but became marginalized as things went on. In the end his most palpable contribution is the set of four small sculptures on the north side of the bridge – consolation prizes, almost. Caro was never discounted: as he put it, 'I was there when all the important decisions were made'.[194] But a tendency soon set in for the art to come apart from the bridge and turn into an add-on, against the rhetoric of integration.

Chris Wise is one of the stream of exceptional, architecture-loving engineers groomed and turned out by Arups, with a streak of romance to balance his structural finesse. The evening before the first meeting with Caro and Foster, Wise has written, he sat down in a wine bar with a colleague, Roger Ridsdill Smith. After essaying many possibilities for the bridge, they finally agreed to 'try to make the structure from cables stretched as tightly as possible between the two banks, and then walk on them'. The formulation is confusing; there had of course to be a deck. That evening, Wise says, they already thought of 'putting the cables outside the deck, so that they would act as outriggers to give some measure of torsional stability'.[195] So it was a suspension bridge they were talking about, but one with such a flat sag-to-span ratio that the 'pylons' atop the piers in the river could be all but ignored by the eye, and deck and cables construed together as a single bounding line of extreme tautness.

Metaphor is a vital crutch in the conceptual and publicity stages of any major building project today, because the true language of modern architectural form is so abstract. Next morning, the Wise and Ridsdill Smith concept won favour over the only alternatives presented, some striking impossibilities developed by Caro 'from his series of long steel table sculptures'.[196] The engineers, it seems, offered for their idea the image of 'a guitar string pulling the banks together'. But soon (it may have been then or later) the alternative of a knife-blade emerged. 'St Pauls on axis!', a Foster sketch is annotated. 'The bridge is tensile – stretched taut – polished – ephemeral – polished – a blade . . . At night the bridge could be a blade of light!'[197]

Architects like to seize an image and work it hard. Buttressed by a fluke aerial photograph showing the bridge-site bathed in weak sunbeams while its environs were in shadow, the 'blade of light' phrase got into the

341, 342, 343. Millennium Bridge, alternatives for the southern abutment. A, stepped extravaganza in steel planes, largely by Caro. B, the circular ziggurat. C, eye of the needle. A simplified version of C was eventually built.

winning submission, percolated into the fund-raising literature and assumed the title of Sudjic's commemorative booklet. Large architectural firms have become adept at publicity. The well-oiled machinery at Fosters was to be crucial in translating the Millennium Bridge from idea into reality.

But Norman Foster is more than a very experienced front-man. With the blade of light idea went the notion of St Paul's on axis. For this, the urbanistic aspect of the project, the architects took the lead. Choosing the exact location and 'footfall' of the bridge was not straightforward. Planning permission depended on alertness to the context of St Paul's. The cross-axis of the cathedral is imprecisely aligned on the narrow open vista towards the Thames which the City planners had wrested from the flotsam to its south;

while the commanding central tower of the power station opposite, as symmetrical as St Paul's itself, is well off-centre from the dome. There was no perfect answer to the conundrum. Some competitors tried cranking their bridges, landing themselves with clumsy transitions. Picking up from Wise's taut string or blade, Foster saw that clarity was the right response. The line of the bridge should shoot straight from the gap below St Paul's, and land on the south side where it fell.

The southern abutment (ills. 341–3) is the main disappointment of the Millennium Bridge. Had the Tate authorities been well disposed towards a direct entrance from the bridge, all might have been resolved. That was never encouraged, leaving the power station's grand outline to be more gratuitously disfigured by other hands. For their submission, the winners came up with the imaginative answer of a bridge that changed as it landed into a gigantic stepped Caro sculpture. This riverside feature – an hors d'oeuvre for what the visitor could expect within the gallery – might wonderfully have renewed the heritage of sculptural abutments for suspension bridges. There were practical difficulties involved, and it would have been expensive. But what most told against it was the modernist insistence upon integration. An essay in the lightest structural engineering was to butt straight up against grave, moody planes of rust-red steel: that was too much like the old aesthetic of contrast. So the judges told the winners to go away and knit the sculpture better with the architecture.

For a time Fosters and Caro toyed with a 'circular ziggurat' that combined ramps with a little central café. Then that too fell away, leaving the architects to devise versions of the 'eye of the needle' solution adopted, with a plain reversal of direction for the access ramps, contradicting the line of the 'blade'. Quieter is the northern abutment, where the pedestrian walks straight on and off and the cables tuck underneath into an anchorage between a cascade of steps down to the water. It is on this side that Caro was given the sad little sculptures – explicitly distinct works of art – in recompense for his lost abutment.

Architects and engineers now expect to work intimately together on projects like the Millennium Bridge. As they shift from outline design to execution, that intimacy is ever more widely shared. The bridge had never truly been a collaboration between just three individuals. During the construction phase, the numbers involved have been estimated at some 5,000. Such figures depend on whom you care to count. Builders, suppliers, sponsors and publicizers all have their claim, putting paid to smug simplifications about creativity. But the bulk of the professional time developing the design was spent at Arups, where some seventy people were co-ordinated by Sophie Le Bourva. Chris Wise having left to form his own firm in 1999, his place on the design side was taken by David Kaye, with Ridsdill Smith as project manager. The team from Fosters was much smaller but always at the engineers' elbow, one of the architects, Catherine Ramsden, having her own work station in the Arups office. The looks of the smallest thing mattered, because in a pedestrian bridge the public eye falls on almost every detail from close quarters. Sudjic: 'There is no room for camouflage, or for a general unfolding of ideas, like a novel. A bridge is more like a three-dimensional poem. Everything is visible, nothing is hidden. It is as beautiful as the smallest bolt head, and as the springing curves of the suspension cables.'[198]

Then came the opening and the wobble, of which enough has been made. What the professionals knew but the general public did not was that London's Millennium Bridge was not the only new bridge so affected. In Paris, there was an uncanny parallel in Marc Mimram's Passerelle Solferino, opened in December 1999, afflicted by sway and slippery surfaces, then briefly closed for remedial works the following autumn, to much grumbling.[199] Since Mimram was joint architect and engineer, there could be no buck-passing there. A potentially graver embarrassment arose with the Erasmus Bridge, Rotterdam – a cable-stayed road bridge crucial to the economic regeneration of that great port's whole southern sector, the Kop van Zuid, opened to much celebration in 1996, then hastily shut after the cables vibrated in rain and wind.

The causes and solutions for the problems were different in each case, and it is not the business of this book to follow them through.[200] But what linked Rotterdam with London was that the Erasmus Bridge too had been fronted, for reasons of prestige and publicity, by an architect. It was therefore again from the architect that the press and the city expected immediate comment and technical reassurance. Ben van Berkel was more completely the 'designer' of the Erasmus Bridge than Foster was of the Millennium Bridge.[201] He had been brought in in the first place to advise one of Holland's best–known structural engineers, Arie Krijgsman, on the aesthetics of the bridge. Then, in limited competition, he had come up with an original concept of his own, involving a cranked pylon, which the Rotterdam City Council endorsed as a symbol of the renewed city and proceeded to develop and build, despite its greater expense. On scrutiny, van Berkel's victory had as much to do with the bridge's approaches and alignment as with its arresting profile. Here is another parallel with Foster's role in London. All was fine while the Erasmus Bridge was acclaimed. But when the crisis came, though the young and not very experienced architect had briefly spent time with Calatrava, he could not provide the technical answers. Van Berkel did not desert his colleagues. But for a moment he looked foolish. It was to the problem-solving engineers that the bridge then reverted.

Image is a hunger of our hasty age. Those that feed it to satisfy our eyes and souls do us service, and we are keen to praise and reward them. In so doing let us not deny equal honour to those who use their ingenuity to make the world safer and better. Always, differences of talent stimulate and enrich. Reflecting on the Millennium Bridge, Foster wrote: 'It is vital to remain open to new ideas and to recognize that they can come from anywhere – and probably won't come from the hermetic world of one's own creative discipline.'[202] The remark implies that engineering and architecture need one another but are not the same. If there is a moral to the story of the bridge, it is that the strands of art and engineering run parallel, often intertwine creatively, but in the last analysis are distinct. They should remain so and be seen to be so.

5

RECONCILIATION

1. Britain

ARUP

For over fifty years now a semblance of peace has descended on the relations between architects and engineers. Among individuals, the greatest of the peacemakers has been Ove Arup (ill. 345). Through personal example, through the firm he fathered and nurtured, this now-consecrated figure stands for a fresh pattern of partnership between architect and engineer. Was he symbol or cause; instigator, philosopher or 'umbrella man'? A little of everything. The facets of his career offer a starting point for tracking the convergence.[1]

Because Ove Arup was passionate about learning, it took him time to settle down. Half Danish and half Norwegian by birth, he spent his first twelve years in Hamburg. In 1913, at the age of eighteen, he went to Copenhagen University to study philosophy and mathematics, only to find the subjects absorbing but too dry for him. He thought of studying architecture instead, but 'wasn't sure of being artistically good enough',[2] so switched to engineering and spent six years at Denmark's Polytechnisk Laereanstalt. None of those early enthusiasms evaporated: Arup turned out to be a fair mathematician, a better designer than he feigned, and a true moralist. Experience added a businessman's shrewdness.

In 1922 Arup joined the Danish engineering contractors Christiani and Nielsen. The firm was among those that prospered and waxed multi-national with the spread of reinforced concrete. After three years back in Hamburg, Arup was dispatched to London as the firm's chief designer there. Though concrete was the order of the day in advanced structural engineering, Britain lagged behind the rest of Europe in familiarity and ease with its use. 'Reinforced concrete was looked on with the greatest suspicion by the majority of potential clients', wrote Arup.[3] It is too glib to explain why in terms of a national reluctance to embrace modern styles. In her penetrating study of the British building industry, Marian Bowley pointed to deeper institutional mechanisms: the 'steel-mindedness' of engineers, their links with steel companies, and shortcomings in British engineering education. Though price-fixing may also have been a factor, Bowley attached the ultimate blame to failings in technical initiative among professionals: 'The lack of development must . . . be attributed primarily to designers.'[4]

Here was an opening for imported expertise; Christiani and Nielsen with their young designer availed themselves of it. They were not alone. Even before the Nazi emigrations, there was a small influx of civil engineers into Britain. Danes made a disproportionate impact on British construction at this time: as early as 1922 a society of Danish civil engineers existed in London.[5] Besides Christiani and Arup, the names of Oscar Faber (English-born but the son of a Dane), Olaf Kier and Peter Lind may be mentioned. Possibly Danes were able to introduce advanced German techniques of construction which might have been resisted under a German name.

Most of what Arup drew up for Christiani and Nielsen was industrial and maritime. Docks and silos made him fluent in a range of concrete building techniques. Concealed

FACING PAGE
344. Hongkong and Shanghai Bank, Hong Kong. Foster Associates, architects, Ove Arup and Partners, engineers, 1981–6.

345. Ove Arup on the completed ramps of the Penguin Pool, 1934.

framed construction was all that most architects wanted from structural concrete at that time, but it offered few challenges. Then in 1932 two things happened. Arup acquired the job of designing a café and shelter on Canvey Island; and Berthold Lubetkin asked Christiani and Nielsen for help with the first commission awarded to his new firm of Tecton, the Gorilla House at London Zoo.

Later, Arup was disposed to mock his Canvey Island café (ill. 346) as a folly which deployed every Corbusian cliché, confirming his suspicions that he lacked the flair to be an architect: 'architecture on the cheap by an amateur architect employed by a contractor, and a client with no money to spend, is not a good way in which to achieve perfection.'[6] Yet he was in earnest at Canvey Island. The building showed Arup's employers how imagination might extend their business; and at a time when Britain had little to show for all the talk about functionalism and a concrete architecture, it staked his claim as a rare modernist among British engineers. Soon he joined the MARS (Modern Architecture Research) Group – only to find himself baffled by what architects meant by research.

As for the approach from Lubetkin, it marked the start of the real thing, of a 'new kind of collaboration': so it has been said in hindsight.[7] Subjectively, the engineer was to identify with what the architect imagined, to throw himself into it heart and soul, while exploring, stretching and if need be challenging its implications. A self-denying role, perhaps: or so Arup came to present it. His ideal became to help bring to fruition what the art-architect conceived. That meant a willingness on the engineer's part not just to foster tolerance of the architectural imagination, but to mate with it. In time, the sense of loving attachment surfaced in Arup's language. The true partnership of architect and engineer he saw as 'a very intimate affair'.[8] 'I didn't know you cared', he joked when given

his gold medal from the RIBA in 1966; and his speech ended with the old pledge: 'until death us do part'.[9]

There were compensations for this deference of technique to art. After all, wasn't woman created out of man's rib, in Joseph Brodsky's words, 'neither to be loved nor to be loving, nor to be judged, but to be "a judge of thee"'?[10] Arup's ideal of collaboration could only work if, for his part, an architect endeavouring to do something new showed an intelligent respect towards technology. Here reinforced concrete's flexibility was vital. Traditional masonry construction might be reliable and look good, but it had structural limitations. Steel was better, but at this date its use ran only to a few tasks like framing and roofing a building. Moreover the separation between manufacturers, steel erectors and contractors hindered creative partnerships between architects and engineers. Concrete on the other hand could do almost anything. In situ or pre-cast, it could take a protean variety of forms: columns, beams, panels or slabs; heftily modelled walls or thin shells; structure or cladding. It followed from the range of concrete's uses that the arrangements for building in it were also looser. An architect who wanted to get the best from it needed specialized help at the start of the design. It was not good enough, as with steel, to wait till the design had been made and then ask an engineer to make it stand up. Concrete technology obliged him to consult early on, and to heed the engineer's judgement. 'I could only work with Lubetkin,' Arup once teased, 'because, as an engineer, I had the last say, which Lubetkin had to respect'.[11]

Lubetkin, for his part, knew plenty about concrete. During his roving youth he had taken lessons from an expert on the material at the Berlin-Charlottenburg Technische Hochschule, and refreshed his knowledge with further technical classes in Paris.[12] There he had attended Perret's atelier, where concrete construction in frame and infill

346. Canvey Island café. Ove Arup, designer, for Christiani and Nielsen, engineer-contractors, 1932.

347. Gorilla House, London Zoo.
Design drawing and analysis by
Lubetkin and Tecton, architects,
c.1932–3. Such drawings were intended to emphasize Tecton's scientific
rigour, and may sometimes have been
produced retrospectively.

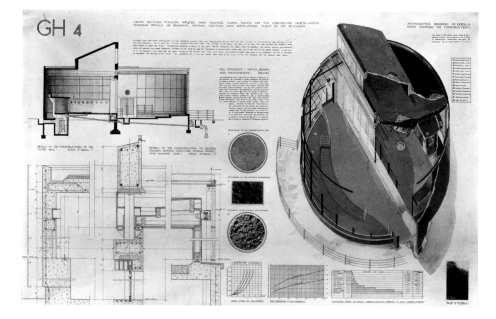

was de rigueur. Just before moving from Paris to London, he built a smart apartment
block in the Avenue de Versailles (1930–1). Its concrete frame was conventional and he
probably needed little help with it. But though the early Tecton projects were cheaper
buildings, structurally they were less straightforward. Taking off as they did from con-
structivism and Le Corbusier's villas, their blend of geometry and slightness called for
a technical midwife. Arup's first role in Tecton's structures at the London Zoo, says
Lubetkin's biographer John Allan, was 'to advance the structural application of reinforced concrete in building from a column and beam medium – a hangover from the
orthodox steelwork tradition – to a panel and slab technique'.[13] In the process, he
opened up fresh avenues of design to Lubetkin and his young colleagues. In exchange,
Arup learnt to loosen up. The dash and sweep of the Gorilla House (ill. 347) and
Penguin Pool eclipsed the Canvey Island café. 'My first real teacher of architecture was
Lubetkin', he said. 'He . . . taught me that "sensible building" must be modified to satisfy the claims of aesthetics.'[14]

These were special, one-off projects, carried out in the 'honeymoon' period between
Arup and Lubetkin. Could so intense a style of joint working broaden into a general
methodology of design? It was harder to apply to building-types where efficient production counted for more than art. High housing blocks were a preoccupation of the 1930s,
a decade when many on the Left believed that new technology held the key to fresh patterns of living and hence to social progress. Lubetkin was an ardent socialist, Arup an
easier-going fellow-traveller. Accordingly, housing proved the next arena for the burgeoning mutual exploration between the professions.

There were problems. Arup was not a consulting professional but an engineering contractor's employee. Christiani and Nielsen had built the Canvey Island café (though Arup
grumbled he had been able to get out of the office to the site only once), the Gorilla
House and the Penguin Pool. When it came in 1934 to Tecton's first block of tall flats,
Highpoint One, the firm declined to tender. No doubt the hours their designer had
already lavished on the project, plus the challenge of what he and the architects were
devising, seemed unlikely to translate into profits. So Arup, committed to architecture
and waxing independent, parted company with Christiani and Nielsen. Through his
friend Olaf Kier he switched to the London affiliate of another Danish-based firm, J. L.
Kier. Kiers erected the concrete structure for all Tecton's famous buildings thereafter up
to the outbreak of war, and through Arup took on other British modernist projects.

By remaining with a contractor, Arup was able to influence both the design of the structure and the means of building it – things that had to be thought about together if building methods and productivity were to improve. Notably, he introduced 'climbing shuttering' to Highpoint, a technique absent from architectural projects hitherto though familiar in engineering work (ill. 348).[15] But his style of partnership between architect and engineer meant a partnership also with the contractor. That raised the vexed question of the 'nominated' contractor or subcontractor. If co-operation and joint development were the way forward for building technique, how could competition enter into the equation? At its worst, competition meant fragmentation and the loss of joined-up thinking. But at its best, it could offer economies and the stimulus of fresh ideas.

For a little while Arup tried to have it all ways. When he finally struck out on his own in 1938, he set up no less than three parallel companies: a design-and-build operation, like Kiers or Christiani and Nielsen; a design firm; and an engineering consultancy to work with architects.[16] In the ensuing circumstances of war, only the design-and-build service flourished. Then, with the return of peace, Arup in 1946 plumped for consultancy. At the age of 51 he founded the company which today bears his name. Now it was to be architect and engineer together, with the contractor doing their bidding.

Ove Arup and Partners was a rapid success. Many of his pre-war collaborators came back to him, among them the Tecton architects. More than ever, social projects were the order of the day. When Tecton picked up their pre-war schemes for housing in

348. Highpoint One erecting, 1934. Lubetkin and Tecton, architects, Ove Arup, engineer, for J. L. Kier and Co.

349. Brynmawr Rubber Factory: beneath the domes in the years of its dereliction. Architects Co-operative Partnership, architects, with Ove Arup and Partners, engineers, 1948–52. The factory never proved economic and despite many attempts to find new uses it was finally demolished.

Finsbury, Arup too developed former ideas; he perfected the 'egg-crate' or cross-wall system for framing high flats (soon to become a standard in British housing) and he improved the climbing shuttering. On the face of it there was entire continuity with the former way of working.

Yet there were subtle differences. As a consulting engineer, not a contractor's agent, Arup now stood at one step removed from the building process. In the uncompetitive conditions of post-war Britain, that did not make much difference; in the long run it would do so. And as regards Lubetkin, the 'honeymoon' intimacy had evaporated. As early as 1945, before Arup's new firm had been set up, Lubetkin was lamenting the loss of old comradeship over their Finsbury projects, the future Spa Green and Priory Green flats:

> We used to be very friendly, and used to collaborate professionally with good results for both sides. Then as time went on, these two functions seemed to drift further and further apart . . . In the original stages of our collaboration . . . we found in you sufficient understanding of our methods, which, coupled with enthusiasm, made you welcome all the trouble involved in the purification of the schemes. Later on, I maintain, this attitude changed. Hence, instead of getting from you the constant stimulus to improvement, we got to feel very much the same as we no doubt should have felt from the beginning with any other consulting engineer, i. e. that we should just be a damned nuisance if we changed anything.[17]

Lubetkin was often combative, and one feels the wisdom in Arup's reply: 'You probably think that your explanation is the right one, but I wonder whether you are a good judge of your own subconscious reactions. You are not given to introspection, as I am.'[18] Yet for all his blandishments there did take place a parting of the ways, which the hypersensitive architect had sensed almost before it happened. Thereafter the bulk of the Finsbury housing work fell within Arup's firm to his lieutenant and future partner Peter Dunican. In arduous post-war circumstances Spa Green and especially Priory Green proved very painful to construct; after them Lubetkin worked with Arup no more.[19]

John Allan, who quotes this exchange, judges that the cooling of relations reflected no 'architectural or artistic differences'.[20] Yet the technology of high-rise housing could hardly be conducted as an affair of the heart. Arup, his firm fast expanding, had to face that as much as Lubetkin. In the event through Dunican's drive, skills and political contacts, Arups developed a powerful interest in industrialized housing.[21] It was the first of many profitable research specialisms that the firm was soon spawning. Such endeavours never deeply touched Arup himself. He still hungered for emotion to permeate any partnership.

Take for instance Arup's style with the Architects Co-operative Partnership. Untried young men entrusted in 1948–52 with the building of a giant rubber factory at Brynmawr in the Welsh valleys, they had started out with a vague idea of a roof like the one Le Corbusier had intended for the Palace of the Soviets. Arup explained to them that this would never work. Instead, he and his partner Ron Jenkins suggested the shell-concrete domes devised by Dischinger and Finsterwalder for the Zeiss-Dywidag company in Weimar Germany. The nine shallow domes that became the core of the Brynmawr Rubber Factory (ill. 349) helped to spread the architectural craze for shell concrete. Essentially they were an extrapolation from one of the Zeiss-Dywidag diagrams, designed by Jenkins and calculated with much pains in pre-computer days by John Henderson on the architects' behalf. But what the latter remembered was the avuncularity of Arup himself, altruistic in his keenness to further their virgin endeavour.[22]

Sydney

Among many such stories, the most gripping is that of the Sydney Opera House (ill. 350) – a turning-point in the affairs of Ove Arup and Partners. So much ink has been spilt upon this famous saga that one hesitates adding to it.[23] There have been phases in the building's historiography. After the first excitement over Jørn Utzon's exhilarating design of 1957 came his resignation in the throes of the work in 1966: hence recriminations and disillusion, not by any means quelled when the opera house opened at last in 1973. Since then there has been forgiving and forgetting. That is bound to occur as any spectacular building ages and endears itself to new generations, who need not bother with its birthpangs. At the time of writing, Utzon lives on. Latterly the New South Wales authorities have had the

350. Sydney Opera House.

largeness to welcome him back and consult him. And fresh enthusiasts have rehabilitated and exonerated a great talent whose career was all but shipwrecked by far-away events.

Now the pendulum has swung back a little. The revisionists' aim was to vindicate the quality of Utzon's imagination. If they laid any blame, it was not on Ove Arup and Partners, as Utzon at first did when he resigned, but on the New South Wales Government. Yet renewed reverence for Utzon in an age of star-architects easily implied an underestimate of Arups' contribution. Recently, the engineer's biographer Peter Jones has returned to his forthright defence, marshalling fresh documentation. No such new evidence is presented in what follows. My excuse for adding to what has been written is that the Sydney Opera House advanced, deepened and maybe darkened the new style of collaboration. Its history adds to the psychology as well as the narrative of the present theme.

Ove Arup always looked upon Utzon as exceptional. 'He is certainly unusually gifted,' he told London colleagues as the storm gathered in 1965,

> and he is very, very much an Architect, one who masters his architectural media and who is very sensitive to space, form, colour and texture, and to aesthetic logic or consistency – if one can talk of such a thing. He has also a very good structural sense, and is quick to learn. And he combines a steadfastness of purpose – call it stubbornness if you like – with complete flexibility of mind . . . He is very good at explaining what he is after and uses every opportunity to demonstrate his views on Architecture or Aesthetics. This is of course a requisite for fruitful collaboration. It is an education in itself and it enables one to offer rather more helpful advice.[24]

Here was the mode of partnership that Arup most eagerly embraced.

It had come about indirectly. When the opera house competition took place in 1957, the jurors included no engineer-assessors and sought no structural advice – an oversight made good in later large competitions. Nor for his hastily concocted entry did Utzon consult any engineer, trusting to his instinct and feeling for structure (his father had been a naval architect). Arup put a diplomatic spin on the outcome:

> It certainly required great courage to back a scheme like this which contained hardly any evidence of its structural feasibility. Had the panel of Assessors included an Engineer it might have meant the loss of one of the great buildings of the world – but I expect it was reassured by the then prevailing faith amongst Architects in the omnipotence of shells.[25]

Expressive concrete shells, developed from research by German engineers (pp. 409–10), were indeed then in high fashion. Among the jurors, Utzon's breathtaking outline for Sydney specially excited Eero Saarinen, boldest exponent of shells among American architects. Saarinen worked with the engineer Fred Severud (pp. 412–14) and knew their complexity. So he and his fellow-juror Leslie Martin prevailed upon their chairman to induce Utzon to work up the winning scheme with concrete engineers of the standing of Arups or Christiani and Nielsen, fellow-Danes. As a next step, Saarinen and Martin introduced Arup and Utzon to one another in London. 'Mr Arup was very enthusiastic for the shell construction of the Opera House and for the constructive clearness of the entry as a whole', they reported.[26] Utzon did not demur. Arups were duly appointed in November 1957 – not as Utzon's consultant, but by direct contract to the New South Wales Government. In time, as Françoise Fromonot summarizes matters, 'Arup was enthroned as administrator of the contract'.[27] For this supervisory work the firm was paid no extra money.

With hindsight, that was one of several early mistakes. It subverted the chain of command then usual in great architectural projects. It is easy to see why it was done. Utzon lived far from Sydney and had built little. Arups' track-record represented an insurance policy, hedging the client against the risk involved in accepting the jurors' radical choice. For

his part, Arup had a dream: that of fostering Utzon's mesmeric creativity, giving the younger man freedom to turn his sketches into a masterpiece, while his own young men sorted out the structure. Development was sorely needed on all sides. Almost everything about the design needed radical research: the spanning of the substructure, the layout of the halls, the acoustics, the fenestration, the ceilings, above all, the shells. Yet the Australians insisted on starting on site in 1959, against both architect's and engineer's advice.

Stage I, the foundations and podium, fell in large measure to Arups, while Utzon's small and changeable team in Denmark (ill. 351) grappled with issues of brief and super-structure. Despite its striking span of prestressed and folded concrete beams roofing the podium, designed by Arup himself with Povl Ahm, it was not well built, and its costs escalated as changes proliferated, partly in response to evolving ideas about the super-structure. Neither architect nor engineer yet had a permanent office in Sydney. Long-range working in those days meant tardy flights, costly phonecalls (Utzon had only one phone) and no fax or e-mail. So long as the project was at design stage, the intimacy Arup was trying to coax could just be sustained. Getting a building out of the ground in Australia by this method was different. Utzon did not really want to go to Sydney, while Arups were worried that they could not afford an office there.

The concrete 'sails' or shells of Stage II were the great challenge.[28] Here expression was inseparable from structure: Utzon and Arups *had* to work together. Utzon's competition designs had shown 'geometrically undefined' (that is to say, freehand) shapes – uplifting but unbuildable (ill. 352).[29] A consistent geometry was inevitable. Shell concrete appealed because of its sleek, thin profile; Utzon first thought of both the outsides and their undersides as eggshell surfaces, and the engineers loyally acquiesced. The shells changed utterly three times between 1958 and 1961 without a solution being found. There were problems with structural stability, not to mention the formwork, the costliest element for shells of in situ concrete. Arup's team under Ron Jenkins, expert in 'membrane action', came up with a double skin for the main shells, strengthened by an internal web. Yet no one was happy. The engineers were worried about the bending moments, and Utzon disliked the double skin and the profiles. Arup poured in staff time and

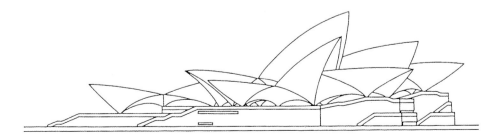

353. Arup and his team: Michael
Lewis, Ove Arup and Jack Zunz at
Sydney, 1966, not long after Utzon's
reignation.

354. All smiles: Jørn Utzon shows a
detail of the ribbed scheme for the
Sydney Opera House shells to Ove
Arup, 1961.

money, spending, so it was said, 375,000 man-hours and 2,000 computer-hours on the
research up to 1961.[30] One of the first uses of computers to solve an architectural problem, it was a proud but pyrrhic episode in the firm's history. Testing the model failed to
confirm the mathematical analysis; Jenkins had reached a dead end.

The breakthrough came in two stages: engineer first, followed by architect. Perhaps,
thought Arup, the shells could be built up on triangulated ribs, perhaps the surfaces too
could be prefabricated in sections. He put Povl Ahm on the job. They came up with a
simpler approach, 'more holistic than analytical, more empirical than theoretical'.[31] The
shells were ceasing to be true shells any more. Utzon at first demurred. It took several
trips to Denmark by Arup and Ahm to woo him away from his clean surfaces. Eventually
he began to play with the ribbed scheme. The shell surfaces posed formidable complexities of standardization and fabrication. These Utzon, not Arups, solved. 'Mr Arup has
been here with us this weekend, we have found a very ingenious and marvellous way of
producing these shells, and they are, finally, as we want them', exulted Utzon in
September 1961.[32] Segments were now to be taken from common spheres, sliced and
translated into a grid of ceramic tiles over concrete ribs.

It was a brilliant but by no means simple solution, found by an architect of first-class
geometrical attainments. Somehow the engineers had missed it. But it had arisen out of
a courageous reappraisal started in London.[33] The solution of the shell problem marked
the high point in the intimate partnership between Arups and Utzon (ill. 354). Later
came simplified propaganda in the architect's favour. In these imagery played its part.
The earliest, a rhapsodic essay by Sigfried Giedion ('a new chapter of *Space, Time and
Architecture*') spoke of three wooden globes sent to him by Utzon, 'from which he had
sliced the different segments of his vaults' and showed pictures of graceful hands laid on
a shallow model from which the segments were extracted (ill. 355).[34] Architecture likes to
portray the visible or organic image as the jumping-off point for creativity, with the technologists moving in dutifully behind to make things work. Things are seldom just like
that.

The solution had the clarity and consistency that Utzon sought, and Arup swung behind it. The jettisoning of four years of work precipitated a damaging row in his firm. Arup now appointed Jack Zunz to take the project forward, along with his friend and fellow-South African, Mick Lewis on site (ill. 353). Utzon and Zunz explained the spherical principle on Australian television in 1962. Yuzo Mikami, who had been Utzon's chief assistant up to that point and had some background in engineering, joined Arups to help work up the shells.

The crisis over the shells had ended in a victory for Arup's vision of an art-architecture supported by engineers, not without cost to himself. A serious illness in 1962 caused him to lose some of his personal grip on the project.[35] Better relations and faster progress ought to have followed after Arups opened an office in Sydney under Lewis, and Utzon finally moved out there in 1963. But from that point relations deteriorated until Utzon, in symbolic reproof to Arup's ideals, bricked up the door between the architect's and engineer's site offices.[36]

No explanation of why that happened can separate personalities from technique and management. Over time, Arups' Sydney staff developed a less forbearing frame of mind towards Utzon and his retentive way of working than their London counterparts. The design, fabrication and tiling of the shells themselves (which saw the début of the architect-friendly Peter Rice as Arups' site-engineer) made good progress (ill. 356). Other items led to antagonisms: the timber ceilings, the fenestration and the unresolved planning and acoustics of the halls. By the time the New South Wales Government changed in 1965 and started pressing Utzon for completion, interrelated problems were piling up. From then events ran on until his resignation.

The proximate cause of that resignation early in 1966 was confrontation with a Minister of Public Works less insistent on the building's integrity than on getting it finished, and declining to pay Utzon's fees until evidence of faster progress could be shown. But matters were never so simple as art versus philistinism, good architecture versus the crude deadline.[37] At the time Utzon walked out, a row was going on over the ceilings. Rival solutions having been proposed, Arup was again asked to choose. This time he backed his own staff against the architect.

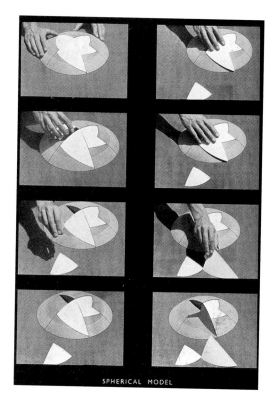

SPHERICAL MODEL

355. Segments of a sphere giving the concept of the adopted design for the Sydney shells. From an article by Sigfried Giedion in *Zodiac*, 1965.

> I have not seen the drawings of our scheme except for the cross-section shown in our report, but as far as I understand it, it gives exactly the same outward appearance as your scheme. But it weighs much less and can be built and costs less. So what is so frightfully wrong? You say that the situation is similar to that which occurred when we changed the shells to a ribbed construction without louvre walls. But that is not so. Then I disagreed with some of my people about the feasibility of building the scheme as you wanted it – with inside ribs and without louvre walls. I thought it could be done, and I also preferred this solution from an architectural point of view. So, in spite of the disruption it caused, I supported you against some of my own people. But this time there is no difference in the architecture, as far as I know, and I have absolute faith in Mick [Lewis] and Co.[38]

In reply, Utzon accused Arups of bad faith and sharp practice, charges redoubled after his resignation. He also picked up a gaffe ('no difference in the architecture') from Arup's letter:

> You have in other stages to a great extent destroyed the architect's position . . . by your whole attitude of dividing structure and architecture and minimising the importance of the architect in this work.[39]

356. Construction of a 'shell', Sydney Opera House, showing the latticed ribs before the addition of the ceramic cladding.

Here was a point of principle. One plank of Arup's philosophy was that architects liked to display structure without knowing much about it – sometimes as a substitute for understanding it. That he found vain and rhetorical.[40] His hope always was that a collaborative handling of structure could liberate architecture to do other things. But if architects loosened their grip upon structure, they might well decline into decorators of what engineers built for them. That was the instinctive fear of Utzon – 'very, very much an Architect'.

Behind this bickering lay the issue of control. As the fitting-out and finishing of the opera house came to the fore and time-pressures piled on, the status of the structural engineers as project managers to the New South Wales Government had become harder for Utzon to bear. The operational arrangements gave them the right to question the architect's consultants and veto his choice of subcontractors. In the interests of progress they were wise to do so, for Utzon was a self-confessed perfectionist, with scant heed for budgets or timetables: 'not a single pipe, dimension or type of material has up till now not been defined completely by me', he trumpeted in 1965.[41] Artistic integrity and reasonableness seldom coincide.

Utzon demanded absolute control as a precondition for his reinstatement. That was never going to happen. It was left for a team of Australian architects under Peter Hall to sort out the brief for the interior, and finish off the building as best they could in partnership with Arups. The task was thankless. They found the designs further backward than anticipated. The Arups team had chosen to conceal that from their clients. Some thought that the engineers too should have resigned. Ove Arup's response was robust: 'It would have been quite wrong to resign and then publicly support Utzon in a matter where right was not on his side.'[42] As for the replacement architects, they would be judged by Utzon's standards, while their very ability to finish the building betrayed their inability to meet them. In the event the biggest cost overruns took place in the years after the resignation. That reveals less about either Utzon or his successors than about the poor procurement arrangements set up from the start.

Costs, delays and obstacles lose their emotional charge, as the making of any great enterprise recedes. Engineers tend to focus more than architects upon construction costs. For his part, Ove Arup never worried unduly about money, either for himself or in building. Years afterwards, Jack Zunz reflected that the total cost of the opera house was about the same as that of a new frigate for the Australian navy. Yet he could still not forgive Utzon:

> He left the job, he let down many of his friends and closest allies, he split not only the architectural profession, but also the community as a whole, and he left the project in chaos.[43]

That animosity had its roots in the lightning recoil with which Utzon had turned upon a firm that felt it had gone out of its way to support him. When Arup's ideal of intimacy went wrong, it left a sense of betrayal. Maybe it also rankled with Zunz that the 'inventive' engineers were fast being forgotten in favour of the 'creative' architect. The eye has decreed that Utzon, not Arup, has the ear of posterity. That was predicted back in 1968:

> In twenty years' time the names of all the honest, able and in some cases brilliant men who worked on the building will have been forgotten and Utzon's name alone will remain.[44]

As to the psychology of why he so captiously resigned, leaving the project of a lifetime in tatters, Peter Murray has hazarded that Utzon felt he had solved all the problems of design, and was never much interested in the actual building.[45] For many architects design is all, construction only an aftermath.

THE POMPIDOU CENTRE

Sydney did Arups no harm. Its international profile brought new jobs and marked a milestone in the firm's great growth. When Ove Arup died in 1988, it had 3,500 staff with offices in forty countries.[46] Could his passion for a shrewdly constructed art-architecture maintain itself and pass to younger engineers? Somehow it did so. Management, social change, fashion, technique and personal charisma all had a hand in the transformation.

An early token of that propagation was the forming of an architectural team within Arups. Back in 1953 Philip Dowson, a Cambridge-trained engineer who had switched over to architecture, joined the firm. A mixed 'building group' of architects, engineers and quantity surveyors soon emerged. Starting with industrial buildings (ill. 357), they moved on to smarter commissions – largely university laboratories and concert halls. In 1963 they became defined as Arup Associates, behaving and trading like a stand-alone

357. Horizon Factory, Nottingham.
Arup Associates, architects, for John
Player Ltd, cigarette-makers.
Photograph of 1972.

FACING PAGE
358. Centre Pompidou. Piano and
Rogers, architects, with Ove Arup
and Partners, engineers, 1972–6.

firm of architects.[47] In their heyday Arup Associates were counted among Britain's leading architects, only better geared up than most in respect of technology and procurement. Their organization stood for much that Arup had dreamed of. Architects and
engineers worked mixed up together in groups, not in separate layers.[48] At the same time
American engineering companies, often with many offices and an international clientele,
were spawning divisions of architects with varying measures of autonomy, and a few
large architectural concerns did the reverse (pp. 394–5). But the blending of the work-
forces was seldom so complete.

Between 1966 and 1977, as its founder withdrew from day-to-day work, Arups introduced a co-partnership constitution for sharing profits and ownership. The change, led
in large measure by Peter Dunican, marked an end to the ethic of liberal paternalism.
But it had a wider symbolism too, germane to the psychology of collaboration. A residue
of class-feelings had attached to relations between the British professions, abetted by the
places and ways in which architects and engineers were taught. Architecture had
formerly been idealized as an art, not even a profession, and barely a money-making
business; engineering was often regarded as a doubtful science, perhaps just a trade. The
lingering of class-attitudes in inter-war Britain helps explain why Owen Williams preferred to work independently (pp. 261–4), and why the new style of partnership began
with immigrants. Two world wars and rampant technology whittled prejudice away but
did not wipe it out. Class-implications still bothered C. P. Snow, watching post-war
Cambridge riddled with hapless divisiveness between the arts and sciences.

In one domain Ove Arup had done his bit to bridge the gap, in the garb of practical
philosopher-priest. Then in the 1960s the barriers broke down. Priests were no longer
needed; partnerships between British architects and engineers could count on social
equality. Continuing differences between them would be rooted in their disciplines, not
assumptions of class. The exhilaration of equal standing helped father a fresh wave of
energy, not least within Arups. The best instance of the change is the making of the
Centre Pompidou in Paris (ill. 358).

The winning team for the Pompidou Centre, explains Nathan Silver in his spirited
study of the project,[49] originated in Arups – to be precise, in 'Structures 3', one of four
divisions on the firm's structural side. The formation of Arup Associates had not lessened

the Arups engineers' engagement with architecture: that was what they were famous for. But most initiatives started with architects, who would approach one of Arups' structural divisions for advice over a specific commission. Even if they consulted early, that still cast the engineers in a passive role.

In 1971 Structures 3 under Povl Ahm and Ted Happold was bristling with energy and independence. Happold was the team's animator (ill. 360). A latecomer to engineering, he had risen through the ranks at Arups, unfolding as a natural leader with an original mind and a passion for architecture. Engagements in Saudi Arabia with Frei Otto, the guru of lightweight structures (pp. 422–3), had emboldened Happold.[50] That encounter took Structures 3 off on fresh adventures, culminating in an experimental group to investigate lightweight structures. Peter Rice was the group's chief designer, Ian Liddell its expert on suspended structures and space-frames. Steel, long out of architectural favour in Europe, was making a come-back. So too were international competitions, few for a time after Sydney.

Structures 3 liked competitions because they offered the chance to experiment. Reading about the French Government's competition for a cultural centre on the Plateau Beaubourg, Happold sent off a cheque for the brief. But he had to find an architect, as the competition was open only to architects or 'teams directed by an architect'.[51] His choice fell upon Richard Rogers, whom he had encountered over an abortive project for a football stand.[52] Along with Norman Foster, Rogers had in the 1960s been a member of the short-lived Team 4 (pp. 385–6). Partial to smart technology and the updating of American steel-building traditions, the engineers knew, he would be amenable to lightweight structures and space-frames. He had just linked up with Renzo Piano, but neither had much on. If you were going in for a major competition, it made sense to find partners with time on their hands. Yet Happold had to press Rogers hard to take part, until he was persuaded by his wife Su and by Piano. None of them had any experience of procurement in France.

The entry submitted by the team in June 1971 was a true joint design between the small Piano and Rogers team and Structures 3. Much of it was invented around Happold's kitchen table. It included some socio-architectural notions, imbued with anarchic or egalitarian intent; and some outspoken structural preferences, for space-frames, steel castings

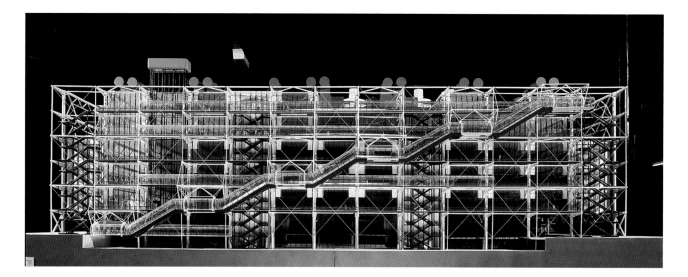

359. Centre Pompidou. Model of
the final design emphasizing the circulation system and services.

(a recent hobbyhorse of Happold's) and movable floors. 'The logic of the structure',
recalled Happold, 'was that we would build exposed steel scaffolding – like a Victorian
structure, Crystal Palace, or a petro-chemical refinery.'[53] The team was in revolt against
the ponderous, concrete-wrapped structures of the post-war years. But much of the
design lay in a limbo between the disciplines. In an architectural project that has to be
built, the engineer must ensure that it will stand up; for that, precision is needed. But in
architectural competitions, he mainly needs to provide a broad guarantee. So Happold
and Rice agreed to present only an 'attitude' towards the structure, says Silver: 'some sort
of trenchant approach to the problem that would make the jury realize, without having
full details, that the structure too was to be a clear expression of the performance of the
building'.[54] To that extent, the engineers behaved like architects. Competitions have contributed to professional convergence.

The euphoria of winning is lovingly conveyed in Silver's book.[55] A detail from the
team's official audience with President Pompidou bears repeating. The architects wore
clothes of affected casualness, but Happold had on 'his civil engineer's suit' with 'a
depressing tie to match'. That allowed Pompidou to pick him out as 'the capitalist of the
group'. Everyone laughed; but it symbolized safety. 'Thank God we are all right', someone is said to have exclaimed when the names of the winning partners were read out to
the competition jury and Arups' participation stood revealed. The Pompidou Centre was
as risky as Sydney. Once again the guarantee of established engineers helped make
unknown architects acceptable. But this time it was Arups rather than Piano and Rogers
who lost out in the sequel. The reasons are a reminder that even if there were international styles of architecture by the 1970s, there were still only national styles of getting
things built.

The first big battle concerned control of the design (ill. 359). If there had to be such a
battle, better to have it at the outset not the middle, as at Sydney. In Paris the organization, under the resolute Robert Bordaz, was tighter. At the outset, the French administrators naturally assumed that Piano and Rogers would take the drawings up to a certain
stage, and then hand them over to an independent bureau d'études, which would detail
them further and oversee the project through construction.

A short digression on the French tradition of the bureau d'études, several times mentioned in this book, may help to highlight how variations between national procurement
practices can affect buildings.[56] The term originated as the name for the design section
of a French engineer-contractor's office. Thus it was in the respective bureaux d'études
of Eiffel and Hennebique that those firms' creative work was done (pp. 164, 221). There

had long been a preference in France for the contractor to sort out the technical ramifications of a design, without a consultant engineer. The Perrets, controversially architect-contractors, also employed such a bureau (pp. 238–9). But only in the reconstruction period after 1945 did the bureau d'études system colonize French architecture proper. Reacting to the massive tasks before them, the Fourth Republic's administrators (frequently graduates of the Polytechnique) allotted the technical direction of state building largely to engineers. The method was adopted mainly to get fast housing completions in a period of acute overcrowding and homelessness. Architects would do the outline designs of housing projects and then pass them on to 'bureaux d'études techniques'. Sometimes independent, sometimes attached to building or engineering firms, these outfits would detail the designs and supervise them on site. That took much of the donkey-work off architects' backs. 'Far from being the builder's *auteur*,' quips Silver, '. . . the architect in France was the person who went home early' – and with less money in his pocket.[57] Still, some architects – Fernand Pouillon is the egregious example – saw recourse to the 'B.E.T.' as a base abandonment of responsibilities and fought to manage the great post-war housing projects themselves.[58]

The bureau d'études system had been drawn into architecture by the post-war housing crisis. Because it delivered buildings cheap and fast, the civil service came to see it as one that might be generally applicable to state building projects. But it had drawbacks for anything out of the ordinary. The system limited the input of consulting engineers at the creative stage, tied the engineering to the construction process, and encouraged buildings with standard detailing. It was wholly unsuitable for a prestigious project like the Pompidou Centre. Had that road been taken, it would have curtailed Piano and Rogers and cut out Arups almost entirely.

In the event, Bordaz and his administrators knew that their duty was to procure a great cultural monument. So the normal system, due for reform in any case, was all but waived. But it left residues and expectations. The timetable pressed; the building was wanted in 1975, which with a bureau d'études might have been feasible. Meanwhile in the interests of quality, not only Piano and Rogers but also Arups were dragging their feet. In the preliminary, six-month development of the project, the design changed radically. Here the engineers' role seemed to the French needlessly detailed and slow. When it came to signing an agreement, anxieties about cost and time were mounting. So the partners were offered a contract which, in return for full control over the design, tied them to a sliding scale of fees linked to their estimate of the future cost of the building: the more it exceeded the estimate, the less they would get. Since the project was shot through with experiment, the risks were great. The Arups management prevaricated, then refused to sign, fearing large losses; but Piano and Rogers had little to lose and signed up.

As a result Piano and Rogers inherited sole legal charge of the project from Spring 1972. The Structures 3 engineers, after overseeing the foundation contract, reverted to orthodoxy as consultants employed and paid by the architects. Having fashioned its image, the Piano and Rogers team put in so much work to make the Pompidou Centre a reality that their rights to the intellectual property of its design brook no challenge. 'We, the engineers – Ted Happold and I – were not the main stars: that was the role of the architects', wrote Peter Rice.[59] Yet to lay everything upon the latter misconstrues not just

360. Lennart Grut (left), Ted Happold (centre) and Peter Rice (right) in the office of Ove Arup and Partners, 1971.

361. Peter Rice pulling on a cable truss at the Grandes Serres, La Villette.

how the design began, but how the partnership panned out down to the building's completion. One token was an agreement that Arups would take just over half (52 per cent) of the designers' fee, should their estimates prove right. That was a measure of responsibilities, not design input. Nevertheless although disputes ensued between the two firms over fees and demarcation, a spirit of creative partnership persisted on the design side after the contractual arrangements had been made.

Happold bowed out in 1973; having 'ruffled a number of Gallic feathers' he was sent home 'desolated' by the Arups management, to take up the lightweight structures work.[60] So the onus of maintaining the collaboration in Paris fell largely upon Peter Rice (ill. 361), forming a bond at this time with Renzo Piano. After Arup himself, Rice the 'improbable engineer', as Bryan Appleyard calls him,[61] best symbolizes reconciliation between the professions. His was a special kind of balancing act. He had grown up within the Arup organization, maturing into a virtuoso with a blend of analytical skills and design instincts. Rice always insisted that he was an 'inventive' engineer, not a 'creative' architect. He did most of his major work in outwardly orthodox professional relationships.[62] Yet he behaved and talked like an architect. That turned him into the architect's ideal engineer, in perfect tune with the aim that Arup had defined – 'trying to conjure forth that mystical spiritual quality which is the essence of art'.[63]

Those words are quoted in Rice's posthumous book, *An Engineer Imagines* (1994). The primary appeal of that text is emotive. It borrows the layout, the organic imagery, the feeling for analogy and the rhetoric of the architect; and it is prefaced by the picture of a primrose. Like Arup, Rice knew the value of optimism and uplift. One essay, 'The Role of the Engineer', warns his fellow-professionals against the 'Iago mentality', against undermining 'romantic and artistic creativity' with plausible rationalization.[64] It celebrates the structural engineer as a scientific adventurer who must help his comrade-architect explore the properties of materials, innovate with them, stretch them, play with them and imbue them, says Rice elsewhere, with a 'feeling of contact and warmth between the person looking and the maker'.[65]

Here is an echo of Ruskin and the Arts and Crafts Movement. Not just Rice's work at the Pompidou Centre and after, but the whole British-based 'high-tech' alliance between architecture and engineering, can be read as the updating of an older passion for craft and detail. Structural particulars are picked out, honed and polished, pointed to and hallowed as significant; components and connectors are fetishized and lovingly pictured, from the first sketch through manufacturing, assembly and installation to the unveiling. The image of the identifiable, photogenic particular stands in for the baffling complexity of the whole.

Rice was conscious of the historical dimension. 'There is a belief', he once explained,

which comes from the past, that the building process is a craft process, a process by which you design and define something by understanding how it works. But this type of architecture, or architectural engineering, seldom happens today. This type of design is about trying to intervene in, and to control, at least a small part of the processes and procedures of building so that we may bring back a direct, tactile quality to the buildings themselves.[66]

He referred his own love of the particular to a lesson he had learnt from Utzon at Sydney – 'the importance of detail in determining scale'.[67] Yet the focus upon detail had started at a lowlier level, with the prefabricated or systems approach to architecture favoured after 1945, in which component design preceded and all but dictated building design. When Silver says of the Pompidou Centre that 'the elevation only emerged after the details were clear',[68] he might have been speaking of the prefabricated housing kits

362. Floor trusses of the Centre Pompidou in construction showing the gerberettes in place.

363. Gerberettes for the Centre Pompidou in foundry.

of Konrad Wachsmann with their totemic universal connectors, or the Hertfordshire schools of post-war austerity. Now in an image-conscious age component design acquired a sacred aura. High-tech architecture can show many instances. Most famous from Rice's own practice are the ferro-cement ceiling-louvres of the Mesnil Museum, Houston; and the archetype of them all, the cast-steel gerberettes of the Pompidou Centre.

Queer in both name and nature, the gerberettes (ills. 362, 363) are the most photographed details of the Pompidou Centre. Plump, pivoting, cantilevered castings exposed outside the building envelope, they connect the 48-metre-span main beams of

the floors with the main columns and, beyond them at the outermost edge, with vertical tension rods. They do a structurally critical job and are therefore engineers' details. They were designed, developed for manufacture and styled in the main by the Structures 3 team: by Rice, Lennart Grut and Johnny Stanton, with help from Laurie Abbott, the practical workhorse within Piano and Rogers.

Structural cast steel was an experimental material, preselected for exploration at competition stage by the engineers and architects together. Looking back to the detail of Victorian cast ironwork as an inspiration, the engineers thought of castings as one way of restoring the individualism that industrial steelwork had lost. They could 'reflect the "human hand" reasonably cheaply since the moulds could be hot wire cut from polystyrene', recollected Happold.[69] The evolution of the gerberettes has been graphically described by both Rice and Silver. The sequences of trial and error repeated the pattern of nineteenth-century iron buildings. The detail of their connections and finishes was worked out between Arups and the foundry contractors (Pont à Mousson, an affiliate of Krupps, who made the main beams). With a delicacy that might have touched Ruskin, every gerberette was even ground to a different finish. It was the same later, wrote Rice, with the ceiling louvres at the Mesnil Museum:

> Once the leaves had been cast, they were rubbed by hand with marble sand. One man wiped each piece up and down; all the rest of the men wiped their pieces from side to side. In the finished building, some of the pieces in the ceiling appear to have slight dirt marks, this is actually the change in reflectivity caused by this one man's technique.[70]

To repeat, these are engineers' details, yet they are invested with weighty architectural significance. It is easy to see why. Belying its flamboyant exterior, much of the Pompidou Centre was recessive to the point of puritanism. The architects tried to prevent themselves from 'falling in love with the details', one of them recalled.[71] Quantities of time were spent trying to reconcile their search for flexibility and indeterminacy with the needs of the different users. The building's interior ended up bland. But its guts, especially its services, were flung exuberantly outwards. That gesture put the spotlight on structure and mechanical servicing and those who provided them. Hitherto in conventional big buildings, engineers had proffered half-hidden support to architects who designed what was seen. The Pompidou Centre reversed the pattern. While the architects were busy with scene-changing, the engineers were strutting about front of house. Was it not up to them to supply refinement and focus?

That is to omit the architects' co-ordination and control of the different engineering interests, so as to make the building rigorous and consistent. Every element of the design, not least the gerberettes, bounced back and forth within the building-team. Procedures of that kind, unrecorded and unattributable, lie at the heart of true partnerships. Nevertheless engineering, both structural and mechanical, had been thrust into the limelight. That raised anew old questions about the boundary between structural imagery and structural reality.

HIGH-TECH AND ITS CRITICS

The Pompidou Centre's opening in 1976 did not start the high-tech movement in architecture, but it helped spread and validate it. 'High-tech' is a label that no one ever cared for or could define. A stab at it was made by Colin Davies: in paraphrase, an architecture of metal and glass aspiring to honest expression, keen on manufacturing styles of production, plundering other industries than construction for its imagery and technology, and promoting flexibility of use.[72] Martin Pawley was blunter. Architects, he argues, ran out of 'functional' ideas round about 1970 and began plagiarizing engineering.[73] Yet to turn structure to expressive purposes is the oldest of architectural tricks. In this fresh alloy, what were the relations between the elements?

364. Reliance Controls. Team 4, architects, Anthony Hunt, engineer, 1965–6. Braced shed bays in fore-ground, unbraced watertower behind.

Peter Rice believed that architects like Rogers accepted that the discipline provided by the engineer was the best framework in which to conduct architecture. On that reading an avuncular Arup, Felix Samuely (of whom more shortly), Ted Happold in his 'capitalist' suit, or Rice and his brave band of analysts and researchers offered stability and objectivity to the unruly creativeness of their architect-partners. But that could be turned round. Maybe it was the engineers who were being guided, their skills and instincts coaxed, challenged and sometimes distorted by their agile, articulate partners.

Cross-currents of that kind can be caught in the infancy of high-tech and the short heyday of Team 4 (1963–7), the early partnership of Norman and Wendy Foster, Richard and Su Rogers, and their friends. Arups were never Team 4's engineers. That role was filled by Tony Hunt, later Norman Foster's regular consultant.[74] Hunt had learnt the flexible frame of mind needed for modern 'architectural engineering' in the small firm of F. J. Samuely and Partners. He set up on his own about the same time as Team 4. The relations between Hunt and Team 4 exemplified the caste-busting collaborations that erupted in the Britain of the 1960s. They were all young, they were friends as well as col-leagues, they had tiny offices and concurred in casual habits and enthusiasms. Hunt had once hoped to be an architect.

In the masonry and concrete houses that were Team 4's early jobs, the engineer's main task was to sort out the architects' structural naiveté with sense and reticence. Then came the Reliance Controls factory at Swindon (1965–6). Looking back, Hunt saw this plain, sleek shed as 'the building that allowed us to do what we wanted to do'.[75] So pervasive had been post-war concrete that it was the first thing any of them – Hunt included – had designed in steel. Not only was steel logical for a single-storey electronic components factory, but it also fitted the architects' emotional preference: lightweight, open, democratic, Californian even. Farewell wetness and stodginess. 'We had found our style', recalled Rogers.[76]

The goal at Reliance Controls was to present a simple steel structure, meticulously refined. The design boiled down to the choice and assembly of a few prefabricated and available ('found') components. That entailed an overlap between architects and engineer; they were crowding over the same few items. Foster found himself developing an engineer's instinct for precision, Hunt an architect's relish for aesthetics: 'it's very difficult, now, to remember who made which decision or how'.[77]

On the other hand, the engineering aesthetic drew out the underlying values of the two disciplines. Reliance Controls flaunted dynamic but mostly superfluous diagonal bracing round the shed's perimeter, a feature insisted upon by Foster (ill. 364). On top of that, Rogers slung in a detached water tank on stilts, copied from the Smithsons' school at Hunstanton. The architects were unfazed by these departures from claims to minimalism and honesty of structure. Hunt's reaction was equivocal:

> The point that I always have to answer for now is the multiple cross-bracing, not only along the sides that would have required some diagonal support anyway, but also along the two elevations . . . that did not require it at all. I am still a little embarrassed: it is not a 'pure' structure, so the engineer in me can never be entirely satisfied. The designer in me, however, tends to agree that it makes the building look better. The real irony – and for me a far more difficult problem – was that Norman, who had used all his charm to persuade me to accept multiple cross-bracing for the building, then decided the water tower would be better without it. As a very tall, very slender portal frame this really did present some problems![78]

As so often in the past, therefore, where it suited the high-tech architects they set engineering logic aside in favour of pseudo-structural gesture. It was a practice against which Arup had always protested. It put in question the nature and equality of the partnership between architects and engineers, because it invited the latter to deviate from their principles. A gesture that makes a building more thrilling to someone who enjoys structural imagery may discomfort an engineer. Just as architects loathe the larding-on of decorative features which a lay public may enjoy, so engineers deprecate the exaggeration or faking of structural features. Different backgrounds teach the eye different ways of seeing.

Thoughts like these lay behind the subtle critique of British high-tech architecture developed by Frank Newby, the engineer under whom Tony Hunt served his apprenticeship. Since Felix Samuely's death in 1959, Newby had been the main creative figure in F. J. Samuely and Partners. That firm had long come second only to Arups among London engineers in serving and stimulating architects: indeed it was the firm that many of them preferred. Samuely and Newby were both 'engagé' about architecture, but in a quiet, backroom style; they kept their office small.

Trained like so many fine structural technicians at the Berlin-Charlottenburg TH, Felix Samuely had come to Britain in 1933. Almost his first job was to calculate the daring ramps of the Penguin Pool at London Zoo for Arup.[79] Thereafter he built up his own practice, always on lines of principle. Whereas, Arup remembered, his own approach tended to be 'How can I build this thing?', Samuely would ask, 'How can I make an elegant structure which does the job with the least material?'.[80] He was 'a bad administrator and organizer', a colleague remembered, but a fine theorist: 'really and truly a scientist'.[81] Catholic in his curiosities, Samuely made his name with welded-steel structures. Then like most engineers he plunged into concrete after the war, preferring pre-cast to in situ solutions and soon discarding shells in favour of folded slabs. For years he taught structures at the Architectural Association, intriguing his baffled pupils and building up fruitful contacts with the post-war generation of architects. It was in the first place through teaching that Samuely wielded his influence.

Newby was Samuely's right-hand man all through the 1950s. Both men are best remembered now for the precocious tensile structures they devised for exhibitions – the Skylon at the Festival of Britain, the British pavilions at the Brussels Expo and the exhibition-style Aviary at London Zoo. Those spectacularities can distract. The core of Newby's work, as the freedoms of the 1960s unfolded, came to be about supporting the bolder British architects in inhabited buildings. He believed that though their bright ideas deserved first-rate back-up, they seldom needed structural gymnastics.

> In architecture the structure is only a part of the whole; it is the architect who is creating the image and the environment. He may or may not welcome or accept ideas from the engineer. However the fine tuning and sculpting of exposed structural elements and joints are often left to the engineer and the quality of such details comes from his experience and flair.[82]

In sustaining risk-taking practices like Stirling and Gowan, he came to feel that 'what I really did was give architects confidence that they could design structures'.[83]

Arup might have said such things, but Newby went further. His doubts about structural rhetoric went back to his struggles with the American Embassy in London (1957–60), where the architect, Eero Saarinen, insisted on exposing the concrete, inside and out (ill. 365): 'this gave me my first insight into the architectural and philosophical problems this poses'.[84] He became wary of equating virtue with virtuosity, and ready to criticize the buildings of the revered Pier Luigi Nervi (whom he knew well) as 'mediocre architecture with jewels of structure within them'.[85]

It was not to be expected that high-tech would escape Newby's censure. Such architecture, he argued, made a fuss about aspects of a building that ought often to recede in favour of usefulness or art. Structure was in danger of driving out culture, he worried –

365. United States Embassy, London, in construction. Eero Saarinen, architect, F. J. Samuely and Partners, engineers, 1957–60. The 'space-frame' concrete ceiling, awkwardly interrupted by piers, was evidently modelled on Louis Kahn's Yale Art Gallery (ill. 378).

366. Hongkong and Shanghai Bank. Foster Associates, architects, Ove Arup and Partners, engineers, 1981–6.

and Newby was a highly civilized man. Moreover, even if one took high-tech on its own terms, the engineering novelties it claimed were seldom great:

> Architects just started using structure as decoration and, because it carried load, we had to deal with it. It made a lot of extra work. But if you are asking me whether high-tech architecture advanced technology in any way, I would have to say no. It made no contribution at all. There was nothing in all of it half as testing as the problems of a long span bridge or a decent sized stadium.[86]

These remarks run counter to Peter Rice's ebullience in *An Engineer Imagines* over the links between high-tech architecture, engineering, technology and production. What about the gerberettes, for instance? Newby could be warm about the exquisite detailing of Rice's work. Nor was he against exposing structure in principle or the occasional redundancy of feature for appearance's sake exemplified by the bracing at Reliance Controls. But as Samuely's successor he was painfully aware of such devices ('Why does the Eiffel

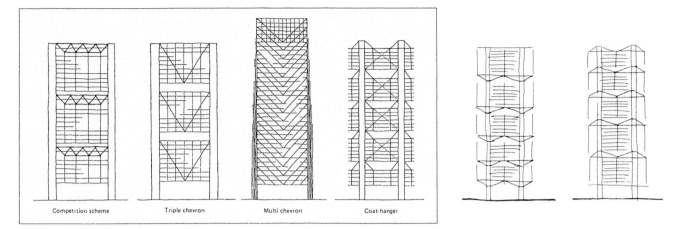

Competition scheme   Triple chevron   Multi chevron   Coat-hanger

Tower have four legs instead of three?').[87] What irked him was the pretence that high-tech architecture was rational, or that it advanced the progressive science of structures at the heart of his own profession. On architects' rights to experimentalism and even irresponsibility however, Newby was a staunch liberal. In the case of the Hongkong and Shanghai Bank, he and a colleague had fun adding extra drawings to the architects' sketches of alternatives for the famous exposed trusses, showing how a simple reversal would have been more logical yet duller (ill. 367):

> it is manipulation of the expressed structure by Foster Associates, be it conscious or unconscious, that is most intriguing. It could be argued that the design clearly illustrates the fact that the essential difference between buildings and nature is that buildings are man-made. The engineer follows natural laws in his designs but in this case the expressed structure follows anti-natural form to create excitement, a quality seen in the other prominent High-Tech buildings. In the case of the Hongkong and Shanghai Bank the architect appears to have upturned the engineer's building and in so doing hopefully heralds a new interpretation of structural form.[88]

HONG KONG

That lavish landmark, built in 1982–5 to the designs of Foster Associates with Ove Arup and Partners as their engineers, can stand here as our final example of far-flung British undertakings, created in the aftermath of empire. The subject of an in-depth book by Stephanie Williams,[89] the Hongkong and Shanghai Bank's headquarters (ills. 344, 366) reveals the architect-engineer partnership built up in the Arup mould developing yet also partly unravelling. The reasons are clear. As buildings became more complex and procurement more competitive, the need to manage and service major projects tended to disrupt intimacy between architect and structural engineer. For all Arup's idealism, there had always been many more than two parties in the construction relationship. Rivals were now barging back into the picture.

The Sydney Opera House programme had been organized around the two ingredients basic to any major building: overall design and structure, the respective provinces of architect and engineer. That was one reason why it went wrong. No one thought enough until too late about how the many other services and skills central to so intricate a project were to be commissioned, integrated and managed.

The means of co-ordinating separate skills and trades in construction have seldom been stable or uniform. Every country and century has developed its own models for procuring different building-types. But between about 1800 and 1950 a common system emerged for most large commissioned buildings in industrialized countries. A contractor hired and co-ordinated most of the trades required to carry on a project from start to finish. The architect furnished the drawings that described the design in outline. It was then

367. Hongkong and Shanghai Bank, ideas for the structural design and the façades, from *Architectural Review*, April 1986. Left, sketches by Birkin Haward junior showing the development stages of the Fosters design, culminating in the 'coathanger' scheme Above, sketches by Frank Newby illustrating 'the engineer's building (logical transfer of loads)' and 'the architect's building – upside down and irrational in engineers' terms'.

practicable for him to oversee the job on the client's behalf, because he dealt largely with the contractor.

Procurement on those lines worked so long as the technology was stable and comprehensible, and most of the building's costs went on basic construction. The system was elastic enough to allow a few consultants and specialists – decorators and electricians, for example – to perform their tasks with a measure of independence from the architect and the main contractor – as 'nominated subcontractors', to use the modern term. But once specialized techniques penetrated the core of the building, and services made appreciable inroads as a proportion of total costs, the single-contractor style of procurement came into question.

Examples from earlier chapters illustrate the point. At London's Houses of Parliament, more than a century before the Sydney Opera House, the decision to construct roofs and floors of prefabricated ironwork led to a subcontractor designing and installing whole hunks of the building, with little input from the architect (pp. 95–6). So extensive too was the heating and ventilation system that the consultant could press for radical changes to the architecture, causing confusion and overspending; the architect disagreed with him but had neither the powers nor the expertise to control him. Later, architect-contractor management was further eroded when proprietary systems of reinforced-concrete construction swept the board after 1900 (pp. 221–2). To take economic advantage of them, the architect had to tailor his building to the system and tolerate a specially licensed subcontractor on site dealing with the core of the construction, in parallel with the main builder.

The chief theme of this chapter – the consultant structural engineer's emergence in architectural projects and his rise to near-parity of status with the architect – follows that pattern of challenge to an obsolescent system. Thus it was that Arups came to be project managers at Sydney. Yet changes in structure ultimately had lesser effects on procurement than changes in the non-structural ingredients of buildings – 'services' for short.

Light, air, water and power have always shaped building form. A history of architecture might be written in terms of a struggle back and forth between the suppression or expression of services. A history of their economic role in construction would be clearer. Since the Industrial Revolution, a mounting proportion of budgets for major buildings has been spent on services and fittings – in other words, on comfort – and an ever-smaller proportion on their carcases. At the time of Sydney, architects had not yet grasped what such changes implied. That neglect was compounded by their obsession with the image of structure. So late as 1980, Norman Foster was referring his Hong Kong bankers by way of model to the Eiffel Tower and the Golden Gate Bridge – projects in scant need of services.[90]

Nevertheless architects were waking up to the clout of services. Louis Kahn's notion of 'served' and 'servant' spaces (p. 406) was one symptom; Reyner Banham's engaging book, *The Architecture of the Well-Tempered Environment* (1969), another. Early reactions tended to be rhetorical. In high-tech buildings like the Pompidou Centre and Lloyd's, service elements like ducting and lifts played unruly counterpoint with structure. Showing them off may be fun, but it does not really make them easier to maintain and replace. As mechanical engineers readily point out, most services are a nuisance if they are not tidied up and out of the way. What matters is how they work, not how they look. To make a wanton display of them is the services equivalent to Ove Arup's 'structural fallacy'. Better to find a balance whereby air, light and power blend discreetly with structure and architecture so as to improve a building's looks and performance.

While the mushrooming of services has added to the overall technical or 'engineering' content in major building projects, it has also reduced the value and status of structure. The diversification of Arups and other big engineering firms into a series of technical divisions is one token of that; acousticians, lighting experts and others now cluster around the original core skill. Instead of a single major consultant engineer, architect and client liaise with a raft of consultants and fabricators to whom the various elements are devolved. Sometimes they are chosen by nomination, sometimes by competition according to price or 'performance specification'. How to orchestrate these players in any great project is very exacting. It has led to the rise of the professional project manager – one of the bogeys of architects today. Yet under that style of procurement the architect, if he plays his cards right, can stay ahead, while the structural engineer may be reduced to one of a range of consultants and subcontractors. That is roughly what happened at the Hongkong and Shanghai Bank, where Fosters declined 'the purely passive role implied by the performance specification procedure'.[91]

When the building was mooted, Arups were an international concern with 3,000 staff in 22 countries.[92] That made it more than a hundred times the size of the Bank's architects, Foster Associates. At the heart of the firm still lay its élite of structural engineers, renowned for supporting inventive architecture. But already most of the Arups divisions carried out other tasks – in communications, transportation, geotechnics, fire safety, acoustics, quantity surveying, and so on. In a large project the divisions sometimes took on near-conflicting roles. So it was at Hong Kong.

For the Bank as for the Pompidou Centre, the engineers were on the scene before the architects. But here they were in reactive rather than initiating mode. Arups' fast-growing office in Hong Kong, established in 1976, specialized in geotechnics, since local foundation conditions were difficult.[93] One of its jobs was the new Mass Transit Railway. In this the British-born oligarchs heading the mighty Hongkong and Shanghai Banking Corporation were deeply involved. When the Bank ran a feasibility study for new headquarters in 1978–9, it was natural for Arups to be retained as structural consultants.

They had no hand in the choice of firms for the limited architectural competition that followed. For that the Bank took advice from Gordon Graham, the British architect who had guided the selection process for the Lloyd's building in London. Graham persuaded them to include Norman Foster's small outfit on the shortlist, and he served on the jury that picked the firm in October 1979. Later he came back and joined Fosters in order to push the project through when the going got hard.

When Roy Munden of the Bank did the rounds of the candidates and their buildings, he was impressed by Foster and his team but less taken by their architecture:

The Willis, Faber Dumas building [in Ipswich, finished in 1974] presents an extraordinary appearance with its tinted glass walls with no supports. In my opinion it is quite unsuitable to the town . . . The building itself has several serious defects. Blinds have had to be installed to make it inhabitable in sunlight (there were none proposed in the original design) and the air-conditioning is still inadequate. The ground floor is largely wasted and the placing of a swimming pool on that floor is crazy. The pool is hardly used.

Conclusions: If we use Foster we shall have to be tough. The benefit of his brilliance and good management could be outweighed by the lack of practicality of his designs.[94]

Up till then Tony Hunt had been the engineer for all of Foster Associates' buildings. Yet for the Bank competition he was set aside in favour of Arups. Foster and Hunt had never built high together; indeed Willis Faber Dumas was about their only effort above a single storey. In some ways the scale of the Bank was easier for Foster's small office (numbering some twenty architects when the project started) to contemplate than for Hunt's. While designers can be quite light on their feet, engineers who stand closer to the business of construction need local knowledge – about soil conditions, building regulations, labour traditions and so on. In any case, following the feasibility stage, the Bank had already decided to employ Arups Hong Kong on the final building. Prudently therefore, Foster beat a path to Jack Zunz in the same firm's London office.

Architects love competitions because they favour concept over object: images, ideas, words. Foster trounced the other competitors for the Bank (many experienced with sky-scrapers) not because of the precision of his design but because his concept was incisive: 'his written submission was judged to be of extraordinarily high quality,' says Williams.[95] At the Pompidou Centre, where the architects and engineers started out on an equal footing, the formal role of the Arups team at competition stage had been to guarantee that the ideas were practicable. So also with the Bank, where Zunz's men were subordi-nate. In the first instance they were asked whether a tall building could be built around and on top of the old banking hall, which the Bank thought to keep. Having answered in the affirmative, they needed only to ensure that Foster's concept would pass muster with the consultants in Hong Kong – notably the local branch of their own firm.

Once the job was in the bag, the game changed. Exactitude of structure and pro-curement now came into the picture. Engineers like Arups reckon to be good at things like that. Most architects profess to be good at them too. Yet often they like to keep things fluid and even contradictory as long as they can. Along with a rhetoric about industrial precision, Foster preached flexibility and practised time-consuming perfectionism. When boundaries between art and business had to be sorted out, Arups were well placed to act as mediators. The toing-and-froing over the structure now began. It wove the same intri-cate course as at Sydney and the Pompidou Centre, between architect and engineer and between London and Hong Kong. The design took shape in London, where Ken Shuttleworth on behalf of Fosters liaised with a team appointed by Zunz under Mike Glover, holed up in a 'dark, hot and badly ventilated' basement at Arups nearby.[96] Collating the Bank's requirements and sorting out the construction happened in Hong Kong, where Arups found space for a team from Fosters under Spencer de Grey.

First the collaborators had to advance from the concept sketch to a definite set of pro-posals. It was at this juncture that Foster, obsessed with expressing the structural system for the building, dangled before the Bank unserviced structures like the Eiffel Tower. More pertinent were the 'exoskeletal' skyscrapers then coming out of Chicago (see pp. 400–1), giving the team examples to latch on to and refine. Out of many trials came Fosters' favourite 'chevron' scheme. Rejected by the client because of its negative feng-shui, it would also have been impracticable for the internal frames and therefore an insincere guide to the structure as a whole. There followed the 'coathanger' scheme, worked up by Shuttleworth with Tony Fitzpatrick from Arups and approved in January 1981. In parallel went a raft of joint investigations and escapades. Some were rational, like a fact-finding mission to see what the Japanese steel industry might come up with for the Bank; others were jaunts to kindle the relationships. An outing to watch the Concorde built and the Army throwing bridges across a river at breakneck speed was laid on by the architects as 'a mind-expanding exercise', a spur to resourcefulness.[97]

Fosters stayed in command, even when the imperfect practicality, predictability and accountability feared by Munden had revealed itself. As the detailing and construction of the Bank wore on, Arups remained crucial to the enterprise's morale as well as it struc-ture, not least when another firm of engineers delivered a damning audit on progress and conduct in 1983. But they were never in charge. Hong Kong was not Sydney.

Construction arrangements for the Bank were settled in outline during 1980, while Fosters and Arups were still busy with the core of the design. The big headache was con-trol of the subcontracts, which ran from off-the-cuff arrangements with local firms to negotiated deals with nominated suppliers. Neither Fosters nor even Arups could cope with their complexity. The Bank sought to appoint a conventional project manager, who would co-ordinate everything and preside over negotiations between client and architect. The alternative, pressed by Foster, was a management contractor who would run the sub-contracts but not obstruct lines of communication between architect and client or indeed between architect and subcontractors. After divided counsels within the Bank, the sec-ond option was agreed; a consortium from two contracting firms, John Lok and Partners of Hong Kong and George Wimpey International, took on managing the multiplicity of contracts. Their position was not so far from the role assumed by Arups at Sydney. Nor were the frictions between Fosters and the Bank during construction without a resonance of old ambiguities with Utzon. While John Lok/Wimpey exercised full sway in Hong Kong, they did not always know what Fosters (or for that matter Arups) were up to in London.

Meanwhile Foster himself was nurturing his own romantic dream of procurement, whereby partnership and cutting-edge methods of industrial production would converge to create a perfect building. As regards the cladding, the dream was realized. Poised midway between structural and services engineering, cladding has grown steadily more ingenious and ambitious. Because of its visibility, the need for good liaison between cladding engineers and architects is mutually comprehended. Here Fosters linked up with Cupples of St Louis, classiest of the American cladding designers. They had already procured the tricky curtain walls for Skidmore, Owings and Merrill's exoskeletal John Hancock Center in Chicago, forerunner of the super-skyscrapers (p. 402), so they knew all about threading structural members through and past glass. For the cladding therefore, the intensive joint working between Fosters and Arups of 1980 repeated itself a year later in London and St Louis, under the buoyant Phil Bonzon for Cupples and Roy Fleetwood for Fosters (ill. 368). A similar, less heroic procedure brought together architects and the Japanese consortium supplying the fully serviced modules for air-conditioning plant and lavatories. But the industrial fantasy of an entire module that could be plugged in and then unplugged for total replacement proved as impracticable as that of completely pre-fabricated floors.

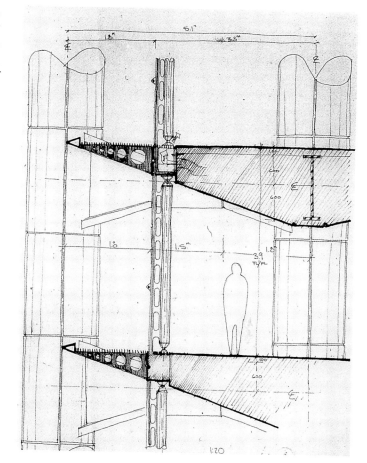

368. Sketch by Phil Bonzon of Cupples Products, showing schematic relations between structure, floors, ceilings and cladding for the Hongkong and Shanghai Bank, March 1981.

Such partnerships between architects and nominated subcontractors, echoing Lubetkin and Arup's style of working fifty years before, were now making inroads into a building world of multiple specialists. Revolutionizing the relations between architect and suppliers would, it was hoped, bring order to the ingrained chaos of construction. Where design was at the centre of relations between architect and subcontractors, that could work well. But it was often a myth when it came to fabrication and installation. There a tougher style of management had to prevail, not just on site but in manufacture as well. At the Bank, the collaborative ideal broke down over the steel supply for the main frame. Following changes in design, the costs got so far out of control that the job nearly collapsed in 1982–3. In the end well over half the total expense went on structure and cladding – a staggering proportion for a building of its day. I. M. Pei's rival Bank of China, built just afterwards on a site close to the Fosters building, included half as much floor area again and is alleged to have come in at one fifth of the cost.[98] That was not just a failure of the architects; Arups as well as Fosters were implicated. But it exposed the limits of passionate partnership.

In the end John Lok/Wimpey pulled things together. In a spirit of frenzy and exhaustion the Hongkong and Shanghai Bank got finished. The sheer spectacle of the completed building proved a triumphant turning-point in Foster's career. For Arups the results were less clear-cut. Their Hong Kong branch had been in on the job before

Fosters; their best engineers in London had contributed deeply to the Bank's creation. Zunz's men had worked with their wonted commitment to supporting the architects and bringing the best analytical methods and structural imagination to the task. Yet in the shadow of the spotlight shone on the architects, the engineers were left looking less like the near-equals they had been at Sydney or the Pompidou Centre and more like the first among many consultants and subcontractors.

Despite their size and breadth of skills, Arups today do not quite enjoy their former unchallenged prestige. That is partly because there are many other British structural engineering firms of note who build things in the creative, collaborative spirit that they originated. But it is also because in the centrifugal world of specialized procurement, other partners besides structural engineers have entered the relationship with architects.

2. *America*

SOM

Partnerships between architects and engineers have become commonplace in modern construction. They take many forms. Among the variables are the type of building in question and the partners' temperaments. To these must be added the nature of the building industry and the patterns of training in each country.

The rise to architectural notice of Ove Arup and his successors and the style of collaboration they fostered might never have happened, had it not been for the poor skills permeating Britain's construction industry, from site operations through to the casual training in 'technics' offered in its leading architectural schools. Elsewhere in Europe, a polytechnic mode of education begat a crop of graduates who permeated the national building industries. In those circumstances, the cross-over between architects and engineers could sometimes make them indistinguishable. Not that that always led to more creative partnerships: confrontation can sharpen up ideas. Broadly, there have developed two styles of collaboration between architects and engineers: an integrative model, and a dialectical one. A good place to watch both at work is the United States, where values vary so widely that different habits of education and management can co-exist.

In 1988 the sociologist Robert Gutman published his *Architectural Practice, A Critical View*. The book depicted an American profession healthier than it believed itself to be, but turning corporate. During the decade 1972–82, the number of small architectural firms (with under twenty employees) had actually increased. There were still over fifty times as many small firms as there were big ones (with over fifty employees), but that proportion was in sharp decline. Moreover their profitability and share in the overall value of construction was declining. Over the same years, the small firms had lost 7.5 per cent of total architectural receipts, while the big firms had gained by 7.8 per cent.[99]

Large firms of engineers, in the sense of engineer-contractors, have a history as old as modern infrastructure projects – canals, railways and the like. Large private firms of architects are more recent, even in the United States. McKim, Mead and White in New York were exceptional in having 89 employees in 1909 and something like an internal organization.[100] Massive architectural projects like Grand Central Terminal and the Rockefeller Center were shared out between several firms ('Associated Architects'), less on account of different specialisms than of limited organizational capacities.

Corporatism began to permeate the American professions after the Second World War, when big architectural firms came about in part through encroachment upon architecture by engineers, so as to create 'EA' or 'AE' partnerships.[101] Often the process was gradual. Large engineers and builders might keep an architect or two in work on appropriate parts of projects. By increase of staff they would turn into a division set up so that it could trade semi-independently and conform to the rules of the professional architectural bodies, which tried to keep design aloof from construction. The emergence of Arup

Associates out of the Arup partnership in Britain took the same route. Such joint concerns were few at first in number. But among American 'design firms' they earned far more by 1968 than those that practised architecture alone.[102]

The rise of Skidmore, Owings and Merrill (SOM), doyens of corporate American architecture, fits this pattern of growth and amalgamation. Yet SOM did not start out as an 'EA' or even entirely an 'AE' outfit. In a management diagram of the firm in 1957, the sundry branches of engineering – civil, structural, mechanical-electrical and 'architectural' – were relegated to 'production', halfway between 'design' and 'construction'.[103] That configuration harked back to the firm's idealistic beginnings. In 1936 the brothers-in-law Louis Skidmore and Nathaniel Owings joined forces. Unambiguously architects but with skills in construction and management (Skidmore had been in the Air Corps Construction Agency during the First World War), they shared the New-Deal dream of making architecture more useful by building in series, not just one-offs. They hoped to become 'master-builders', so Owings claimed afterwards:

> to offer a multidisciplined service competent to design and build the multiplicity of shelters needed for man's habitat . . . We were not after jobs as such. We were after leverage to influence social and environmental conditions . . . To work, we must have volume . . . Volume meant power. We would try to change men's minds.[104]

The interdisciplinary ideal was confirmed when an engineer, John Merrill, became the third partner in 1939. In that year Skidmore set up a New York office supplementing the original one in Chicago, so as to contribute to the New York World's Fair. For a firm of architects to have offices in more than one city was still a rarity; but a firm of architects it essentially was.

There followed in 1943–6 the building of Oak Ridge, the town where Uranium 235 was made for the Manhattan Project.[105] That New-Deal-style enterprise, target population 12,500, took form on virgin Tennessee territory in secrecy and at high speed. Oak Ridge helped to confer upon SOM its sense of scope, while underlining that its values were architectural. In overall command was the Manhattan Engineering District of the US Corps of Engineers, which laid out the plant and even ran the town. The town-plan had first been entrusted to a builder-engineer, but Wilbur Kelly of the Corps was appalled by the quality of the housing proposed. So he contacted Skidmore, already working with the Corps on wartime interceptor stations. Once guaranteed the job, SOM drew in engineering expertise to help them. Secrecy and haste favoured improvisation. The six architects who travelled to Oak Ridge to run site operations under John Merrill had no idea where they were to be taken. Most of the design had to be done from maps and photographs alone, while the town's sociology was based on intuition. 'We did not dare (being experts) admit to this nonscientific formula', remembered Owings.

> We clutched our briefcases and changed the subject whenever the army asked for any information. We hinted at satellite towns, talking of concentric systems of planning, ribbon-type cities as developed by foreign-named people like Hilbersheimer.[106]

A builder-engineer might have been no more scientific, nor might less alcohol have been consumed in the frenzy of construction. But the quality of the 'controlled hysteria' would have been different.

After the war SOM's centre of gravity swung back towards Chicago, the premier office and the largest one. That took time. It was SOM New York that raised the famous Lever House (1950–2), earliest of the 'Miesian' office towers to blunt the Manhattan skyline. Fresh branches sprang up, the first in San Francisco and Portland, Oregon. Their existence underscored the need for Chicago to co-ordinate. That office began with a mixed bag of commissions, including military work and housing. But after 1955 it was chiefly the Chicago-centred output that earned SOM its profile as smart architects to corporate America. About this time the firm became a haven for the 'Mieslings', a link

having been forged with Mies van der Rohe's architectural course at the Illinois Institute of Technology.

The organization of SOM was horizontal, not pyramidal; as in a medical group practice or law firm, the partners distributed the jobs. The handful of dominant architects worked in their own ways and with their own teams beneath the corporate umbrella. They shared a passion for structure and an ease with construction and co-operation, often going back to lessons learnt in the war. But their relationship with engineering and engineers, beneath or beyond the umbrella, differed appreciably. From about 1960 the status of structural engineers within the firm improved and their numbers mounted in SOM Chicago and San Francisco – though not in New York. It was because engineers were present in-house, believed Hitchcock in 1963, that SOM's work didn't 'freeze into fixed channels'.[107]

Walter Netsch was one of those ensconced in the firm before Mies's influence pervaded it.[108] A practical architect with a penchant for geometry, Netsch served during the war with the Corps of Engineers, then shifted into SOM in 1947. His earliest building of note was the US Air Force Academy at Colorado Springs (1954–64). For his austere placing of the main buildings on a wide-open site (ill. 369), he made tentative use of the mathematics of field theory – an obsession in later Netsch projects. But in the chapel added afterwards he changed tack and waxed expressionist, manipulating the fashionable geometry of tetrahedra into seventeen heaven-pointing, tubular splinters. The structural engineering was all done in-house. Over the academy itself, the SOM engineers needed only to be supportive, but for the chapel Netsch relied much on his colleague Ken Nasland. Some geometrical impulses in architecture entail a more intensive technical input than others.

Myron Goldsmith was the eminence grise within SOM Chicago. He was that rare type, an architect-engineer evenly balanced between the callings. His qualifications from Armour Institute of Technology – predecessor to the Illinois Institute of Technology – embraced both subjects. It was during Goldsmith's last year at Armour that Mies arrived to direct the architecture course (pp. 475–7); and it was to Mies's office that he returned after war service. From the Miesian aesthetic of structure Goldsmith took ideals of scale, proportion and natural analogy which he synthesized in an influential thesis of 1953 on tall buildings (ills. 371–2). Perhaps because of his ease with engineering, he seems not to

have worried about the equivocation in Mies's architecture between truth and refinement. Goldsmith spent two years in Europe, mainly with Nervi, before joining SOM in 1955. At first he worked mainly as an engineer, but in the 1960s he branched out into architectural ventures of his own. Some, like the Kitt Peak telescope and certain bridge projects, had a strong engineering component, but not all. Neither the ethos of the firm nor his private romanticism contented him with a supporting role:

> Although I worked as a structural engineer for many years, it was precisely this realization, that the sensitive problems were not only technical, that caused me to leave this field and return to architecture . . . I hope I bring to the solution of visual problems the objectivity of the structural engineer, for here is a field where one must really examine cause and effect.[109]

If Goldsmith's is the integrated face of interdisciplinary working within SOM,

371, 372. Proposals for exoskeletal structures from Myron Goldsmith's Illinois Institute of Technology thesis on tall buildings, 1953. A. 80-storey tower with concrete frame on a waterfront. B. Alternatives for 60-storey tower in steel, diagonally braced for lateral stability and without internal columns.

Gordon Bunshaft of the New York office presents the opposite pole: the pure architect drawn into creative challenge with outsiders, as occasion demands.[110] Bunshaft 'strained the theory of anonymity for us all', remembered Owings.[111] He had been trained, like Netsch, in formal ideals at MIT and admired the work of Auguste Perret, whom he met in Paris while on war-service. Perret meant concrete. But on returning to Skidmore (whom he had joined in 1937), Bunshaft saw how the land lay: 'America is largely a steel-building country'.[112] He and his colleagues therefore ran with steel. The exposed frame and curtain wall found precocious expression in Lever House, the building that made Bunshaft's name (ill. 370). Towers of that modest size required refinement from their architects, but few novelties of structure. So SOM New York, which had no division of engineers, operated in the old way: after the lines of a design had been set, the drawings were mailed to Chicago or subcontracted out within the city for engineers to work up.

That could carry on so long as architecture was premised on the familiar culture of steel. But Bunshaft always hankered for concrete. European commissions found him helping himself to that material, at the Istanbul Hilton (1951–5) and the Banque Lambert in Brussels (1959–65) – the latter an early effort at rendering corporate opulence in concrete.[113] If he were to do the same in America, his outfit had first to change its structural habits. 'We knew we could do all sorts of things with concrete,' Bunshaft remembered, 'but we needed education.'[114]

Their mentor was Paul Weidlinger, one of the handful of men born and trained in Europe who cut through the complacency of American steel engineers of the 1950s and 60s, just as Arup and Samuely had shaken up British engineering. Weidlinger had strong cultural affiliations, having studied with Moholy-Nagy and Le Corbusier.[115] Always an

373. Heinz corporate campus, Hayes, London. Skidmore, Owings and Merrill (New York), Gordon Bunshaft, project architect, with Matthews, Ryan and Simpson, 1962–5.

independent consultant, he first worked with Bunshaft on the Banque Lambert. Then in a run of concrete buildings starting with the Emhart corporation headquarters at Bloomfield (1961–3), he helped SOM New York create a fresh image. Weidlinger regarded Bunshaft as the most rational of all the architects he had worked with, but sometimes had to dissuade him from his fixation on concrete. In Bunshaft's memory, the engineer was never a front-running member of the design team: 'Sometimes Paul would say, "You can't do this", but he never made design suggestions,' he recalled.[116] The fruits of their partnership were largely corporate statements in office parks, bland in manner so as to signal discipline, dignity and wealth (ill. 373). The concrete is detailed with unfailing suavity – detoxified almost. Here the gravity of Perret is updated to flatter and soothe the American businessman.

Eventually SOM New York took in-house engineers into its bosom. By the 1980s, when the making of that firm's Worldwide Plaza for William Zeckendorf junior was portrayed in Karl Sabbagh's exemplary television series and book *Skyscraper*, the workforce

FACING PAGE
375. Chestnut-De Witt Apartments. The first of the fully framed tubes: a contrast and challenge in concrete to the familiar steel frame used by Mies for his Lake Shore Drive apartments nearby. Skidmore, Owings and Merrill (Chicago), Bruce Graham, project architect, Fazlur Khan, project engineer, 1964–6.

374. Towards the framed tube: plan of the Brunswick Building. A low-rise structure prefaces a 37-storey concrete-framed tower. Closely spaced perimeter columns are connected to the central core by waffle-slab floors without intermediate supports. Skidmore, Owings and Merrill (Chicago), Bruce Graham and Myron Goldsmith, project architects, Fazlur Khan, project engineer, 1961–5.

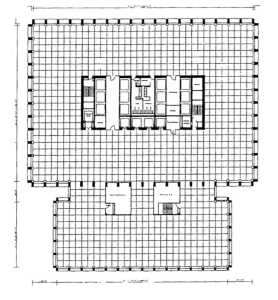

there had risen to some 400 strong, including a hefty division of engineers. They were not always an advantage, as SOM's job manager for Worldwide Plaza, Rob Schubert, explained:

> Structural engineers in the New York office are a pretty new phenomenon – we used just to go out and hire consultants. If I hired a consultant and he screwed up I could call him up and scream at him. Now I've got people who are our own employees who I don't have the same clout with.[117]

Mightiest of the collaborations within SOM was that which brought Bruce Graham and Fazlur Khan together.[118] It was Graham, the Chicago office's strong man, who did most to take the structural ideal of high-rise architecture worked out by Mies and the Mieslings and transform it for grosser buildings and slicker corporate tastes. As in the 1880s, raising the taller towers the property moguls now wanted required fresh technological imagination. That is chiefly why the engineering divisions within SOM began to grow. The Inland Steel Building, Chicago (1957–8), precursor of exoskeletal skyscrapers, symbolized changes to come. But the great leap forward came with the advent of Fazlur Khan.

A Bengali from Dacca, Khan had started out as a government engineer in East Pakistan, then won a scholarship to write a doctorate at the University of Illinois. After a stint with SOM he returned home, only to come back to the firm for good in 1960, when he began working with Myron Goldsmith. Experience both inside and outside SOM showed Khan that both architects and engineers were often content with mutual isolation. 'Any structure can be made to work with many engineers gladly willing to play with their computers and come up with the answers to hold up the building', he noted.[119] It seems to have been Goldsmith who prompted Khan to broader reflection. Graduating from a technologist's narrow passion for structures, he became inquisitive about cities and the mutual relationship of tall buildings, and started reading about aesthetics. That helped catalyse a close partnership between Graham, intuitively good with structure and committed to showing it; Khan, the brains behind the new ideas; and his number two, Srinivasa Hal Iyengar.

During the 1960s Fazlur Khan devised and developed a range of inventions for building economically to greater heights while also counteracting the effects of wind. They entailed discarding conventional frames running regularly through high buildings in favour of so-called tubes, whereby structure was concentrated on the perimeter in the form of closely spaced columns. Sometimes the columns were complemented at the perimeter by major load-bearing beams (the framed tube), sometimes they were diagonally braced (the trussed tube), sometimes they were connected to a second such system within the building (the tube in tube). Tube systems threw everything into the air. They offered better resistance to shear forces than the regular steel or concrete frame. They dramatically brought down the costs of building above thirty storeys, freed up interiors, cut back or eliminated the massive lift cores in the centre of skyscrapers, and eroded the dominance of the curtain wall. They also required radical revision of foundation technology, into which Khan plunged with the same brilliance and courage he brought to superstructures.

It is telling that Graham and Khan's earliest ventures in Chicago deployed in situ concrete, creeping into favour for high buildings in America around 1960. As concrete did not need fireproofing it could be candidly expressed at the perimeter, unlike steel which needed cladding.[120] First came the Brunswick Building (1961–5), which still had a core and was therefore not a proper tube structure (ill. 374). The

376. Fazlur Khan and Bruce Graham
with a model of the final design for
the John Hancock Center, *c.*1966.

true prototype was a block of Chicago apartments, the 43-storey Chestnut-De Witt Building (1964–6) (ill. 375). In these mid-sized skyscrapers the box-like outline of the post-war generation of towers as yet remained. But soon under the prompting of tube structures the sober rectangular profiles began to drop away.

There followed the most spectacular collaborations between Graham and Khan (ill. 376). The trussed tube made its flamboyant debut in the hundred-storey John Hancock Center (1965–70). Though there was no need to bare the frame and bracing 'exoskeletally', Graham wanted it so: 'It was as essential to us to expose the structure of this mammoth as it is to perceive the structure of the Eiffel Tower, for in Chicago, honesty of structure has become a tradition.'[121] At the even loftier Sears Tower (1972–4) clusters of tubes were so disposed as to create enough rigidity for different portions to be topped off at different heights. Predictably, Graham invoked San Gimignano. Here at any rate died the classic model of skyscraper.

The coarsening of subsequent tall buildings can hardly be laid at Khan's door. A logical theorist as well as a natural constructor, he disliked irrational novelties. David Billington believed Fazlur Khan responsible for the 'wresting of art out of almost pure structure',[122] while according to the Boston engineer William LeMessurier, 'he understood the construction process, and everything that was done had an enormous practicality, unlike the case of that Hong Kong monster'.[123] A mark of the breadth and honour of the man was that when his great work on skyscrapers was done, he quietly turned his talents elsewhere (pp. 424–6).

LOUIS KAHN

Khan or Kahn is a name that resonates in American architecture. Turning in this tour d'horizon of post-war architects' relations with engineers to the best-known of those who have borne it, Louis Kahn of Philadelphia, we encounter a pattern at the opposite pole from the corporate model. Kahn of course was a 'star'. Distinguishing the way such architects work from his own firm's collaborative culture, Richard Keating of SOM defined the star system as one in which 'architecture is delivered as artifact . . . and the subsequent reality is a byproduct'.[124] If that is so, Kahn's dealings with other building professionals can be of scant interest. Happily the truth is less clear-cut.

Louis Kahn was an architect's architect: a teacher, mystic and 'poet' whose goal was spiritual uplift through a monumental vision of beauty. A merely rational response to his oeuvre would misapprehend its point. Yet having visions is less hard than building them without losing the visionary quality. When that is ventured, poetry must come to terms with everyday life. The compromises and exasperations involved in that sometimes turn architecture into tragi-comedy. Kahn blended a religious gravity about his calling with an awareness of the absurd which, now and then, he could direct against himself. Accordingly, his relations with colleagues seesawed between inspiration, egotism, bluff, dependence and rare self-abasement.

Kahn's closest compeer in engineering was August Komendant. *18 Years with Architect Louis I. Kahn*, Komendant's memoir of their relations, sets out how it was to assist the great man: the high ideals, the skills, the charm, the gnomic utterances, the perfectionism, the crablike way of working, the second thoughts and delays, the almost wilful technical blindness, the emotional blackmail, the ultimate neglect.[125] A note of resentment is made up for by the recognition that genius is just about worth being close to. But Komendant does not tell quite the entire tale. Eighteen years back from Kahn's death in 1974 takes us only to 1956, when his career had already taken off. Moreover Komendant represented only half of what Kahn wanted from engineering. He was a brilliant technician, a constructor at the forefront of concrete technology, whom Kahn used (so he thought) solely to help him get his buildings built. To stimulate his actual designing, Kahn also needed another type of engineer to give him ideas and to crib from – to galvanize him, not to work with him.

The story starts with Kahn's lover and long-term inspirer, Anne Tyng, an assistant (on and off) in the firm of Stonorov and Kahn from 1945. Tyng, who had geometrical gifts, had spent a short time in the New York office of the German-born engineer and guru of modular systems, Konrad Wachsmann.[126] In 1949 she attended a two-day lecture at the University of Pennsylvania ('Penn') by Buckminster Fuller. The charismatic Fuller was just then tightening the grip he came to exert on schools of architecture around the world.[127] A production engineer by origin, he had progressed from low-cost housing to inventing the Dymaxion car and the geodesic dome. In time Fuller started to peddle his personal version of a technological utopia. While pouring scorn on architects' style-mongering and image-making, in architectural students he found idealists eager to hang on to his gush of words. They were also thrilled to join him in making and testing his domes, based on a simple tetrahedral geometry. Ingenious devices with many applications for shelter, Fuller's domes solved only a limited range of problems, whether in engineering or in architecture. Tyng, though captivated, saw that straight away:

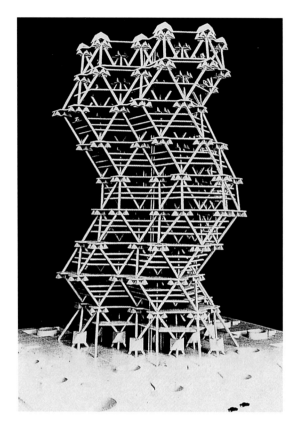

377. City Tower project: Louis Kahn and Anne Tyng, architects. Model of 1956–7, going back to a first project of 1952 when Buckminster Fuller's tetrahedral geometries were being pressed upon Kahn by Tyng. The structure was meant to be of concrete.

> His forms were purely mathematical, and there was a great gap between what he was doing and what I felt was an architectural expression. But these forms had a powerful appeal . . . I was not satisfied with the pure symmetry of the dome form, and was determined to find principles of asymmetry to extend these shapes in different ways.[128]

In qualified homage to Buckminster Fuller, Tyng went on to design first an unbuilt elementary school and then a house for her parents, who insisted against her wishes on a pitched roof. In revenge, she gave it a gawky tetrahedral frame.

The house coincided with the beginnings of a wildly speculative City Tower project by Tyng and Kahn, reliant on a vertical build-up of tetrahedra (ill. 377),[129] and with the building in 1952–3 of Kahn's first major commission, the Yale Art Gallery. On her account, it was Tyng who dissuaded him from a conventional structure for the gallery and drew him into tetrahedral and octahedral geometries.[130] Buckminster Fuller had been proposing trussed floors of this type in various studios he had taught and, as it happened, was visiting critic at Yale School of Architecture in 1952. He and Kahn spoke separate languages and didn't connect well. Yet another engineer was also hovering in the background. That was the Frenchman Robert Le Ricolais, the instigator of the modern space-frame during the run-up to the Second World War.[131] In search of opportunity Le Ricolais moved in 1951 to America, at first to the University of Illinois. There he built an agricultural hangar with a single-layer roof of tetrahedra that was perhaps the first practical space-frame. Yale students got to hear of it. By April 1953 Le Ricolais and Kahn were in touch, and soon teaching together at Penn.[132] Later, Komendant joined them to make up a formidable trio of teachers. He judged Le Ricolais 'a quiet creative man, not a promoter, therefore his important work and achievements are not widely known'.[133]

The Yale Art Gallery ceilings (ill. 378) are a classic instance of an architect taking an

378. Yale Art Gallery ceiling. Louis Kahn, architect, 1952–3. The deep pseudo-space-frame ceiling was used to unify and solemnize the gallery spaces.

engineering invention and warping it into something elegant that flouts its original purpose. Space-frames were devised to span large widths without columns; yet the spans of the gallery did not need to be very wide. They were also conceived for steel. But this was the time of the Korean War and steel shortages; in any case, Kahn had become a committed late-Corbusian – a heavy-concrete man. Excited by the depth of the open tetrahedra which gave him shadows and space for the lighting, he therefore designed a kind of pretence space-frame in concrete. His structural consultant, the sage Henry Pfisterer, Professor of Architectural Engineering at Yale, affected not to approve. It was not a proper space-frame at all, he grumbled, just a set of 'fireproofed metal' beams braced with ribs to form triangular openings; it would be costly, in reinforcing steel especially, and anyway it violated the New Haven building code. Still, Kahn charmed everyone, critically the Yale president, Whitney Griswold, and got his way. A host of helpers got sucked into the adventure of 'making theoretical ideas work': Pfisterer; a pair of young engineers, Nicholas Gianopulos and Thomas Leidigh, from the Philadelphia firm of William Gravell; Kahn's New Haven associate, Douglas Orr; and an excellent builder, C. Clark Macomber. They all did Kahn's bidding with meticulous regard for what he labelled '"architectural" concrete', but none was just a passive partner. Pfisterer actually varied the truss dimensions, so as to express the structure more economically and honestly. After this had been accepted, at the last moment Kahn standardized the sizes, partly in order to reinstate 'the ambiguity of the ceiling's directionality'. Once the queer new ceiling was complete, he could lay on the metaphors: 'It's beautiful and it serves as an electric plug . . . and a lung. It breathes.'[134]

August Komendant did not cross Kahn's path till after the Yale gallery was finished. An engineer of Estonian birth, like Kahn himself, he had imbibed advanced German concrete technique and worked on rebuilding bridges in Europe before emigrating in 1950.[135] On the strength of a textbook on concrete, he was recommended to Kahn for

379. Richards Laboratories,
University of Pennsylvania. Louis
Kahn, architect, August Komendant
with Keast and Hood, engineers,
1957–61. Photograph by Barnabas
Calder.

an abortive competition by Gravell's successors, Keast and Hood. Then Kahn took his
students from Penn to visit a state-of-the-art plant for precasting and prestressing concrete designed by Komendant. They were shown how concrete mixes should vary
according to the nature of the job, and how vital it was to vibrate and test them individually. Enthralled, Kahn delivered an ecstatic homily.[136] It was the prelude to an on-off
partnership that began with the Richards Medical Laboratories at Penn and ended only
with Kahn's death.

The Richards Laboratories (1957–61) (ill. 379) are famous for articulating a principle
which, as mentioned earlier (p. 390), had always subsisted in architecture but came to the
fore as services took up a mounting proportion of building budgets: the concept of
'served' and 'servant' spaces. Thomas Leslie has pointed to recent precedents including

SOM's Inland Steel Building, which 'displayed a very clear hierarchy among its vertically finned service tower, its outboard structure, and its clear-span office space'.[137] But after Kahn made the served and servant motto the key to his treatment at Richards, it resonated round the architectural world. He drew back from it in his later buildings, in part because the original laboratories proved hard both to build and to work in. Komendant's task was to liaise with the contractors so that they could try out his techniques of precasting and prestressing for the aboveground structure. The foundations, service towers, stairs, fumestacks were left to Keast and Hood. The technical triumph at Richards, in concept and execution alike, was the wall surface – the relation and jointing between structure, brick panel and glass. But the mechanical installation was a muddle, leaving much work to be redone: 'in Kahn's later work he did not advocate exposed mechanicals any more', says Komendant.[138] The crux of the served and servant idea was not any solution for displaying or corralling up services, but the way Kahn used it to affirm the supremacy of the architecture over the engineering, structural or mechanical. 'Served' was architecture, 'servant' was engineering; appearances made clear which mattered more. Komendant again: 'Engineering, in Kahn's mind, was servant to the architecture.'[139]

Yet like every master Kahn could not function without his servant. Anyone who relishes the former's architecture should also enjoy the latter's mordant commentary on their many joint projects after Richards. Specially fraught was the saga of the Salk Institute, La Jolla (1959–65), where the fit between engineering design, the science of how materials behave, the process of construction and ultimate architectural success was frictionally close. Rows would sometimes break out between the obdurate partners: and Komendant reconstructs for the reader a rant he directed at Kahn's arrogance and ignorance, after he had objected to the office façade at Salk and been told he knew nothing about architecture.[140] Kahn's government buildings at Dacca (started in 1962) also come under fire. Komendant felt they neither respected climatic conditions nor stood 'the test of reason', the forms chosen bearing little relation to the means of construction.[141] Dacca caused a temporary parting of ways between the stormy couple.

One of Komendant's chapters concerns the Kimbell Art Museum, Fort Worth (1967–72) (ill. 380).[142] Since the Kimbell has been much studied, we can get deeper into the professional intercourse that brought it to birth. In June 1966 Louis Kahn was chosen from various architects recommended to the Kimbell Art Foundation by its director, Richard Brown. It was agreed that he would work with a local architect, Preston M. Geren, who had been associated with an adjacent museum building and had his own engineering consultants in tow. Such arrangements are common in a country where distances are great and an architect must have a licence in each state if he is to build there under his sole recognizance. The Texans were to sort out the building permissions, do the working drawings, advise on servicing consultants, look after the tenders and supervise construction.

That left Kahn's small office free to focus on the design. He soon came up with the 'parti' of a series of parallel concrete vaults, gashed open at their crown for toplighting with a reflector underneath. The idea was simple; as always with Kahn, its quality would depend on the light and the detail. Following his usual meditations and second thoughts, it took the best part of a year before a detailed model got made. It revealed an angular, folded-plate roof and proved too big and expensive. A scaled-down design with shallow, curved shells followed: and in the late summer of 1967 a former assistant of Kahn's, Marshall Meyers, rejoined the practice to work it up.

At that point memories diverge. According to Komendant, Kahn had imagined the concrete shells of the vaults in his preliminary design as old-fashioned arches, when really they would be beams; he had 'misunderstood the carrying action of a shell . . . the roof design was dishonest and the elegance of a shell system was entirely missing.'[143] That affected not just their means of construction but their very shape. Komendant

380. Kimbell Art Museum, exterior.
Louis Kahn, architect, August
Komendant, engineer, 1967–72.

claimed that it was he who substituted for segmental vaults the rare cycloid profile which is now reckoned one of the triumphs of the Kimbell. It conveys (so one of its admirers feels) 'metaphorical movement . . . a movement of tremendous logical depth because it is so purely mathematical'.[144]

From the architects' side, Meyers is circumstantial: it was he who came up with the cycloid curve, in response not to technicalities but to Richard Brown's apprehension that the original roofs were too lofty and would make the building look too monumental. Meyers tells us that he pulled off his shelves a book by Fred Angerer, *Surface Structures in Building*, and found in it various cross-sections for barrel shells, including the cycloid. He liked it and asked their local engineer if it could be built. When he added that there was to be a slot at the crown of the shell, Meyers was advised that the analysis would be difficult, and to 'Call Gus':

> Komendant greeted us, puffing on his perpetual pipe, and looked at our drawings. Then he opened a large German book, pointed to a drawing of a cycloid and, in his abrupt manner of speaking just said 'yup' (translated 'yes'). And he confirmed that the shell could be only four inches thick (we had intuitively drawn it that way) and he added upright curbs on either side of the skylight opening. He also added what he called a 'diaphragm' of concrete at the open end of the cycloid, two feet thick and a constant one foot deep which followed the contour of cycloid.[145]

Here we witness architectural instinct mating with engineering technique at a critical creative moment. It was Komendant's pride, not so much in getting Kahn's buildings built as in having made unacknowledged contributions to their looks, that made him prickly and perhaps led him to exaggerate his initiative at the Kimbell. Kahn himself was marginal to the change. But he could see it was an improvement and accepted it right

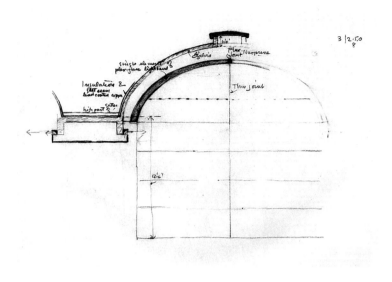

381. Kimbell Art Museum, sectional drawing by Louis Kahn.

382. Kimbell Art Museum, interior of a gallery, showing reflectors under central opening in cycloid vault, and open strip of daylight above end wall.

away. His version of events was typically foggy: 'Komendant actually made the design of our sense of the vault. We were the instigator of it, we wanted the cycloid right away, and we gave him the cycloid, with the problem of all the parts.'[146]

When it came to developing the ends of the shells, further interaction took place. Komendant wanted to thicken up his 'diaphragm' at the crown, where reinforcement was concentrated. That was against the architects' instincts; it broke the regularity of the cycloid profile, and they wanted thinness at the top, where there was least weight. A stand-off ensued between the two principals, eased (so Meyers recalls) by Aquavit and rival stories of their boyhoods.[147] Kahn grumbled but gave in: 'When he says no, it never becomes yes. But the engineer who always says "tell me what you want and I'll give it to you" is useless.'[148] To reflect the change, Kahn deftly reshaped the window strips on the end walls (which are there, in an architect's gesture, not for light but to prove that these walls do not support the roofs) so as to reflect the cycloid curve on the lower side and the curve of Komendant's diaphragm on the upper side. It was a simple touch; but the feature has come to be revered for its grace and candour (ills. 381, 382).

As construction loomed, relations between the architects deteriorated. Kahn's office was unpredictable and dilatory; Geren and the contractors showed signs of being unable to cope with the roofs, and at one stage appalled everyone by suggesting a flat slab instead. In the midst of things Geren died, to be succeeded by a son less biddable or indulgent of unauthorized consultants' bills. The day was saved by the commitment

of those closest to the actual making of things: by the contractors' engineer, by Geren's site superintendent, and by Komendant, who was sucked back into the process. By taking the shells through working-drawing stage, sorting out the sequencing of the roof operations and monitoring on site the first episodes of critical reinforcement and post-tensioning, Komendant believed he warded off a lawsuit. He had bargained for none of this when first casually consulted over the Kimbell. He was naturally sore to find himself omitted from draft press releases for the opening that highlighted Kahn's genius alone. He declined to attend, and had yet to visit the museum when he wrote his memoir:

> It was typical of Kahn not to give credit to any one of his associates . . . But in this respect Kahn was not an exception, it is common enough to almost all architects . . . Only very secure persons, with ability and progressive views, teach and develop their successors to one day carry on their ideas and work. Kahn was not such a person.[149]

3.   *Wide Spans, Natural Structures, Broad Horizons*

It is high time in this last section to drop the basis of separate professions in separate countries on which this book has been founded. At the topmost level of construction, at least, that gradually ceased to make sense over the course of the twentieth century. Since Ove Arup came to London in 1925, many architects and engineers have shifted between countries, under the pressure of events or for advancement's sake.[150] Since the Sydney Opera House leading individuals and firms, creative and commercial, have operated around the world almost as of right. As we saw in Hong Kong (pp. 392–3), the making of major structures has become internationalized. In that process, projects still get caught on the snags of local law, skill and labour. But a plethora of global contractors, subcontractors and managers has smoothed out many differences. Meanwhile – this has been the chief burden of the chapter – high-class architects and engineers have learnt to work intimately if quixotically together, spurring one another on in the quest for the grail of innovation.

THE SHELL

One theme that illustrates the internationalization of these processes during the twentieth century is that of buildings with wide-span internal spaces or expressive roof-coverings. Two have already been touched on: Sydney, whose uplifted 'sails' mark the peak of fashion for roofs of shell concrete; and the Pompidou Centre, whose floors span unsupported from one side of the building to the other. It may help to set them in broader context.

Concrete came first. From about 1910 the concrete arch and vault, initially in German hands, developed stirring capacities for elasticity and beauty. The pioneering monument was the Centennial Hall in Breslau (now Wroclaw) of 1911–12, a dome of 213-foot diameter designed by the architect Max Berg in tandem with the specialist concrete engineering contractors Dyckerhoff and Widmann (ill. 383). After the First World War concrete-arched or vaulted buildings ran amok, waxing internationally fashionable for churches, halls and swimming baths. A craze developed for the parabolic arch and vault. But the outstanding interwar innovations with concrete spans came from Dyckerhoff and Widmann, whose brilliant engineers Franz Dischinger and Ulrich Finsterwalder embarked in the 1920s on a research programme in partnership with the firm of Zeiss. The upshot was a growing mathematical grasp of membrane action in shells, and a repertoire of shell-concrete domes, vaults and other roofing forms, exploited under the Zeiss-Dywidag patents. They were first sensationally deployed for a series of wide-span market halls in Weimar Germany, at Frankfurt and Leipzig. Most later concrete shells

383. Breslau Centennial Hall.
Max Berg, architect, Dyckerhoff
and Widmann, engineer-contractors,
1911–12.

drew on the Zeiss-Dywidag inventions. The domes of the Brynmawr Rubber Factory
have been mentioned (p. 371). In America, the development of shells relied on Anton
Tedesko, a Dyckerhoff and Widmann employee who emigrated in 1930.[151]

The leading post-war champion of concrete shells in their pure form was the Spanish-
Mexican architect-engineer, Felix Candela.[152] Spanish architects tend to be technically
well-trained; Candela was no exception. He was acquainted with the leading Spanish
engineer of his day, Eduardo Torroja,[153] whose concrete spans of the 1930s took up from
the Zeiss-Dywidag patents. He would have gone to study with Dischinger had it not been
for the Spanish Civil War. Instead he went to Mexico, there to win his spurs from 1949
with a run of expressive shell structures, often graceful hyperbolic paraboloids or 'hypars'
of double curvature, which now came into architectural vogue for a generation. Candela
worked as an engineer-contractor, usually in a dominant relationship with a local architect. He claimed that his shells were simple to calculate, cheap to construct and therefore

generalizable. But most covered only small spans; and they were built with plentiful labour and in a clement climate (ill. 384).

Candela appealed to architects because his shells of double curvature looked spectacular and he seemed to promise a return to the mythical master-builder. Engineers were less convinced. When in 1954 he tried to argue that the mathematization of engineering and the theory of elasticity had atrophied structural thinking, he was set on by American engineers. Mario Salvadori warned that shells were trickier to calculate than Candela made out; Paul Weidlinger found his reasoning 'primitive, imprecise and oversimplified'. A third engineer pooh-poohed Candela's whole advocacy of double-curvature shells: 'It is true that in nature . . . flat surfaces are practically non-existent, but since we are not equipped to walk on the ceiling, life within a modern community of necessity requires that most living functions be exercised on horizontal planes.'[154] This last criticism emanated from another exponent of expressive spans – the pioneer of modern cable-hung structures, Fred Severud.

384. The hyperbolic paraboloid as art-form: open-air restaurant at Xochimilco. Felix Candela, architect-engineer, 1957–8.

385. Sesquicentennial arch, St Louis. Eero Saarinen, architect, Severud-Elstad-Krueger, engineers. Conceived in 1948, built after Saarinen's death in 1962–5.

FACING PAGE
386. State Fair Livestock Judging Pavilion (now Dorton Arena), Raleigh. Matthew Nowicki and William H. Deitrick, architects, Severud-Elstad-Krueger, engineers, 1952–3.

Severud and Saarinen

Severud is something of a dark horse among the pack of Germans and Scandinavians who transformed British and American structural engineering.[155] Norwegian by birth, he emigrated to the United States after graduating in 1923. By the post-war period he was in a fair way of practice. It will be simplest to deal with his activities and collaborations before reverting to his thoughts.

In 1948 Severud helped Eero Saarinen at the moment of his breakthrough, when the latter won the competition for a monument in St Louis to commemorate the Louisiana Purchase. The giant arch that dominated Saarinen's entry was cribbed fairly blatantly from a design by Adalberto Libera for the exhibition that never was, Mussolini's planned Espozione Universale Romana of 1942. It was not in the event raised until 1963–5, after Saarinen had died, summing up half a century of fashion for catenaries and parabolas (ill. 385). If the engineer's account is trustworthy, the precise line of the curve chosen for the competition came from his firm, Severud-Elstad-Krueger:

> The architects proposed a profile for the arch with suggested span and height. From this we worked out mathematically a curve which would place the pressure line as near the center of the arch as possible and at the same time give the architects an arch that was pleasing to the eye.[156]

Not that the idea of the arch was anything but Saarinen's. Doubtless it is his memory of events that is reflected by Hélène Lipstadt: 'it was aesthetics and not his [Severud's] very minimal input that dictated the change in the profile of the arch.'[157] As usual, the engineer's role in the competition was largely that of guarantor. Later came the meat of the collaboration, whereby alternative catenary formulae and profiles were set out by Severud's staff, worked up into models by Saarinen's office and put before the architect for him to choose between, like any designer-product. That is how Saarinen liked to work, selecting at the critical moment from copious variants.

Following the St Louis competition, Severud was drawn into the project for which he is now best remembered: the Dorton Arena, formerly the State Fair Livestock Judging Pavilion, at Raleigh, North Carolina (ill. 386). This remarkable structure revolutionized wide-span enclosures.[158] The concept consisted of two low-slung catenary arches set in contrary directions – like two men in a tug of war, quipped Severud[159] – intersecting at the foot, and acting as edge-beams for a cable-hung, saddle-shaped roof in between. Its

387. Raleigh State Fair Pavilion, preliminary sketch by Nowicki, from *Architectural Forum*, October 1952.

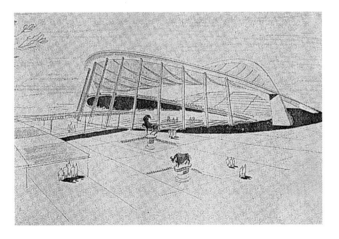

388. Ingalls Ice Rink, Yale. Eero Saarinen, architect, Severud-Elstad-Krueger, engineers, 1956–9.

first author was a brilliant young émigré architect from Poland, Matthew Nowicki, who had come to teach at the Raleigh School of Design in 1948. He cannot have taken it far before consulting Severud.[160] Within months of the design's creation in 1949 Nowicki died in a plane crash, leaving the arena to be sorted out and built in 1952–3 by Severud and the architect William H. Deitrick. The budget was always short. Modifications to the purity of Nowicki's sketch (ill. 387) ensued, by way of guyropes and additional supports beneath the arches around the perimeter, which was closed in with ugly glazing. Though the Raleigh pavilion was a breathtaking idea, it had severe 'shortcomings . . . in refinement'.[161] The acoustics were indescribably bad. As for the roof, it deflected up to 10° in high winds: 'all these additional clumsy, costly measures defeated the potential economy of the new form without wholly solving the problem of vibration and flutter.'[162] Pioneer structures can often be crude. For that kind of enclosure to progress more work would have to be done, technical and artistic.

About the time of the Raleigh design, Nowicki had been working informally with Eero Saarinen. After his death Saarinen's big, expressive roofs took up from where his friend left off. The first to be built, the Kresge Auditorium at MIT (1950–5) and the Ingalls Ice Hockey Rink at Yale (1956–9) were planned with Severud's firm. But in the later spans, both for airports, the great hall at Dulles (1958–63) and the smaller TWA terminal at Kennedy (1959–62), Saarinen's collaborators were another New York firm of engineers, Ammann and Whitney, who came to work regularly with him on other projects. At Dulles, Ammann and Whitney were first on the scene, drawing the architect in behind them.[163]

Eero Saarinen's output often has an enigmatic quality. His expressive structures are no exception. He seems to have chosen a pragmatic path with his structural engineers, midway between the managed collaboration of SOM's corporate highway and the bumpy track in high mountain air taken by Louis Kahn. Compressed into a dozen or so mete-

oric years of practice, his buildings were always meticulously conceived. In designing them, he drew on a studio craftsmanship acquired from his Finnish roots and his father's teaching. Modelling furniture out of new materials and into new profiles came naturally and joyfully to Saarinen, as it did to his friend Charles Eames. But repeated crafting in the studio did not guarantee the same refinement in buildings of scale and structural complexity.

The design impulse behind all four famous Saarinen spans is the same – that of fashioning monuments out of the box-breaking engineering techniques for wide spans then entering architectural currency. To achieve that, it hardly mattered to him whether he employed the shell dome or the cable-hung roof. The Kresge auditorium has a concrete shell roof resting by a mannered tour de force on three points. At the Ingalls rink (ill. 388) the span is suspended lengthwise, with the cables hung from a humped, reptilian rib. While that project was in progress, Saarinen helped choose Utzon's design for the Sydney Opera House. That revelation touched his two later airport terminals. Structurally they again divide between the cable-hung roof of the great hall at Dulles and the sculptured shells of the miniature TWA terminal (ills. 389, 390). Unmissable at TWA, even if it can be ignored at Ingalls, is the blatant hint at representation: the allusion to flight, the aeroplane and the American eagle. Here Saarinen ventured on to the vexed ground of symbolism.

Saarinen's spans split the critics. Henry-Russell Hitchcock adduced them to rebut those who believed that 'architects, as such, are unnecessary and that, if those who are responsible for buildings will only *costruire correttamente*,

389. TWA Terminal, Idlewild (now Kennedy) Airport. Eero Saarinen, architect, Ammann and Whitney, engineers, Grove Shepherd Wilson and Kruge, contractors, 1959–62. The working out of the building presented a massive task for the contractors.

390. TWA Terminal, interior model, with Eero Saarinen looking on.

we can have architecture without architects'. At Dulles, argued Hitchcock, Saarinen had 'beaten the engineers at their own games (needless to say with plenty of engineering advice) and yet achieved an architecture that no Nervi or Candela entirely on their own could rival'.[164] Puritans and technicians demurred. The Saarinen spans were attacked for their impurity, their cost, their showiness, their ponderous symbolism, their sheer vulgarity. Nervi himself thought the Kresge Auditorium 'invaded by extravagance, and . . . deprived of all justification'; Vincent Scully savaged the 'exhibitionism, structural pretension, [and] self-defeating urbanistic arrogance' of the Ingalls ice rink.[165] Certainly the shells were indulgent. To shift from the cardboard models of the TWA terminal to the formwork on which the concrete was poured was to move from the exhilaration of studio design to a nightmare of harrowing complexity and cost.[166] Apprehensions over formwork were among the reasons for delay at Sydney. They help explain why concrete shells fell from favour after their short spell in the architectural sunshine.

The Natural Analogy

Saarinen was not alone in his neo-symbolism. Even if Utzon denied it, the instinct to represent patently underlay the 'sails' of the Sydney Opera House. Here were symptoms of a broader urge among those who designed post-war structures of wide span to respect and reflect the natural world, and pay homage to the principles of creation. Architects and engineers were equally drawn to analogy. But they tended to experience it differently.

Fred Severud was among the first to express it. 'Turtles and Walnuts, Morning Glories and Grass' is the title of an article he wrote for the architectural press in 1945 (ill. 391) . It makes the rounds of various natural phenomena with striking structural properties – those of the title, eggs, spiders' webs, bamboo stems and so on – and urges designers to learn from them. But it concludes with a caution. Analogy should be used only as a point of departure and scientific enquiry:

> It is inconvenient to live and work on surfaces which are not flat and level, and in modern life it is often absolutely necessary to superimpose one such surface on top of another. We cannot, therefore, completely abandon right angles and straight lines in favour of Nature's curves. This fact alone would prevent any romantic attempts to slavishly mimic Nature's designs.

That, Severud later felt, was Candela's error. He goes on:

> What we can and should do is to understand that her use of curves is merely the expression of a principle of structural continuity; and this latter quality is what distinguishes her designs. We should study her principles, not attempt to copy her shapes. These are rich and bountiful, displaying everywhere the means whereby architects and engineers can achieve beautifully efficient form.[167]

The story of the 'natural analogy' in architecture is involved.[168] As we belong to the natural world, we can hardly say where compulsion ends and imitation begins. Engineering, in Tredgold's definition, is 'the art of directing the great sources of power in Nature for the use and convenience of man'.[169] Physical forces must be studied in order to be harnessed, and the technologist's job is to apply a law of nature revealed by science. But there is a spiritual side to the question as well. Since it is a human instinct to live in harmony with nature, there have always been attempts to derive architecture's principles and forms from nature. The two strands of thought can both be found in Vitruvius. They can be set against one another, if it is suggested that engineering is there to counteract or withstand nature, while architecture is there to express it. Or they can be reconciled, if engineering is construed in a kinder guise, as the beneficial direction of natural forces. Both insights are valid: man is simultaneously at peace and at war with his environment.

Both strands affected the development of modern building spans. Take the analysis of the line of forces in an arch or vault. That investigation, mathematical in the main, had been undertaken in the late seventeenth century by Hooke in England and De La Hire in France. They concluded that the ideal line of force for an arch was a catenary. Visible architectural form might have been expected to follow. But though the analysis did something for the streamlining of masonry bridges (p. 295), few explicitly catenary or parabolic arches were designed till concrete challenged masonry in the early twentieth century. Just at that time Gaudí, always keen to integrate natural and architectural form, made his famous model for the Guell Colony Church, hanging weighted sacks from a network of strings, so as to derive the catenary line of his church vaults from the inverted model and thus to construct a vault using as little material as possible. Typically for an architect, Gaudí's point of departure was a physical model rather than calculation.[170] Only in the next phase, during the era of concrete arches and vaults, did engineers develop a livelier sense of the relationships between architectural form and the forces at work in spans.

If architecture draws on the expression of inanimate physical forces, it has drawn equally on respect for the forces which define life. That is the core of what is called an organic architecture. It can be traced back to the mystical concept of 'Vitruvian man'. It by no means vanished with the advent of a rationalistic approach to building. When French engineers and architects of the Enlightenment studied efficiency in man-made structures, they compared them to the skeletons of animals (p. 297). Later, the enterprise of classifying natural history and its patterns had an incalculable impact on architectural thought. The first worked-out formulation of the 'form follows function' doctrine, for instance, came from Horatio Greenough, who argued that his compatriots should adapt the European styles of architecture to fit the American conditions in which they found themselves, just as animals and plants adapted to their context.[171] And yet most man-made artefacts went on defying the lessons of biology. How to get life – literally, life – back into buildings was the transcendent architectural puzzle of the nineteenth century. It was the great cause for Ruskin and for Sullivan alike, whose writings and buildings respectively offer the highest guides to the struggle.

Modernism gave fresh force to the biological analogy. Stimulated by the Scottish biologist Patrick Geddes and others, architects and planners became persuaded of the organic nature of cities and settlements. At the level of the building, the breakdown of the styles tempted some architects to mimic natural structures. The result could be ingenious, beautiful and naïve at the same time – as with the famous 'lilypads' used by Frank Lloyd Wright to carry the roof of his Johnson Wax Building. After the Second World War, the hunger for analogy intensified. Suddenly *On Growth and Form* (1917), an old biological textbook by D'Arcy Thompson, a colleague of Geddes, came into fashion among English-speaking designers. Only a few pages of that exhaustive work, replete with images of bones, leaves and shells, touch on man-made structures.[172] But Thompson linked beauty, natural form and functional efficiency suggestively enough for his book to be revered in British and American schools of architecture after 1945.[173] *On Growth and Form* drew the two aspects of the analogy together: on the one hand the analysis of physical forces, on the other the appeal of life-forms.

So much, cursorily, of background to the natural analogy. As the structural options opened out after 1945, most engineers remained wary of too forthright an appeal to nature. A few certainly embraced it in ways of their own. Buckminster Fuller, for instance, took the line

391. 'Turtles & Walnuts': from Fred Severud's article in *Architectural Forum*, September 1945, pressing the natural analogy on the architectural profession.

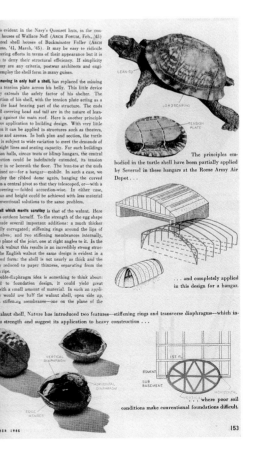

that nature's geometries were an infinite source of technological inspiration, if only they could be unleashed. He also thought that natural structures should be used as a teaching tool to overcome the fear of geometry and number – in other words that analogy was a spur to creativity for those allergic to calculation, as many architects are.[174] Since Fuller was never an orthodox structural engineer, his example might be discounted were it not for the impact of his teaching on schools of architecture. His charisma emboldened young architects to think resourcefully about the infinite scope of structures as tools for living.

Severud, to whom we now return, was more typical of adventurous practising engineers in his generation. Though he delighted in natural structures and dallied with metaphor and symbolism, he saw them mainly as jumping-off points for the designer.[175] We have seen that he was quick to point out their drawbacks. As a practitioner Severud was bold, open-minded but ultimately conventional. His limitations emerge from the unhappy history of one of his major projects and the opprobrium heaped on it by an architect who started out as his admirer.

FREI OTTO AND SUSPENDED STRUCTURES

Like the great nineteenth-century constructors, the post-war engineers who assisted architectural audacity took risks. Working on the frontiers of calculation and building technology, they could not always guarantee the result. Occasionally, as at the Congress Hall in West Berlin (1956–7), the upshot could be ignominious.

The Congress Hall (ill. 392) was a politically charged endeavour. Sited provocatively close to the border with East Berlin, it was a symbol of liberty and reconstruction, mainly paid for from American funds. The designers too were American: the Boston architect Hugh Stubbins working in tandem with Fred Severud, who used the occasion to further the cable-hung, saddle-shaped roof of wide span. So the Congress Hall promoted the latest American knowhow in design and structural technology along with the American version of freedom. But its charmless monumentality earned the so-called 'pregnant oyster' few friends among cynical Berliners, while technicians were inclined to quibble with Severud's roof structure. For the sceptics there was belated satisfaction when one of its heavy ring beams collapsed dramatically yet without casualties in 1980 (ill. 393). In the reconstruction, the roof system was changed entirely.[176]

After the Congress Hall opened, a debate about it took place in Berlin. The participants were Stubbins, Severud and a local man just starting to make a name: Frei Otto. The young architect had had high hopes. On a short visit to Severud's New York office some years before, the design of the Raleigh arena had thrilled him and helped confirm him in a resolve to make lightweight structures his life's work.

But Otto was grievously disappointed with the Congress Hall and felt no compunction in telling its architect and engineer so. Technically the saddle roof was a fraud, he felt, betraying the promise of Raleigh; much of it was supported by the auditorium walls and hidden compression beams connected with the abutments, instead of being wholly hung from the edge beams, as the eye led one to expect (ill. 394). But what Otto loathed the most was the hall's symbolic ponderousness.

> Otto: As a shape, the building dominates the skyline . . . From a distance, the simplicity of the original concept is apparent. But close up, everything becomes too heavy: arches, columns, railings, even the flag poles. It is hard for anyone to speak freely in the shadow of such overpowering forms.
>
> Stubbins: What we are talking about here is monumentality. Even though humanism is at the roots of modern architecture, man should not be deprived of the true and eternal qualities of contrast in scale. Perhaps the principal goal should be stated here: the development of an environment which would enhance deeply the processes of thought and communication which are to occur in it . . .

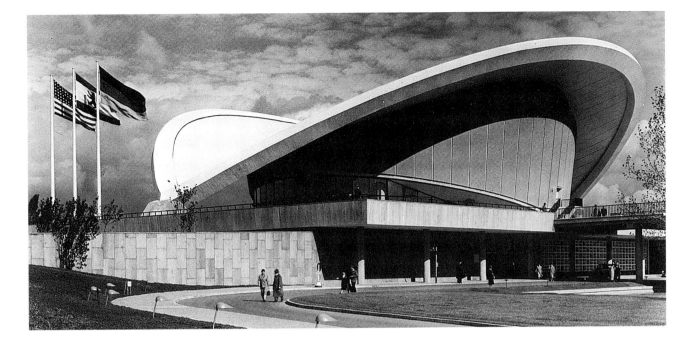

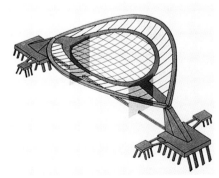

392. Kongresshalle, Berlin. Hugh Stubbins, architect, Severud-Elstad-Krueger, engineers, 1956–7.

393. The Kongresshalle after failure of a ring beam, 1980.

394. Diagram of original structure for the Kongresshalle roof.

Otto: In reality, 'free speech' is something that cannot be built but a building housing free speech can. Man often symbolizes ideas as a warning, a reminder, a lesson
. . . But when art tries to teach a lesson, it ceases to be art. Of course we need symbols. But can a suspended roof be a symbol of free speech? I believe that we are past
the time of the Statue of Liberty or the Goddess of Victory. Is there any difference
between a symbol taken from nature or a symbol taken from a new form of construction? . . . People today are full of fear. They also fear us architects – not without a certain justification. But in architecture we must be daring because we have very
important problems to solve. Shells, light framework, and suspended roofs were developed to keep the weight of a building to a minimum, to save money and energy, and
especially to increase interior space. They were developed to free us from the rigid old
building forms and to obtain flexibility for today's quickly changing circumstances. Of
these methods, the thin net especially allows buildings with a fraction of previous
efforts. If the simplest space enclosure for a certain building program is to be achieved,
only one structural solution is especially suited. The shape of the building develops
during an intensive search. The more intense this search is and the less preconceived
the architectural ideas are, the greater is the chance of finding a form of the highest
plastic quality and therefore symbolic expressiveness. I will go so far as to say that
buildings should not be designed – an architect can only be of assistance to their taking shape. A Congress Hall designed in this way would look quite different. It would
be more modest and less conspicuous. And it would be more fitting to the purpose of
this building.[177]

Behind Otto's passion for technical adventure, this torrent of thoughts betrays, lay an
idealist for whom innovations were means to higher ends. The building was not the main
point. First came human needs, natural forces.

Frei Otto (ill. 395) came from a family of sculptors.[178] He took a youthful interest in
gliders and aircraft, but had already enrolled to study architecture by the time he flew for
the Luftwaffe. As a prisoner of war repairing bridges in France, he found that well-
deployed tension members could reduce the need for scarce materials. At the Berlin-
Charlottenburg TU he then acquired a thorough technical grounding – the strength of
the German polytechnical model of architectural education. His visit to Severud's office
(made at Eero Saarinen's suggestion) occurred during a study trip to America in 1950–1
taking in buildings, structures and urban developments of all kinds.

395. Frei Otto around 2000.

396. German pavilion, Montreal
Expo, 1967. Rolf Gutbrod, architect,
Frei Otto, Peter Stromeyer, and Fritz
Leonhardt and Partners, engineers.

After graduating Otto wrote a thesis on suspended roofs, published as *Das hängende Dach* (1954). He began designing and building canopies, tents and temporary spaces, notably with the tentmaker-engineer Peter Stromeyer who helped finance his early career. Partnership and shared knowledge were instincts for Otto. In due course he was invited by the veteran engineer Fritz Leonhardt to set up the Institute of Lightweight Structures in Stuttgart. Wider fame came following his designs for the German pavilion at the Montreal Expo (1967) and his contribution to the roofs for the Munich Olympics (1972). A stream of experimental structures and collaborations has followed, in many technical categories. In most the point of departure is a roof of arresting grace, impermanent or at least anti-monumental, and in notional harmony with natural forms and forces.

Frei Otto cannot be pigeonholed; he is both architect and engineer and yet not quite either. Contractually his usual status is that of a consultant sandwiched between the two. But the design – at least of the roof – normally starts with him. On the Montreal pavilion (ill. 396), for instance, his province lay between colleagues from Stuttgart. The invitation to join in on a competition entry came from the vivacious architect Rolf Gutbrod (afterwards Otto's coadjutor on projects in Saudi Arabia), while Leonhardt calculated what could be calculated, giving the authorities the assurances they needed. But devising the cable-net structure and polyester membrane and its tireless testing and refinement, first in model form and then on site, belonged largely to Otto, Stromeyer and their assistants.[179]

At Munich (ill. 397), Otto never controlled the overall concept. The sensation of Montreal was fresh in architects' minds when the competition for the Munich stadium took place in 1967, so several bold cable-net roof designs were submitted. The winning entry, by Behnisch and Partner, was guaranteed by the respected Swiss engineer Heinz Isler. But the Olympic authorities doubted its technical feasibility for such extensive roofs. Otto came to the rescue with a solution that involved a mixture of external masts and

397. Munich Olympics, 1972, aerial view of stadium and tents. Behnisch and Partner, architects, Fritz Leonhardt and Partners, engineers, assisted and inspired by Frei Otto.

internal supports for the cable-nets. Working that out proved controversial, not least because the growing use of computers in calculating design by the engineers (Leonhardt's firm) threatened to marginalize Otto's method of physical modelling and full-scale site testing (ill. 398). He distanced himself from the result. 'We saw . . . our light-weight structures, developed as material saving architecture . . . distorted into a craze for the huge and spectacular', he grumbled. To blame was the 'euphoric enthusiasm of the architects and engineers, who were making their debut and aware of their abilities, who took the risk of working in this area that was unfamiliar to them in order . . . to outdo everything that had ever been'. Otto's hatred of monumentalism had not abated.[180]

Otto's foray with Rolf Gutbrod into Saudi Arabia led to a fast friendship with Ted Happold of Structures 3 at Arups, who had been appointed engineers for the work there. Happold, like Otto, had been entranced by Fred Severud's early cable-hung work and spent eighteen months in his office as a young man. Years of blithe interchange between Otto and Structures 3 ensued.[181]

On one occasion three of the Structures 3 team, Happold, Peter Rice and Lennart Grut, sat down together (ill. 360) to try and pin down the Frei Otto secret. The key, they agreed, was his 'integrated use of models, both as an inventive design and finally an analysis tool'. Experiments with soap bubbles typically came first, then small canvas models and lastly full structural models, sedulously load-tested and recorded. Like Buckminster Fuller's, the models were made in a mood of creative enthusiasm using cheap or free student labour. The engineers were intrigued by such painstaking, physical methods, almost quaint in an age of structural analysis by computer. Otto's models, Grut thought, were 'partly to explain to others and partly to himself . . . Just to materialize the idea he has to have a model.' Rice felt the soap bubbles helped Otto to formulate a design, while the later models were just for testing. Happold had the last word: 'I do not see how these structures could ever be achieved without his cycle of model tests. They are the only protection against the heavy hand of the engineers.' Architect, engineer and inventor were terms equally descriptive of Otto, he added:

> Unlike most architects his approach to form springs from a knowledge of structure rather than a knowledge of sculpture. In this sense he is entirely different from Utzon for example. Utzon seems to know the sculptural form he wants and then sets about, or sets others about, finding how to achieve it. Frei starts with an extensive armoury of techniques for producing new forms which have structural simplicity (like most great architect/engineers he is always striving for direct force structures) and a great curiosity to explore more forms. He would never 'force' the form to achieve an architectural effect. This is at the root of his work and his basic architectural philosophy is built up on it.[182]

If what Happold calls 'direct force structures' – letting nature take the strain – lies at the root of Otto's philosophy, it took time to mature. At first he accepted something of the positive technological mindset deep-rooted in American engineering. For his early

398. Munich Olympics, measurement model of one of the roofs. Frei Otto in centre, looking down.

engagement with inflatable structures or 'pneus' he was indebted, for instance, to the American engineer-inventor Walter Bird.[183] Then in the 1960s Otto started working with biologists as well as engineers and thinking in terms of harnessing life's structures. A fan of D'Arcy Thompson, he came to cast himself as a midwife guiding the creative processes of nature to birth – not an 'arranger' but a 'discoverer', to use his own categorization of architects.[184] Hence the status he accorded to the soap bubbles and kindred experiments. Needless to remark, the midwife was expected to be a technician, not a mystic.

In a resource-conscious world, the allure of an architecture relying on natural forces and minimal materials has given Otto exalted standing today as a master of 'natural design' as well as lightweight construction. But his invocation of nature is best regarded as an idealist's critique of normal construction conditions. Natural design, as Severud saw in 1945, is never going to transform the modern urban habitat. The uplifting, stand-alone structures of Frei Otto and his disciples solve only some of the issues engineers are charged with, and even fewer that architects must tackle. They end, as they begin, with abnormal spans and spaces.

In one country however the renaissance of the tent under Otto's inspiration had particular application and resonance. That was Saudi Arabia, where he worked many times with Gutbrod, Happold and others.[185] Sometimes what he planned was built, often not. But the finest tribute to Otto's liberating impact came in a Saudi project with which he had nothing to do. It can stand as a postscript to several themes that intersect in this chapter: to the careers not only of Frei Otto but, surprisingly also, of Fazlur Khan, inventor of the tube-framed skyscraper, and Gordon Bunshaft of SOM New York; to the intensifying internationalism of smart construction; and to the great spans that have been the topic of this section. That is the Haj Airport Terminal at Jeddah (1981–3).

After the oil crisis of 1973–4 the demand for tall buildings in America fell away for a time. So a shift took place in Fazlur Khan's career within SOM (pp. 400–2).[186] He found himself travelling more on foreign assignments, especially to the Middle East. Khan, a Bengali, had never repudiated his Islamic roots and now plunged into just those issues of culturally appropriate building that skyscraper design ignored. In the process he turned into an architect-planner as much as an engineer. Two Saudi projects absorbed his talents: a university campus, never built, on which he worked with Hassan Fathy and others; and the Haj Terminal.

The terminal had its origins in a more than tenfold rise in Muslim pilgrims travelling on the Haj to Mecca, following the opening in the 1960s of an airport at Jeddah nearby. It had become impossible to cope with the ephemeral influx. In 1974 the Saudi Government held an international competition for improving the traditional makeshift tented accommodation on the Meena plateau outside Mecca (ill. 399). Otto like many others entered, suggesting infrastructural changes and a new type of pilgrim's tent.[187] Nothing came of the idea for some years. Meanwhile SOM had been retained as consultants for the airport. With the growth in numbers a new building became inevitable, however brief its annual usage. On arrival the pilgrims needed to rest, prepare, change their clothes and take refuge from the heat. On their way home, everyone would arrive together and there could be quite a long wait for planes. All this indicated an ample, part-time terminal offering a simple style of shelter.

The Haj project began in SOM's small Washington office, whence it passed to New York in about 1976. Raul de Armas, the architect in charge, suggested a series of concrete umbrellas. This his superior Roy Allen capped with the notion of a tensile structure with membrane roofs. Architects had long enthused over Frei Otto's experiments. But until the 1970s the failures of membranes in waterproofing and fireproofing had inhibited their use in permanent buildings – never Otto's prime interest. By dint of intensive exchanges between American manufacturers and designers the problem had recently been solved.

The prospect of the Haj Terminal roof now caught Gordon Bunshaft's imagination; once involved, he secured Fazlur Khan's services. For both it was almost their last job. But Bunshaft and the New York architects at first drew complex structures with supports 450 feet apart and great sails of fabric in between. When Khan turned his Chicago engineers to the job, he saw that the design cried out for simplification. Model-making and physical analysis à la Frei Otto offered the key. So a series of fabric models shuttled between New York and Chicago. The result was a more practicable plan. It consisted of repeated open modules 150 feet square, with semi-conical translucent tent roofs on each, shaped so as to keep the air moving and designed with regard for the erection procedure and sequence. There were to be 210 units, though not all were built. It was, said de Armas, a high-tech roof for a low-tech terminal.

The profiles, materials and connectors for the Haj Terminal roof entailed to-ing and fro-ing between many parties, including Birdair Structures and Geiger-Berger, the American engineering firms which had done most to bring permanence to membrane roofs. As at the Hongkong Bank, built at the same time, procurement had become internationalized; the pylons came from Japan, the cables from France and the fabric from the United States, while the main contractor was German. Collectively, the pavilions made up much the largest roof then constructed (ill. 400). It paved the way for later membranes over airport and conference halls.

The experimentation and partnership that went into building the Haj Terminal roof apart, its connotations can hardly be ignored. Bunshaft the SOM corporate modernist's reading was pragmatic: 'it's nice to tell them that it was inspired by nomad tents. But it's just a coincidence . . . it came from empirical engineering.'[188] Fazlur Khan did not dispute that. But as an engineer who had grown into architecture, also as a Muslim, he perceived that the team had, almost by accident, created a monument. Nor, *pace* Frei

399. Tent city during the Haj.

400. Haj Terminal, Jeddah. Skidmore, Owings and Merrill (New York), architects, with Fazlur Khan of the SOM Chicago office and Geiger-Berger Associates, 1979–82.

400. Haj Terminal, Jeddah. Skidmore, Owings and Merrill (New York), architects, with Fazlur Khan of the SOM Chicago office and Geiger-Berger Associates, 1979–82.

Otto, had they been wrong to do so. And so Khan was proud to evoke the memory of the teeming faithful, encamped outside Mecca:

> When we started testing making models, the structural models, we were really seeing the infinite possibility of creating forms simply by pushing and pulling at different locations to achieve two-directional curvature. The fabric's form from an intellectual point of view could be millions of forms. As long as we created a negative-positive surface curvature, we could build it. At that point, the cultural heritage comes in – the culture and the response in terms of what is Haj itself, what ambiance and environment are; and then you visualize the millions of tents on the plains of Meena.[189]

Here it was the engineer who looked deeper. In March 1982, not long after these words were written and as the Haj Terminal neared completion, Fazlur Khan died of a heart attack in Jeddah. He was not quite 53.

No Gap to Bridge?

Almost all the post-war histories in this narrative portray a growing interdependence between high-class architects and engineers. They had become intimate with one another. But intimacy can never be assumed. To avoid things going wrong, couples must keep talking and sharing their feelings.

In 1989 the Building Arts Forum hosted a conference about the relations between the two professions, called 'Bridging the Gap'. Academics and practitioners of standing from both sides of the Atlantic converged on New York to exchange their tales of uplift or woe. Mostly it was the educators who sounded the note of frustration and division. Mario Salvadori, a veteran structural engineer from Weidlinger's office who had taught a long stint in the school of architecture at Columbia, opened the proceedings. Architects and engineers were different beasts, he proposed. Teaching maths to architects was hopeless, since most of them could only understand structures physically. Instead, they needed technicians whom they had to learn to respect and understand 'in their own physical way'. Salvadori did not think architects were stupid; it was just that the business of architecture was 'much more complex . . . than post-Newtonian physics'. David Billington, expert on Maillart and revered teacher of engineering to architects at Princeton, came next. He had long concluded that there was an art of engineering distinct from that of architecture; this he rehearsed anew. Tom Peters, the Swiss-American historian of engineering, gloomily diagnosed a 'mutual feeling of inferiority'.[190]

But when the practising architects and engineers arose, the mood lightened. Not that they agreed with one another or for that matter crossed swords; mostly they just exchanged habits and experiences. There was easy unanimity about the need for teamwork. Only when Richard Rogers claimed that architects ought to be the team-leaders even in bridge-building did a frisson arise. Architects and structural engineers might be different beasts (though Calatrava and LeMessurier, both present, had track-records in both disciplines), but that seemed to pose no problem to their working harmoniously together. Not a single practitioner diagnosed acute difficulties of communication. Obstacles lay elsewhere, in bringing service engineers into equally fruitful partnerships, or in talking to clients and the construction industry. Then there was education. Yes, there lay a problem. Maybe education was the cause of the problem, or why a problem was perceived. Kenneth Frampton pointed to Delft (he might have added Zurich) as one of the few places where architects got a long enough, good enough technical training. But nobody took that up. Training was forgotten, as architects and engineers in the higher echelons of practice basked in the sensation that there was hardly any longer a gap to bridge, or rather, no particular point in bridging it.[191]

Why education should be fraught is reserved for the last chapter. Here it need only be restated that behind the need for the dialogue and exchange chronicled in this chapter lay impersonal factors bound up with the growing ambition and complexity of building technology and design. All that might have forced the professions further apart. That that did not happen is owed most of all to two generations of cultured and broad-minded engineers, who championed partnership while yet maintaining their profession's integrity. Ove Arup is their great exemplar. Much is also due to those who dismantled or ignored professional demarcations, and treated the idea of designing and building as without borders. Frei Otto, one such, was thoroughly trained. On the other hand Buckminster Fuller and Robert Le Ricolais, the engineers who did most to excite post-war American architects about light, deployable or mass-produced types of structure, both had irregular educations which helped them to leap fences.

The final question for this chapter is whether in the contemporary alliance between architecture and engineering one or the other discipline dominates. In the public mind today, the architects undoubtedly do so. They attract more glamour, more students and more media coverage – to the frustration of the engineers.[192] One response to that has been for the engineers to reconfigure themselves as artists, and for those who write about

engineering to follow suit.[193] That tendency goes some way back. From his time with Lubetkin onward, Arup was keen to present himself as a servant of art-architecture – to get close to, participate in and promote 'that mystical quality which is the essence of art' (p. 382). Art is now more commodified than it was in Arup's day, and its allure even stronger. It enjoys a unique status in the developed world, above and beyond any profession.

Nevertheless the calling of engineers cannot be wholly identified with the making of art, as it is commonly understood. Take the relation between structure and expression in architecture. Arup thought it misguided pedantry on the part of modernist architects to insist on expressing the means of structure. But he also accepted that the further a structure departs from logic and economy, the less reasonable, objective and truly dialectical becomes the relationship between architect and engineer. That can happen when the instinct for art or expression is unqualified. When the structure of a building is distorted or hidden to comply with a designer's imaginative bidding, as in the architecture of Gehry (ill. 401), Hadid, Koolhaas or Libeskind, a resourceful engineer can find creative challenges galore to address. For Cecil Balmond, the Arups engineer who has worked closely with several of these neo-expressionists, 'the computer opens a door and gives unparalleled freedom to explore – the result is a bewildering and mind-bending free-for-all where anything goes'. Accordingly, the relationship between architect and engineer becomes about the end of 'subservience' and the ever more adventurous 'writing of new stories' – one artist going hand in hand with another, or perhaps egging the other on.[194]

Does an engineer in experimental support of the irrational find freedom, or does he forego a portion of his identity and responsibility? That question has arisen with increasing frequency since the Sydney Opera House. It is stimulating for engineering as a whole when a few engineers find absorbing enough work on the margins of architecture or the arts to rebrand themselves as artists. But it can hardly be doubted that the present art-tendency in engineering reflects a measure of professional anxiety and envy.

What about the other side to the coin? Since the Industrial Revolution architects have often feared that engineers were taking over from them. The fear resurfaces from time to time. Big, emotive public projects like stadia, transport interchanges and bridges deploying structure as their natural medium of expression, are now so widespread that it is sometimes argued that engineering is the true architecture of today. The title of a handbook about the movement for tensile and membrane structures by the artist Tony Robbin puts that claim with a dash of indefinition: *Engineering A New Architecture*. The qualification is wise. Previous chapters have shown that in some specialized or spectacular building-types – fortifications, bridges, railway stations, exhibition pavilions – engineer-constructors commonly displace or limit the role of designer-architects. That never spelt the end of architecture as an independent discipline, any more than the supremacy of the master-architect in a library like the Bibliothèque Sainte Geneviève or an art museum like the Kimbell threatened the demise of the servant-engineer.

For years now, the demand for special building-types at the ends of the constructional spectrum has been constant enough in the developed world for either profession to nurture its independence along with an undertow of rivalry. In times of war or urgency, the balance tips towards the engineer. Then come peace, plenty, leisure and luxury, and the art of architecture swells in confidence. Culture is the reward of affluence. Countries that are poor or in long-term turmoil do not boast a powerful architectural profession.

What cannot be gainsaid is that during the last luxurious half-century of peace among the rich nations of the world, the lines between the two professions have drawn closer, at least at the high end of building ambitions. But they will never meet. Engineering is too focussed, too careless of context, too indelicate, ever to be *the* new architecture. Architecture is too wanton, too irrational, too marginal, too distractable, to swallow engineering. We need a modest gap.

6

A QUESTION OF UPBRINGING

1.   School Culture 1750–1914

A RIOT AND ITS REASONS

The instant the Professor opened his mouth to deliver his inaugural lecture, a crowing of cocks, trumpeting of elephants, roaring of lions, clucking of hens, braying of asses, hinnying of horses, miaowing of cats, howling of tigers, yelping of foxes and yapping of dogs erupted in the packed amphitheatre of the Ecole des Beaux-Arts. For a time Eugène Viollet-le-Duc stood imperturbably at the rostrum. The Comte de Nieuwerkerke, superintendent of the Beaux-Arts, tried in vain to stem the student hubbub. At length he abandoned the struggle and swept out, Viollet[1] and cultural dignitaries of the Second Empire in tow. Harried all the way by jeers, out filed the crestfallen party through the Beaux-Arts courtyard (not without a recriminatory backward gesture from Nieuwerkerke), to disperse in ignominy. The date was 29 January 1864.[2]

Viollet-le-Duc (ill. 403), newly elected Professor of the History of Art and Aesthetics, was not a man to be cowed by opposition. Next week he was back on the podium. Half the audience walked out, so he was able to deliver his lecture to the two hundred or so remaining. But the recalcitrants, realizing that absence was a tactical mistake, trickled back for the rest of the series and renewed their interruptions. A scheme to make students sign in aborted. Comprehensively insulted at his seventh and final lecture on 18 March, Viollet the selfsame evening gave notice of his resignation. He was fifty now, he wrote: why preach any more to ill-behaved boys? Other duties called him, and he would be out of Paris for some days.[3] His departure was symbolic. The imperial campaign to reconstruct the Ecole des Beaux-Arts, most illustrious of Europe's institutions for teaching the visual arts, had met with a setback from which it was not to recover.

Why such antagonism to lectures of anodyne content by an architect-scholar of repute?[4] Politics holds the key. Resentment had been mounting in France against the quasi-autocracy of Napoleon III. Outright opposition being treasonable, hostility battened upon marginal aspects of the régime's policies like the arts, and upon the imperial court's favourites. One token of the mood abroad was the setting-up in 1863 of the famous Salon des Refusés, unnerving the Académie des Beaux-Arts, parent-body to the Ecole. More vociferous activity started with theatre-riots, then percolated into the lecture-hall.[5] In both venues it relied on the half-sanctioned licence and animal spirits of Paris's student population.

Nieuwerkerke and Viollet-le-Duc were natural targets of anti-imperial resentment. A sculptor of sorts, Nieuwerkerke was rumoured to be the lover of the Emperor's cousin, Princesse Mathilde, and to have picked up his title and post at the Beaux-Arts in consequence. Viollet was of sterner stuff. Forthright, methodical, intellectual, he had hewn his way to eminence outside the closed shop of Beaux-Arts education and patronage. Lately he had filtered into imperial circles through the mediation of his politic crony, Prosper Mérimée, whom some suspected as the puppeteer behind what ensued.

Viollet saw government initiative as the key to modernizing the official French system for the arts, which he abhorred. He began in 1862 by publishing four polemical articles, 'The Teaching of the Arts: Something Must be Done'.[6] From the imperial seat of

FACING PAGE

402. The customary mixture of studiousness and larking about in the Beaux-Arts atelier of Julien Guadet, *c*.1894. This was the atelier attended by Auguste and Gustave Perret.

403. Viollet-le-Duc in 1860,
photograph by Marville.

Pierrefonds, which he was restoring, he wrote next year to
Nieuwerkerke: 'a big, a very big game is afoot. Mérimée
has already draughted the report.'[7] A decree followed in
November 1863. It aimed to take control of the Ecole des
Beaux-Arts, hitherto a 'State within the State'[8] run by
professors appointed by the Académie, and make it
accountable to Government. It abolished the self-select-
ing system for election to the Académie, lowered the
maximum age at which prize-winning students from each
of the three art-disciplines – painting, sculpture and
architecture – could go to the government's finishing
school for artists in Rome, and threatened a shake-up of
teaching and competitions in the Ecole des Beaux-Arts
itself. 'The revolution has begun and like all revolutions it
can't now be stopped', Viollet exulted. 'The best thing is
to direct it.'[9]

In the protests that ensued, disparate interests combined. The academicians and professors feared for their
privileges. Teachers on the fringes of the school foresaw
a loss of income caused by a drop in the number of competitions, for which intensive training was customary.
Older students saw their chance to win the coveted years
in Rome and a guaranteed start to their career snatched
away. Formalist architects objected to Viollet's insistence
on approaching design through a rigorous study of construction. Last but not least were those who opposed the
régime on principle. In the event, the government dared
not face down this ramshackle alliance. The reforms had already been watered down by
the time Viollet suffered his public basting. So his resignation acknowledged a wider failure. It also left him diminished, resentful and averse to any further official role in the politics of French culture and architecture.

Just what did Viollet object to in the Beaux-Arts system, and what did he want in its
place? Long-term aims in education are never easy to disentangle from the circumstances
and means of its delivery and the politics of the moment. Reaction to the Beaux-Arts
reforms hinged upon personalities and interests. Luckily, Viollet was a writer of rare clarity who allows us to extricate the deeper issues. His polemics of the 1860s afford a handy
introduction to the quandaries of architectural training.

Viollet-le-Duc brought to education an architect's instincts and agenda. But the
Académie des Beaux-Arts served the visual arts as a whole. It united the duties of two of
the many academies founded under Louis XIV. The school over which the Académie
presided could be traced back to the establishment of those two academies, for painting
and sculpture in 1648, for architecture in 1671.[10] They were the first and the last of the
Sun King's foundations, the fine arts having proved easier to discipline and delimit than
those pertaining to construction.

In the form which it had taken since a major reorganization in 1819,[11] the Ecole des
Beaux-Arts was split between two constituent sections that perpetuated the old distinction between the academies – one for painting and sculpture, the other for architecture.
Lectures and competitions being all that was officially offered, the real teaching went on
in ateliers attached informally to the school. These ateliers varied in character according
to the discipline. Nevertheless, the Beaux-Arts remained a single school and academy for
the three leading visual arts. The same structure of lectures, private ateliers, competitions
and, for the top alumni in each field, promotion to Rome and a government pension,
obtained for all three disciplines. In 1863–4 most of the argument focussed on the Beaux-

404. Salle des Etudes Antiques, Ecole des Beaux-Arts.

Arts system as a whole: on its irresponsibility, narrowness and waste of talent, or its traditions, Frenchness and international prestige, according to viewpoint.

Everyone, Viollet-le-Duc included, agreed that architecture was an art, just like painting and sculpture; often it was claimed as first among the arts. It did not follow that architects at large learnt their trade in the same way as painters or sculptors. Most up to that time had acquired their skills through apprenticeship and on the job, supplemented by institutional teaching where that was available. In the 1860s that was still the case even in France, the country that furnished the most opportunities for a formal architectural education, aloof from practice. Nor did the existence of schools of architecture offer 'protection of title'. Any Frenchman could call himself an architect, irrespective of his 'formation', until 1942.

Only an elite of students applied to study architecture at the Ecole des Beaux-Arts in Paris or even attached themselves to one of its satellite ateliers (ills. 402, 404). And as is often remarked, only a fraction of that elite won through to Rome, topped the greasy pole and acquired, years later, the slow-moving but career-sustaining government commissions which had been the goal of the whole arcane system since Louis XIV. What the elect – the Labroustes, the Ducs and the Dubans – built could be superb. They created monumental exemplars of high civility. But few of these were far from Paris, or impinged upon everyday tasks of construction. The charge against the Ecole des Beaux-Arts has been repeatedly levelled since against architectural education: it was directed to the happy few, absorbed with its internal agenda, and inculcated few of the technical and commercial skills that an architect needed in practice. Back in 1820, there had been complaints that some Prix de Rome winners were 'taught so little of construction that they could hardly be entrusted with the smallest buildings'.[12] That had been addressed in the best of the ateliers during the 1840s and 50s. It was not the main thing amiss with the Beaux-Arts. The whole school was autonomous, superior, smug. This it was that irked Mérimée and Viollet-le-Duc the most and spurred them to action.

Viollet and his reforming friends had first to decide what the whole edifice of the Beaux-Arts was for. Was it a liberal, learned body (corps savant), or was it a government service (service de l'Etat)? Naturally it played at being both, the academicians deploying whichever profile seemed likelier to preserve their privileges. With national needs in mind, the reformers plumped for the latter.[13] That choice having been made by the decree of November 1863, things would be clearer, they felt.

The question of what state and country require is at the heart of Viollet's polemical articles of 1862. Painting and sculpture are straightforward, he believes. Their role is noble but circumscribed. Artists need to be an elite, whose educational structure should therefore be opposite to that for technologists or the armed services:

> The idea of running a single régime, a system of exact teaching suitable for bringing young people up to an equal level as in institutions like the Ecoles Polytechniques or Saint-Cyr is entirely plausible. A certain number of engineers or officers are *needed*; these nurseries are *required* to furnish the country with a contingent every year. But the number of painters and sculptors is not important: it is their quality which is essential. An officer of fairly mediocre ability can still render important services, since not everyone can be generals. But a feeble artist renders none; he is a useless cog. In painting and sculpture as in poetry or music, there is only one grade; you must be a general or nothing.[14]

Engineers, then, can be turned out en masse; artists should be limited to a privileged few.

But what about architects? Does the modern world need generals or middle-ranking officers – cogs in a machine? That conundrum Viollet addresses in his third article. Pinning down what skills an architect requires is problematic, he begins. Most people suppose them to be about 'a higher direction exercised over workmen who look after the details of execution, a direction which anyone can get the knack of picking up'.[15] But Philibert de l'Orme in the sixteenth century was already aware of how much an architect had to know. Since then, what with travel, archaeology, industrialization and the explosion of building-types, the field has expanded out of recognition. The Ecole des Beaux-Arts has reacted to none of this, and is still teaching architecture in the old formalized, artistic way.

In a modern economy, pursues Viollet, the state should exercise a regulatory role over architectural education, distinct from that of turning out exemplary government buildings:

> Architects are called not just to contribute to the magnificence of the State, but more often to look after private interests of major importance. However innocent his intentions, an incompetent or inexperienced architect can put his client's money at risk and bring disaster on a family. It would not be out of kilter with the duty of protection for all its citizens which the State takes upon itself up to a certain point, if it insisted that every architect should give proofs of his capacity and if it intervened directly in his teaching just as it intervenes in the teaching of engineers, sailors and so on, and if it insisted that this teaching should keep up to the mark of contemporary knowledge.[16]

As things stand, even the construction course in the Ecole is repetitive and inadequate:

405. Vignettes of the Ecole Centrale des Arts et Manufactures in 1884, the last year of its original home in the Marais, now occupied by the Musée Picasso.

Given the materials supplied by modern industry, given the courageous experiments of our engineers, who are experts rather than artists; given the need of a society which sets fresh programmes daily and does not admit the impossible, what is the construction course at France's imperial Ecole des Beaux-Arts? . . . The work of the best pupils leaving the school shows patently and every day that the school's construction course fails to meet the needs of our time.[17]

All that gave notice of this supposed mediaevalizer's modernizing stance, filled out when the first volume of Viollet's famous *Entretiens* (or *Discourses on Architecture*) appeared in 1863. Technical education was to come first in the training of architects, since most were to be useful officers for the state and reliable servants for their private clients – cogs in the machine, perhaps. The star system, as we now call it, might apply in Viollet's mind to painters and sculptors, but not to architects. No wonder backs in the Beaux-Arts had been put up.

Architectural education is seldom straightforward, however. The logic of these opinions is not quite borne out by Viollet's own teaching record. As a young man, he had taught composition of ornament in the Ecole des Arts Décoratifs, one of several schools in Paris to offer architectural topics as adjuncts to other disciplines. Then in 1857 he opened a short-lived atelier and ran a lecture course on the fringes of the Beaux-Arts, only to be blocked, so Viollet claimed, by interests within the school. That incident intensified his antipathy to the Beaux-Arts and prompted him to embark upon the *Entretiens*.[18]

In neither the Arts Décoratifs nor the atelier, nor for that matter in the *Entretiens* or his ill-starred Beaux-Arts lectures of 1864, did Viollet depart from celebrating architecture primarily as high art. As a devotee of the great Gothic churches, how could he do so? The furthest he went was to insist that architects base their designs on a rational analysis of materials and construction, ancient and modern, fit them to the task in hand and express the means of construction with vigour and candour. At the core of that analysis lay historical study. So architecture in Viollet's view was a matter not just of technique but of culture, and its relation with technology one not just of science but of selection and interpretation.

The point is reinforced if we turn to his relations with other Parisian establishments sympathetic to the view that modern society required properly equipped private architects. Take the Ecole Centrale des Arts et Manufactures (ill. 405), founded in 1829 to supply engineers and technicians for French industry. Architecture was thoroughly taught there from the 1830s under Charles-Louis Mary, and once the school secured government funding it turned out useful and sometimes eminent alumni – the centraliens – geared to the needs of the nineteenth century.[19] Viollet approved. He pointed out that the Ecole Centrale went on site visits, analysed monuments and looked at processes, which Beaux-Arts students never did. When his new flèche was hoisted on to Notre Dame, fifty centraliens asked to witness the event; no one bothered from the Beaux-Arts. At that rate the Ecole Centrale would soon become the true Ecole des Beaux-Arts, at least for architects, he predicted.[20] Yet he never had a connection with it.

After his humiliation at the Beaux-Arts, Viollet did briefly help foster a school of architecture: the Ecole Centrale d'Architecture, founded in 1865. That tiny, confusingly named establishment (soon changed to the Ecole Spéciale d'Architecture, the name it bears today) was in part an offshoot of the Arts et Manufactures; centraliens, engineers and businessmen figured among the original shareholders. The prime mover was Emile Trélat, an architect-engineer whose father had been an expert in hygiene and minister of public works in the radical government of 1848. The objectives of the school were in character: to train well-informed technicians who would help modernize the nation. Viollet was among a handful of architect-shareholders, while his disciple Anatole de Baudot and his son both served on the faculty.[21]

The connection proved short-lived. In 1868 Viollet was complaining that the new school was 'falling into the bad ways of the Ecole des Beaux-Arts, and we are paying insufficient attention to the application of technical knowledge'. Another teacher worried that 'studies verge in the particular direction of creating artists which is not what the shareholders intended'.[22] Soon the Viollet faction withdrew, leaving Trélat to struggle on. That same year saw the appearance of Viollet's fourteenth *Entretiens*, on the teaching of architecture. It repeated all his former charges against the Beaux-Arts, but in a bleaker tone tinged maybe by this fresh setback.[23] Things did not improve during his lifetime. Under the Third Republic, keen to bury the heritage of the Second Empire, the Académie des Beaux-Arts got all its old privileges back.[24]

French Schools and Their Purposes

Much comes together in the story of Viollet-le-Duc's failure to reshape French architectural education. There is the persistence and adaptability of official academic institutions in France, compared with governments. There is the combative individualism of Viollet himself, a better critic than teacher, and a subtler writer than designer. Brilliant at pinpointing faults and absurdities, he was less able to meet fellow professionals on equal terms or foster successful institutions.[25] Nor in the last analysis was his recipe for training architects precise enough to be persuasive. The ambiguities of architecture set snares for the clearest minds.

But there is a broader factor, warranting lengthier exploration. The type of education which Viollet professed to champion for architects when the Beaux-Arts controversy erupted was already in ample supply in Paris: in part through the Ecole Centrale des Arts et Manufactures, which he respected; but also through institutions of higher authority, the Ecole Polytechnique and one of the several 'écoles d'application' to which it sent its pupils, the Ecole des Ponts et Chaussées. Behind Beaux-Arts exclusivity, the failures of 1863–4 and Trélat's faltering little school loomed the spectre of a second system of education whose alumni could turn out buildings better built and not much less handsome. France's most thoroughly trained architects were engineers, and had been for a century.

Nor, if the record is impartially examined, were the Beaux-Arts school and the school of the Académie d'Architecture from which it sprang so laggardly in technical training as Viollet made out. Design is the glamorous side of architecture. It always hogs the lime-light. Just as buildings are judged on looks alone, so students of architecture often resent or repress the lessons in construction and elementary engineering which belong to their lot. Among the legion who have done so was Le Corbusier (p. 270). That does not mean the subjects are ignored or badly taught.

In the system of the Beaux-Arts and its predecessor, the ideal competition designs instituted around 1700, codified in 1717 and continued throughout the nineteenth century were the focus of architectural values and passions. Since preparing for them was the main goal of the informal ateliers to which Beaux-Arts students attached themselves, design and seductive rendering – skills not so different from what the painters learnt – dominated atelier training. But if what the school offered by way of official lectures is examined, another picture emerges.

Though much about the early years of the French academy school is obscure, we know something of its professors and their lectures.[26] Its first director and professor (1671–87), François Blondel, normally defined as an engineer and a latecomer to architecture (pp. 25–6), certainly lectured on design but spent more time on mathematical and technical topics.[27] His successor, Philippe de la Hire (1687–1718), was 'even more the mathematician and engineer', says Robin Middleton, and his lectures seem to have been largely of that ilk. Even after the competition system and Prix de Rome had got into their stride, mathematics, construction and stereotomy were steadily taught through good and bad periods for the school, balancing the cultural sides of the course. Technique and culture can never be wholly divided in architecture; and in an era when the technology of materials was restricted and skill in design rested upon law-like, proportional systems, there was much continuity between them. Nevertheless competitions, stimulants to the young imagination, always draw design away from construction. This the lecture courses endeavoured to redress.

The pairing of the academy school for architecture with its counterparts for the fine arts was a product of the Revolution. Combined with the paucity of building opportunity for architects in the turbulence between 1790 and 1815, that threw it back upon design. After the chops and changes of that period, the decisive reform of 1819 established the Ecole des Beaux-Arts as Viollet knew it. With the return of peace, it was again time to build. The architecture section was allotted four professors, one each for theory, history, construction and mathematics. So technique and culture received equal weighting. There was no official instruction on the design side – hence the growth of the ateliers.

The first professor of construction under the 1819 reform was Soufflot's old assistant from the Panthéon, Rondelet. He was blunt about the purposes of architecture:

> Architecture is not an art like painting and sculpture whose sole purpose is to please and in which the artist in carrying out his own work can abandon himself entirely to the heat of his imagination. It is a science whose essential aim is to construct solid buildings which deploy the finest of forms and the aptest of dimensions to unite all the parts necessary for their purpose.[28]

Rondelet wanted architectural students in Beaux-Arts to be proficient in mathematics and construction before they advanced to the higher classes. He instituted four competitions in construction each year, of which one always dealt with 'serrurerie', the new materials of iron and glass. After his retirement in 1825, construction was undoubtedly downgraded in the school. From about this time the results of the major competitions began regularly to be published, attracting lopsided attention and perverting the school's balance. Viollet had legitimate bones to pick with its disdain for technique away from the drawing-board and lecture hall. By the 1860s, redress was needed in the architectural section of the Beaux-Arts. But the basis of a balanced system existed, and had done since the school started under Louis XIV.

Both the idealism and the instability persistent in architectural education feed off obscurities attending the subject's aims and the public utility of its practitioners. With engineers there is less uncertainty. They are thoroughly trained because they are found useful. That, baldly, is the background to the Ecole Polytechnique and the Ecole des Ponts et Chaussées, to which we now turn.

The foundation date of the Ecole des Ponts et Chaussées is usually pinpointed as 1747, while that of the Polytechnique is fixed as 1795. The former date marks a departure point for the campaign to endow the growing body of engineers recruited into the civil side of French government service with preliminary training in the skills they would need in the field, among them architectural design. Thereafter the Bureau des Dessinateurs de Paris, the central draughting office that made maps and plans for the Corps des Ponts et Chaussées, evolved into a school under Jean-Rodolphe Perronet's guidance.

The second date, 1795, stands for the moment when under the pressure of extreme events the French state recast its system for training technicians. Surrounded by foes, the revolutionary government sorely needed middle-ranking expertise, especially for its military forces (85 per cent of Polytechnique graduates went into the armed services between 1806 and 1813[29]). It was resolved to distinguish between an establishment where recruits could acquire basic technical knowledge, the Ecole Polytechnique, and a series of higher, more specialized 'écoles d'application', of which the established Ecole des Ponts et Chaussées became one. The Polytechnique offered an introductory course (three years at first, soon reduced to two) in principles founded on classroom instruction, after which many students left. These arrangements, hammered out amidst endless altercation and instability but fixed by the start of the new century, compromised between the administrators who saw the Polytechnique as a cheap means to immediate ends, and the 'savants' who looked upon the principles of knowledge as superior ends in themselves – as academics tend to do.

Institutionally, the Ponts et Chaussées and Polytechnique schools were there to turn out servants for the various arms of state administration. Like the first of their imitators in other countries, they derived from the state services which they fed. Uniforms were worn; after 1804 when Napoleon militarized the Polytechnique, it was mostly under the Ministry of War and its students were theoretically confined to barracks.[30] All that is easily forgotten. Construction tends to be task-related, discontinuous and vagrant; a job once completed, the team moves on. In comparison a technical school is static, self-renewing, articulate. It can easily come to believe in its own autonomy, and in its ability to dictate the relation between what gets taught and what gets made.

In reality, technical schools and colleges have a habit of drifting into existence. Decrees and dates of foundation, later hallowed, regularize arrangements and formalize classes already half in being. First come exams or entry requirements, set by a body wishing to restrict or raise the standard of entry into its ranks. Classes then arise in response. From the programme of the parent body may be deduced the slant of the school.

The history of training for the two great construction corps of the Ancien Régime, the Génie and the Ponts et Chaussées, illustrates this. The Génie or military corps was the senior service (pp. 19–24). Recruitment to its ranks proceeded piecemeal up to 1692, when a new war minister, Le Peletier, insisted that Vauban scrutinize all applicants. Pressed with commitments, the latter took on an examiner – the first of a series all belonging to the Académie des Sciences.[31] In 1720 the examiner of the day was told that those wishing to enter the Génie should be able to 'elevate and draw the plan of works of fortifications and buildings, make sections and details of them, arrange estimates and breakdowns, and make plans of an area'.[32] One such candidate spent two years with a coach in Paris learning maths and draughtsmanship before being examined. Those who passed the exam went straight to the various provincial directors of the Génie, learning the rest of their trade on the job.

At that time architecture was a fully constituent part of a Génie officer's skills. Its role
diminished once a formal school was set up.
That came about through the energy of
Chastillon, the provincial director at Mézières.
From 1748 Mézières became the centre to which
most recruits were sent once they had passed the
examination, which continued to be held in Paris
under an academician of mathematical bent.[33]
Over the next quarter-century Mézières shifted
its focus towards scientific and theoretical
prowess. Civilian teachers replaced practising
engineers; Gaspard Monge the mathematician,
founder of the Polytechnique, began his career
there. Architecture still had a restricted place in
the curriculum. In 1772 it was classified as one of
twelve subjects of study; and architectural skills
were limited to furnishing clear sets of geometrical drawings for civil and military buildings and
fortifications. A sense of remoteness from the site
and from craft was now perceptible. The shift
corresponded to the waning in military construction after France's setbacks in the Seven
Years War. Guns now meant more to the army,
fortresses less. As the Génie lost effectiveness, its
school drifted into a theoretical and autonomous
stance.

Meanwhile on home territory the Corps des
Ponts et Chaussées, fed by a school of its own,
was by now handling all manner of architecture
and construction. The record of the Ponts et
Chaussées school has been masterfully addressed
by Antoine Picon. The rise of the Corps des
Ponts et Chaussées from its junior status to the
Génie came about, he explains, not just because
French military expansionism withered, but also
because it contrived to lock itself into the supply
system of government finance.

OPTIMO VIRO ET CLARISSIMO CIVI JOANNI RODOLPHO PERRONET,
Regiæ Scientiarum Academiæ Parisiensis Sodali,
et à Viis, Pontibus et Ædificiis Publicis Galliæ conficiendis Architecto et Præfecto,
offerebant et consecravère Institutori, Amico, Patri,
testes Virtutum assidui et Beneficiorum memores Alumni.

ANNO M.DCC.LXXXII.

406. Jean-Rodolphe Perronet.
Frontispiece to his *Description des projets
et de la construction des ponts*, 1782.

The success of the Ponts engineers can be largely ascribed to the support afforded
them by the financial administration to which they belonged, to the detriment of the
war department on whom the Génie engineers depended.[34]

One of the great French civil servants, Daniel-Charles Trudaine, intendant to the Ponts
et Chaussées from 1734, drew the Corps into centralized planning. Trudaine saw the supply of surveying and mapping skills as a preliminary towards the Colbertian dream of
transforming the country's communications (pp. 28, 293). He therefore set up a special
drawing bureau in 1744. It was this Paris office, taken on three years later by Perronet,
which turned by stages into a school. The name Ecole des Ponts et Chaussées became
fixed only in about 1756.

Perronet's blend of skills allowed him to become the greatest of eighteenth-century
bridge-builders (pp. 296–304) and architectural educators alike (ill. 406). The cast of his
early career shows how technicians could still move then between different organs of the
state and what we think of as different disciplines in construction. He had a mathematical uncle, and had hoped to enter the Génie. But he was sidetracked into the office of the

Paris city architect, where he worked on sewers and quays before transferring to the Ponts et Chaussées and making his mark as a road-builder. He also studied manufacturing at first hand (his report on the stages of pin-making in Normandy has been hailed as 'one of the landmarks in the study of the division of labour'[35]). So he knew not just about design but about process and the sequencing of operations – trickier in the procurement of roads and canals (later, railways too) than it is even for buildings.

Quite soon the Ecole des Ponts et Chaussées inculcated the gamut of skills, from designing monuments to managing infrastructure. In its infancy it did not stand alone. Perronet, half an architect, enjoyed good relations with J.-F. Blondel, whose Ecole des Arts, started in 1740, was the stablest of several private academies offering elements of design education to aspirant young architects in the French capital. Blondel's 'pépinière' compensated for the decadence of the Académie d'Architecture school at that moment. At first he took in some employees from the Ponts et Chaussées drawing office, and trained others for entering it. As the Corps' own school matured, architectural skills – drawing, the elements of design and a grasp of the orders – were nurtured by Perronet and his staff, along with construction, surveying, geology, calculation and administration. All these skills the recruits would find of value when they took up their provincial posts as state employees in the Corps. For though its main output consisted of bridges, roads, river, canal, port and drainage schemes, some buildings also fell to its engineers, as public or private commissions (pp. 31–2).

At the outset Perronet's school was locked into the Corps des Ponts et Chaussées. Between a third and a half of pupils' time was taken up by practical tasks for the Corps, says Picon, mainly the map-making fundamental to the 'aménagement du territoire'. Another third was spent 'en campagne', learning the practicalities of the engineer's trade and helping out colleagues.[36] On-site training persisted in the Ecole des Ponts et Chaussées. Stand-alone architectural schools today often aspire to it, but seldom enjoy the institutional backing to make it regular.

Soon though, a liberal, semi-independent atmosphere developed. Practical mapping jobs for the Corps dwindled. By 1770 Perronet could tell his pupils, 'our insistence on several years' study of architecture is more for your own sake than for the service.'[37] That is partly because only a minority of alumni during the Ancien Régime went on to make their careers as engineers in the Corps. Adding in those who went as official engineers to the pays d'état like Burgundy and Languedoc, where the writ of the pre-revolutionary Corps des Ponts et Chaussées did not run, the number who became engineers in state service rises to over half. But alternative career structures also beckoned. Some alumni became geographers, others builder-developers, others again architects, as a sprinkling of independent architects – though hardly yet independent engineers – found they could subsist in the private economy.[38]

A convergence also took place between the Ponts et Chaussées school and that of the Académie d'Architecture after Blondel took over the latter in 1762. Both drew in pupils who did not end up in state service, and both were increasingly seduced by the glamour of design competitions. Though present from early on in the Ecole des Ponts et Chaussées, these intensified after 1775, reaching a climax in the handsomely rendered schemes of the decade before the Revolution. But the Ponts et Chaussées competitions, unlike those of the Académie, were mostly geared to priorities in government policy. Debates of the 1780s about modernizing France's ports, for instance, were reflected in ideal designs which Picon christens 'prototypes of engineer's space'.[39]

An Unravelling

Here then is a pre-revolutionary moment of equilibrium, offering an ideal of professional training which in a dislocated design-world eighty years later Viollet-le-Duc would seek in vain. Architects and engineers, still scarcely distinguishable, were taught on similar lines, learning to match technique with art in the service of national development, yet given much freedom. At bottom these privileges derived from a long-term commitment to France's infrastructure by enlightened if over-taxing administrations. But combining usefulness with art pervaded architectural thought. It can be seen, for instance, in Pierre Patte's famous ideal section through an urban street, where the servicing is accorded equal value with the architecture. Likewise, Blondel in his *Cours d'architecture* insisted that his students communicate with society. The Ecole des Ponts et Chaussées and other schools for design training in the Paris of the 1780s stood for more than the pragmatic unity between architect and engineer inherited from earlier centuries. They offered a consciously enlightened attitude to construction in which art and need were in balance.

That equilibrium could not survive the instability of the Revolution. In its aftermath, the royal academies were dissolved. The academy school of architecture suffered a prolonged purgatory of hand-to-mouth arrangements. Spatchcocked to the fine arts, architecture bumped along until the Restoration. Sorting out the engineering side was more urgent. The Ecole du Génie, dissolved as a nest of aristocrats, had hastily to be re-established to address the challenges of invasion. The more meritocratic Ecole des Ponts et Chaussées scraped along until it too was swept up into the new structure of the 1790s for technical education.

The ambiguities of the Ecole Polytechnique helped make it the world's most influential technical school. The arrangements of 1795, whereby students destined to be military, civil or mining engineers undertook a common preliminary course together, allowed technicians to be prepared for state service fast and cheaply. That way, the mediocre goats could be weeded out from the fewer, cleverer sheep who would go on from the Polytechnique to further years of intensive and practical study in the 'écoles d'application'. The economy of mere classroom teaching, as opposed to the model of instruction half in Paris and half 'en campagne', was to interest those German states who in due course fused the Polytechnique and the écoles d'application into the single cursus of their Technische Hochschulen.

Another reading of the Polytechnique's establishment sees it as the child of a cluster of far-sighted 'savants', the new priests of the Enlightenment: and pre-eminently of the mathematician-turned-Jacobin Gaspard Monge.[40] Though Monge wanted the new school to be open to all and to serve practical ends, he followed Condorcet in believing that practical human progress depended upon applying a fundamental grasp of scientific and mathematical principles. That is the reading that academics naturally prefer – one seemingly borne out by the growing scientific tenor of engineering education. It was certainly upheld by Monge's ally Gaspard Riche de Prony, first professor in mechanics and analysis at the Polytechnique, and head of the reformed Ecole des Ponts et Chaussées from 1798 until his death in 1839. A figure of public authority in Napoleonic France, Prony was Perronet's chosen successor. Yet he was 'not a man for design',[41] notes Picon, but a teacher of enlightened classroom principle.

In his 1837 'éloge' of Navier, an ideal of the intellectual style of engineer who flourished under the reformed system, Prony summed up his philosophy of technical education:

> The complex of different areas of knowledge that at the present day go to make up the science of the engineer (in which architecture should also be considered as included) embraces almost all the different branches of the physical and mathematical sciences in all their aspects, whether theoretical, experimental or practical. This complex has never been nearly as enlarged as it is today; but if you compare the monuments erected in successive periods, judging them not by their size or magnificence or the amount of labour they have involved but by the level of education required for these projects to be executed, you can observe a clear relationship between the progress of science in the human spirit and the art of construction. The range of knowledge which Vitruvius believed an architect needed and which he sets out was most probably not possessed by the men who built the pyramids of Egypt, yet now they would amount to a poor level of expertise.[42]

407. Typical diagram from J. N. L. Durand's *Précis des leçons*, 1809 edition, showing how students can develop a square courtyard building by starting from a simple grid. Durand's text reads: 'An ensemble has been specified consisting of a courtyard ... together with four rectangular rooms and four square rooms at the corners. With this abstract proposition clearly in mind, set down your mental image of the model in an initial croquis, which, by means of signs placed in an appropriate relation to each other, will fix the idea and allow its author to examine it and to judge whether it is truly what he intended; if it is satisfied, he can then realize it by indicating the axes of the walls that are to enclose the ensemble and the parts of the building. The number, shape, and situation of these parts being once indicated, a second croquis is made in order to gain an idea of their relative sizes' (translation by David Britt).

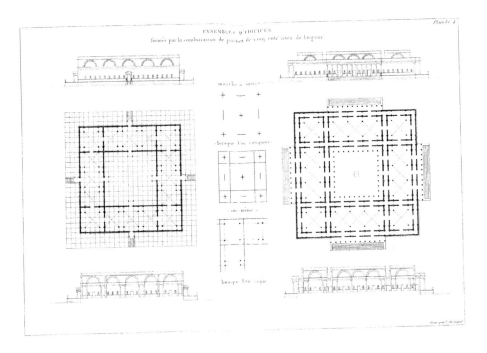

Where does this leave architecture, so strong a strand in Perronet's philosophy of public works? In Prony's frame of reference it is relegated to a parenthesis, its pedigree brushed aside as an obsolescent afterthought. It can be admitted only as one of many skills in the progressive arts of construction which make up the 'science of the engineer'.

Understanding of architectural teaching for engineers in this new era habitually focusses on the rigorous figure of Jean-Nicolas-Louis Durand, professor of the subject at the Polytechnique from 1797 until 1833.[43] From his texts alone, it is easy to read the minimalism of Durand's teaching as determined solely by an ideology of design. That is to strip it of context. Like the Polytechnique itself, Durand's was never a self-sufficient enterprise. It just expressed what could and could not be taught in the classroom over two years to young men destined for a range of careers.

In his earlier years, most of Durand's students would be hurrying on to the Army. Nonetheless in due time some might have the opportunity to build, and on a large scale. All this he states with pith and realism at the outset of his famous textbook, the *Précis des leçons d'architecture données à l'Ecole Polytechnique*.

Architects are not alone in being required to erect buildings: so, frequently, are engineers, both civil and military. It might even be said, speaking of engineers, that they have more opportunities to carry out large undertakings than do architects proper. The latter may well build nothing but houses all their lives; but the former, aside from being frequently called upon to do the same in those remote provinces where architects are rare, find themselves professionally required to construct hospitals, prisons, barracks, arsenals, magazines, bridges, harbours, lighthouses: a host of buildings of the first importance; and so knowledge and talent in architecture are at least as necessary for them as they are for architects.

But youthful students embarking on the profession of engineering, whether civil or military, or on some other branch of the public service, have very little time to spare for such a study, whether at the Ecole Polytechnique, or at the special schools to which they proceed on leaving it, or even when they have attained the rank of engineer. For such students, therefore, it has been necessary to make their study of architecture, although extremely brief, nonetheless fruitful.[44]

SCHOOL CULTURE 1750–1914   443

Durand's is a fallback position. Never an engineer, all architect, the times had deprived this former student of Boullée of opportunities to build, as they had his master. Like many architects in such circumstances, and almost all architect-teachers, he retreated to the abstraction of design as the device to keep his cherished subject alive and pertinent. 'What is architecture?', Boullée had asked at the start of his own treatise. 'Am I to define it, with Vitruvius, as the art of building? No, Vitruvius's definition contains a flagrant error. He takes the effect for the cause. To execute you must first conceive . . . It is that production of the mind, that creation, which constitutes architecture.'[45] So the core of architecture, and by implication of architectural teaching, becomes a methodology of design. That was Durand's position too. Only Durand was operating not in Boullée's isolation, but in an official institution charged with turning out useful technicians.

Like the best teaching, Durand's was memorably simple (ill. 407). If its synthesis and dogmatism were new, its contents were pragmatic. He accepted the Orders, drew upon earlier authors for his analysis by elements and types, and for their assembly into big compositions upon the competitions of the old Ponts et Chaussées and recent Napoleonic projects for new towns. All that helped familiarize things for his students. Drawings were confined to the plainest: no perspectives, no washes, no shading, no hint of the 'descriptive geometry' dear to Monge. The building-types Durand covered were conventional. There was nothing about industrial buildings, and little about new materials such as iron.[46] Though he taught budding engineers, structure played no part in his lessons, because it drew students away from analysis and composition to the next stage of their skills, which the écoles d'application would teach. Nor does the *Précis* dwell on context. The character of a building is developed and expressed not from its setting but from within, by its own type and plan-form. All this smacks of not overwhelming the tiro confined to the classroom with material he can take in later.

The value of Durand's unvarnished message can be gauged by the elemental, neatly set-out public buildings that proliferated over France and beyond during the first half of the nineteenth century, on the strength of his diagrams. Lowest-common-denominator architecture can be pleasing, if it is well built.[47] Designed by architects and engineers alike, it reaffirmed a level of unity between the professions. As for teaching, the example set by Durand at the Polytechnique ran on into architectural instruction at the Ecole des Ponts et Chaussées, and beyond into the technical and design schools that sprang up all over Europe after 1820. His method was also tacitly welcomed in the Ecole des Beaux-Arts. For despite the lip-service he paid to economy and utility, Durand had been 'constantly concerned to preserve the autonomy of the discipline of architecture'.[48] By abstracting it from the field and defining its scope within the classroom, he had rescued and reconfigured his beloved subject.

Against the ideal of science upheld by Monge and Prony, however, Durand's efforts look like a holding operation to keep architecture aboard the vessel of progressive engineering. Not architecture alone but the whole empiricism of building construction was in retreat within the Ecole des Ponts et Chaussées during the first half of the nineteenth century. State-trained engineers now acquired their own enhanced position, buttressed by the technocratic authority conferred by their education, and the demands with which a development-led society bombarded them.[49]

The polytechnical élite never had things all its own way. On the engineering side, a body of French opinion always looked on its training methods as impractically abstract. In 1825 the authorities of the Ecole du Génie at Metz complained that the mathematics and mechanics taught at the Polytechnique were too complex, and that students who came on to them arrived 'fed up with analysis and so exhausted by analytical calculations that they had a horror of everything involved in them.'[50] The failure of the state-sponsored schools to adjust and turn out the types of engineer, civil or mechanical, which the private sector needed, led to the foundation of the Ecole Centrale des Arts et

Manufactures in 1829; while the alleged practical deficiencies of Ponts et Chaussées training elicited a swipe from Balzac in *Le Curé de Village* (1839).

Throughout these years there was much peeking over shoulders at Britain, which seemed to be racing ahead in engineering without lifting a finger about technical education. From that perspective the Polytechnique often looked out of step. Take, for instance, Monge's great pedagogical innovation, 'descriptive geometry'.[51] The tradition of setting out masonry from which Monge had developed this lucid method of analysing and depicting structure was losing its grip. Yet in the illusion that it would help with industrial and technical drawing, descriptive geometry got rammed down the throats of generations of Frenchmen who as yet were being taught little about steam engines. British engineers remained, in Picon's words, 'disdainfully ignorant' of the whole method.[52] A contrary disdain for dirty hands, and a preference for the mollycoddling fellowship of the academy, recur in French engineering education. So late as 1899, reported the director of the Ecole Centrale, when students have 'a course on bridges, roads, railroads, architecture, they say – "that is meant for masons and workers" and we have to propagandize them for months in order to make them understand that they cannot make a living on algebra.'[53] And once they got on site, engineers often found the science they had learnt in the schools of limited help, as a study of building the Suez Canal attests.[54]

An academic, science-based ideology; a diminished status for architecture; and a need for state-trained engineers to tackle a wider variety of infrastructural projects in the national interest: such then was the background to teaching at the Ecole des Ponts et Chaussées in the first half of the nineteenth century. But though architecture had lost its gloss since Perronet, that did not prevent it from being well taught there, in alliance with construction. Students coming in from the Polytechnique took a general construction course with an architectural bias until the 1830s, after which it was broken up into specialities: roads and bridges, navigation, railways, etc. There was also a separate architecture course, taught for most of the Durand years by Charles-François Mandar, whose pragmatism and stress on materials made up for the abstractions of the Polytechnique.[55] Architectural competitions continued, more realistic now in topic and scope if less inspired in their rendering.

In the 1840s architectural training for French state engineers rediscovered some of its lustre under Léonce Reynaud, professor at the Polytechnique from 1837 and at the Ponts et Chaussées from 1842, later head of the latter school. Reynaud's background in Saint-Simonianism (he had been thrown out of the Polytechnique in youth for political activity), plus his hands-on experience of big public projects (his first Gare du Nord is described on pp. 118–20) led him to blend practical authority with vision. He sketched out for his students the avant-garde role he conceived them to possess in constructing the modern nation. In a way that his predecessors had never quite dared or thought to do, he admitted the claims of imagination and inspiration in the making of architecture and indeed in the great social tasks ahead. Engineers were not just technicians, Reynaud implied; they were also artists. Here was a vision more enlarged and uplifting than that of Durand or Prony.

Reynaud was still exalting the priesthood of the all-round expert in construction to his students when Viollet-le-Duc embarked on his hapless démarche over the Ecole des Beaux-Arts in 1863–4. Why then did the latter refer only briefly to the record of architecture in the Ecole des Ponts et Chaussées? Personalities and stylistic preferences aside, the main reason was the smallness and inwardness of the engineering school. The arrangements of the 1790s had strengthened the link between the écoles d'application and their Corps, to which it was expected all successful graduates would proceed, at least for a time. The Ponts et Chaussées school was taking in between twenty and thirty students per year from the Polytechnique during the first half of the century. It had also started admitting a few foreign students after about 1830, just before the Ecole des Beaux-Arts did the same. But compared to Perronet's pre-revolutionary school or the new Ecole Centrale, it was a closed shop. It might offer the best-grounded architectural training, but it still aimed at producing engineers for the state, not architects or civil engineers for private practice.

With the rise of the railways, the private sector in civil engineering grew hungry for trained personnel. Half a century later than in Britain, the era of the consultant engineer opened out in France. Thereafter, once they had served their time in the Corps, Ponts et Chaussées graduates habitually shifted into lucrative private work. But nearly all became engineers, not architects. There lay the fame, the money and the large scale of challenges. Over and above that, the polytechnical 'formation' had been stamped upon them. For all Reynaud's efforts at redress, science had triumphed over art. The botched reform at the Ecole des Beaux-Arts confirmed how hard the fissiparous societies of the nineteenth century found it to keep the two together – even in the country that had first fostered formal training in the arts and sciences of construction as a continuum.

Towards the Technische Hochschule

On the face of things, the pattern of French education for architects and engineers between 1750 and 1850 is one of divergence. The systematic application of mathematics and science to technology and the growth of subdisciplines within engineering both pointed to separate development, shielding its alumni from the messy empiricism of building construction and the confusions and squabbles of art. So deeply did the polytechnical élite embed a scientific structure for technical education into French government policy that it remains in place today. At best, architecture played a bit-part in that historic endeavour, clinging on by its coat-tails.

On a different reading, the two subject-areas behaved erratically, veering apart only to converge. During Perronet's heyday in the 1780s, architecture and engineering came closer than they had been when Trudaine took hold of the Corps des Ponts et Chaussées. Later, making up for the famine of real projects during the era of revolution and educational restructuring, a glut of Durandesque buildings pointed to a renewed identity or at least an ease of exchange between the disciplines. Later again, the rhetoric of both Reynaud, eager to draw art back into engineering, and Viollet-le-Duc, intent on grounding the art of architecture in construction, stand for fresh efforts at redress or reconciliation between the 'two-culture' poles of the Polytechnique and Beaux-Arts.

Nevertheless the overall slant of the nineteenth century lay towards the culture of science and engineering, based on a material understanding of utility. Wherever governments had a hand in it, the main trend in training technicians for construction therefore was to start with engineers and move outwards, with architects holding on for the ride but breaking free when they could. That is best illustrated from German-speaking states, on the rise throughout the century and less entrenched in their educational dispositions than France had become by the 1860s.

Like other initiatives of the revolutionary years, the creation of the Polytechnique sent out waves of energy into the German states. None was large or rich enough to emulate the French two-tier system. Instead, omnium-gatherum central technical schools became the norm, small at the outset but quick to grow. Most began with infrastructure and building technique, though mechanical engineering's claims were not neglected.

Prussia, readiest to modernize, reacted first by founding its Bauakademie in 1799.[56] Here from the start tension arose between Berlin's art-architects, fostered in the new school by the scholar-artists of the established Akademie der Bildenden Künste, and the technicians, upheld by the state building service or Oberbaudepartement. The Oberbaudepartement enjoyed the asset of continuity, but lacked the independence and status of a French corps. It saw the school as training officials for state service, took an interest in skills and outcomes, and managed to splice into the curriculum mechanical engineering and other technical subjects that did not concern the artists. The state service largely had the better of the argument, especially after the charismatic young architect Friedrich Gilly died in 1800. Under the Bauakademie's first director, Eytelwein, its engineering side became dominant. Nonetheless the school maintained classical-humanistic entry requirements. Its critics felt that the school had 'unwarranted academic pretensions' coupled with 'a teaching level too advanced for most students, lax discipline, and ultimately technical incompetence'.[57]

In the next Prussian phase, the critical figure was Schinkel's friend and partial mentor, Peter Beuth. An expert on tax and finances, when Beuth became head of the state office for economic development and technical education in 1820, he tried to focus all government-supported training upon national development. In no state had so cohesive a programme been articulated or implemented before. But it implied intolerance of the dominant humanistic slant within German higher education. The resulting tensions long bedevilled the national culture.

At the lower level of technical training, Beuth set up a network of trade schools culminating in the central Gewerbeinstitut in Berlin, which offered artisans elements of scientific instruction coupled with good workshop instruction. At the higher level, he addressed the conflicts within the Bauakademie by renaming the school the Allgemeine Bauschule and splitting it into two departments, one under the trade ministry, the other under the fine arts academy. 'As a consequence of Schinkel's considerable influence at the Bauakademie', claims Ulrich Pfammatter, always keen to assert harmony between architect and engineer, 'this differentiation remained illusory.'[58] The truth seems rather to be that the architects tried and failed to establish a separate identity. According to Vincent Clark,

> The engineering department prospered, while the department of architecture suffered from low enrollment and was later closed. Architecture was soon revived and reunited with engineering, but in the Academy of Building the engineering emphasis remained.[59]

Following Beuth's retirement, a vocal minority of architects managed after 1845 to secure a separate cursus to official employment within the Prussian building bureaucracy and hence a renewed department within the school for themselves. The old name now returned. But so long as the main goal of pupils at the Bauakademie remained state employment, interdisciplinary rivalries kept resurfacing. Only the rise of a sustainable private sector in architecture, as Prussia's economy grew and liberalized itself, allowed architects and engineers to establish their own patterns of training on a distinct and secure basis. Architects, it seemed, could only control their own education when there were enough of them, and they no longer relied on state patronage.

The *Deutsche Bauzeitung* was founded in 1866 in part to press for this legal division of the professions and their separate training. 'With force they want to make us simultaneously into architects and engineers', complained one architect of the Bauakademie. 'And what do they achieve other than that we are in truth neither architects nor engineers but dilettantes in both fields, merely administrators with technical knowledge.'[60] Not everyone agreed even then. Bickering about whether architecture and engineering should be taught together or apart was still rumbling on when in 1879 the Bauakademie was swept up into the grandest of the German Technische Hochschulen, the Königliche Technische Hochschule, Berlin-Charlottenburg.

The nomenclature of the Technische Hochschulen was imposed after German unification, to link the various higher training institutions founded in the princely states on the polytechnical model – and for the most part first christened 'polytechnikum'. What these schools did for the level and repute of German technological training can hardly be overstated. Some were broader in range than the Bauakademie, but so far as our subjects go the overall pattern was similar. In Berlin's wake, most of the early examples, like Karlsruhe, Württemberg and Hanover, leant initially towards the main branches of engineering and recruitment into local state service. If architecture was never excluded (there was occasional talk of doing so), its proponents had often to plead its special orientation and needs. For example at Karlsruhe, the outstanding early foundation (1825) – in effect a fusion of Friedrich Weinbrenner's tiny architectural Bauschule with a larger engineering school – Weinbrenner had to fight against putting the architectural student through

advanced mathematics, 'since he already has so much to learn that he cannot make this science his chief study'.[61]

These technical schools, strong in engineering and mostly weaker in architecture, were the only ones available to the local student unless he ventured abroad. Nowhere in Germany was there a Beaux-Arts system wherein to take refuge. Perhaps for that reason, Weinbrenner sent his best students to complete their training in Paris.[62] Many a young German architect therefore acquired a deeper technical training in the theory of construction than he preferred – until a twentieth-century reaction set in. But though engineering might loom, it did not always crush, since the German tradition of self-governance in universities extended to the polytechnics as they matured, and indeed within them to the departments themselves. Karlsruhe's pre-history in two fledgling schools proved a precedent, shaping the continued independence of the architecture and engineering departments. As relations with the state building corps loosened and officialdom ceased to be the students' certain destiny, self-regulating departments of architecture managed to forge a tentative identity, without straying far from the umbrella of engineering.

The prime example of that was the Eidgenössische Polytechnikum at Zurich, later the ETH Zurich – earliest and sincerest tribute from a neighbour-country to the new-minted German model of technical education, and to this day one of the world's foremost schools for both engineering and architecture. Autonomous the two disciplines have always been at the ETH; but the relations between them are all but inextricable.[63]

From its foundation in 1855, the key to the success of Zurich's polytechnic lay in its federal funding and status, envisaged under Switzerland's modernizing constitution of 1848. Exempt from cantonal control, relieved from undue interference by the various state corps, it could promote national economic development while taking – and in due course teaching – the best from abroad. In the words of one admirer, the British naval engineer John Scott Russell, its founders trawled

> the annals of pure philosophy and applied science, for the names of those men who were best known for science, skill, and the love of teaching; and these men from every country they selected, and entreated to come and teach their children, considering only how they could best make it agreeable and convenient to them to become the teachers and patterns of Swiss youth.[64]

On the humanities side, not neglected, among the original luminaries were Jakob Burckhardt, Francesco De Sanctis and F. T. Vischer; in engineering, Karl Culmann, bridge-expert and inventor of the graphical analysis of structures; in architecture, Gottfried Semper, designer of the Polytechnikum's building (ill. 408), theorist, and knotty author of *Der Stil*.

Did the two eminent German professors, Culmann and Semper, connect? Though Culmann is said to have taught in the architecture department,[65] there seems no evidence they did so. Here lies the enigma of their subjects and indeed of the Zurich Polytechnikum's thrust. There were at first to have been only three main faculties, civil engineering, mechanical engineering and chemistry. Architecture was an afterthought, after the authorities had been persuaded that Switzerland afforded just enough opportunities for it. Though it was tiny – between nine and twenty students per year during Semper's time – the fact that it enjoyed a faculty of its own distinguished it from, say, the lectures of Burckhardt, which were a cultural add-on.

The architectural course was a three-year one (Semper argued in vain for a fourth year), with a fiercely technical Vorkurs to make up for the unevenness of Swiss schooling. In the first year, there were five hours of calculus per week, and four of descriptive geometry. 'Composition' featured, but 'designing' came in only in the second year and predominated only in the last. That was typical of polytechnical courses, which tended to postpone what would now be called studio teaching until students were technically grounded. Yet to judge the faculty by its physical first-fruits, including Semper's Swiss buildings – not least the school itself – the impression is of a conventional art-academy, remote from the progressive thrust of the Polytechnikum.[66]

As an art-architect, Semper was out of tune with the conditions he took on at Zurich. A good mathematician, versed in the nature and behaviour of materials, he insisted that architects should have a thorough technical understanding. But he was vehemently against technology determining architectural form or, worse, presuming upon its overall direction and destiny which, so he believed, grew out of habit and custom. 'My only business here is theoretical teaching and lecturing,' he complained to Henry Cole, 'which is neither my chief inclination nor the best method of forming efficient pupils'.[67] Since his ideas were too hard for juveniles (as he found his students), much of his instruction was delegated to assistants. Semper had never greatly wanted the Zurich professorship in the first place, thought often about leaving before he decamped to Vienna in 1871, and spent much of his time writing or drumming up architectural jobs for himself while bemoaning the stinginess of his Swiss patrons.

All this is rehearsed not for criticism's sake, but to convey the difficulty of reconciling speculative thought in architecture with rigorous technical training. As for engineering, with the brilliant Culmann and his assistants hard by, Semper's obduracy against exposed iron structures must surely be lamented. One clue to the gap between them is Semper's rejected design of 1861 for the main station in Zurich, which prefaces a wayward, timber-trussed trainshed with an essay in opulent streetscape. Later, Culmann was drawn in as a consultant on the sensible iron roof erected.

Yet the long-term trajectory of the so-called Semperschule cannot be written off as reactionary. Without any sudden change of direction, the ETH Zurich matured over time into one of the few centres of architectural teaching where technical competence and flair for design have managed to co-exist on calm and equal terms. Many factors had a hand: Semper's art-idealism; the technical grounding inculcated; the overlap but underlying dialectic between architecture and engineering; and the school's overall scale and internationalism, balanced against its appointed role in Switzerland's economic and social development. The ability of teachers and architects from Zurich to reconcile technical discipline with art and to proselytize that union has long been the envy of other schools of architecture. Its success surely proceeded from its original institutional context. Zurich's legacy is that of the TH or polytechnic movement at its best, whereby architecture flourishes and even cavorts a little under the protective skirts of engineering.

The Polytechnical Model in America

The history of continuous academic education for the construction professions in the United States starts with West Point. Though drawing schools and similar add-ons to builders', craftsmen's or architects' offices can be traced before then, they tended to be bound up with pupillage. Few lasted long.[68]

Not that West Point offers a simple date of departure. Like the French official schools from which it borrowed men and methods, it arose pragmatically out of the creation of a government corps – the US Corps of Engineers – and turned into a formal school by stages.[69] When the original corps was set up during the War of Independence, there was talk of making it a 'school of engineers'. But there was no money or time to sort out a structure before the corps was disbanded at the end of the war (pp. 51–2). In 1794 came tentative renewal, when West Point was designated the headquarters of a 'Corps of Artillerists and Engineers'. Some training was given there by French veterans still lingering in America. Though the arrangements soon collapsed, the germ of a military academy had been sown. Gradually it triumphed over ingrained American resistance to a standing army. Latrobe made a plan at one point. The idea kept surfacing during the presidency of John Adams, to be implemented by Jefferson when he took over at the end of 1801.

It has often been supposed that the far-sighted Jefferson saw the refounding of West Point (1802) as a way of putting American science and engineering on an institutional basis and setting a technical élite of the new permanent armed forces, if they had to exist, to the service of his young country's civil development. Theodore Crackel, West Point's latest historian, disputes this. Early instruction was feeble and, wrote one cadet, 'barely sufficient to excite a desire for military enquiries and of military pursuits.'[70] Only the trauma of a second war with Britain in 1812–14 promoted renewal, under West Point's second superintendent, Alden Partridge. An academic department for engineering now appeared in embryo. In 1816 the polytechnician and French army engineer Claude Crozet (one of several Bonapartist technicians to take their engineering skills to America after Napoleon's defeat) arrived to teach descriptive geometry.[71]

West Point really took off only in 1817 after Partridge had been controversially ousted in favour of Sylvanus Thayer, 'Father of the Military Academy'.[72] Thayer had been two years in Europe at his government's expense, studying fortifications and methods of education. Naturally the Ecole Polytechnique occupied the top of his agenda. Under Thayer and Crozet, West Point's technical programme became unequivocally French and polytechnical in inspiration. Mathematics was the staple of the early years along with the French language, since most engineering textbooks were still in French.[73] Later in the course came military engineering, the 'science of war', rhetoric, and moral and political sciences.

Civil engineering was spliced in some years later, against Thayer's advice. Architecture as a stand-alone subject never got taught at West Point. But the civil skills, once present, proved invaluable; they were what the young country most needed. As in France, openings in canal-building and, soon, railway-building lured graduates into private work once they had done their statutory term of military service. As American development depended upon free enterprise, it was the natural career path to follow. The Army would sometimes release its officers early for such tasks, so subsidising the private sector. An example is 'Whistler's father', Major George Washington Whistler, one of the leading American railway engineers of the pioneer years. That brilliant West Point alumnus was loaned by the War Department in the late 1820s to the Baltimore and Ohio Rail Road and ended up laying out the St Petersburg to Moscow line for Nicholas I.[74]

West Point's contribution to American engineering was not confined to feats of construction by individuals or even by the Corps of Engineers as a whole, much though that body shaped the national infrastructure. It moulded the country also through the teachers it trained and the methods it passed on. As early as 1819, Thayer's disgraced predecessor, Alden Partridge, had set up a short-lived 'American Literary Scientific and Military Academy' in Vermont. By various mutations this was to turn into Norwich University, a specialist engineering college. Other such academies, most of them ephemeral, were started by Partridge or his pupils.[75] Some states created their own military colleges, notably the Virginia Military Academy, where Crozet became the first professor of engineering in 1839. Likewise some of the burgeoning scientific and engineering colleges in major towns and cities selected their original professors from West Point teachers. Such were David Douglass at New York University (1832), and H. L. Eustis at Lawrence Scientific School, sister establishment to Harvard (1849).[76]

Even where West Point personnel could not be had, the French model of engineering college it aped soon got adapted to the private sector. An early example was the Rensselaer Polytechnic Institute. Remote from the cities in up-state New York, the Rensselaer School (its original name) arose in 1824 out of the Erie Canal's impact on local farming territory. Stephen Van Rensselaer, shareholder in and long-time president of the Erie Canal Commission, hired a self-taught surveyor and science teacher, Amos Eaton, first to give lectures on science subjects in settlements along the canal, and then to start a college in the town of Troy. Land-surveying and hydraulics were the original subjects, reflecting the students' regional and agricultural destinies. Civil engineering soon muscled in, and by 1835 those enrolled in its one-year degree course were described as 'of the Engineer Corps'. Their curriculum featured 'Mechanical powers', along with 'construction of Bridges, Arches, Piers, Rail-Roads, Canals, running circles for Rail-Ways'.[77]

Architecture tucks easily under the coat-tails of such subjects, and we can catch it trying to insinuate itself with a new director, Franklin Greene. Resolved to raise the status of his small-town college, Greene changed the name to 'polytechnic institute' in 1849–50. Rensselaer's objects were now declared to be 'the education of architects and civil, mining and topographical engineers, upon an enlarged basis and with a liberal development of mental and physical culture'. A shared first-year curriculum took polytechnical guise, founded on 'the common scientific basis of the four professional courses'. Alas for Greene and for architecture, only the engineering specialisms could be afforded (or were demanded) in rural Troy. Greene made the best of things by stating that Rensselaer stood midway between the Polytechnique and the Ecole Centrale: 'It claims no other resemblance to these celebrated and *richly endowed* institutions. To its peculiar mode of study there is *no known counterpart*'.[78]

An architectural course at Rensselaer had to await an era of ampler private funding for American colleges. By the time it was inaugurated in 1929, engineering had diversified beyond the imagination of Greene's days. The restitution of his dream had to be presented in terms of modest redress for a national culture too far in thrall to technology:

> Every course previously established had, up to this time, a beneficial effect upon those previously existing and there was no reason to think otherwise of the architectural course. As architecture is a science as well as an art, it was appropriate that such a course should be available in a school of science and engineering, particularly the latter, since criticisms of engineering works on account of their want of architectural beauty were then, as now, common and often quite just. As a result most of the engineering curriculums now contain courses in architecture.[79]

Rensselaer's prematurity is telling. Until the United States had developed a quantum of riches, leisure and urban fabric, room could not be found for stand-alone instruction in architecture. France's different history had endowed it with parallel systems of training in the arts of construction. One side, opaque in methodology, was rooted in the demand for prestigious buildings; the other, broader but subdivided into specialisms, put scientific principle to the service of security, infrastructure and growth. Besides the luxury of art, the Beaux-Arts system also connoted class hierarchies which to a democracy intent on physical development had scant appeal. So it was to the French polytechnical model, to West Point and its homegrown derivatives, and to the image of discipline and efficiency conveyed by a 'corps', that ante-bellum statesmen and educators turned when they thought about training personnel for constructing the nation.

That viewpoint could only be reinforced as the German states took up and recast poly-technical education. With Central Europeans migrating apace after 1848, not least to the frontier regions where infrastructural needs were greatest, the skills which German-speaking technicians imported started to be valued, copied and diffused. So much is con-firmed by the Land-Grant Act of 1862 and its consequences. That far-sighted legislation prompted new foundations for higher education in almost every state, endowed from the sale of federal lands. The flavour of the resulting colleges tended to be German more than French. Agriculture, engineering and home economics were the subjects specifically promoted under the terms of the Act. Soon though, a few schools began to offer courses in architecture also. At first, most of these infant departments ran on practical and poly-technical lines. But as the Civil War receded and the nation prospered and relaxed, it was to be only a matter of time before they worked free, proliferated, and explored what art could offer.

SCHOOLS OF ARCHITECTURE: THE AMERICAN PIONEERS

Each of the first four stand-alone academic programmes for architectural education in the United States originated in land-grant institutions founded in the decade following the Civil War.[80] All – at the Massachusetts Institute of Technology, Cornell, Syracuse and the University of Illinois – were at first very small.[81] Like Rensselaer, the last three were in small inland towns. In the first instance they taught the unvarnished principles of science and construction, intending that their alumni would find local employment.

The early history of the Illinois course, in landlocked Urbana-Champaign, is that of a miniature.[82] In 1867 an architectural degree programme was bravely announced in the 'polytechnic' division of the 'Illinois Industrial University', alongside civil, mechanical and mining engineering. No takers turned up until 1870, when a mature student, the 27-year-old Nathan Ricker, enrolled – part-time perhaps, as he vanished to Chicago now and again to earn a living. He was taught drawing by an engineering graduate from Michigan, and technique by a Swedish-born member of the university's engineering staff, Harold Hansen, who had spent time at the Bauakademie. After Hansen left, Ricker taught himself along with three new students. On graduating, he was offered the post of instructor in architecture if he would travel to Europe to enrich his skills. To the Bauakademie and to Vienna he duly went for six months, returning in September 1873 to promote the Illinois course.

Over the years Ricker doggedly drew in students and teachers. Though defining him-self as an architect, he believed in the polytechnical virtue of locking architecture together with engineering. Technical skills came first; drawing started only in the second year, a design studio only in the fourth. The textbooks he taught (and translated) were mostly German. In 1892 he added a four-year course in 'architectural engineering', a term seemingly invented by Ricker that soon spread. Its justification, says Roula Geraniotis, was that 'since few students were equally competent in design and construc-tion, an architecture school should provide for the specialized training of each'.[83] It was doubtless also the school's response to the innovations in construction then going on 135 miles to the north in Chicago. That city then led the world in architecture, yet possessed no special school for the subject.[84] Ideas and techniques travelled from Chicago to Urbana, not the other way round. Perhaps the most telling contribution made by Ricker and the early University of Illinois School of Architecture to their subject's development was neither technical nor artistic but restrictive. In 1897 the existence of the school's final exam as a criterion of competence allowed Illinois to become the first American state to protect the title of architect by law.[85] Here is food for thought about the purposes of architectural schools.

MIT's school of architecture (ill. 409) differed from the other pioneers because it was in a long-settled city. With industrial and infrastructural development accumulating around Boston, the new institute was dedicated to equipping New England with

409. Classroom scene at the MIT
School of Architecture, *c*.1900. The
atmosphere of the art school is
unmistakeable.

'technologists'. The main model was Karlsruhe. But by 1868, when teaching began with
four students,[86] a trickle of demand for architecture and architects *per se* was perceptible
in America's cities. Boston had got far enough beyond mere building to boast an architectural identity of its own. That helped the new department to diverge from the name
and connotations of its parent body and take a separate, artistic direction. As an independent discipline, architecture relies on concentration and affluence: which is why
architects like talking about cities.

In search of a method, MIT under its first director William Ware (ill. 410) identified
a tentative model in the Ecole des Beaux-Arts. As yet the applicability of Beaux-Arts
training to American circumstance was untested. As admirers of Viollet-le-Duc, Ware
and his architectural partner Henry Van Brunt will have known his detestation of that
school, at its height when MIT was founded. But what alternative was there? Like
Thayer of West Point before him and Ricker after him, Ware took a European tour to
find out. For all the vigour of its Victorian architecture, which he admired, Britain could
furnish nothing; German-speaking culture suggested only the polytechnical model,
which would have thrown Ware back upon engineering.[87] Only the Beaux-Arts promulgated architecture as an adjunct to the fine arts, offering the lure of social aspiration
and personal expression.

How the Beaux-Arts colonized American architectural ideology is a story that has
often been told. It started with two wealthy charismatics, Richard Morris Hunt and

H. H. Richardson, who travelled to Paris to buy themselves an art-education in architecture and brought its methods and message back. Their success suggested that a living and a name might be earned in American cities from the message that architecture was an art.

Richardson's trajectory illustrates the change in ideals common around the time of the Civil War. Proficient in mathematics, he had dreamed of West Point, only to be ruled out on the grounds of a boyhood speech-impediment. Instead he went to Harvard in 1856, meaning to become an engineer. There he drew pleasure enough from good company and the good life to shift his sights to architecture. It was his step-father, says James O'Gorman, who then pointed him towards the Ecole des Beaux-Arts, writing:

> a good architect, if he is industrious, cannot help but succeed, and in order before you come out to New Orleans to pursue the architectural business, I have thought that six or nine months in London and Paris . . . will do you more good than three times the time spent in N.O.[88]

It was the gradual revelation of Richardson's genius in built form, not a commitment on his part to a style or method of design, let alone teaching, that hinted to his admirers the value of time spent at the Beaux-Arts.

Hunt's case, a decade earlier, had been different. He too first thought of West Point on the strength of his good mathematics and his 'schoolboy's knowledge of Mechanics, Hydrostatics, Pneumatics, Astronomy, and Bookkeeping . . . the sum total of all my *Gumption*'.[89] He too might have gone like his brothers to Harvard, had not his mother decamped with her younger children to Europe. In Geneva he studied with an architect who had been at the Beaux-Arts. There he managed to enroll at the end of 1846, aged nineteen – the first American to do so. Unlike Richardson, Hunt had no prior degree. The force of the art-idealism he imbibed during almost a decade in Paris may therefore have been the fiercer, making up for the mediocrity of his own imagination. What he did possess was enthusiasm and charm. He returned home in 1855, 'accredited as an ambassador of art from the abounding wealth of the old world to the infinite possibilities of the new', wrote Van Brunt.[90]

In the little atelier Hunt set up in New York round about 1858, a coterie mainly of 'university men' foregathered for the pleasure of artistic comradeship, drawing and composition. Ware and Van Brunt, recent Harvard graduates, were typical of those who attended. 'Mr. Hunt's élèves', wrote Ware, 'were, of course, utterly without the systematic instruction by lectures in history, science and construction, which, to a certain extent, supply in Paris the deficiencies of mere atelier education.'[91] Nevertheless, he added later, 'we left him with our imaginations no longer sterilized by partisanship, but enlightened by his influence' and eager 'to hand on to others the light that we were receiving.'[92] Here architecture is configured as for the personal development of its authors in the service of art, not that of clients, users or material progress. America was growing more cultured but also more self-regarding.

The MIT programme grew out of a copycat, Hunt-like atelier started by Ware when he set up practice in Boston. The lineage was clear; and indeed Ware soon hired a young Beaux-Arts graduate, Eugène Letang, to help teach design. Nevertheless the programme

410. William R. Ware. From Moses King's *Notable New Yorkers*, 1899.

was bolted on to an institute of technology in which all the students had to take a number of joint courses. Ware explained in 1867:

> The architectural classes are part of a general school of applied science – the only part which touches the domain of fine art; though drawing, as a useful accomplishment and a natural language, is taught in all the departments. In the exercises of these departments the architectural students may learn what they require of mathematics, physics, chemistry, engineering construction, mechanics, and the modern languages. All this work is taken off our hands. Whatever we need for our students can here be supplied. Our own work thus relates only to our own profession and our own art.[93]

So the independence he claimed was equivocal. Ultimately, Ware admitted, he envisaged the eventual 'protection of title'. On the one hand, the high purposes of art; on the other, a training technical enough to shoo rivals out of the corner of construction which architects felt belonged to them. The aims meshed together.

In theory, full-time instruction at MIT took four years: two of general followed by two of technical education. That was an ideal programme, set out before architectural teaching started. In practice, most students came for as long as they cared or could afford to, dropping into the courses they needed. Louis Sullivan, for instance, attended for a few months in 1872–3; he was well grounded by the 'Tech' in calculus and mechanics, but bored by Ware's lectures.[94] The course as built up was therefore a compromise, slow to attract degree-candidates.

In 1881 Ware was poached from MIT to start up New York City's first full school of architecture at Columbia. That foundation's origins were pragmatic and social. Keen to improve housing in New York, the original benefactor, F. A. Schermerhorn, sought to link the teaching of architectural design with sanitary engineering, 'as I was sensible that the architect's profession was mainly deficient in this particular branch as well as in Engineering knowledge'. For want of funds, the new foundation was linked to Columbia's school of mines, whose faculty taught the technical subjects. Despite the watchful Schermerhorn, Ware gradually worked in more freedom: 'little by little the chemistry and physics, the botany and hygiene, the sanitary engineering and the economic geology . . . were crowded out and dropped.' Nevertheless mathematics and mechanics remained strong. For the first decade of the Columbia school's existence design was not taught in the first year.[95]

After 1890 stand-alone schools of architecture in America's urban universities mushroomed in number, enrolment and prestige. They also turned their face from utility towards the glamour of the Beaux-Arts. Ware, whose tastes in architecture were quite catholic, came under pressure to make the Columbia course mimic Paris more slavishly and acquire a Frenchman to head the design teaching – like Letang, only at a higher level. MIT became the first school to take that route when it hired Désiré Despradelle in 1893.[96] By the time Columbia followed suit after Ware's retirement in 1903, even an up-country college like Cornell had gone Beaux-Arts and was employing a Frenchman.

Scholars of the Beaux-Arts era in American architecture have tended to stress the design teaching – the side of things Ware meant by 'our own work', 'our own profession', 'our own art'. But it never stood alone. Because American schools of architecture were part of wider colleges or universities and beholden to larger neighbouring departments, they could call in both technical and humanistic help with an ease unavailable to the Ecole des Beaux-Arts itself, much though that institution broadened its scope after the blistering confrontations of the 1860s.

A survey of the leading American schools by the English architect Arthur Cates in 1900 reveals both their uniform devotion to the Beaux-Arts and the plethora of technical courses offered by non-architects.[97] Columbia, for instance, claimed that its architects were being taught by

411. Casts in Robinson Hall, Harvard, in blatant homage to the Salle des Etudes Antiques in the Ecole des Beaux-Arts.

the Professors of Chemistry, of Physics, of Mining, of Analytical Chemistry and Assaying, of Mineralogy, of Geology, Adjunct of Mining, of Electrical Engineering, of Civil Engineering, of Mechanics, the Instructors in Civil Engineering, and in Chemical Philosophy and Chemical Physics, the Tutor in Civil Engineering, and the Assistant in Mineralogy.[98]

In construction, advanced courses in the schools were often now gathered under Ricker's rubric of architectural engineering. If some were superficial, others taught by academic engineers gave the students a fair grasp of, or at least a sense of ease with, steel and concrete construction.

The balance of courses at Harvard, most humanistic of these schools, has been assessed by Anthony Alofsin. His conclusion may surprise those who picture the American Beaux-Arts tradition as lofty or effete. Harvard architects of the 1920s and early 1930s, pronounces Alofsin, were 'trained with a rigor in construction technology that equaled and usually surpassed the level of expertise attained by students trained under the modernists of the GSD' (the successor school).[99]

As elsewhere, architectural instruction at Harvard started on the fringes of engineering, within the Lawrence Scientific School, the college's semi-separate sister institution. Several first-rank architects, among them Jenney, McKim and Ware, spent time at Lawrence. As specialisms proliferated after the Civil War, its engineering department got larger and baggier.[100] Then Lawrence lost its independence to Harvard in 1890. Hence new cross-disciplinary groupings across the arts and sciences, among them the architectural course set up in 1893–4.

'Collaboration between the Lawrence Scientific School and the Department of Fine Arts was crucial for the new program in architecture,' says Alofsin. That is not to gainsay that it was driven from the arts and humanities side (ill. 411), chiefly by H. Langford Warren, the architect-scholar who started it up, and by Charles H. Moore in Fine Arts, who feared that subordinating the architects to the engineers 'would be very undesirable and even harmful'.[101] Although the course came under the science side of the university, Moore's point of view was not controversial. After 1906, when the Lawrence name disappeared, it became part of the Department of Fine Arts within the Graduate School of Applied Science. Institutions that teach architecture are used to occupying such curious interstices.

Schools of architecture that hitch their wagons to the star of design and composition easily grow remote from the world beyond the studio. Several factors prevented that from going to the extreme at Harvard. One was that it became largely a graduate school. So students had already taken a measure of other disciplines (as Richardson, Ware and McKim had all done) before they turned to design. A second was the rise of landscape architecture – interpreted in terms of city-planning – as a sister discipline within the school. At a time when improving the physical design of cities seemed to hold out their salvation, the analytic and social bent of urbanism balanced the self-absorption which goes with art. Finally, there was the technical manure dug in deep around the course's roots. If these considerations applied to such a school as Harvard, superficially averse to technology, they did so in equal or greater measure throughout the other American Beaux-Arts schools of architecture as the twentieth century took its course.

2.   Shop Culture at Bay: Britain 1750–1914

THE PUZZLE

The spectre of Britain haunts the topic of technical education for architects and engineers. On the face of things, the country that had taken the lead in material development daydreamed its way through the nineteenth century, heedless of how to train its technologists and professionals. Even after 1870, when warnings about national nonchalance over technical skills reached cacophony, remedial action can look sluggish. The University of Oxford possessed no professor of engineering before 1908; no full-time architectural course existed anywhere in the kingdom until 1895. How could a nation advanced in commerce, unvanquished in war, let that happen?

If formal education were crucial, Britain should have suffered a disastrous deficit in technical skills. But though it had lost its overall industrial lead by 1914, it had not fallen back irrecoverably. Some experts believe that the empire artificially sustained the home country, while lulling industry and invention into complacency. Others argue that a loss of primacy after 1850 was inevitable and that in many respects Britain did as well thereafter as could be expected.[102] Others again, Michael Sanderson chief among them, contend that British tardiness and sluggishness over technical education have been overdrawn.[103] Their studies distinguish between different industries, technologies and activities, the types of training they needed, and the response in each case.

The spectre emerges from another line of argument. Persistent in the British tradition has been a suspicion that the usefulness of academic education is overrated, and that the best place to learn skills is on the job. The thought is variously expressed: 'learning by doing'; 'sit next to Dolly'; 'muddling through'; or, as the academics put it, a preference for 'shop culture' over 'school culture'. Outmoded though that faith has become in the teaching of technology, it lurks in the background still, as a sullen reproach to the educational edifices which other nations look to have raised more readily and funded more liberally. In certain activities, the suspicion lingers that the academic backdrop is decorative, and that the show would go on as well without it. National scepticism about institutional training has often bolstered ignorance and smugness. But it has also made the British debate about technical education singularly impassioned and robust.

Inherently non-repetitive, site-specific, hard to manage, construction has been among the activities where a disdain for the classroom has endured. Let us hear 'learning by doing' articulated from experience, first by an engineer, then by an architect. Here is the Royal Engineer Francis Fowke telling an investigative committee of 1861 about his first practical experience of construction, the building of a barrack in Bermuda for which his training had not prepared him:

> I was put to construct that barrack without any assistance, and I found that it was necessary to instruct myself, and I took every means of doing so, by first of all picking up as much as I could from books, and also from actual observation, construction having always been rather the bent of my inclination . . . I believe that to have been the beginning of all the instruction that I have had in practical work, the actual doing of the work without any assistance; and I believe you will find that this has been the experience of many other officers of the engineers who have been thrown on their own resources. I have frequently heard officers say in private conversation that that has been their experience; I refer to officers who are good constructors.[104]

Now the Arts and Crafts architect and future professor Edward Prior, speaking in 1891 about the instruction he received as an articled pupil:

> He [Prior] was also anxious to learn something about old buildings, but in that desire he received little encouragement from his master [Norman Shaw], although he gave him Viollet-le-Duc's book to read. He also felt a very strong desire to know something of construction, for he felt that behind the drawings upon which he was engaged was the mystery of construction. His master rather pooh-poohed this idea, and said that it would come in time. One day, however, his master suddenly told him that he was to go down and stay for two or three years on a building as a clerk of works. He said he did not know what a clerk of works' duties were; but he was told to go and find out. He went, and then he found that the idea of wonderful construction was all an imposture: there was no science of construction, but there was an experience of construction to be gained by the man who worked with his own hands, and that the real artist was the man who worked with his hands, and not the man who made the drawing.[105]

Fowke and Prior would have disagreed over the aims of architecture. While the engineer sought technical efficiency, the architect dreamt of penetrating to the social secrets of his art. But they shared an impatience with formal instruction, and a passion for making things over an abstracted science of design.

It was to prove easier to corral engineers than architects into the British classroom. The science disciplines offer a home base for engineering, securing the various branches of the subject. Their principles must be learnt and understood before the natural world can be mastered. But science and technology are for architecture just two components in a construct of many parts. Its overall progress, though often presented as logical, is as wayward as that of humanity itself.

In addition, no state can ignore the tool for material welfare offered by engineering, in all its power and diversity. Here lie the roots of the long British campaign for better technical education. Skills must be improved, argue the reformers, so as to raise national performance and living standards. If they dwell on quality, it is because quality affects prosperity. A machine must be well designed and assembled if its makers are to be profitable; a fabric needs to be pretty in order to appeal to the market. Ultimately, design matters because it sells the product.

Calls for reforming technical education in nineteenth-century Britain therefore fastened upon activities affecting employment and wealth. Infrastructure – transport projects, mines, telegraphy, sewers – were the loins of the nation. That meant engineering. So too did heavy manufacturing, which generated jobs and exports. Likewise, industrial design and the crafts were perceived as vital, which is why the applied arts figured alongside engineering in calls for educational and technical reform.

Architecture, though, was largely missing. Architects do not sell anything obvious. The material improvements they make to buildings – the 'value-added' element – lie at the margins. They create few jobs and, even today, few exports. The idea that the architectural design of cultural 'icons' begets economic activity through tourism is recent, and did not occur to the nineteenth century. Besides, though it has always been appreciated that architecture confers prestige, uplift and identity, it has never been certain that the ability to purvey such qualities can be systematically taught.

In the end, the following pages show, British patterns of formal education for engineers and architects did not depart far from those set by other European countries and the United States. Much was copied from elsewhere. Yet throughout there ran a strain of reluctance. In time engineers settled into a system responsive enough to the multiple public calls for technological skills. For their part, British architects moved towards a style of training that gave them the freedom they craved from outside constraints and disciplines, along with a measure of status and a marginal academic recognition. But they proved able to discard shop culture only in part, and at a price. Unanimous only about the privileges due to their art, and fearful that too exacting a training might cost them their independence and their magic, they created a ramshackle, strangely confused style of schooling.

The Military

As in France and the United States, the first concerted school-training in construction available to British engineers came out of preparation for the Army. For years it was small-scale and haphazard. The Royal Engineers date back to 1716, and by 1761 had become effectively separate from the Royal Artillery.[106] The service boasted just 44 officers in 1800, rising to 262 at the height of the Napoleonic wars, then dropping back to 193 in 1819. From 1741 to 1812 cadets for the engineers and artillery acquired their instruction together at the Royal Military Academy, Woolwich, described by John Weiler as 'essentially a militarized public school'; recruits came in as young as thirteen and stayed up to five years. Excellent mathematicians and scientists taught at Woolwich, but 'there was no specific training in engineering or architecture'.

The impetus to supplement Woolwich came from a single individual, the Royal Engineer Captain Charles Pasley. Wounded at Walcheren, Pasley spent his convalescence penning an essay critical of British military organization.[107] In 1811 he improvised a course in Plymouth at his own expense, training non-commissioned officers and men to help the engineer officers in the field (ill. 412). So useful did this prove in the Peninsular War that his course rapidly found a permanent footing at Chatham. Here, in the new Royal Engineer Establishment, young officers just out of Woolwich were taught a year's-worth of construction from 1812 alongside Pasley's 'military artificers'. But the Army would not pay for a master. So the training was hand-to-mouth, teach-yourself stuff, reliant on textbooks by Pasley filletted from civil authorities. All this at the acme of Britain's industrial and military expansion.

In 1822 the Royal Engineers took over the design and construction of barracks at home and abroad, hitherto procured from civilians. As that entailed a modicum of architecture, Pasley added a new course and textbook. A civilian instructor was even hired. But the classroom teaching still relied on rote-learning and copying. Thus things drifted on at Woolwich and Chatham. Attempts at reform were obstructed on the grounds that practice, not principle, was the key to better construction. During intensive criticism of

412. Working dress of the Royal Sappers and Miners, 1813, around the time Charles Pasley inaugurated training in construction for officers of the Royal Engineers. From T. W. J. Connolly, *History of the Royal Sappers and Miners*, 1857.

the Royal Engineers in 1860, the Inspector-General of Fortifications, Sir John Burgoyne, fended off complaints of its educational policy thus:

> I have the impression that too much time should not be given to what is called the architectural course; for the practical purposes we require, it is scarcely susceptible of being learned by book and theory; the proportions of details of buildings and constructions, and to define proportion and put together several materials are essential items, and will be more readily acquired by closely witnessing the actual practice and operations, and studying by experience and effects how to gain strength with the smallest means, and therefore chiefly to be learned when employed on great works.

As has been seen, Francis Fowke felt likewise. Though the rudiments of theory were certainly taught at Woolwich, the constituency for reforming the classroom component in the Royal Engineer Establishment was weak.

Yet out in the yard at Chatham, and wherever the Royal Engineers built once Pasley had got a grip on their training, experimentation was rife. Despite their low status, the continuity they enjoyed allowed them to do things for which consulting civil engineers and contractors seldom had the time. They could take ideas on from one structure to the next, test and authorize materials, inspect and publish. Under Pasley, for instance, cements were investigated and assessed (pp. 212), while in a famous run of dock buildings in the 1840s and 50s, Chatham-based engineer-architects advanced the spanning and framing of enclosed structures. Because learning by doing was an ideal, it is hard to draw the line between education and experimental building at the Royal Engineer Establishment. The formal poverty in the school did not inhibit creativity in the shop.

Elsewhere in Europe and the United States, we have seen, state corps of engineers and the training they spawned spread out to the general development of architecture and engineering. That hardly happened in Britain. Some cross-over did occur. The Royal Engineers sometimes helped on the fringes of urgent national tasks like the great exhibitions, or adjudicated in technical enquiries; the railway inspectorate, for instance, was

413. Faculty at the Royal Indian
Engineering College, Coopers Hill,
*c.*1882.

staffed by engineer-officers. A few major Victorian prisons and hospitals were designed
by REs. Henry Cole during his reign at South Kensington also employed them, maintaining that they were more efficient than private architects (pp. 135–7). In the 1870s Cole
wrote a memo urging that Royal Engineer officers should be regularly seconded to the
civil service and made inspectors of public buildings, but it came to nothing.[108] Within
Britain, RE architect-engineers usually stuck to building-types within their bailiwick like
forts, docks, barracks and military hospitals. Even here there was always a threat of
reversion to designs by civilian professionals.[109]

Beyond the home-country things were different. Obliged to make shift with what construction skills they could find, colonizers willingly embraced military methods and training. India is the famous example. In 1809 the East India Company established Britain's
first engineering college, the Addiscombe Military Seminary. Addiscombe trained most
of the engineers active in British India before the Mutiny of 1857. But it could not long
survive the abolition of the company entailed by that event, and the subsequent expansion of the Indian Public Works Department. The upshot was the Royal Indian
Engineering College, Coopers Hill, started in 1871.[110] As Mountstuart Grant-Duff,
under-secretary for India, put it to Parliament:

> The experience of eleven years had taught them [the India Office in London] that our
> scientific training in England, like so much of the rest of our training, was in a chaotic,
> and indeed, contemptible state. They had learned that an engineer is trained in
> England as a barrister is trained, by what some people are pleased to call 'the practical
> system' – that is, by rule of thumb . . . Do you suppose that if India had belonged to
> Germany, had belonged to France, had belonged even to Switzerland, it would have
> been necessary for its representatives to set up an Engineering College of their own?[111]

In the event Coopers Hill elicited only minimal government funding and lukewarm
support from India itself. Still, it offered a three-year course in civil engineering under
four professors, three instructors and a draughtsman (ill. 413). No British university could

then offer that much. Once they could, a clamour arose about the cost of Coopers Hill. Down it duly closed in 1903, despite its record in training most of the construction professionals who joined the Public Works Department and created the impressive infrastructure and buildings of late-Victorian India.[112]

The Universities and Engineering

Here then is the spectacle of a nation willing to pay for the technical training of its soldiers and colonizers only if it was basic, cheap and impinged little on civil society. So long as national development ran on, construction could look after itself. Such was the attitude usual before 1850 and common afterwards. Even the great professional societies, at their mid-Victorian zenith, were nonchalant about education. Both the Institution of Civil Engineers and the Royal Institute of British Architects discussed it now and then, but neither did much about it until late in the century. For most of their members professionalism was bound up with the apprenticeship system, from which they received fees. Controlled examinations sometimes interested the societies, as routes to restrict future entry into their professions and promote 'registration of title' – the closed shop. But the nature and contents of training drew scant response.[113]

It was through the efforts of those acquainted with other countries that education for engineers trickled into British universities. In the vanguard was Scotland, where high literacy rates in the Lowlands plus four established universities and a proud intellectual tradition helped prompt demand for a modern higher education.[114] Glasgow established its chair in civil engineering in 1840. Its first tenant was the young and versatile Lewis Gordon, who had learnt his trade under Marc Brunel before studying mining at Freiburg.[115] Gordon practised as an engineer before, during and after his stint at Glasgow. He taught mainly mechanics, amidst the jealousy endemic in universities. Told not to encroach on natural philosophy and mathematics, he asked for a place where he could show models and apparatus. The chemists denied him the use of their room, so he ended up in the law classroom. No apparatus was supplied, nor was a degree conferred for years. At least it was a start.

Edinburgh followed suit in 1868, electing to its engineering chair Fleeming Jenkin, another supple practitioner with foreign experience, able to range across the civil, mechanical and emergent electrical fields.[116] Again there was no laboratory; classroom instruction had to suffice. Jenkin devoted his inaugural lecture to 'the Education of Civil and Mechanical Engineers in Great Britain and Abroad'. After a glowing review of provision in France and Germany, he recommended a middle course, with engineers moving on from an academic training in principles to practical apprenticeship. Both were needful, he believed; and he trusted to the elasticity of British universities to adapt. So it turned out.

Jenkin came to Edinburgh from part-time teaching at University College London. UCL and its rival, King's College London, were the first English university-level institutions to dabble in engineering and architecture. Their early courses from the 1840s catered for whoever cared to come along and pay. 'Engineering and the Application of Mechanical Means to the Arts' were among the subjects adumbrated in the original UCL prospectus of 1827, but it was long before the subject emerged from beneath the wing of classroom physics.[117] Even in London, where industry, commerce and the professions most powerfully coincided, the gap between practice and theory took years to bridge. The technical education lobby, findings its voice as Britain's share of world markets declined and the great exhibitions highlighted shortcomings in its products, did its best. But it was split between those who saw artisan training and social welfare as its chief aim, and the partisans of scientific innovation by a theoretically trained élite. While both ends were pursued, engineering found itself caught in the middle. Nor did institutions wishing to teach on the cheap scruple to argue for the superiority of theory, so as to save money on apparatus for demonstration and experiment.

Cambridge is a case in point.[118] Hostility towards professional education and practical training has persisted in England's two ancient universities, and has as persistently to be smashed. In respect of technology the erosion began the sooner in Cambridge, because of that university's repute in mathematics. Many great technologists such as Lord Kelvin started out as Cambridge mathematicians. Brilliant lectures on engineering principles 'with their practical applications to Manufacturing processes, to Engineering and Architecture' were also on offer after 1837 from Robert Willis, professor of natural and experimental philosophy, and historian of Gothic structure (indeed in Pevsner's eyes 'England's greatest architectural scholar in the nineteenth century'). Willis favoured a professional training for engineers in London, where he taught at the Royal School of Mines, and promulgated a special kit for 'lecturers and experimenters in mechanical philosophy'. Yet he obstructed the setting-up of an engineering school in Cambridge, as recommended by a Royal Commission in 1852.

After Willis's death in 1875, James Stuart was appointed to a new chair in 'mechanism and engineering'. Stuart, who had come to the fore as an extension lecturer teaching engineering to working men, threatened to draw Cambridge headlong down the path of practice, setting up a smithy and workshop at his own expense behind his lecture room and fostering contacts with industry. That ran counter to the classroom and collegiate system. His work having been pronounced 'too practical', 'not yet integrated into the rest of the university', Stuart was frozen out and failed to establish a degree in engineering. It was left for his emollient successor to tack back from craft towards science and secure what became known as the Mechanical Sciences Tripos in 1892.

In his great study *The Universities and British Industry*, Michael Sanderson has described how the English knocked their applied higher education into shape at the end of the nineteenth century, creating a network of civic 'university colleges' where science bedded down with technology, and overtaking the Scottish universities.[119] Because of the challenge from American and German manufacturing, the burgeoning branches of engineering were to the fore. Awareness of the boost supplied to German-speaking countries by the Technische Hochschule system also helped promote a second tier of locally supported polytechnical and technical schools, in which engineering and craft training ran parallel with social aims; vestiges even of architectural teaching can be found. Imperial College London, by contrast, was an attempt to transplant the TH system wholesale into a grand technological university.

Historians now tend to think that the Technische Hochschulen were excessively revered or feared.[120] Falling over one another to update technical training, every nation came up against the rigidity of its own traditions and institutions. By 1900, despite specific shortcomings, engineering education in British universities and colleges was solidly based at last, with an expanding range of specialisms responding to shifts in technology. The detail of its development needs detain us little longer. Formal architectural education, on the other hand, was only just then emerging from its chrysalis. The oddities and hesitancies of its autonomous existence, and the evasive attitude it took to technical training, in Britain and beyond, will dominate the remainder of this chapter.

BIRTH OF THE BRITISH ARCHITECTURAL SCHOOL

Architecture is absent from the Victorian campaign for technical education, I have argued, because its economic impact was weak. Few denied that it affected art and welfare alike, or that it required many skills and years to master. But it presented no pressing needs. Apprenticeship had not evidently failed to deliver professionals who could design sturdy, practical buildings. If there were structures that an architect could not or would not design, an engineer or a builder could take his place. From the national standpoint that sufficed.

The clamour for better architectural education in Britain therefore arose from within: from apprentices ('articled pupils') who felt ill-instructed by their masters, and architects who wanted to protect their patch against competitors – chiefly builders, surveyors and engineers. These interests coincided in a protracted campaign, from the 1850s until 1938, to devise recognized examinations which could both meet student aspirations and serve as agreed entry qualifications to the profession. The staple of the British debate on architectural education is about examinations and 'registration of title'. Against that backdrop schools of architecture emerge, multiply and puzzle over what should be taught and how.

Until after 1900, it was assumed that a formal supplement to office-training could benefit young architects but need take up no more than an afternoon or evening or two a week. Lectures long predominated. In London a burst of initiative took place in 1840–1, when T. L. Donaldson started his courses on architecture at University College, William Hosking did the same at King's, and C. R. Cockerell revived the custom of professorial lectures to students at the Royal Academy. (The Academy has some claim to priority among Britain's schools of architecture, but hitherto its instruction had been remote and intermittent.) All three had their own slants, ranging from Hosking's technical bias to the art-inspiration of Cockerell. Donaldson's lectures, part of a polytechnic-style 'Vorkurs' running across the sciences, seem to have been modelled on Durand.[121]

These departures whetted demand. In 1847 young architects and students banded together to found an educational club. That was the Architectural Association, slowly to evolve into a unique, self-governing school of architecture, always vulnerable to swings in architectural fashion. It began with lectures and debates, but there was also a fortnightly 'class of design', building on the drawing side which was the core of office training. That practical side grew and intensified from the 1860s.[122] The Royal Academy also instituted evening classes in design after 1870 with the Beaux-Arts trained Richard Phené Spiers as its rather hands-off master, and architectural academicians dropping in as visiting critics.

These drawing and design classes were successful, popular but unsystematic. 'I do not remember', wrote Charles Reilly about the AA as late as 1900, that 'we had any definite teacher. I suppose our designs were hung up and criticized by someone, but I am sure we learnt mostly from each other.'[123] The lecture or taught class, relaying definite information, technical or cultural, was still the main mode of instruction. In this vein half-remembered figures like T. Roger Smith at University College (1880–1903) and Banister Fletcher at King's (1890–9) addressed large, mixed audiences of budding architects and engineers. There was little inkling of full-time education or of the dominant 'studio' of today. The same held true beyond London, where classes grew out of civic architectural associations on the AA model, or on the fringes of technical or art schools.

Such part-time arrangements subsidised established architects, who still pocketed the fees of articled pupils while training them less. Disquiet about that helped tip the balance towards a proper school-model for British architectural training. But other forces were at work. The 1890s witnessed a first abortive attempt to make architecture a closed profession by examination. It elicited a confused ('rather stupid', Reilly thought[124]) controversy as to whether architecture was 'a profession or an art'. Those who successfully repressed the infant registration movement claimed that only technical and business topics could be formally taught and examined, whereas the art of architecture – design and its relation to building – depended on intuition and experience alone. That argument, while upholding architecture's autonomy, consigned schools to the margins.

Yet the opponents of registration accepted, and regretted, that a division had taken place between the arts of design and of building, one deepened by new construction techniques. There were two possible responses, the younger generation came to realize: leave construction to the engineers and builders, and get on with design as the best guarantee of independence and status; or try and regroup the skills of construction under architectural leadership. Here was a broader argument to fight out in the classroom. Should they be constructors or designers? It was against the backdrop of that issue that the formal training of British architects finally took shape.

414. W. R. Lethaby, plaster relief by
Gilbert Bayes, 1923.

The constructors, led by W. R. Lethaby (ill.
414), hoped to convert the part-time schools of
architecture into part-time schools of building.
In powerful, Ruskinian homilies, Lethaby
asserted that architecture as a stand-alone subject was a modern aberration. All the good art of
the past had come out of skilled construction, he
argued. Therefore:

> We must learn about building in schools –
> schools for practice and theory, experiment and
> research. Building schools exist in several conti-
> nental cities, and it seems to me that such
> schools, as representing a very large and impor-
> tant industry in big towns, should be established
> or assisted out of public funds. In a big London
> institution I should like to see all the building
> crafts carried on side by side, where experi-
> ments might be made in brick-arching, stone-
> cutting, timber-framing, and so on, with due
> supply of apparatus and testing-plant. Here also
> the mechanics of construction should be taught
> mechanically and demonstrated in models . . .
> Then I would have planning and normal
> arrangements for given purposes taught. Even
> taste, the objecting negative taste of a good
> critic, might be allowed, but not a word on 'art,'
> and 'design,' and the styles, in the usual accept-
> ance of those words. In such a school we might
> hope to bring together the different craftsmen,
> builders and architects, all studying together the
> true art of building and evolving a reasonable
> architecture . . . It would have been well if we

could have been ready with a scheme in which all might join a dozen years ago, when
Technical Education was first being practically dealt with; but I fear unless we are less
vague in our aims nothing will be done for a further dozen years, and that I feel would
be a calamity. But some day, pleasant, natural, living architecture, will be refounded on
common building – it can stand on nothing else.[125]

Here was shop culture restored and enthroned. The best hope for such a programme,
Lethaby believed, lay not in the half-fledged academies with their dry lectures and
evening 'classes of design' attended by young architects, but in the hands-on schools of
craft and technology that had sprung up in the 1890s. Steps were indeed taken along
these lines at London technical schools subsidised by the London County Council, which
Lethaby served as art inspector. At the Central School of Arts and Crafts, the Regent
Street Polytechnic and the Brixton School of Building (ill. 415), traditional building construction proceeded by practice rather than lecture, while attempts were made to draw
in new materials and structural mechanics. Advanced building fabrication also took place
at Brixton, where Beresford Pite headed an architectural department from 1906.[126]

Such courses took place mostly in modest institutions whose success at integrating
architecture with construction was later forgotten. Where architects were taught on their
own, that ideal fared less well. At the AA, it flowered briefly between 1895 and 1898
under Lethaby's disciple Owen Fleming, who promoted an 'AA School of Design and
Handicraft', in other words a practical construction course based on working with mate-

rials. But it was always resisted, and after Fleming resigned it petered out. As one recalcitrant put it, 'skill in a trade, the price of long years in constant work, carries with it no inevitable faculty for design, nor is the power of composition denied to those who possess no manual dexterity.'[127] In those terms, the split between design and building was an accomplished fact; architects should stick to the drawing board and leave construction to engineers and builders.

Class-issues underlay the failure to integrate the training of British architects with that of engineers, builders and craftsmen. But there were other currents. As in Germany, an independent education for architects could only make headway once there was sufficient and steady demand for their specific services. Once that was secure, full British schools of architecture arose after 1900. But in their heedlessness of technology, they differed from their counterparts throughout Europe and America, which having started at an earlier phase of economic and cultural development had grown out of and maintained contacts with schools of engineering. The Ecole des Beaux-Arts was the one important exception.

Another factor that came into play was the resurgence of classicism around 1900. Teaching creative design might be a mystery. But the classical orders had long offered a rule-based language which could be taught and learnt as objectively as technical topics. The drawing and understanding of the orders, inculcated in offices and lectures alike, now presented itself anew as the basic technical discipline of the autonomous architectural school. The Beaux-Arts offered a model, familiar in itself and from American adaptations, for building upon that basic training by means of an atelier system. Pursuing that route, the loose British 'classes of design' turned into homegrown versions of the Beaux-Arts-style atelier or studio. Upon the rock of classicism architectural training in Britain thus built the semblance of an objective technique.

The final ingredient was the drawing of British universities into educating architects. That was not a foregone conclusion. In France and Germany, architecture had confined itself to technical and art schools (the system still obtains in France today). Only in the

415. Clay modelling class at the Brixton School of Building, May 1914. The school was predominantly artisan in character but included instruction in drawing and architecture.

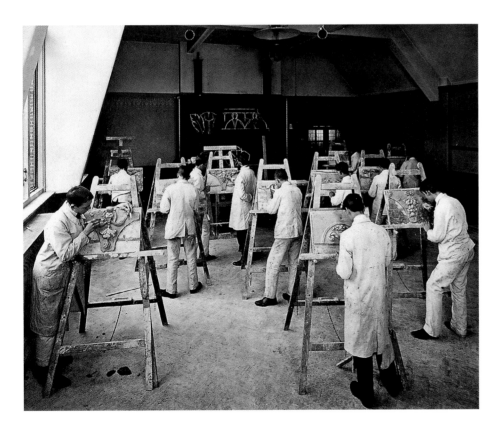

United States had university and technical colleges become inextricable. In British universities, we have seen, engineering had often been resisted on the grounds that it was a
practical, professional subject. It had penetrated in the end because of national pressures.
Architecture became a university subject for other reasons: chiefly, perhaps, because the
academic ideal suited those who configured it as a pursuit of polite learning rather than
as a professional or technical service. Here again class enters the picture.

Two examples, from a new and an old university, illustrate the point. Liverpool was
the first British university to offer a full course in architecture, in 1894–5. It grew out
of supplementary training classes for articled pupils, held under the auspices of a local
architectural association. These were then adopted by academics keen to temper the
technological bias of Liverpool's 'university college', as it matured into an independent university. So the first director, F. M. Simpson, was hired from the art end of the
professional spectrum. Most of the pupils came in part-time from offices, though there
was a full-time option. Linked with the architectural section was an equally small art
school, which started out on a crafts trajectory before veering off towards the chic of
the fine arts.[128]

The Liverpool school took off after Simpson gave way in 1904 to Charles Reilly (ill.
416).[129] The son of a London surveyor-architect, Reilly had been among the first to study
the new Mechanical Sciences Tripos at Cambridge. He excelled, but felt himself 'a black
sheep in taking engineering', reacted against the 'purely technical education' he had
received there,[130] and espoused architecture as an art. Reilly's career abounds in para-
doxes. He might have made an excellent engineer, yet spent his early years dressing up
power stations and substations in classical clothes. A sharp operator, he brought to
Liverpool resolve, dash, and a talent for improvisation.

In time Reilly worked out that the way to curry respect for architectural education in
Britain was to mimic the American university schools. It was the Beaux-Arts side of
American instruction that he chose to stress, coupled with its civic agenda. By taking
advantage of a benefaction he was able to house his architects alongside the university
school of art and a new department of civic design in their own building, aloof from the
technologists. Its atelier arrangement boosted the school's confidence and independence.
The figure with the best academic background in engineering of any British architect up
to that time thus became the prime champion of an aesthetic university education in

architecture. Nevertheless Reilly's Liverpool was no mere school culture, contemptuous of office experience. The ordinary architectural degree instituted in 1906 could not be taken until three years in an office had followed two years' coursework. Nor was construction (of a conservative kind) ignored in the school.

At Cambridge, a town too small to generate the local demand for training that underlay the start of Liverpool, architecture was thrust upon a reluctant university from within – by its classicists.[131] The prime mover, the American-born Charles Waldstein, had been educated at Columbia and Heidelberg before running through a plethora of posts and activities centred upon classical archaeology and Cambridge. Drawn towards architecture by developments in art history, Waldstein dallied with the notion of an intellectual 'university student of art', midway between artist and amateur.[132] By 1904 the idea was abroad that architecture might be taught not as professional training but as part of a humanistic education. Waldstein decided to see what might be done in Cambridge. He consulted William Ware, not long retired from Columbia, who told him of the difficulties he had had in winnowing out architecture from the technical courses taught there.

The scheme for an examination – no more – that issued in 1908 after much wrangling within the university differed from what Waldstein had envisaged. Containing some technical study, some history and a little design, it pleased nobody. It even brought down ire from a constituency which might have been expected to support it – the few senior architects who, having taken Oxbridge degrees, had submitted to long professional apprenticeships. Among these was a figure active in the intricate politics of training, Reginald Blomfield. He and his allies opposed architecture in universities, most of all in Oxford and Cambridge, on the grounds that their purpose was to offer a general education: 'a technical school at a University which set out "to provide a liberal education" was a contradiction in terms', he told Waldstein and his coadjutors.[133]

A technical school was not what had been proposed, but the truth is that Cambridge had no clear aims. Such as it was, the tiny new school had a bias towards history and 'the study of art'. Awarded no teaching funds, it could not operate until Waldstein injected some of his private money in 1911–12 and resigned his professorship in favour of Edward Prior, the Arts and Crafts architect who had once declared that the science of construction was an imposture (p. 457). Whereas engineering, once legitimized, developed apace at Cambridge, 'architectural studies' remained a marginal discipline, held at arm's length from professional training. Somehow the frail plant clung on, but it was often touch and go.

A turning point for the training of British architects came in 1903–5, when the RIBA set up a Board of Architectural Education to advise on and validate the multiplying courses. In casting its recommendations the main voices came from Lethaby and Blomfield. Though they collaborated respectfully, the drift of events and opinion was from the former's constructive passions to the latter's forthright art-idealism. 'We want to train expert constructors, engineers of building, men of initiative and daring', Lethaby had urged. But his emphasis on workshop or laboratory training vanished from the Board's final advice, to which after genuflection to construction this rider was added:

> The student should however be taught that architecture is something more than engineering, that while the engineer is concerned with scientific instruction, it is the architect's province to see that his construction is not only scientifically sound but aesthetically beautiful. The students should therefore be trained in the selection of form, and in the study of mass and proportion.[134]

That formulation guaranteed architects the independence and status they sought but gave little guidance for teaching. Nor did the Board tackle the shocking standards of entrants to the schools. At Liverpool, Simpson had done his best to broaden his students' minds by urging them to take extra courses in 'the subjects ancillary to architecture which are daily growing in importance', instancing 'mathematics, mechanics, French,

417. Hermann Muthesius sets off on his bicycle from Hammersmith to pursue his researches. His wife Anna bids him goodbye.

etc.'[135] 'Pupils, owing to their imperfect general education, do not know how to learn', pronounced Blomfield and his allies at the time of the Cambridge controversy.[136]

The problem was bound to surface once schools of architecture aspired to do more than just supplement office training. So broad is the reach of their subject that architects benefit from a wider general education than technologists. Achieving this with the time and resources available is always a stumbling block. The schools' dilemma was not only what best to teach but also in what order. Should a sound knowledge of the humanities come first, and design and technique follow later, as Blomfield believed? Or should drawing and mathematical competence precede design, as in polytechnical training? No sooner had schools of architecture emerged than the cry arose for their courses to be longer or, as in some American schools, for the topic to be studied only as a graduate degree. That was unlikely in the frailer British context. Along with that went a paradox. Breadth or depth of knowledge does not necessarily add to ability in design. Yet while vaunting themselves as specialists in design, the schools were tempted by the quality of their recruits to postpone design until a raft of other subjects had been mastered.

Reviewing the state of architectural education in Britain and abroad in 1910, Alfred Cross reached unpalatable conclusions.

In comparing the results obtained in the architectural schools of France and America with those of similar teaching institutions in our own country it appears to be quite evident . . . that the educational facilities provided for the students of architecture in this country are wholly inadequate either for his present or his future needs . . . the root of the evil lies in the insufficiency of the educational tests instituted by that body [the Board of Examiners of the RIBA] . . . Among other defects its mathematical test is of a phenomenally low standard, whilst the papers in Geography and History are ridiculously easy . . . As to the questions in Elementary Mechanics and Physics, they are so elementary that they can scarcely be regarded as being in any way in the nature of a searching test of the candidate's knowledge of these subjects . . . If we compare the strenuous lives led by the young architects of the Ecole des Beaux-Arts, try to realise the vast amount of work involved in the successive examinations which step by step have to be grappled with in their entirety, for a long terms of years, and compare the courses of study at the French and American schools with those necessary to satisfy the examiners in our Intermediate and Final Examinations, we are compelled to come to the conclusion that the education of the artist is regarded very seriously in France and other countries other than our own.[137]

Cross focussed upon art, the ideal upon which the Edwardian architectural schools had chosen to fix. But following the watering down of Lethaby's advice, the majority of them had also ducked the pressing issues of construction. Having plucked students out of the offices, they had drawn them inwards to the narcissism of the design class instead of outwards to the challenge of workshop or site. Architects had not just failed to grasp the techniques of 'common building', Lethaby came to believe; they had shut their eyes to engineering. 'Although all these modern activities frighten me', he wrote in 1918,

and I would rather be dealing with rubble and thatch than with concrete and steel . . . I have seen much which causes one to look again, in great bridges spanning a valley like a rainbow; in roofs meshed across with thin threads of steel; in tall factory chimneys, great cranes and ships; or even in gasometers . . . I have no love for modernism as such, and fain would hide my head in the sands of the past, but I cannot help seeing that the courageous mind will shape even seemingly hopeless materials to its purpose.[138]

English diffidence and nostalgia apart, here are the sentiments of Le Corbusier and other twentieth-century rhapsodes bidding architects board the good ship technology – welcome now within the armada of art.

The means to address such structures had to start as much in the classroom as on site or in the office. Lethaby half-admitted that when he called for 'high mechanical training, wide practical experience, and great geometry'.[139] Whether they construed architecture as an art-destiny with rules of its own, or as a means to civilize the industrial environment, the skills which young British architects needed could no longer be acquired without the schools. By the end of the First World War, shop culture still hung on in architectural education, in the form of office experience. But it had lost its allure. To be revived as an ideal, it needed a new guise.

3.   Triumph of the Art School

BEFORE THE BAUHAUS

When Hermann Muthesius (ill. 468) visited England in 1896, he was following in the footsteps of Schinkel and Beuth seventy years previously (pp. 86–7). Both visits were sponsored by the Prussian Government, both involved an architect, and the charge both times was to report back to the government on British prowess in design and industry. Schinkel and Beuth stayed three months, looking mainly at engineering and infrastructure. Muthesius stayed over six years. His brief was broad and liberal. John Maciuika has defined it as 'to present coherent evaluations of English technical and cultural trends to the Prussian Ministry of Commerce and Trade and the Prussian Ministry of Public Works, especially where these could inform the development of Prussia's infrastructure and economic-development policies.'[140] But in effect he represented the transformed German nation as a whole. Soon, Muthesius was being asked to stop sending back 'tedious engineering reports', like those he wrote on the railway system.[141] Instead training became his main official concern, as he looked into schools of design and progress in the applied arts. In addition, out of personal enthusiasm he became a devotee of British architecture and the chronicler of the English house.

On the merry-go-round of technical education, nineteenth-century nations chase one another. Schinkel and Beuth hope to learn how Britain has managed its industrial revolution and to suggest strategies which, in respect of design and technology, can allow Prussia to follow suit. Then in 1837 Britain, apprehensive about French skills and budding German improvements in textiles, ceramics and other manufactures, creates a Government School of Design to stimulate the applied arts. Taken up by Henry Cole and his circle after the Great Exhibition, that develops into a national network of design-school training, the so-called South Kensington system. Berated by Ruskin for its preoccupation with pattern-making, the system does at least exist, unlike schools of architecture; its alumni find posts with manufactures in which pattern is to the point. As funding swells for technical education in the 1890s and the English Arts and Crafts Movement finds its voice, the next generation reacts against 'South Kensington'. But it is something to which a unified Germany is alert, as its centrally steered trade and industry thrust forward. The carousel has turned again: hence the presence of Muthesius in London.

Why Muthesius and, given his brief, why an architect? Like Schinkel, at the time of his mission he was a state official.[142] His training had been intense: apprenticeship under his father, a builder in Saxony; modern secondary schooling; a year at university studying philosophy and art history; and the four-year architectural course at the Königliche Technische Hochschule, Berlin-Charlottenburg. He next spent some years with the Berlin architects Ende and Böckmann, including a spell in Japan helping supervise public buildings they were designing there.[143] In 1895 Muthesius joined the Prussian Ministry of Public Works as a government architect. Such positions were constraining but not all-consuming. The Prussian state, Maciuika has written, 'had a long tradition of supporting architects and engineers in their efforts to develop private practices outside of their government posts'.[144] Muthesius could follow his nose, so long as he carried out official assignments.

Of the two ministries which sent Muthesius to Britain, the Ministry of Commerce was in the ascendant. In 1884 it had won a demarcation dispute with the Ministry of Education over control of the applied arts and their teaching. That lay at the start of a concerted commercial-industrial policy throughout Germany towards design and production in this sphere, with international competitiveness in mind. The dispatch to London of a young civil servant expert in the arts to report on how things were managed there marked a step in that process. Not that Muthesius was a pawn. He had written about British architecture before he went,[145] and may have suggested the mission himself.

The choice of an architect to look into teaching in applied arts and design echoes arrangements again going back to Beuth, Schinkel and the origins of the Berlin Gewerbeakademie. The claim of architects to leadership over the range of the visual arts had a record of support in the German states. When a school of art and industry opened at Offenbach in 1868, for instance, architects had priority as teachers over painters and sculptors.[146] Muthesius was the harbinger of their growing involvement in directing such applied art or design schools – the Kunstgewerbeschulen. A few of them – Walter Gropius of the Bauhaus is the most famous – were to head such schools.

Architects justified this by means of a dogma hallowed in English Arts and Crafts ideology. 'Almost all artisanal activity only finds its true existence in the context of or against the background of architectonic intentions', proclaimed Fritz Schumacher, one of the movement's leaders, in a typical formulation.[147] The attempt by Lethaby and his allies to draw architecture into the crafts schools, which Muthesius studied, rested on that questionable belief. In Germany it was matched by a wider opportunity that reformers like Gropius and Schumacher would eagerly seize. Investment was starting to funnel into the Kunstgewerbeschulen, with commercial ends in mind. But most such schools taught architecture only at the margins, if at all; that was reserved for the established Technische Hochschule system, where architects typically learnt their trade in a dry and regimented way, alongside engineers.[148]

It became the reformers' aim to promote architecture or at least architectural ideas and leadership in the Kunstgewerbeschulen, while also refreshing its teaching in the Technische Hochschulen with the competitive forces driving the applied arts. In that way architectural education could slip the yoke of engineering, tighten its grip over other art-disciplines, and claim relevance to an industrialized economy. That process culminated in the glamour of the Bauhaus. Under the guise of modernization and the quest for 'form' (that compelling concept in German culture) through 'design', the teaching of architects would shed its drudgery, and the bondage of the engineering school give way to the freedoms of the art school.[149]

That is to anticipate. What Muthesius brought back from London in 1903 was an affection for homespun English ideals of craft production and building, coupled with a resolve to update them for the furtherance of Germany's ambitions. His own private love continued to be architecture, in which he built up a fair practice. But the Ministry of Commerce, to which after his return he remained loosely attached, saw that there was more to be gained from his advocacy of design reform within the home and civic environment, by bringing it to bear on trade and on teaching in the Kunstgewerbeschulen. The impact of improved training for apprentices could already be seen. Germany's applied-art workshops were on a roll, winning plaudits at international exhibitions and making inroads in sales even to that art-proud nation, France.[150] If their energy could be channelled into an alliance between enlightened industrialists, high-class designers and the schools, everyone would be a winner. Hence the inauguration of the Deutscher Werkbund in 1907.

The Werkbund's purpose, as put by its most articulate founder, Friedrich Naumann, was to create fresh, practical and marketable forms for an industrialized culture by means of design, in order expressly 'to extend Germany's economic power'.[151] Attached to that crusade came artists and architects, keen to endue that technical-commercial directive with a veneer of art-idealism. Peter Behrens is an example. Appointed through Muthesius's influence in 1903 to head the Düsseldorf Kunstgewerbeschule, Behrens was explicit about his new role: 'What pleases me greatly', he told Muthesius, 'is the opportunity it extends to be able, in a direct sense, to be subservient to the interests of the state.'[152] Yet Behrens was a designer to his fingertips. A painter by training, he had gravitated through the workshop movement into the applied arts, and then into architecture. Possessing an instinct for form and a knack of collaborating without loss of command, he could create a poster, a light fitting, cutlery or the carcase for a turbine factory with equal fluency. It was for the veneer of memorable, marketable design which he could impose on disparate products, not for his technical abilities, that Behrens was valued within the Werkbund and by those that hired him, notably the great electricity combine AEG.

In the early days of the Werkbund the primacy of the applied arts was never in doubt. 'We designed artistic ashtrays and beermugs, and in that way hoped to work up to the great building', remembered Gropius.[153] Architecture stood in an oblique relationship with the Werkbund's public purposes. It enjoyed unique authority among the arts; architects were eager to participate in the movement; and building projects helped advertise the reform programme. Yet though architecture can publicize and inspire, it can seldom be put into direct production. Efforts to close the gap were certainly made. It was at this time that the young Gropius first embarked on his life-long crusade for prefabricated housing, in an effort to turn buildings into industrial commodities that could be bought and sold.[154] In the event, the dream of houses from the factory was to prove a constant struggle against the architectural instinct for uniqueness. Nor did Werkbund styling run deep in buildings. In the exhibition halls and factories that became its public symbols, the architects were commonly confined to the design of the façades behind which, as of old, engineers took over.[155] What they offered was an image of modernity, a rhetoric that consecrated technology while claiming also to transcend it. Writing on 'the problem of form in engineering construction', Muthesius admitted as much.[156]

If that was to be the twentieth-century condition, architects did not need to learn their trade in the painstaking depth which Muthesius, for instance, had plumbed. The pedantry of the Technische Hochschulen on the cultural and technical sides alike had long been resented: 'the whole method of training suppressed the individual', recalled Gropius later.[157] Behrens had leapt into the front rank among German architects without any architectural education; other leading lights of the Werkbund, Henry van de Velde, Bruno Paul and Richard Riemerschmid, did the same. Maybe it was enough to inculcate the talents of a Behrens as a designer and ringmaster, co-ordinating modern trades and skills. If so, that could best be done via the Kunstgewerbeschulen.

In a symptom of pre-war Germany's febrility, the tensions between art and commerce broke at the Werkbund's Cologne conference of June 1914. The organization's focus had never been art and architecture *per se* but the design and production of industrialized goods. That priority Muthesius and his official friends reasserted one more time at Cologne. For their part the artist-architects had tasted influence in government through the Werkbund. In the shift of its offices to Berlin, in Muthesius's creeping autocracy, and in his insistence that members should focus on types, they feared not only for their freedom but also for their grip on the Kunstgewerbeschulen. The onset of war put paid to these bickerings. But Gropius, siding with the artists, did not forget. It was in a shattered post-war Germany, with industrial production close to zero, that he resurfaced to shape the Weimar Kunstgewerbeschule into the Staatliches Bauhaus.

The Bauhaus

It is a curiosity that the most famous twentieth-century school of architecture taught that topic for less than half its brief and restless existence, between 1927 and its closure in 1933. From its foundation in 1919, and even after the formal addition of architecture, the Bauhaus was largely a radicalized Kunstgewerbeschule. Yet in each of its phases instruction took place under architectural overlordship. All three of its directors were architects. From the moment when Gropius (ill. 418) became heir apparent to Henry van de Velde at Weimar, there was talk of teaching architecture in whatever combined school might replace its pre-war Kunstgewerbeschule and its fine-arts academy. 'The ultimate aim of all visual arts is the complete building', proclaimed Gropius in the original Bauhaus programme.[158] Throughout its vicissitudes, within the school architecture was always upheld as the grail.

To the official world, things were presented otherwise. External funding for such schools depended on their turning out people and products for economic regeneration. So long as architecture did not obviously contribute to that, support for teaching it was lukewarm. Adept politically, Gropius varied his formulations to cover the discrepancy. Amidst the misery and unemployment of 1919 he appealed to utopianism, idealism and the unity of the arts under architectural command. These were the years of art-school speculation and free experiment. When the economy picked up, he reverted in 1923 to the Werkbund ideal of art and technology uniting to boost trade. The pretext for omitting architectural classes now was that students needed first to learn about craft and elementary form. They started therefore in the craft workshops. Internally, their function was understood to be to give play to the imagination, but Gropius explained them as 'really laboratories for working out practical new designs for present-day articles and improving models for mass production'.[159] Those who proved talented architectural 'journeymen' could then go on to 'complete these studies at technical universities and polytechnic institutes'.[160] As yet the Bauhaus lacked even an architectural workshop, Oskar Schlemmer lamented. Gropius's small private office formed a finishing school for a handful of students, in line with the old apprenticeship model.

By 1925 the Thuringian Government doubted it would ever see the profits promised from the Bauhaus workshops for the regional economy. Divorce from Weimar ensued. Though politics came into the dispute, most of the locals sought reform not closure; in fact another modernist architect, Otto Bartning, took over the Weimar school from Gropius and ran it until the Nazi era. With the move to Dessau at the end of 1926, an endowment of exemplary new buildings, and big housing programmes coming on stream in Germany's renascent cities, an architectural course at the Bauhaus could be delayed no

418. Walter Gropius in 1923.

longer. It began under Hannes Meyer the following spring. Housing was its natural focus, since Dessau now entrusted the Bauhaus with an estate of workers' flats to build; it was in housing that Meyer had won his spurs.

Housing enjoys an ambiguous status in architecture. Individual houses like those that Muthesius and his English and German contemporaries built fire the imagination of architects, and had helped focus the pre-war reform movement. But the wider goal and disciplines of mass housing have seldom been truly susceptible to their control. Weimar Germany proved a fleeting exception. Briefly in the 1920s architects took command of housing provision, notably in Berlin and Frankfurt. They did so by drawing on the pre-war ideal of production and turning it to social ends. That entailed a depth of technical investigation never demanded of Behrens and his Werkbund comrades. But it impinged on social habits and equipment more than on building technology. The outstanding innovation in the Frankfurt housing programmes, for instance, was the planning and equipment of the 'Frankfurt kitchen'. Despite some experiments with building materials and production, most of the Neues Bauen housing tended to be conventionally constructed. As in pre-war days, architects found objects within the house more amenable to industrialization than the buildings themselves. Despite Gropius's interest, prefabrication, for instance, played only a small part in the German housing movement till after the Crash of 1929, when large manufacturing firms sought new markets for their products.[161]

That background is germane to the way in which architectural teaching was presented under Meyer at the Bauhaus. At first sight it is awash with rhetoric about ruthlessly embracing a scientific outlook, industrial production and new technologies. The article that became Meyer's passport into the Bauhaus captures the tone:

Building is a technical not an aesthetic process, artistic composition does not rhyme with the function of a house matched to its purpose. Ideally and in its elementary design our house is a living machine. Retention of heat, insulation, natural and artificial lighting, hygiene, weather protection, car maintenance, cooking, radio, maximum possible relief for the housewife, sexual and family life, etc. are the determining lines of force. The house is their component.[162]

And much more to the same effect. The liberation of technique embraces modern life as a whole, not construction alone.

Since everything was new, the technical teaching under Meyer at the Bauhaus attempted no more than to scratch the surface. It was enough to render the form-seeking student alert and open to all novelties. Certainly there were changes in emphasis. Bauhaus students could now start on architecture right away, instead of waiting till after the welter of preliminary art-courses. They also took lessons on psychology, sociology and economics, all pertinent to the housing programmes. On the technical side, they continued to have workshop sessions to hone their manual skills. Engineering was now taught for the first time under Alcar Rudelt, who continued on till the school closed. But it took no pride of place. Even under Meyer's materialistic direction (1928–30), the Bauhaus remained a school of art-form, not of technology.[163]

The story of Bauhaus architectural teaching under Mies van der Rohe (1930–3), the school's last director, is one of sharp retraction of horns. After Hannes Meyer's dismissal and the collapse of the housing programmes, modernist architects could only assert their usefulness and create their visions of future cities, while drawing back to lick their wounds within the cave of art. A non-political leader such as Mies was the only chance of survival for an institution that had overreached itself before Nazism contrived its downfall. In the contracted Bauhaus's architectural section, there was little theory or collaboration. Mies simply tried to produce high-class art-architects from his half-dozen or so pupils by the oldest of atelier design-teaching methods: iteration, obedience and attention to detail. For one student he was 'the worst educator one could imagine'.[164] Others felt otherwise. One of several young Americans who participated in the last phase of the Bauhaus, in this case from an engineering background, remembers Mies saying: 'step out of the automobile and draw me a freehand sketch of what you see.'[165] Another summed up his lesson thus:

> The teaching of architecture, or of any other art, for that matter, involves the tediously slow awakening in the student of aesthetic insight; it has relatively little to do with the training of the intellect, however important that may be. It is concerned with the development of wisdom rather than the accumulation of knowledge. Mies van der Rohe had this profound understanding.[166]

Such was the intuitive vision that Mies would take to America.

The impact of architectural teaching at the Bauhaus was not very immediate. It trained few architects and, except Marcel Breuer, few notable ones. Within Weimar Germany it was one of a number of updated Kunstgewerbeschulen and Technische Hochschulen to breed modernist architects. Not the least of its legacies was its ability to thrive on publicity, challenge and instability – all features of the modern liberal art school. If that strategy involved some public posturing in order to win it attention, at least we know more about the Bauhaus than about its rivals or successors, and can better judge its endeavours.

To attract funding the school had to present itself as economically useful and inventive, like the pre-war Kunstgewerbeschulen. To that end, the supersession of craft by technology in the workshops was exaggerated. Once architecture came into the picture, the same overdrawn publicity applied. Had the housing programmes lasted longer, the technical teaching might have settled and deepened. But the ideal of social and economic utility ran less deep in the Bauhaus than the instinct for control through the mantra of creative form. Far from reconciling art with technology, or architecture with engineering, the school kept them apart behind a smokescreen of words and images, so as to bolster the status of the architect. Though the cloak of its inclusive name seemed to fulfil Lethaby's dream of a school of building, it did so at best for a short time only. In truth, the Bauhaus prefigured the modernist art school as a refuge for those who champion freedom of imagination in architectural training over depth of technique.

SINCE THE BAUHAUS

Words, as much as buildings or designs, have become the output of schools of architecture. Needing to define, defend and promote themselves, and having hitched themselves to the academic wagon, they vie less now in teaching than in density of publication. Much of that writing engages in issues about educating design professionals or reveals the slant of the school in question. Yet there have been few objective studies of how any particular school has operated over time. That makes these final pages on how architecture and technology have fared in schools since the Bauhaus episodic and tentative. They limit themselves to remarks on a few schools, drawn from a handful of studies.

Harvard has a unique link with the Bauhaus.[167] It was there that Gropius transferred by way of England in 1937, welcomed into the American architectural firmament with signal grace and ease. That welcome drew upon the custom of importing design-leadership into American schools of architecture, coupled with the sense of crisis and thirst for change that gripped the United States in the 1930s. The combination prompted Harvard to look to Germany and an architecture that seemed to marry art and technology rather than France, which symbolized art alone. In Germany meanwhile, Gropius and his compeers were being discredited as materialists who had betrayed their nation.

The reforms that turned the Harvard's School of Architecture into its Graduate School of Design (GSD) had nothing to do with Gropius, Anthony Alofsin reminds us. They emanated from Joseph Hudnut, appointed dean in 1935. Early in his career Hudnut had worked for Werner Hegemann, a city-planner, editor and moderate modernist who partitioned his career between Germany and America and did much to transmit urban understanding between the two. If Hegemann expanded Hudnut's outlook, it was the slump that provoked him to breadth of vision as to what a school of architecture might stand for. At Columbia, where he taught and was briefly dean before his Harvard appointment, Hudnut expressed 'an exigent desire to improve the environment of the human race'. He came to Harvard as a reformer who would draw three rather stale departments – architecture, landscape and civic design – into a fresh unity. But architecture was always the senior partner in the GSD. There Hudnut planned specifically, says Alofsin, 'to teach the science of construction particularly in its abstract principles and general processes, and to teach the economic facts and intellectual forces that affect students' lives and work'. Art was by no means excluded: but it had to proceed from the social function.[168]

These sentiments meshed with Gropius's thinking and led to his selection when a new head for Harvard's department of architecture was sought in 1936. The only other serious candidates were both Europeans. Oud the Dutchman turned out to be uninterested; Mies van der Rohe was much favoured but had poor English, possessed little social philosophy or track record as an educator, and took quite a casual attitude to the post. He was already being wooed by Armour Institute of Technology in Chicago.[169]

The department Gropius inherited at the GSD could hardly have differed more from the Bauhaus. Combining the character of academy and professional training school in the American way, it had been turning out its small quota of architects since the 1890s (pp. 445–6). All its pupils were graduates, and it was locked into Harvard's structure at staff and student level. What brought its sister-departments – Regional Planning and Landscape Architecture – together with architecture within the GSD was the shared

419. Walter Gropius with students at
the Harvard GSD, 1946. Harry
Seidler, with tie, stands to Gropius's
left; I. M. Pei, with waistcoat, has his
back to the camera.

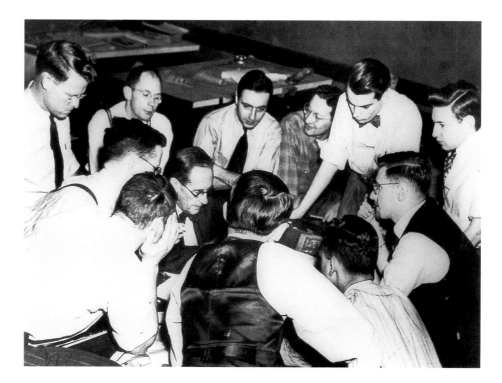

419. Walter Gropius with students at the Harvard GSD, 1946. Harry Seidler, with tie, stands to Gropius's left; I. M. Pei, with waistcoat, has his back to the camera.

rubric of drawn design, not the workshop-based practices of the Bauhaus. If the school's former reputation had rested upon architecture as an art, the technical side of the teaching under the long-serving engineer Charles Killam had also been thorough.

> The theses of architectural students in the later 1920s and mid-1930s contained more detailed engineering calculations than those of the late 1930s or 1940s, and their knowledge of construction matched or surpassed that of late students. The older generation had used models to study structure, and their abilities to render light, texture and color provided tools that better communicated at least the visual appearance of reality.[170]

In the event, the Hudnut-Gropius régime presided over a reduced level of formal teaching, cultural and technical alike. Gropius notoriously saw little virtue in teaching history to architects. Yet despite cuts on the cultural side, which Hudnut did his best to mitigate, no equivalent increase took place in the time devoted to technique. Wartime students and a visiting committee both articulated the view that engineering and construction had been neglected since Killam's departure. Instead everything focussed on the master-studio (ill. 419), called in 1940 'the crown of our system of education'.[171] If under Gropius and Breuer studio methodology had a social and analytic flavour, it was naturally also supposed to stimulate design and composition. Yet some post-war students found the arrangements too rigid to foster creativity. That may have been a by-product of reducing formal teaching and forcing instruction into the studio.[172]

In response to the size and conditions of modern projects, the GSD promoted 'teamwork' as the new architectural ideal. What that meant depended on whom you talked to. Hudnut intended by it primarily collaboration between the three design disciplines of the GSD; Gropius thought of teamwork as between mutually supporting architects. Sometimes engineers were drawn into the team concept, sometimes not. What it seldom implied was equal partnership across the construction industry, since that could not be nurtured in a school of 'design'. The name and operation of Gropius's post-war practice, The Architects Collaborative, give the flavour of his own interpretation. The archi-

tect was always to lead the team – on the basis of status and special access to form, rather than a superior grasp of technique or management.[173]

The GSD enjoyed a decade or so of zest and popularity. After that, the differences between Hudnut and Gropius deepened and the school suffered. Hudnut became less radical as he aged and drew back towards cultural ideals. Unwilling to accept his authority, Gropius pressed increasingly for an autonomous department of architecture in which the abstraction of form became the focus of a Bauhaus-style Vorkurs. The stand-off ended with both men's departure, Gropius by resignation, Hudnut by retirement. It was left to José Luis Sert to take on their combined jobs and steady the ship.

Gropius's fame depends on his educational work more than his designs. Yet his contribution to training American architects, praised and publicized in its day, has stood the test of time less well than that of Mies van der Rohe, lured to teach at the Armour Institute of Technology, Chicago, from 1938.[174] That Armour was less prestigious than Harvard stood Mies in good stead, since its structures were more open to change. Though it had taught architecture since the 1890s, its graduate programme was of recent origin. Behind the search for a new leader in architecture lay a wider campaign by its engineer-president Henry Townley Heald, to turn Armour into a more thoroughly scientific school. No doubt to Heald and his colleagues Mies's kind of architecture presented a more scientific image than the Beaux-Arts method previously peddled. Like Gropius, Mies took the post on condition that he could introduce a new curriculum and bring with him some of his former Bauhaus colleagues. That was the more needful, since Mies's teaching experience was limited and he wanted not just to teach but to build.

The new order only got going after the war. Armour meanwhile had merged with another college to become the Illinois Institute of Technology (IIT). With a fresh campus to design (ill. 420) and no rival to challenge his direction, Mies was less exposed and constrained than Gropius. To those advantages were added Chicago's size and strength as a centre of pragmatic, structure-based architecture. Technology was enshrined in the name of IIT and could not be avoided. Its engagement with architecture became a defining aspect of the school. Mies laid down a set of dogmas about how it was to work:

Some people are convinced that architecture will be outmoded and replaced by technology. Such a conviction is not based on clear thinking. The opposite happens.
Wherever technology reaches its real fulfillment, it transcends into architecture.[175]

Vatic though such utterances may be, they help students know what they are supposed
to believe. In reality, beneath the 'science-engineering sequence' and the 'construction
sequence' of the IIT curriculum lay the old, polytechnical style of teaching developed
across America since the Civil War. That eased creative cross-over between the professions. But Mies's unremitting ideal of raising structure to art shielded the course from
technical banality. There were drawbacks. By focussing its methodology on one man,
and on structure in preference to social enquiry, the post-war IIT school gave birth to an
architecture that too often mimicked the master, along with the cluster of alumni known
as the Mieslings. In comparison, the GSD of Hudnut and Gropius was broad-minded
and pluralist. Yet on balance IIT until Mies's retirement 1958 offered the more dynamic
environment, yielding stronger outcomes in Chicago and beyond (pp. 396–400). To flourish, a school of architecture needs some belief-system but not a complete one.

The brief take-over of the University of Texas's School of Architecture at Austin in
the 1950s by the so-called Texas Rangers has been fully and exuberantly chronicled by
Alexander Caragonne.[176] It belongs to the next phase of architectural training, when the
socio-technical certainties of the pioneer modernists lapsed and complexity hove on to
the horizon. But there were continuities too. Schools of architecture took another step
along the path of the autonomous art-school, reconfiguring themselves now as havens for
personal development – 'helping the students to find themselves and happiness through
creativeness', as one of the Texas initiative's originators wrote.[177] That was Bernhard
Hoesli, a Swiss: once again too, Europeans were setting the American agenda.

The reaction was the sharper at Austin because architecture there had been a constituent part of the College of Engineering until 1950, in line with the University of
Texas's origins as a land-grant college. Indeed it shared space with the engineers
throughout the Texas Rangers episode. The school inherited by the new director,
Harwell H. Harris, had found freedom but not direction. 'Formerly', he recalled, 'the faculty's common enemy – the College of Engineering – had united the architects by their

422. The Architecture Faculty, University of Texas in 1954.

opposition to it. Now that engineering was no longer a menace, the architects [ill. 422] fought each other.'[178] But Harris was no Hudnut. The process whereby the main authors of the new direction converged upon Austin had more to do with his forceful wife, Jean Harris, than himself.[179] Those authors were Hoesli, trained at ETH Zurich, and Colin Rowe, an acerbic English intellectual (ill. 421). Harris just permitted, moderated and for a time defended what the pair proposed.

The programme they set out in 1954 under Harris's name was premised on the belief that the purposes of architectural education were 'not alone to train a student for professional occupation, but . . . above all to stimulate his spiritual and intellectual growth . . . Any educational program of a School of Architecture cannot be based on the mechanics of the professional occupation but only on the intellectual content of architecture'. That responded to an old teaser: since students arrived with limited general education, should not the school offer them more than technique? What was new, and came specially from Rowe, was the idea that architects should be groomed to stand aloof from the compact which professionals made with modern society. Thus (so Caragonne paraphrases the line of thought) Louis Sullivan was construed as having contracted 'a shotgun marriage of economics and poetry', whose consequences led schools of architecture to 'function more or less as trade schools providing solid technical training leading to gainful employment in the building industry'. Likewise the amalgam of social endeavour and suspect studio methodology concocted by the Bauhaus and its alleged offshoot, the GSD, was to be condemned as 'a flawed enterprise'. Frank Lloyd Wright, on the other hand, stood for the type of creative critic whose independence might be admired, though his architecture was not to be copied. By disengaging from the pressures of the Zeitgeist, students could now embark on 'personal voyages of exploration and discovery'.[180]

That philosophy turned the Texas school – and others later – into art-school versions of 'the academy': strongholds of formal experiment, aloof from the constraining impositions of brief and technique. In the hands of charismatic teachers like Hoesli and Rowe, studio procedures begat a subjective discipline engendering intense loyalties and intimacies. In its focus on the studio Texas differed little from the GSD or IIT, merely replacing socio-technical guidelines with cultural ones. Once alumni were out in practice, however, disillusion could ensue. Nor did the new formalism bring concord or stability. Unrest at Austin rumbled on; by 1958 Harris, Hoesli and Rowe had all departed. A milder version of the method found wider acceptance under their successors. Rowe moved on to penetrate the schools at Cornell and Cambridge. Hoesli returned to the Zurich ETH, but in a bigger school, armed with the full panoply of construction skills, he could not make the same personal impact.[181] That breadth and connection the Texas school had abjured when architecture elected to go it alone.

A difficulty confronts design schools that embrace any external theory, cultural, social, mathematical or technical: theories cannot avail students at the moment they compose. In the days when originality in art had been construed as the inflection of common rules of style via common techniques of drawing and construction, it was not so hard to start off a design. Once these fell away and abstractions about space and form prevailed, the gap between architectural ideas and the act of design widened. This it was that Bauhaus pedagogy tried to address. If most of its teaching veered away from the specific tasks of architecture, at least it set out a methodology supposed to assist the actual process of design. That was why with sundry modifications it came to be adopted almost everywhere, including the three American schools mentioned above. All found it hard to translate ideas coherently into designs. IIT had the least difficulty, because Mies was indifferent to ideas, or to questioning the obscure process of design. Resistant to verbalization, that process is basically irrational; yet the IIT course, because of the value it set upon clarity of structure, is often dubbed rational.

This tour of some architectural schools after the Bauhaus ends with the British one that tried hardest to reconcile architecture with the full range of modern applied sciences – an attempt that failed, because it could not bridge the gap between these rational and irrational forces. That is the Bartlett School, at University College London.[182]

While the Texas school turned its back on its engineering inheritance, conversely the Bartlett, in appointing Richard Llewelyn-Davies as its new head in 1960, reacted against the watered-down Beaux-Arts training that had long been purveyed there. Llewelyn-Davies personified the technocratic mandarin: administratively assured, personally uncreative. Like Charles Reilly, he had trained as an engineer at Cambridge before transferring to architecture. But instead of rejecting that background, as Reilly had done, he built on it. Llewelyn-Davies was one of many in post-war Britain who felt that the art of architecture should be used to develop the public realm by turning technology to social purposes. When he took over the Bartlett, it was the reverse of an art school or disengaged academy that he envisaged.

Llewelyn-Davies' inaugural lecture at the Bartlett is a manifesto of exceptional lucidity.[183] It begins by reviewing the history of architectural education, deploring the division between design and construction, and blaming it on the nineteenth-century withdrawal from the Vitruvian ideal of universal knowledge. The private cubicles in which students at the Ecole des Beaux-Arts matured their designs become a symbol for the architects' isolation from society; and Viollet-le-Duc is honoured as the first to condemn that baleful monasticism. By contrast Llewelyn-Davies proclaims the Bauhaus as his lodestar: but 'its theoretical basis is often misunderstood,' he says, 'and its lessons have never been fully exploited.'

Next, he passes in review two subjects which link academic education with professional training: medicine and engineering. These, he notes, start with a long theoretical training before the practical sides of the subject are addressed, usually at post-graduate level.

Architectural education presents a striking contrast. Postgraduate teaching and research hardly exist. The main strength lies in the training it gives in the intuitive and creative process; i.e. design work in the studio. Lectures, with their attendant examinations, are regarded as necessary but rather irritating interruptions. The principal lecture courses are given in three main subjects: history of architecture, structural engineering, and building construction. There is little attempt to separate theory from application, or to ensure that theory is taught first . . . Some schools of architecture are thought of as traditional, and some as revolutionary. But the difference seems to rest in the flavour imparted to the instruction by the teachers, and not to any differences in the form and content of the courses.

Salvation, argues Llewelyn-Davies, lies in reverting via the sciences to universality. But these must be construed as the human sciences as a whole, not structural technology alone.

The idea that these sciences are related to architecture is fairly new and we have still to work out how best to teach them. In doing so we shall be greatly helped by the development of research. There are already several examples which point the way. One is the study of natural lighting in buildings, which involves architecture, physics and psychology . . . There is also important work on the functional requirements of certain types of buildings – schools, hospitals and laboratories – by research groups which include people from many disciplines working as a team. We must encourage the growth of these studies within university departments of architecture. They provide the essential link between teaching, theory and advanced practice, without which professional education stagnates.

The difficulty, as before, was how to squeeze new areas of knowledge into the curriculum without stripping others out. Like Gropius, Llewelyn-Davies thought history now less necessary. The studio he dared not touch; there it was that the skills were supposed to come together. How then to teach the sciences? The more you put in, the more superficial the teaching was bound to be. In respect of building technology, principles would have to suffice, thought the new professor.

An architect without a strong intuitive sense of the play of forces in a structure has one hand tied behind his back. It is of the utmost importance to get this part of his education right, but there are very great difficulties in doing so. Very few architects have managed to acquire this sense, and few engineers have it either. At the conclusion of his course, an architect can carry out some of the calculations necessary to analyse what is likely to happen to a given structure subjected to a given force. But ask him to imagine a structure appropriate to a particular set of circumstances, or to guess imaginatively the pattern of stress in a given structure without performing his calculations, and, most often, he will not be able to do it. How are we to teach this? First we must face the hard fact that the concepts of the theory of structure are mathematical. There is no easy way of imparting a sense of structure – it can only be done through mathematical understanding. We will therefore need to give more attention to our teaching to the fundamental, mathematical side of engineering design, and to bring the student, through a mastery of theory, to an intuitive grasp of structure. At the same time I hope we can drop some of the teaching in applied structures. No architect today designs his own steel frame, and no point is served in trying to train him to do so.

Likewise with building construction:

We must give up the attempt to pursue in an academic course the application of technology in every field of building. These applications will be better learnt by the young architect during his early years of practice, after leaving the school. Our task is to educate a man who can master these problems when he meets them. He needs to be sufficiently literate in physics and chemistry to relate materials and methods of construction to the needs of any particular job. He needs to be able to read and understand discussions of a scientific character and to know how to work with and learn from experts. We should have something to learn from the training of doctors in this matter. The medical profession has also had to face the transition from the days when the doctor mixed his potions to the circumstances of today, when he has to prescribe medicine made by processes outside his experience and whose chemical structure he could not describe.

This document has been quoted at length because of its clarity, and because it fore-shadows the stampede of many schools of architecture since the 1960s towards 'research'. Yet the programme it laid out proved ill-starred. Llewelyn-Davies certainly transformed the Bartlett, bringing in technologists, planners and boffins in the human sciences to provide the missing expertise above and beyond the studio. But quite soon a bitter reaction had set in against these changes. No figure in post-war British architecture is now more despised than Llewelyn-Davies. His crime in his critics' eyes was simple but heinous: he challenged the studio's autonomy. The experts he appointed proved helpless when the students embarked upon the design-process; those with creative powers felt themselves stifled. Instead of backing and reinforcing design, the effect of applied science had been to undermine it. Once again architects realized that if they valued freedom and glamour, they should pursue art not science. And so in the 1980s the Bartlett swung back, on the design side, towards the Arabian nights fantasy of the art school.

REFLECTIONS

'Arabian nights' is harsh. Readers who have persevered with this chapter may have been repelled by the illiberalism of its concluding pages. What is architecture for, they may say, if not for enjoyment, imagination, protest against materialism? As affluence spreads, should we not rejoice if design freedoms can now be explored, and utility and vulgar pro-gressivism resisted? On that reading, the parting of the ways for the training of architects and engineers marks progress. Young men and women now have the chance for a wider education in design instead of – or at least before – becoming office-fodder for corpora-tions in thrall to the bleak logic of contractual service-provision.

There is a subtler argument, inherent in the thoughts of Hannes Meyer and Llewelyn-Davies and indeed the whole Bauhaus tradition, against too much 'technics' in architec-tural training. Technologies have become so various and change so fast that it makes little sense for a young architect to enter into any of them deeply. The understanding of struc-ture is just one such strand, and one maybe of lessening importance. In many buildings an architect handles there will be no special structural challenge to master; in others the service element today may approach as much as fifty per cent of the budget – far more than the structure of the building.[184] Who is to pull all these activities together if not the architect? In Fritz Schumacher's maxim quoted earlier (p. 471), technology might today replace craft: 'Almost all technological activity only finds its true existence in the context of or against the background of architectonic intentions.'

Whether in its original or in this amended form, the claim is the same: only the archi-tect has the right to be the team-leader. But why that should be is not self-evident. Some say because only the architect is a generalist, others because only the architect is an artist. But as this book shows, many engineers too have been remarkable generalists and some have been remarkable artists. Nor, in many useful or exciting buildings and structures, have architects been the team-leaders. Leadership must be earned: it depends on cir-cumstances and merit. In construction, as in all else, it requires a blend of personality, skills and experience. Formal education can do little for personality, nor can the school stand in for the shop. What it can teach is skills. That the technical sides of these skills are many and changeable should not allow architectural training to slide into self-absorp-tion or bluff.

There are other reasons for limiting architects' tendency to appoint their own educa-tional agenda. Since in most countries their formal training is subsidised, there is an expectation of the uses to which their skills will be put. The same goes for engineers, doc-tors and others who undertake a lengthy training for professions which society deems useful and which may or may not prove lucrative. Architecture seldom brings large finan-cial rewards. Those who favour an education aloof from technique and practice may argue that this expectation of low earnings justifies an experimental, imaginative peda-gogy. Nevertheless some anticipation of usefulness attaches to architectural training: less than in engineering, more than in the fine arts. Just these issues were raised by Viollet-le-Duc at the time of the Beaux-Arts controversy in the 1860s (pp. 434–5). He found it harder to judge how architects should be trained than engineers or artists, or to say how many there needed to be. But he never doubted that a public interest was at stake.

Since special schools for teaching architecture took shape, the supply of those keen to create special buildings has always exceeded the demand for jobs that satisfy their aspirations. Today many more architects are trained than in Viollet's time. Fewer than ever follow through to a long-term career in the profession. Against that, architectural graduates now often move into a cluster of occupations which include a design element and profit from their synthesis of skills. In addition, architecture in some countries (Italy and Japan are examples) has become a popular first degree combining practical, visual and intellectual attainments. Here is a fulfilment of Charles Waldstein's Cambridge dream a century ago, of architecture as a liberal training for the affluent classes. Maybe we should not be over-prescriptive about the link between architectural training and subsequent careers.

But there is a second aspect to resources. Just as to be a proper doctor you must treat the sick, so to be a proper architect – or structural engineer – you must build. That, rather than design or research, is the object of the architectural process. The temptation in the schools is to forget or deny this. Part of the reason is practical; building demands time and money. Though schools of architecture often experiment in making buildings, such initiatives seldom endure.[185] Of the many institutions mentioned in this chapter, the Ecole des Ponts et Chaussées most successfully combined design with construction, because it was locked into an organization that built. Under later conditions of divided labour and separate schools, the trick proved impossible to pull off for long, as the record of Lethaby's English programme and of the Bauhaus showed.

Furthermore, however difficult design may be as an art, it is free and individual; construction is expensive and collective. A design belongs to its creator in the moral and intellectual sense, but the building which it adumbrates does not. To fulfil their destinies, architects must spend others' money, cater to others' needs and tastes, and work with others' skills. The challenges of their calling are all about balancing their own perceived rights and instincts with those of others. All that is hard to impress in the schools, where there is neither money nor time for building. Studio teachers feel they have done their best if they nurture fluent designers, without wrestling with such perplexities. But a consequence is to foster the belief that the rights in a work of architecture belong to architects alone.

The burden of this chapter has been that the teaching of architectural skills has grown up entwined with that of engineering skills. That will not surprise readers familiar with the ETH Zurich, with the prestigious Dutch schools of Delft and Eindhoven, with the tenor of training in Spain, Germany (despite the distractions of the Bauhaus) and with many other countries, where in line with the polytechnic tradition a unity or parallel between the subjects has been upheld to this day. In schools of that background, the sense that construction contributes to social and technical progress continues without undue friction to rub along with the notion of architecture as an art.

Alongside that, the chapter has traced the growth of efforts to escape the technical aspects of construction or to subordinate them to the idea of design. That process has often been mired in myth. Its true roots lie in the growing specialization of demand. It is fair that a number of schools should focus on the art side of architecture, just as the Ecole des Beaux-Arts once did. But it is surely an immature reaction if the unique excitement and uplift derived from public works of art is allowed to distract from or distort the techniques required for a rounded professional training.

Students endowed with a good technical or mathematical grasp are not necessarily less creative than those who are more intuitive. Nor need design always be taught in an autonomous world of its own, remote from formal instruction. That is not the way that engineers, among others, understand the term and learn the skill of design. If architectural design often looks like something that is unique, that is mainly because institutions have isolated it for the purposes of professional protection. Too often, the modern architectural studio has become an enlarged version of the Beaux-Arts cubicle condemned by Llewelyn-Davies. In truth architects, engineers and others contribute in subtly different ways to the design and making of buildings. They should do so on equal terms.

CONCLUSION

It is time to pull threads together and see what these long pages of narrative, anecdote and analysis come to. Three plain questions will suffice to shape the conclusion. Were architects and engineers once more or less indistinguishable, as has often been alleged? If so, how and why did they separate? And how far-reaching has been the reconciliation of the twentieth century diagnosed in Chapter 5? The patient reader will find that the answers proffered tend to undermine the questions. A short coda rounds things off.

WERE ARCHITECTS AND ENGINEERS ONCE THE SAME?

In the later middle ages, major western buildings were built by craftsmen plying their separate trades. Masonry and carpentry being the primary media of building, technical co-ordination and design were normally exercised by men deeply versed in one of them. Most scholars are now happy to call them architects, though they are seldom so called in the documentation. So the term is at least a little misleading. By training and experience they had graduated from practised craftsmen into 'masters' or 'master builders' – a phrase that has resounded from that day to this. Nevertheless the term 'architect' was familiar from Vitruvius. In mediaeval times it referred sometimes to technicians, sometimes to patrons or clients. We are familiar with that latter extended usage today, when 'architect' often means the moving force behind any scheme of complexity.

In the same years the term 'ingeniator' (not, incidentally, a term used in classical Latin) meant in the first place an expert adept at machinery, usually in connection with water management or the arts of war, offensive and defensive. Here too there has been continuity of usage. In common English-language understanding today (not wholly shared in continental Europe, where the term has acquired a professional-scientific flavour), an engineer is someone who deals with machinery. In *Tess of the D'Urbervilles*, for instance, Thomas Hardy describes an itinerant technician with a 'strange northern accent' who plies his steam-threshing machine around Wessex in reaping time: 'If any of the autochthonous idlers asked him what he called himself, he replied shortly, "an engineer"'.[1] Likewise the surname Engineer, quite common in India, comes from the men who drove steam engines for the British Raj.

If the ideas of architect and engineer were once notionally distinct, as they are now, how did they get muddled together? Two answers can be suggested. Firstly, major structures need machinery. The skills involved in designing and constructing the means of making a mediaeval cathedral were as remarkable as those that went into the finished edifice. At some points they might even be indistinguishable. Therefore many men skilled with machines came out of or mingled with the building trades, the carpenters especially. Someone with the capacity to design a building might also design the equipment that helped make it too. In the same way the arts of war, siege warfare in particular, involved both building structures, permanent or temporary, and machinery, fixed or mobile. In that vital activity the gifted engineer – whether soldier, mathematician or craftsman by background – could wield authority over the skilled craftsmen and labourers.

The other answer is bound up with the vexed question of what happened during the Renaissance. According to the usual account, the crafts suffered a decline in prestige. Conception in building projects came to be valued over execution; drawing and 'disegno' (an elusive term) began to exert a hold over the attention of princes and scholars – people who increasingly controlled the flow of information through printed books – and consequently over the wider public. From that change can be dated the modern idea of the architect as an artist and conceptualist, first clearly set out in Vasari. Some diagnose a divorce between architects and engineers as originating out of the disegno revolution. That will only work if the architect is identified with the artist-conceiver of a project, and the engineer with its practical builder. But there is little evidence for that kind of division in Renaissance times. Architects and engineers were already distinct in principle; disegno in itself cannot explain the fundamentals of their separation.

A better argument runs like this. As craft control over buildings weakened, experts from various backgrounds were able to concentrate more and more on the conceptual stage of construction – what in the broadest sense we call design. In the first place that conduced to the integration of design skills in individual personalities, not their separation. It hardly matters whether we label the great building-designers of the Italian Renaissance artists, architects or engineers. On the whole, the terminology of the time applied not to the person but to the job he was doing. If you designed secular or religious buildings and their adornment, you were likely to be called an architect; if you designed forts, walls, towns, ports, canals or machines for war and peace you were likely to be called an engineer. As the training of these designers varied, so too naturally did their talents. Some succeeded in specializing. But more moved between fields, if only because the constraints of the times meant that they could not spend a lifetime in a single specialism. Whatever their titles, they belonged to a growing class of experts who had moved beyond handicraft into more conceptual and managerial activity.

Permanent professional hierarchies, that is to say sizeable bodies of experts in fairly continuous employment by the state on the conceptual side of construction projects, begin to appear in sixteenth-century Venice. But they first fully articulate themselves in the France of the *grand siècle*, in response to monarchical power and purposes. Since French state organizations for construction descended directly from mediaeval royal administration, they sharpened and strengthened old demarcations. The engineers of the Corps du Génie took charge of war and infrastructure, the architects of the Bâtiments du Roi housed the king and helped articulate his magnificence, his generosity and his piety. Though there was some interchange between these two august institutions, they differed in size and style. While Louis XIV and Vauban lived, the engineers were the more numerous, prestigious and productive. Not that their construction techniques varied much; when building permanent structures, both drew on the same masonry technology. So Vauban employed people we think of as architects; and once the Corps des Ponts et Chaussées got going, its first chief engineers were 'architectes du roi'.

To sum up so far: a workable distinction existed between the 'professions' right through from the middle ages to the Enlightenment. But there was also interchange and overlap. Between about 1400 and 1700, architect and engineer in Europe were differentiated more by the hierarchies they belonged to and the tasks they carried out than by the building technologies they used or the design skills they deployed. These differences prevailed beyond the stratified conditions obtaining in France. In Britain, where state institutions were looser and weaker, until well into the eighteenth century engineers were thought of either as military experts or as consultants who advised on machinery and water management.

How and Why did Architect and Engineer Separate?

If then the two callings already enjoyed their own institutions and tasks before 1700, we must rephrase our second question. The division that obtains today between architects and engineers is commonly ascribed to the broad epoch covered by the Enlightenment and the Industrial Revolution. Why is it to these years, shall we say between 1750 and 1900, that a divorce between the professions is particularly assigned?

A clear and concise explanation for the alleged professional break in this period was set out in a textbook by the Swiss scholar Hans Straub in 1949, at the height of modernist certainties. Straub based his case for separation on the advent of the 'new materials', iron and steel with their special properties, and of a 'rational' approach to design. It became 'impracticable to adapt the historical building styles to construction in iron', argued Straub. Moreover another 'new fact' had

> a decisive and, to some extent, a fateful influence on architecture at large. The scientific solution of the structural problems of statics and strength permitted a more rational design of the structures, and thus made it possible to cope with extensive and difficult structural tasks in an economic way, without prejudice to safety requirements. At the same time, however, it was now possible to design structures according to two points of view, different in principle: the one emphasizing the engineering aspect, i.e. structural analysis and calculation, and the other stressing the architectural aspect, i.e. the aesthetic appearance. All according to the nature of the task, the one or the other of them prevailed: the engineering aspect in the case of utility buildings, and the architectural aspect in the case of monumental buildings. With certain engineering works such as bridges in towns and the like, both aspects must be reconciled as far as possible.[2]

Let us look closer at this argument. Firstly, there is the coming of 'new materials', iron and steel, and of a 'rational' approach to design. Are we supposed to equate these two phenomena? Though they became connected during the nineteenth century, they did not arrive hand in hand. Britain pioneered the liberal use of iron for constructional purposes. There were always good reasons for using iron – advertisement, economy, its supposed fireproof properties, and sheer novelty among them. But structurally speaking, it would be hard to claim that the ways in which iron was at first deployed were rational in a mathematical or scientific sense, if that is what is meant. Nor did engineers of the consulting or official variety lead or monopolize in the structural use of iron. Businessmen, architects and fabricators were equal initiators in iron along with engineers until the railway age.

Crossing the Channel to France, we encounter an allegedly rational approach to design making itself felt during the Enlightenment, long before the candid use of iron had any impact upon the country's building culture. In buildings it took two paths, both linked to masonry construction. The study of stereotomy – the jointing, analysis and drawing of advanced masonry structures – led on to the techniques of descriptive geometry championed by Gaspard Monge, first at the Ecole du Génie and then at the Ecole Polytechnique. Descriptive geometry became the tool of choice for teaching the principles of building construction to engineers. Along with that went a revival of minimalism and clarity in masonry structures. Going back to the great French mediaeval churches, this tradition found fresh life in the hands of classical architects like Perrault and Soufflot and in the propaganda of the Abbé Laugier.

So there was an engineering thread and an architectural thread to the rationalizing movement in French construction – both originally more concerned with masonry structures than with new materials. The two strands correspond to the already separate institutional paths of official French engineers and architects. Nevertheless there was still overlap. For instance, the career of Perronet, founding father of the Ecole des Ponts et Chaussées, can well be interpreted as a campaign to hold architecture and engineering together within a socially rationalistic framework. Like Ove Arup in later times, Perronet regarded breadth of skills and outlook as the key to a holistic, civilized approach to construction. He trained his engineers to be competent architects: so they were in all but name.

The fashion for structural rationalism during the Enlightenment, therefore, did not promote a sharper separation between architects and engineers than already obtained as a result of their differing tasks and career-paths. If anything, the ideological thrust after Perronet was towards integration. An educational system, radiating in modified form from France to the rest of Europe and America during the nineteenth century, endeavoured in principle to keep the teaching of architecture and engineering together. The polytechnical movement dominated construction training in continental Europe and the Americas from 1800 until as late as the Second World War. Polytechnical education believed in the fundamental unity of the arts of construction. The integrative traditions and impulses of the movement have not been exhausted today.

Nevertheless it is fair to remember that from Prony onwards many engineering teachers in the polytechnics saw architecture as an irrelevance, while many architects chafed under engineering leadership and did their best to escape them. That brings us back to Straub's proposition. Even if the linking of new materials with the so-called rationalist movement was not the original cause for the separation of the two professions, common sense compels us to admit that the gap between them widened during the course of the nineteenth century. Now for the first time this had less to do with career structures and tasks than with skills – specifically, as Straub says, the interrelation of new materials and calculation. As the manufacture and properties of iron and later steel were more intensively investigated, their deeper understanding and better application passed to specialists of one kind or another – to metallurgists, to fabricators, to production engineers, ultimately to those whom we now call structural engineers. That made the integrative ideal of polytechnical training hard to maintain. Institutions were obliged to adapt to the fissiparous forces at work in the industrialized economies.

The change should neither be overstated nor dated too early. Bridges were the structural type first affected. Until about 1830 major bridges were often built by men classified as architects. Thereafter they passed into the hands of engineers. The suspension bridge had a lot to do with this. But the early examples were not especially 'rational'. James Finley and Samuel Brown were clever amateurs. Telford, starting out as a mason-architect, drew on experience. Only after Navier's codification was there a body of technical principle for later specialists to draw on. Even in France many of the nineteenth-century suspension spans were the work of a single empirical building firm. With tall buildings in America, the separation took place much later. As late as the 1880s George Post and John Root still reckoned to design the iron frames of their skyscrapers themselves, as part of their overall architectural remit. The take-over by structural specialists of the foundations and framing of tall buildings took place only in the last decade of the nineteenth century.

It is plausible, then, to ascribe the widening gulf between architectural and engineering skills to more complex materials and structures and the need for specialized calculations. But there is also a simpler overall way of viewing the processes described by Straub. Specialisms arise to meet demand. It was because Louis XIV and his ministers persistently required particular services from both that the Génie could be so different from the Bâtiments du Roi. What made nineteenth-century conditions new was the steadiness of demand for construction projects of many differing types. To an extent the designers of the Renaissance could not have dreamed of, men in private as well as public employment were able to specialize in some conceptual aspect of construction. Specialists before had been few in number. Now they were a professional class. And as those with common skills and interests always do, they banded together in their clubs, their societies and their professional institutions, to learn from one another and to protect or even to corner their share of the ramifying market in construction. Whatever they may say in public, such institutions feed the tendency to fission.

The supreme exponent of the division of labour has been Adam Smith. It is regrettable that Smith never turned his great intellect concertedly to construction. His main instances of the process came from fixed manufactures. It was famously through analysing the productivity that accrues when they are organized in separate, specialized stages that he illustrated the benefits of dividing labour. In every kind of economy, construction is a critical and sensitive species of manufacture. But it is also an exception among industries: hard to classify, even harder to reform. Being site-specific, it lacks the full advantages of fixed machinery and a constant labour force. The principles of divided labour in construction are time-honoured and have changed little over the centuries. Carpenters do not lay bricks. Yet how to make construction efficient, let alone civilized, has taxed brains from Smith's day to the present.

Perhaps Smith's ideas can illuminate the design side of construction. As demand grows, specialisms arise to meet particular needs, finding their place in the chain of production. How they interrelate will be governed by the nature of the end-product. The railways, for instance, elicit sub-classes of engineers to create the locomotives, the rails, the earthworks, the bridges and the stations. To take each of these items through from conception to completion they must work with others who have different skills, some more specialized and educated than others. Architects come into the picture in respect of railway bridges and stations. The two require different treatment. Stations are multi-functional and need a variety of rooms and scales, which architects tend to be better than engineers at addressing. Bridges do not need rooms. But if they are conspicuous they call for architectural dignity, as do stations. In the end all these different skills have to be put together to make one thing. As Straub revealingly remarked, for 'bridges in towns and the like', the engineering and architectural aspects of the project must be 'reconciled as far as possible'.

Have the Professions Been Reconciled?

And so we are led to the last of our questions. The urge for reconciliation, co-ordination or unity within a given job must always be there to counteract the tendency to professional fragmentation, if the result is to turn out satisfactory. Different aspects of a project are broken down into separate parts at the conceptual stage and assigned to specialists. They must then be reassembled into a whole. Major nineteenth-century building projects are constantly criticized for lacking a sense of unity. Often the criticism is naïve; not all multi-functional buildings can or need be seen as a whole or as of a single nature. But the call represents the instinct for wholeness crying out against the productivity of fragmentation.

A broadbrush way of describing the change in relations between architects and engineers over the past two centuries might be to say that they used to work on different projects but have similar skills, whereas they now work on the same projects but have different skills. That is far too sweeping. Though customs vary between countries, the old separate spheres of interest remain. Architects are marginal to most infrastructural work, while on small domestic jobs engineers do little except check foundations and roof-structures. There are also plenty of early examples of engineers and architects, so denominated, yoked together on a single project, as when the younger Mansart's architects were called in to embellish the gates of Menin and Neuf-Brisach for Vauban's engineers. Nevertheless the spotlight has increasingly fallen on the relationship and the rivalry that arise when the professions collaborate, as they have been forced to do since techniques and materials changed and the range of things a designer needed to know for a major building outran the scope of all but a few exceptional all-rounders.

What happens when architects and engineers work together? The most familiar scenario today is that of the consulting engineer answering to an architect at some stage in the design, the earlier the better. As a fixed model, that emerges only in the twentieth century. Before then, and often since, the relationship has been fluid. This book has traced many varieties of working patterns between architects, engineers, fabricators and contractors. Often indeed the contractor has looked like a 'third person in the marriage', invalidating the idea of special intimacy between architect and engineer. Yet ultimately the late-maturing of the consulting engineer supplied legitimacy and even romance to the link.

Private civil consulting engineers, in the sense of independent professionals who devise infrastructural projects yet stand aloof from taking contracts for construction, appear to be a British invention, going back to Smeaton or a little before and proliferating with the railways. When they first work with architects they seem to do so only in a lowly and intermittent way, sizing components, or assessing competing tenders and methodologies from rival engineer-contractors, as at the Kew Palm House. From about 1860 consultants begin to advise architects about structures more often. Sir Gilbert Scott seems to have been the first British architect regularly to use engineers in that way, iron always being involved. But the engineering consultant who consistently helps the architect work out the whole structure of his design comes in only around the time of the switch from iron construction, first to steel and then to reinforced concrete. Though there had been some sniff of it in earlier exhibition structures, the Galerie des Machines of 1889, designed by a Beaux-Arts architect, Dutert, with an intensive input from his structural engineer, Contamin, best exemplifies the change. It also makes a neat, symmetrical contrast with the contemporary Eiffel Tower, where an architect, Sauvestre, inflects an engineering design by Eiffel's lieutenants Nouguier and Koechlin.

Even after that things did not quite settle down. It is sometimes said that since steel was hard and reinforced concrete harder for architects to understand, they were forced to employ independent consultants. After 1900, masonry-fronted structures with steel frames were certainly farmed out to experts, often after the architectural design was well advanced. But in respect of concrete, though most architects certainly had to seek help, their original collaborators were seldom consulting engineers. Reinforced concrete was first successfully marketed in France and beyond by engineer-contractors such as Hennebique and his licensees. While the concrete engineering firms of those years – approximately, 1890–1914 – did indeed take much of the structural effort away from architects, they were not consulting professionals.

As the patents for the early concrete systems lapsed, independent consulting engineers who advised on building in concrete emerged. But the link between advanced designing in reinforced concrete and engineer-contractors remained unbroken between the world wars. Auguste Perret designed and built not with a consulting engineer but in partnership with the contracting firm run by his brothers. Pier Luigi Nervi too operated in a contracting partnership. Likewise in his inter-war works in partnership with Lubetkin and others, Ove Arup was not officially a consultant but the agent of engineer-contractor firms. When it came to creating a modern concrete architecture in which a building's expression followed from its materials and structure, the consulting and the contracting paths developed side by side. Both proved to have pros and cons.

After 1945 a regular mode of collaboration between independent professionals for ambitious buildings at last firmed up in Britain and the United States – though not to the same extent in France, where the bureau d'études system tackled the means of translating architectural design into construction via a different path. Much was due to a new, humane brand of structural engineer. Some were independent consultants like Arup, Bodiansky, Samuely, Severud and Weidlinger who had roots in European culture, cared about art and architecture and dreamt of imaginative buildings rationally built. Others like Fazlur Khan settled for anonymity within interdisciplinary firms of architects and engineers like Skidmore Owings and Merrill.

What role architects played in this new sense of common purpose is harder to specify. Years of modernist propaganda had deprived the old styles of their strength and dinned into their heads that they should aim for the honest expression of structure. They had been fed the line that works of engineering with simple functions like the Crystal Palace, the Eiffel Tower, the bridges of Maillart or the authorless silos of the prairies were the best the built environment had come up with over the past century. All that helped architects to listen to, learn from and trust their engineers. For years they persisted with their own ideas about expressing structure – ideas that their engineer-collaborators seldom took seriously. That mattered less than the lesson that they ought to be more like, and therefore more responsive to, engineers. They needed the engineers in any case. The more imaginative their idea, the earlier they needed them, even if it was just to prove to the clients that the design could be built.

The post-war consensus was never quite stable. For a start the partnership was seldom an affair of equals. Most architectural jobs come to architects first; the tenor of collaboration is set in advance by that fact. Arup's ideal was always to subordinate himself to the architectural concept, because he fervently believed that art mattered. Sometimes the ideal went wrong, the flagrant instance being the Sydney Opera House; often it did not fit the circumstances of the case. Meanwhile technology and procurement moved on. Demand grew further, and with it the fragmentation of conceptual skills, threatening the intimacy between architect and structural engineer. Engineering was already split between many sub-disciplines. In the later twentieth century the junior branches dealing with construction flexed their muscles. Showing structure began to be emulated by a fashion for showing services. If seldom prudent, at least it reflected the growing proportion of building budgets spent on such items. The services engineer could now demand a presence earlier in the design process. In complex projects the structural engineers began to look less like uniquely privileged collaborators at the design stage, more like first among a litter of equal consultants and subcontractors.

By the end of the twentieth century, in buildings of high prestige another change had penetrated the architect-engineer relationship. With the expression of structure a waning ideal, some architects turned to buildings of flagrant imagery, showiness and daring. How to build such designs was as challenging and maybe more fun for engineers to sort out than the austerer equivalent forty years before. But now the relation between structure and architecture stood in danger of losing some of its dialectical discipline. In an art-obsessed world, the architect had dragged the engineer out of the temple of reason and beguiled him to worship in the temple of art. This tour d'horizon of architect-engineer relations concludes with a quick peek inside that hallowed structure.

ART, USEFULNESS AND MARRIAGE

Like architecture, engineering is unquestionably an art. Throughout western history, art as a matter of making has always fundamentally meant skill – *techne*. The Greeks knew no other word. The skills or techniques which a proficient engineer brings to his tasks are as palpable as those an architect employs, and rather easier to pin down. It has been argued that the skills a pre-industrial engineer brought to bear in above-ground construction did not greatly differ from those deployed by an architect, even if they tended to be employed on different tasks. On that reasoning if the skills are the same, then the art must be the same. A similar conclusion might emerge if we looked at art from the standpoint of the perceiver, as something man-made that evokes an emotion. Clearly engineering structures can evoke emotions as much as architectural ones. Once again, if the emotion is the same, then the art must be the same. That side of the question could doubtless be infinitely qualified. However it is not the perceiver we are concerned with here, but the maker.

With the steady separation in skills during the nineteenth century, the two professions develop their own techniques, the mathematico-scientific or rational strain in engineering skills becoming more pronounced. Do they therefore become less artistic? If we hold on to the idea of art as skill, plainly not. Findings in other areas of engineering frequently confirm that the skills of applied technology cannot be deduced from science. In his study of aeronautics, for instance, Walter Vincenti quotes an engineer who said back in 1922:

> Aeroplanes are not designed by science, but by art in spite of some pretence and humbug to the contrary . . . there is a big gap between scientific research and the engineering product which has to be bridged by the art of the engineer.[3]

Many scholars, eminently David Billington, have championed an understanding and appreciation of structural design as an art-form distinct from that of architecture. What that argument seems to amount to is that since the skills and methodology of a fine engineer are differently deployed, they should be otherwise analysed and interpreted than those of a fine architect. Nobody could disagree with that.

Difficulties arise only when it is not the skills but the object created that is, by time-honoured extension, described as art. At that point we move over from what is involved in making something to what is involved in experiencing it. Only when that happens do we meet arguments to the effect that engineering is less artistic than architecture, either inherently or because in most instances it lacks some aesthetic quality. Or, to the contrary, it is claimed, as Le Corbusier and others did, that engineering structures have caught an emotional quality which the architecture of the styles has missed, and therefore point the way forward for art. The subjectivity and mutability of such perceptions should warn us not to abandon the firm territory of skills for the soft footing of experience. Every creative designer creates an aura that is sui generis. But whether that experience follows from our understanding of his skills is questionable. There seems no reason to insist that the effect of a bridge by Maillart belongs in an aesthetic category of its own – that of engineering-art.

Attempts to claim that the objects as well as the skills of engineering amount to art have their roots in the desire for engineers to be allotted the same high status as architects, and perhaps also in the instinct for unity in face of the fragmentation of skills. Even if we accept that engineering-art differs from architecture-art, they come together in the overriding category of art. But if an engineer becomes so bound up with the art-object side of the question as to sideline the rationalizing skills and efficiencies of his discipline, it may be questioned whether he is actually doing his bit for construction as a whole.

One last strand to the question of art and engineering remains to be touched on. That has to do with utility. Engineering is valued and rewarded by governments and communities at large because it is useful, practical, progressive. Utility and organization lie at the heart of Tredgold's famous definition of engineering as 'the art of directing the great sources of power in Nature for the use and convenience of man'.

Architecture too claims to be useful. Normally it fulfils specific human needs. It also has a broader utility, concisely defined by Wren as architecture's

> political Use; publick Building being the Ornament of a Country; it establishes a Nation, draws People and Commerce; makes the People love their native Country, which Passion is the Original of all great Actions in a Commonwealth.[4]

All the same, architecture lacks the linearity of engineering, that quality of consistent serviceability and of the creative skill subordinating itself to the practical end at issue. For some commentators, it is architecture's freedom from the manipulations of utility that makes it a true art – and makes engineering not an art or, at best, a handmaid to art.

On that view, once engineering and engineering education were earmarked by governments to further national and industrial ends, once they became narrowly numerate in method and pragmatic in scope, they began to lose the quality of art. Durand, the teacher who turned classical architecture into a system that could be manipulated by the engineers, is the greatest of the bogey-figures for those who think in that way. That interpretation is allied to a reading of western history which argues that some form of cultural 'dissociation' took place during the Enlightenment, after which the arts and sciences went their separate ways, never to be reunited. For architecture and engineering, the case for dissociation has been best set out by Alberto Perez-Gomez in his well-known book *Architecture and the Crisis of Modern Science* (1983).

All such arguments seem to be based on texts, not the experience of building buildings. Once we enter into what happens when a structure is actually assembled in any age, we find designing and making, architecture and engineering, art and science muddled up together so constantly and utterly that a once-and-for-all process of dissociation in an age of reason or enhanced technology appears implausible. People have always struggled between the claims of the ideal and of the real, and always will. It is because I respect that implacable struggle and, even more, the long human struggle for material improvement that the present book has eschewed texts in the main and stuck to the nobler chronicle of building practice.

It is our instinct to configure great issues and thought-patterns in opposing, dialectical terms, yet to try and resolve the dialectic. Art and science; mind and body; reason and emotion; invention and creation: the list could go on. Carlyle has a good one: 'Priest and Prophet to lead us heavenward; or Magician and Wizard to lead us hellward'. Always crude, inexact and misleading, the dualities and the parallels between them are also compelling and recurrent. In the world of construction, the eternal duel between architect and engineer plays the same role; the two stand for contrasted facets of our common, riven humanity. If it is proper to regard art as a consolation for the stresses and damage imposed upon us by material progress, then the wayward pleasures of architecture can stand as a compensation for the compunctionless efficiencies of engineering.

Shortly after I began this investigation, a friend suggested the idea of sibling rivalry as a way to describe the jostling relationship between the two professions. I thought it a good one, and have adopted it for my subtitle. Ove Arup went further: he saw the relations between architect and engineer as marital, with all the potential for harmony, conflict, perpetuity and fruitfulness which marriage implies. The comparison is tempting but fraught. It would be a rash person who assigned genders to the partners.

Better perhaps to leave the final thought about dualities to Andrew Marvell, just as my preface ended with Marvell's friend Milton. In his dialogue charting the consuming struggle between soul and body, Marvell perceived that the rivalry was transforming, for good and bad alike. His last words are given to body:

> What but a soul could have the wit
> To build me up for sin so fit?
> So architects do square and hew,
> Green trees that in the forest grew.

ABBREVIATIONS

ANB   *American National Biography* (New York, 1999)

BDCE   *A Biographical Dictionary of Civil Engineers in Great Britain and Ireland, Volume 1: 1500-1830* (London, 2002)

Colvin   Howard Colvin, *A Biographical Dictionary of British Architects 1600–1840*, 3rd edition (London, 1995)

DAB   *Dictionary of American Biography* (New York, 1928-58)

EP   *Encyclopédie Perret* (Paris, 2002)

LFP   *Les frères Perret: l'oeuvre complète* (Paris, 2000)

ODNB   *Oxford Dictionary of National Biography* (Oxford, 2004)

TNS   *Transactions of the Newcomen Society*

NOTES

1 'IMPERIAL WORKS AND WORTHY KINGS'

1 Among the many French books on Vauban, the best now is Anne Blanchard, *Vauban* (Paris, 1996). Valuable also are Michel Parent and Jacques Verroust, *Vauban* (Paris, 1971); Robert Bornecque, *La France de Vauban* (Paris, 1984); Bernard Pujo, *Vauban* (Paris, 1991); *Vauban Réformateur* (Association Vauban, Actes du Colloque Vauban Réformateur 15–17 Dec. 1983, Paris, 2nd edn, 1993). Source-material is supplied in Rochas d'Aiglun (ed.), *Vauban, sa famille, ses écrits, ses oisivetés et sa correspondance* (Paris, 1910, 2 vols.). In inter-war England, Daniel Halévy's study was translated as *Vauban Builder of Fortresses* (London, 1924), and Reginald Blomfield wrote his robust, readable but dated *Sebastien le Prestre de Vauban* (London, 1938). There has been no monograph in English since, but much can be gleaned from Christopher Duffy, *The Fortress in the Age of Vauban and Frederick the Great* (London, 1985), Ch. 2; Robert W. Berger, *A Royal Passion: Louis XIV as Patron of Architecture* (Cambridge, 1994), Ch. 13; and above all, Janis Langins, *Conserving the Enlightenment: French Military Engineering from Vauban to the Revolution* (Cambridge, Mass., 2004).
2 Halévy, *op. cit.*, pp. 10, 162–3.
3 For Vauban and Louvois, see André Corvisier, *Louvois* (Paris, 1983), pp. 359–74.
4 See Geoffrey Parker, *The Military Revolution: Military Innovation and the Rise of the West 1500–1800* (Cambridge, 1988), esp. Ch. 1.
5 Langins, *op. cit.*, p. 21, perhaps borrowing from the Italian phrase 'sogno tecnologico' see Paolo Galluzzi, *Gli ingegneri del Rinascimento* (Florence, 1996), pp. 11–97 (e.g. p. 49).
6 Horst de la Croix, 'Military Architecture and the Radial City in Sixteenth-Century Italy', *Art Bulletin* 42 (1960), pp. 263–90, esp. pp. 279–80; Martha D. Pollack, *Turin 1584–1640* (Chicago, 1991), pp. 19–20.
7 Parker, *op. cit.*, p. 12; Langins, *op. cit.*, p. 39.
8 Compare the phrase 'artista-ingegnere', employed for the mechanical gurus of the Quattrocento by Paolo Galluzzi, *op. cit.*, and others.
9 J. R. Hale, *Renaissance Fortification: Art or Engineering?* (London, 1977), pp. 18–19, notes that at Ferrara, Biagio Rossetti is referred to 'almost in alternate documents as architect and engineer'. As late as 1666 Guarini, though a priest, was appointed by the Duke 'our Engineer for the said Chapel of the Most Holy Shroud' in Turin. Guarini was of course an accomplished mathematician; a treatise on fortification was published over his name in 1677. H. A. Meek, *Guarino Guarini and his Architecture* (New Haven and London, 1988), pp. 61, 145–7.
10 For a brave attempt to define engineers from historical usage, see Hélène Vérin, *La gloire des ingénieurs; l'intelligence technique du XVIe au XVIIIe siècle* (Paris, 1993), pp. 19–42; for architects during the Renaissance, see the essays by Leopold Ettlinger and Catherine Wilkinson in Spiro Kostof (ed.), *The Architect: Chapters in the History of a Profession*, (New York, 1977), pp. 96–160.
11 M. E. Mallett and J. R. Hale, *The Military Organization of a Renaissance State: Venice c.1400–1617* (Cambridge, 1984), pp. 87–96, 409–28; and for Sanmicheli, the essays by Ennio Concina, Giuliana Mazzi, Jordan Dimacopoulos and Kruno Prijatelj in H. Burns, C. L. Frommel and L. Puppi, *Michele Sanmicheli: architettura, linguaggio e cultura artistica nel cinquecento* (Milan, 1995), pp. 196–203, 204–9, 210–21, and 222–7.
12 Pollack, *op. cit.*; Pietro Marchesi, *Fortezze veneziane 1508–1797* (Milan, 1984), pp. 61–83.
13 Mallett and Hale, *op. cit.*, p. 485.
14 Parker, *op. cit.*, p. 62.
15 D. J. Buisseret, 'Les Ingénieurs du Roi au temps de Henri IV', *Bulletin de la Section de Géographie, Ministère de l'Education Nationale, Comité des Travaux Historiques et Scientifiques*, 77 (1964), pp. 13–84, esp. pp. 15–19; Anne Blanchard, *Les ingénieurs du 'Roy' de Louis XIV à Louis XVI* (Montpelier, 1979), pp. 51ff.; Anne Blanchard in Philippe Contamine (ed.), *Histoire militaire de la France, vol. 1, Des origines à 1715*, (Paris, 1992), pp. 459ff.; Blanchard, *Vauban*, pp. 104–8.
16 Blanchard in Contamine, *op. cit.*, pp. 471–2.
17 Duffy, *op. cit.*, pp. 65–6.
18 *Ibid.*, p. 29.
19 For attempts to analyse Vauban's 'architecture' and its power, see e.g. Michel Parent in Parent and Verroust, *op. cit.*, pp. 147ff.; Bornecque, *op. cit.*, pp. 29–30; Claude Parent in *Vauban réformateur*, pp. 137–44. Comments by Grodecki, Hautecoeur and Lavedan are relayed in Peter, *op. cit.*, pp. 280–1. Blomfield, *op. cit.*, pp. 51–6, is characteristically direct, calling Vauban 'not an architect' and his gateways 'costly affairs rather clumsily designed . . . The entrances to Vauban's great forts seem to me to be the worst things about them.' Older French architectural critics tended to dismiss Vauban's major ensembles as unsophisticated, e.g. Grodecki, *loc. cit.*: 'l'effort architectural de Vauban à Toulon a été nul'.
20 Berger, *op. cit.*, p. 165.
21 Hale, *op. cit.*, pp. 58–9.
22 For an exemplary study of how disegno theory can falsify the nature of the skills and relations attending construction, see William E. Wallace, *Michelangelo at San Lorenzo: the Genius as Entrepreneur* (Cambridge, 1994).
23 In his article on radial cities, De la Croix, *op. cit.*, concludes (p. 289) that 'the military planner trimmed the radial plan of all its symbolic and most of its aesthetic qualities. He accepted an objet d'art and transformed it into a functional tool.' This seems to be an art-historical way of saying that radial plans proved impractical. The first attempted, Villefranche-sur-Meuse of the 1540s by Girolamo Marini, was never completed. In the most famous, the Venetians' Palmanova of the 1590s, the radial arrangement was soon compromised.
24 René Descartes, *Discourse on Method*, introduction to section 2. Langins, *op. cit.*, p. 145, adds the point that in the making of fortifications, management was the crucial engineering skill. Architects were subordinate to engineers in that field because they were regarded as technicians not managers.
25 For Alphand and Belgrand see, for instance, David P. Jordan, *Transforming Paris: The Life and Labors of Baron Haussmann* (New York, 1995), pp. 270–2, 280–3. Haussmann thought of architecture as a secondary matter ('l'architecture n'étant autre chose que l'administration . . .'): quoted *ibid.*, pp. 159–60, 392.
26 Berger, *op. cit.*, p. 177.
27 Langins, *op. cit.*, p. 192.
28 Blanchard, *Ingénieurs*, pp. 436–50.
29 Blanchard, *Vauban*, p. 72.
30 James Fergusson, *History of the Modern Styles of Architecture* (London, 1862), p. 485.
31 Concina in Burns, Frommel and Puppi, *op. cit.*, pp. 199–203.
32 Pujo, *op. cit.*, p. 112; see also Berger, *op. cit.*, p. 176, Halévy, *op. cit.*, pp. 80–1, Blomfield, *op. cit.*, p. 55. Vauban also had an argument with Louvois about the cost of a garrison church at Strasbourg.
33 Quoted in Philippe Truttmann, *Fortification, architecture et urbanisme aux XVIIe et XVIIIe siècles* (Thionville, [1975]), p. 35.
34 Pierre Clément (ed.), *Lettres instructions et mémoires de Colbert* (Paris, 1868), vol. 5, p. 227. See also Philippe Deshayes in *Actes du Colloque Vauban Réformateur*, p. 111, and for French barracks generally, François Dallemagne, *Les casernes françaises* (Paris, 1990). For the arsenal, a building-type which was not standardized, see Philippe Jacquet, 'Les arsenaux de Vauban', in André Lanotte (ed.), *L'Arsenal de Namur 1692/1982* (Brussels, 1984), pp. 19–27.

35 J.-H. Mansart was brought in to help with gates at Menin and Neuf Brisach: see Bornecque, *op. cit.*, pp. 24–5, Berger, *op. cit.*, p. 169; also Rochas d'Aiglun, *op. cit.*, vol. 2, p.195.

36 Bernard de Bélidor, *La science des ingénieurs* (Paris, 1729), Book 4, Ch. 5, p. 37. This chapter on military buildings includes, among later projects, plans from Neuf-Brisach and other Vauban designs, e.g. 'magasin de poudre selon M. de Vauban'.

37 Cited in Pujo, *op. cit.*, p. 180, and Blanchard, *Ingénieurs*, p. 104.

38 Quoted in Halévy, *op. cit.*, p. 27.

39 Vauban to Louvois, quoted in Blanchard, *Vauban*, p. 134.

40 Quoted in Langins, *op. cit.*, p. 154. By the Director General is meant Vauban's administrative superior, at that time Le Peletier, not Vauban himself.

41 This paragraph is largely derived from Bornecque, *op. cit.*, pp. 13–15.

42 Blanchard, *Ingénieurs*, p. 84.

43 *Ibid.*, pp. 92–3, 483; Blanchard, *Vauban*, pp. 359–60.

44 A vignette of how Vauban-style fortification worked in a place of smaller scale, Ath in Hainault (1668–74), is given in Robert Mousnier, *The Institutions of France under the Absolute Monarchy 1598–1789, vol. 1, Society and the State* (Chicago, 1979), pp. 723–9.

45 For Lille, see Blanchard, *Vauban*, pp. 129–34, 428; Paul Parent, *L'architecture civile à Lille au XVIIe siècle* (Lille, 1925), pp. 153–61; Alain Demangeon, *Lille, portrait de ville* (Paris, 1993), pp. 12–14.

46 Jean-Louis Harouel, *L'embellissement des villes: l'urbanisme français au XVIIIe siècle* (Paris, 1993), p. 84.

47 Demangeon, *op. cit.*, p. 13.

48 For Lalonde, see Langins, *op. cit.*, pp. 43–4.

49 Louvois to Vauban, 5 January 1671, cited in Rochas d'Aiglun, *op. cit.*, vol. 2, p. 40.

50 Louvois to Vauban, 30 Dec. 1680, cited *ibid.*, p. 195; see also pp. 174–5.

51 Vauban to Louvois, 17 July 1674, cited *ibid.*, p. 109.

52 The paragraphs on Toulon are based on Jean Peter, *Vauban et Toulon* (Paris, 1994); Jean Meyer, 'Vauban et la Marine: milieu ambiant et invention', in *Vauban Réformateur*, (Paris, 1993), pp. 276–89; and Blanchard, *Ingénieurs*, pp. 98–100.

53 For Blondel see Placide Mauclaire and Charles Vigoureux, *Nicholas-François de Blondel, ingénieur et architecte du roi* (Paris, 1938), a rather partial study; and Louis Hautecoeur, *Histoire de l'architecture classique en France*, vol. 2 Part I (Paris, 1948), pp. 468, 511–14.

54 From Naudin's *L'ingénieur françois* (1695), quoted by Langins, *op. cit.*, p. 42. Mauclaire and Vigoureux, *op. cit.*, p. 240, say there is no record of personal contact between Vauban and Blondel. But Vauban and Louvois were forced to accept alterations to the Aqueduc de Maintenon suggested by Blondel in 1685 shortly before the latter's death (pp. 127–8).

55 For the Rochefort dockyard, which involved Blondel, Clerville, Le Vau and others, see Monique Moulin, *L'Architecture civile et militaire au XVIIIe siècle en Aunis et Saintonge* (La Rochelle, 1972), and Josef W. Konvitz, *Cities and the Sea: Port City Planning in Early Modern Europe* (Baltimore, 1978), pp. 81–8, 105–14.

Konvitz interprets the final plan for Rochefort as pragmatic: 'the grid plan was taken to its dullest and most economical extreme' (p. 110).

56 Quotations from Peter, *op. cit.*, pp. 26, 67.

57 Meyer, *op. cit.*, pp. 278, 282, calls this project 'among the most prescient for industrial history . . . Albertian voluptas has had to give way to commoditas; compromise is no longer possible.'

58 Clément, *op. cit.*, vol. 5, pp. 230–1.

59 Blanchard, *Vauban*, pp. 264–9. Technical help for the original Canal du Midi or Canal des Deux-Mers, the great project of Pierre-Paul Riquet, was given in some measure by Vauban's predecessor Clerville, who supplied a site engineer: Jean Petot, *Histoire de l'administration des Ponts et Chaussées 1599–1815* (Paris, 1958), pp. 76–7, 97. Robert W. Berger, *A Royal Passion: Louis XIV as Patron of Architecture* (Cambridge, 1994), p. 86, says that J.-H. Mansart was dispatched by Colbert in 1673 to help with the canal. According to Vauban, Riquet was 'ny conduit ni aidé'; hence the canal's deficiencies. But he esteemed Riquet for his vision and felt that the one thing lacking was a statue to the founder, now ubiquitously commemorated along its length. Among many books on the canal, L. T. C. Rolt's delightful *From Sea to Sea: the Canal du Midi* (London, 1973) may be singled out.

60 Langins, *op. cit.*, pp. 58–62.

61 See the entry on Bélidor by Charles C. Gillispie in *Dictionary of Scientific Biography* (New York, 1970), vol. 1, pp. 381–2; and Langins, *op. cit.*, pp. 224–34. His first book was the *Nouveau cours de mathématique à l'usage de l'artillerie et du génie* (Paris, 1725).

62 Bélidor, *op. cit.*, Book 4, Ch. 13, p. 88, says he will be brief on private buildings, 'parce qu'ils ne font partie de mon ouvrage qu'autant qu'un Ingénieur, sans vouloir être Architecte du premier ordre, ne peut ignorer les proportions qu'il faut donner aux parties d'un Bâtiment pour être commode et gratieux'.

63 Langins, *op. cit.*, p. 86. Uniforms came in in 1732.

64 For the paragraphs on the Bâtiments du Roi see Jeanne Laurent, *A propos de l'École des Beaux-Arts* (Paris, 1987), pp. 21–6; Hautecoeur, *op. cit.*, pp. 413–25, 462–91; Myra Nan Rosenfeld, 'The Royal Building Administration in France from Charles V to Louis XIV', in Spiro Kostof (ed.), *The Architect: Chapters in the History of the Profession* (New York, 1977), pp. 173–9; and the summary by Alden Gordon in *Macmillan Dictionary of Art* vol. 20, p. 133, under 'Maison du Roi'.

65 Berger, *op. cit.*, pp. 5–6.

66 For the historical background see the rich and discursive Charles Woolsey Cole, *Colbert and a Century of French Mercantilism* (New York, 1939), 2 vols.

67 'Colbert ne toléra pas l'indépendance des villes': Hautecoeur, *op, cit.*, p. 424.

68 Richard L. Cleary, *The Place Royale and Urban Design in the Ancien Régime* (Cambridge, 1999).

69 Cleary, *op. cit.*, pp. 25–6, 256–7; Harouel, *op. cit.*, pp. 60, 120, 219.

70 Blanchard, *Ingénieurs*, pp. 436–50; Harouel, *op. cit.*, pp. 60–3, 77, 91–2, 97–9, 104, 121, 285, 300, 309.

71 For this and the following paragraphs see Petot, *op. cit.*, pp. 67–138; Antoine Picon, *L'invention de l'ingénieur moderne* (Paris, 1992), pp. 29–33, 39–47; Hautecoeur, *op. cit.*, p. 423.

72 For François Le Vau's repair work on roads and bridges and his unexecuted plans for Rochefort, see Robert W. Berger, *The Palace of the Sun: The Louvre of Louis XIV* (University Park, 1993), pp. 17–18.

73 Petot, *op. cit.*, pp. 67–8.

74 *Ibid.*, p. 77 (discussion of procurement under Colbert, pp. 73–80).

75 For J.-H. Mansart's part in the Pont Royal, Pont de Moulins (unexecuted), Canal de la Marne and Aqueduc de Maintenon, see Hautecoeur, *op. cit.*, pp. 621–3, and Pierre Bourget and George Cattaui, *Jules Hardouin Mansart* (Paris, 1960), p. 128.

76 Petot, *op. cit.*, p. 125.

77 For Gabriel and the Ponts et Chaussées, see Michel Gallet and Yves Bottineau, *Les Gabriel* (Paris, 1982), pp. 44–7; for Boffrand, *Germain Boffrand 1667–1754: l'aventure d'un architecte indépendant* (Paris, 1986), pp. 116–18.

78 'Ingénieurs de la Finance', a phrase used by the eminent Génie engineer Le Michaud d'Arçon in a publication of 1789, is quoted in Picon, *op. cit.*, p. 226.

79 Petot, *op. cit.*, pp. 259–318.

80 For the Languedoc service, *ibid.*, pp. 297–318, and F. de Dartein, *Etudes sur les ponts en pierre remarquables pour leur décoration antérieurs au XIXe siècle* (Paris, 1912), vol. 3, pp. 1–21.

81 Arthur Young, *Travels in France during the Years 1787, 1788 & 1789* (ed. Constantia Maxwell, Cambridge, 1950), p. 51.

82 Picon, *op. cit.*, pp. 39–47; Antoine Picon, *French Architects and Engineers in the Age of Enlightenment* (Cambridge, 1992), pp. 150–3, 346–7; Anne Cochon, 'Road Construction in Eighteenth Century France', in *Proceedings of the Second International Congress on Construction History* (Cambridge, 2006), vol. 1, pp. 791–7.

83 Michel Yvon, 'Pierre-Marie-Jérôme Trésaguet, ingénieur des Ponts et Chaussés 1716–1796', in *Les routes du sud de la France* (Paris, 1985), p. 301.

84 Picon, *L'invention . . .* pp. 211–14, 225–7; and for Cherbourg, Langins, *op. cit.*, p. 402. Another symptom of the loss of prestige suffered by military construction after 1760 was the dismantling of some French city walls and gates by the Ponts et Chaussées and others to help build roads or supply cheap stone for urban renewal projects: see André Guillerme, *The Age of Water: The Urban Environment in the North of France, A.D. 1300–1800* (College Station, 1988), pp. 210–11.

85 For these paragraphs see, Moulin, *op. cit.*, pp. 27–8, 38, 46–52, 59–7, 80–7 and 145–9.

86 For Toufaire or Touffaire in Paris, see Michel Gallet, *Les architectes parisiens du XVIIIe siècle* (Paris, 1995), pp. 463–4; and for his industrial work, W. H. Chaloner, 'The Brothers John and William Wilkinson and their Relations with French Metallurgy, 1775–1786', in D. A. Farnie and W. O. Henderson, *Industry and Innovation: Selected Essays by W. H. Chaloner* (London, 1990), pp. 19–32, originally published in *Annales de l'Est*, Mémoire 16: *Actes du colloque internationale 'Le fer à travers les âges: hommes et tech-*

niques' (University of Nancy, Faculté des Lettres, 1956), pp. 285–99.

87 Decree quoted in Moulin, *op. cit.*, p. 147; a dead letter, Picon, *op. cit.*, p. 214.

88 Helen Rosenau, *Social Purpose in Architecture: Paris and London Compared, 1760–1800* (London, 1970).

89 For Frézier see Blanchard, *Les ingénieurs . . .*, p. 517; Langins, *op. cit.*, pp. 234–7; Robin Middleton, 'The Abbé de Cordemoy and the Graeco-Gothic Ideal', *Journal of the Warburg and Courtauld Institutes* 25 (1962), pp. 287–90.

90 Jacques Heyman, *Coulomb's Memoir on Statics: An Essay in the History of Civil Engineering* (Cambridge, 1972), esp. pp. 190–8 (quotation, p. 194).

91 René Taton, 'L'Ecole Royale du Génie de Mézières', in Roger Hahn and René Taton, *Ecoles techniques et militaires au XVIIIe siècle* (Paris, 1986), pp. 559–615.

92 Figure given By E. Ingress Bell in 'The Modern Barrack: Its Plan and Construction', *Transactions of Royal Institute of British Architects 1880–81* (1881), p. 16. Austin Woolrych, *Britain in Revolution 1626–1660* (Oxford, 2002), p. 702, calculates Cromwell's home-based English army during the Protectorate at 13,500. For the small size of Charles II's army at the end of his reign, see his p. 789.

93 Lord Grey and Sir Thomas Palmer, quoted in J. R. Hale's spirited essay, 'The Defence of the Realm 1485–1558', in *The History of the King's Works*, vol. 4 (London, 1982), p. 395.

94 For this paragraph see the above essay by Hale (pp. 367–401): quotations from pp. 369 ('the one scheme . . .' from B. H. St J. O'Neil) and 374.

95 H. M. Colvin in the preface to *The History of the King's Works*, vol. 4, p. x.

96 The preface to BDCE vol. 1 consists of an essay on 'The practice of civil engineering 1500–1830'. Fortunately there are entries on many military engineers including Lee, Rogers, de Gomme, Romer etc., but the coverage in this respect is uneven.

97 King's Order in Council, 26 April 1667, quoted in Andrew Saunders, *Fortress Builder: Bernard de Gomme, Charles II's Military Engineer* (Exeter, 2004), p. 241. Note that the Order just preceded the Medway-Thames raid.

98 For this paragraph see Howard Tomlinson, 'The Ordnance Office and the King's Forts, 1660–1714', *Architectural History* 16 (1973), pp. 5–25; H. C. Tomlinson, *Guns and Government: the Ordnance Office under the late Stuarts* (London, 1979); Frances Willmoth, *Sir Jonas Moore: Practical Mathematics and Restoration Science* (Woodbridge, 1993), pp. 138–57; Nigel Barker, 'The Building Practice of the Board of Ordnance', in John Bold and Edward Chaney (eds.), *English Architecture Public and Private: Essays for Kerry Downes* (London and Rio Grande, 1993), pp. 199–214; and Saunders, *op. cit.*, pp. 237–66.

99 Charles II as recorded by Pepys, Sept. 1680, quoted e.g. in Saunders, *op. cit.*, p. 309.

100 Saunders, *op. cit.* I have called de Gomme Dutch rather than Flemish because he was born in Terneuzen, then as now just in the Netherlands, and served with the forces of the United Provinces. The nomenclature of the time was loose.

101 Quoted in Tomlinson, 'The Ordnance Office', p. 17. For Hull and the Fitches see also Barker, *op. cit.*, pp. 205–7.

102 Quotation of 1692–3 from Tomlinson, *Guns and Government*, p. 49. It was to be many years before Britain got over an inferiority complex about its military engineers. In 1746, after witnessing an abortive attack on L'Orient, David Hume could write to his brother: 'It has long been the Misfortune of English Armies to be very ill serv'd in Enginiers; and surely there never was on any occasion such an Assemblage of ignorant Blockheads as those which at this time attended us.' Ernest Campbell Mossner, *The Life of David Hume* (Oxford, 1980), p. 199.

103 Romer and Lilly: NDNB; Romer also in BDCE, pp. 585–6.

104 Bodt/Bott: Colvin, p. 136; Dubois: Colvin, pp. 323–5; Winde: Colvin, pp. 1064–8 and NDNB. For the pattern of army officers becoming domestic architects, see also Rolf Loeber, *A Biographical Dictionary of Architects in Ireland, 1600–1720* (London, 1981), pp. 4–5.

105 James Douet, *British Barracks 1660–1914* (London, 1998), pp. 29–30; Paul Kerrigan, *Castles and Fortifications in Ireland 1485–1945* (Cork, 1995), pp. 130ff.; and Edward McParland, *Public Architecture in Ireland 1680–1760* (London, 2001), pp. 123–41, quotations from p. 125. For Burgh, an Irish Protestant who acted as an engineer at the siege of Namur and became an important architect in Dublin, see Loeber, *op. cit.*, pp. 31–9, and McParland, *op. cit.*, pp. 145–6.

106 Douet, *op. cit.*, pp. 22–4; Geoffrey Stell, 'Highland Garrisons 1717–23: Bernera Barracks', *Post-Medieval Archaeology* 7 (1973), pp. 20–30. Fort Augustus, started 1729, and Fort George, built after the '45 Rebellion, follow similar patterns.

107 For the work of William, John and Robert Adam for the Board of Ordnance see Colvin, pp. 49, 51, 62, and BDCE, pp. 5–9; also John Fleming, *Robert Adam and his Circle* (London, 1962), pp. 52, 334–5. For General Wade, see BDCE, pp. 751–3. Out of some forty bridges under Wade's direction, the exception in ambition was William Adam's Bridge of Tay, Aberfeldy (1732–5).

108 For the Ravensdowne Barracks, see Richard Hewlings, 'Hawksmoor's "Brave Designs for the Police"', in John Bold and Edward Chaney (eds.), *English Architecture Public and Private: Essays for Kerry Downes* (London and Rio Grande, 1993), pp. 215–29.

109 Barker, *op. cit.*, pp. 210–14. For Richards, see BDCE, pp. 577–8; for Jelfe, a significant figure who had a hand in many of the biggest contracts of his age, from the Highland garrison forts to Westminster Bridge, see particularly the ODNB account by Richard Hewlings.

110 Douet, *op. cit.*, p. 56, and for an illustration of the Dubois scheme, *Transactions of Royal Institute of British Architects 1880–81* (1881), f.p. 15.

111 J. A. Bennett, *The Mathematical Science of Christopher Wren* (Cambridge, 1982).

112 Wren's obvious but less wide-ranging French rival is Claude Perrault: see Antoine Picon, *Claude Perrault, 1613–1688: ou la curiosité d'un classique* (Paris, [1989]).

113 From Wren's second tract on architecture, reprinted in Lydia M. Soo, *Wren's 'Tracts' on Architecture and Other Writings* (Cambridge, 1998), pp. 159–61.

114 James Fergusson, *History of the Modern Styles of Architecture* (London, 1862), pp. 272–3.

115 The best account of the rebuilding plans is still T. F. Reddaway, *The Rebuilding of London after the Great Fire* (London, 1940), pp. 40–67. For the way in which later architects and critics twisted Wren's sketch plan for London from an idea into a tragic reject, see his pp. 311–12.

116 I am grateful to Gillian Darley for help on Evelyn. She pointed me to the importance of his *The State of France* (1652), and *The Character of England* (1659), and to the excellent article by Mark Jenner, 'The Politics of London Air: John Evelyn's *Fumifugium* and the Restoration', *Historical Journal* 38, 3 (1995), pp. 535–51.

117 Quoted by Michael Hunter, *Science and Society in Restoration England* (Cambridge, 1981), p. 40.

118 Lisa Jardine, *On A Grander Scale: The Outstanding Career of Sir Christopher Wren* (London, 2002), pp. 234–9, 259–63, gives evidence for the Evelyn-Wren alliance from 1662 but, as often, overstates the case.

119 Margaret Whinney, 'Sir Christopher Wren's Visit to Paris', *Gazette des Beaux-Arts* 51 (1958), pp. 229–42. For what was then built in Paris and what was only in prospect, see Robert W. Berger, *A Royal Passion: Louis XIV as Patron of Architecture* (Cambridge, 1994), Chs. 2, 4, 7.

120 Wren's letter from Paris to 'a friend', construed by Jardine (pp. 239–47) as Evelyn, was first printed in *Parentalia* but is now conveniently found in Soo, *op. cit.*, pp. 103–6: 'the old reserved Italian', p. 105.

121 From Wren's report on a new dome for St Paul's, 1 May 1666, in *Wren Society* 13 (1936), p. 17. The part of the Louvre that Wren inspected was probably Louis Le Vau's inner south wing, not the east wing which was still under discussion. At the time of Wren's visit, Colbert had suspended Le Vau from the works. It was assumed Bernini would carry on, at least with the east wing. That of course did not happen.

122 Edward Browne, quoted in Jardine, *op. cit.*, pp. 245–6.

123 For the quays, see Soo, *op. cit.*, pp. 100 and 269, n. 44; Pierre Lavedan, *Histoire de l'urbanisme à Paris* (Paris, 1975), pp. 258ff.; and for the earlier embanking work under Henri IV, Hilary Ballon, *The Paris of Henri IV, Architecture and Urbanism* (New York, 1991), Ch. 3. The bridge planned by Le Vau was realized more than a hundred years later in the form of the Pont des Arts between the Institut (as the Collège des Quatre-Nations became) and the Louvre. I am grateful to Martin Meade for advice on the complex issue of the Paris quays.

124 Reddaway, *op. cit.*, pp. 200–43.

125 Wren's 'letter to a friend' on the church-building commission of 1711 is reprinted from *Parentalia* in Soo, *op. cit.*, pp. 112–18: quotations from pp. 112, 113.

126 For these two paragraphs see Berger, *op. cit.*, pp. 45–52; Antoine Picon, *Claude Perrault, 1613–1688: ou la curiosité d'un classique* (Paris, [1989]), pp. 197–223; Roger Hahn, *The Anatomy of a Scientific Institution: The Paris Academy of Sciences* (Berkeley, 1971), Ch. 1; J.-A. Dulaure, *Histoire physique, civile et morale*

de Paris (Paris, 1839), vol. 3, pp. 217–20.

127 This summary of Greenwich Observatory follows Willmoth, *op. cit.*, pp. 178–95, in stressing the role of Jonas Moore. See also Derek Howse, *Greenwich Time and the Discovery of Longitude* (Oxford, 1980), pp. 20–44; Jardine, *op. cit.*, pp. 307–15; and John Bold, *Greenwich: An Architectural History of the Royal Hospital for Seamen and the Queen's House* (London, 2000), pp. 19–24.

128 Lisa Jardine, *Ingenious Pursuits: Building the Scientific Revolution* (London, 1999), p. 160.

129 Wren to Dr Fell, 3 Dec. 1681, quoted e.g. in Jardine, *On A Grander Scale . . .* , p. 315.

130 For a previous fantastical observatory, compare Tycho Brahe's Uraniborg at Hven: Victor E. Thoren, *The Lord of Uraniborg: A Biography of Tycho Brahe* (Cambridge, 1990), pp. 144–91.

131 Bold, *op. cit.*, p. 21.

132 The attempts by Wren and Hooke to plan the Monument and appropriate parts of St Paul's for zenith telescopes and other scientific observations, for which see Jardine, *op. cit.*, pp. 317–21 and 424, confirm this. Neither proved suitable, but the thought was there.

133 Quoted in T. S. Willan, *River Navigation in England 1600–1750* (London, 1964), p. 79.

134 On the early growth of the English surveying profession, see F. M. L. Thompson, *Chartered Surveyors* (London, 1968), pp. 19–38; and Andro Linklater, *Measuring America* (London, 2002), Ch. 1, with a summary of 'the invention of property' and the importance of Edmund Gunter's measuring chain, first promulgated in 1623.

135 L. T. C. Rolt and J. S. Allen, *The Steam Engine of Thomas Newcomen* (Ashbourne, 1997), pp. 24–30, quotation p. 27; Mike Chrimes in BDCE, pp. 594–5.

136 Mike Chrimes in BDCE, pp. 641–5. Another major waterworks engineer of this generation was John Hadley of Worcester, for whom see Skempton in BDCE, pp. 289–91.

137 For Lombe's Derby mill see Anthony Calladine, 'Lombe's Mill: An Exercise in Reconstruction', *Industrial Archaeology Review* 16 (1993), pp. 82–99.

138 *Ibid.*, p. 644.

139 See Mark Girouard, *The English Town* (London, 1990), pp. 75–100; P. J. Corfield, *The Impact of English Towns 1700–1800* (Oxford, 1982), pp. 168–85.

140 This is the core of the interpretation of Pope's mid-period poetry in Howard Erskine-Hill, *The Social Milieu of Alexander Pope* (London, 1975). See, besides the remarks on the *Epistle to Lord Burlington*, Erskine-Hill's comments (pp. 8–41) on 'The Man of Ross' in the *Epistle to Lord Bathurst*, and (pp. 205–40) on John Allen of Bath. Pope met Allen after he wrote these poems, but praised him in 1740 as 'the Most Noble Man in England' because of his schemes of public benevolence.

141 John Butt (ed.), *The Poems of Alexander Pope* (London, 1963), p. 586.

142 For the ancestry and development of this argument, see Jules Lubbock, *The Tyranny of Taste* (London, 1995).

143 Butt, *The Poems of Alexander Pope*, pp. 594–5 (lines 177–204).

144 *Ibid.*, p. 595.

145 This and the following paragraph are based on Samuel Wilson jr, 'Colonial Fortifications and Military Architecture in the Mississippi Valley', in Francis McDermott (ed.), *The French in the Mississippi Valley* (Urbana, 1965), pp. 103–22; and James Pritchard, *In Search of Empire: The French in the Americas, 1670–1730* (Cambridge, 2004), pp. 42–3, 104. The sneer about the Spanish and the English is from the Abbé Delaporte (1769), quoted by Wilson. See also John W. Reps, *The Making of Urban America: A History of City Planning in the United States* (Princeton, 1965), pp. 78–84. The same author's *Cities of the American West: A History of Frontier Urban Planning* (Princeton, 1979), p. 57, adds information about La Salle's expedition further west in 1684, when he founded Fort St Louis near the present Port Lavaca, Texas.

146 According to Pritchard, *op. cit.*, p. 115, the French seventeenth-century colonial grids at Fort Royal (Fort-de-France, Martinique) and Le Cap Français, (Cap Haïtien, Haiti), had both been laid out by French military engineers. Louisbourg on Cape Breton Island, founded a little before New Orleans in 1712, certainly was, and boasted European-style fortifications: see Reps, *The Making of Urban America*, pp. 65–8.

147 Reps, *Cities of the American West*, pp. 35–40 (quotation, p. 37); see also his *The Making of Urban America*, pp. 29–36.

148 Valerie Fraser, *The Architecture of Conquest: Building in the Viceroyalty of Peru 1535–1635* (Cambridge, 1990), pp. 36–40, 49, 73, 101–7: quotation from pp. 106–7.

149 Reps, *Cities of the American West*, pp. 57–84 for the Texan story, with plans of presidios, pp. 61–4; and pp. 87–115 for the Californian story, with an 'expeditions engineer' first mentioned in connection with the founding of Monterey in 1770 (p. 89) and of Branciforte in 1796 (pp. 102–3).

150 Early Portuguese forts in Asia were by no means geared just against other European powers: see C. R. Boxer, *The Portuguese Seaborne Empire* (London, 1969); A. J. R. Russell-Wood, *A World on the Move: The Portuguese in Africa, Asia and America, 1415–1808* (Manchester, 1992); and Timothy J. Coates, *Convicts and Orphans: Forced and State-Sponsored Colonizers in the Portuguese Empire* (Stanford, 2001), which at p. 73 quotes the Portuguese viceroy in Goa in 1653 as saying that he urgently needed military engineers who were being 'sent to the hinterlands of Brazil, but not here in spite of the very great need for men practiced in that art'.

151 Juliet Barclay, *Havana: Portrait of a City* (London, 1995). pp. 41–50, quotation p. 44.

152 Danes: Christianborg (1661–70); Dutch: Coenraadsburg (1662–6); English: Cape Coast Castle (*c.*1674), Commenda (1686), Dixcove (1692); Prussians: Gross Friedrichsburg (1683). Elsewhere in Africa may be added Cape Town Castle (1666–79), and the English works at Tangier (1667–83). Dates etc. from Derek Linstrum in Dan Cruickshank (ed.), *Sir Banister Fletcher's History of Architecture* (20th edn, Oxford, 1996), pp. 1171–5.

153 Marcus Whiffen and Frederick Koeper, *American Architecture 1607–1976* (London, 1981), pp. 38–41; Daniel L. Schodek,

Landmarks of American Civil Engineering (Cambridge, Mass., 1987), pp. 269–71. For the history of St Augustine see Reps, *The Making of Urban America*, pp. 32–5.

154 The different ways in which permanent corps of engineers emerged in different countries before the Napoleonic Wars (after which they became widespread) affected the authorship of city layouts and other public works. The evolution of these corps has not to my knowledge been studied on a comparative basis. Each state had its unique pattern, linked to political events, reform of the armed forces and the founding of academies. In Spain and its colonies, the development can be traced to the Bourbon succession in 1700. It is usually said to proceed from the establishment of the military academy of mathematics in Barcelona under Prospero de Verboon following decrees of 1710–11: see Alfredo Vigo Trasancos, *Arquitectura y urbanismo en El Ferrol del Siglo XVIII* (Segovia, 1984), pp. 11–12; Josep Maria Montaner, *Barcelona: A City and its Architecture* (Cologne, 1997), pp. 16–20. In respect of Latin America, the invaluable Ramon Gutierrez, *Arquitectura y urbanismo en Iberoamerica* (Madrid, 1983), notably Ch. 13 (pp. 343–50), explains that Spain did not allow independent military academies in its colonies, though Portugal permitted one in Brazil from 1792. The case of Portugal is specially intriguing because of the rebuilding of the Lisbon Baixa from 1755 under José I's veteran chief military engineer, Manuel da Maia. The new layout and building typology, and thus the creation of the austere, so-called 'Pombaline style' fell largely to da Maia's protégé Eugénio dos Santos (d.1760) and Carlos Mardel (d.1763), whose previous careers had straddled military and civil architecture. Though there had been an official course in fortification and military architecture since 1735, a corps of engineers seems not to have been formalized by the time of the Lisbon earthquake, while in Portuguese colonies the town-planning regulations adopted by the Spanish hardly existed. See José-Augusto França, *Une ville des Lumières: La Lisbonne de Pombal* (Paris, 1965); José-Augusto França, *A reconstrução de Lisboa e a arquitectura pombalina* (Lisbon, 1981); John Bury, *Arquitectura e arte no Brasil colonial* (São Paolo, 1991), pp. 215–17; Kenneth Maxwell, *Pombal, Paradox of the Enlightenment* (Cambridge, 1995), pp. 24–35, 102–5.

155 Even when English colonies were official or licensed, they did not often go in for much formal planning. Joseph Konvitz, *Cities and the Sea: Port City Planning in Early Modern Europe* (Baltimore, 1978), pp. 64–5, remarks: 'On the whole, the English had no sense that city planning had anything to do with sea power.'

156 Reps, *The Making of Urban America*, pp. 157–74. It has been hazarded that Holme (1625–95) had been a navy captain under Penn's father Admiral Penn, but this appears to be a guess. As he had land-surveying skills, he is more likely to have served in some army.

157 These paragraphs are largely based on the documents and commentary in Paul K. Walker, *Engineers of Independence: A*

Documentary History of the Army Engineers in the American Revolution, 1775–1783 (Office of History, US Army Corps of Engineers, Washington, 1981). See also *The History of the U.S. Army Corps of Engineers* (Washington, 1998), pp. 17–19.

158 Walker, *op. cit.*, p. 10.

159 H. Paul Caemmerer, *The Life of Pierre Charles L'Enfant, Planner of the City Beautiful, The City of Washington* (Washington, 1950, repr. New York, 1970), pp. 34–41.

160 Janis Langins, *Conserving the Enlightenment: French Military Engineering from Vauban to the Revolution* (Cambridge, Mass., 2004), pp. 155–8, 215–22.

161 The above quotations are from Walker, *op. cit.*, pp. 35–6 and 37 (bis) (capitalization and punctuation changed).

162 *Ibid.*, p. 106.

163 *Ibid.*, p. 352.

164 *Ibid.*, p. 341.

165 *Ibid.*, p. 358 (capitalization, punctuation and some spelling changed).

166 Theodore J. Crackel, *West Point* (Lawrence, Kansas, 2002), pp. 32–3. Crackel's narrative emphasizes the discontinuities at West Point between 1794 and 1817. He also notes that Alexander Hamilton procured a plan for an academy from du Portail in 1798. See also *The History of the U.S. Army Corps of Engineers* (Washington, 1998), pp. 25–7.

167 Caemmerer, *op. cit.*, p. 268.

168 Jonathan Williams, quoted in Todd Shallat, *Structures in the Stream: Water, Science and the Rise of the US Army Corps of Engineers* (Austin, 1994), p. 2.

169 But Crackel, *op. cit.*, pp. 46–50, argues that the idea of a peacetime role for the Corps was not decisive in 1802 and carried little weight until 1817.

170 The Topographical Section of the War Office was created during wartime in 1813, subsumed as the US Topographical Bureau within the Corps of Engineers in 1818, separated out into the US Corps of Topographical Engineers in 1838, and merged again in 1863. See Frank N. Schubert (ed.), *The Nation Builders: A Sesquicentennial History of the Corps of Topographical Engineers, 1838–1863* (Office of History, US Army Corps of Engineers, Fort Belvoir, Virginia, 1988).

171 Quoted in Lois Craig (ed.), *The Federal Presence: Architecture, Politics, and Symbols in United States Government Building* (Cambridge, Mass., 1978), p. 49.

172 Shallat, *op. cit.*, p. 42.

173 Joseph Totten in a letter of Feb. 1824, quoted in Shallat, *op. cit.*, p. 75.

174 These paragraphs on L'Enfant and Washington rely chiefly on Caemmerer, *op. cit.*; John W. Reps, *Monumental Washington: The Planning and Development of the Capital Center* (Princeton, 1967); Dumas Malone, *Jefferson and the Rights of Man* (Boston, 1951), pp. 371–87; Pamela Scott, '"This Vast Empire": The Iconography of The Mall, 1791–1848', in Richard Longstreth (ed.), *The Mall In Washington, 1791–1991* (2nd edn, Washington, 2002), pp. 37–58; and André Corboz, *Deux capitales françaises: Saint-Pétersbourg et Washington* (Gollion, Switzerland, 2003).

175 Caemmerer, *op. cit.*, p. 113, from the *Massachusetts Magazine*, June 1789.

176 *Ibid.*, pp. 127–30 (L'Enfant to Washington, 11 September 1789).

177 For Federal Hall, see Louis Torres, 'Federal Hall Revisited', *Journal of the Society of Architectural Historians* 29 (Dec. 1970), pp. 327–38. On the promise, Caemmerer, *op. cit.*, pp. 214–15, quotes Washington to L'Enfant, 28 February 1792, as to 'the compliment which was intended to be paid you in depending alone upon your plans for the public buildings instead of advertising a premium for the person who should present the best'.

178 Corboz, *op. cit.*, pp. 74–6, reproducing at Fig. 60 a plan for a 'ville imaginaire' of 1784 from the Ponts et Chaussées mapping competitions, with grids and diagonal axes of the Washington variety – but fortified, as Washington was not. In his absorbing essay, Corboz is sceptical of the influence of Versailles, out of fashion in the late eighteenth century, on the Washington plan. Pamela Scott in her essay, *op. cit.*, makes a case for the influence of the layout at Marly. But neither of these authors takes account of process in city-planning; the L'Enfant layout is treated as a static work of art, not as a blueprint from which it was necessary to build.

179 Caemmerer, *op. cit.*, pp. 145–9. The maps Jefferson sent were of Frankfurt-am-Main, Karlsruhe, Amsterdam, Strasbourg, Paris, Orleans, Bordeaux, Lyons, Montpellier, Marseilles, Turin and Milan.

180 *Ibid.*, p. 146.

181 For the Ellicotts, see DAB and ANB, and Caemmerer, *op. cit.*, pp. 206–7, 220, 343. Joseph Ellicott subsequently surveyed three million acres of New York State and laid out the town of Buffalo.

182 Malone, *op. cit.*, p. 375, quoting an undated note from L'Enfant.

183 *Ibid.*, p. 22.

184 The serviceable Andrew Ellicott, after making the first alterations to L'Enfant's plan on Jefferson's advice, was likewise sacked by the Commissioners at the end of 1793.

185 Roger G. Kennedy, *Orders from France: The American and the French in a Revolutionary World, 1780–1820* (New York, 1989), p. 97. This rather chaotic book is full of fascinating details about Franco-American relations in this period.

186 Caemmerer, *op. cit.*, p. 213 (L'Enfant to Washington, 28 Feb. 1792).

187 For Roberdeau see DAB and ANB; Caemmerer, *op. cit.*, pp. 198–201; Shallat, *op. cit.*, pp. 61–3.

188 Albert E. Cowdrey, *A City for the Nation: The Army Engineers and the Building of Washington, D.C., 1790–1967* (Office of the Chief of Engineers, Washington, 1979), pp. 11–14.

189 For Mills in Washington see John M. Bryan, *America's First Architect: Robert Mills* (New York, 2001), pp. 10–36, 216–319.

190 Agnes Addison Gilchrist, *William Strickland, Architect and Engineer, 1788–1854* (New York, 1969), pp. 15, 100–3; Bryan, *op. cit.*, p. 304; Antoinette J. Lee, *Architects to the Nation: The Rise and Decline of the Supervising Architect's Office* (New York, 2000), pp. 33–4. According to Cowdray, *op. cit.*, p. 15, Andrew Humphreys, best known for his later flood defences along the Mississippi, was also involved with Strickland on the House of Representatives plan, while Humphreys and another officer, Campbell Graham, had worked with Mills on an unexecuted War Department building, presumably the predecessor to the Strickland-Abert plan. Mills certainly had close relations with the Corps of Engineers, designing a building at West Point in 1839.

191 Quoted in Lee, *op. cit.*, pp. 44–5; previous quotation, p. 43. For the Bureau of Construction and the Supervising Architect, see Lee, pp. 39–72, and Craig, *op. cit.*, pp. 91, 99.

192 Cowdrey, *op. cit.*, p. 18. For the aqueduct see also Shallat, *op. cit.*, pp. 179–84; and for the famous Cabin John Bridge, Daniel L. Schodek, *Landmarks in American Civil Engineering* (Cambridge, Mass., 1987), pp. 112–14.

193 For this paragraph see Dean A. Herrin, 'The Eclectic Engineer: Montgomery C. Meigs and his Engineering Projects', and Martin K. Gordon, 'The Engineers and the Architects: Whose Profession Shall Build for the Government?', in William C. Dickinson, Dean A. Herrin and Donald R. Kenner (eds.), *Montgomery C. Meigs and the Building of the Nation's Capital* (Athens, Ohio, 2001), pp. 3–20 and 49–54; William C. Allen, *The Dome of the United States Capitol: An Architectural History* (Washington, 1992), pp. 36–47. 'Technologically feasible': Herrin, p. 8.

194 These paragraphs are based on Cowdrey, *op. cit.*, pp. 23–34, 35, 45; Craig, *op. cit.*, p. 188; Lee, *op., cit.*, pp. 174–5; Gordon in Dickinson, Herrin and Kennan, *op. cit.*, pp. 51–3; Reps, *Monumental Washington*, pp. 58–93; *The History of the U.S. Army Corps of Engineers* (Washington, 1998), pp. 61–3.

195 Cowdrey, *op. cit.*, p. 27.

196 Reps, *op. cit.*, p. 75.

197 For this paragraph: Cowdrey, *op. cit.*, pp. 56–8; Craig, *op. cit.*, pp. 412, 425.

198 Arthur Young, *Travels in France during the years 1787, 1788 & 1789* (ed. Constantia Maxwell, Cambridge, 1950), pp. 40–1.

2 IRON

1 For classical antiquity, see S. B. Hamilton, 'The Structural use of Iron in Antiquity', TNS 31 (1957–9), pp. 29–47, casting doubt on W. B. Dinsmoor's earlier claims that the Greeks made use of solid sections of iron; and for the mediaeval period, John Reeves, Gavin Simpson and Peter Spencer, 'Iron Reinforcement of the Tower and Spire of Salisbury Cathedral', *Archaeological Journal* 149 (1992), pp. 385–406, citing Aachen, the Sainte Chapelle, Canterbury and Westminster Abbey for the use of iron as flitched ties and plates. The bracing of the lantern at Salisbury, probably of the early fourteenth century, seems its most extensive known use in Gothic architecture. Another example was the continuous iron 'chain' in the upper triforium of the chevet at Bourges Cathedral (*c*.1200–25): Robert Branner, *La cathédrale de Bourges et sa place dans l'architecture gothique* (Paris/Bourges, 1962), p. 83.

2 For the proliferation of minor uses of iron in Venetian building, see Antonella Zanussi, 'L'impiego del ferro nell'edilizia storica veneziana', in Claudia Conforti and Andrew Hopkins, *Architettura e tecnologia:*

acque, tecniche e cantieri nell'architettura rinasci-mentale e barocca (Rome, 2002), pp. 126–39.

3   Frank D. Prager and Gustina Scaglia, *Brunelleschi: Studies of his Technology and Inventions* (Cambridge, Mass., 1970), pp. 33–8, 55–9; Rowland Mainstone, 'Brunelleschi's Dome', *Architectural Review* 162 (1977), pp. 157–66; Howard Saalman, *Filippo Brunelleschi: The Cupola of Santa Maria del Fiore* (London, 1980), p. 72. Prager and Scaglia point out that the word 'catena' has several meanings. The primary elements of the 'chains' in the Florence dome were timber (chestnut) and sandstone. How continuous the iron ties were remains unclear. But the use of metal was, in their view, innovative: 'The combination of stone and iron members constituting such a tie ring had long been forgotten . . . The reintroduction of metal-connected tie rings in a cellular vault fabric by Brunelleschi at the end of the Gothic age appears to us a major part of his "renewal of Roman masonry" ' (p. 59). This claim might not stand up to a careful comparison of Brunelleschi's usage with Gothic examples.

4   Deborah Howard, 'Renovation and Innovation in Venetian Architecture', *Scroope*, 6 (1994–5), pp. 66–74; Manuela Morresi, *Jacopo Sansovino* (Milan, 2000), pp. 165–9. The extent of Sansovino's iron girdling of the dome of the Ascension at St Mark's may have been obscured by nineteenth-century restorations. Vasari speaks briefly of 'catene di ferro'; but Morresi quotes Temanza as mentioning in 1778 'a great girdle of iron, in several pieces, notched and well tightened, with wedges and ties of the same metal'. Howard is confident that the encircling 'twisted ropes of wrought iron' found when the dome was opened up were indeed Sansovino's. When he took on this job, Sansovino as yet had little technical expertise in construction. Very likely he depended heavily on ironworkers from the Arsenale.

5   Saalman, *op. cit.*, pp. 214–17; Filippo Baldinucci, *The Life of Bernini* (University Park and London, 1966), p. 100.

6   James Campbell and Robert Bowles, 'The Construction of Wren's Cathedral', in Derek Keene, Arthur Burns and Andrew Saint, *St Paul's, The Cathedral Church of London* (London and New Haven, 2004), pp. 215–17. The structural ironwork was supplied mainly by Thomas Robinson, not, as is sometimes said, by Jean Tijou.

7   E.g. the Cappella dei Principi, Florence (early seventeenth century): Marzia Marandola, 'La Cappella dei Principi: un cantiere secolare', in Conforti and Hopkins, *op. cit.*, pp. 92–4; and the Mansion House, London, *c.*1739–40: Sally Jeffrey, *The Mansion House*, (Chichester, 1993), pp. 72–4, mentioning other English examples. For a tantalizingly obscure allusion to chains suggested by Ammannati for the repair of Siena Cathedral in 1558, see Saverio Sembranti, 'Bartolomeo Ammannati a Siena tra il 1558 e il 1559', in Niccolò Rosselli Del Turco and Federica Salvi, *Bartolomeo Ammannati scultore e architetto 1511–1592* (Florence, 1995), p. 360.

8   Rowland J. Mainstone in Robert W. Berger, *The Palace of the Sun: The Louvre of Louis XIV* (University Park, Pennsylvania, 1993), pp. 69ff; Robin Middleton, 'Architects as Engineers: The Iron Reinforcement of Entablatures in Eighteenth-Century France', *AA Files* 9, 1985, pp. 54–64.

9   Simon Thurley, *Hampton Court: A Social and Architectural History* (London, 2003), pp. 169–70; Howard Colvin, 'The Building', in David McKitterick (ed.), *The Making of the Wren Library, Trinity College Cambridge* (Cambridge, 1995), p. 43 & fig. 23. In the Trinity Library, the weight of the book-stacks determined the need for iron ties.

10 The best text of Wren's 1668 report on Salisbury Cathedral is in Lydia M. Soo, *Wren's 'Tracts' on Architecture and Other Writings* (Cambridge, 1998), pp. 62–78; an abridged version occurs in Wren's *Parentalia*. For an authoritative discussion of this report and the mediaeval work at Salisbury, see Reeves, Simpson and Spencer, *op. cit.*

11 Reeves, Simpson and Spencer, *op. cit.*, p. 400.

12 Mainstone in Berger, *op. cit.*, p. 72.

13 There are certainly counter-examples. John Rennie, for instance, was a fabricator-manufacturer in the mechanical field as well as a consulting engineer; Charles Fox started off as a junior railway engineer under the Stephensons, then moved into contracting.

14 Barrie Trinder, *The Industrial Archaeology of Shropshire* (Chichester, 1996), p. 57.

15 *The History of the King's Works*, vol. 5 (London, 1976), pp. 402–4. This augments and corrects an article by John Harris, 'Cast Iron Columns 1706', in *Architectural Review* 130 (July 1961), pp. 60–1.

16 *Op. cit.*, p. 403.

17 John Tijou, *A New Book of Drawings* (1693), Plate 53. The drawing was plagiarized by Batty Langley in his *City and Country Builder's and Workman's Treasury of Designs* (1741), while the whole book was reissued in France in 1722–3: see Eileen Harris, *British Architectural Books and Writers 1556–1785* (Cambridge, 1990), p. 456. For Tijou's great influence on English ironwork, see the excellent books by Raymond Lister, *Decorative Wrought Ironwork in Great Britain* (London, 1957), and *Decorative Cast Ironwork in Great Britain* (London, 1960).

18 *The History of the King's Works, loc.cit.*

19 Ted Ruddock, *Arch Bridges and their Builders 1735–1835* (Cambridge, 1979), pp. 132–6; Barrie Trinder, 'The First Iron Bridges', *Industrial Archaeology Review* 3, no. 2 (1979), pp. 112–21, reprinted in R. J. M. Sutherland (ed.), *Structural Iron 1750–1850* (Aldershot, 1997), pp. 247–56; Julia Ionides, *Thomas Farnolls Pritchard of Shrewsbury, Architect and 'Inventor of Cast Iron Bridges'*, Ludlow, 1999, pp. 247–70.

20 Trinder, 'The First Iron Bridges', *op. cit.*, p. 116, reprinted in Sutherland, *op. cit.*, p. 251.

21 Neil Burton tells me that the pulpit at St George's Bloomsbury (*c.*1730) had an iron core.

22 Terry Friedman, *Church Architecture in Leeds 1700–1799*, Thoresby Society 2nd series no. 7 (Leeds, 1997), pp. 52–6. The gallery columns were removed in 1867.

23 David Nortcliffe, 'A Preliminary Report on the Kirklees Iron Bridge of 1769 and its Builder', Yorkshire Archaeological Society, 1979.

24 The surveyor-builder was Cuthbert Bisbrown, who was buried at St James's. Frank Kelsall, note on St James's Toxteth for English Heritage, n.d.; and Colvin, p. 127.

25 Notably St John's, Hanley, Stoke on Trent (1788–90), where cast iron is used not just for the columns supporting the galleries but also for the window tracery and tower crenellations.

26 In a case beyond easy reach of the iron-founders, Tetbury Church, Gloucestershire (1777–81), a core of iron hidden in the nave piers allowed Francis Hiorne to raise a high and airy Gothic arcade, but the style was still furnished by a timber casing.

27 *Survey of London*, vol. 35, 'The Theatre Royal, Drury Lane, and the Royal Opera House, Covent Garden' (London, 1970), pp. 49–57; Ian Donaldson, 'New Papers of Henry Holland and R. B. Sheridan, (i) Holland's Drury Lane, 1794', *Theatre Notebook* 16 no. 3 (1962), pp. 91–6.

28 Unpublished researches kindly communicated by Frank Kelsall.

29 *Survey of London, op. cit.*, pp. 91–3.

30 John Foulston, *The Public Buildings erected in the West of England as designed by John Foulston, F.R.I.B.A.* (London, 1833), pp. 18–20 and Plates 36–42.

31 R. F. Mould, *St George's Church, Everton: The Iron Church* (Liverpool, 1977), p. 17.

32 British patent no. 3761, 1813, for 'Facing Walls of Brick etc.', which refers to various features for 'Churches or other buildings of pure Gothic Design'. Cragg had also acquired a patent in 1810 for 'Certain Improvements in the Casting of Iron Roofs for Houses, Warehouses, and other Buildings'.

33 [John Cragg], *Remarks on the Gothic Style of Building, with Some Observations on the Practicability of Restoring Ecclesiastical Architecture in England* (Liverpool, 1814), pp. 15–16. A copy of this rare pamphlet is in the RIBA Library.

34 Information and quotations in this and the following paragraph come from the diaries of Thomas Rickman in the RIBA Library.

35 See Colvin, p. 812.

36 M. H. Port, *Six Hundred New Churches* (Reading, 2nd edn, 2006), pp. 152–4, and for the rest of this paragraph, pp. 59–64, 145–9. I am grateful to Michael Port for advice on the role of the Crown architects.

37 Notably the famous (and climatically horrendous) iron church by William Slater for the Ecclesiological Society, published in *Instrumenta Ecclesiastica*, 2nd series (London, 1856).

38 See especially the article by Eric Robinson on Thomas Yeoman, 'The Profession of Civil Engineer in the Eighteenth Century', in A. E. Musson and Eric Robinson, *Science and Technology in the Industrial Revolution* (Manchester, 1969), pp. 372–92.

39 For Albion Mills see John Mosse, 'The Albion Mills, 1784–1791', *TNS* 40 (1967–8), pp. 47–60.

40 For dock buildings in London, see Malcolm Tucker, 'Warehouses in Dockland', in *Dockland, An Illustrated Historical Survey of Life and Work in East London* (London, 1986), pp. 21–30; and A. W. Skempton, 'Engineering in the Port of London', *TNS* 50 (1978–9), pp. 87–108, and 53 (1981–2), pp. 73–96.

41 A. W. Skempton and H. R. Johnson, 'The First Iron Frames', in *Architectural Review*, 131 (March 1962), pp. 175–86, reprinted in R. J. M. Sutherland (ed.), *Structural Iron 1750–1850* (Aldershot, 1997), pp. 19–44; H. R. Johnson and A. W. Skempton, 'William Strutt's Cotton Mills, 1793–1812', TNS 30 (1955–7), pp. 179–205.

42 W. G. Rimmer, *Marshalls of Leeds, Flax-Spinners 1788–1886* (Cambridge, 1960), p. 49.

43 'The Boulton and Watt drawings of engines they installed always contained a considerable amount of structural detail because of the interrelation betwen engine and building': A. W. Skempton in discussion of paper by Mosse, *op. cit.*, p. 59. But earlier, Matthew Boulton had been unwilling to be drawn into construction, writing in 1776: 'It is not compatible with my business nor Mr Watt's to attend the erection of buildings'. Quoted in R. H. Campbell, *Carron Company* (Edinburgh, 1961), p. 43.

44 Mike Williams and D. A. Farnie, *Cotton Mills in Greater Manchester*, (Preston, 1992), pp. 78–9.

45 *The Life of Sir William Fairbairn, Bart*, ed. A. E. Musson (Newton Abbot, 1970), pp. vii–ix, 111–17.

46 Skempton and Johnson, *op. cit.*, p. 178.

47 *Ibid.*, p. 179, and in Sutherland, *op. cit.*, p. 27.

48 Ron Fitzgerald, 'The Development of the Cast Iron Frame in Textile Mills to 1850', *Industrial Archaeology Review* 10, no. 2 (1988), pp. 127–45, reprinted in R. J. M. Sutherland (ed.), *Structural Iron 1750–1850* (Aldershot, 1997), pp. 45–63.

49 Keith A. Falconer, 'Fireproof Mills – the widening perspectives', *Industrial Archaeology Review* 16, no. 1 (1993), pp. 11–26; Colum Giles and Ian H. Goodall, *Yorkshire Textile Mills* (London, 1992), pp. 67–9.

50 Frances H. Steiner, *French Iron Architecture* (Ann Arbor, 1984), p. 6; and for the background to continental metallurgy, David S. Landes, *The Unbound Prometheus* (Cambridge, 1969), pp. 174–84.

51 W. H. Chaloner, 'The Brothers John and William Wilkinson and their Relations with French Metallurgy, 1775–1786', in D. A. Farnie and W. O. Henderson, *Industry and Innovation: Selected Essays by W. H. Chaloner* (London, 1990), pp. 19–32, originally published in *Annales de l'Est*, Mémoire 16: *Actes du colloque internationale 'Le fer à travers les âges: hommes et techniques'* (University of Nancy, Faculté des Lettres, 1956), pp. 285–99; Steiner, *op. cit.*, pp. 19–20; Louis Bergeron, *Le Creusot: une ville industrielle, un patrimoine glorieux* (Paris, 2001).

52 Robin Middleton, 'Architects as Engineers: The Iron Reinforcement of Entablatures in Eighteenth-Century France', *AA Files* 9 (1985), pp. 54–64.

53 Pierre Patte, *Mémoires sur les objets les plus importants de l'architecture*, (Paris, 1769); Middleton, *op. cit.*, p. 54. Most scholarship about structural iron in French architecture before 1840 relies on the publications of Patte and Rondelet. Archaeology might reveal a different pattern of diffusion.

54 Middleton, *op. cit.*, p. 62.

55 J. Rondelet, *Traité théorique et pratique de l'art de bâtir*, vol. 4, Part 2 (Paris, 1817), pp. 75ff.; Steiner, *op. cit.*, pp. 23–6. The Louvre work of 1789, in 'the salon next to the Grande Galerie', is attributed to J.-B.-A. Renard, working under his father-in-law Guillaumont, says Steiner. Again, this history of early wrought-iron trusses derives almost exclusively from Rondelet, who worked with Soufflot, and may not give the full story.

56 *Mercure de France*, March 1770, p. 44; section of Moreau's Opéra in Musée Carnavalet, Topo G C X. I am grateful to Sophie Descat for these references: further on Moreau, see her 'Deux architectes-urbanistes dans l'Europe des Lumières: Pierre-Louis Moreau et George Dance à Paris et à Londres', Thèse de Doctorat, Université de Paris I, 2000. It should be added that Soufflot's theatre at Lyons (1754–6) had apparently featured a folding iron 'chassis' between auditorium and stage: Erich Schild, *Der Nachlass des Architekten Hittorffs* (Cologne, 1958), pp. 285–8.

57 Rondelet, *op. cit.*, p. 123, Plate 173.

58 A coherent but evangelistic account of early hollow-pot development is given in C. L. G. Eck, *Traité de construction en poteries et en fer* (Paris, 1836), pp. 1–7.

59 Steiner, *op. cit.*, p.25.

60 Eck, p. 50, where it is also asserted that Victor Louis had used the same method for his Bordeaux theatre 'une année avant', i.e. in 1785. This seems confirmed in *Victor Louis 1731–1800: dessins et gravures* (Bordeaux, 1980), p. 86.

61 For the Louvre trusses and rooflights see Rondelet, *op. cit.*, p. 123, Plate 174.

62 See the remarks by J. G. James in his 'Thomas Paine's Iron Bridge Work 1785–1803', TNS 59 (1987–8), pp. 193–9, 219–20.

63 Steiner, *op. cit.*, p. 27; Antoine Picon, *L'invention de l'ingénieur moderne* (Paris, 1992), pp. 169–70.

64 This account follows Mark K. Deming, *La Halle au Blé de Paris 1762–1813* (Brussels, 1984), pp. 167–97; Dora Wiebenson, 'The Two Domes of the Halle au Blé in Paris', *Art Bulletin* 55 (1973), pp. 262–79; Jean Stern, *A l'ombre de Sophie Arnould: François-Joseph Bélanger, Architecte des Menus Plaisirs* (Paris, 1930), vol. 2, pp. 201–12, 233–8, 245–52; Steiner, *op. cit.*, pp. 28–32; and Bertrand Lemoine, *L'architecture du fer: France, XIXe siècle* (Paris, 1986), p. 155. Donald David Schneider, *The Works and Doctrine of Jacques Ignace Hittorff 1792–1867* (New York, 1977), vol. 1, pp. 47–56, gives a few extra details.

65 Kenneth Woodbridge, 'Bélanger en Angleterre: son carnet de voyage', *Architectural History* 25 (1982), pp. 8–19. Woodbridge believes the date of the visit was c.1778, later than Stern had believed.

66 Wiebenson, *op. cit*, p. 264; Deming, *op. cit.*, p. 174. Wiebenson says Deumier was to be specially responsible for the copper covering.

67 Legrand had structural expertise, having trained as an engineer in the Ponts et Chaussées but abandoned that career because he found it 'too narrow': Deming, *op. cit.*, p. 175.

68 Stern, *op. cit.*, vol. 2, p. 208: quotations from Becquey's report of 31 July and the Commission's conclusion of August 1807.

69 F. Brunet, *Dimensions des fers qui doivent former la coupole de la Halle aux Grains* (Paris, 1809), p. 3.

70 Brunet describes himself on the title page of his book as 'ancien entrepreneur des batiments, et controleur des travaux de la Halle'. He had submitted his own 'fireproof' scheme for a new roof in 1806: Wiebenson, *op. cit.*, p. 270. She follows Eck in calling him an engineer (p. 279).

71 Bergeron, *op. cit.*, pp. 34–5. Le Creusot iron-work was also used for thousands of pipes, partly rejects from cannon, for the Paris water supply in 1808, and for a famous Paris fountain by Girard of 1811; Denis Woronoff, *L'industrie sidérurgique en France pendant la Révolution et l'Empire* (Paris, 1974), pp. 424–7.

72 Eck, *op. cit.*, p. 53.

73 Stern, *op. cit.*, vol. 2, p. 236.

74 Steiner, *op. cit.*, p. 31.

75 Sigfried Giedion, *Building in France, Building in Iron, Building in Ferro-Concrete* (Santa Monica, 1995), p. 104 (translation of Giedion's *Bauen in Frankreich . . .* of 1928).

76 Deming, *op. cit.*, p. 195.

77 Pierre-François-Léonard Fontaine, *Journal 1799–1853* (Paris, 1987), p. 375 (24 November 1813).

78 Steiner, *op. cit.* p. 27. Napoleon also urged Fontaine to investigate the idea of reinforcing the shaky dome of Ste Geneviève (by then the Panthéon) with iron columns and for the columns at the Panthéon: *ibid.*, p. 28.

79 *Ibid.*, p. 8.

80 This section on Prussia relies chiefly on Werner Lorenz, *Konstruktion als Kunstwerk: Bauen mit Eisen in Berlin und Potsdam 1797–1850* (Berlin, 1995). See also Ursula Ilse-Neuman, 'Karl Friedrich Schinkel and Berlin Cast Iron, 1810–1841', in Derek E. Ostergard (ed.), *Cast Iron from Central Europe 1800–1850* (New York, 1994), pp. 54–73.

81 Karl Friedrich Schinkel, *The English Journey*, ed. David Bindman and Gottfried Riedmann (New Haven and London), 1993.

82 Lorenz, *op. cit.*, pp. 79–86.

83 On Persius and iron see the essays by Sabine Bohle-Heintzenberg, 'Die Dampfkraft in der Parklandschaft', and Andreas Meinecke, 'Gestalt und Konstruktion – Persius und die Schutzkuppel von St. Nikolai' in *Ludwig Persius – Architekt des Königs: Baukunst unter Friedrich Wilhelm IV* (Potsdam, 2003), pp. 73–9 and 80–7. Bohle-Heintzenberg argues that Persius's romantic steam-engine houses at Klein-Glienicke, Sanssouci and Babelsberg were more rational and collaborative than in Schinkel's prototype at the Charlottenhof waterworks. Meinecke claims that Persius had more to do with the outer iron shell of the dome at St Nikolai, Potsdam, than is allowed by Lorenz, who gives most of the credit to Borsig.

84 Lorenz, *op. cit.*, pp. 87–93.

85 For Schinkel's influence on the change in roof design of Leo von Klenze's Valhalla, Regensburg, from a barrel vault to wrought iron, c.1836, see Werner Lorenz and Annegret Rohde, 'Building with Iron in Nineteenth Century Bavaria – The Valhalla Roof Truss and its Architect, Leo von Klenze', *Construction History* 17 (2001), pp. 55–74. This article shows that Klenze, like Schinkel, had a largely 'architectural' attitude towards iron construction. But like many articles of engineering scholarship it labours the point, as if anyone coming out of the Bauakademie ought to have taken a more rational attitude towards iron.

86 An example is Tuscany, where in 1834 Leopold II invested in a foundry at the coastal town of Follonica for exploiting the Elban iron ore that was one of the grand duchy's assets. Experimental buildings were commissioned from the Florentine architect-professor Carlo Reishammer, 'Commissario Regio delle Strade Ferrate': notably a church in Follonica itself and a new city gate for Leghorn. In both, Reishammer integrated structural iron with the arched masonry of Tuscan tradition. San Leopoldo, Follonica (1838) has a double-height cast-iron portico and a wealth of iron fittings, yet the core of the church itself is of masonry. As a church built to advertise a foundry's virtuosity, it makes a pendant to the earlier Liverpool churches of Cragg and Rickman. See Romano Jodice, L'architettura del ferro: l'Italia (1796–1914) (Rome, 1985), pp. 124–40.

87 For this section on Russia see R. H. Campbell, Carron Company (Edinburgh, 1961), pp. 74–5, 144–53; Eric Robinson, 'The Transference of British Technology to Russia, 1760–1820', in Barrie M. Ratcliffe, Great Britain and her World 1750–1914 (Manchester, 1975), pp. 12–15; Anthony Cross, By the Banks of the Neva (Cambridge, 1997), pp. 242–60; J. G. James, 'Russian Iron Bridges to 1850', TNS 54 (1982–3), pp. 79–104, reprinted in R. J. M. Sutherland (ed.), Structural Iron 1750–1850 (Aldershot, 1997), pp. 348–72; Roger P. Bartlett, 'Charles Gascoigne in Russia', in A. G. Cross (ed.), Russia and the West in the Eighteenth Century (Newtonville, Mass., 1983), pp. 354–67; Sergey G. Fedorov, 'Matthew Clark and the Origins of Russian Structural Engineering, 1810–1840s', Construction History 8 (1992), pp. 69–88, reprinted in Sutherland, op. cit., pp. 103–22; Sergey G. Fedorov, 'Early Iron Domed Roofs in Russian Church Architecture: 1800–1840', Construction History 12 (1996), pp. 41–66; Sergey G. Fedorov, Wilhelm von Traitteur: ein badischer Baumeister als Neuerer in der russichen Architektur, 1814–1832 (Berlin, 2000); ODNB sub Charles Gascoigne (Eric Robinson).

88 Dimitri Shvidkovsky, The Empress and the Architect (London, 1996), pp. 25ff.

89 Ibid., pp. 239–40; Bartlett, op. cit., p. 359.

90 Architectural Magazine 2 (March 1835), p. 122, in an account of his Russian visit two decades before, in 1813–14.

91 Fedorov, 'Early Iron Domed Roofs', pp. 41–5.

92 For an analysis of the slightly later iron roofs over the Hermitage, similar in type, see Bernhard Heres, 'The Iron Roof Trusses of the New Hermitage in St Petersburg', in Proceedings of the Second International Congress on Construction History (Cambridge, 2006), vol. 2, pp. 1555–68.

93 Fedorov. 'Matthew Clark', p. 74 (Sutherland, op. cit., p. 108).

94 Rondelet in his revised edition added that the trusses of the Bourse 'hold the first rank among works of this kind': Traité théorique et pratique de l'art de bâtir (7th edition, Paris 1834), vol. 3, p. 318. While Rondelet gives the credit exclusively to the architect, Charles Eck in his Traité de construction en poteries et en fer (Paris, 1836), p. 49, once again

makes a point of mentioning the serrurier Albouy, 'depuis longtemps au nombre de nos plus habiles practiciens'.

95 Eck, op. cit., plates passim. Important examples shown: the Bourse, 1823, architect for the roof, Labarre, serrurier, Albouy; Théâtre de l'Ambigu-Comique, 1827–8, architects Hittorff and Lecointe, serruriers Roussel and Baudrit; Théâtre des Nouveautés, 1829, architect, Debret, serrurier, Mignon. For Hittorff's later work at the Panorama and the Gare du Nord, see pp. 122–4.

96 Bertrand Lemoine, L'architecture du fer: France, XIXe siècle (Paris, 1986), p. 68. A curiosity was the steep iron roof erected over Chartres Cathedral following the fire which destroyed the ancient timber charpente above the stone vault in 1836. This concealed superstructure was of cast iron and by the serrurier Mignon, who had already inserted iron into the roof of the chapel in the Palais Royal for Fontaine. In some rival proposals for the roofs at Chartres, serruriers or their architects were cowed by the context into proposing a measure of Gothic detailing, but Mignon and the cathedral architect, Emile Martin, shunned that as pointless. See ibid., pp. 70–2, 173–5. Hidden iron roofs over French churches were quite common thereafter.

97 L'art de l'ingénieur, ed. Antoine Picon (Paris, 1997), p. 369.

98 John Summerson, 'Records of an Iron Age', Official Architect 8, no. 5 (May 1945), p. 235.

99 John Summerson, The Life and Work of John Nash Architect (London, 1980), pp. 17–18, 27–8, 35–6, 101–9. For Nash's use of iron at Corsham see Frederick J. Ladd, Architects at Corsham Court (Bradford-on-Avon, 1978), pp. 95–111; and for his ingenuity with timber, Jonathan C. Clarke, 'Cones, Not Domes: John Nash and Regency Structural Innovation', in Proceedings of the Second International Congress on Construction History (Cambridge, 2006), vol. 1, pp. 717–39.

100 Quoted in Clifford Musgrave, Royal Pavilion (Brighton, 1951), p. 62.

101 The fabricators of the structural iron for the Brighton Pavilion are largely unrecorded, though Views of the Royal Pavilion, ed. Gervase Jackson-Stops (London, 1991), p. 76, states that the bamboo staircases were made by William Stark of London and put in place in 1815. Andrew Barlow of the Royal Pavilion kindly drew my attention to a sentence in the Sussex Weekly Advertiser, 28 Sep. 1818: 'The immense iron work used has been supplied from the foundries in the neighbourhood of Lewes, and part from London'.

102 The History of the King's Works, vol. 6 (London, 1973), pp. 137, 267, 274–6.

103 Quoted from appendix to second report of Select Committee on Windsor Castle and Buckingham Palace, October 1831, in R. J. M. Sutherland, 'The Age of Cast Iron 1780–1850: Who Sized the Beams?', The Iron Revolution: Architects, Engineers and Structural Innovations 1780–1850, ed. Robert Thorne (London, 1990), p. 30, reprinted in R. J. M. Sutherland (ed.), Structural Iron 1750–1850 (Aldershot, 1997), p. 72.

104 Quoted in The History of the King's Works, vol. 6 (London, 1973), p. 276.

105 Early Smirke country houses where iron beams are recorded include Cirencester Park (c.1810) and Worthy Park (c.1816).

106 Ibid., pp. 425–6; J. Mordaunt Crook, The British Museum (Harmondsworth, 1972), pp. 132, 140–2; Sutherland in Thorne, op. cit., pp. 26–7, reprinted in Sutherland, Structural Iron, pp. 67–9. For Smirke's use of iron generally, see entry on him by Mike Chrimes in BDCE, pp. 628–9; and for iron in his bombed Inner Temple Library and Harcourt Buildings, John Summerson, 'Records of an Iron Age', Official Architect 8, no. 5 (May 1945), p. 235.

107 See entry on Rastrick by Mike Chrimes in BDCE, pp. 544–7.

108 R. J. M. Sutherland in Thorne, op. cit., pp. 24–33, reprinted in Sutherland, Structural Iron, pp. 65–75.

109 See entry on Tredgold by L. G. Booth in BDCE, pp. 716–22; also L. G. Booth, R. J. M. Sutherland and N. S. Billington, 'Thomas Tredgold (1788–1829): Some Aspects of his Work', TNS 51 (1979–80), pp. 57–94.

110 Booth, Sutherland and Billington, op. cit., p. 93.

111 Francis Fowke, 'On Civil Construction', Reports on the Paris Universal Exhibition (1856), Part III, p. 259.

112 Lawrance Hurst, 'The Age of Fireproof Flooring', in Thorne, op. cit., pp. 34–9. The earliest such British system, invented by H. H. Fox in the 1830s, only took off after his partner James Barrett substituted wrought iron beams for cast ones after 1851. For France, Lemoine, op. cit., pp. 50–1, lists a plethora of floor patents dating from 1846–51, of which the most popular at first was the Vaux system. Later the patents of Fernand Zorès of Besançon became dominant. A study of one Burgundian foundry shows that iron for floors and other structural uses during the 'Haussmannization' of Paris in the 1860s could amount to a high proportion of a firm's output: Serge Benoit, 'Croissance de la sidérurgie et essor de la construction métallique urbaine au siècle dernier: le cas de Châtillon-Commentry', in Frédéric Seitz (ed.), Architecture et métal en France: 19e–20e siècles (Paris, 1994), pp. 99–129, esp. pp. 119–20.

113 This account follows primarily M. H. Port (ed.), The Houses of Parliament (London, 1976), notably Michael Port's chapters on the construction, pp. 97–121 and 142–94, and Denis Smith's chapters on technique and servicing, pp. 195–231.

114 Port, op. cit., p. 106. Meeson is here described by Charles Barry junior as 'long my father's chief assistant in his office', suggesting that he worked out the primary structural implications of Barry senior's designs. Later, Meeson was involved in the main building for the International Exhibition of 1862, and in the re-erection of parts of it at the first Alexandra Palace, Muswell Hill, 1873. He also made the drawings of the ironwork for the Langham Hotel (1863–5) on behalf of the architects John Giles and James Murray: The Builder, 7 Dec. 1867, p. 898.

115 On the roofs, see Smith in Port, op. cit., pp. 198–200. See Skempton in TNS 35 (1959–60), pp. 57–78.

116 *Survey of London*, vol. 35, 'The Theatre Royal, Drury Lane, and the Royal Opera House, Covent Garden' (London, 1970), pp. 80–3, 98–105; Andrew Saint, 'Covent Garden's Three Theatres', in *A History of the Royal Opera House Covent Garden 1732–1982* (London, 1982), pp. 26–31; Richard Leacroft, *The Development of the English Playhouse* (London, 1988 edn), pp. 220–30.

117 The poor acoustics of iron and glass architecture soon put paid to the concerts.

118 On the long-span roof, see notably R. J. M. Sutherland, 'Shipbuilding and the Long-Span Roof', TNS 60 (1988–9), pp. 107–26, reprinted in R. J. M. Sutherland (ed.), *Structural Iron 1750–1850* (Aldershot, 1997), pp. 123–42. Workshop and factory roofs have been less coherently discussed, though there are helpful remarks in Keith A. Falconer, 'Fireproof Mills – the Widening Perspectives', *Industrial Archaeology Review* 16, no. 1 (1993), pp. 11–26, and Colum Giles and Ian H. Goodall, *Yorkshire Textile Mills* (London, 1992), pp. 71–2. The 55-foot cast-iron span of Henry Maudslay's roof at his works in Lambeth (1826), also comes into this story: see Mike Chrimes's entry on Maudslay in BDCE, pp. 455–7.

119 These paragraphs rely much on Edward John Diestelkamp, 'The Iron and Glass Architecture of Richard Turner', Ph.D., University College London, 1982, pp. 6–17, and Melanie Louise Simo, *Loudon and the Landscape* (New Haven, 1988), pp. 111–18.

120 For Loudon as an architect see Colvin, pp. 623–4, and as an engineer, BDCE, pp. 412–13 (Tess Canfield).

121 The date of the Bicton hothouse is unknown but is ascribed to around 1825.

122 John Summerson, *The Life and Work of John Nash Architect* (London, 1980), p. 40.

123 Neil Burton, 'Thomas Hopper, 1776–1856: The Drama of Eclecticism', in Roderick Brown (ed.), *The Architectural Outsiders* (London, 1985), pp. 117–19.

124 Simo, *op. cit.*, p. 264, points out that Loudon could state that plant houses had only made progress when freed from the domination of 'mansion architects', yet also advocate conservatories decorated with columns or pinnacles.

125 Stefan Koppelkamm, *Glasshouses and Wintergardens of the Nineteenth Century* (London, 1981), pp. 17, 48–54.

126 Koppelkamm, *op. cit.* p. 17, quoting from Charles Macintosh, *The Greenhouse, Hothouse and Stove* (1838).

127 Colvin, p. 378.

128 Robert Thorne, *Covent Garden Market, its History and Restoration* (London, 1980), pp. 15–18, and for the later roofs pp. 28–32; Charles Fowler, 'Description of the Metal Roof at Hungerford Market', *Transactions of the Institute of British Architects*, 1, 1837, pp. 44–6; Gavin Stamp, 'The Hungerford Market', *AA Files* 11 (1986), pp. 62–6.

129 In discussion of a paper on Horace Jones's metropolitan markets: *Transactions of the R.I.B.A.*, 1877–8, pp. 128–9.

130 Lemoine, *op. cit.*, pp. 130–2.

131 Diestelkamp, *op. cit.*, p. 89.

132 These paragraphs are largely based on Edward Diestelkamp's dissertation, 'The Iron and Glass Architecture of Richard Turner', Ph.D., University College London,

1982, and on his two articles: 'Richard Turner and the Palm House at Kew Gardens', TNS 54 (1982–3), pp. 1–26, reprinted in R. J. M. Sutherland (ed.), *Structural Iron 1750–1850* (Aldershot, 1997), pp. 77–102, and 'The Design and Building of the Palm House, Royal Botanic Gardens, Kew', *Journal of Garden History* 2 no. 3 (1982), pp. 233–72.

133 Edward J. Diestelkamp, 'Fairyland in London: The Conservatories of Decimus Burton', *Country Life* 19 May 1983, p. 1343.

134 Quoted in G. F. Chadwick, *The Works of Sir Joseph Paxton 1803–1865* (London, 1961), p. 94. Chadwick's assignment of responsibilities for the Great Stove to Paxton alone is hardly borne out by the evidence he cites, pp. 72–98.

135 *Ibid.*, p. 79.

136 Quoted in Diestelkamp, 'Richard Turner', p. 4 (Sutherland, p. 80).

137 *Ibid.*, p. 6 (Sutherland, p. 82).

138 Diestelkamp, 'Richard Turner', p. 20 (Sutherland, p. 96).

139 Both quotations *ibid.*, p. 19 (Sutherland, p. 95).

140 In some countries engineer meant the man who drove the train. But the term always denoted some level of superior status.

141 F. R. Conder (ed. Jack Simmons), *The Men Who Built Railways* (London, 1983), pp. 27–8 (edited reprint of Conder's *Personal Recollections of English Engineers*, 1868).

142 R. S. Fitzgerald, *Liverpool Road Station Manchester, An Historical and Architectural Survey* (Manchester, 1980), pp. 14–15.

143 *Ibid.*, pp. 17–24; and for Water Street, which involved William Fairbairn and Eaton Hodgkinson, see Michael Bailey, 'The Early Years', in Michael R. Bailey (ed.), *Robert Stephenson – The Eminent Engineer* (Aldershot, 2003), pp. 26–7.

144 R. H. G. Thomas, *The Liverpool and Manchester Railway* (London, 1980), pp. 98–9.

145 Fitzgerald, *op. cit.*, p. 52.

146 Robert E. Carlson, *The Liverpool and Manchester Railway Project 1821–1831* (Newton Abbot, 1969), pp. 185–7; Thomas, *op. cit.*, pp. 108–10.

147 Thomas, *op. cit.*, p. 119; he cites Cunningham as the architect for the rear part of the station, and Samuel and James Holme as the builder-designers of the roof, with other builders also involved.

148 *Ibid.*, pp. 120–2; *The Builder* 17 Feb. 1849, pp. 77–8. For Turner's part and Locke's opposition, see Edward Diestelkamp, 'The Iron and Glass Architecture of Richard Turner', Ph.D., University College London, 1982, pp. 261–8.

149 The best account of Euston is given in John Summerson, *The Architectural History of Euston Station* (London, 1959), printed by the British Transport Commission but never distributed because of embarrassment about the impending demolition of the Euston Arch. For Charles Fox's role in the Euston extension, see Mike Chrimes, 'Building the London & Birmingham Railway', in Michael R. Bailey (ed.), *Robert Stephenson – The Eminent Engineer* (Aldershot, 2003), p. 243; and James Sutherland, 'Iron Railway Bridges', *ibid.*, pp. 306–9.

150 The early 'Projet Bourla' for a northern station for Paris (1838) took its cue from Euston, with an arch detached from the

working parts of the station: see Karen Bowie (ed.), *Les grandes gares parisiennes du XIXe siècle* (Paris, [1997]), pp. 96–9.

151 A. W. N. Pugin, *An Apology for the Revival of Christian Architecture* (London, 1843), pp. 10–11 and Plate III.

152 Summerson, *op. cit.*, p. 10.

153 The theme of the open head station is developed in Carroll L. V. Meeks, *The Railroad Station: An Architectural History* (New Haven, 1956). For the endurance of this tradition in France, see Lemoine, *L'architecture du fer: France, XIXe siècle* (Paris, 1986), pp. 143–54.

154 Georges Tubeuf, quoted in *All Stations* (London, 1981), p. 25: originally *Le temps des gares* (Paris, 1978).

155 On the early London and Birmingham Railway (1835–8), 'Architects designed the station buildings excluding the platforms and train-shed roofs', says Mike Chrimes, 'Building the London & Birmingham Railway', in Michael R. Bailey (ed.), *Robert Stephenson – The Eminent Engineer* (Aldershot, 2003), p. 243. Chrimes reports that besides Philip Hardwick's work at London and Birmingham, George Aitchison designed the intermediate stations, Daniel Bagster (an engineering draughtsman?) organized the platform layouts, while Francis Thompson was brought in to design goods stations after December 1838.

156 Christian Barman, *An Introduction to Railway Architecture* (London, 1950). The classification comes from the division of his illustrations.

157 Quoted in Michael Freeman, *Railways and the Victorian Imagination* (New Haven, 1999), p. 73.

158 Quoted by Martin S. Briggs in his article on Tite, *The Builder* 20 January 1950, p. 95. In claiming precedence for Gothic at Carlisle, Tite was forgetting or ignoring the first Bristol Temple Meads and Bath stations. Elsewhere he talks about trying 'to introduce some degree of beauty' into an iron-built railway shed at Perth: *The Builder* 18 Jan. 1851, p. 39.

159 There were similar figures, e.g. David Mocatta, Francis Thompson and Sancton Wood. A summary of how architects were chosen by Robert Stephenson is given by Mike Chrimes and Robert Thomas, 'Railway Building', in Michael R. Bailey (ed.), *Robert Stephenson – The Eminent Engineer* (Aldershot, 2003), pp. 283–5.

160 Obituary of Tite in *The Builder* 3 May 1873, pp. 337–9.

161 Victoria Haworth, 'Inspiration and Instigation: Four Great Railway Engineers', in Denis Smith (ed.), *Perceptions of Great Engineers: Fact and Fantasy* (London 1994), p. 69.

162 Haworth, *ibid.*; obituary of Tite in *The Builder* 3 May 1873, pp. 337–9; details of timber roof at station in Rouen in *The Builder* 2 Feb. 1850, p. 52; and for the two stations at Rouen, *The Builder* 20 January 1950, p. 95. For British involvement in the Paris–Le Havre line, made infamous by Zola's *La bête humaine*, see David Brooke, 'William Mackenzie and Railways in France', *Construction History* 13 (1997), pp. 17–28; and David Brooke (ed.), *The Diary of William Mackenzie the First International Railway Contractor* (London, 2000). Mackenzie's

inability to spell Tite's name correctly (*Diary*, pp. 133, 170–2) suggests that they were not intimate. There was resentment about British involvement in French railway-building. A note in the architectural press about the building of the Rouen–Le Havre line complained that 'Les travaux d'art [a term no British railway-builder would have used of bridges, viaducts and stations] sont mal exécutés et avec de mauvais matériaux.' Brick only was being used, alleged the writer: 'De ces deux espèces de briques, les unes sont jaunes, les autres rouges. La première semble n'avoir été cuite qu'aux rayons du soleil, à l'instar des Égyptiens, tant à cause de sa couleur que de sa friabilité.' So there was some relish when the Barentin viaduct collapsed. See *Revue de l'architecture et des travaux publics* 5 (1844), cols 555–6; Marc Saboya, *Presse et architecture au XIXe siècle* (Paris, 1991), p. 149. According to Frances Steiner, *French Iron Architecture* (Ann Arbor, 1984), pp. 40, 210, many of the original stations on the Paris–Le Havre line were designed not by Tite but by Alfred Armand, who had been responsible for stations on the original Paris (St Lazare)–St Germain-en-Laye line from which the Paris–Rouen–Le Havre route diverged. This seems likely, but she is wrong in attributing the layout of the line to Allard and Buddicom, partners of Mackenzie and Brassey who, David Brooke makes clear, mainly contributed locomotives and equipment.

163   This account is based on John Addyman and Bill Fawcett, *The High Level Bridge and Newcastle Central Station: 150 Years across the Tyne* (Newcastle, 1999). See also Thomas Faulkner and Andrew Greg, *John Dobson, Architect of the North East* (Newcastle, 2001), pp. 101–6.

164   L. T. C. Rolt, *George and Robert Stephenson* (London, 1960), p. 287.

165   Quoted e.g. in L. T. C. Rolt, *Isambard Kingdom Brunel* (London, 1957), pp. 231–2.

166   Francis Conder gives this character-sketch of Brunel: 'His conscientious resolve to see with his own eyes, and to order with his own lips, every item of detail entrusted to his responsibility, brought on him an enormous amount of labour, which, on another and a more easy system, would have been borne by subordinates, perhaps with equal advantage to the public. His exquisite taste, his knowledge of what good work should be, and his resolve that his works should be no way short of the best, led rather to the increased cost, than to the augmented durability, of much that he designed and carried out . . . He was highly impatient of insubordination, or of contradiction. Very much more was to be done with him by following his course, than by attempting to take one's own.' F. R. Conder, (ed. Jack Simmons), *The Men Who Built Railways* (London, 1983), pp. 125–7 (edited reprint of Conder's *Personal Recollections of English Engineers*, 1868).

167   John Binding, *Brunel's Bristol Temple Meads* (Hersham, 2001).

168   Thomas Foster of Bristol: Binding, *op. cit.*, p. 41.

169   A. W. N. Pugin, *An Apology for the Revival of Christian Architecture* (London, 1843), p. 11.

170   Binding, *op. cit.*, p. 79.

171   Steven Brindle, *Paddington Station: Its History and Architecture* (London, 2004), p. 32. This book is the source for most of my account of Paddington.

172   Matthew Digby Wyatt, 'Iron work and the principles of design', *Journal of Design* 4 (1850), pp. 77–8, quoted in Brindle, *op. cit.*, p. 40.

173   Quoted in Brindle, *op. cit.*, p. 43.

174   For the Gare du Nord, see especially Karen Bowie (ed.), *Les grandes gares parisiennes au XIXe siècle* (Paris, [1987]), pp. 95–115; also the preface by Béatrice de Andia, pp. 13ff. I am grateful to Karen Bowie for her thoughts on the complex issue of responsibility for the early Paris termini on which, she tells me, much evidence is lost.

175   Georges Ribeill, 'Les fondations stratégiques des grandes gares parisiennes', in Bowie, *op. cit.*, pp. 28–38; Frank Dobbin, *Forging Industrial Policy: The United States, Britain, and France in the Railway Age* (Cambridge, 1994), pp. 105–16, and for opposition to the Ponts et Chaussées monopoly from Perdonnet and the Ecole Centrale, p. 122.

176   For the intellectual side to Reynaud's career see David Van Zanten, *Designing Paris* (Cambridge, Mass., 1987), pp. 45–59; and Karen Bowie, 'Les polytechniciens et l'architecture métallique', in Bruno Belhoste, Francine Masson and Antoine Picon (eds), *Le Paris des polytechniciens* (Paris, 1994), pp. 203–5.

177   Meeks, *op. cit.*, pp. 61–2; Bowie, *Les grandes gares . . .*, pp. 100–3.

178   Passengers were not allowed on to the platforms until their trains were ready in early French stations, so waiting rooms were important.

179   *The Builder* 6 April 1850, p. 159. The writer presumably means that only comparatively short lengths of timber were then obtainable in Paris.

180   Quoted by Ribeill in Bowie, *Les grandes gares . . .*, pp. 36–8, from *Journal des chemins de fer*, 25 Oct. 1845.

181   Van Zanten, *op. cit.*, pp. 117–18.

182   Lemoine, *op. cit.*, pp. 143–54; foremost among these engineers was Eugène Flachat.

183   Thomas von Joest, 'Hittorff et la nouvelle Gare du Nord', in *Hittorff: un architecte du XIXème* (Paris, 1986), p. 270.

184   Meeks, *op. cit.*, p. 61.

185   Bowie, *Les grandes gares . . .*, pp. 104–15, based on her thesis: K. Bowie, 'L'eclecticisme pittoresque et l'architecture des gares parisiennes au XIXe siècle', Université de Paris I, 1985. Lejeune, she says, began work on the station, to be joined by Ohnet in 1858 when planning was already under way. Further on Hittorff and the second Gare du Nord, see Donald David Schneider, *The Works and Doctrine of Jacques Ignace Hittorff 1792–1867* (New York, 1977), pp. 613–82; and Joest, *op. cit.*, pp. 266–77. The obscurity about the Gare du Nord is complicated by the fact that Haussmann and Hittorff had bad relations.

186   In Bowie, *Les grandes gares . . .*, p. 21.

187   For de Baudot's critique see *Encyclopédie d'architecture* 1863, pp. 190–1, and Schneider, *op. cit.*, pp. 663–71. 'Confectionery': in 1956 Carroll Meeks could still compare the front of the Gare du Nord to 'the stucco-covered

confectionery of a world's fair' (*op. cit.*, p. 67).

188   Lemoine, *op. cit.* p. 146, says that Couche conceived the original space of the Gare du Nord. Some older accounts say that Reynaud was involved in the engineering of the second Gare du Nord though not in the architecture, but there seems no evidence for this.

189   Some earlier sheds had involved collaboration between architects and engineers. Karen Bowie, 'Les polytechniciens et l'architecture métallique', p. 205, mentions the case of Alexis Cendrier, the architect who designed the iron-and-timber train sheds for the first Gare de Lyon (1847–53, demolished) under the supervision of Adolphe Jullien of the Ponts et Chaussées, as well as the station at Lyon-Perrache (surviving).

190   Joest, *op. cit.* p. 272 (for 'Gomlay' read Gourlay).

191   *Victor Laloux: L'architecte de la gare d'Orsay* (Paris, 1987), pp. 21–41.

192   This section is based largely on Aart Oxenaar, *Centraal Station Amsterdam: Het paleis voor de reiziger* ('s-Gravenhage, 1989): quotation from p. 17. See also H. Romers, *De Spoorwegarchitektuur in Nederland 1841–1938* (Zutphen, 1981), pp. 99–110.

193   Quoted in Oxenaar, *op. cit.*, p. 91, from meeting of 19 April 1887.

194   This account derives primarily from: Carl W. Condit, *The Port of New York*, vol. 2 (Chicago, 1981), pp. 1–10, 54–100; Deborah Nevins (ed.), *Grand Central Terminal, City within the City* (New York, 1982); Kenneth Powell, *Grand Central Terminal, Warren and Wetmore* (London, 1996); Kurt C. Schlichting, *Grand Central Terminal: Railroads, Engineering and Architecture in New York City* (Baltimore, 2001); and Peter Pennoyer and Anne Walker, *The Architecture of Warren and Wetmore* (New York, 2006), pp. 78–107.

195   For Pennsylvania Station, see William Couper, *History of the Engineering Construction and Equipment of the Pennsylvania Railroad Company's New York Terminal and Approaches* (New York, 1912); Steven Parissien, *Pennsylvania Station, McKim, Mead and White* (London, 1996); Hilary Ballon, *New York's Pennsylvania Stations* (New York, 2002).

196   Ballon, *op. cit.*, p. 22.

197   *Ibid.*, pp. 69–73 for the distortions of the concourse. But Ballon is mistaken in attributing the structural engineering at Penn Station to the Westinghouse firm. For Purdy and Henderson's authorship see Paul Starrett, *Changing the Skyline* (New York, 1938), p. 144, where he says they designed 'that graceful archwork of steel within the station which has excited the admiration of architects and artists ever since'. Starrett claimed to have recommended Purdy and Henderson to William Mead as 'the best steel designers I know' after seeing 'frightfully ugly' steelwork designs for the trainshed by a rival firm.

198   Edward Hungerford, article from *Harpers Weekly* quoted approvingly in John A. Droege, *Passenger Terminals and Trains* (New York, 1916), p. 6.

199   *American Architect and Building News*, 1904, quoted in Condit, *op. cit.*, p. 61.

200   Quoted in Parissien, *op. cit.*, p. 35.

201   Amidst the huge literature on the Crystal

Palace, the best starting points for the story of its evolution and construction now are John McKean, *Crystal Palace: Joseph Paxton and Charles Fox* (London, 1994), and J. R. Piggott, *Palace of the People: The Crystal Palace at Sydenham 1854–1936* (London, 2004).

202 Also important to Paxton in the preparatory stage of his design was the mathematician and engineer Peter Barlow: Elizabeth Bonython and Anthony Burton, *The Life and Work of Henry Cole* (London, 2003), p. 135.

203 Piggott, *op. cit.*, pp. 8–9: quotation from the *Art Journal Illustrated Catalogue* for the 1851 Exhibition.

204 Michael Darby and David Van Zanten, 'Owen Jones's Iron Building of the 1850s', *Architectura* 1 (1974), p. 54; Carol A. Hrvol Flores, *Owen Jones: Design, Ornament, Architecture, and Theory in an Age in Transition* (New York, 2006), pp. 81–4.

205 *The Builder* 18 Jan. 1851, p. 39, a phrase repeated in Cockerell's Royal Academy lecture reported *ibid.*, 6 March 1851, p. 149: 'The Iron order was essentially Britannic . . . Iron structures were susceptible of a similar classification to that of Doric, Ionic, and Corinthian in the Masonic order; for they might be governed by similar proportions to those of the palm, the cane or the reed . . . A Gothic cathedral in this material might be rendered tenfold more astonishing in its effect.' Cockerell showed similar enthusiasm the last time he gave his Royal Academy lectures, arguing that iron could give architecture 'that characteristic style which we had long been whining about, as the desideratum and deficiency of our age . . . Iron might be termed the osteology of building . . . Now our buildings would have bones, giving unity and strength which never before existed.' *Ibid.*, 12 Jan. 1856, p. 13.

206 The main executed examples were St James's Hall (1856–8), the Crystal Palace Bazaar (1857–8), and Osler's Shop, Oxford Street (1858–60), all in the West End of London. In each of these projects, the structural ironwork was qualified and partly hidden by rich plaster and coloured glass. See Flores, *op. cit.*, pp. 130–47, and Darby and David Van Zanten, *op. cit.*, pp. 53–75.

207 John Physick, *The Victoria and Albert Museum – The History of its Building* (Oxford, 1982), pp. 22–6; Bonython and Burton, *op. cit.*, pp. 172–5.

208 *Survey of London*, vol. 38, 'The Museums Area of South Kensington and Westminster' (London, 1975), p. 98. C. D. Young and Co. also erected the hall for the Manchester Art Treasures Exhibition of 1857, designed by the architect Edward Salomons. Here an architectural competition won by Owen Jones was set aside because the Executive Committee had already chosen Youngs to manufacture and erect the building: see Flores, *op. cit.*, pp. 133–4.

209 Physick, *op. cit.*, p. 24.

210 *Ibid.*, p. 26.

211 See M. H. Port, *Imperial London: Civil Government Building in London 1851–1915* (New Haven and London, 1995), Chapter 11, 'The Golden Age of Competition', and Chapter 12, 'Rival Modes of Selection'.

212 See David Van Zanten, *Designing Paris* (Cambridge, Mass., 1987), pp. 121–35.

213 Physick, *op. cit.*, p. 27; *Survey of London*, vol. 38, 'The Museums Area of South Kensington and Westminster' (London, 1975), pp. 86–9, 102–3. Grover, along with R. M. Ordish, played an important role in designing the roof for the Royal Albert Hall: *ibid.*, p. 188.

214 Physick, *op. cit.*, pp. 97ff.

215 *Ibid.*, pp. 47–56; *Survey of London*, vol. 38, pp. 104–6.

216 *Survey of London*, vol. 38, p. 106.

217 *Building News* 24 Feb. 1865, pp. 130–1.

218 For the Natural History Museum see *Survey of London*, vol. 38, 'The Museums Area of South Kensington and Westminster' (London, 1975), pp. 200–16; Mark Girouard, *Alfred Waterhouse and the Natural History Museum* (London, 1981).

219 Sir Henry Hunt, quoted in *Survey of London*, vol. 38, p. 210.

220 Jonathan Clarke's report for English Heritage, 'Early Structural Steel in London Buildings' (2000), highlights Waterhouse's place in this story. His earliest significant building in this respect was the National Liberal Club (1884–7), which has flooring and roofs of W. H. Lindsay's patent steel decking.

221 Quoted in Physick, *op. cit.*, p. 47.

222 Frederick O'Dwyer, *The Architecture of Deane and Woodward* (Cork, 1997), p. 160. Paddington may be relevant, as Philip Pusey was an MP and must have travelled up regularly to London from Berkshire via the station.

223 *Ibid.*, p. 258. The following account derives largely from O'Dwyer's chapter on the Oxford Museum, pp. 152–283, notably his section on the roof, pp. 257–75.

224 'The Work of Iron, in Nature, Art and Policy', published in *The Two Paths* (1859). Ruskin's views on iron continue to be caricatured in much of the modern literature about him.

225 Quoted in O'Dwyer, *op. cit.*, p. 261.

226 Charles Grey to the Queen, 1864, quoted in Peter Howell, 'Francis Skidmore and the Metalwork', in Chris Brooks (ed.), *The Albert Memorial* (London, 2000), p. 268. A full account of Skidmore's career is given in this essay, pp. 253–85.

227 Skidmore's lecture 'On the Use of Metals in Church-Building and Decoration' is reported in *The Ecclesiologist* 17 (1856), pp. 221–2, 333–8, with comments from Scott, Matthew Digby Wyatt and others. On the ecclesiologists and iron, see further Stefan Muthesius, 'The Iron Problem in the 1850s', *Architectural History* 13 (1970), pp. 58–63.

228 O'Dwyer, *op. cit.*, p. 265. It was presumably on the basis of Skidmore's involvement in the Slater design, never fabricated, that he felt confident in tendering to build the Oxford Museum roof.

229 Fairbairn's report, 10 April 1858, kindly communicated by Peter Howell.

230 *Building News* 18 Feb. 1859, p. 161.

231 George Gilbert Scott, *Remarks on Secular & Domestic Architecture, Present and Future* (London, 1857), p. 113.

232 For iron in the Foreign Office see Ian Toplis, *The Foreign Office, An Architectural History* (London, 1987), pp. 189–90.

233 Quoted in a modern article on Norman Shaw's 180 Queen's Gate, *Country Life*, 30 August 1956.

234 Gavin Stamp, 'The Hungerford Market', *AA Files* 11 (1986), p. 61, where Scott's recollections are quoted: 'The work was constructed on principles then new. Iron girders, Yorkshire landings, roof and platforms of tiles in cement, and columns of granite being its leading elements . . . I ought, too, to mention the advantage of constant reference to Mr. Fowler's working drawings, some of the best and most perspicuous I have ever seen.'

235 Robert Thorne, 'Building the Memorial: the Engineering and Construction History', and Peter Howell, 'Francis Skidmore and the Metalwork', in Chris Brooks (ed.), *The Albert Memorial* (New Haven & London, 2000), pp. 134–59, 253–85.

236 George Gilbert Scott and F. W. Sheilds, 'Salisbury Cathedral: Reports on the Tower and its Sustaining Piers', August 1865, Salisbury Cathedral Archives, kindly communicated by Claudia Marx. Ten years previously, in 1855, the railway engineer Eugène Flachat had been brought in by the Ministère des Cultes to save the crossing tower of Bayeux Cathedral, following mistakes by the architect Victor Ruprich-Robert and a shockingly vandalistic report by Viollet-le-Duc: see Wim Denslagen, *Architectural Restoration in Western Europe: Controversy and Continuity* (Amsterdam, 1994), pp. 133–8. That was not a case of an engineer working under an architect as his consultant, but of a competent engineer replacing an incompetent architect. The job at Bayeux was underpinning and iron was not a primary component of Flachat's solution, whereas both Sheilds at Salisbury and Ordish in the Westminster Abbey Chapter House were specifically employed to bring ironwork to the rescue of wide mediaeval spans.

237 For Ordish see the splendid article by Jonathan Clarke, 'Like a Huge Birdcage Exhaled from the Earth: Watson's Esplanade Hotel, Mumbai (1867–71), and its Place in Structural History', *Construction History* 18 (2002), pp. 37–78: biographical section on Ordish, pp. 58–61, and notes on Ordish's work for Scott and quotation from *The Engineer* obituary, p. 75. Another job for which Scott might have asked for engineering help was Brill's Baths, Brighton, built in 1866 by Jackson and Shaw, the same builders as the Midland Grand Hotel, with a high elliptical dome over the pool carried on iron ribs sheathed in oak: *Civil Engineer and Architect's Journal* Nov. 1866, p. 313. The baths were demolished in 1926.

238 See Simon Bradley, *St Pancras Station* (London, 2007), p. 67, and Jack Simmons, *St Pancras Station* (revised edition by Robert Thorne, London, 2003), pp. 33–4. I am grateful to Robert Thorne for discussions about Sheilds, Ordish and Scott. For Moreland's floors in the hotel, see *Building News* 22 May 1874, p. 554. For the relationship between Barlow and Scott at St Pancras see Gilbert Herbert, 'St Pancras Reconsidered: A Case Study in the Interface of Architecture and Engineering', *Journal of Architectural and Planning Research* 15, no. 3 (Autumn 1998), pp. 197–224. Herbert concludes, p. 221: 'There is much architecture in the train shed, deriving from both

Barlow and Scott. Conversely, there is evidence of both structural and mechanical engineering in the hotel.'

239 Jonathan Clarke, 'Early Structural Steel in London Buildings', report for English Heritage, 2000.

240 Parliamentary Commission on the Paris Exposition of 1855, cited in Steiner, *op. cit.*, p. 7. The view that France was leaping ahead of Britain in iron building and architecture is corroborated by the expert findings published in *Reports on the Paris Universal Exhibition* (London, 1855). See Warington Smyth, 'On Mining and Metallurgical Products', Part II, pp. 24–48; William Fairbairn, 'On Machinery in General', Part II, pp. 63–81 and 133–5 (with a damning section on the progress of British architecture in relation to iron, pp. 78–81); and Francis Fowke, 'On Civil Construction', Part III, pp. 259–75. Smyth remarks that France has increased her production of coal-made iron one hundred and fifty fold since 1819, and reduced the price of pig iron by one third since 1830: 'with their wits sharpened by the high price of coal, the French have exceeded us in some branches of economy' (*ibid.*, pp. 44–6). It seems likely that calculation went further in iron structures in France before 1850 because the cost of iron required more efficient sections and truss-forms. For improvements in French metallurgy between 1815 and 1850 and for the exporting of French engineering skills, see Rondo E. Cameron, *France and the Economic Development of Europe 1800–1914* (Princeton, 1961), pp. 66, 70, 98–102.

241 Helene Lipstadt, *Architecte et ingénieur dans la presse: polémique, débat, conflit* (Paris, 1980); and her 'Early architectural periodicals', in Robin Middleton (ed.), *The Beaux-Arts and Nineteenth-Century French Architecture* (London, 1982), pp. 51–7; Lemoine, *op, cit.*, p. 256.

242 Meredith Clausen in her *Frantz Jourdain and the Samaritaine* (Leiden, 1987), p. 82, says architects became obliged 'to turn to engineers, or more demeaning yet, to ordinary building contractors, for assistance'. Indeed they turned to building contractors, but these were hardly 'ordinary'.

243 Whether Louis-Napoleon (later Napoleon III) really said 'Du fer, du fer et rien que du fer' in connection with the second competition for Les Halles in 1853 is however doubtful. The phrase comes from Haussmann's memoirs, written in retrospect and intended to suggest that a discussion between the Haussmann and his master had been decisive at this stage of the planning of Les Halles. See Bertrand Lemoine, *Les Halles de Paris* (Paris, 1980), pp. 150–1, where the passage from Haussmann is quoted in full.

244 Emile Zola, *Le ventre de Paris* (1873), p. 240.

245 Not all infrastructure was inartistic: to take an obvious instance, Adolphe Alphand, the Ponts et Chaussées engineer who was Baron Haussmann's right-hand man in replanning Paris, was involved in designing parks, gardens and iron items of 'équipement urbain'.

246 François Loyer, 'Horeau et la technique', *Hector Horeau 1801–1872* (Paris, n.d.), p. 177.

247 Lemoine, *op. cit.*, pp. 173, 258, 311. The subjects ranged from conservatories in the 1820s to railway stations and churches in the 1840s and 50s.

248 *Revue générale d'architecture* 2 (1841), col. 553 (full article, cols. 551–63); see also C.-P. Gourlier et al., *Choix d'édifices publics projetés et construits en France depuis le commencement du XIXe siècle* (Paris, 1825–50), 8th section, pp. 33–4; and Donald David Schneider, *The Works and Doctrine of Jacques Ignace Hittorff 1792–1867* (New York, 1977), vol. 1, pp. 432–56 and vol. 2 pp. 254–64, which I have largely followed for this account of the Panorama.

249 *Revue générale d'architecture* 2 (1841), col. 558.

250 *Ibid.*

251 I have used mainly the following sources: René Plouin, 'Henry Labrouste, sa vie, son oeuvre', Thèse de Doctorat, Faculté des Lettres et Sciences Humaines, Université de Paris, 1965, pp. 263ff.; Neil Levine, 'The Romantic Idea of Architectural Legibility: Henri Labrouste and the Neo-Grec', in Arthur Drexler (ed.), *The Architecture of the Ecole des Beaux-Arts* (London, 1977), pp. 334ff.; David Van Zanten, *Designing Paris* (Cambridge, Mass., 1987), pp. 89, 98, 236–44; Robin Middleton, 'The Iron Structure of the Bibliothèque Sainte-Geneviève as the Basis of a Civic Décor', *AA Files* 40 (1999), pp. 33–52.

252 Van Zanten, *op. cit.*, pp. 121–35.

253 Middleton, *op. cit.* p. 44, where further information is given on Calla, who had made fountains designed by Hittorff for the Champs-Elysées. See also *Dictionnaire de biographie française* sub Calla.

254 Veugny's Marché de la Madeleine of the 1820s, for instance, put Giedion in mind of the sketchy columns in Pompeian wall paintings: Sigfried Giedion, *Building in France, Building in Iron, Building in Ferro-Concrete* (Santa Monica, 1995), p. 105 (translation of Giedion's *Bauen in Frankreich* . . . of 1928).

255 Achille Hermant in *L'Artiste*, Dec. 1851, quoted in Van Zanten, *op. cit.*, p. 236.

256 Wolfgang Herrmann, *Gottfried Semper: In Search of Architecture* (Cambridge, Mass., 1984), p. 179.

257 Quoted from an article by Garnier on 'Architecture en Fer' in *Le Musée des Sciences*, 11 Feb. 1857, by Van Zanten, *op, cit.*, p. 239. Garnier's views on iron have been as unjustly caricatured by modernists as Ruskin's. He set them out clearly in his book *A travers les arts* (Paris, 1869; reprinted with an essay by François Loyer, Paris, 1985), which includes a chapter on iron and some acute remarks on the architect-engineer brouhaha in his discussion of 'Le style actuel' (1985 edition, pp. 93–5). He used iron liberally in the Opéra, worked with Eiffel over the observatory at Nice, and latterly accepted exposed structural iron so long as it was designed with artistry, which he felt was not the case at the Eiffel Tower. He therefore lent his name to the 'protest of the artists' against the tower on grounds of practice, not principle.

258 Other readings abound about the 'meaning' of the Bibliothèque Ste Geneviève and its successor, the reading room at the old Bibliothèque Nationale. Attractive if fantastic is the idea put forward by Michael Graves and filled out if not quite endorsed by Van Zanten, to the effect that they are 'groves of academe'. The Ste Geneviève

vestibule and the reading room ceiling in the Nationale thus become inside-outside spaces – a conservatory or a sky above the garden. If that is so, it suggests that despite his determination to show iron in public buildings, Labrouste felt the urge to compensate for its dumbness. See Van Zanten, *op. cit.*, pp. 98, 241–3.

259 Bertrand Lemoine, *L'architecture du fer: France, XIXe siècle* (Paris, 1986), p. 238.

260 For Boileau and St Eugène see *L'église Saint-Eugène à Paris* (Paris, 1856); Bruno Foucart, 'La cathédrale synthétique de Louis-Auguste Boileau', *Revue de l'Art* 3 (1969), pp. 49–66; Robin Middleton and David Watkin, *Neoclassical and 19th Century Architecture* (New York, 1980), pp. 366–71; Lemoine, *op. cit.*, pp. 178–82; and Bernard Marrey (ed.), *La querelle du fer: Eugène Viollet-le-Duc contre Louis-Auguste Boileau* (Paris, 2002).

261 Quoted in Marrey, *op. cit.*, p. 16.

262 *Ibid.*, p. 36.

263 Different versions are given of Boileau's appointment. Robin Middleton in Middleton and Watkin, *op. cit.*, p. 369, seems the most reliable. He says that Louis-Adrien Lusson, another architect, took Boileau's published design for a church in the Chaussée d'Antin and proposed something like it to the Archbishop of Paris for St Eugène. When Boileau heard this, he protested and was given the job. See also *L'église Saint-Eugène à Paris* (Paris, 1856), pp. 9–10; Lemoine, *op. cit.*, p. 178; and Pascal Étienne, *Le Faubourg Poissonnière: architecture, élégance et décor* (Paris, 1986), pp. 295–6.

264 A note in *Revue générale de l'architecture* 13 (1855), pp. 82–3, written before the church is finished, is already disparaging and declines to name the architect. *The Builder* 21 Feb. 1857, pp. 106–7, is more respectful, while calling St Eugène 'not a perfect realization of M. Boileau's system'.

265 After Viollet-le-Duc's death, Boileau prided himself that his work had started the former on the road towards a more positive view of iron architecture: see L. A. Boileau, *Histoire critique de l'invention en architecture* (Paris, 1886), pp. 145–6.

266 *L'église Saint-Eugène à Paris* (Paris, 1856), pp. 15–16. Most of the cast iron was supplied by Grebel; the wrought-iron 'charpente' was begun by Liandier and completed by Bertrand.

267 According to Boileau, *op. cit.*, pp. 141–2, in 1866–7 Viollet-le-Duc and he were each invited to propose models for cheap churches for the Paris suburbs. His own proposals were judged too costly and the matter dropped.

268 Middleton, *op. cit.*, p. 370, though Boileau, *op. cit.*, pp. 136ff., claims the vaults of these later churches answered better to the properties of metal. The list of Boileau's later iron churches given there, supplemented by Marrey, includes Notre Dame, Rochefort-sur-Mer (1858–60); St Germain de Marencennes (1861–2); Ste Marguerite, Le Vésinet (1863–5), for which see pp. ••; St Paul Montluçon (1863–7); Notre Dame de France, Leicester Square, London (1866–8); and churches at L'Abergement-Clémencia (1868), Juilly (1869) and Le Gua (1869). Since Le Vésinet, Montluçon and the

Leicester Square church shared the same structural 'ossature', an analysis of the dimensions and contracting arrangements for those churches that survive might be revealing.

269　The limits of this study preclude investigation of the architect-engineer relationship in Belgium. But it may be remarked that the flowering of Belgian architecture at the end of the nineteenth century owed much to the combination of its flourishing iron industries and its French traditions of professional training. For the architectural side of the subject, see Richard Vandendaele, 'Le métal dans l'architecture au XIXe siècle', in *Poelaert et son temps* (Brussels, 1980), pp. 78–100, and for the Flemish and Gothic dimension, Bart de Kayser, *De Ingenieuze Neogotiek: Techniek en Kunst, 1852–1925* (Leuven, 1997), pp. 15–30, 34–7, 77–90. The role of Alphonse Balat, master of Victor Horta and architect for the original portions of Leopold II's great glasshouses at Laeken, is of special interest. Horta himself was an excellent technician who persistently looked beyond a merely rational approach to metal construction: see David Dernie and Alastair Carew-Cox, *Victor Horta* (London, 1995), and Françoise Aubry and Jos Vandenbreeden, *Horta: naissance et dépassement de l'Art Nouveau* (Brussels, 1996). Horta's Maison du Peuple (1896–9, dem. 1965) revealed the limits of his individualism. A fascinating attempt to make structural iron politically representative of the industrial working classes, it nevertheless proved costly and possibly incoherent; it should not be judged a masterpiece just because it was shamefully demolished. See Franco Borsi, 'Victor Horta et la Maison du Peuple de Bruxelles', in *Maisons du Peuple* (Brussels, 1984), pp. 10–31.

270　For Gouin, see Rondo E. Cameron, *France and the Economic Development of Europe 1800–1914* (Princeton, 1961), pp. 98–102.

271　D. Ramée, quoted by Karen Bowie, 'Les polytechniciens et l'architecture métallique', in Bruno Belhoste, Francine Masson and Antoine Picon (eds.), *Le Paris des polytechniciens* (Paris, 1994), p. 206.

272　Many authors have written about the Bon Marché, giving contradictory dates and information about its architects and phases. This account generally follows Bernard Marrey, *Les grands magasins* (Paris, 1979), pp. 68–83, confirmed by Lemoine, *op. cit.*, pp. 194–7, except in relation to Eiffel, who appears not to have been involved until 1879. See also Steiner, *op. cit.*, pp. 59–62; Meredith Clausen, *Frantz Jourdain and the Samaritaine* (Leiden, 1987), p. 198. For the business history of the shop, see Michael B. Miller, *The Bon Marché* (London, 1981), where the account of the reconstruction is not accurate. For Eiffel's role, see Michael Carmona, *Eiffel* (Paris, 2002), pp. 157–8. How far L. A. Boileau (Boileau père) was involved in the early phases is confusing, some alleging that he worked with Laplanche. Boileau fils says he has profited by the experience of previous architects for the Bon Marché, 'notamment . . . mon père': *Encyclopédie d'Architecture* 5 (2nd series) (1876), pp. 120–2. Marrey, *op, cit.*, p. 71, suggests this just means that Boileau fils benefited from his father's experience with iron.

273　Lemoine, *op. cit.*, p. 194.

274　*Encyclopédie d'Architecture* 5 (2nd series) (1876), pp. 120–2 and 9 (2nd series) (1880), pp. 183–5.

275　*Ibid.*, 5 (2nd series) (1876), p. 120. The prescient sentence about voids rather than solids runs: 'Je dirai que ce point de vue devra consister à envisager non plus les pleins de l'édifice, mais bien le vide qu'il enveloppe.'

276　*Ibid.*, 9 (2nd series) (1880), p. 184 ('un seul réseau de fer').

277　*Id.*

278　For Moisant see Madeleine Fargues, *Armand Moisant: de l'architecture métallique aux fermes modèles tourangelles* (Sainy-Cyr-sur-Loire, 2004), with a brief account of the Bon Marché, pp. 19–21, giving the dates for Moisant's work as 1871–2.

279　For the Printemps see Marrey, *op. cit.*, pp. 96–109; Lemoine, *op. cit.*, pp. 194, 197–9; and *Encyclopédie d'Architecture* 4 (3rd series) (1885), pp. 1–35.

280　See profile of Sédille in *La Construction Moderne* 23 May 1891, pp. 387–8.

281　Marrey, *op. cit.*, p. 104, where it is added that Sédille designed the monument to Eugène Schneider, owner of the firm, in the main square of Le Creusot, also in 1878.

282　*Encyclopédie d'Architecture* 4 (3rd series) (1885), pp. 2–3.

283　Lemoine, *op. cit.*, p. 299.

284　*Ibid.*, pp. 13, 20–22.

285　*Gazette des Beaux-Arts* March 1883, pp. 239–53.

286　Meredith Clausen, *Frantz Jourdain and the Samaritaine* (Leiden, 1987). Further on the Samaritaine, see Arlette Barré-Despond and Suzanne Tise, *Jourdain* (New York, 1991), pp. 145–203; Marrey, *op. cit.*, pp. 122–39, Lemoine, *op. cit.*, pp. 198–202.

287　*L'Architecture* 1890, pp. 446–7. For previous comments on 'ingénieurophobie', see pp. 398–402 and 423–5 following a series of articles by Roux.

288　Clausen, *op. cit.*, pp. 258–9, notes that the construction of the three-hinged arches over the glazed courts of Samaritaine No. 2 quoted on a smaller scale from Dutert's Galerie des Machines of 1889.

289　Zola's researches for *Au Bonheur des Dames* included prolonged visits to stores, and eliciting information from the management of the Bon Marché and other establishments. The Printemps was rebuilding while the novel was being written and no doubt influenced Jourdain's notes as well as Zola's text. The first section was opened in March 1883, just as the novel finished serialization and appeared as a whole. The store in the novel also opens in March. Jourdain reviewed *Au Bonheur des Dames* in *Le Phare de la Loire*, 19 March 1883. See Clausen, *op. cit.*, pp. 20–2; Henri Mitterand, *Zola* vol. 2 (Paris, 2002).

290　Marrey, *op. cit.*, pp. 125–6; Barré-Despond and Tise, *op. cit.*, p. 151. For Schwartz and Meurer, see Lemoine, *op. cit.*, pp. 304–5.

291　Samaritaine No. 1 building, rebuilt in 1912 on similar lines to No. 2, stood to the latter's east on the angled site occupied by the old shop.

292　The Brussels stores of Old England (by Saintenoy, 1899) and L'Innovation (Horta, 1901) and the Tietz store in Berlin (Sehring, 1900) are commonly cited.

293　*Ibid.*, p. 248. Jourdain's main artistic collaborators on the Samaritaine were his son Francis Jourdain (enamels), Eugène Grasset (frieze and tympani), Janselme (panelling), Alexandre Bigot (ceramics) and Edouard Schenk (copperwork).

294　For Eiffel I have consulted mainly the following: Bernard Marrey et al., *Gustave Eiffel et son temps* (Paris, 1982–3); Bertrand Lemoine, *Gustave Eiffel* (Paris, 1984); Henri Loyrette, *Gustave Eiffel* (Fribourg, 1985); Bernard Marrey, *Gustave Eiffel: une entreprise exemplaire* (Paris, 1989); Bertrand Lemoine, *La tour de Monsieur Eiffel* (Paris, 1989); Bertrand Lemoine, 'L'entreprise Eiffel', in Robert Thorne (ed.), *Structural Iron and Steel, 1850–1900* (Aldershot, 2000), pp. 247–59 (originally in *Histoire, Economie et Société*, 14 (1995), pp. 273–85); Michel Carmona, *Eiffel* (Paris, 2002).

295　Lemoine, 'L'entreprise Eiffel', p. 248 (original version, p. 274).

296　The metal frame of the synagogue in Rue des Tournelles, Paris (Varcollier, architect, 1867) and the roof-trusses of Notre-Dame des Champs, Boulevard Montparnasse (Ginain, architect, 1866–7) are early examples of Eiffel's involvement in Parisian architectural projects.

297　*Encyclopédie d'Architecture* 5 (2nd series) (1876), pp. 75–6 and Plates 387–8, where however Sauvestre's name is not mentioned.

298　For Sauvestre see *1889: La Tour Eiffel et l'Exposition Universelle* (Paris, Musée d'Orsay, 1989), p. 251; Frédéric Seitz, *L'Ecole Spéciale d'Architecture 1865–1930* (Paris, 1995), pp. 102, 145, 152, 166, 174; and Carmona, *op. cit.*, pp. 143, 145, 238, 394–5. Carmona reports a suggestion that Sauvestre's first connection with Eiffel may have been over detailing the roof of the station at Pest in Hungary, a project of 1874–7 entailing complex collaboration between Seyrig, Eiffel and various other engineers and architects. Much later, in 1905 6, Sauvestre collaborated with the reinforced-concrete specialist Armand Considère to build a large addition to the famous Meunier chocolate factory at Noisiel-sur-Marne, connected to a 44.5-metre-span concrete bridge (le 'Pont Hardi').

299　Lemoine, *Gustave Eiffel*, p. 86. In Lemoine's *La Tour de Monsieur Eiffel*, pp. 26–7, he adds that Sauvestre at this early stage also designed the masonry bases to the tower, put arcades round the glass hall, and set a glass roof on the apex of the tower in the form of a bulb.

300　Loyrette, *op. cit.*, p. 171, speaks of 'the engineers' definitive victory'.

301　Eiffel's answer to the artists' protest, quoted in Loyrette, *op. cit.*, p. 176.

302　For the Galerie des Machines, I have followed mainly Marie-Laure Crosnier Leconte, 'La Galerie des Machines', in *1889: La Tour Eiffel et l'Exposition Universelle* (Paris, Musée d'Orsay, 1989), pp. 164–95; and John W. Stamper, 'The Galerie des Machines of the 1889 Paris World's Fair', in Robert Thorne (ed.), *Structural Iron and Steel, 1850–1900* (Aldershot, 2000), pp. 261–84, originally in *Technology and Culture* 2 (1989), pp. 330–53.

303　'Lorsque l'idée de construire une grande ferme a été donnée par l'architecte . . .':

report of Alphand, 17 Jan. 1887, quoted by Crosnier Leconte, op. cit., p. 167.

304 J. W. Schwedler's Unterspree Bridge, Berlin (1864–5), and his station roofs at Ostbahnhof, Berlin (1867) and Frankfurt (1887) were probably the critical exemplars. See Stamper in Thorne, op. cit., p. 271 (original version p. 340). French bridge design, even in great spans like the Garabit, had confined itself to two-pin structures.

305 The final decision on grounds of cost to build the trusses of the Galerie des Machines from wrought iron rather than steel, with the changes in design which that entailed, probably rested also with Contamin, consulting with Alphand: see Crosnier Leconte, op. cit., p. 172. Incidentally, it is often stated by modernists that the Eiffel Tower and the Galerie des Machines were exemplary early steel structures, when in fact both were largely of wrought iron.

306 Quoted by Crosnier Leconte, op. cit., p. 192.

307 Building News 21 March 1890, p. 403. Some however remained unconvinced, including the great Belgian engineer Arthur Vierendeel: 'the girder is not balanced . . . it has no base . . . the eye is not reassured.' Quoted in Stamper, op. cit., p. 272 (original version p. 341).

308 Building News 21 March 1890, p. 402: Eugène Hénard took a similar line. In a literary parallel, Huysmans depicted the Eiffel Tower as a vulgar steeple for the church of industrial capitalism and consumption represented by the 1889 Exhibition as a whole. The Gothic overtones of the Galerie des Machines, on the other hand, lifted his spirits: 'La forme de cette salle est empruntée à l'art gothique, mais elle est éclatée, agrandie, folle, impossible à réaliser avec la pierre, originale avec ses pieds en calice de ses grands arcs'. J. K. Huysmans, 'Le Fer', in Certains (Paris, 1889), pp. 169–81.

309 Building News 2 May 1890, p. 642.

310 The Papers of Benjamin Henry Latrobe, Volume 3, 1811–1820 (New Haven, 1988), pp. 121–3. This letter is partly excerpted in Talbot Hamlin, Benjamin Henry Latrobe (New York, 1955), p. 564, and in Lois Craig (ed.), The Federal Presence: Architecture, Politics and Symbols in United States Government Buildings (Cambridge, Mass., 1978), p. 38. Its recipient, the radical Constantin Volney, had escaped France in 1796 to America, where he met Latrobe. Volney soon returned home but grew disillusioned under Napoleon and had retired from public life by the time of Latrobe's letter.

311 For overviews of Latrobe's career, see Hamlin, op. cit., supplemented and updated by Darwin H. Stapleton, The Engineering Drawings of Benjamin Henry Latrobe (New Haven, 1980), pp. 3–71, and Mike Chrimes, BDCE, pp. 394–6. In his preface, Hamlin says he started out thinking of Latrobe primarily as an architect, but grew fascinated with the 'industrial' side of his career as he pursued the subject. Stapleton treats Latrobe exclusively as an engineer: see also Darwin H. Stapleton, The Transfer of Early Industrial Technologies to America (Philadelphia, 1987), pp. 35–6, 59–71, 126–7.

312 The dates given by Stapleton (Engineering Drawings, pp. 5–6) and Chrimes for

Latrobe's time with Smeaton and Jessop correct those given in Colvin, p. 601.

313 Quoted in Stapleton, Transfer, p. 35.

314 Letter of 4 Nov. 1804 quoted in Hamlin, op. cit., p. 149.

315 Letter of 23 Jan. 1812 quoted ibid., p. 148.

316 Letter of 13 Dec. 1812 quoted in Stapleton, Engineering Drawings, p. 38.

317 Hamlin, op. cit., pp. 156, 442, 561.

318 Stapleton, Engineering Drawings, p. 29, and for the bulk of information in this paragraph, pp. 28–41. When Robert Fulton began developing steamboats from 1806, it is notable that in the first instance he imported Boulton and Watt engines.

319 Ibid., p. 41.

320 Infomation in this paragraph is from Stapleton, Engineering Drawings, pp. 49–52, and Hamlin, op. cit., pp. 192, 348.

321 For Finley, see Eda Kranakis, Constructing A Bridge (Cambridge, Mass., 1997), pp. 17–37. It is not clear when Finley started working on his suspension bridges; Kranakis thinks in the 1790s while he was in Philadelphia, though the first one was not built till 1801. For Strickland (a pupil of Latrobe), see Agnes Addison Gilchrist, William Strickland Architect and Engineer, 1788–1854 (New York, 1969 edn), pp. 7–8, 30, 61, 79.

322 Margot Gayle in Kenneth T. Jackson (ed.), The Encyclopedia of New York City (New Haven, 1991), p. 186, where the date 1825 is given for New York's first iron shop-front. Sarah Bradford Landau and Carl W. Condit, Rise of the New York Skyscraper (New Haven, 1996), p. 403 n. 3, say 1830. Meanwhile in Pottsville, Pennsylvania, Haviland in 1829–30 produced a two-storey bank front entirely clad with cast-iron plates in imitation of ashlar: Turpin C. Bannister, 'Bogardus Revisited: Part I, The Iron Fronts', in Robert Thorne (ed.), Structural Iron and Steel, 1850–1900 (Aldershot, 2000), pp. 52–3, previously in Journal of the Society of Architectural Historians 15 no. 4 (1956), pp. 15–16.

323 Landau and Condit, op. cit., p. 41; Gayle in Jackson, op. cit., p. 186.

324 Badger's most impressive surviving work is the all-iron Watervliet Arsenal storehouse, built adjacent to the Erie Canal in 1859, with cast-iron columns and wall panels, and versions of the Fink wrought-iron truss spanning the 50-foot centre and galleried aisles: see Daniel L. Schodek, Landmarks in American Civil Engineering (Cambridge, Mass., 1987), pp. 272–4. Further on Badger, see the reprint of his catalogue of 1865 in W. Knight Sturges (ed.), The Origins of Cast Iron Architecture in America (New York, 1970).

325 Margot Gayle and Carol Gayle, Cast-Iron Architecture in America: The Significance of James Bogardus (New York, 1998), p. 157. This lifetime's work replaces earlier work by Margot Gayle and is the principal source for the following paragraphs. It in part supersedes the articles by Turpin C. Bannister in Thorne, op. cit., pp. 49–68, previously in Journal of the Society of Architectural Historians, 15, no. 4 (1956), pp. 12–22 and 16 (1957), pp. 11–19.

326 Sturges, op. cit., reprint of James Bogardus, Cast Iron Buildings: Their Construction and Advantages (New York, 1856), p. 4 (text by John Thomson).

327 For Jackson see William J. Fryer, 'A Review of the Development of Structural Iron', in

A History of Real Estate, Building and Architecture in New York City (New York, 1898, repr. 1967), pp. 456–7.

328 For the ornamental elements of the Laing Stores (1849) the founder was William Miller.

329 David G. Wright, 'The Sun Iron Building', in James D. Dilts and Catharine F. Black (eds.), Baltimore's Cast-Iron Buildings and Architectural Ironwork (Centreville, Md, 1991), pp. 22–33. This paragraph is based on Wright's article together with Gayle and Gayle, op. cit., pp. 96–102.

330 See Landau and Condit, op. cit., pp. 97–8, where Hatfield's Seamen's Bank for Savings (1870–1) is analysed and illustrated.

331 For the Harper Building see Jacob Abbott, The Harper Establishment (New York, 1855, reprinted 1956); Gayle and Gayle, op. cit., pp. 136–51; Wright, op. cit., p. 29; Landau and Condit, op. cit., pp. 47–50.

332 The rapid shift in building technique at this time is well illustrated by the practice of the US Treasury Department, which by 1858 was advocating wrought-iron construction in its customs houses, court houses and post offices as standard practice. In that year Secretary to the Treasury James Guthrie had plans and specifications of such buildings distributed to colleges and learned societies, writing: 'The introduction of wrought iron beams and girders in these Edifices, instead of groined arches as formerly used, is, I believe, wholly new, and this improvement will, it is hoped, prove interesting and useful to you, or to those who, through you, may have the opportunity of inspecting them'. Quoted in Antoinette J. Lee, Architects to the Nation: The Rise and Decline of the Supervising Architect's Office (New York, 2000), p. 61.

333 Bannister, 'Bogardus Revisited: Part II, The Iron Towers', in Thorne, op. cit., pp. 60–8, previously in Journal of the Society of Architectural Historians 16 (1957), pp. 11–19.

334 See articles by William J. Fryer in The Builder 3 July 1869, pp. 529–30: 'In the city of New York there are a greater number of entire cast-iron building fronts in process of erection than ever before at any one period of time'; and 'A Review of the Development of Structural Iron', in A History of Real Estate, Building and Architecture in New York City (New York, 1898, repr. 1967), pp. 458–9: 'The cast-iron front business in New York reached its greatest proportions in the early seventies'. For Fryer, 'constructor in iron', see Landau and Condit, op. cit., pp. 106, 120–1, 166, 218.

335 Bannister, 'Bogardus: Part I', reprinted in Thorne, op. cit., pp. 54 and 59 n. 63 (previously pp. 17 and 22). He notes that architects are named for 174 of Badger's 419 published projects in New York and Brooklyn, and suggests that 'some [architectural] practitioners took the easy course of securing design service, in addition to production, from Badger's firm'.

336 George H. Johnson (1830–79), English-born, learnt his skills with builders in Manchester. He is said to have worked for Badger between 1852 and 1862, then briefly had his own architectural office in New York before moving after the Civil War to Richmond, then Buffalo and in 1871 back to

New York with frequent visits on fireproofing business to Chicago. The best résumé of his career is in Bannister, 'Bogardus: Part II', reprinted in Thorne, *op. cit.*, p. 63 (previously p. 14). Typically, Johnson is given different professional status by different authors, contemporary and modern. Fryer, in his 'Review of the Development of Structural Iron', in *A History of Real Estate, Building and Architecture in New York City* (New York, 1898, repr. 1967), p. 476, refers to him as a 'civil engineer' connected with Badger; Bannister, 'Bogardus: Part I'), reprinted in Thorne, *op. cit.*, p. 54, previously p. 17, calls him 'manager of [Badger's] new architectural department, a young and energetic English builder'; Carl W. Condit, *The Chicago School of Architecture* (Chicago, 1964), p. 24, refers to him as 'a designer on the staff of Daniel Badger's Architectural Iron Works' and 'a practical builder of unusual inventive ability'. Further on Johnson see Landau and Condit, *op. cit.*, pp. 28, 48–9. Those interested in the evolution of framing have drawn attention to the two almost identical grain elevators designed by Johnson for Badger: the US Warehousing Grain Elevator Building, Brooklyn, and the almost identical Pennsylvania Railroad Elevator Building in Philadelphia. Their importance was noted by P. B. Wight in 1892: see G. R. Larson and R. Geraniotis, 'Towards a Better Understanding of the Evolution of the Iron Skeleton Frame in Chicago', in Thorne, *op. cit.*, pp. 300–1, originally in *Journal of the Society of Architectural Historians* 46 no. 1 (March 1987), pp. 40–1.

337 The outstanding partnership of the 1850s between an architect and a New York ironfounder was between Thomas U. Walter of Philadelphia and Janes, Beebe and Co. (later Janes, Fowler, Kirtland and Co.) at the Capitol in Washington. Together they built the 'fast-track' iron-framed Library of Congress (1852–3) after the fire of 1851, and the framing of Walter's high dome for the Capitol (1859–63), the latter in the teeth of busybodying interventions from Montgomery Meigs. Walter had worked previously with Janes, Beebe at Girard College, Philadelphia, and clearly enjoyed a close relationship with one of the partners, Charles Fowler. See William C. Allen, *The Dome of the United States Capitol: An Architectural History* (Washington, 1992), and *History of the United States Capitol* (Washington, 2001). In designing the framing for the Capitol dome, Allen highlights the role of Walter's assistant, the German-born and trained August Schoenborn.

338 Landau and Condit, *op. cit.*, passim; and Sarah Bradford Landau, *George B. Post: Picturesque Designer and Determined Realist* (New York, 1998).

339 Paul R. Baker, *Richard Morris Hunt* (New York, 1980), pp. 99, 106, 483. For the NYU course see the chapter on 'The College of Engineering' in Theodore Francis Jones (ed.), *New York University 1832–1932* (New York, 1933), pp. 305ff. David Douglass, previously an important teacher at West Point, was an original member of the NYU faculty in 1832 as 'Professor of Natural Philosophy, Architecture and Civil Engineering', but taught only intermittently.

Regular courses in engineering and architecture began from 1854.

340 Landau and Condit, *op. cit.*, p. 66 (discussion of the Equitable Building, pp. 62–75).

341 *Ibid.*, pp. 116–25.

342 Quoted in Theodore Turak, 'Remembrances of the Home Insurance Building', *Journal of the Society of Architectural Historians* 44 (March 1985), p. 64.

343 Quoted in Frank A. Randall, *History of the Development of Building Construction in Chicago* (Urbana, 1949), p. 23.

344 Sanford E. Loring and W. L. B. Jenney, *Principles and Practice of Architecture* (Chicago, 1869), pp. 50–1.

345 Theodore Turak, *William Le Baron Jenney: A Pioneer of Modern Architecture* (Ann Arbor, 1986), p. 348 n. 14, says that 'an injury caused Loring to leave his profession'. But Michael Stratton, *The Terracotta Revival* (London, 1993), pp. 147–52, shows Loring actively promoting terracotta in construction from 1870 at the latest.

346 Turak, *op. cit.*, pp. 75–112 and 342 n. 36; Daniel Bluestone, *Constructing Chicago* (New Haven, 1991), pp. 48–52, 215 n. 44.

347 Theodore Turak, 'Remembrances of the Home Insurance Building', *Journal of the Society of Architectural Historians* 44 (March 1985), pp. 60–1. The honour was offered by the Bessemer Steamship Company in 1896.

348 Sullivan's judgement on Jenney comes from Louis H. Sullivan, *The Autobiography of an Idea* (Chicago, 1924), p. 203. It is quoted along with Wight's comment in Theodore Turak, 'Jenney', pp. 260–1. Both men were writing in their old age and had axes to grind. Turak, from whom these paragraphs on Jenney are largely derived, has defended Jenney's contribution to the skyscraper against detractors and argued that his technical competence relied heavily on his education in France. This view is endorsed rather uncritically in Ulrich Pfammatter, *The Making of the Modern Architect and Engineer* (Basel, 2000), pp. 166–77. For further commentary on Sullivan's remark, see Robert Twombly and Narciso G. Menocal, *Louis Sullivan: The Poetry of Architecture* (New York, 2000), pp. 81–2.

349 Quoted in Turak, *Jenney*, p. 74, from Jenney's autobiographical sketch.

350 According to Thomas J. Misa, *A Nation of Steel* (Baltimore, 1995), p. 64, 'So proficient was Jenney at turning abandoned cotton gins, church pews and chimneys into bridges that a southern newspaper stated that it was useless for the retreating armies to burn the railroad bridges because Sherman evidently carried with him a full-line of ready-built bridges'. This surely ascribes too much to Jenney personally.

351 Laura Wood Roper, *FLO: A Biography of Frederick Law Olmsted* (Baltimore, 1973), pp. 221–2; Elizabeth Stevenson, *A Life of Frederick Law Olmsted, Parkmaker* (New Brunswick, 1977/2000), p. 236.

352 Turak, *Jenney*, pp. 75–6.

353 Later at least, Olmsted felt some equivocation about Jenney's talents. In 1876 he declined to support him for a professorship of architecture at the University of Michigan, stating that when he knew him at the end of the 1860s Jenney was 'more in the condition of feeling his way, than a thor-

oughly disciplined designer working with sure hand and fixed principles': quoted in Turak, *Jenney*, p. 112.

354 Lawrance Hurst, 'The Age of Fireproof Flooring', in Thorne, *op. cit.*, pp. 34–9; Lemoine, *op. cit.*, pp. 50–2; Landau and Condit, *op. cit.*, pp. 26–30; Sarah Wermeil, 'The Development of Fireproof Construction in Great Britain and the United States in the Nineteenth Century', in Thorne, *op. cit.*, pp. 69–92, previously in *Construction History* 9 (1993), pp. 3–26. For an early overview of the subject, see J. K. Freitag, *Architectural Engineering, with Special Reference to High Building Construction* (New York and London, 1895), Ch. 2.

355 For Johnson, see above, n. 336, and for his fireproofing work, going back to at least 1869, Landau and Condit, *op. cit.*, pp. 28–9; Wermeil in Thorne, *op. cit.*, p. 82, previously p. 16; and Bannister, *Bogardus* (Part II), *ibid.*, pp. 63 and 67, previously pp. 14 and 18. Johnson's hollow floor voussoirs were first used in Van Osdel's Kendall Buildng (1872–3). Bannister notes that Van Osdel and Johnson had worked together on cast-iron buildings in Chicago supplied by Daniel Badger in the 1850s, and that Johnson went to Paris to study flooring systems in 1871. Fryer, 'Review', *op. cit.*, pp. 474–6, attributes the original Johnson patent of 1871 exclusively to the manufacturer, Balthasar Kreischer, and states that it was finally voided 'for want of originality' as far as flat hollow-brick floor arches were concerned. The Johnson patents and inventions were exploited after his death by the Pioneer Fireproof Company, founded in 1880 and managed by his son.

356 Sarah Bradford Landau, *P. B. Wight: Architect, Contractor, and Critic, 1838–1925* (Chicago, 1981), pp. 44–7; Turak, 'Jenney', pp. 161–2, 348 n. 14; Stratton, *op. cit.*, pp. 147–51;. The exact relation between Wight and Loring is obscure.

357 Quoted in Landau, *Wight*, p. 44: other information in this and the next paragraphs from same source, pp. 9–59.

358 Donald Hoffmann, *The Architecture of John Wellborn Root* (Baltimore, 1973), p. 6.

359 The best general introduction to foundation technology in Chicago is in Freitag, *op. cit.*, Ch. 10, pp. 171–200.

360 I have avoided the term 'hardpan', a word favoured for Chicago foundations but used by different writers to mean sometimes the trustworthy ground just below grade, sometimes lower levels. Paul Starrett refers to the top layer of Chicago soil as 'gumbo'.

361 Quoted from *Chicago Times* 2 Nov. 1879, in Hoffmann, *op. cit.*, p. 90.

362 Sullivan, *op. cit.*, p. 245.

363 This theme is explored by Roula Mouroudellis Geraniotis, 'An Early German Contribution to Chicago's Modernism', in John Zukowsky (ed.), *Chicago Architecture 1872–1922* (Munich, 1987), pp. 90–105, with reference to Baumann, pp. 92–4, & 105 n. 5. For his contribution to the German-led sphere of brewery design, see Susan K. Appel, 'Brewery Architecture in America from the Civil War to Prohibition', in John S. Garner (ed.), *The Midwest in American Architecture* (Urbana, 1991), pp. 196–9. Beyond Chicago, Baumann was executant

architect for the colossal Soldiers and Sailors Monument, Indianapolis, designed by the German architect Bruno Schmitz.

364   Larson and Geraniotis in Thorne, *op. cit.*, p. 306 (original version, p. 46).

365   Hoffmann, *op. cit.*, p. 91; Robert Twombly, *Louis Sullivan, His Life and Work* (Chicago, 1986), p. 93.

366   A clear exposition of Baumann's argument is given by Joseph M. Siry, *The Chicago Auditorium: Adler and Sullivan's Architecture and the City* (Chicago, 2002), pp. 151–2.

367   Though Freitag, *op. cit.*, p. 173 says that Baumann's method was 'brought to a high degree of perfection by the engineers of Chicago', this appears to refer to the general idea of isolated footings, not to his specific solution.

368   Hoffmann, *op. cit.*, pp. 24–7, 41–2, 47, 68, 73, 133–7, 146, 164; see also Harriet Monroe, *John Wellborn Root: A Study of his Life and Work* (Boston, 1896), pp. 96–103.

369   A transitional case is Holabird and Roche's Tacoma Building (1886–9), where the subsoil was test-bored to 50 feet and partly filled with concrete below the reinforced footings: Condit, *op. cit.*, p. 117.

370   For Sooy Smith and his son Charles Sooysmith [sic], see *Dictionary of American Biography* vol. 17 (1935), pp. 367–8, 397. Tom F. Peters, *Building the Nineteenth Century* (Cambridge, Mass., 1996), p. 403 n. 114, cites a later Sooy Smith bridge for the first American use of pneumatic caissons.

371   Siry, *op. cit.*, pp. 154–8. As Siry explains, the serious problems later experienced with the Auditorium's foundations arose from the directors' insistence on replacing brick with stone for the elevations after the footings were already in position.

372   Twombly, *op. cit.*, pp. 296, 302, 316. For polite criticism of Sooy Smith's recommendations to pile to bedrock, see Freitag, *op. cit.*, pp. 194–7, mentioning the Chicago Public Library (1891–7) as among the buildings which involved him. Siry, *op. cit.*, p. 159, cites Adler and Sullivan's unexecuted Odd Fellows Temple design (1890–1), rising to twenty-one storeys, as the first skyscraper conceived with caisson foundations throughout. An elliptical sentence in Sullivan's autobiography, *op. cit.*, p. 311, refers to 'an invention of English origin, an automatic pneumatic ejector, which rendered basement depths independent of sewer levels', as leading the way to deep foundations. For caisson foundations in New York, where bedrock was closer to the surface, see Landau and Condit, *op. cit.*, pp. 23–5, 178–82, 222–4, and 423; also Fryer, 'Review', *op. cit.*, pp. 479–80. Charles Sooysmith, son of William Sooy Smith, was the foundation engineer for the first important New York skyscraper with caisson foundations, the Manhattan Life Building (1893–4). Landau and Condit point out that pneumatic caissons were more regularly used in New York than Chicago.

373   Sullivan, *op. cit.*, p. 312.

374   Thomas J. Misa, *A Nation of Steel* (Baltimore, 1995), pp. 45–89.

375   As Sullivan said with the peremptory judgement of old age, 'the squabblings as to priority are so much piffle'.

376   Misa, *op. cit.*, p. 61.

377   *Ibid.*, p. 65.

378   For Whitney, see *ibid,*, p. 64. He later worked with or for Burnham and Root and then founded the short-lived Whitney and Starrett Construction Company with Theodore Starrett.

379   Freitag, *op. cit.*, p. 39.

380   Siry, *op. cit.*, provides a magnificently full and illuminating account of this great project, but curiously does not mention Strobel or Marburg.

381   Condit, *op. cit.*, p. 76 (quoted in Misa, *op. cit.*, p. 300 n. 8).

382   For Mueller see pp. ••. His memorandum of 1925, at the time of a lawsuit over the Auditorium, is printed in Edgar Kaufmann jr., *9 Commentaries on Frank Lloyd Wright* (New York & Cambridge, Mass., 1989), pp. 42–62. The phrase 'Man Friday' for Mueller occurs in Wright's *Genius and the Mobocracy*.

383   Misa, *op. cit.*, pp. 47, 64–5, 71–2. Randall, *op. cit.*, pp. 28–30, gives an overview of Strobel's career. Though born in Cincinnati, he received his technical education in Stuttgart. He appears to have been a Carnegie employee 1878–85, and a consultant to the firm 1885–93. From 1905 to 1926 he ran the Strobel Steel Construction Company.

384   Starrett, *op. cit.*, pp. 37–8.

385   Shankland supervised the structural side of the World's Columbian Exposition in 1893 and was a partner in D. H. Burnham and Co. till 1900: Charles Moore, *Daniel H. Burnham* (New York, 1921), vol. 1, pp. 82–3.

386   Misa, *op. cit.*, p. 64.

387   Freitag, *op. cit.*, p. 43, says that to allow a manufacturer to prepare the complete details of the frame 'is not consistent with the best results, in the judgment of the writer'.

388   Cited in Schuyler's article on Adler and Sullivan in Montgomery Schuyler, *American Architecture and Other Writings*, ed. William H. Jordy and Ralph Coe, (Cambridge, Mass., 1961), vol. 2, p. 381.

389   Freitag, *op. cit.*, pp. 1–8.

390   The general contract was not of course unheard of in the 1880s, but it was a relatively late development in the United States. An article in *Architectural Record* 24 (1908), pp. 231–6 states that it had then been in use for about fifty years.

391   See Paul Starrett, *op. cit.* The careers and interrelationships of the four remarkable Starrett brothers, Theodore, Paul, William (author of *Skyscrapers and the Men Who Build Them*, 1928) and Goldwin, and the rivalry between the George A. Fuller and Thompson-Starrett companies, are too complex to enter into here. For Paul and William, see *American National Biography* vol. 20 (1999), pp. 578–81.

392   An account of Fuller's career is given by Sara E. Wermiel, 'Norcross, Fuller, and the Rise of the General Contractor in the United States in the Nineteenth Century', in *Proceedings of the Second International Congress on Construction History* (Cambridge, 2006), vol. 3, pp. 3304–9.

393   Starrett, *op. cit.*, p. 70.

394   Robert Bruegmann, *The Architects and the City: Holabird and Roche of Chicago, 1880–1918* (Chicago, 1997), pp. 81–2, 484 n. 63.

395   See the article on Purdy by Sara Wermiel in *American National Biography* vol. 17 (1999), pp. 943–4. Another engineer of similar type was Louis E. Ritter, a graduate of the Case School who had experience with the Erie Railroad before working for Jenney and Mundie in the 1890s.

396   The structure of Holabird and Roche's first big success, the Tacoma Building of 1886–9, was apparently drawn out by the obscure Karl Seiffert: Bruegmann, *op. cit.*, pp. 80, 483 n. 55. According to one source Purdy and Henderson were engineers for the Tacoma Building, but that firm was not yet then in existence; it seems more likely that Seiffert was working for George A. Fuller. William Holabird was trained at West Point and competent in engineering. Holabird and Roche's regular employment of Purdy (Bruegmann, p. 39) after 1892 no doubt corresponds with the shift from iron to steel.

397   Starrett, *op. cit.*, p. 144.

398   *Building News* 12 Dec. 1902, p. 819.

399   Sullivan, *op. cit.*, pp. 287–8.

400   Monroe, *op. cit.*, pp. 117–78; most of this extract is also quoted in Carl Condit, *The Chicago School of Architecture: A History of Commercial and Public Building in the Chicago Area 1875–1925* (Chicago, 1964), p. 51.

401   Starrett, *op. cit.*, p. 131.

402   Cited in Rochelle S. Elstein,' The Architecture of Dankmar Adler', *Journal of the Society of Architectural Historians*, Dec. 1967, p. 242. This article (pp. 242–9) covers Adler's career as a whole and mentions many less-familiar projects, but touches only on appearances. See also Twombly, *op. cit.*, p. 97.

403   Elstein, *op. cit.*, p. 247; Twombly, *op. cit.*, pp. 322–6.

404   About his Civil War experience with the 'Topogs', following two years in an artillery unit and a wound, Adler was less specific than Jenney: 'I made as good use of my time and was [as] well equipped as if my studies had been prepared at home'. Quoted from the autobiography in Charles E. Gregersen, *Dankmar Adler: His Theatres and Auditoriums* (Athens, Ohio, 1990), p. 2.

405   It is usually agreed that before Sullivan, first John Edelmann and then Francis Whitehouse were allowed to design much of the 'architectural embellishment' for Adler: Gregersen, *op. cit.*, pp. 47–8.

406   'I must confess that I cannot agree with those who place the matter of structural design as first in importance . . . The first requisite to the successful occupation of any premises for use as offices . . . is light and air.' From an article by Adler quoted by William H. Jordy, 'The Tall Buildings', in Wim de Wit (ed.), *Louis Sullivan: the Function of Ornament* (New York, 1986), p. 80.

407   This paragraph is based mainly on Gregersen, *op. cit.*, pp. 9–29, and Michael Forsyth, *Buildings for Music* (Cambridge, 1985), pp. 236–53, where the limitations of Adler's method and of the Auditorium's acoustics for music are explained.

408   Sullivan, *op. cit.*, pp. 246–50.

409   Twombly, *op. cit.*, p. 138.

410   Louis Sullivan, *Kindergarten Chats and Other Writings* (New York, 1947), pp. 137–8.

411   Twombly, *op. cit.*, p. 320.

412   Joseph Siry, *Carson Pirie Scott: Louis Sullivan*

and the Chicago Department Store (Chicago, 1988), p. 109. This book has been my main source for the paragraphs that follow.

413 Though it is interesting to note that as late as the 1903 section of the store, cast-iron columns had to be used in place of steel because of delivery problem: Siry, op. cit., p. 105.

414 The Brickbuilder 12 May 1903, p. 101, quoted in Siry, op. cit., pp. 170, 172–3.

415 Quoted ibid., p. 148.

416 Ibid., p. 155.

3   CONCRETE

1 Karl Sabbagh, Skyscraper: The Making of a Building (London, 1989), p. 78. Though the potted history of concrete that follows in Sabbagh's book is a little awry, his account of the factors determining its strength is clear and exemplary.

2 For the paragraphs on concrete and cement before Smeaton my principal sources have been C. W. Pasley, Observations on Limes (London, 1838, republished 1997), pp. 1–2, 44ff.; Encyclopaedia Britannica (11th edition, 1910), articles on cement and on concrete, pp. 653–9, 835–40; Norman Davey, A History of Building Materials (London, 1961), pp. 97–104; J. Mordaunt Crook, 'Sir Robert Smirke: A Pioneer of Concrete Construction', TNS 38 (1965–6), pp. 10–12, reprinted in Frank Newby (ed.), Early Reinforced Concrete (Aldershot, 2001), pp. 6–8; Major A. J. Francis, The Cement Industry, 1796–1914: A History (Newton Abbot, 1977), pp. 19–20; André Guillerme, Bâtir la ville: révolutions industrielles dans les matériaux de la construction (Paris, 1995), pp. 147–70; B. L. Hurst, 'Concrete and the Structural Use of Cements in England before 1890', Proceedings of the Institution of Civil Engineeers 116, nos. 3–4 (1996), pp. 2–6, reprinted in Newby, op. cit., pp. 22–6; and Cyrille Simonnet, Le béton: histoire d'un matériau (Marseilles, 2005), pp. 7–15.

3 For the 1748 works at Toulon see Bernard de Bélidor, Architecture hydraulique (Paris, 1790 edn), Tome 2, Part II, Livre III, Sec II, p. 187 (section on 'béton', pp. 178–90); and for the 1777 works, Simonnet, op. cit., p. 13. The latter followed the publication of experiments by the chemist Faujas de Saint-Fond.

4 For Smeaton and cements see BDCE, pp. 621–2 (Denis Smith), Guillerme, op. cit., p. 158, and Francis, op. cit., p. 20: 'overset the prejudices', Pasley, op. cit., p. 5.

5 For this paragraph see Francis, op. cit., pp. 26–159, and A. W. Skempton, 'Portland Cements, 1843–1887', TNS 35 (1962–3), pp. 129–31, reprinted in Newby, op. cit., pp. 73–5.

6 Skempton, op. cit., p. 147 (Newby, op. cit., p. 91).

7 For Vicat and French development of cements see Antoine Picon, L'invention de l'ingénieur moderne: L'Ecole des Ponts et Chaussées 1747–1851 (Paris, 1992), pp. 364–71; Guillerme, op. cit., pp. 161–90, 284; Simonnet, op. cit., pp. 21–38. Saint-Léger, whose relationship to Vicat seems to have been close, had been to London and taken out a patent there as early as 1818: see Pasley, op. cit., pp. 12–13. According to Simonnet, op. cit., p. 22n., Saint-Léger's was

not the first artificial cement factory. That was installed at Nemours in 1818 by an architect-entrepreneur, Giraut, but soon failed because good natural ingredients were available in the area.

8 Picon, op. cit., p. 368.

9 For this paragraph see Crook, op. cit., pp. 5–24: 'the merit of introducing', p. 13, quoting from Pasley, op. cit., p. 16. A note of 1825 on St Mark's Church, Myddelton Square, Clerkenwell, names the expert who prepared the pioneering foundation at the Millbank Penitentiary as the otherwise obscure Durant Hidson (Church of England Record Centre, file 18196). I'm grateful to Peter Guillery for drawing this to my attention.

10 For Cointeraux and Treussart, see Peter Collins, Concrete: The Vision of a New Architecture (London, 1959), pp. 21–4; Guillerme, op. cit., pp. 188, 284–5.

11 For Lebrun, see Collins, op. cit., pp. 24–6; Guillerme, op. cit., pp. 176, 188, 280; Simonnet, op. cit., pp. 34–5.

12 Picon, op. cit., pp. 368–9.

13 The Builder 18 Sep. 1847, p. 442, in a debate on the construction of breakwaters.

14 Ranger was the builder of Charles Barry's St Peter's, Brighton (1824–8). Some accounts claim that there is concrete in this church, but in what form is not clear.

15 For Ranger see BDCE, p. 543 (Mike Chrimes); Colvin, p. 792–3; Diana Burfield, Edward Cresy, 1792–1858, Architect and Civil Engineer (Donington, 2003), pp. 125–8. Crook, op. cit., p. 9 (Newby, op. cit.. p. 6), states that Cooper, George Ledwell Taylor and Ranger produced concrete blocks – whether together or apart is not clear. I have seen evidence only for Ranger as an initiator. For a proposal by Taylor to use 'Ranger's stone' for nave columns at Holy Trinity, Sheerness, see Michael Port, Six Hundred New Churches (2006 edn), pp. 137, 268.

16 Pasley, op. cit., p. 99.

17 John Michael Weiler, 'Army Architects: The Royal Engineers and the Development of Building Technology in the Nineteenth Century', Ph.D., University of York, 1987: for Pasley on concrete and cement, see pp. 60–80, 171–9.

18 Cited ibid., p. 64.

19 For concrete pipes of 1801 see Simonnet, op. cit., pp. 18–19. The early productions of Joseph Monier, described in Jean-Louis Bosc etc., Joseph Monier et la naissance du béton armé (Paris, 2001), pp. 65–98, consisted mostly of equipment for the farm and park; structures were a sideline.

20 For Lascelles see Andrew Saint, Richard Norman Shaw (London, 1976), pp. 165–71 and the works cited there; he is mentioned in Collins, op. cit., pp. 42–3.

21 For Coignet see Collins, op. cit., pp. 27–35; Bernard Marrey and Franck Hammoutène, Le béton à Paris (Paris, 1999), pp. 21–5; Bosc, op. cit., pp. 31–41; Simonnet, op. cit., pp. 41–6.

22 Cited in Collins, op. cit., p. 32.

23 Paul Séjourné, Grandes voûtes, vol. 1 (Bourges, 1913), pp. 210–9. The authorship of the Yonne viaduct design is ascribed there to the engineers Eugène Belgrand and Edmond Humblot. Other accounts add the name of Charles (sometimes called

Edmond) Huet. Dates here follow Séjourné: it is striking that the viaduct was built during the turbulence that attended the collapse of the Second Empire. The Vanne scheme as a whole started in 1867 and was completed in 1874: see George Atkinson, 'Eugène Belgrand (1870–1878): Civil Engineer, Geologist and Pioneer Hydrologist', TNS 69, no. 1 (1997–8), pp. 107–9.

24 Quotations from letter from L. A. Boileau to Le Moniteur des Architectes 1 Dec. 1867, cols. 207–14. Coignet replied, ibid., 1 Feb. 1868, cols. 19–25, without referring to issues of design; Boileau wrote again, 1 May 1868, cols. 67–72, giving the additional information that the iron 'ossature' for the Le Vésinet church had been fabricated by the Fonderie Boignes-Rambourg. Collins, op. cit., pp. 33–4, naturally offers a pro-Coignet reading of the argument.

25 Tom F. Peters, Building the Nineteenth Century (Cambridge, Mass., 1996), pp. 66–9.

26 The best short account of Hyatt's work is in W. K. Hatt, 'Genesis of Reinforced Concrete Construction', American Concrete Institute, Proceedings of the Twelfth Annual Convention, 1916, pp. 25–30, reprinted in Newby, op. cit., pp. 235–40.

27 Skempton, op. cit., pp. 149 (Newby, op. cit., p. 93), and for information in this paragraph, ibid., pp. 137–45 (Newby, pp. 81–9).

28 For this paragraph see O. Kohlmorgen, 'The Historical Evolution of Reinforced Concrete in Germany', Concrete and Constructional Engineering 1 (Nov. 1906), pp. 325–9 ('extensive experiments', p. 326); Simonnet, op. cit., pp. 71–3 (a particularly clear account); Bosc, op. cit., pp. 108–13; Collins, op. cit., pp. 60–1; Peters, op. cit., p. 77; Delhumeau p. 65; and www.wayss-freitag.de. For the Victoriastadt, Berlin, using Tall's system, see Heike Hinz, 'Frühe Betonhäuser in Berlin in der Victoriastadt ab 1871', Master's thesis, Berlin TU, 2002.

29 Bosc, op. cit., p. 113. For Wildegg Bridge, Hatt, op. cit., p. 35 (Newby, p. 245). For the nature of the early structures see Ludwig Hess, 'The Historical Evolution of Reinforced Concrete in Austria', Concrete and Constructional Engineering 2 (Sep. 1907), pp. 265–72. Peters, op. cit., p. 386 n. 156 identifies the Leipzig Krystallpalast of 1887, built by Wayss to the designs of Arwed Rossbach, as among the earliest German buildings largely of reinforced concrete.

30 For Ransome see Carl W. Condit, 'The First Reinforced-Concrete Skyscraper', Technology and Culture 6, no. 1 (Jan. 1968), pp. 7–9, reprinted in Newby, op. cit., pp. 260–2; John Snyder and Steve Mikell, 'The Consulting Engineer and Early Concrete Bridges on California', pp. 2–3, reprinted in Newby, op. cit., pp. 294–5; Collins, op. cit., pp. 61–3; Reyner Banham, A Concrete Atlantis (1986), pp. 32–8, 65–79; Daniel L. Schodek, Landmarks in American Engineering (Cambridge, Mass., 1987), pp. 145–6, 285. For the US and early reinforced concrete: Captain Sewell, 'Reinforced Concrete in the United States', Concrete and Constructional Engineering 1 (May 1906), pp. 79–87; Schodek, op. cit., pp. 283–5; Amy E. Slaton, Reinforced Concrete and the Modernization of American Building, 1900–1930 (Baltimore,

2001), with references to Ransome, pp. 17, 138, 144–7, 199–200.

31   For the following paragraphs on Hennebique see Gwenaël Delhumeau, *L'invention du béton armé: Hennebique 1890–1914* (Paris, 1999); statistics for personnel are given on p. 102; for the remark by Gubler, see p. 15. For an English summary, Gwenaël Delhumeau, 'Hennebique and Building in Reinforced Concrete', in *Rassegna* 49 (1992), *Reinforced Concrete: Ideologies and Forms from Hennebique to Hilberseimer*, pp. 15–25; and for a photographic record of the Hennebique legacy with essays by Delhumeau, Gubler and others, see also Delhumeau *et al.*, *Le béton en représentation* (Paris, 1993).

32   Delhumeau, *L'invention*, p. 18.

33   Apart from the earlier intermittent efforts by Monier, the first entrepreneur to exploit reinforced concrete successfully in France was Edmond Coignet, son of François Coignet. His patents date from 1889 and his first works in Paris (mainly pipes and conduits) from 1892, effectively the start of practical reinforced concrete in the French construction industry. Cottancin's patent also dates back to 1889, before Hennebique's. For an early overview of the French story by a collaborator of Edmond Coignet's, see N. de Tedesco, 'The Historical Evolution of Reinforced Concrete in France', *Concrete and Constructional Engineering* 1 (July 1906), pp. 159–70.

34   R. Nelva and B. Signorelli, *Avvento ed evoluzione del calcestruzzo armato in Italia: il sistema Hennebique* (Milan, 1990), p. 19. For early reinforced concrete in Switzerland, see F. Schüle, 'The Historical Evolution of Reinforced Concrete in Switzerland', *Concrete and Constructional Engineering* 2 (May 1907), pp. 91–7; and Hans-Ulrich Jost, 'The Introduction of Reinforced Concrete in Switzerland (1890–1914): Social and Cultural Aspects', in *Proceedings of the Second International Congress on Construction History* (Cambridge, 2006), vol. 2, pp. 1741–53. For Mörsch at Wayss and Freytag, Kohlmorgen, *op,. cit.*, pp. 327–8; for a review of the changing editions of his book, Peters, *op. cit.*, p. 475.

35   For L.-C. Boileau and Hennebique, see Delhumeau, *op .cit.*, pp. 232–5, 248–53; and for Boileau's articles on 'Le ciment armé', *L'Architecture* 2 , 9, 16 and 30 Nov. and 7 Dec. 1895, pp. 369–74, 380–2, 389–91, 402–4, 411–12. In the fourth article (30 Nov., p. 403) comes the question: 'Quelles formes l'architecture donnera-t'il aux organes d'une construction de ciment armé?' In the last article, on fresco, Hennebique is forgotten as Boileau waxes ecstatic on the possibilities for surface decoration.

36   *L'Architecture* 13 Jan. 1906, pp. 12–14.

37   For de Baudot and Cottancin I have largely followed Marie-Jeanne Dumont's article 'The Philosopher's Stone: Anatole de Baudot and the French Rationalists' in *Rassegna* 49 (1992), *Reinforced Concrete: Ideologies and Forms from Hennebique to Hilberseimer*, pp. 37–43; also her contributions to *Rassegna* 68 (1996), *Anatole de Baudot 1834–1915*, pp. 7–13, 46–9. For a clear English account of Cottancin's system with useful diagrams see G. J. Edgell, 'The

Remarkable Structures of Paul Cottancin', in *Structural Engineer* 6, no. 7 (July 1985), pp. 201–7, reprinted in Newby, *op. cit.*, pp. 169–86. See also Delhumeau, *L'invention*, pp. 244–6, 253–5; Collins, *op. cit.*, pp. 113–17; and the entries on de Baudot and Cottancin in *Encyclopédie Perret* (Paris, 2002), pp. 72–4.

38   Quoted from *Le Ciment*, 25 Nov. 1896, in Simonnet, *op. cit.*, p. 50.

39   C. F. Marsh, quoted in Collins, *op. cit.*, p. 115. The remark refers to the constructive process, not to the Cottancin system or the design. For the dates of construction and the religious side of the story, see the excellent booklet and complementary pamphlet on sale at Saint-Jean de Montmartre.

40   From the list of works in Edgell (Newby, *op. cit.*, p. 174), and an article in *Building News* 29 April 1904, pp. 613–14, it seems that Cottancin had employment in France, Belgium, Portugal, Algeria, Tunisia, Morocco, England and the United States about this time. In 1903 he and his partner Garcin had successfully tendered to construct the first stage of Albert Ballu's cathedral at Oran (see p. 235–6). Not having Hennebique's well-organized licensing system, he may have over-extended himself. Another factor may have been an accident to a Cottancin-system structure in Santiago, Chile, in October 1904: Simonnet, *op. cit.*, p. 104. Cottancin's main English work was St Sidwell's Methodist Church, Exeter (1902): see the analysis of its 'extraordinary structure' in *Structural Engineer*, 17 Jan. 2006, pp. 19–20. According to the *Building News* article, Cottancin was introduced to the UK via an English architect working in Paris, Arthur Vye-Parminter.

41   Dumont begins her 'Philosopher's Stone' article, *op. cit.*, p. 37, with the thought that de Baudot and Cottancin represented an 'intellectual concrete' as opposed to the 'heroic concrete' of the engineers and the 'pragmatic, prosaic concrete' of the entrepreneurs.

42   For information in this paragraph see Delhumeau, *op. cit.*, pp. 257–9, 289–91; Simonnet, *op. cit.*, pp. 84–5, 93–4, 194. Rabut's pupils included Freyssinet and Limousin.

43   Anatole de Baudot, *L'architecture: le passé – le présent* (Paris, 1916), p. 164: whole passage, pp. 147–65.

44   Quoted in Simonnet, *op. cit.*, p. 96.

45   Patricia Cusack, 'Agents of Change: Hennebique, Mouchel and Ferro-Concrete in Britain, 1897–1908', *Construction History* 3 (1987), pp. 61–74, reprinted in Newby, *op. cit.*, pp. 155–68.

46   *Concrete and Constructional Engineering* 5 (Oct.1910), p. 703, quoted in Patricia Cusack, 'Architects and the Reinforced Concrete Specialist in Britain 1905–1908', *Architectural History* 29 (1986), pp. 183–96, reprinted in Newby, *op. cit.*, pp. 217–30, from which much of this paragraph derives (quotation p. 187, Newby p. 221).

47   See besides Cusack, *op. cit.*, pp. 185–6 (Newby pp. 219–20), the obituary of Dunn in *RIBA Journal* 24 Feb. 1934, p. 418.

48   For Sachs see David Wilmore (ed.), *Edwin O. Sachs, Architect, Stagehand, Engineer and Fireman* (Summerbridge, 1998), especially pp. 18–20

(part of a memoir by his son Sir Eric Sachs, with minor errors of date); Cusack, 'Architects and the Reinforced Concrete Specialist', pp. 189–92 (Newby, pp. 223–6); and entry in ODNB (Saint).

49   In a memo of 1914 the invariably precise Sachs dates the founding of the Concrete Institute to 1896: 'I was founder and first Chairman of this Institute in 1896. When the Chairmanship was abolished, I was elected as Vice-President . . . I no longer take any active part in its work' (papers in author's possession). But there is no published evidence for the Concrete Institute's work before the announcement of its inauguration in *Concrete and Constructional Engineering* 2 (Jan. 1908), p. 423. Possibly Sachs had the idea in 1896 and instituted a few private meetings but could not get others interested until 1908. For a report of the inaugural lunch see *ibid.*, 2 (Sep. 1908), pp. 261–4 and for the first open meetings, *The Builder* 28 Nov. 1908, pp. 584–6, and 26 Dec. 1908, pp. 701–2.

50   For the inconclusive theoretical answers to this question offered from 1900 onwards, see Peter Collins, *Concrete* (London, 1959), pp. 112–49.

51   The section on the Perrets is drawn unless otherwise stated from two invaluable productions: *Les Frères Perret: L'oeuvre complete* (Paris, 2000), hereafter LFP, and *Encyclopédie Perret* (Paris, 2002), hereafter EP, more particularly the latter.

52   Claude-Marie Perret was the masonry contractor for the 1876–8 section of the great glasshouses built at Laeken to the designs of Alphonse Balat.

53   For the nomenclature of the firms see Guy Lambert in EP p. 46.

54   LFP pp. 374–7.

55   Concrete at Saint-Malo: LFP, p. 82; EP p. 16, and photo of Claude-Marie Perret on the site, p. 28.

56   Though the highly glazed rear façade of the block in the Avenue de Wagram has points in common with the fenestration at the Rue Franklin, and portions of the floors at the block in the Avenue Niel were of reinforced concrete, procured as at Rue Franklin from Latron and Vincent: EP pp. 71, 194.

57   For 25 bis rue Franklin, besides EP pp. 74–5, 184–5 and LFP pp. 88–91, see *Rassegna* 28 (1986): *Perret: 25 bis Rue Franklin*, with articles on construction by Paul Poitevin and on decoration by Hélène Guéné.

58   Quoted by Gwenaël Delhumeau in EP p. 71. For Latron et Vincent see this entry, and Delhumeau, *L'invention du béton armé: Hennebique 1890–1914* (Paris, 1999), pp. 299–300.

59   LFP pp. 92–4: Auguste Perret's recollection of the job, EP p. 16.

60   It is run close by the house for Dr Carnot on the Avenue Elisée Reclus, Paris (1906-8), designed by the Perrets' close friend Paul Guadet and built by Perret Frères with an internal concrete frame: EP p. 287.

61   For Guy Lambert's assessment of the roles of the brothers, EP p. 29; for Nitzchké's statement and comments, *ibid.* p. 51.

62   For Oran Cathedral see EP pp. 74–8 and LFP pp. 332–5.

63   The full story is set out in EP pp. 78–84 (Claude Loupiac), and pp. 307–9

(Christophe Laurent). For an earlier summary see Bernard Marrey, 'Qui est l'architecte du Théâtre des Champs-Elysées?', *L'architecture d'aujourd'hui* 174 (July/Aug. 1974), pp. 115–25.

64 EP pp. 65–6.

65 *Ibid.*, pp. 59–65 (Guy Lambert).

66 EP pp. 51–4, 66: quotation in this paragraph from p. 54.

67 EP pp. 96–7; LFP pp. 124–9; and for an English appreciation, Andrew Saint, 'Notre-Dame du Raincy', *Architects' Journal* 13 Feb. 1991, pp. 26–45.

68 Quoted in EP p. 54.

69 Quoted e.g. in LFP pp. 22, 109: original from a speech given in May 1933, printed in *Revue d'art et d'esthétique* nos.1–2, June 1935, p. 48. For the Perrets' airship hangar designs of 1917–19, see LFP pp. 108–9, 381; Gellusseau's involvement in the first of these, which was of steel, is recorded p. 381.

70 From a speech of 1958 by Freyssinet entitled 'Constructeur, pas professeur': Eugène Freyssinet, *Un amour sans limite* (Paris, 1993), p. 143: quoted also in EP p. 100.

71 For Albert Kahn my main sources have been: Grant Hildebrand, *Designing for Industry: The Architecture of Albert Kahn* (Cambridge, Mass., 1974); W. Hawkins Ferry, *The Legacy of Albert Kahn* (Detroit, 2nd edn, 1987); Federico Bucci, *Albert Kahn, Architect of Ford* (New York, 1993); and Brian Carter (ed.), *Albert Kahn: Inspiration for the Modern* (Ann Arbor, 2001).

72 Hildebrand, *op. cit.*, p. 24, n.14.

73 *Ibid.*, p. 17.

74 Family details mainly from Hildebrand. Moritz Kahn worked for Julius from 1906 to 1923 and then joined Albert. For Felix Kahn see Bucci, *op. cit.*, p. 68, n.12.

75 For Julius Kahn see the memoir by O. W. Irwin in *Transactions of the American Society of Civil Engineers* 72 (1945), pp. 1742–7. I have followed dates in that account: e.g Irwin says Julius Kahn spent only one year in Japan in 1900 and joined Albert early the following year, whereas Albert in his memoir in *Architectural Forum* (see below) says Julius had spent 'several years' in Japan.

76 Hildebrand, *op. cit.*, p. 72.

77 Albert Kahn, 'Industrial Architecture', *Architectural Forum* Feb. 1939, pp. 131–2. The text is from a talk given at a lunch for the Detroit building industry, 21 Dec. 1938.

78 Its first publication took the form of an article by Julius Kahn in *Railroad Gazette* 16 Oct. 1903, pp. 734–6.

79 The sources disagree as to whether the University of Michigan's Engineering Building (now West Engineering Building) was the first Kahn-system structure, or whether its unsatisfactory construction using an earlier method caused Julius Kahn to devise his system. It seems possible that it was constructed in two stages of which the second, not finished till 1904, involved using the new system for the first time. To compound the confusion, O. W. Irwin's obituary of Julius Kahn states that the Agricultural Building of the University of Michigan was the first to use the system, but there was no such building at Ann Arbor. I am very grateful to Rebecca Price of the University's Art, Architecture & Engineering Library for help on these points.

80 Probably at first on Roosevelt Hall, before being extended to other buildings. The Army War College, a pet project of Theodore Roosevelt's, was started under Sewell in 1903: see Captain Sewell, 'Reinforced Concrete in the United States', *Concrete and Constructional Engineering* 1 (May 1906), pp. 79–87. It may be noted that one of Julius Kahn's employers in the period after he graduated was the US Navy Bureau of Yards and Docks.

81 See e.g. Moritz Kahn, *The Design and Construction of Industrial Buildings* (London, 1917).

82 Quoted in Hildebrand, *op. cit.*, p. 67, n.5. The assertion that Julius Kahn made reinforced concrete skyscrapers possible is of course false. The first such structure was the sixteen-storey Ingalls Building, Cincinnati (1903), constructed by the engineer Henry N. Hooper of the Ferro-Concrete Company of Cincinnati using an evolved version of the Ransome system: see Carl W. Condit, 'The First Reinforced-Concrete Skyscraper', *Technology and Culture* 6, no. 1 (Jan. 1968), pp. 1–33, reprinted in Frank Newby (ed.), *Early Reinforced Concrete* (Aldershot, 2001), pp. 255–87; Daniel L. Schodek, *Landmarks in American Engineering* (Cambridge, Mass., 1987), pp. 283–6. Cincinnati was notably advanced and experimental in early reinforced concrete.

83 Condit, *op. cit.*, pp. 24–6 (Newby pp. 260–2).

84 *Architectural Forum* Feb. 1939, p. 132.

85 In a paper of 1924 reviewing the development of reinforced concrete, Albert Kahn also ascribed the European advances in the exposed external use of the material to labour costs: 'With labor costs much lower and careful workmanship more general than here, it was only natural that they should produce results quite impossible in this country': Albert Kahn, 'Reinforced-Concrete Architecture These Past Twenty Years', *Proceedings of the American Concrete Institute*, 20 (1924), p. 109.

86 Amy E. Slaton, *Reinforced Concrete and the Modernization of American Building, 1900–1930* (Baltimore, 2001), p. 227 n.41.

87 Reyner Banham, *A Concrete Atlantis* (1986), p. 84.

88 I have omitted from my narrative the George N. Pierce automobile factory in Buffalo (1906). This confusing job has fascinated historians because it was entirely of reinforced concrete, while the various shops were all on one storey, prefiguring the horizontal production line. Perhaps because it was built in one rapid burst, it was divided between several concerns. Albert Kahn's involvement seems to have been as consultant for Julius's firm, the Trussed Concrete Steel Company. There is no evidence that he designed the overall layout, and to ascribe to him the most generous space in the complex, the Assembly Building, where development of the Kahn beams allowed bay widths in trabeated concrete to stretch to 61 feet, is guesswork. Hildebrand, *op. cit.*, pp. 34–43, gives the fullest description of the George N. Pierce plant. Lockwood, Greene and Company, the Boston engineer-builders who feature prominently in Banham's *A Concrete Atlantis*, were employed there in an unknown capacity. A local architect, George

Cary, designed the Administration Building, while according to Banham, *op, cit.*, pp. 86–7, the Body Building on the south side of the complex was built by the Aberthaw Construction Company, licensees of the Ransome system in the north-eastern states (for the Aberthaw Company, see Slaton, *op, cit.*, pp. 157–83). Banham, who is generally pro-Ransome and Lockwood, Greene and anti-Kahn, identifies the 'production shop and some other structures immediately to its north and west' as revealing 'exactly the mean-minded rationality and gracelessness that characterize Packard 10'. He says nothing about the Assembly Building.

89 Albert Kahn, 'Reinforced-Concrete Architecture These Past Twenty Years', *Proceedings of the American Concrete Institute*, 20 (1924), p. 118.

90 Betsy Hunter Bradley, *The Works: The Industrial Architecture of the United States* (New York, 1999), p. 168, dates the marketing of Julius Kahn's steel sashes to 1910. However she seems to be mistaken in saying that the earlier sashes at Highland Park had been supplied by the Detroit Steel Products Company, since Albert Kahn definitely states that they were imported from England.

91 For Fiat-Lingotto see Marco Pozzetto, *La Fiat-Lingotto: un'architettura torinese d'avanguardia* (Turin, 1975); Banham, *op. cit.*, pp. 236–53 (an emotional paean); and Grant Hildebrand, 'Beautiful Factories', in Carter, *op. cit.*, pp. 16–27.

92 Exceptions include the Body Plant for the Studebaker Corporation, South Bend, Indiana (*c.*1922–3), of six storeys high on a concrete frame: Albert Kahn, 'Reinforced-Concrete Architecture These Past Twenty Years', *Proceedings of the American Concrete Institute* 20 (1924), p. 119.

93 *Ibid.*, p. 109.

94 Hildebrand, *op. cit.*, p. 124.

95 These details are from Hildebrand, *op, cit.*, pp. 59–61.

96 Henry Jonas Magaziner, 'Working for a Genius: My Time with Albert Kahn', *APT Bulletin* 32 nos. 2/3 (2001), p. 61.

97 Cited by Hildebrand, *op, cit.*, p. 156 (from *Architectural Forum* Dec. 1940, p. 501).

98 Frank Lloyd Wright, *An Autobiography* (London, 1945 edn), p. 144. This section is an abbreviated version of my article 'Frank Lloyd Wright and Paul Mueller: the architect and his builder of choice', *arq (Architectural Research Quarterly)* 7 no. 2 (2003), pp. 157–67.

99 From Mueller's testimony of 1925 on the Chicago Auditorium, printed in Edgar Kaufmann junior, *9 Commentaries on Frank Lloyd Wright* (New York, 1989), p. 42 (whole document pp. 42–62).

100 Kaufmann, *op. cit.*, p. 37, says that Mueller was with the architect J. L. Silsbee before joining Adler and Sullivan. That is not mentioned in Mueller's witness statement, *ibid.*, pp. 42–3, and may be wrong. There Mueller says he joined Adler and Sullivan in 1883, probably a misprint for 1886. Wright was with Silsbee for most of 1887 and seems definitely first to have met Mueller when he went to prospect for a job with Adler and Sullivan: see Wright, *op. cit.*, pp. 84–7.

101 *Ibid.*, p. 54.

102   Kaufmann, *op. cit.*, p. 48.

103   Quoted in Jack Quinan, *Frank Lloyd Wright's Larkin Building: Myth and Fact* (New York, 1987), p. 129.

104   See e.g. the 'monolithic bank', probably of mass concrete, in Terence Riley (ed.), *Frank Lloyd Wright, Architect* (Museum of Modern Art, New York, 1994), p. 123. The first mention of reinforced concrete in his work seems to occur in 1904 in connection with beams for the plant house of the Darwin D. Martin house, Buffalo: Bruce Brooks Pfeiffer (ed.), *Frank Lloyd Wright: Letters to Clients* (London, 1987), p. 13.

105   Letter from Darwin D. Martin, quoted in Brendan Gill, *Many Masks: A Life of Frank Lloyd Wright* (New York, 1987), p. 161.

106   In an impressionistic essay on Wright's technology in Riley, *op. cit.*, p. 60.

107   Joseph Siry, *Unity Temple: Frank Lloyd Wright and Architecture for Liberal Religion* (Cambridge, 1996), the main source for the following paragraphs.

108   Frederick Gutheim, *Frank Lloyd Wright, In the Cause of Architecture: Essays by Frank Lloyd Wright in Architectural Record 1908–1952* (New York, 1975), p. 141.

109   Quoted in Siry, *op. cit.*, p. 146.

110   Gutheim, *op. cit.*, pp. 205, 208.

111   Wright, *op. cit.*, p. 145.

112   *Ibid.*, p. 159. For Midway Gardens see Anthony Alofsin, *Frank Lloyd Wright: The Lost Years, 1910–1922* (Chicago, 1993), pp. 137–50, and Paul Kruty, *Frank Lloyd Wright and Midway Gardens* (Urbana, 1998).

113   Kruty, *op. cit.*, p. 33.

114   Alofsin, *op. cit.*, p. 359 n.68.

115   Kruty, *op. cit.*, p. 40, says that Mueller borrowed money from Seipp to get Midway Gardens finished. This led to an inevitable quarrel.

116   Bruce Brooks Pfeiffer, quoted in Kaufmann, *op. cit.*, p. 62.

117   R. K. Reitherman, 'The Seismic Legend of the Imperial Hotel', *AIA Journal* 69 (1980), pp. 42–7, 70.

118   Cited *ibid.*, p. 46.

119   Julius Floto, 'The Imperial Hotel', *Architectural Record* 55 (1924), pp. 119–23. Reitherman, no doubt thinking of Floto as Japanese, misreads him as 'Hoto'.

120   Bruce Brooks Pfeiffer, *Frank Lloyd Wright: Letters to Architects* (London, 1987), pp. 39–40.

121   For Mueller at Ocatillo and San Marcos in the Desert see Kaufmann, *op. cit.*, p. 62; for Westhope, Tulsa, see Meryle Secrest, *Frank Lloyd Wright* (London, 1992), pp. 371–2.

122   Vincent Scully, 'Wright versus the International Style', essay of 1954 reprinted in *Modern Architecture and Other Essays* (Princeton, 2003), p. 60.

123   The larger Usonian houses profited from an idealist-builder in the shape of Harold Turner: see John Sergeant, *Frank Lloyd Wright's Usonian Houses* (New York, 1984), pp. 40, 62, 78, 118–17.

124   Jonathan Lipman, *Frank Lloyd Wright and the Johnson Wax Buildings* (London, 1986).

125   Robert Silman, 'Fallingwater: Solving Structural Problems', in Susan Macdonald (ed.), *Preserving Post-War Heritage: The Care and Conservation of Mid-Twentieth-Century Architecture* (Donhead St Mary, 2001), pp. 186–94; and for a fuller account of the concrete cantilevers at Fallingwater, Franklin

126   Toker, *Fallingwater Rising* (New York, 2003), pp. 210–23.

126   For Williams I have largely followed David Yeomans and David Cottam, *The Engineers' Contribution to Contemporary Architecture: Owen Williams* (London, 2001). See also David Cottam, *Sir Owen Williams 1890–1969* (London, 1986).

127   At Hayes, Middlesex, for the Gramophone Company, later EMI. A. C. Blomfield junior was the architect (Yeomans and Cottam, *op. cit.*, p. 14). In his first job, working for the Indented Bar Company, 1911–12, Williams probably met another engineer of future note, Oscar Faber, who was then also working on one of the Gramophone Company's buildings at Hayes: John Faber, *Oscar Faber* (London, 1989), p. 9. I assume the buildings are different.

128   For the use of concrete at Wembley, besides Yeoman and Cottam see *Concrete and Constructional Engineering* 17 (Nov. 1922), pp. 695–703 and 19 (July 1924), pp. 410–19; remarks by Oscar Faber in *Architectural Review* 55 (June 1924), pp. 218–21; and Sir E. Owen Williams, 'Concrete and its Uses in the Construction of the British Empire Exhibition', *Journal of the Institution of Municipal and County Engineers*, 12 Aug. 1924, pp. 212–32 with discussion, pp. 233–7. I am grateful to Elain Harwood for communicating these and for advising on Wembley.

129   *Concrete and Constructional Engineering* 16 (Oct. 1921), pp. 629–30.

130   See ODNB entry on the McAlpine family by Iain F. Russell. McAlpine had built concrete cottages in Lanarkshire as far back as the 1870s.

131   *Architects' Journal* 30 April 1924, pp. 744–5.

132   Simpson and Ayrton's Clare College Mission Church, Rotherhithe, London, of 1911, still extant, was of lightly reinforced concrete with roughcast render.

133   C. H. Reilly, 'First Impressions of the Wembley Exhibition', *Architects' Journal* 28 May 1924, pp. 893–4.

134   *Journal of the Institution of Municipal and County Engineers* 12 Aug. 1924, p. 220.

135   Harry Barnes, in *Architectural Review*, cited by Yeomans and Cottam, *op. cit.*, p. 26.

136   *Journal of the Institution of Municipal and County Engineers*, 12 Aug. 1924, p. 234.

137   *British Engineers Export Journal*, July 1924, cited in Yeomans and Cottam, *op. cit.*, p. 30. For a discussion of Ayrton's and Williams' views on collaboration at this stage with further citations, see *ibid.*, pp. 29–31.

138   O. Maxwell Ayrton, 'Modern Bridges', *RIBA Journal* 38 (April 1931), pp. 479–88; remarks by Williams, pp. 490–3. See also Yeomans and Cottam, *op. cit.*, pp. 56–7.

139   Recounted *ibid.*, pp. 58, 62–7.

140   'Great whitewashed barn' from an anecdote told to David Cottam, *ibid.*, p. 64; 'period opulence' from the account of the Dorchester Hotel episode in J. M. Richards, *Memoirs of an Unjust Fella* (London, 1980), pp. 49–52. Richards worked as an architectural assistant for Williams on the Peckham Health Centre as well as the Dorchester.

141   Cited in Yeomans and Cottam, *op. cit.*, p. 79.

142   See Susie Barson and Andrew Saint, *A Farewell to Fleet Street* (London, 1988), with mention of Williams' activities and Gallannaugh's role, pp. 45–9. For an analy-

sis of Williams' intricate work at the Daily Express, see Yeomans and Cottam, *op. cit.*, pp. 67–79.

143   Peter Bak (ed.), *J. Duiker Bouwkundig Ingenieur* (Delft, 1982); J. Molema, *Ir. J. Duiker* (Rotterdam, 1989).

144   Ida Jager, 'Bouwkunst-schepper door constructie en techniek', in Rainer Bullhorst, Kees van Harmelen and Ida Jager, *Duiker in Den Haag* (The Hague, 1999), pp. 10, 17.

145   Ida Jager, *Willem Kromhout Czn.*, (Rotterdam, 1992), pp. 24–5.

146   Cited in *Het Nieuwe Bouwen: Voorgeschiedenis* (Delft, 1982), pp. 46–7.

147   *Ibid.*, p. 47.

148   Ida Jager, 'De Haagse Bijvoet en Duiker', in Bullhorst, van Harmelen and Jager, *op. cit.*, pp. 19–47.

149   For most of what follows on Wiebenga see Jan Molema and Peter Bak (eds.), *Jan Gerko Wiebenga, Apostel van het Nieuwe Bouwen* (Rotterdam, 1987).

150   *Ibid.*, pp. 48–9.

151   *Ibid.*, pp. 59–61.

152   *Ibid.*, pp. 58–9; Ronald Zoetbrood, *Jan Duiker en het Sanatorium Zonnestraal* (Amsterdam, 1985).

153   Molema and Bak, *op. cit.*, pp. 70–7; Bullhorst, van Harmelen and Jager, *op. cit.*, pp. 64–79.

154   Cited in Molema and Bak, *op. cit.*, p. 77: see also Jager in Bullhorst, van Harmelen and Jager, p. 14.

155   From the preface to Le Corbusier, *Oeuvre complète 1910–1929*, eds. Boesiger and Stonorov (Paris, 1929), pp. 10, 13. It is typical of Le Corbusier's vagueness about engineers that in this preface he names Cottancin, not Contamin, as the engineer of the Galerie des Machines.

156   The following paragraphs depend much on H. Allen Brooks, *Le Corbusier's Formative Years* (Chicago, 1997). I have used Le Corbusier's post-1919 pseudonym throughout, rather than calling him Charles-Edouard Jeanneret for this early period.

157   Brooks, *op. cit.*, p. 71.

158   *Ibid.*, p. 135.

159   *Ibid.*, pp. 156–7, quoting from Le Corbusier's memoir on Perret from *L'Architecture d'aujourd'hui*, October 1932.

160   *Ibid.*, p. 175.

161   Translated from Marie-Jeanne Dumont (ed.), *Le Corbusier: Lettres à Charles L'Eplattenier* (Paris, 2006), p. 187 (letter of 22 November 1908). The excerpt in Brooks, *op. cit.*, p. 153 is misleading at this point.

162   Quoted in Dumont, *op. cit.*, p. 163.

163   *Ibid.*, p. 197.

164   Brooks, *op. cit.*, pp. 191–2.

165   *Ibid.*, pp. 326, 339.

166   *Ibid.*, pp. 428–9, 448–9, with figs. 377–8. The Zurich engineers were Terner and Chopard. Brooks reports that in the lawsuit the client alleged that the use of a concrete frame had added to the expense by 20,000 Swiss francs. Expert witnesses agreed with the figure but disputed the allegation that the frame had no technical or aesthetic benefits.

167   *Ibid.*, p. 471. My account of Le Corbusier and Du Bois largely follows Brooks's pp. 382–9 and 471–3 (a biographical note on Du Bois appears on p. 383). I am grateful to Tim Benton for cautioning me against

accepting the Du Bois point of view uncritically.

168   Le Corbusier, *Oeuvre complète 1910–1929*, eds. Boesiger and Stonorov (Paris, 1929), p. 12.

169   Quoted in Gregh, pp. 66, 80. In the original, the second sentence reads: 'C'est hors la construction, dans la construction.'

170   *Ibid.*, p. 71.

171   Colin Rowe, *As I Was Saying* (Cambridge, Mass., 1996), vol. 1, p. 31.

172   These abattoir designs were probably influenced in part by Tony Garnier. Le Corbusier had visited Garnier's half-built cattle market and abattoir at Lyons in 1914 and wrote to him afterwards, asking for photographs: Brooks, *op. cit.*, pp. 371–2, 408–9. Garnier's notorious designs for a 'cité industrielle' were finally published in 1917, though made long before.

173   Quoted *ibid.*, p. 486.

174   Tim Benton, 'From Jeanneret to Le Corbusier: Rusting Iron, Bricks and Coal and the Modern Utopia', *Massilia: annuaire d'études corbuséennes* 2003, article 17, pp. 28–39.

175   Stanislaus von Moos, 'Industrie', in *Le Corbusier: une encyclopédie* (Paris, 1987), pp. 190–9.

176   Quoted *ibid.*, p. 196.

177   This sketch of Pierre Jeanneret is mostly based on *Le Corbusier: une encyclopédie*, pp. 212–14. During the Second World War, unlike his cousin, Jeanneret was a resister.

178   For a table of the villa-projects with the names of the various participating contractors, including Summer, see Tim Benton, *The Villas of Le Corbusier 1920–1930* (London, 1987), p. 220. Summer's name recurs throughout this book, but little is said about his skills or habits; as Benton says, p. 12, 'Little work has yet been done on the Corbusian équipe.'

179   In personal communication, 2006. I am grateful to Tim Benton for his patience with my (for him) unpalatable ideas about Le Corbusier. This section has been adjusted and chastened after he kindly read it, but our approaches remain at variance.

180   The fullest study of Pessac etc. now is Tim Benton, 'Pessac and Lège Revisited: Standards, Dimensions and Failures', *Massilia: annuaire d'études corbuséennes* 2004, article 34, pp. 64–99. See also Brian Brace Taylor, *Le Corbusier at Pessac* (Cambridge, Mass., 1972), notably pp. 11–17, and Gilles Ragot and Mathilde Dion, *Le Corbusier en France* (Paris, 1992 edn), pp. 135–9.

181   Quoted by Benton, 'Pessac and Lège', p. 96: in French, 'd'une insipidité stupéfiante'.

182   Tim Benton remarks that the cement guns 'did not give the precise machine finish that Le Corbusier wanted and which a craftsman applying cement render "à la taloche" on breeze block walls could achieve. Also, the Solomite compressed straw panels onto which the concrete was to be sprayed could not be delivered in the right sizes. Thirdly, it was difficult to direct the spray into re-entrant corners, such as those formed by the semi-circular "chais" or stores' (personal communication, 2006).

183   See T. Benton, 'Villa Savoye and the Architect's Practice', in Le Corbusier, *Villa Savoye and Other Buildings and Projects, 1929–1930* (New York, 1984), pp. ix–xxxi;

Brian Brace Taylor, *Le Corbusier, The City of Refuge Paris 1929/33* (Chicago, 1987), notably pp. 83ff.; Jacques Sbriglio, *Le Corbusier: L'Unité d'Habitation de Marseille* (Marseilles, 1992), notably pp. 121–43; and Eduard F. Sekler and William Curtis, *Le Corbusier at Work: The Genesis of the Carpenter Center for the Visual Arts* (Cambridge, Mass., 1978), notably pp. 171–7, 201–23.

184   Taylor, *The City of Refuge*, p. 103.

185   The best summary of Bodiansky's career is by Marion Tournon Branly in *Architectural Design* 35, no. 1 (1965), pp. 25–8.

186   For ATBAT see Marion Tournon Branly, 'History of ATBAT and its Influence on French Architecture' in *Architectural Design* 35, no. 1 (1965): pp. 20–4. 'Largely unpaid': Tim Benton in his account of the Marseilles Unité, *Le Corbusier Architect of the Century* (London, 1987), p. 222.

187   This account of the Marseilles Unité follows Sbriglio, *op. cit.*, especially pp. 45–50 and 121–43. A list of the main architects and engineers who participated is given on pp. 47–8.

188   Quoted *ibid.*, p. 122.

189   ATBAT-Afrique was particularly the vehicle of Georges Candilis and Shadrach Woods, both of whom had worked on the Marseilles Unité.

4   THE BRIDGE

1   Ivo Andric, *The Bridge on the Drina* (London, 1959). The original version was published in 1947.

2   Michael Sells, 'The Saddest Eyes I've Seen', and 'Visegrad and Andric', 1996, on www.haverford.edu/relg/sells/postings; *The Guardian*, 11 August 2005, G2, pp. 1–3.

3   Piero Gazzola, *Ponti romani* (Florence, 1963), vol. 2, p. 12.

4   Colin O'Connor, *Roman Bridges* (Cambridge, 1993), pp. 38–42.

5   Gazzola, *op. cit.*, p. 16.

6   H. J. Hopkins, *A Span of Bridges* (Newton Abbot, 1970), pp. 33, 35.

7   For the Blenheim bridge see Howard Colvin and Alistair Rowan, 'The Grand Bridge in Blenheim Park', in John Bold and Edward Chaney (eds.), *English Architecture Public and Private: Essays for Kerry Downes* (London and Rio Grande, 1993), pp. 159–75. The pretty Pont des Belles Fontaines on the Paris–Fontainebleau road at Juvissy (1728) is the French equivalent.

8   For details and nomenclature of the bridges in Rome see Samuel Ball Plattner and Thomas Ashby, *A Topographical Dictionary of Ancient Rome* (London, 1929), and Eva Margareta Steinby (ed.), *Lexicon Topographicum Urbis Romae*, vol. 4 (Rome, 1999).

9   Mark S. Weil, *The History and Decoration of the Ponte S. Angelo* (University Park, 1974).

10   For imaginative reconstructions of the Castel and Ponte Sant'Angelo between 1450 and 1700 see George Kunoth, *Die historische Architektur Fischers von Erlach* (Düsseldorf, 1956).

11   L. B. Alberti, *On the Art of Building in Ten Books*, trans. Joseph Rykwert, Neil Leach and Robert Tavernor (Cambridge, Mass., 1988), pp. 262–3.

12   For the Rialto Bridge, besides generous advice from Deborah Howard, I have consulted Bruce Boucher, *Andrea Palladio: The Architect in his Times* (New York, 1994), pp. 216ff.; Donatella Calabi and Paolo Morachiello, *Rialto: le fabbriche e il Ponte 1514–1591* (Turin, 1987), pp. 173–300; Deborah Howard, *Jacopo Sansovino* (2nd edn, London, 1987), pp 52ff.; Giangiorgio Zorzi, *Le chiese e i ponti di Andrea Palladio* (Milan, 1967), pp. 223–64; William Barclay Parsons, *Engineers and Engineering in the Renaissance* (Cambridge, Mass., 1939), pp. 507–38; Manfredo Tafuri, *Venice and the Renaissance* (Cambridge, Mass., 1989), pp. 137, 161–6, 179.

13   Frank Brangwyn and Walter Shaw Sparrow, *A Book of Bridges* (London, 1920).

14   See Peter Murray and Mary Anne Stevens (eds.), *Living Bridges: The Inhabited Bridge, Past, Present and Future* (London, 1996).

15   Murray and Stevens, *op. cit.*, pp. 58–61; Parsons, *op. cit.*, pp. 553–77; Jocelyne Van Deputte, *Ponts de Paris* (Paris, 1994), pp. 114–37.

16   Benedetto Agnello, quoted in Calabi and Morachiello, *op. cit.*, p. 221.

17   To come.

18   Parsons, *op. cit.*, pp. 488, 510; for other interpretations of the timber bridges, see Boucher, *op. cit.*, pp. 205–10, and Donata Battilotti, 'I ponti vicentini di Palladio della teoria al cantiere', in Claudia Conforti and Andrew Hopkins (eds.), *Architettura e tecnologia: acque, tecniche e cantieri nell'architettura rinascimentale e barocca* (Rome, 2002), pp. 108–25.

19   Boucher, *op. cit.*, p. 228.

20   I have mainly used Jonathan Scott, *Piranesi* (London, 1975); Nicholas Penny, *Piranesi* (London, 1978).

21   Quoted in Scott, *op. cit.*, p. 49.

22   Manfredo Tafuri, *Venice and the Renaissance* (Cambridge, Mass., 1989), p. 137.

23   *Ibid.*, p. 179.

24   E.g. Morachiello in Calabi and Morachiello, *op. cit.*, pp. 288–300, with the ambivalent conclusion that da Ponte had 'repudiated the principle of authority in architecture'.

25   John Ruskin, *The Stones of Venice*, Book 3, Venetian index *sub* Rialto.

26   For the Ponte Santa Trinità I have used Paolo Paoletti, *Il Ponte a Santa Trinità: com'era e dov'era* (Florence, 1987); Michael Kiene, *Bartolomeo Ammannati* (Milan, 1995), pp. 124–8; Daniela Lamberini, 'Bartolomeo Ammannati: tecniche ingegneristiche e macchine da cantiere', in Niccolò Rosselli Del Turco and Federica Salvi, *Bartolomeo Ammannati: Scultore e Architetto 1511–1592* (Florence, 1995), pp. 349–55; Amedeo Belluzzi, 'Il cantiere cinquecentesco del ponte a Santa Trinità', in Conforti and Hopkins, *op. cit.*, pp. 28–43; Parsons, *op. cit.*, pp. 539–46.

27   There is also a theory that Michelangelo made a sketch for the arches which Ammannati carried out: see Kiene, *op. cit.*

28   Belluzzi, *op. cit.*,: 'quasi del continuo' cited on p. 30.

29   For French masonry bridges the outstanding and indispensable authority remains F. de Dartein, *Etudes sur les ponts en pierre remarquables par leur décoration antérieurs au XIXe siè-*

cle (4 vols., Paris, 1907–12).

30   For the Pont Royal see Dartein, *op. cit.*, vol. 2 (1907), pp. 77–86; Deputte, *op. cit.*, pp. 158–63, and Bernard Marrey, *Les ponts modernes 18 et 19 siècles* (Paris, 1990), p. 27.

31   Quoted in Deputte, *op. cit.*, p. 163.

32   It is interesting that Arnold de Ville and Rennequin Sualem, the engineer and contractor for the famous Machine de Marly which raised the water for Versailles, of 1681–4 and therefore almost contemporary with the Pont Royal, also came from the Southern Netherlands – in their case Liège.

33   According to Gauthey, Romain was brought in only after difficulties with the first pier at the Pont Royal, but Dartein (p. 77) says the evidence is that he was paid from the start. He succeeded Libéral Bruand in control of the generality of Paris for the Ponts et Chaussées from 1695, but was sent elsewhere on 'missions spéciales', perhaps to do with hydraulic works. He died in 1735. No design work after Maastricht seems attributed to him.

34   On Moulins see Marrey, *op. cit.*, pp. 46ff., and Pierre Bourget and Georges Cattaui, *Jules Hardouin Mansart* (Paris, 1956), pp. 127–9.

35   For Blois, see Dartein, *op. cit.*, pp. 91–106; and for other bridges by Jacques V. Gabriel see Michel Gallet and Yves Bottineau, *Les Gabriel* (Paris, 1982), pp. 44–7.

36   For the confusing de Régemorte (or Regemortes) family, see Jean Petot, *Histoire de l'administration des Ponts et Chaussées 1599–1815* (Paris, 1958), pp. 128–9, and Janis Langins, *Conserving the Enlightenment: French Military Engineering from Vauban to the Revolution* (Cambridge, Mass., 2004), p. 90. Jean-Baptiste, the head of the family, is said to have had a Dutch background. He was the general contractor for Vauban at Neuf-Brisach, served as an engineer in the War of the Spanish Succession in Alsace and also built roads there. He was contrôleur of the bridge-works at Blois, and inspector of the 'turcies et levées' of the Loire above Orléans from 1720.

37   For these bridges see Dartein, *op. cit.*, pp. 117–32; Marrey, *op. cit.*, pp. 37–52; Antoine Picon, *L'Invention de l'ingénieur moderne* (Paris, 1992), pp. 65–70; Antoine Picon, *French Architects and Engineers in the Age of Enlightenment* (London, 1991), pp. 156–9. For an earlier English analysis see Edward Cresy, *An Encyclopaedia of Civil Engineering* (London, 1847), pp. 250ff.

38   For an overview of Perronet as a bridge-builder see Dartein, *op. cit.*, pp. 1–53; also Picon, *op. cit.*, pp. 69–80, and his *French Architects and Engineers in the Age of Enlightenment* (London, 1991), pp. 160–8. His principal book is J. R. Perronet, *Description des projets et de la construction des ponts* (Paris, 3 vols., 1782–3).

39   R. D. Middleton, 'The Abbé de Cordemoy and the Graeco-Gothic Ideal: A Prelude to Romantic Classicism', Part II, *Journal of the Warburg and Courtauld Institutes* 26 (1963), p. 111.

40   Dartein, *op. cit.*, pp. 7–8.

41   Quoted in Mae Mathieu, *Pierre Patte: sa vie et son oeuvre* (Paris, 1940), pp. 400–1.

42   W. Hosking, *Essay and Treatises on the Practice and Architecture of Bridges* (London, 1843), p. 33.

43   Quoted in Dartein, *op. cit.*, p. 19 (memorandum of March 1774).

44   For the Pont de la Concorde, see Dartein, *op. cit.*, vol. 2, pp. 20–3, 197–244; Marrey, *op. cit.*, pp. 64–7; Deputte, *op. cit.*, pp. 168–9. Picon, *L'Invention . . . .*, pp. 73–6.

45   Sir John Soane (ed. David Watkin), *The Royal Academy Lectures* (Cambridge, 2000), p. 139.

46   Quoted in Dartein, *op. cit.*, p. 205.

47   Etienne-Louis Boullée, *Architecture: Essai sur l'art*, ed. J. M. Pérouse de Montclos (Paris, 1968), pp. 145–6.

48   For Gauthey see Dartein, *op. cit.*, vol. 4 (1909); and Anne Coste, Antoine Picon and Francis Sidot (eds.), *Un ingénieur des lumières: Émiland-Marie Gauthey* (Paris, 1993), notably the essay by Picon, pp. 207–36.

49   Dartein, *op. cit.*, p. 49.

50   Christine Lamarre in Coste, Picon and Sidot, *op. cit.*, p. 36, discussing the Pont sur la Thalie by Gauthey's predecessor as chief engineer in Burgundy, Thomas Dumorey.

51   Paul Séjourné, *Grandes voûtes*, vol. 5 (Bourges, 1914), p. 113.

52   Christopher Chalklin, *English Counties and Public Building, 1650–1830* (London and Rio Grande, 1998), Chapters 5 & 6.

53   See R. J. B. Walker, *Old Westminster Bridge* (Newton Abbot, 1978); Ted Ruddock, *Arch Bridges and their Builders, 1735–1835* (Cambridge, 1979), pp. 3–18.

54   Howard Colvin and Alistair Rowan, 'The Grand Bridge in Blenheim Park', in John Bold and Edward Chaney (eds.), *English Architecture Public and Private: Essays for Kerry Downes* (London and Rio Grande, 1993), pp. 159–75.

55   Nicholas Hawksmoor, *A Short Historical Account of London Bridge: with a Proposition for a new Stone Bridge at Westminster* (London, 1736): quotations from pp. 36 and 46.

56   ODNB (Roger Bowdler) and BDCE (Ted Ruddock).

57   Quoted in Walker, *op. cit.*, p. 68.

58   Walker, *op. cit.*, pp. 168, 238–41.

59   *Observations on Bridge Building, and the Several Plans Offered for a New Bridge* (London, 1760), p. 29.

60   Ruddock, *op. cit.*, pp. 64–79; Roger Woodley, 'Robert Mylne, 1733–1811: The Bridge between Architecture and Engineering', Ph. D., University of London, 1998; Mylne in BDCE (Ruddock).

61   *Observations . . .* , p. 39. This pamphlet has often been assumed to be Mylne's, but Woodley, *op. cit.*, does not accept the attribution.

62   *Ibid.*, p. 12.

63   *Ibid.*, pp. 24–5.

64   E.g. *Observations sur Londre et ses environs, par un Athéronome de Berne* (Paris, 1777), pp. 6–7; Thomas H. Shepherd and James Elmes, *Metropolitan Improvements of London* (London, 1847 edn.), p. 164.

65   Ruddock, *op. cit.*, p. 91.

66   *Ibid*, pp. 178–84.

67   The seductive hypothesis that the young C. R. Cockerell, who later married Rennie's daughter, may have helped with the detailing will not wash. He was away on his Grand Tour during the years Waterloo Bridge was designed and built. But there were certainly early connections between the Cockerell and Rennie families.

68   Shepherd and Elmes, *op. cit.*, p. 158.

69   Thomas Telford and Alexander Nimmo, 'Bridges', in *Edinburgh Encyclopaedia* (Edinburgh, 1830), p. 521. The opinion is confirmed by Southey's notes on his Scottish tour with Telford in 1819: 'I learnt in Spain to admire straight bridges; but T. thinks there always ought to be some curve, that the rain water may run off, and because he would have the outline look like the segment of a larger circle, resting on the abutments.' Robert Southey, *Journal of a Tour in Scotland in 1819* (London, 1929), p. 116.

70   For Jolliffe and Banks see BDCE *sub* Sir Edward Banks (David Brooke).

71   A. W. Skempton and Esther Clark Wright, 'Early Members of the Smeatonian Society of Civil Engineers', *Transactions of the Newcomen Society* 44 (1971–2), p. 36.

72   See the lines on the Caledonian Canal, reprinted as the epigraph to L. T. C. Rolt's *Thomas Telford* (London, 1979 edn), and for the record of their travels, Robert Southey, *Journal of a Tour in Scotland in 1819* (London, 1929).

73   For the poetry and utterances of Pope, Steinman and Roebling see Alan Trachtenberg, *Brooklyn Bridge: Fact and Symbol* (Chicago, 1965).

74   David Billington, *The Tower and the Bridge: The New Art of Structural Engineering* (Princeton, 1983).

75   Tom F. Peters, *Building the Nineteenth Century* (Cambridge, Mass., 1996), pp. 178, 413.

76   For Finley and the early suspension bridge see Eda Kranakis, *Constructing A Bridge* (Cambridge, Mass., 1997), Chapters 1–3.

77   Quoted in Kranakis, *op. cit.*, pp. 36–7.

78   BDCE *sub* Samuel Brown (Tom Day). Brown's original idea was to use his chains for ships' rigging but this was not taken up. For his bridges see Emory L. Kemp, 'Samuel Brown: Britain's Pioneer Suspension Bridge Builder', in A. Rupert Hall and Norman Smith (eds.), *History of Technology*, 2nd annual volume (London, 1977), pp. 1–38; and Thomas Day, 'Samuel Brown: his Influence on the Design of Suspension Bridges', *History of Technology*, 8th annual volume (London, 1983), pp. 61–90, reprinted in R. J. M. Sutherland (ed.), *Structural Iron 1750–1850* (Aldershot, 1997), pp. 181–210.

79   Rolt, *op. cit.*, pp. 131–2: and for Telford's suspension bridges, BDCE (Roland Paxton), pp. 686–7.

80   Charles Stewart Drewry, *A Memoir on Suspension Bridges* (London, 1836), gives a valuable account of most of the major British examples: Union Bridge, pp. 37–41; Menai and Conway, pp. 54–68; Brighton chain pier, pp. 69–74; Réunion (Ile de Bourbon), pp. 75–81; Hammersmith, pp. 82–8.

81   For the Seguins, see Seguin ainé, *Des ponts en fil de fer* (Paris, 1826); Michel Cotte, 'Seguin et Cie (1806–34): du négoce familial de drap à la construction du pont suspendu de Tournon-Tain', *History and Technology* 6 (2/3) (1988), pp. 95–144; Bernard Marrey, *Les ponts modernes 18 et 19 siècles* (Paris, 1990), pp. 114ff.

82   Illustrated in R. A. Paxton, 'Menai Bridge (1818–26) and its Influence on Suspension Bridge Development', TNS 49 (1977–8), p. 94, fig. 8.

83 For the Union Bridge see Drewry, *op. cit.*, pp. 37–41, and Kemp, *op. cit.*, pp. 12–16. The attribution of the masonry elements to Rennie is supplied by Charles Dupin.

84 Drewry, *op. cit.*, p. 148; and Denis Smith, 'The Works of William Tierney Clark (1783–1852)', TNS 63 (1991–2), pp. 181–207. For the Széchenyi bridge see Gyöngyvér Török (ed.), *The Széchenyi Bridge and Adam Clark* (Budapest, 1999), with many interesting essays. Besides its arch-portals this bridge also formerly boasted under the Buda approaches a fine tunnel in granite with flanking Doric columns, much influenced by Waterloo Bridge. It is illustrated in an article by the Franco-Hungarian architect Joseph Vago, *Journal of the Royal Institute of British Architects* 18 Sep. 1939, pp. 975–9.

85 Richard Beamish, *Memoir of the Life of Sir Marc Isambard Brunel* (London, 1862), pp. 178–83. A single span was for Saint Suzaine; a double span, with the A-frame towers, was for Rivière du Mât. Neither survives. The early-looking masonry towers of the present suspension bridge at Rivière de l'Est are later. Many thanks to Rosemary Boyne for her on-site investigations and report.

86 See Richard G. Carrott, *The Egyptian Revival: Its Sources, Monuments and Meaning, 1808–1858* (Berkeley, 1978), pp. 103–8 and 124. Carrott quotes *Architectural Magazine*, 1, 1834, p. 246, to the effect that Egyptian architecture was 'peculiarly applicable to engineering work, particularly for the piers, and engineering of suspension bridges'.

87 Karl Friedrich Schinkel, '*The English Journey': Journal of a Visit to France and Britain in 1826*, ed. David Bindman and Gottfried Riemann (London, 1993), p. 188. In the year that Schinkel crossed the Menai there were two serious episodes of wind damage. Further damage in 1836 and, most gravely, 1839, led to modifications of the deck. Before Menai, there had been damage to the Dryburgh Bridge in 1818 and to the Union Bridge in 1821.

88 Sergej G. Fedorov, *Wilhelm von Traitteur: ein badischer Baumeister als Neuerer in der russischen Architektur 1814–1832* (Berlin, 2000). In building elegant iron bridges in St Petersburg, Traitteur had been preceded by William Hastie, who built five such in the period 1804–16; only the Red Bridge of 1814 survives. See Anthony Cross, *By the Banks of the Neva* (Cambridge, 1997), p. 306.

89 Quoted in Fedorov, *op. cit.*, p. 142, from Traitteur's *Plans . . . des ponts en chaînes exécutés à Saint-Pétersbourg* (1825).

90 For Menai and Conway see William Alexander Provis, *An Historical and Descriptive Account of the Suspension Bridge Constructed over the Menai Strait, in North Wales: with a Brief Notice of the Conway Bridge* (London, 1828); Rolt, *op. cit.*, pp. 133–41; Paxton, *op. cit.*, pp. 87–110.

91 Drewry, *op. cit.*, p. 64.

92 John Rickman (ed.), *Life of Thomas Telford, Civil Engineer, written by Himself* (London, 1838), p. 233.

93 For the Pont des Invalides, see Kranakis, *op. cit.*, Chapters 4–6; and Antoine Picon, 'Navier and the Introduction of Suspension Bridges in France', *Construction History* 4 (1988), pp. 21–35, reprinted in Sutherland *op. cit.*, pp. 211–24.

94 Quoted in Kranakis, *op. cit.*, p. 170.

95 Stephen G. Buonopane and David P. Billington, 'Theory and History of Suspension Bridge Design from 1823 to 1940', *American Society of Civil Engineers (Structural Division), Journal of Structural Engineering* 119 no. 3 (March 1993), p. 957. The typical sag-to-span on a modern, tall-towered suspension bridge is 1 : 10, say the authors.

96 For French suspension bridges between 1825 and 1850 see Picon, 'Navier', pp. 34–5; Marrey, *op. cit.*, pp. 114–35; and Dominique Barjot, 'From Tournon to Tancarville: The Contribution of Civil Engineering to Suspension Bridge Construction', *History and Technology* 6 no. 3 (1988), pp. 177–202.

97 W. Hosking, *Essay and Treatises on the Practice and Architecture of Bridges* (London, 1843), pp. 45, 246–7.

98 For Clifton, see L. T. C. Rolt, *Isambard Kingdom Brunel* (London, 1957), pp. 51–64, 103–4, 314–15; Geoffrey Body, *Clifton Suspension Bridge* (Bradford on Avon), 1976; Adrian Vaughan, *Isambard Kingdom Brunel: Engineering Knight-Errant* (London, 1991), pp. 37ff., 270.

99 Quoted in Rolt, *op. cit.*, p. 57.

100 Bazaine's design of 1825 is illustrated in Fedorov, *op. cit.*, pp. 170–1. It was promptly published in Paris and so could easily have been known to the Brunels. Marc Brunel himself had been commissioned by Alexander I to send designs for a Neva bridge in about 1818, but they were eventually rejected, ostensibly for want of money. One of his designs, for a trussed timber arch with central opening, is illustrated in Beamish, *op. cit.*, pp. 158–64.

101 *Ibid.*, p. 58.

102 The quotation is from Elizabeth B. Mock, *The Architecture of Bridges* (New York, 1949), pp. 56–7.

103 For Hungerford, see Vaughan, *op. cit.*, pp. 57, 146–7; Gavin Stamp, *The Changing Metropolis* (Harmondsworth, 1984), pp. 155–7.

104 *London Observed: A Selection from the Letters of Jacob Burckhardt* (Hayloft Press, Birmingham, 1997), p. 9.

105 For Britannia Bridge: Peters, *op. cit.*, pp. 159–78; Edwin Clark, *The Britannia and Conway Tubular Bridges* (London, 2 vols., 1850); G. Drysdale Dempsey, *Tubular and Other Iron Girder Bridges* (London, 1864); James Sutherland, 'Iron Railway Bridges', in Michael R. Bailey (ed.), *Robert Stephenson – The Eminent Engineer* (Aldershot, 2003), pp. 301–35; and for Thompson's other collaborations with Stephenson including further bridges, Mike Chrimes and Robert Thomas, 'Railway Building', *ibid.*, pp. 284–5.

106 Peters, *op. cit.*, p. 168.

107 Montgomery Schuyler, *American Architecture and Other Writings*, ed. W. H. Jordy and R. Coe (Cambridge, Mass., 1961), vol. 2, p. 358 (original of 1900). Nevertheless some critics have found the Britannia Bridge beautiful: see e.g. the remarks by Joseph Husband in the interdisciplinary debate on the aesthetics of bridges reported in *Minutes of the Proceedings of the Institution of Civil Engineers*, 145 (1901), pp. 134–244: 'a greater success has seldom been achieved' (p. 165). In the discussion that followed, several architects and engineers agreed with that view. One who did not was Benjamin Baker, the designer of the Forth Bridge (pp. 206–9).

108 Among many studies see Alan Trachtenberg, *Brooklyn Bridge: Fact and Symbol* (Chicago, 1965); David McCullough, *The Great Bridge* (New York, 1972); Daniel L. Schodek, *Landmarks in American Civil Engineering* (Cambridge, Mass., 1987), pp. 132–42.

109 An early romantic reference to the Brooklyn Bridge occurs in Chapter 13 of Helen Hunt Jackson's Californian novel *Ramona*, written in a New York hotel just after the bridge opened, in 1883–4. 'There had been no crises of incident, or marked moments of experience such as in Felipe's imaginations of love were essential to the fulness of it growth. This is a common mistake on the part of those who have never felt love's true bonds. Once in those chains, one perceives that they are not of the sort full forged in a day. They are made as the great iron cables are made, on which bridges are swung across the widest water-channels, – not of single huge rods or bars, which would be stronger, perhaps, to look at; but of myriads of the finest wires, each one by itself so fine, so frail, it would barely hold a child's kite in the wind: by hundreds, hundreds of thousands of such, twisted, re-twisted together, are made the mighty cables, which do not any more swerve from their place in the air, under the weight and jar of the ceaseless traffic of two cities, than the solid earth swerves under the same ceaseless weight and jar. Such cables do not break.' I am grateful to my daughter Lily for making this fine novel known to me and giving me a copy of McCulloch's *The Great Bridge*.

110 Hamilton Schuyler, *The Roeblings: A Century of Engineers, Bridge-Builders, and Industrialists* (Princeton, 1931), pp. 12–13; Trachtenberg, *op. cit.*, p. 41; McCulloch, *op. cit.*, p. 42. The formulation 'Royal Polytechnic Institute' goes back to the account of Roebling's life given in Charles B. Stuart, *Lives and Works of Civil and Military Engineers in America* (New York, 1871), pp. 301–26; it refers to the later title of the Bauakademie after reforms in 1849. Hamilton Schuyler says that Roebling specialized there in foundations and bridge-building under J. F. W. Dietleyn, and dyke-construction under the better-known J. A. Eytelwein, head of that side of the Bauakademie from 1824.

111 Quotations from the paragraph by the family friend cited in Hamilton Schuyler, *op. cit.*, pp. 12–13, repeated by Trachtenberg and McCulloch.

112 *Ibid.*, p. 54, in a letter of 1840 soliciting employment on Ellet's Schuylkill bridge: for Roebling's interest in a suspension bridge at Bamberg, *ibid.*, pp. 13–14. Roebling's early career is now covered by Andreas Kahlow, 'Johann August Röbling (1806–1869): Early Projects in Context', in *Proceedings of the Second International Congress on Construction History* (Cambridge, 2006), vol. 2, pp. 1755–76.

113 Gene D. Lewis, *Charles Ellet junior: The Engineer as Individualist* (Urbana, 1968), p. 17.

114 Hamilton Schuyler, *op. cit.*, p. 50. This seems to have been the wire rope developed for

Government mines at Clausthal in the 1830s by Wilhelm Albert and freely communicated by him for 'the good of humanity': it led also to submarine cables. See [Thomas Constable], *Memoir of Lewis D. B. Gordon, F.R.S.E.* (Edinburgh, 1877), pp. 46–7; and Gillian Cookson and Colin A. Hempstead, *A Victorian Scientist: Fleeming Jenkin and the Birth of Electrical Engineering* (Aldershot, 2000), pp. 31–2.

115 Quoted in Hamilton Schuyler, *op. cit.*, p. 61.

116 Schodek, *op. cit.*, pp. 84–9.

117 *Journal of the Franklin Institute* 10 (1865), pp. 306–9, quoted in www.buildingtechnology.com/bcba/bridges/articles/PA2-07 (Bruce S. Cridelbaugh). See also the description by Lindenthal in *Transactions of the American Society of Civil Engineers*, 12 (Sep. 1883), pp. 354–5.

118 Letter of Roebling, June 1860, quoted in Hamilton Schuyler, *op. cit.*, p. 118.

119 See Göran Werner, 'John August Roebling – The Niagara Railway Suspension Bridge', in *Proceedings of the Second International Congress on Construction History* (Cambridge, 2006), vol. 3, pp. 3315–31.

120 For Cincinnati-Covington (the John A. Roebling Bridge) see Schodek, *op. cit.*, pp. 115–18.

121 As Lewis Mumford put it: 'If the architectural elements of the massive piers have perhaps too much the bare quality of engineering, if the pointed arches meet esthetic betrayal in the flat solidity of the cornices, if, in short, the masonry does not sing as Richardson alone perhaps could have made it sing, the steel work itself makes up for this, by the architectural beauty of its pattern; so that beyond any other aspect of New York, I think, the Brooklyn Bridge has been a source of joy and inspiration to the artist.' *Sticks and Stones* (New York, 1955 edn), pp. 51–2. But a few like the English architect W. D. Caroe have felt that the crudity of the towers counteracted the beauty of the cables' curve and made the bridge's architectural qualities 'irredeemable': *Minutes of the Proceedings of the Institution of Civil Engineers*, 145 (1901), p. 180.

122 Eidlitz had contacts on the Brooklyn Bridge Design Review Committee of 1869. His design of 1871 for an unexecuted terminal for the Manhattan end of the Viaduct Railway which crossed the bridge took a decidedly Gothic stance. I am grateful to Kathryn Elizabeth Holliday, author of the doctorate 'Leopold Eidlitz and the Architecture of Nineteenth Century America' (University of Texas at Austin, 2003), for this information.

123 Montgomery Schuyler, *American Architecture and Other Writings*, ed. W. H. Jordy and R. Coe (Cambridge, Mass., 1961), vol. 1, pp. 49ff. (editors' introduction), and 154–5 (Schuyler's account of the Eidlitz offer).

124 *Ibid.*, vol. 2, pp. 331–44: original in *Harpers Weekly*, 26 May 1883.

125 David P. Billington, 'History and Esthetics in Suspension Bridges', *Proceedings of the American Society of Civil Engineers, Journal of the Structural Division* 103, ST8 (Aug. 1977), pp. 1655–72, seems to me unfair on Schuyler. He attacks Schuyler's criticism of the Brooklyn Bridge towers on the grounds that he was 'essentially a literary man'

(p. 1656) who did not appreciate that the Roeblings' towers had to be heavy to sink the caissons beneath and stiff to stabilize the stays attached to them. But Schuyler criticized the towers for their clumsiness of detail, not of material, weight or form, apart from the issue of cross-bracing. Later in the article (p. 1663), Billington uses the same argument against an engineer, when he criticizes Ammann for failing to distinguish between masonry towers as used at Menai and at Brooklyn, when he invoked both to support masonry cladding for his George Washington Bridge: 'the latter [Brooklyn Bridge] represented an engineering solution by using the heavy masonry to push the deep caissons into the resistant bed of the East River as well as to serve as stiff supports for the stays, whereas the early English towers were simply built before metal towers had ever been tried.' This last point is not strictly correct.

126 For Lindenthal I have relied largely on Henry Petroski, *Engineers of Dreams* (New York, 1995), pp. 122–216.

127 According to Gregory F. Gilmartin, *Shaping the City: New York and the Municipal Art Society* (New York, 1995), p. 124, Lindenthal 'lied on his résumé when he claimed to have studied engineering at the Polytechnic Institutes of Brno and Vienna. (His brother had.)'

128 For the Smithfield Street Bridge see G. Lindenthal, 'Rebuilding of the Monongahela Bridge at Pittsburgh, Pa.', in *Transactions of the American Society of Civil Engineers* 12 (Sep. 1883), pp. 353–92; Petroski, *op. cit.*, pp. 125–30; Schodek, *op. cit.*, pp. 129–31; David P. Billington, *The Tower and the Bridge* (Princeton, 1983), pp. 123–4.

129 From an article in *Engineering News*, 4 Feb. 1888, pp. 78–9, quoted in Petroski, *op. cit.*, pp. 139–40.

130 A controversial view put to the author in the Museum Tavern by Irénée Scalbert.

131 See *Minutes of the Proceedings of the Institution of Civil Engineers*, 145 (1901), pp. 134–244. The original, rather banal paper by Joseph Hubbard on 'The Aesthetic Treatment of Bridge Structures', elicited a confused discussion most notable for the fact that architects and engineers took an equal part in it. The frequent reference to the Pont Alexandre III indicates how large a part it played in stimulating debate on bridge design at this date.

132 Carl W. Condit, *The Port of New York: A History of the Rail and Terminal System from the Beginnings to Pennsylvania Station* (Chicago, 1980), p. 256.

133 Thomas Hastings in Henry Grattan Tyrrell, *Artistic Bridge Design* (Chicago, 1912), pp. 1–8.

134 Henry Van Brunt in J. A. L. Waddell, *De Pontibus* (New York, 1898), pp. 43–4.

135 Montgomery Schuyler, *op. cit.*, vol. 2, p. 346.

136 For the internecine politics of the New York bridges covered in the following paragraphs, see the fascinating account in Gilmartin, *op. cit.*, pp. 21–4, 76, 121–32.

137 For Williamsburg Bridge, see Petroski, *op. cit.*, pp. 158–64. Buck had already reconstructed parts of Roebling's Niagara Gorge Bridge and later replaced it with an arched bridge.

138 Robert A. M. Stern, Gary Gilmartin and John Massengale, *New York 1900: Metropolitan*

Architecture and Urbanism 1890–1915 (New York, 1983), p. 28.

139 Montgomery Schuyler, 'Bridges and the Art Commission', *Architectural Record*, Dec. 1907, pp. 469–75. For Hornbostel, see Walter C. Kidney, *Henry Hornbostel: An Architect's Master Touch* (Pittsburgh, 2002), pp. 20–7. According to Gilmartin, p. 76, he got the job through the recommendation of A. D. F. Hamlin of Columbia. Hornbostel added some touches to the closing stages of the Williamsburg bridge but was too late to influence its design.

140 For Manhattan and Queensboro Bridges, see Petroski, *op. cit.*, pp. 165–82; Gilmartin, *op. cit.*, pp. 121–7, 132; and Stern, Gilmartin and Massengale, *op. cit.*, pp. 51–4. The normally comprehensive Petroski does not mention the involvement of either Buck or Hornbostel on Queensboro Bridge.

141 Montgomery Schuyler, 'New York Bridges', *Architectural Record*, Oct. 1905, pp. 259–60. Schuyler also remarks here that the Paris-trained Hornbostel had revised the outside spans of the Queensboro Bridge on the lines of Résal's recent Pont Mirabeau.

142 Kidney, *op. cit.*, p. 23.

143 Gilmartin, *op. cit.*, p. 123.

144 *Ibid.*, p. 124.

145 For Hell Gate Bridge, see Petroski, *op. cit.*, pp. 182–91; Condit, *op. cit.*, pp. 336–41; Billington, *The Tower and the Bridge* (Princeton, 1983), pp. 125–8.

146 Quoted in Condit, *op. cit.*, p. 336.

147 For Ammann, see the essay in Petroski, *op. cit.*, pp. 217–319; and Darl Rastorfer, *Six Bridges: The Legacy of Othmar H. Ammann* (New Haven, 2000).

148 For the background to the George Washington Bridge, see Jameson W. Doig, *Empire on the Hudson: Entrepreneurial Vision and Political Power at the Port of New York Authority* (New York, 2001), pp. 120–57; also the essays by Jameson W. Doig, 'Joining New York City to the Greater Metropolis: The Port Authority as Visionary, Target of Opportunity, and Opportunist', in David Ward and Olivier Zunz (eds.), *The Landscape of Modernity: New York City 1900–1940* (Baltimore, 1992), pp. 76–105; and by Jameson W. Doig and David P. Billington, 'Ammann's First Bridge: A Study in Engineering, Politics, and Entrepreneurial Behavior', *Technology and Culture* 35, no. 3 (July 1994) pp. 537–70.

149 Quoted in Doig, *Empire . . .* , p. 128 (also in Doig and Billington, *op. cit.*, p. 548, Petroski, *op. cit.*, p. 251, and Rastorfer, *op. cit.*, p. 15).

150 For the Delaware River Bridge (engineer, Ralph Modjeski), see Petroski, *op. cit.*, pp. 201–6; H. D. Ebelein, 'The Delaware River Bridge', *Architectural Record* 61 (Jan. 1927), pp. 1–12; and Elizabeth Greenwell Grossman, *The Civic Architecture of Paul Cret* (Cambridge, 1996), pp. 157–8. For Cret's musings on collaboration, see Paul Philippe Cret, 'The Architect as Collaborator with the Engineer', *Architectural Forum* 49 (July 1928), pp. 97–104.

151 Quoted in Doig, *Empire . . .* , p. 139.

152 From a paper of 1933 to the American Society of Civil Engineers quoted by Billington, 'History and Esthetics', p. 1662.

153 Quoted in an essay by Mary Beth Betts in Margaret Heilbrun (ed.), *Inventing the Skyline:*

The Architecture of Cass Gilbert (New York, 2000), p. 156.

154   From *When the Cathedrals were White* (1936), quoted in Rastorfer, *op. cit.*, pp. 41–3. The book derived from Le Corbusier's lectures during his American trip of 1935, which he often started with an evocation of the George Washington Bridge: see Mardges Bacon, *Le Corbusier in America: Travels in the Land of the Timid* (Cambridge, Mass., 2001), pp. 75, 149, 167, 265–9.

155   For Ammann's reactions, see his paper of 1933 to the American Society of Civil Engineers quoted by Billington, 'History and Esthetics', p. 1664; and for the inaccurate article in the *New Yorker*, 2 June 1934, claiming that he hoped masonry would never be added, *ibid.*, p. 1670.

156   For Embury, see Rastorfer, *op. cit.*, pp. 59, 105, 109, 165, 169, 173, and Robert A. Caro, *The Power Broker: Robert Moses and the Fate of New York* (New York, 1975), pp. 365, 371–2, 384, 486–7, 665, 676, 1085.

157   Quoted in Rastorfer, *op. cit.*, p. 125.

158   For Tacoma Narrows and the deepening of the decks, see Petroski, *op. cit.*, pp. 291–308; Rastorfer, *op. cit.*, pp. 32–4; F. Leonhardt, *Brücken/Bridges* (London, 1982), p. 290; Buonopane and Billington, *op. cit.*, pp. 972–3; and Billington, 'History and Esthetics', p. 1665–7.

159   Quoted in David P. Billington, *Robert Maillart: Builder, Designer and Artist* (New York, 1997), p. 23.

160   Billington's final synthesis is *Robert Maillart: Builder, Designer and Artist* (New York, 1997); see also his *Robert Maillart's Bridges: The Art of Engineering* (Princeton, 1979); and his *Robert Maillart and the Art of Reinforced Concrete* (New York, 1990).

161   For Stauffacher, see Billington (1997), pp. 13–14.

162   Perhaps most to the point was the Pont Coulovrenière, Geneva, of 1895–6, in concrete with masonry cladding, designed by the municipal engineer Constant Butticaz with 'decoration' by the local architect Bouvier: see Paul Séjourné, *Grandes voûtes*, vol. 4 (Bourges, 1913), pp. 81–6.

163   For Zuoz and Tavanasa, see Billington (1997), pp, 17–22 and 38–40.

164   Quoted in Billington, *ibid.*, p. 17.

165   *Ibid.*, p. 41.

166   *Ibid.*, pp. 58–69.

167   Leonhardt, *op. cit.*, p. 216.

168   Jose A. Fernandez Ordóñez, *Eugène Freyssinet* (Barcelona, 1979 edn), p. 88.

169   From an essay of 1936 entitled 'Moyens mécaniques, émotion esthétique', published among the selection of Freyssinet's writings edited by Henri Lemoine and Pierre Xercavins as Eugène Freyssinet, *Un amour sans limites* (Paris, 1993), pp. 147–8. A translation of the essay appears with a short introduction by the present author in 'Eugène Freyssinet on the Sublime'. in *arq* (*Architectural Research Quarterly*) 5, no. 3 (2001), pp. 249–53.

170   For the bridges referred to in this paragraph see Fernandez Ordóñez, *op. cit.*, pp. 37–89.

171   Séjourné, *op. cit.*, vol. 2 (Bourges, 1913), pp. 135–44. The other bridges are at Castelet on the Ariège, and Antoinette on the Agoût.

172   Pont Adolphe: *ibid.*, vol. 2, pp. 60–82; Pont Catalan (des Amidonniers): vol. 1, pp.

188–207. A third notable Séjourné bridge was the Pont Sidi Rached at Constantine, Algeria (1908–12): vol. 2, pp. 107–14.

173   *Ibid.*, vol. 5 (1914), pp. 67–9.

174   *Ibid.*, vol. 1, p. 207.

175   Quotations in this paragraph are from *ibid.*, vol. 5, pp. 98–100.

176   But Séjourné's range, unlike Dartein's, is international. His foreign coverage, though uneven, is at its best on Switzerland and Austria.

177   For Villeneuve-sur-Lot and Tonneins, see Fernandez Ordóñez, *op. cit.*, pp. 94–104, 302–15.

178   Séjourné, *op. cit.*, vol. 6, pp. 211–77.

179   Fernandez Ordóñez, *op. cit.*, p. 104.

180   Quoted in Billington (1997), p. 216.

181   See e.g. Gilmore D. Clarke, 'The Parkway Idea', in W. Brewster Snow (ed.), *The Highway and the Landscape* (New Brunswick, 1959), pp. 33–55; James Drake, H. L. Yeadon and D. I. Evans, *Motorways* (London, 1969), pp. 30–4; Geoffrey Hindley, *A History of Roads* (London, 1971), pp. 114–76; Christopher Finch, *Highways to Heaven: The Auto Biography of America* (New York, 1992), pp. 77–82; Bruce Radde, *The Merritt Parkway* (New Haven, 1993), pp. 8–11.

182   See Radde, *op. cit.*, pp. 40–67, for an extended analysis of the styling of the Art Deco-ish bridges on the Merritt Parkway, 'uneven in their pursuit of delight', by George Dunkelberger, 'a trained architect'.

183   For the autobahns and their bridges, see Leonhardt, *op. cit.*, passim, esp. p. 10; Christoph Hölz, 'Verkehrsbauten', in Winfried Nerdinger, *Bauen im Nationalsozialismus: Bayern 1933–1945* (Munich, 1993), pp. 54–97; Iain Boyd White in *Art and Power: Europe under the Dictators* (Hayward Gallery exhibition catalogue, London, 1995), pp. 267–9; Frederic Spotts, *Hitler and the Power of Aesthetics* (London, 2002), pp. 386–95. For the role of Bonatz see Roland May, 'Paul Bonatz and the Search for an Art-Form for Motorway Bridges', in *Proceedings of the Second International Congress on Construction History* (Cambridge, 2006), vol. 2, pp. 2139–58; also *Paul Bonatz 1877–1956* (Stuttgarter Beiträge, Heft 13, 1977), pp. 29–30.

184   For the Italian background, see Rees Jeffreys, *The King's Highway* (London, 1949), pp. 123–7; Hindley, *op. cit.*, p. 27; George Charlesworth, *A History of British Motorways* (London, 1984), pp. 11–13.

185   Quoted in Elizabeth B. Mock, *The Architecture of Bridges* (New York, 1949), p. 7.

186   Harry Tour, a TVA 'architectural engineer', quoted in Walter L. Creese, *TVA's Public Planning: The Vision, The Reality* (Knoxville, 1990), pp. 177–8. Creese's Chapter 4, 'The Imagery of Structure Triumphant' (pp. 147–238), discusses the uses to which professionals were put in the TVA. As in the autobahn programme, engineers were dominant; architects were employed alongside them for styling and propaganda. Their purpose was to create a fresh and independent image from that of the excellent previous dams and ancillary buildings, designed by the US Corps of Engineers and the Bureau of Reclamation (which built the first important 'architectural' dam project of the Reconstruction era at Boulder on the

Colorado River in 1931–6). Roland Wank, the TVA chief architect appointed in 1933, was not a pure modernist. In the early TVA projects, his architects largely streamlined what the engineers had designed. After 1937, with the support of Albert Kahn as consultant architect, there was more give and take. Tour recalled that when a new dam and ancillary buildings were contemplated, one of the designers was deputed 'to sit in the conference with the engineers and the site planners and others and get their reaction to what kind of a project we were going to have, how big it was going to be and what its physical size was going to be and all that. Then he was the one that would sit down and make the initial sketches leading up to the adoption of the final design and work with the engineers in the evolution of this design . . . It was Rudolph Mock for a while. It was Roth for a while. Then it was Joe Passoneau for a while. Different ones would stay for a few years and they were probably the most important members of the staff' (*ibid.*, p. 173).

187   Leonhardt, *op. cit.*, pp. 103, 111, 135–6.

188   *Evening Standard*, 12 June 2000. This was the leading story of the day.

189   Except where indicated, this account follows Deyan Sudjic, *Blade of Light: The Story of London's Millennium Bridge* (London, 2001). A subsequent article by Sudjic, 'Quando i ponti tremano/The Bridge that Wobbled', in *Domus* 847 (April 2002), pp. 92–103, is a valuable supplement.

190   *Architects' Journal*, 6 July 2000, p. 17: 'Lord Foster sat quietly to one side'.

191   I am most grateful to Sophie Le Bourva who co-ordinated the Arups team for generous information and counsel on the Millennium Bridge,

192   Peter Murray and Mary Anne Stevens (eds.), *Living Bridges: The Inhabited Bridge, Past, Present and Future* (London, 1996).

193   Sudjic, *op. cit.*, p. 26.

194   Quoted *ibid.*, p. 32.

195   Quotations from essay by Chris Wise, 'The Birth of London's Millennium Bridge', on www.expedition-engineering.com.

196   *Ibid.*

197   Reproduced in Sudjic, *op. cit.*, p. 62.

198   *Ibid.*, p. 39.

199   See Françoise Fromonot, *Marc Mimram: Passerelle Solferino/Solferino Bridge* (Paris, 2001), with references (p. 21 n. 6) to commentaries on the difficulties in *Le Monde*, 28 and 29 May 2000. The Passerelle Solferino, sponsored by the Etablissement Public du Grand Louvre, was a classic French state-funded project involving an equilibrium between various arms of the administration which broke down when problems arose (p. 12). The diverse procurement patterns of the Millennium Bridge and the Passerelle Solferino, otherwise similar in type, timing and teething troubles, reflect eternal differences between the cultures and administrations of London and Paris.

200   For an analysis of the Millennium Bridge's problems, see the explanation by David Newland in Sudjic, *op. cit.*, pp. 88–93. They were largely caused by the unanticipated effect of simultaneous compensatory *horizontal* (i.e. side to side) footfall when the deck was crowded and there was a slight sway in

the wind – an effect exacerbated by the bridge's narrowness and tautness. The feature of simultaneous *vertical* footfall on suspension bridges had of course long been known and guarded against.

201   For the Erasmus Bridge see *De Brug/The Bridge: Geschiedenis, Architectuur en Kunst/History, Architecture and Art* (three boxed booklets, Rotterdam, 1996); and Aaron Betsky in *Ben van Berkel and Caroline Bos: UN Studio Unfold* (Rotterdam, 2002), p. 11. Details given here are mainly taken from the booklet *De Brug/The Bridge: Licht en Techniek/Light and Engineering*, produced by the Centrum Beeldende Kunst of the Nederlands Architectuurinstituut.

202   Quoted in Sudjic, *op. cit.*, p. 63.

5   RECONCILIATION

1   For Arup's career and personality, see now Peter Jones, *Ove Arup: Masterbuilder of the Twentieth Century* New Haven and (London, 2006). For his firm and its work, see *Ove Arup and Partners: 1946–1986* (London, 1986); Degenhard Sommer, Herbert Stöcher and Lutz Weisser, *Ove Arup and Partners* (Basel, 1994); Robert Thorne, 'Continuity and Invention', in David Dunster (ed.), *Arups on Engineering* (Berlin, [1999]), pp. 234–61. Other enlightening discussions may be found in John Allan, *Berthold Lubetkin: Architecture and the Tradition of Progress* (London, 1992), pp. 205–7; Peter Rice, 'Ove Arup the Man', in *An Engineer imagines* (London, 1994), pp. 67–70; and 'Ove Arup Recalled', *Architects' Journal* 23 Feb. 1995, pp. 35–50. For Arup's own writings on the architect-engineer theme see the collected articles in *Arup Journal* 20, no. 1 (Spring 1985), pp. 2–47; also *RIBA Journal* 72 (April 1965), pp. 176–83, and 73 (Aug. 1966), pp. 350–9.

2   *RIBA Journal* 72 (April 1965), p. 176.

3   *Id.*

4   Marian Bowley, *The British Building Industry: Four Studies in Response and Resistance to Change* (Cambridge, 1966), pp. 39–52: quotation from p. 51.

5   John Faber, *Oscar Faber: His Work, His Firm and Afterwards* (London, 1989), p. 6.

6   From a witty account of the Canvey Island café in *Arup Journal* 20, no. 1 (Spring 1985), p. 41 (original text of 1972).

7   Thorne in Dunster, *op. cit.*, p. 237. For an overview of the collaborative position at this time see Alejandro Bernabeu Larena, 'Origin of the Collaboration breween Engineers and Architects in Great Britain in the Thirties', in *Proceedings of the Second International Congress on Construction History* (Cambridge, 2006), vol. 1, pp. 357–77.

8   *RIBA Journal* 72 (April 1965), p. 180.

9   *RIBA Journal* 73 (Aug. 1966), pp. 350, 359.

10   Joseph Brodsky, 'In the Shadow of Dante'. The reference and quotation 'a judge of thee' are from a poem from Robert Frost's New Hampshire period.

11   Quoted in Martin Pawley, 'The Secret Life of Engineers', *Blueprint*, March 1989, p. 36.

12   Allan, *op. cit.*, pp. 50, 68, and 94 (n.17). Lubetkin told John Allan that during the design of Highpoint and other projects of the 1930s he often referred to *Der Eisenbetonbau*, a textbook of 1912 by Kersten,

his Berlin teacher. Unless he used a later edition, it must have been out of date by then.

13   *Ibid.*, p. 207.

14   *RIBA Journal* 73 (Aug. 1966), p. 353 (quoted in Allan, *op. cit.*, p. 206).

15   For an analysis of the technical innovations in the Tecton housing from Highpoint onwards see D. T. Yeomans and D. Cottam, 'An Architect/Engineer Collaboration: the Tecton/Arup Flats', *Structural Engineer* 67 (May 1989), pp. 183–8.

16   Thorne in Dunster, *op. cit.*, p. 238.

17   Quoted in Allan, *op. cit.*, pp. 442–3.

18   *Ibid.*, p. 443.

19   For analysis of the practical difficulties in building these estates, see the forthcoming *Survey of London* volume on North Clerkenwell (vol. 47).

20   Allan, *op. cit.*, p. 443. Lubetkin's random reminiscences, dictated to his daughter in old age and now in the RIBA's collections, show much retrospective bitterness towards Arup, whom he accuses of being mercenary and insincere.

21   For valuable articles by Dunican on the technology and economics of housing, see *Architects' Journal* 5 June 1952, pp. 702–9, and 28 Jan. 1954, pp. 137–41.

22   For the Brynmawr Rubber Factory see *AA Files* 10 (Autumn 1985), pp. 3–12 (John Henderson and the German domes, p. 8); Victoria Perry, *Built for a Better Future: the Brynmawr Rubber Factory* (Oxford, 1994).

23   The main sources I have used for Sydney are Michael Baume, *The Sydney Opera House Affair* (Sydney, 1967), laying out much documentation at an early stage; John Yeomans, *The Other Taj Mahal: What Happened to the Sydney Opera House* (London, 1968); Jack Zunz, 'Sydney Revisited', in *Arup Journal* 23, no. 1 (Spring 1988), pp. 2–11; Peter Rice, 'Sydney', in *An Engineer Imagines* (London, 1994), pp. 59–66; Françoise Fromonot, *Jørn Utzon – The Sydney Opera House* (Corte Madera, 1998); Philip Drew, *Utzon and the Sydney Opera House* (Annandale, 2000); Yuzo Mikami, *Utzon's Sphere: Sydney Opera House – How it was Designed and Built* (Tokyo, 2001); Richard Weston, *Utzon* (Hellerup, 2002), pp. 112–201; Peter Murray, *The Saga of Sydney Opera House* (London, 2004). For the reader seeking balance, Murray's book is the best.

24   Quoted in Baume, *op. cit.*, p. 121, from an address by Arup to the Prestressed Concrete Development Group.

25   *Ibid.*, p. 120.

26   Fromonot, *op. cit.*, pp. 63–4.

27   *Ibid.*, p. 64.

28   The best accounts of the Sydney shells' evolution are given in Mikami, *op. cit.*, pp. 59–69, and Murray, *op. cit.*, pp. 30–2.

29   Zunz, *op. cit.*, p. 5: Weston's phrase, *op. cit.*, p. 122, is 'geometrically unregulated'.

30   Fromonot, *op. cit.*, p. 133; Zunz, *op. cit.*, pp. 5–6.

31   Mikami, *op. cit.*, p. 60.

32   Quoted in Fromonot, *op. cit.*, p. 85.

33   A summary by Arups of the fresh initiative on the shells is given in their report of 1964–5, quoted in Baume, *op. cit.*, pp. 98–108, notably p. 102.

34   *Zodiac* 14 (1965), pp. 36–63, quotation from p. 43. Weston, *op. cit.*, p. 132, is among those who insists that the solution to the shells came from Utzon alone: 'the breakthrough

came, and in Helleback, not London'. The best account of Utzon's experiments with spheres, which involved a red rubber ball in the bath before proceeding to specially made wooden spheres, is given by Mikami, *op. cit.*, pp. 64–5: 'I disbelieve other dubious stories told years later . . . of orange peels and so forth as the origin of the spherical idea, though they might appeal to laymen and be easier to understand.'

35   Murray, *op. cit.*, pp. 36, 41–2; Zunz in *Architects' Journal*, 23 Feb. 1995, pp. 44–5; Weston, *op. cit.*, p. 138.

36   Baume, *op. cit.*, p. 26.

37   The best account of the resignation from that standpoint is in Weston, *op. cit.*, pp. 172–6.

38   Quoted in Baume, *op. cit.*, p. 41.

39   Quoted *ibid.*, p. 42.

40   See Arup's famous article, 'Modern Architecture: The Structural Fallacy', in *Arup Journal* 20, no. 1 (Spring 1985), pp. 19–21: originally a radio talk published in *The Listener*, 7 July 1955.

41   Baume, *op. cit.*, p. 64.

42   Quoted in Murray, *op. cit.*, p. 151.

43   Zunz, *op. cit.*, p. 7, in one of several resentful pronouncements.

44   Yeomans, *op. cit.*, p. 216.

45   Murray, *op. cit.*, p. 139, in a brilliant summary of the issues ('Ars Longa, Vita Brevis'), pp. 136–56.

46   Thorne in Dunster, *op. cit.*, p. 259; Sommer, Stöcher and Weisser, *op. cit.*, p. 27.

47   For Arup Associates see Thorne in Dunster, *op. cit.*, pp. 244–5; and *RIBA Journal* 88 (Aug. 1981), pp. 59–65 (presentation of the RIBA Gold Medal to Philip Dowson).

48   Sam Price, 'Different Models of Interdisciplinary Collaboration', in Robin Spence, Sebastian Macmillan and Paul Kirby (eds.), *Interdisciplinary Design in Practice* (London, 2001), pp. 78–85.

49   Nathan Silver, *The Making of Beaubourg* (Cambridge, Mass., and London, 1994). For the Centre Pompidou I have relied much on this source, supplemented by Bryan Appleyard, *Richard Rogers – A Biography* (London, 1986), pp. 153–226; Peter Rice, 'Beaubourg', in *An Engineer Imagines* (London, 1994), pp. 24–46, with a corrective review by Ted Happold in *RSA Journal* 143 (1995), pp. 85–6; Derek Walker and Bill Addis, *Happold: The Confidence to Build* ([London], 1997), pp. 50–3; and the special issue of *Architectural Design* 47 (Feb. 1977), pp. 86–151, including articles by Ted Happold and Dennis Crompton and a critique by Alan Colquhoun.

50   For Happold's early career and work at Arups see Walker and Addis, *op. cit.*, pp. 37–49; entry on Happold in ODNB (Mike Chrimes); and for Happold, Otto and Structures 3, Christian Brensing, 'Frei Otto and Ove Arup: A Case of Mutual Inspiration', and Michael Dickson, 'Frei Otto and Ted Happold: 1967–1996 and Beyond', in *Frei Otto Complete Works: Lightweight Construction, Natural Design* (Basel, 2005), pp. 103, 111–13.

51   Silver, *op. cit.*, p. 13.

52   The ambitious Rogers had approached Frei Otto, whom he did not know, out of the blue to work on the football stand. Otto passed Happold's name on to him.

53 Happold in *RSA Journal* 143 (1995), p. 85.

54 Silver, *op. cit.*, p. 29.

55 Information and quotations in this paragraph: *ibid.*, pp. 42, 48, 53, 53; Appleyard, *op. cit.*, p. 165.

56 I am grateful to Nick Bullock, Anne-Marie Châtelet, Charlotte Ellis and Martin Meade for discussions and advice on the bureaux d'études. For a post-war British account of one such bureau, see *The Builder* 12 April 1963, pp. 755–755c; and for Arups' analysis of the system, Michael Barclay, Edmund Happold, John Martin and Brian Watt, 'Working in France', *Structural Engineer* 52 (January 1974), pp. 3–16.

57 Silver, *op. cit.*, p. 72.

58 For Fernand Pouillon on the bureau d'études system, see his *Mémoires d'un architecte* (Paris, 1968), pp. 223–6 and 284–5; and for his alternative system of designing and managing large-scale housing projects, 'la Set', invented for his first great housing project at the Vieux Port, Marseilles, pp. 103–5. This vainglorious book must be the most gripping of architectural autobiographies. It starts with Pouillon's escape from prison hospital and ends with his trial, while offering many insights into post-war French construction and politics, as into Pouillon's bizarre personality and talents. Further on Pouillon, see Bernard Félix Dubor, *Fernand Pouillon* (Milan and Paris, 1986); Jacques Lucan (ed.), *Fernand Pouillon, architecte* (Paris, 2003).

59 Rice, *op. cit.*, p. 27.

60 Walker and Addis, *op. cit.*, p. 51.

61 Appleyard, *op. cit.*, p. 202.

62 That excepts the work Rice did after 1982 through the Paris-based RFR (Rice Francis Ritchie), 'an engineering group but with the involvement of architects giving it design aspirations': Rice, *op. cit.*, p. 183. A short-lived partnership with Piano did not work out.

63 Quoted *ibid.*, p. 68, in Rice's essay 'Ove Arup the Man'.

64 *Ibid.*, pp. 71–80.

65 *Ibid.*, p. 30.

66 Peter Rice, 'Building as Craft, Building as Industry', in *Bridging the Gap: Rethinking the Relationship of Architect and Engineer* (New York, 1991), p. 88.

67 Rice, *An Engineer Imagines*, p. 28.

68 Silver, *op. cit.*, p. 95.

69 From Happold's review of Rice's book in *RSA Journal* 143 (1995), p. 85. Happold seeks in this review to correct the impression conveyed by Rice that the steel castings at the Pompidou Centre were the latter's idea. He implies that the 'trace de la main' was as much in his own mind as it was in Rice's.

70 Rice, 'Bridging the Gap', p. 94.

71 Silver, *op. cit.*, p. 183.

72 Colin Davies, *High Tech Architecture* (New York and London, 1988), p. 6. See also Angus Macdonald, 'The Aestheticisation of the Steel Framework: the Contribution of Engineering to a Strand of Modern Architecture that Became Known as High Tech', in *Proceedings of the Second International Congress on Construction History* (Cambridge, 2006), vol. 2, pp. 2037–54.

73 Martin Pawley, 'The Secret Life of Engineers', *Blueprint*, March 1989, p. 36.

74 For Hunt, see Angus Macdonald, *The Engineer's Contribution to Contemporary Architecture: Anthony Hunt* (London, 2000). For

Team 4 and Reliance Controls, see also Ian Lambot (ed.), *Norman Foster, Team 4 and Foster Associates, Buildings and Projects Volume 1, 1964–1973* (London, 1991), esp. pp. 74–87 (Rowan Moore) and 145–6 (Hunt).

75 Quoted in Macdonald, *op. cit.*, p. 52.

76 *Id.*

77 Hunt, quoted in Lambot, *op. cit.*, p. 145.

78 *Ibid.*, p. 146.

79 A full study of Samuely's career is wanted. The best sources remain Malcolm Higgs's commemorative essay in *Architectural Association Journal* 76 (June 1960), pp. 2–31, and Frank Newby's tribute in *Architects' Journal* 12 March 1959, pp. 451–2. See also Gregory Hardie, 'Felix Samuely: Teacher, Innovator, Engineer', M.St. thesis for the IDBE course, Cambridge, March 2005. I am grateful too for reminiscences from the late Colin Boyne.

80 Hardie, *op. cit.*, p. 7.

81 A. V. Pilichowski, quoted in Higgs, *op. cit.*, p. 27.

82 Frank Newby, 'High-tech or Mys-tech?', *RIBA Transactions* 6 (vol. 3, no. 2, 1984), p. 22 (article pp. 18–27).

83 Quoted from interview between Frank Newby and Martin Pawley, *Architects' Journal* 22 Jan. 1986, pp. 30–1.

84 'Engineers and Architects: Newby and Price', *AA Files* 27 (1994), p. 26 (article pp. 25–32).

85 *RIBA Transactions* 6 (vol. 3, no. 2, 1984), p. 20.

86 Quoted in Martin Pawley, 'The Secret Life of Engineers', *Blueprint*, March 1989, p. 36.

87 *RIBA Transactions* 6 (vol. 3, no. 2, 1984), p. 26.

88 Frank Newby and Tom Schollar in *Architectural Review* 179 (April 1986), p. 110.

89 Stephanie Williams, *Hongkong Bank: The Building of Norman Foster's Masterpiece* (London, 1989). The rarity of readable but relatively objective studies of recent buildings at any length makes Williams' book, like Silver's on the Pompidou Centre, of special value. It is complemented by a full account of the bank in *Architectural Review* 179 (April 1986), pp. 35–117, with contributions by Colin Davies and Frank Newby.

90 Williams, *op. cit.*, pp. 101–2.

91 Colin Davies in *Architectural Review* 179 (April 1986), p. 82.

92 Williams, *op. cit.*, p. 91.

93 Robert Thorne, 'Continuity and Invention', in David Dunster (ed.), *Arups on Engineering* (Berlin, [1999]), pp. 258–9.

94 Williams, *op. cit.*, p. 43.

95 *Ibid.*, p. 49.

96 *Ibid.*, p. 94.

97 *Ibid.*, p. 126.

98 According to Charles Jencks the Bank's structure and cladding came in at 58.4% of the total cost: interview with Jack Zunz in *Architectural Design* 57, no. 11/12 (1987), p. 45. Zunz's not untypical response was that people worried too much about money. For the comparative costs of the two banks see Carter Wiseman, *The Architecture of I. M. Pei* (London, 1990), pp. 292–3; and for comments on cost by Frank Archer of the Bank, *Architectural Review* 179 (April 1986), p. 107.

99 Robert Gutman, *Architectural Practice, A Critical View* (New York, 1988), pp. 3–22 and 115–7.

100 Bernard Michael Boyle, 'Architectural Practice in America 1865–1965 – Ideal and

Reality', in Spiro Kostof (ed.), *The Architect: Chapters in the History of a Profession* (New York, 1977), p. 313 (essay, pp. 309–44).

101 Gutman, *op. cit.*, p. 7; Andrew Saint, *The Image of the Architect* (London, 1983), p. 154.

102 Giovanni Brino, *La professione dell'architetto in USA* (Turin, 1968), pp. 54–62.

103 Boyle in Kostof, *op. cit.*, p. 329.

104 Nathaniel Alexander Owings, *The Spaces In Between: An Architect's Journey* (Boston, 1973), p. 66.

105 *Ibid.*, pp. 83–97; Charles W. Johnson and Charles O. Jackson, *City Behind a Fence: Oak Ridge, Tennessee, 1942–1946* (Knoxville, 1981).

106 Owings, *op. cit.*, p. 94.

107 Henry-Russell Hitchcock, in *Architecture of Skidmore, Owings and Merrill, 1950–1962* (London, 1963), p. 10. Further and later on SOM, see Christopher Woodward, *Skidmore, Owings and Merrill* (London, 1970); *Architecture of Skidmore, Owings and Merrill, 1963–1973* (Stuttgart, 1974); Albert Bush-Brown, *Skidmore, Owings and Merrill, Architecture and Urbanism, 1973–1983* (London, 1984); and 'SOM, A Legend in Transition', *Architecture*, Feb. 1989, pp. 52–9.

108 For Netsch and the US Air Force Academy see 'Walter Netsch interviewed by Detlef Mertins', *SOM Journal* 1 (2001), pp. 136–51.

109 Myron Goldsmith, *Poet of Structure* (Centre Canadien d'Architecture, Montreal, 1991), p. 30. For Goldsmith see also Alfred Swenson and Pao-Chi Chang, *Architectural Education at IIT 1938–1978* (Chicago, 1980), pp. 155–6, 170–7; *RIBA Journal* 73 (June 1966), pp. 252–7; and many passing references in Phyllis Lambert (ed.), *Mies in America* (New York and Montreal, 2001), mostly based on taped interviews between Goldsmith and Kevin Harrington held at the Centre Canadien d'Architecture.

110 For Bunshaft see Carol Herselle Krinsky, *Gordon Bunshaft of Skidmore, Owings and Merrill* (Cambridge, Mass., 1988). For information about SOM New York I am grateful for an interview with Tom Killian.

111 Owings, *op. cit.*, p. 75.

112 For a later confirmation of this bias from the perspective of New York labour practices, see Dick Rowe of SOM as quoted by Karl Sabbagh, *Skyscraper: The Making of a Building* (London, 1989), p. 41: 'A composite building of steel plus concrete is extremely difficult to achieve in New York City. It's done very easily in other parts of the world, but it's extremely difficult to do here. The two trades – steel and concrete – have very different working requirements and so they don't want to work in close proximity to each other.'

113 For a Bunshaft building in concrete in the UK, the Heinz offices at Hayes, see Elain Harwood, 'Prestige Pancakes', in *Twentieth-Century Architecture* 1 (1994), pp. 81–3.

114 Krinsky, *op. cit.*, p. 138.

115 *L'Art de l'ingénieur* (Centre Pompidou, Paris, 1997), pp. 543–4.

116 Krinsky, *op. cit.*, p. 138.

117 Sabbagh, *op. cit.*, p. 39. Sabbagh's study is a superb introduction to the human issues involved in complex building procurement, where respect between partners must be balanced by discipline and clear lines of command. It should be compulsory reading or viewing in schools of architecture.

118 See *Bruce Graham of SOM* (Milan, 1989); Mir M. Ali, *Art of the Skyscraper: The Genius of Fazlur Khan* (New York, 2001); Yasmin Sabina Khan, *Engineering Architecture: The Vision of Fazlur R. Khan* (New York, 2004).

119 Ali, *op. cit.*, pp. 50–1.

120 Concrete frames had been expressed earlier on several high-rise American buildings, notably the Kipps Bay Plaza apartments, New York, by I. M. Pei with August Komendant (1959–61): see Yasmin Sabina Khan, *op. cit.*, pp. 100–2. They had of course also featured in large European slab structures, the most famous being Le Corbusier's Unités d'Habitation. Neither manifestation is connected to Khan's tubular framing.

121 *Bruce Graham of SOM*, p. 46.

122 David Billington, 'The New Art of Engineering', in *Bridging the Gap: Rethinking the Relationship of Architect and Engineer* (New York, 1991), p. 17.

123 Ali, *op. cit.*, p. 156. No mention of LeMessurier and skyscrapers can pass without allusion to the notorious case of the Citicorp Building, New York (1974–7), for which Hugh Stubbins was architect and LeMessurier's firm were engineers. The lower stages of the tower were supported in a daring manner which involved cantilevering the corners at lower levels. After completion LeMessurier came to believe that its stability was doubtful in certain wind conditions. Remedial works were discreetly undertaken without the true cause being given. The scare did not become generally public until it was revealed in an article by Joe Morgenstern in the *New Yorker*, 29 May 1995. LeMessurier's conduct then became celebrated as an example of engineering ethics at its best. Eugene Kremer has since questioned that reading: *arq (Architectural Research Quarterly)* 6, no. 3 (2002), pp. 269–76.

124 Richard Keating, 'Collaboration and the Culture of SOM', in *Bridging the Gap: Rethinking the Relationship of Architect and Engineer* (New York, 1991), p. 75.

125 August E. Komendant, *18 Years Working with Architect Louis I. Kahn* (Englewood Cliffs, 1975).

126 Thomas Leslie, *Louis I. Kahn: Building Art, Building Science* (New York, 2005), p. 36. Leslie's book, which I saw at a late stage, is much the best text for fully understanding the construction of Kahn's buildings.

127 Fuller's prolixity is rivalled by the numerous texts about him, mostly written by acolytes. Lloyd Steven Sieden, *Buckminster Fuller's Universe* (Cambridge, Mass., 1989), is perhaps the best overall account of his life and work.

128 Quoted in Alessandra Latour, *Louis I. Kahn, l'uomo, il maestro* (Rome, 1986), pp. 49, 285; see also Anne Gryswold Tyng (ed.), *Louis Kahn to Anne Tyng: The Rome Letters 1953–1954* (New York, 1997), p. 39.

129 David B. Brownlee and David G. De Long, *Louis I. Kahn: In the Realm of Architecture* (New York, 1991), pp. 60–1; Leslie, *op. cit.*, pp. 84–6.

130 Latour, *op. cit.*, p. 49; Leslie, *op. cit.*, pp. 60–2.

131 For Le Ricolais see *L'Art de l'ingénieur* (Centre Pompidou, Paris, 1997), p. 264; and *AA Files* 39 (Autumn 1999), pp. 56–60.

132 Tyng, *op. cit.*, p. 47.

133 Komendant, *op. cit.*, p. 177.

134 Information in this paragraph is from Patricia Cummings Loud, *The Art Museums of Louis I. Kahn* (Durham, N.C., 1989), pp. 52–99, with quotations from pp. 72, 73, 82–4; and from Leslie, *op. cit.*, pp. 48–89, with quotation from p. 67.

135 Leslie, *op. cit.*, pp. 96–7.

136 Komendant, *op. cit.*, pp. 1–6.

137 Leslie, *op. cit.*, p. 102, in his discussion of the Richards Labs, pp. 92–127.

138 Komendant, *op. cit.*, p. 23.

139 *Ibid.*, p. 24.

140 *Ibid.*, pp. 91–2.

141 *Ibid.*, p. 87.

142 *Ibid.*, pp. 115–31. See also Leslie, *op. cit.*, pp. 178–221; Loud, *op. cit.*, pp. 101–70; and *La costruzione del Kimbell Art Museum/The Construction of the Kimbell Art Museum* (Brescia, 1998).

143 Komendant, *op. cit.*, pp. 116–17.

144 Michael Benedikt, *Deconstructing the Kimbell* (New York, 1991), p. 65.

145 Marshall D. Meyers, 'Making the Kimbell: A Brief Memoir', in *La costruzione del Kimbell Art Museum/The Construction of the Kimbell Art Museum* (Brescia, 1998), p. 19.

146 Quoted in Loud, *op. cit.*, p. 148.

147 Meyers, *op. cit.*, pp. 19–21.

148 Quoted in *Architectural Review* 155 (July 1974), p. 332.

149 Komendant, *op. cit.*, pp. 130–1.

150 Just to take the major figures mentioned in this chapter, the following were all refugees or émigrés: Arup, Calatrava, Candela, Louis Kahn, Fazlur Khan, Komendant, Le Ricolais, Lubetkin, Nowicki, Eero Saarinen, Salvadori, Samuely, Severud, Tedesko and Weidlinger.

151 This paragraph abbreviates information in Andrew Saint, 'Some Thoughts about the Architectural Use of Concrete', *AA Files* 22 (1991), pp. 7–13. For the fetish of the parabolic arch and vault see Francis S. Onderdonk, *The Ferro-Concrete Style* (New York, 1928, reprinted Santa Monica, 1998), pp. 186–221.

152 For Candela see Colin Faber, *Candela: The Shell Builder* (London, 1963), esp. pp. 11–15.

153 For Torroja see *L'Art de l'ingénieur* (Centre Pompidou, Paris, 1997), pp. 506–7.

154 *Progressive Architecture* June 1954, an issue entitled 'Towards New Structural Concepts', pp. 83–125: article by Felix Candela, 'Stereo-Structures', pp. 84–93. Letters occur in the same issue from Salvadori, pp. 16–22, from Weidlinger, p. 22, and from Severud, pp. 181–2.

155 For a summary of Severud's career see *L'Art de l'ingénieur* (Centre Pompidou, Paris, 1997), pp. 453–4, where it is noted that he became a Jehovah's Witness in 1935 and after his retirement gave himself up to religious interests.

156 Quoted in Jayne Merkel, *Eero Saarinen* (London, 2005), p. 198 (account of the St Louis arch, pp. 194–203).

157 Hélène Lipstadt, 'The Gateway Arch', in Eeva Liisa Pelkonen and Donald Albrecht (eds.), *Eero Saarinen: Shaping the Future* (New Haven and London, 2006), p. 226. In her essay (pp. 222–9) Lipstadt notes that there was much debate about the arch at the Cranbrook Academy, the Saarinen stronghold, and that the sculptor Car Milles is alleged to have suggested the change from a square to a triangular section.

158 For the Raleigh pavilion see *Architectural Forum* October 1952, pp. 134–9 and 162, and April 1954, pp. 130–4. Nowicki had originally wanted a stressed skin or membrane roof, but according to the engineering analysis by Igor Voshinin of Severud-Elstad-Krueger, *ibid.*, October 1952, p. 162, such a membrane could not then be fabricated 'as thin as theoretical computations allow'. The cable-hung modification seems to have been agreed before Nowicki's death. There were a few precedents, going back to Hittorff's Panorama des Champs Elysées of 1838–9 (p. ••). One of the buildings at the Chicago Exhibition of 1933 had a catenary roof. For the later history of the pavilion see Ernest Wood, 'A Radical Settles Down in Raleigh, N.C.', *AIA Journal* Sep. 1980, pp. 54–61, mentioning that at one stage it was proposed to hang 13,000 Confederate flags from the roof to solve the acoustic problems, and that when the glazing was renewed 32 bullet holes were found.

159 Fred N. Severud and Raniero G. Corbeletti, 'Hung Roofs', *Progressive Architecture* March 1956, p. 100.

160 See the articles by Lewis Mumford, 'The Life, the Teaching and the Architecture of Matthew Nowicki', *Architectural Record* June 1954, pp. 139–49; July 1954, pp. 128–35; Aug. 1954, pp. 169–76; and Sep. 1954, pp. 153–9. The August 1954 article is informative about the collaboration between Nowicki and Eero Saarinen over designs for Brandeis University, which involved several experimental roof forms, and more briefly so about the Raleigh arena.

161 Paul Rudolph, 'The Great Livestock Pavilion Complete', *Architectural Forum* April 1954, pp. 130–4.

162 Lawrence Lessing, 'Suspension Structures', *Architectural Forum* Dec. 1957, p. 137.

163 The paragraphs on Eero Saarinen are mainly drawn from Merkel, *op. cit.*: collaboration with Nowicki (on Brandeis University), pp. 107–8; Kresge Auditorium, pp. 113–20; Ingalls Ice Hockey Rink, pp. 123–30; TWA Terminal, pp. 205–13; Dulles Airport, pp. 216–29, for the role of Ammann and Whitney, who also brought in Saarinen at Athens Airport, pp. 213, 245. See also scattered remarks in Pelkonen and Albrecht, *op. cit.*

164 Henry-Russell Hitchcock, 'American Architecture in the Early Sixties', *Zodiac* 10 (1962), p. 16.

165 Quoted in Merkel, *op. cit.*, pp. 117, 210. Candela too took exception: 'his reaction[s] to projects like Niemeyer's, or the TWA terminal at Idlewild, or the Sydney Opera House are somewhat vitriolic': Faber, *op. cit.*, p. 10.

166 'Shaping a Two-Acre Sculpture', *Architectural Forum* Aug. 1960, pp. 118–23; Pelkonen and Albrecht, *op. cit.*, pp. 199–220. Ammann and Whitney's job engineer for the TWA terminal was Abba Tor: *ibid.*, p. 78.

167 Fred N. Severud, 'Turtles and Walnuts, Morning Glories and Grass', *Architectural Forum* Sep. 1945, pp. 149–62.

168 For an exploration of analogical tendencies in architecture see Philip Steadman, *The Evolution of Designs: Biological Analogy in Architecture and the Applied Arts* (Cambridge,

1979); and for a summary of the many ideas about the relation between architecture and nature, Adrian Forty, *Words and Buildings* (London, 2000), pp. 220–39.

169 Quoted by L. G. Booth on Tredgold in BDCE, p. 720.

170 George R. Collins, *Antonio Gaudí* (London, 1960), pp. 23–5, dating the model to around 1908. The relations between Gaudí's forms, the Catalan tradition of masonry vaulting on which he drew, and the shell-concrete designs of Torroja and Candela is obvious in principle but hard to pin down.

171 Horatio Greenough, 'American Architecture', reprinted e.g. in Lewis Mumford (ed.), *Roots of Contemporary American Architecture* (New York, 1952), pp. 32–56.

172 Mainly Chapter 16, 'On Form and Mechanical Efficiency', pp. 670–718 in the original edition of 1917. The Forth Bridge is mentioned, and Thompson had mastered Culmann's graphical analysis of engineering structures.

173 E.g. at the IIT school, where Mies was an enthusiast: see Detlef Mertins in Phyllis Lambert, *Mies in America* (New York and Montreal, 2001), pp. 611–12. Mies probably came across the book through Moholy-Nagy, who (Assimina Kaniari kindly told me) appears to have known *On Growth and Form* before the 1942 edition made it more widely accessible. In England it seems to have been Herbert Read who introduced Thompson to the inter-war art-world. The text was well known at the Architectural Association in the 1940s. It has been a fixture on architectural-school reading lists ever since.

174 Lloyd Steven Sieden, *Buckminster Fuller's Universe* (Cambridge, Mass., 1989), pp. 9–10, 33–5.

175 An example of an architect-engineer who uses natural form in this way, as an aid at the conceptual stage of design, is Santiago Calatrava: see Bryan Lawson, *Design in Mind* (Oxford, 1994), pp. 21–34.

176 See *Berlin Baut 2: Die Kongresshalle* (Berlin, 1987), issued to celebrate the reconstruction of the building.

177 From 'The Congress Hall Debate', *Architectural Forum* Jan. 1958, pp. 115–21, 170–2. The debate was also published in German in *Bauwelt* 1958, no. 1, pp. 13–16. In Severud's fairly brief contributions, he brushed aside Otto's suggested remedies of tension wires or nets.

178 The authoritative source now for Otto is Winfried Nerdinger (ed.), *Frei Otto Complete Works: Lightweight Construction, Natural Design* (Basel, 2005), the book accompanying the Frei Otto exhibition in Munich that year: for biographical data see pp. 368–9.

179 *Ibid.*, pp. 227–36. Gutbrod's assistant Hermann Kendel was Otto's main architectural collaborator in Montreal. Otto and Kendel had previously worked together on an important but unbuilt auditorium roof at Stuttgart-Hohenheim in which, said Kendel, 'Otto guided and taught me through every step of the project': *ibid.*, pp. 219, 364–5.

180 *Ibid.*, pp. 260–9 (quotations from p. 269).

181 For Happold and Severud, see Derek Walker and Bill Addis, *Ted Happold: The Confidence to Build* ([London], 1997), pp. 11–12, 55; and for Happold and Otto, Michael Dickson, 'Frei Otto and Ted Happold, 1967–1996 and Beyond', in Nerdinger, *op. cit.*, pp. 111–22.

182 *Architectural Design* 3 (1971), pp. 137–67, quotations from pp. 144, 150, 162.

183 The aviation engineer Walter Bird developed non-metallic inflatable shelters at the Cornell Aeronautical Laboratory in the 1940s. For his company, Birdair Structures, and their 1950s 'radomes' see Tony Robbin, *Engineering a New Architecture* (New Haven, 1996), pp. 10, 13; Nerdinger, *op. cit.*, p. 23; and Walker and Addis, *op. cit.*, p. 63.

184 In the essay by Irene Meissner, 'In Harmony with Nature and Technology' in Nerdinger, *op. cit.*, pp. 56–63, Otto is quoted (p. 61) as dividing architects into 'arrangers, discoverers and thieves'. For Otto and nature see also in the same book Rainer Barthel, 'Natural Forms – Architectural Forms', pp. 16–31, and Ulrich Kull, 'Frei Otto and Biology', pp. 44–55; for Otto and D'Arcy Thompson, Walker and Addis, *op, cit.*, p. 59.

185 See Nerdinger, *op. cit.*, pp. 240–7, 292–3, 295–8, 300, 312–19, 325, and the note by Bodo Rasch, p. 361.

186 For Khan's later career see Yasmin Sabina Khan, *Engineering Architecture: The Vision of Fazlur R. Khan*, pp. 259–373; and for the Haj Terminal, *ibid.*, pp. 287–313; *Progressive Architecture* Feb. 1982, pp. 116–22; Carol Hershelle Krinsky, *Gordon Bunshaft of Skidmore, Owings and Merrill* (Cambridge, Mass., 1988), pp. 261–8; Robbin, *op. cit.*, pp. 9, 14–15.

187 Nerdinger, *op. cit.*, pp. 293, 312–19.

188 Quoted in Krinsky, *op. cit.*, p. 268.

189 *Progressive Architecture* Feb. 1982, p. 121.

190 *Bridging the Gap: Rethinking the Relationship of Architect and Engineer* (New York, 1991).Salvadori, pp. 1–2; Billington, pp. 3–20; Peters, pp. 23–35 (quotation p. 23).

191 *Ibid.*, Rogers, pp. 139–55, with criticisms in the ensuing discussion, pp. 156–66, from Jörg Schlaich; Frampton on Delft, p. 162. Other speakers included Santiago Calatrava, Richard Keating, William LeMessurier, M. Levy, Peter McLeary and Peter Rice.

192 See Tanya Ross, 'Stepping out of the Shadows: Engineers in the Modern Media', M.St. thesis for the IDBE course, University of Cambridge, 2001.

193 E.g. Alan Holgate, *The Art in Structural Design* (Oxford, 1986); *L'Art de l'ingénieur* (Centre Pompidou, Paris, 1997); David P. Billington, *The Art of Structural Design: A Swiss Legacy* (Princeton, 2003).

194 Cecil Balmond, *Informal* (Munich, 2002), pp. [13–15].

6 A QUESTION OF UPBRINGING

1 This abbreviated form for Viollet-le-Duc is not admitted by French scholars, but Mérimée used it and it has become common in English.

2 My account follows Maxime Du Camp, *Souvenirs d'un demi-siècle* (Paris, 1949), pp. 221–2. Du Camp's colourful account of the 'chahut babylonien' at the Beaux-Arts, written in the 1880s, may be a little embroidered. He incorrectly states that Mérimée was present, though Gautier and Sainte-Beuve were. See Mérimée to Viollet-le-Duc, 1 Feb. 1864, in *Correspondance générale de Prosper Mérimée*, 2nd series, vol. 6, 1864–8 (Toulouse, 1958), pp. 41–3.

3 Geneviève Viollet-le-Duc, 'Viollet-le-Duc et l'Ecole des Beaux-Arts: La bataille de 1863–64', in Eugène Viollet-le-Duc, *Esthétique appliquée à l'histoire de l'art* (Paris, 1994), pp. 140–2, quoting letter to Henri Courmont. See also pp. 143–5 and 152–3 (letters to Mérimée and Sainte-Beuve). This book publishes Viollet-le-Duc's Beaux-Arts lectures of 1864.

4 This section is mainly based on the following sources: Jeanne Laurent, *A propos de l'Ecole des Beaux-Arts* (Paris, 1987), pp. 124–39; Louis Vitet and Eugène Viollet-le-Duc, *Débats et polémiques: à propos de l'enseignement des arts du dessin* (Paris, 1984), with preface by Bruno Foucart (pp. 7–75); Eugène Viollet-le-Duc, *Esthétique appliquée à l'histoire de l'art*, ed. (Paris, 1994), with essay by Geneviève Viollet-le-Duc (pp. 115–56); E. Viollet-le-Duc, *Intervention de l'Etat dans l'enseignement des Beaux-Arts* (Paris, 1864); *Correspondance de Mérimée et Viollet-le-Duc*, ed. Françoise Bercé (Paris, 2001).

5 Du Camp, *op. cit.*, pp. 213–36.

6 Reprinted from the *Gazette des Beaux-Arts* in Vitet and Viollet-le-Duc, *op. cit.*, pp. 105–43.

7 Quoted by Bruno Foucart in his preface to Vitet and Viollet-le-Duc, *op. cit.*, p. 14.

8 Remark of Charles Giraud, quoted in Laurent, *op. cit.*, p. 126.

9 Viollet-le-Duc to Henri Courmont, 21 Dec. 1863, quoted in Geneviève Viollet-le-Duc, *op. cit.*, p. 127.

10 Nikolaus Pevsner, *Academies of Art Past and Present* (Cambridge, 1940), pp. 82–110; Laurent, *op. cit.*, pp. 21–6.

11 1819 is the nearest the nineteenth-century school had to a formal date of refoundation: see Laurent, *op. cit.*, p. 111. The name 'Beaux-Arts' was first formally attached to the school in 1806, following suggestions put forward in 1802. The full name kept changing in small particulars throughout the century; I am grateful to Charlotte Ellis for pointing this out.

12 Letter to the Minister of the Interior, quoted in Laurent, *op. cit.*, p. 123.

13 See Nieuwerkerke's report of 1863, reprinted in Vitet and Viollet-le-Duc, *op. cit.*, p. 148.

14 Vitet and Viollet-le-Duc, *op. cit.*, p. 122 (second article).

15 *Ibid.*, p. 126.

16 *Ibid.*, p. 128.

17 *Ibid.*, p. 133.

18 For Viollet-le-Duc's teaching, see *Eugène Viollet-le-Duc* (Paris, 1965), 'Problèmes de l'enseignement', pp. 140–7; Jean-Michel Leniaud, *Viollet-le-Duc, ou les délires du système* (Paris, 1994), pp. 110–16; and Eugène Viollet-le-Duc (tr. Bucknall), *Discourses on Architecture*, vol. 1 (London, 1875), pp. 5–8.

19 See Ulrich Pfammatter, *The Making of the Modern Architect and Engineer* (Basel, 2000), pp. 106–208; Jean-François Belhoste (ed.), *Le Paris des Centraliens* (Paris, 2004).

20 Vitet and Viollet-le-Duc, *op. cit.*, p. 142; and E. Viollet-le-Duc, *Intervention de l'Etat dans l'enseignement des beaux-arts* (Paris, 1864), p. 26.

21 Frédéric Seitz, *Une entreprise d'idée: L'Ecole Spéciale d'Architecture 1865–1930* (Paris, 1995).

22 *Ibid.*, p. 110.

23 Eugène Viollet-le-Duc (tr. Bucknall), *Discourses on Architecture*, vol. 2 (London, 1881), pp. 140–69.

24 Laurent, *op. cit.*, pp. 133–9.

25 The exception was the impressive Service des Edifices Diocésains, for which see Jean-Michel Leniaud, *Les cathédrales au XIXe siècle: étude du service des édifices diocésains* (Paris, 1993).

26 The following paragraphs are largely based on Chapter 11 of Robin Middleton and Marie-Noëlle Baudouin Matuszek's then unpublished book, *Jean Rondelet: The Architect as Technician*, on 'The Cours de Construction at the Ecole des Beaux-Arts'. I am most grateful to Robin Middleton for allowing me to use this.

27 Claude Mauclaire and C. Vigoureux, *Nicholas-François Blondel: ingenieur et architecte du roi* (Paris, 1938), pp. 120–30, 242.

28 Quoted in Middleton and Baudouin Matuszek, Chapter 11.

29 Bruno Belhoste, *La formation d'une technocratie: l'Ecole Polytechnique et ses élèves de la Révolution au Second Empire* (Paris, 2003), p. 126; and for the rest of the paragraph, in which I have simplified the complex evolution of the Ecole Polytechnique, pp. 47–50, 109–23.

30 *Ibid.*, pp. 48, 123–6. The Polytechnique was under the Interior Ministry 1815–22, but then reverted to the War Ministry.

31 The early Génie examiners, Joseph Sauveur and his successor and nephew François Chevallier, were both members of the Académie des Sciences: see Janis Langins, *Conserving the Enlightenment: French Military Engineering from Vauban to the Revolution* (Cambridge, Mass., 2004), pp. 52, 81, 88; Belhoste, *op. cit.*, p. 312.

32 Quoted in René Taton, 'L'Ecole Royale du Génie de Mézières', in Roger Hahn and René Taton, *Ecoles techniques et militaires au XVIIIe siècle* (Paris, 1986), pp. 562–3.

33 For Mézières see Taton, *op. cit.*, pp. 559–615, and Langins, *op. cit.*, pp. 95–8 and 237–46. Langins gives an outstanding account of the Génie between Vauban and the French Revolution.

34 Antoine Picon and Michel Yvon, *L'Ingénieur artiste: dessins anciens de l'Ecole des Ponts et Chaussées* (Paris, 1989), p. 15. See also Antoine Picon, *L'Invention de l'ingénieur moderne: L'Ecole des Ponts et Chaussées 1747–1851* (Paris, 1992), a masterpiece in the desert of texts on technical education; and his *French Architects and Engineers in the Age of Enlightenment* (Cambridge, 1992: French original, 1988), broader in scope but narrower in chronology.

35 Frank A. Kafker and Serena L. Kafker, *The Encyclopedists as Individuals: Studies on Voltaire and the Eighteenth Century*, 257 (Oxford, 1988), pp. 297–302. This article offers insights into Perronet's personal life and relations with the philosophes. It also makes clear that for the famous pin-making passage at the start of *The Wealth of Nations*, Adam Smith used not Perronet's report on pin-making at Aigle, which though written in 1740 was not published till 1760 (in vol. 4 of the plates to the *Encyclopédie*), but the earlier article on pin-making by Alexandre Deleyre in vol. 5 of the main text. However, Smith was in

Paris for most of 1766 and in touch with Turgot, the future protector of the Ecole des Ponts et Chaussées. He may have met Perronet at this time. See Ian Simpson Ross, *The Life of Adam Smith* (Oxford, 1995), pp. 210–14, 273.

36 Picon, *L'Invention . . .* , pp. 25, 103–7, 111–15.

37 Picon and Yvon, *op. cit.*, p. 28.

38 Picon, *L'Invention . . .* , pp. 145–8, where he points out that some alumni went into alternative professions for some years and then returned to the Corps, like Saint-Far, who designed civil hospitals in the 1780s. Career structures were quite fluid at this time.

39 *Ibid.*, p. 27.

40 The fullest and clearest account of the many currents within the Polytechnique is provided by Belhoste, *op. cit.* For the influence of Condorcet, see his p. 106, and for a summary of the 'Ecole de Monge', pp. 200–3.

41 Picon, *French Architects and Engineers . . .* , p. 128, and for a résumé of Prony's career, pp. 349–53.

42 Quoted in Picon and Yvon, *op. cit.*, p. 20.

43 For Durand I have relied primarily on Jean-Nicolas-Louis Durand, *Précis of the Lessons on Architecture* (Los Angeles, 2000), a translation by David Britt with a luminous introduction by Antoine Picon, pp. 1–68. See also Werner Szambien, *Jean-Nicolas-Louis Durand 1760–1834: de l'imitation à la norme* (Paris, 1984), pp. 64–71, 159–64; and Belhoste, *op. cit.*, pp. 276–7.

44 Durand, *op. cit.*, p. 73 (translation by Britt).

45 Quoted in Picon's introduction to Durand, *op. cit.*, p. 23, translation by Britt with some amendment from Etienne-Louis Boullée, *Essai sur l'art* (Paris, 1968), p. 49.

46 Except for a design for a theatre with a wide-span iron-framed roof, 'a form of covering that would by no means be impracticable': Durand, *op. cit.*, p. 165 and Part III Plate 16.

47 For one among countless examples see Urbain Vitry's abattoirs at Toulouse, chronicled by Marie-Laure de Capella and Rémi Papillault in *Les Abattoirs: histoires et transformation* (Toulouse, 2000), pp. 15–80.

48 Picon in Durand, *op. cit.*, p. 52.

49 The following paragraphs are indebted to Picon, *L'Invention . . .* , pp. 389–577, and Belhoste, *op. cit.*, passim.

50 Picon, *L'Invention . . .* , p. 427.

51 For descriptive geometry see Belhoste, *op. cit.*, pp. 261–73; Joel Sakarovitch, *Epures d'architecture: de la coupe des pierres à la géometrie descriptive* (Basle, 1998); and for an English summary, Peter Jeffrey Booker, *A History of Engineering Drawing* (London, 1963), pp. 86–113.

52 Picon in Durand, *op. cit.*, p. 61, n.68. For the general British reaction see Alexander W. Cunningham, *Notes on the History, Methods and Technological Importance of Descriptive Geometry* (Edinburgh, 1868), pp. 48–55.

53 Paul Buquet, quoted in Robert R. Locke, *The End of the Practical Man: Entrepreneurship and Higher Education in Germany, France, and Great Britain, 1880–1940* (Greenwich, Ct., 1984), p. 45.

54 By Nathalie Montel, cited in Belhoste, *op. cit.*, pp. 299–301.

55 See Antoine Picon, 'Charles-François Mandar (1757–1844) ou l'architecture dans tous ses détails', *Revue de l'Art* 109 (1995), pp. 26–39.

56 This account of the Bauakademie follows principally Pfammatter, *op. cit.*, pp. 223–4; Kees Gispen, *New Profession, Old Order: Engineers and German Society, 1815–1914* (Cambridge, 1989), pp. 27–34, 87–95; and Vincent Clark, 'A Struggle for Existence: The Professionalization of German Architects', in Geoffrey Cocks and Konrad H. Jarausch (eds.), *German Professions, 1800–1950* (New York, 1990), p. 147.

57 Gispen, *op. cit.*, p. 30, paraphrasing views of Peter Beuth.

58 Pfammatter, *op, cit.*, p. 224.

59 Clark, *op. cit.*, p. 147. This view is corroborated by Gispen, *op. cit.*

60 Quoted from *Deutsche Bauzeitung* 1867, p. 443, in Gispen, *op .cit.*, p. 90.

61 Clark, *op. cit.*, p. 148. For Karlsruhe, see Pfammatter, *op. cit.*, pp. 228–36.

62 Mallgrave, *Gottfried Semper, Architect of the Nineteenth Century* (New Haven, 1996), p. 19.

63 For this section I have relied chiefly on J. Scott Russell, *Systematic Technical Education for the English People* (London, 1869), pp. 145–77; Pfammatter, *op. cit.*, pp. 240–61; Harry Francis Mallgrave, *op. cit.* pp. 225–46; and Bruno Maurer, 'Lehrgebaüde – Gottfried Semper am Zürcher Polytechnikum', in *Gottfried Semper 1803–1879: Architektur und Wissenschaft* (Zurich, 2003), pp. 306–13.

64 Scott Russell, *op. cit.*, p. 154.

65 Pfammatter, *op. cit.*, p. 247. For a succinct account of Culmann's importance see Stephen P. Timoshenko, *History of Strength of Materials* (New York, 1953), sec. 43.

66 For a review of early projects and buildings, see Heinz Ronner (ed.), *Arbeitsberichte der Architekturabteilung, Eidgenössische Technische Hochschule Zürich*, (Zürich, 4 vols., 1971), 'A11: Die Bauschule am Eidgenössische Polytechnikum Zürich 1855–1915'.

67 Letter of 3 July 1857 from Semper to Cole quoted in Maurer, *op. cit.*, p. 309.

68 Mary N. Woods, *From Craft to Profession: The Practice of Architecture in Nineteenth-Century America* (Berkeley, 1999), pp. 58–66.

69 For a general history of the Corps of Engineers, see *The History of the US Army Corps of Engineers* (Alexandria, Va., 1998, 2nd edition).

70 Theodore J. Crackel, *West Point: A Bicentennial History* (Lawrence, Kansas, 2003), pp. 46–50, quotation p. 49.

71 *DAB*, *sub* Crozet and Partridge.

72 Crackel, *op. cit.*, pp. 81–105.

73 A graduate of 1828, after civil engineering had been added, remembered grappling with J.-M. Sganzin's cours de construction (1809) and finding it 'very diffuse and with difficulty comprehended': Crackel, *op. cit.*, p. 317 n.51.

74 See Albert Parry, *Whistler's Father* (New York, 1939), whose precision and penetration combined with a light touch put much architectural and engineering history to shame. As is well known, Whistler the painter also attended West Point briefly, but art triumphed over engineering.

75 Palmer C. Ricketts, *History of Rensselaer Polytechnic Institute 1824–1934* (New York, 1934), p. 6; DAB *sub* Partridge.

76 Theodore Francis Jones (ed.), *New York University 1832–1932* (New York, 1933), pp. 305ff.; Hector James Hughes in Samuel Eliot Morrison (ed.), *The Development of*

Harvard University since the Inauguration of President Eliot 1869–1929 (Cambridge, Mass., 1930), pp. 413–27.

77 Ricketts, *op. cit.*, pp. 82–6.

78 *Ibid.*, pp. 92–6.

79 *Ibid.*, p. 150.

80 The point is made in Woods, *op. cit.*, p. 68: this section follows her pp. 66–81. MIT was not strictly a land-grant institution but was boosted by land-grant money.

81 Precise dates are not given here because, as Anthony Alofsin says in his *The Struggle for Modernism: Architecture, Landscape Architecture and City Planning at Harvard* (New York, 2002), p. 272 n.4, 'dates on the founding of schools vary'. Curricula were often set out long before students arrived. Graduation dates are little help, since at first few of the students actually graduated.

82 This account of Ricker at Illinois derives mainly from Roula Geraniotis, 'The University of Illinois and German Architectural Education', in *Journal of Architectural Education*, vol. 38 no. 4 (Summer 1985), pp. 15–26. See also the preface to John S. Garner (ed.), *The Midwest in American Architecture* (Urbana, 1991), pp. ix–xv; and Woods, *op cit.*, pp. 71–3.

83 Geraniotis, *op. cit.*, p. 16.

84 Frank Lloyd Wright's sole formal instruction consisted of two semesters in an engineering course run by the architect-engineer Allan Conover at the University of Wisconsin. This was not untypical of Chicago architects' background in his day. Lessons in architecture were given at the Art Institute of Chicago from 1889, but the Armour Institute of Technology's course, started in 1893, was the city's first full architectural programme; the two courses merged in 1895: see Alfred Swenson and Pao-Chi Chang, *Architectural Education at IIT 1938–1978* (Chicago, 1980), p. 9.

85 Andrew Saint, *The Image of the Architect* (London, 1983), pp. 90–1.

86 Caroline A. Shillaber, *Massachusetts Institute of Technology School of Architecture and Planning, 1861–1961: A Hundred Year Chronicle* (Cambridge, Mass., 1963), p. 12; Paul R. Baker, *Richard Morris Hunt* (Cambridge, Mass., 1980), p. 105.

87 Shillaber, *op. cit.*, p. 9, says Ware visited Britain, France, Italy and the Low Countries; Germany and Switzerland are not mentioned. This may have been a question of languages. If the date of this tour was the winter of 1866–7, when Ware was in Britain, it postdates the preliminary Beaux-Arts-style curriculum, which he published in 1866.

88 Quoted in James F. O'Gorman, *H. H. Richardson: Architectural Forms for an American Society* (Chicago, 1987), p. 11, with minor changes to the orthography.

89 Quoted in Baker, *op. cit.*, p. 24.

90 Quoted *ibid.*, p. 62.

91 William R. Ware, 'On the Condition of Architecture and of Architectural Education in the United States', *Papers Read at the Royal Institute of British Architects, Session 1866–7* (1867), p. 86.

92 Woods, *op. cit.*, p. 65, and for further detail about Hunt's atelier see Baker, *op. cit.*, pp. 100–5.

93 Ware, *op. cit.*, p. 86.

94 Robert Twombly, *Louis Sullivan, His Life and Work* (Chicago, 1986), pp. 28–38.

95 Richard Oliver (ed.), *The Making of An Architect 1881–1981: Columbia University in the City of New York* (New York, 1981), essays by Steven M. Bedford and Susan M. Strauss, pp. 5–12 and 23–48. Quotations here are from Schermerhorn (pp. 6–7) and A. D. F. Hamlin (p. 23).

96 Unless one counts Louis J. Millet, more of a decorator than an architect, who headed the classes at the Art Institute of Chicago from 1889.

97 A general article by Cates on 'Architectural Education in the United States of America' appeared in *Journal of the Royal Institute of British Architects*, 7 (1900), pp. 394–7. It was followed by articles on Columbia, *ibid.*, 8 (1900–01), pp. 16–19; Cornell, pp. 39–44; MIT, pp. 52–4; and Harvard, pp. 96–100. For a British view of some of these schools a decade later, see Alfred W. S. Cross in *The Builder*, 17 Sep. 1910, pp. 306–7 (Cornell); 24 Sep. 1910, pp. 331–2 (Harvard); 8 Oct. 1910, pp. 388–9 (MIT), with comparative remarks on these schools, 15 Oct. 1910, p. 418 and 22 Oct. 1910, pp. 458–9.

98 Cates in *Journal of the Royal Institute of British Architects*, 8 (1900–1), p. 17.

99 Alofsin, *op. cit.*, pp. 13–14.

100 Hector James Hughes in Morrison, *op. cit.*, pp. 413–27.

101 Alofsin, *op. cit.*, p. 19.

102 David S. Landes, *The Unbound Prometheus* (2nd edn, Cambridge, 2003), Chapters 4 & 5; David Edgerton, *Science, Technology and the British Industrial 'Decline', 1870–1970* (Cambridge, 1996).

103 Michael Sanderson, *Education and Economic Decline in Britain, 1870 to the 1990s* (Cambridge, 1999).

104 Quoted in John Michael Weiler, 'Army Architects: The Royal Engineers and the Development of Building Technology in the Nineteenth Century', Ph.D., University of York, 1987, p. 24.

105 *The Builder* 19 Dec. 1891, p. 470. The building where Prior acted as clerk of works was St Margaret's Church, Ilkley; the book Shaw gave him was presumably Viollet-le-Duc's *Dictionnaire de l'architecture française*.

106 Material and quotations in this and the following three paragraphs are from Weiler, *op.cit*, pp. 1–37.

107 C. W. Pasley, *Essay on the Military Policy and Institutions of the British Empire* (London, 1810). A glance at this book leads one to suspect that there was a definite streak of the bore in Pasley. As against that, it should be recorded that he had been a friend of the poet Coleridge and accompanied him on a tour in South Italy in 1805: Richard Holmes, *Coleridge: Darker Reflections* (London, 1998), pp. 39, 51–2.

108 'Extracts from a Memorandum on the Corps of Royal Engineers as Civil Servants of the Crown', in *Fifty Years of Public Work of Sir Henry Cole, K.C.B.* (London, 1884), vol. 2, pp. 322–7.

109 See James Douet, *British Barracks 1600–1914* (London, 1998). The contribution to barracks and forts in the 1880s and 90s of E. Ingress Bell, a civilian architect of some note working within the War Office, is of particular interest.

110 Brendan Cuddy and Tony Mansell, 'Engineers for India: The Royal Indian Engineering College at Cooper's Hill', *History of Education*, 23, no. 1 (1994), pp. 107–23; J. G. P. Cameron, *A Short History of the Royal Indian Engineering College Coopers Hill* (private circulation, 1960). The architectural alterations at Cooper's Hill, previously the property of the notorious Baron Grant, were undertaken by Matthew Digby Wyatt.

111 *Hansard Parliamentary Debates*, 3rd series, vol. 204, 3 March 1871, cols. 1333–8. Both Grant-Duff (sometimes called Duff-Grant) and his superior, the Duke of Argyll, were Scots, which probably made them more favourable to technical training.

112 For insights into the relations between architects, engineers and builders in the PWD, see Gavin Stamp, 'British Architecture in India 1857–1947', *Journal of the Royal Society of Arts*, 129 (1981), p. 359.

113 For the engineers, see George Emmerson, *Engineering Education: A Social History* (Newton Abbot, 1973), pp. 166–94; and R. A. Buchanan, *The Engineers: A History of the Engineering Profession in Britain, 1750–1914* (London, 1989), pp. 161–79. For the architects, see Barrington Kaye, *The Development of the Architectural Profession in Britain* (London, 1960); and Mark Crinson and Jules Lubbock, *Architecture Art or Profession? Three Hundred Years of Architectural Education in Britain* (Manchester, 1994).

114 T. C. Smout, *A History of the Scottish People 1560–1830* (London, 1969), pp. 476–9. There were originally two universities at Aberdeen, so the number is strictly five not four.

115 See [Thomas Constable], *Memoir of Lewis D. B. Gordon, F.R.S.E.* (Edinburgh, 1877); James Small, 'Engineering', in *Fortuna Domus: A Series of Lectures Delivered in the University of Glasgow in Commemoration of the Fifth Centenary of its Foundation* (Glasgow, 1952), pp. 335–55; Crosbie Smith and M. Norton Wise, *Energy and Empire: A Biographical Study of Lord Kelvin* (Cambridge, 1989), p. 30. Gordon's most famous engineering work is the Crumlin Viaduct in South Wales. A reference in the memoir (p. 42) suggests he studied at the Ecole Polytechnique as well as at Freiburg.

116 See Gillian Cookson and Colin A. Hempstead, *A Victorian Scientist and Engineer: Fleeming Jenkin and the Birth of Electrical Engineering* (Aldershot, 2000), esp. pp. 95–112. For his inaugural lecture, see Fleeming Jenkin, *A Lecture on the Education of Civil and Mechanical Engineers in Great Britain and Abroad* (Edinburgh, 1868).

117 Sir Gregory Foster, 'These Hundred Years', in *University College London, Centenary Addresses* (London, 1927), pp. 13–14.

118 These paragraphs on engineering at Cambridge are based on T. J. N. Hilken, *Engineering at Cambridge University 1783–1965* (Cambridge, 1967), with quotations from pp. 52–3 and 74; Michael Sanderson, *The Universities and British Industry 1850–1970* (London, 1972), pp. 43–6, with quotation on p. 44; and the article on Robert Willis in ODNB by Ben Marsden. For Pevsner's judgement of Willis, see *The Buildings of England: Cambridgeshire* (Harmondsworth, 1954 edn), p. 167.

119 Sanderson, *op. cit.*, especially Chapters 2 & 3.

120 For the TH model in Britain, see E. P. Hennock, 'Technological Education in England, 1850–1926', *History of Education*, vol. 19, no. 4 (1990), pp. 299–331.

121 Barrington Kaye, *op. cit.*, p. 93; David Watkin, *The Life and Work of C. R. Cockerell* (London, 1974), pp. 105–32; Crinson and Lubbock, *op. cit.*, p. 49; J. Mordaunt Crook, 'Architecture and History', *Architectural History* 29 (1984), pp. 555–78. On the inspiration of Durand upon Donaldson's course, I am grateful for information from Adrian Forty.

122 John Summerson, *The Architectural Association 1857–1947* (London, 1947), pp. 19–21.

123 C. H. Reilly, *Scaffolding in the Sky* (London, 1938), p. 55.

124 *Ibid.*, p. 115. For the controversy see Crinson and Lubbock, *op. cit.*, pp. 62–4.

125 W. R. Lethaby, 'Education in Building', *RIBA Journal* 8 (1900–1), pp. 393–4.

126 For the Central School, Brixton etc., see notably Crinson and Lubbock, *op. cit.*, pp. 68–71; and for Beresford Pite, Alan Powers, 'Professor Pite', in Brian Hanson (ed.), *The Golden City: Essays on The Architecture and Imagination of Beresford Pite* (London, 1999), pp. 95–103.

127 Arthur Bolton in 1895, quoted in Alan Adrian Robelou Powers, 'Architectural Education in Britain 1880–1914', Ph.D. University of Cambridge, 1982, p. 49. The discussion in this paragraph relies on Chapter 2 of that dissertation, supplemented by Crinson and Lubbock, *op. cit.*, pp. 65–72.

128 Reilly, *op. cit.*, p. 85, ridicules 'the subtle connection between making silver rings and enamelled brooches for one's girl friends with architecture'. For the early years at Liverpool see Powers, *op. cit.*, pp. 65–9.

129 For Liverpool see Reilly, *op. cit*, pp. 114–39; Powers, *op. cit.*, pp. 65–9, 135–54; Myles Wright, *Lord Leverhulme's Unknown Venture: The Lever Chair and the Beginnings of Town and Regional Planning* (London 1982), pp. 48–74; Reilly's letter to *The Times*, 10 Feb. 1908, p. 2f; and the article on the Liverpool school by Alfred W. S. Cross in his series 'Architectural Education at Home and Abroad', in *The Builder* 16 July 1910, pp. 62–3.

130 Reilly, *op. cit.*, p. 31, from chapter on his time at Cambridge, pp. 26–43.

131 These paragraphs on Cambridge are based on research by the author for an unpublished lecture given in March 2006, 'The Cambridge School of Architecture: A Personal View of its History and Meaning'.

132 Charles Waldstein, *The Study of Art in Universities* (London, 1896).

133 Sir Reginald Blomfield, *Memoirs of an Architect* (London, 1932), p. 125. For the public controversy over the proposed Cambridge course, see letters from the various protagonists to *The Times*, 27 Jan. to 22 Feb. 1908.

134 Powers, *op. cit.*, p. 92; Lethaby quotation above, p. 91.

135 Letter from F. M. Simpson in *RIBA Journal* 13 April 1907, pp. 415–16.

136 T. G. Jackson, Reginald Blomfield and Basil Champneys, letter to *The Times*, 27 Jan.1908, p. 8b.

137 *The Builder*, 24 Dec. 1910, p. 773. Cross's series, which proceeds on a weekly basis throughout the second half of 1910, valuably summarizes the state of architectural education in Britain, America and France at that date.

138 *The Builder*, 25 Oct. 1918, p. 261. For further writing in this vein see W. R. Lethaby, 'The Architecture of Adventure', *RIBA Journal* (April 1910), pp. 469–78; W. R. Lethaby, *Architecture* (London, 1912), pp. 248–51; and Godfrey Rubens, *William Richard Lethaby* (London, 1986), pp. 254–8.

139 W. R. Lethaby, 'The Architecture of Adventure', p. 477.

140 John V. Maciuika, *Before the Bauhaus: Architecture, Politics and the German State 1890–1920* (New York and Cambridge, 2005), p. 72. I am most grateful to John Maciuika for the opportunity to use this work and for guidance. I have relied heavily on his Chapter 2, 'The Prussian Commerce Ministry and the Lessons of the British Arts and Crafts Movement', pp. 69–103.

141 *Ibid.*, p. 74: also p. 323, n. 8. This instruction came from Otto March, the senior architect in the Ministry of Public Works. The Public Works Ministry was naturally more interested in architecture, the Trade Ministry in technical education and the applied arts.

142 For Muthesius's early career see Maciuika, *op. cit.*, pp. 17–18, and Stanford Anderson, introduction to his translation of Hermann Muthesius, *Style-Architecture and Building-Art* (Cambridge, Mass., 1994), pp. 2–3.

143 Muthesius may have been introduced to England via Edwin Sachs (1870–1919), a young Anglo-German architect who worked for Ende and Böckmann in Berlin at this time, returning to London in 1892. On various occasions Muthesius reviewed Sachs's work on theatres and fire prevention for German periodicals. See ODNB on Sachs (Saint).

144 Maciuika, *op. cit.*, p. 18.

145 See the bibliography in Muthesius, *Style-Architecture*, p. 107.

146 Rainer K. Wick, *Teaching at the Bauhaus* (Ostfildern-Ruit, 2000), p. 56.

147 Quoted *ibid.*, p. 59.

148 For a bitter memory of teaching at the Berlin-Charlottenburg TH in the 1920s, see Julius Posener, *Heimliche Erinnerungen* (Munich, 2004), Ch. 13, pp. 191–200 ('Verlorene Jahre').

149 The subtitle of the Bauhaus and the main title of its post-war successor at Ulm was 'Hochschule der Gestaltung', echoing a coinage of Schumacher's: 'Hochschule des Gestaltene'.

150 Nancy J. Troy, *Modernism and the Decorative Arts in France; Art Nouveau to Le Corbusier* (New Haven, 1991), pp. 52–102. A little later, between 1909 and 1913, says Troy (pp. 58–9), German sales to France in the applied arts advanced from 378 to 572 million francs, Munich being specially strong as a source of goods.

151 Joan Campbell, *The German Werkbund* (Princeton, 1978), p. 18.

152 Quoted in Maciuika, *op. cit.*, p. 103.

153 From Gropius's address to the Bauhaus students in 1919, in Hans M. Wingler, *The Bauhaus* (Cambridge, Mass., 1978), p. 36.

154 For Gropius's early interest in prefabrication see Gilbert Herbert, *The Dream of the Factory-Made House: Walter Gropius and Konrad Wachsmann* (Cambridge, Mass., 1984), pp. 32–7.

155 At the famous AEG Turbine Factory, Behrens was also responsible for the 'spatial configuration'. But according to Mies van der Rohe, who was working for Behrens at the time, his involvement went deeper in this project than in other buildings, where he was often confined to the façades. For the demarcation at the Turbine Factory between Behrens and his collaborators, the structural engineer Karl Bernhard and the production engineer Oskar Lasche, see Tilmann Buddensieg, *Industriekultur: Peter Behrens and the AEG 1907–1914* (Cambridge, Mass., 1984), pp. 59–66 and 249–55.

156 Maciuika, *op. cit.*, p. 274. Fritz Neumeyer, 'Nexus of the Modern: The New Architecture in Berlin', in Tilmann Buddensieg (ed.), *Berlin 1900–1933: Architecture and Design* (New York and Berlin, 1987), pp. 34–40, discusses the 'transcendent' quality of Behrens's architecture for AEG in the light of Josef-August Lux's book of 1910, *Ingenieur-Aesthetik*.

157 Quoted in Alofsin, *op. cit.*, p. 243. Gropius himself had only a brief TH schooling, in Munich.

158 Quoted in Wingler, *op. cit.*, p. 31. For the negotiations with Gropius to teach architecture in 1915, *ibid.*, p. 22.

159 Walter Gropius, *The New Architecture and the Bauhaus* (London, 1935), p. 36.

160 Quoted in Howard Dearstyne, *Inside the Bauhaus* (London, 1986), pp. 197–8. *The New Architecture and the Bauhaus*, Gropius's retrospective sketch of his creation, offers a similar but longer formulation (pp. 55–7): 'In so far as our curriculum did not provide finishing courses in the theoretical side of the more specialized branches of engineering – such as steel and concrete construction, heating, plumbing, etc. – or advanced statics, mechanics and physics, it was usually found advisable to let the most promising of the architectural pupils round off their studies by attending complementary classes at various technical institutes.' After discussing urbanism and the need for 'a manifold simplicity arrived at by deliberate restriction to certain basic forms used repetitively; and the structural subdivision of buildings according to their nature, and that of the streets they face', Gropius concludes: 'This was at once the limit of our Structural Instruction and the culminating point of the entire Bauhaus teaching.'

161 Herbert, *op. cit.*, pp. 105–59.

162 From 'The New World', 1926, reprinted in Claude Schnaidt, *Hannes Meyer* (Teufen, 1965), pp. 91–5.

163 For the teaching under Hannes Meyer, see Wick, *op. cit.*, pp. 78–81; for Rudelt see Wingler, *op. cit.*, pp. 151–2.

164 Hubert Hoffmann, quoted in Wick, *op. cit.*, p. 84.

165 William Priestley, quoted in Cammie McAtee, 'Alien 5044325: Mies's First Trip to America', in Phyllis Lambert (ed.), *Mies in America* (New York and Montreal, 2001), p. 143.

166 Dearstyne, *op. cit.*, p. 226.

167 For Harvard I have relied on Anthony Alofsin, *The Struggle for Modernism: Architecture,*

Landscape Architecture and City Planning at
Harvard (New York, 2002), supplemented by
Jill Pearlman, 'Joseph Hudnut's Other
Modernism at the "Harvard Bauhaus"',
Journal of the Society of Architectural Historians
56, 4 (Dec. 1997), pp. 452–77; and McAtee,
op. cit., pp. 141–9.

168 On Hudnut see Alofsin, op. cit., pp. 119ff.;
quotations from pp. 120, 121.

169 Ibid., pp. 131–3; McAtee, op. cit., pp. 146–52.
Margret Kentgens-Craig, The Bauhaus and
America: First Contacts 1919–1936 (Cambridge,
Mass., 1999), pp. 194–8, suggests that
Hudnut felt threatened by Mies.

170 Alofsin, op. cit., p. 154.

171 Ibid., p. 178.

172 For a damning critique of Gropian studio-
teaching at the GSD, see Klaus Herdeg, The
Decorated Diagram: Harvard Architecture and the
Failure of the Bauhaus Legacy (Cambridge,
Mass., 1983).

173 For this paragraph see Alofsin, op. cit., pp.
178–9, 240.

174 These paragraphs are based on McAtee, op.
cit., pp. 150–2, 181–5; Phyllis Lambert, 'Mies

175 Immersion', in Lambert, op. cit., pp.
193–221; and Alfred Swenson and Pao-Chi
Chang, Architectural Education at IIT
1938–1978 (Chicago, 1980).

175 Ibid., p. 62.

176 Alexander Caragonne, The Texas Rangers:
Notes from an Architectural Underground
(Cambridge, Mass., 1995). Supplementary
material may be found in Colin Rowe, As I
Was Saying: Recollections and Miscellaneous
Essays, vol. 1 (Cambridge, Mass., 1996), pp.
25–53.

177 Caragonne, op. cit., p. 10.

178 Ibid., p. 17.

179 See the memoir 'Texas and Mrs. Harris', in
Rowe, op. cit., pp. 25–40.

180 Caragonne, op. cit., quotations from pp.
27–8, 78, 137.

181 Ibid., pp. 67–9, 376–8.

182 For a summary of the Bartlett's contribution under Llewelyn-Davies to what they
call the 'official system' in British architectural education, see Crinson and Lubbock,
op. cit., pp. 148–51.

183 RIBA Journal 68 (January 1961), pp. 118–20.

I am grateful to Adrian Forty for drawing
this absorbing document to my attention.

184 See M. J. Long, 'Architect and Engineer', in
Education for the Built Environment (Ove Arup
Foundation, [1991]), pp. 14–22.

185 See William J. Carpenter (ed.), Learning by
Building: Design and Construction in Architectural
Education (New York, 1997).

CONCLUSION

1 Thomas Hardy, Tess of the D'Urbervilles, Ch.
47.

2 Hans Straub, A History of Civil Engineering
(London, 1960), p. 181. The original
German-language edition is of 1949.

3 J. D. North, quoted in Walter G. Vincenti,
What Engineers Know and How They Know It:
Analytical Studies from Aeronautical History
(Baltimore, 1990), p. 4.

4 Lydia Soo, Wren's 'Tracts' on Architecture and
Other Writings (Cambridge, 1998), p. 153.

INDEX

ILLUSTRATION CREDITS

This list acknowledges owners of pictures first, followed where possible by photographers or illustrators. Every effort has been made to contact copyright-holders, but if there are any errors or omissions please write to the publishers, so that corrections can be made in any subsequent editions. Myke Clifford provided much help with photographs and scans, as did Helen Jones. The list aims also to offer a succinct guide to sources. Where those are mentioned in the notes, an abbreviated title is given below. Illustrations out of copyright for which adequate references are given in the captions are mostly not listed below, though additional help in procurement is acknowledged. These include ills. 9, 14, 47, 59, 62, 92, 100–2, 123, 127–8, 131, 135, 138, 139, 170, 179, 180, 183, 189, 197–9, 214, 252, 255–7, 264, 268, 269, 277, 312, 315–8, 338–9, 355, 387, 391, 406, 407 and 410.

Cover, 64, 107, Rheinisches Bildarchiv Köln (Wallraf-Richartz Museum); 1, Department of Engineering, Cambridge; 2, Department of Architecture, Cambridge; 3, Farrells/Nigel Young; 4, Le Corbusier-Saugnier, *Vers une architecture* (1923 edn.); 5, 52, 54, 55, 57, 122, 172, 232, 235, 245, 261, 273, 275, 311, 320, 329, 335, 336, 349, 409, Author; 6, 7, 17, 18, 19, 50, 134, 149, 201, 250, 258, 266, 310, Bridgeman Art Library; 8, 200, Centre des Monuments Nationaux, Paris (Monum); 10, Gino Pavan (ed.), *Palmanova, fortezza d'Europa* (1993); 11, www.italiantourism.com; 12, Galluzzi, *Gli ingegneri del Rinascimento*; 15, Hampden Maps; 16, www.trekearth.com; 21, 249, 253, 270, 290, 295, 322, Charles S. Whitney, *Bridges, Their Art, Science and Evolution* (1929); 23, Moulin, *L'architecture en Aunis et Saintonge*; 24, 51, 91, National Monuments Record; 25, David Davison; 26, Edward Impey; 27, Steen Eiler Rasmussen, *London, The Unique City* (1934); 28, Picon, Perrault; 29, 60, 63, Photothèque des Musées de la Ville de Paris; 30, John Bold, *Greenwich* (Yale University Press); 31, Anthony Calladine/ Tony Berry; 32, Reps, *Making of Urban America*; 33, The Newberry Library; 34, 99, 210, 344, 366, Martin Charles; 35, Library Company of Philadelphia; 36, 38, 41, 42, Library of Congress; 37, 111, 157, 171, 309, 315, 323, New York Historical Society; 39, 40, Reps, *Monumental Washington*; 43, Olin Library, Cornell University; 44, Fine Arts Commission/ Sue Kohler; 45, 84, 103, 104, 116, 118, 119, 129, 130, 142, 168, 174–6, 178, 209, 220–3, 225, 226, 229, 286, 301, 308, Mosette Broderick (Henry-Russell Hitchcock collections); 46, Robert Bowles; 48, Thurley, *Hampton Court* (Daphne Ford); 49, National Portrait Gallery; 52, 280, Sir John Soane's Museum; 53, 78, 120, 121, *Survey of London*; 56, 125, 236, 248, 281, 282, English Heritage, London Division; 58, Skempton & Johnson, *Architectural Review*, March 1962; 61, Steiner, *French Iron Architecture*; 65, Schinkel, *Sammlung architektonischer Entwürfe*; 66, 67, Bildarchiv Preußischer Kulturbesitz; 68, *Karl Friedrich Schinkel: Lebenswerk*; 69, Brandenburgisches Landesamt für Denkmalpflege und Archäologisches Landesmuseum; 70, 71, 292, 300, Sergey Fedorov; 72, Summerson, *Life and Work of John Nash*; 73, 291, The Royal Pavilion and Museums, Brighton & Hove; 74, James Sutherland; 75, 124, Robert Thorne; 76, Parliamentary Archives; 77, Michael Port; 79, Bicton Park Botanical Gardens; 80, 184, Neil Burton; 81, 82, 85–7, 274, Edward Diestelkamp; 83, 117, 380, Gavin Stamp; 88, National Railway Museum/ Science & Society Picture Library; 89, 90, 93, 96, 183, 264, 288, 289, 294, 299, 303, 312, 332, 333, 365, 412, 413, Institution of Civil Engineers (Mike Chrimes/ Annette Ruehlmann); 92, Tim Brittain-Catlin; 94, Faulkner and Greg, *John Dobson*; 95, Laing Art Gallery (Tyne & Wear Museums); 97, University of Bristol Library/Steven Brindle; 98, Westminster City Archives/Steven Brindle; 105, Hector Horeau 1801–1872; 106, 361, Michel Denancé; 108–10, 237–9, 241, Nederlands Architectuurinstituut; 112–5, Avery Library, Columbia University; 126, 133, 314, 404, Agence Roger Viollet; frontispiece, 132, 140, 141, 262, 272, Bibliothèque Nationale/ Anne-Marie Châtelet; 136, Fargues, Armand Moisant; 137, Siry, Carson Pirie Scott; 143, 147, Réunion des Musées Nationaux (Musée d'Orsay); 144, 326, 328, 330, 408, Image Archive ETH-Bibiothek Zurich; 145, 146, Lemoine, *La tour de Monsieur Eiffel*; 148, Écomusée de la communauté urbaine Le Creusot Montceau-les-mines/ Daniel Busseuil; 150, 153, Maryland Historical Society; 151, New Jersey Historical Society, Newark, New Jersey; 152, Sturges, *Origins of Cast Iron in America*; 154–6, Abbott, *The Harper Establishment*/ Gayle and Gayle, *Cast Iron Architecture in America*; 158–60, Sarah B. Landau/ New York Historical Society; 161, 177, 420, Chicago Historical Museum; 162, Landau, *P. B. Wight*; 163–5, Siry, *Chicago Auditorium Building*; 166, Thomas Van Leeuwen, *The Skyward Trend of Thought*; 167, Deutsches Architekturmuseum; 169, Architectural Record, 1892; 173, Art Institute of Chicago; 181, 202–8, 402, Fonds Perret, *Cité de l'Architecture et du Patrimoine*, Centre d'Archives du XXe Siècle; 182, 276, 279, 283, 284, 302, Guildhall Library/Art Gallery; 185, Lewisham Library Services; 186, Ernest Newton, *Sketches for Country Residences* (1882); 187, Simonnet, *Le béton*; 188, 271, Ecole Nationale des Ponts et Chaussées; 190, Bosc, *Joseph Monier*; 191, Stadgeschichtliches Museum Leipzig/ Claudia Marx; 192, Stanford University Archives; 193, *Progressive Architecture*, Sep. 1957; 194–6, Fonds Béton Armé Hennebique, *Cité de l'Architecture et du Patrimoine*, Centre d'Archives du XXe Siècle; 211, 331, 334, Fernandez Ordóñez, *Freyssinet*; 212, 217, 218, Albert Kahn Associates; 213, 216, Moritz Kahn, *Design and Construction of Industrial Buildings*; 215, Henry Ford Museum & Greenfield Village & Henry Ford Motor Co.; 219, Frank Lloyd Wright, *Autobiography* (1932 edn.); 224, Siry, Unity Temple; 227, H.-R. Hitchcock, *In the Nature of Materials* (1942); 228, Lipman, *Frank Lloyd Wright and the Johnson Wax Buildings*; 230, 233, David Yeomans; 231, Brent Archive; 234, C. G. Holme, *Industrial Architecture* (1935); 240, Gemeentearchief Rotterdam; 242–4, Fondation Le Corbusier; 246, *Architectural Design* 35 (1965), p. 20; 247, Musée d'Histoire de Marseille/ Ville de Marseille; 251, Weil, *Ponte Sant'Angelo*/Alinari; 254, The Royal Collection, Her Majesty Queen Elizabeth II; 259, 260, Teresa Sladen; 263, British Library; 265, 268, Perronet, *Description des projets*; 267, 277, 287, Ted Ruddock; 278, Yale Center for British Art/Yale University Press; 285, 296, 297, Eda Kranakis; 293, Ekaterina Shorban; 298, www.bridgemeister.com; 304, Editions du Désastre/ Sempé; 305, Library of Congress (Historic American Engineering Record); 306, 307, Carnegie Library of Pittsburgh; 321, Architectural Record, 1906; 324, Museum of the City of New York, Print Archives; 325, *MTA Bridges & Tunnels*, Special Archives; 327, Baugeschichtliches Archiv, Stadt Zürich Amt für Städtebau; 332, 333, Séjourné, Grandes Voûtes; 337, Radde, *Merritt Parkway*; 340–3, Foster and Partners/ Nigel Young; 345, 348, 353, 357, 360, Ove Arup and Partners; 346, Jason Orton; 347, 414, British Architectural Library/ RIBA; 350, Australian Air Photos/ Ove Arup and Partners; 351, Arne Magnussen/ *Helsingør Dagblad*; 352, Rice, *An Engineer Imagines*; 354, Baume, Sydney Opera House Affair; 356, Max Dupain Associates/ Ove Arup and Partners; 358, 359, 363, Richard Rogers Partnership; 362, *Architectural Design*, Feb. 1977; 364, Norman Foster; 368, Williams, *Hongkong Bank*; 369, *Architecture of Skidmore, Owings and Merrill* (1963); 370, 373, Ezra Stoller "Esto; 371–2, Illinois Institute of Technology, Graham Resource Center, College of Architecture/ Family of Myron Goldsmith; 374–6, 400, Skidmore, Owings & Merrill LLP; 377, 381, Louis I. Kahn Collection, University of Pennsylvania and Pennsylvania Historical and Museum Collection; 378, Yale University Art Gallery; 379, Barnabas Calder; 382, Kimbell Art Museum, Fort Worth; 383, 385, *L'Art de l'ingénieur*; 384, Faber, *Candela*; 386, *Architectural Record*, Aug. 1954; 388–90, Balthazar Korab Ltd.; 392–4, *Berlin Baut 2: Die Kongresshalle*; 395, *Ingenhoven Architekten*, Düsseldorf; 396, 399, ILEK Archive, Institute for Lightweight Structures, University of Stuttgart; 397, Architekturmuseum der Technischen Universität München/ ILEK; 398, Fritz Dressler, Worpswede/ILEK; 401, Gehry Partners LLP; 403, Geneviève Viollet-le-Duc; 405, Belhoste (ed.), *Le Paris des Centraliens*; 411, 419, Anthony Alofsin; 415, London Metropolitan Archive; 416, University of Liverpool, Archives; 417, Sibylle Mutheisius-Boyle/ Museum der Dinge; 418, Bauhaus-Archiv-Berlin; 421, 422, Caragonne, Texas Rangers/Renata Hejduk.